24th International Conference on Marine Technology (SENTA 2024)

IOP Conference Series: Earth and Environmental Science
Volume 1461

Surabaya, Indonesia
31 October - 1 November 2024

ISBN: 979-8-3313-1737-9
ISSN: 1755-1307

Printed from e-media with permission by:

Curran Associates, Inc.
57 Morehouse Lane
Red Hook, NY 12571

Some format issues inherent in the e-media version may also appear in this print version.

Institute of Physics
Dirac House, Temple Back
Bristol BS1 6BE UK

Phone: 44 1 17 929 7481
Fax: 44 1 17 920 0979

techtracking@iop.org

Additional copies of this publication are available from:

Curran Associates, Inc.
57 Morehouse Lane
Red Hook, NY 12571 USA
Phone: 845-758-0400
Fax: 845-758-2633
Email: curran@proceedings.com
Web: www.proceedings.com

TABLE OF CONTENTS

doi:10.1088/1755-1315/1461/1/011001

CONFERENCE
PROCEEDING OF SENTA 2024

Innovation in Marine Technology Toward Sustainable and Resilient Maritime Industry

2024

OCTOBER, 31st - NOVEMBER 1st

The 24th International Conference on Marine Technology and Application (SENTA 24)
Institut Teknologi Sepuluh Nopember
Surabaya, Indonesia

IOP Conf. Series: Earth and Environmental Science **1461** (2025) 011001 doi:10.1088/1755-1315/1461/1/011001

Preface

It is a great honor and privilege to welcome you to the 9th International Conference on Marine Technology (SENTA 2024), organized by the Faculty of Marine Technology, Institut Teknologi Sepuluh Nopember (ITS), Surabaya, Indonesia. With the theme "Innovations in Sustainable Marine Technology for Resilient Maritime Industry," this conference serves as a platform for researchers, professionals, and policymakers to exchange insights and breakthroughs that address challenges in achieving sustainable and resilient maritime solutions.

SENTA 2024 underscores its commitment to advancing knowledge in marine technology, focusing on critical areas such as low-carbon ship design, smart port operations, decarbonization in logistics, and the adoption of renewable energy in offshore engineering. By fostering interdisciplinary collaborations, this event aspires to contribute significantly to sustainable development goals while enhancing the maritime industry's resilience.

We are delighted to host prominent keynote speakers, including **Prof. Osman Turan (Naval Architecture, Ocean and Marine Engineering, The University of Strathclyde Glasgow, UK), Yuen Kum Fai, Ph.D. (Director Maritime Energy and Sustainable Development Center of Excellence), Ruri Indrasari Rachmaputri (Commercial and Technical Director, PT Pelindo Solusi Logistik and Head of Mutiara Pelindo), Lucia Karina (Director of Public Affairs, Communications, Sustainability, Coca Cola Europasicfic Partners)** who will share their expertise on groundbreaking innovations. With over 100 participants and technical paper sessions, SENTA 2024 promises a dynamic exchange of ideas.

On behalf of the organizing committee, I extend my deepest gratitude to all contributors, sponsors, and participants for their invaluable support in making this event a success. We look forward to a fruitful and inspiring conference that drives forward the future of sustainable marine technology.

Sincerely,

Ir. Christino Boyke S.P., S.T., M.T, Ph.D
SENTA 2024 Chairman

IOP Conf. Series: Earth and Environmental Science **1461** (2025) 011001 doi:10.1088/1755-1315/1461/1/011001

Editors

Ir. Tri Achmadi, Ph.D.
Institut Teknologi Sepuluh Nopember, Sukolilo Campus, Surabaya 60111
triachmadi@its.ac.id

Dr. Eng. Ir. I G. N. Sumanta Buana, S.T., M.Eng.
Institut Teknologi Sepuluh Nopember, Sukolilo Campus, Surabaya 60111
buana@its.ac.id

Dr.-Ing. Ir. Setyo Nugroho
Institut Teknologi Sepuluh Nopember, Sukolilo Campus, Surabaya 60111
snugroho@its.ac.id

Ir. Firmanto Hadi, S.T., M.Sc.
Institut Teknologi Sepuluh Nopember, Sukolilo Campus, Surabaya 60111
firmanto@its.ac.id

Dika Virginia Devintasari, S.Si., M.Sc
Institut Teknologi Sepuluh Nopember, Sukolilo Campus, Surabaya 60111
dika.virgina@its.ac.id

IOP Conf. Series: Earth and Environmental Science **1461** (2025) 011001 doi:10.1088/1755-1315/1461/1/011001

COMMITTEE

A. Advisory Board

1.	Rector ITS	Bambang Pramujati, S.T., M.Sc. Eng., Ph.D
2.	Dean of Faculty of Marine Technology	Dr. Eng. Trika Pitana, S.T, M.Sc. Dr.
3.	Vice Dean of the Faculty of Marine Technology	Eng.Rudi Waluyo Prastianto, S.T., M.T.
4.	Head of Dept. of Marine Transportation Engineering	Dr.-Ing.Ir.Setyo Nugroho
5.	Head of Dept. of Naval Architecture	Ir. Wasis Dwi Aryawan, M.Sc., Ph.D.
6.	Head of Dept. of Marine Engineering	Beny Cahyono, S.T., M.T., Ph. D
7.	Head of Dept. of Ocean Engineering	Prof. Herman Pratikno, S.T, M.T, Ph.D

B. Organizing Committee

1.	General Chair	Ir. Christino Boyke, S.T., M.T., Ph.D.
2.	General Co-Chair	Ir. Siti Dwi Lazuardi, S.T., M.Sc.
3.	Secretary	Ir. Maulana Yafie Danendra, S.T., M.Log.
4.	Finance Chair	Ir. Irwan Tri Yunianto, S.T., M.T.
5.	Program Chair	Ir. Firmanto Hadi, S.T., M.Sc.
6.	Program Co-Chair	Ir. Muhammad Riduwan, S.Kom., M.Kom.
7.	Paper & Proceeding Chair	Dika Virginia Devintasari, S.Si., M.Sc.

C. Internal Reviewer SENTA 2024

1. Prof. Ir. Raden Sjarief Widjaja, Ph.D., FRINA
2. Prof. Ir. Achmad Zubaydi, M.Eng., Ph.D.
3. Prof. Ir. I Ketut Aria Pria Utama, M.Sc., Ph.D.
4. Prof. Dr. Ir. Heri Supomo, M.Sc.
5. Prof. Dr. Ir. I Ketut Suastika, M.Sc.
6. Prof. Aries Sulisetyono, S.T., MA.Sc., Ph.D.
7. Prof. Raja Oloan Saut Gurning, S.T., M.Sc., Ph.D.
8. Prof. Drs. Mahmud Musta'in, M.Sc., Ph.D
9. Prof. Dr. Ir. Sardono Sarwito, M.Sc
10. Dr. Ir. Hasan Ikhwani, M.Sc.
11. Dr. Eng. Muhammad Zikra, S.T., M.Sc.
12. Nur Syahroni, S.T., M.T., Ph.D
13. Dr. Eng Yeyes Mulyadi, S.T., M.Sc.
14. Dr. Eng. Shade Rahmawati, S.T., M.T
15. Dr. Dendy Satrio, S.ST
16. Silvianita, S.T., M.Sc., Ph.D
17. Dr. Eng. Trika Pitana, S.T., M.Sc.
18. Dr. Eng. Muhammad Badrus Zaman, S.T., M.T.
19. A.A.B.Dinariyana Dwi P., S.T., MES., Ph.D.
20. Sutopo Purwono Fitri, S.T., M.Eng., Ph.D.
21. Dr. Eng. Dhimas Widhi Handani, S.T.,M.Sc.
22. Sunarsih, S.T. M.,Eng, Ph.D
23. Dr. Nurhadi Siswantoro, S.T., M.T
24. Fadilla Indrayuni Prastyasari, S.T., M.Sc., Ph.D

IOP Conf. Series: Earth and Environmental Science **1461** (2025) 011001 doi:10.1088/1755-1315/1461/1/011001

25. Dr. Emmy Pratiwi, S.T
26. Dr. Ir. Dony Setyawan, S.T., M.Eng.
27. Teguh Putranto, S.T.,M.T., Ph.D
28. Dr. Eng. Yuda Apri Hermawan, S.T., M.T
29. Imam Baihaqi, S.T., M.T, Ph.D
30. Ardi Nugroho Yulianto, S.T., M.T., Ph.D
31. Dr. Ing. Ir. Setyo Nugroho
32. Dr. Eng. IGN. Sumanta Buana, S.T., M.Eng.
33. Dr. Eng. Christino Boyke Surya P, S.T., M.T

D. External Reviewer SENTA 2024

1. AP Cdr. (Rtd) Ts. Dr. Aminuddin Md Arof (UniKL-MIMET)
2. AP Ts. Dr. Mohd Yuzri Mohd Yusop (UniKL-MIMET)
3. Ts. Dr. Wardiah Mohd Dahalan (UniKL-MIMET)
4. Ts. Dr. Khairul Anuar Mat Saad (UniKL-MIMET)
5. Dr. Nurshahnawal Yaacob (UniKL-MIMET)
6. Dr. Roslin Ramli (UniKL-MIMET)
7. Dr. Noorlee Boonadir (UniKL-MIMET)

IOP Conf. Series: Earth and Environmental Science **1461** (2025) 011002 doi:10.1088/1755-1315/1461/1/011002

Peer Review Statement

All papers published in this volume have been reviewed through processes administered by the Editors. Reviews were conducted by expert referees to the professional and scientific standards expected of a proceedings journal published by IOP Publishing.

- **Type of peer review:** Double Anonymous
- **Conference submission management system:** Morressier
- **Number of submissions received:** 83
- **Number of submissions sent for review:** 83
- **Number of submissions accepted:** 65
- **Acceptance Rate (Submissions Accepted / Submissions Received × 100):** 78.3
- **Average number of reviews per paper:** 1
- **Total number of reviewers involved:** 26
- **Contact person for queries:**
  Name: Dika Virginia Devintasari
  Email: dika.virginia@its.ac.id
  Affiliation: Dika Virginia

IOP Conf. Series: Earth and Environmental Science **1461** (2025) 012001 doi:10.1088/1755-1315/1461/1/012001

The effect of damage on single and double hull construction to the ultimate strength of the ship

Muhammad Zubair Muis Alie[1*], Taufiqur Rachman[1], A. Ardianti[2], Alamsyah[3], Nur Indah[1], and Nur Salsabila Aulia[1]

[1] Department of Ocean Engineering, Hasanuddin University, Makassar, Indonesia
[2] Department of Naval Architecture, Hasanuddin University, Makassar, Indonesia
[3] Department of Naval Architecture, Institut Teknologi Kalimantan, Balikpapan, Indonesia

*E-mail: zubair.m@eng.unhas.ac.id

Abstract. Ship construction typically includes single hull and double hull designs, each offering distinct advantages and disadvantages concerning the ship's ultimate strength. Because a single hull does not have an inner hull or inner bottom, it may carry more cargo than a double hull. Compared to a single hull, a double hull offers greater safety during collisions and grounding incidents due to the presence of an additional inner hull and inner bottom. As a result, this influence has to be examined in relation to the ship design and construction specifications. This study aims to examine how damage impacts the ultimate strength of single and double hull ship designs when subjected to hogging and sagging conditions. Both single and double hulls' midship cross sections are considered in the study. One side of the cross section is rotated by a force, while the other side is fixed under restrictions. In this case, just one side of the cross section is subject to the multiple point constraint. It is assumed that the midship cross section will remain flat during the gradual collapse process. The materials have uniform density, yield strength, Poisson's ratio, and Young's modulus, with single hull and double hull ships having identical proportions. An inner hull is incorporated into the simulation for the double hull structure. For simple analysis, longitudinal direction is done in a single frame space. This study employs an analytical method to assess the effects of damage on the structural integrity of single-hull and double-hull ships. The findings reveal that damage has a notable influence on the ultimate strength of both hull types. The moment-curvature relationship is utilized to represent the ultimate strength of these structures. Additionally, the paper illustrates the deformation patterns and stress distributions in single hull and double hull designs.

1. Background

Ship hulls have two types, namely single and double hulls. Both types of hulls have advantages and disadvantages. A single hull offers the benefit of increased cargo capacity but is more susceptible to damage from forces resulting from grounding or collisions. While the double hull has additional strength due to the inner hull, which functions to prevent liquid from directly entering the cargo space, on the other hand, the load carried is reduced due to the additional construction element, namely the inner hull. Damage that may occur to ship hulls is generally asymmetrical. This damage led to a decrease in the ship's strength. Therefore, it is very important to assess the effects of damage to the single and double-hull construction of the ship in the event of a grounding or collision.

1.1 Literature review

Numerous studies have assessed ship structural analysis, with a particular emphasis on ultimate strength. To broaden and deepen the study of hull structures' ultimate strengths, Cui et al. [1] introduced the concept of elastic shakedown and integrated it into the analysis, thereby enhancing the scope and depth of research in this area. Woloszyk et al. [2] proposed a method for assessing the ultimate strength of ship hulls, incorporating advanced corrosion models. Shi and Gao [3] explored the influence of superstructures on the ultimate strength of cruise ship hull girders. Their research examined the combined impact of bending moments and shear forces, evaluated the effectiveness of superstructures, and assessed the applicability of the incremental iterative technique for determining the ultimate strength of cruise ships. Muis Alie et al. [4] evaluated the ultimate strength of a Ro-Ro ship after damage in both hogging and sagging situations. Quispe et al. [5] performed a four-point bending test on a small size hull box girder Geometric slenderness ratios that match a double bottom panel at the midship of a normal Suezmax tanker were used to determine the model's dimensions.

Jagite et al. [6] examined the dynamic ultimate strength of super-large cargo ships under realistic loading scenarios. Their research employed a long-term hydroelastic analysis to compute the cumulative bending moments generated by the application of lateral and localized loads on the hull girder across different cargo loading conditions. Deng et al. [7] on the other hand, utilized both experimental approaches and finite element simulations to study the ultimate strength and buckling collapse behavior of single and double hull girders with large deck openings when subjected to cyclic ultimate bending forces. Ma et al. [8] conducted a comprehensive study on multiple hull girders of a 10,000 TEU container ship, exploring different combinations of geometric length and plate thickness to analyze the effects of bottom lateral pressure on scaled hull girders under combined loading conditions. Chen et al. [9] presented experimental findings from tests on scaled double-hull specimens subjected to localized deformations induced by two distinct rigid indenters. Quasi-static studies were conducted to investigate the crushing behavior of steel structures in ship collision scenarios. Zhao et al. [10] examined the ultimate strength of a ship hull girder model featuring openings, specifically focusing on a model with relatively large apertures in several compartments. The design and testing phases assessed the ultimate bearing capacity of these compartments under pure sagging bending stress.

To enhance the design-focused analysis of collision and grounding resistance, Liu et al. [11] streamlined the required input parameters. Meanwhile, Shi and Gao [12] performed a collapse experiment on a steel model with superstructures inspired by a real cruise ship design. Their experimental results validated critical aspects of the nonlinear finite element analysis, such as initial imperfections and welding residual stresses. Zhang et al. [13] introduced a new method for calculating strength loss based on stiffness loss, using the section modulus as an indicator of strength. Their findings revealed that the reduction in a hull girder's stiffness after damage does not always correspond to a similar reduction in its strength. To validate this concept of the discrepancy between stiffness loss and strength loss (UNSS), they tested thirteen common cross-sectional designs. Wang et al. [14] initially developed and fabricated a scale model of a 10,000TEU cargo ship, ensuring that both geometric and strength similarities were properly considered. In their assessment of the precision and computational demands of two analytical methods under biaxial bending conditions, Kuznecovs et al. [15] examined various ship scenarios, including an undamaged hull, a hull damaged by collision, a newly built hull, and a hull that had aged due to corrosion.

This study will primarily focus on the ultimate strength of single and double hull constructions, considering the impact of damage. Therefore, it presents a unique approach to investigating these effects.

2. Methodology

2.1 Ship particular

This study employs analytical techniques to examine the impact of single and double hull girder designs on ultimate strength. It examines the structures under both hogging and sagging situations, concentrating on the cross-sections of bulk carriers with single and double hull designs. The damage is located at the bottom and is assumed to be symmetrically placed in both single and double-hull construction. The ship's breadth and depth dimensions are 32.2 meters and 19.062 meters, respectively. In the longitudinal direction, the frame space is taken as 5.1 meters. These dimensions are constant for the single and double hull of the ship. The simulation is done by adding the inner hull for double hull construction. The material properties, such as density, yield strength, Poisson's ratio, and Young's modulus, are identical for both single and double hull structures (Table 1).

Table 1. Material properties

Items	Value	Unit
Density	7.89×10^{-9}	N/mm^3
Young Modulus	206000	N/mm^2
Poisson Ratio	0.3	-
Yield Strength	315/355	N/mm^2

2.2 Damaged model, load, and boundary condition

The bulk carrier ship examined in this study is represented by two models: the single hull and double hull structures, as shown in Figures 1 and 2, respectively.

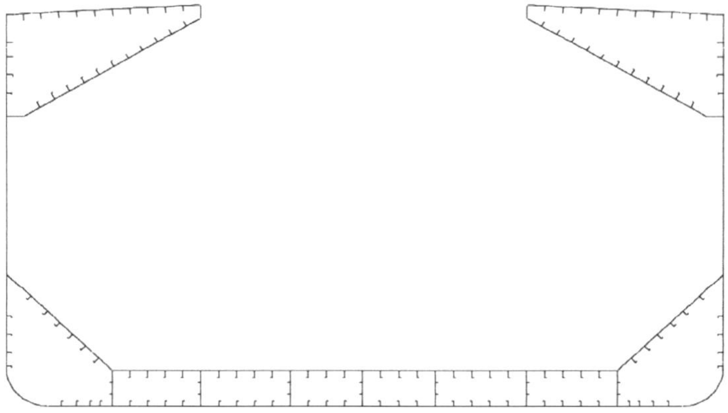

Figure 1. Model-1 Single Hull Construction of Bulk Carrier

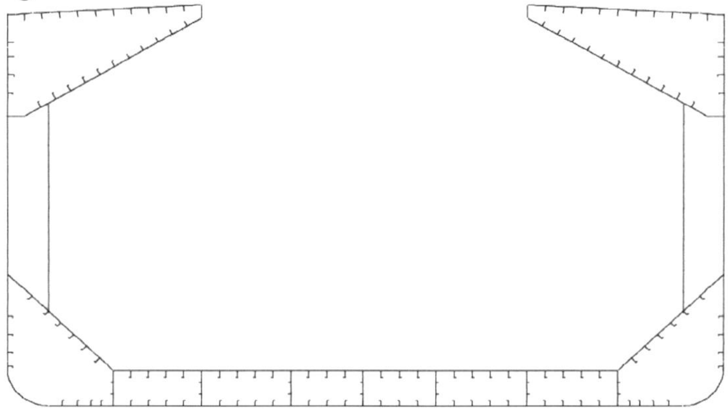

Figure 2. Model-1 Double Hull Construction of Bulk Carrier

The components at the outside bottom section are taken out of it in the event of damage. The damage is assumed to be symmetrical position in the centre part of the bottom area, as shown in Figures 3 and 4, both single and double hull constructions of Bulk Carrier.

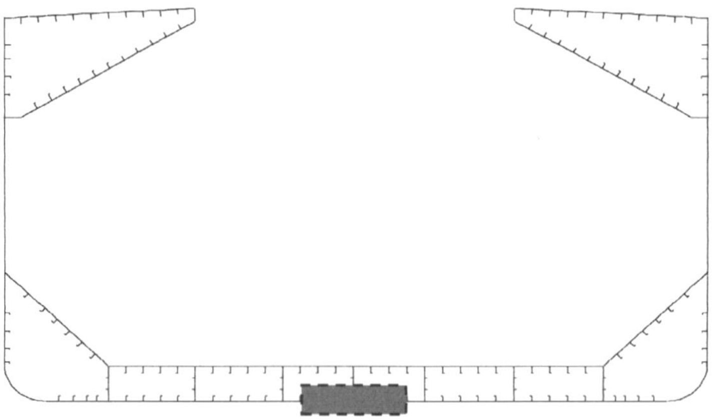

Figure 3. Damage on Single Hull Construction of Bulk Carrier

IOP Conf. Series: Earth and Environmental Science **1461** (2025) 012001 doi:10.1088/1755-1315/1461/1/012001

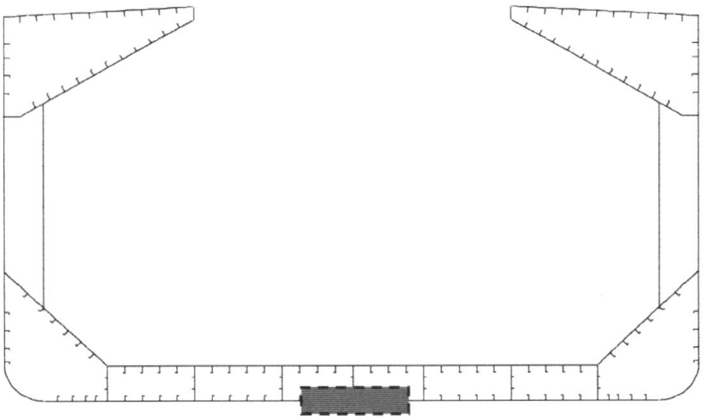

Figure 4. Damage on Double Hull Construction of Bulk Carrier

For simplicity, the damage is considered to be 3 meters from the centerline at the outer bottom section. The HULLST method, created by Yao and Nikolov [16] takes damage to the outer bottom portion into consideration while analyzing the maximum strength of single and double hull structures.

3. Result and Discussion

3.1Single and Double Hull

Single-hull construction has just one watertight outer layer extending across the entire structure. Because there is only one layer, single hull ships present a higher risk to the marine environment in the event of an accident. In contrast, double hull construction incorporates an additional layer, with the space between the two layers serving as ballast tanks. These ballast tank spaces run the full length of the cargo area, providing a significant safety advantage. Single hull designs lack these ballast spaces. However, double hull construction requires more steel, making the building process longer. Ballast compartments in double hull ships are more susceptible to hull fractures and minor damage compared to single hull vessels. Operators of double hull ships frequently report instances of cargo leaking into ballast tanks due to stress, fatigue, or construction defects. The damage to the outer bottom part of bulk carriers with single and double hulls is examined in this research. It is assumed to be symmetrically placed in the center line, as shown in Figures 3 and 4. The neutral axis position for single hull construction of bulk carriers is 7.70 meters. While, for double hull construction of bulk carrier is 7.84 meters. The difference between the two cases indicates that the ultimate strength caused with and without an additional inner hull is also different. This also results in changes in inertia and cross-sectional modulus.

3.2Ultimate strength

The maximum strength obtained through the analytical method is presented in the moment-curvature relationship for both hogging and sagging conditions in this study. Figure 5 presents a comparison of the ultimate strength of bulk carriers with single and double hull design.

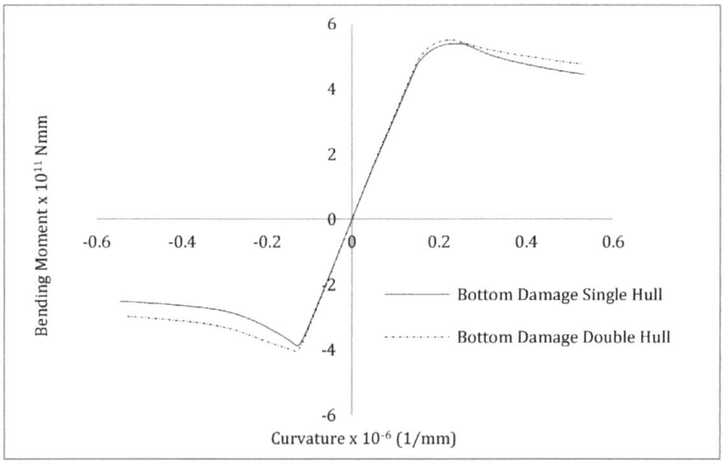

Figure 5. Damage in the Double Hull Structure of a Bulk Carrier

As depicted in Figure 5, the type of hull construction, whether single or double, significantly influences the ultimate strength under both hogging and sagging conditions. These differences are noticeable both at and beyond the ultimate strength (in the post-ultimate strength phase). A double hull with bottom damage has a higher ultimate strength than a single hull. This suggests that even in cases where damage is sustained at the bottom, the inner hull strengthens the ship. The inner hull causes this effect by shifting the neutral axis's location, which impacts the cross-sectional modulus and inertia. However, it should be noted that the shape of the modeled damage is symmetrical and is assumed to be located amidships.

4. Conclusion

This study analyzes the ultimate strength of bulk carriers with both single and double hull designs using an analytical approach. It concludes that a double hulled bulk carrier with an inner hull has a greater ultimate strength than a single hulled bulk carrier. Furthermore, the final strength under hogging and sagging situations is greatly influenced by the kind of hull construction double or single. The rules for recommending ship hull design and construction are the focus of this study's contribution.

5. Acknowledgment

The authors express their gratitude to the Ministry of Research, Technology, and Higher Education of the Republic of Indonesia for their support through Hasanuddin University, under contract numbers 050/E5/PG.02.00.PL/2024 and 02035/UN4.22.2/PT.01.03/2024.

References

[1] Cui H, Chen Z, Hu R and Ding Q 2024 Ultimate strength assessment of hull girders considering elastic shakedown based on Smith's method *Ocean Eng.* **293** 116695
[2] Woloszyk K, Goerlandt F and Montewka J 2024 A methodology for ultimate strength assessment of ship hull girder accounting for enhanced corrosion degradation modeling *Mar. Struct.* **93** 103530
[3] Shi G jie and Gao D wei 2021 Model experiment of large superstructures' influence on hull girder ultimate

strength for cruise ships *Ocean Eng.* **222** 108626

[4] Muis Alie, M. Z, Sitepu G, and Latumahin S I 2018 The Assessment of the Ultimate Hull Girder Strength of RO-RO Ship after Damages *IOP Conf. Ser. Earth Environ. Sci.* **135** 880653

[5] Quispe J P, Estefen S F, Lourenço de Souza M I, Chujutalli J H, Amante D do A M and Gurova T 2022 Numerical and experimental analyses of ultimate longitudinal strength of a small-scale hull box girder *Mar. Struct.* **85**

[6] Jagite G, Bigot F, Malenica S, Derbanne Q, Le Sourne H and Cartraud P 2022 Dynamic ultimate strength of an ultra-large container ship subjected to realistic loading scenarios *Mar. Struct.* **84** 103197

[7] Deng H, Yuan T, Gan J, Liu B and Wu W 2022 Experimental and numerical investigations on the collapse behavior of box type hull girder subjected to cyclic ultimate bending moment *Thin-Walled Struct.* **175** 109204

[8] Ma H, Wang Q, and Wang D 2022 Scaling characteristics of the hull girder's ultimate strength subjected to the combined hogging moment and bottom lateral pressure: An empirically modified scaling criterion *Ocean Eng.* **257** 111520

[9] Chen B Q, Liu B and Guedes Soares C 2022 Experimental and numerical investigation on a double hull structure subject to collision *Ocean Eng.* **256** 111437

[10] Zhao N, Chen B Q, Zhou Y Q, Li Z J, Hu J J and Guedes Soares C 2022 Experimental and numerical investigation on the ultimate strength of a ship hull girder model with deck openings *Mar. Struct.* **83** 103175

[11] Liu B, Villavicencio R, Pedersen P T and Guedes Soares C 2021 Analysis of structural crashworthiness of double-hull ships in collision and grounding *Mar. Struct.* **76** 102898

[12] Shi G jie and Gao D wei 2021 Model experiment of large superstructures' influence on hull girder ultimate strength for cruise ships *Ocean Eng.* **222** 108626

[13] Zhang Y, Guo J, Xu J, Li S and Yang J 2021 Study on the inequivalence between stiffness loss and strength loss of damaged hull girder *Ocean Eng.* **229** 108986

[14] Wang Q, Wang C, Wu J and Wang D 2020 Experimental and numerical investigations of the ultimate torsional strength of an ultra-large container ship *Mar. Struct.* **70** 102695

[15] Kuznecovs A, Ringsberg J W, Johnson E and Yamada Y 2020 Ultimate limit state analysis of a double-hull tanker subjected to biaxial bending in intact and collision-damaged conditions *Ocean Eng.* **209** 107519

[16] Yao, T. and Nikolov, P.I. (1992). "Progressive Collapse Analysis of a Ship's Hull Girder under Longitudinal Bending (2nd Report)," J Soc. Naval Arch. of Japan, 172, 437-446.

IOP Conf. Series: Earth and Environmental Science **1461** (2025) 012002 doi:10.1088/1755-1315/1461/1/012002

Simulation on deck side corner and its effect to the ultimate strength

Muhammad Zubair Muis Alie[1]*, Juswan[1], Nur Salsabila Aulia[1] and Nur Indah[1]

[1] Department of Ocean Engineering, Hasanuddin University, Makassar, Indonesia

*E-mail: zubair.m@eng.unhas.ac.id

Abstract. The deck is one of the ship constructions located above the draft. The deck structure must be designed appropriately, especially at the corner side. This is caused by the impact of water flowing on the deck. As a result of this water flowing, the side corner of the deck must be constructed so that water does not pool on the deck. The aim of this study is to examine how the shape of the deck-side corner affects the ultimate strength of a double-hull ship. Three models of double-hull boats are considered to be investigated with different shapes of deck side corners. The ship's dimensions are 42 meters in width and 20.3 meters in depth. The longitudinal direction with respect to length is considered one frame space of 5 m. The three models ' breadth, depth, and length dimensions are constant. The material properties, such as density, yield strength, elastic modulus, and Poisson's ratio, are assumed to remain constant. The analysis focuses on the cross-section of the double-hull ship. In this study, the influence of the deck side corner shape on ultimate strength is investigated using the Multiple Point Constraint (MPC) method. The position of the neutral axis is determined in advance to allow for the placement of the MPC. The MPC is applied to one side of the cross-section, while the opposite side is fixed securely. A four-node quadrilateral shell element is used in the Finite Element Method to model the entire cross-section of the double-hull ship. The analysis shows that the first model of the double hull ship has the highest ultimate strength, followed by the second model, with the third model having the lowest strength.

1. Introduction

The deck side corner shape of the ship is made for special purposes because water spills onto the deck will run over the deck side of the vessel. The design of the deck side corner shape should take into account the water flow toward this area and how it impacts the overall structural strength. This study has not explored the impact of the deck-side corner shape on the ultimate strength of ship construction. Hence, it is essential to evaluate the effect of the deck side corner on the maximum strength.

1.1 State of the art

Studies on the ultimate limit state of ship hulls have been ongoing for several years. Woloszyk et al. [1] introduced a method for assessing the ultimate strength of a ship hull, which includes advanced corrosion modeling.

Li and Kim [2] created the R–D diagram by applying a Smith-type progressive collapse method to four different types of double-hull oil tankers: VLCC, Suezmax, Aframax, and Panamax. Additionally, Li et al. [3] also introduced a probabilistic method to evaluate the uncertainty in predicting the hull girder's ultimate strength, which arises from key characteristics within the LSCs. Liu et al. [4] provided experimental data for scaled double-hull specimens exposed to locally applied loads. Shi and Gao [5] performed a collapse test on a steel model with superstructures that were designed to replicate those of an actual cruise ship. Babazadeh and Khedmati [6] expanded on previous studies to investigate the impact of cracks on the ultimate longitudinal strength of ship hull girders. They performed a progressive collapse analysis on an ISSC2000 bulk carrier to evaluate its ultimate strength, considering cracking damage at different locations on the hull girder, such as the bottom, double bottom, side, and deck structures.

Muis Alie et al. [7] assessed the ultimate longitudinal strength of the Ro-Ro ship by taking into account the influence of the superstructure. Liu and Soares [8] demonstrated that, under cyclic loading, the ship hull girder may have a lower ultimate strength than when subjected to monotonically increasing bending moments, which is the common assumption in conventional analyses. Wang and Wang [9] performed modeling and analysis on various real ship hull girders from a 10,000 TEU container vessel, utilizing different combinations of geometric length and plate thickness scales. Their research sought to investigate the scaling properties of hull girders' ultimate strength and collapse behavior, aiming to create a more precise scaling criterion for comparing the ultimate strength of a scale model with that of the full-scale ship. Muis Alie et al. [10] assessed the ultimate hull girder strength of the Ro-Ro ship post-damage, concentrating on the ship's cross-sectional structure. Furthermore, Muis Alie and Latumahina [11] examined the Ro-Ro ship by considering the local components that influence its ultimate strength and progressive collapse. Wang and Wang [12] investigated the connection between the ultimate strength properties of the compartment hull girder and the overall hull girder. They focused on the ultimate strength envelope of the global hull girder for a 10,000 TEU ultra-large container ship subjected to combined loading conditions, emphasizing its importance as a crucial factor for evaluating the ship's safety margin.

Xia et al. [13] examined the residual compressive ultimate strength characteristics of ship hull plates under cyclic loading. Zhang et al. [14] proposed a novel approach for calculating strength loss by assessing stiffness reduction, using the section modulus as the strength indicator. Their results revealed that the reduction in stiffness of a hull girder following damage does not necessarily align with an equivalent decrease in strength. The theory explaining the difference between stiffness and strength loss (UNSS) was confirmed through an analysis of 13 typical cross-sectional shapes. Wang and Wang [15] illustrated the torsional collapse behavior of a single-compartment hull girder, which could act as a scaled representation of the mid-ship compartment of a container vessel, through both experimental and numerical analysis. Ma et al. [16] conducted an in-depth study on the impact of bottom lateral pressure on scaled hull girders subjected to combined loads, analyzing various hull girders from a 10,000 TEU container ship with varying geometric lengths and plate thicknesses. Previous studies have highlighted the lack of research on the impact of the deck side corner shape on ultimate strength. Therefore, this study focuses on exploring this influence and its consequences for structural stability.

IOP Conf. Series: Earth and Environmental Science **1461** (2025) 012002 doi:10.1088/1755-1315/1461/1/012002

2. Research Strategy

2.1 Ship dimension and model

This study examines the impact of the deck side corner shape of a double hull tanker under hogging and sagging conditions, using numerical methods. Three different double hull tanker models are analyzed, emphasizing the cross-sectional analysis of the tanker.

The double hull tanker has a length of 42 meters and a breadth and depth of 20.3 meters. In the longitudinal direction, a frame space of 5 meters is considered relative to the length. Dimensions of breadth, depth and length are constant for three models of double hull tankers. Member dimensions like plates and stiffened plates and the other construction elements both longitudinal and transversal direction are similar for three models including their member layout. The three models of double hull tankers are shown in Figures 1-3. Model-1 is the existing condition. While Model-2 and Model-3 are modified on the deck side corner shape. It is mentioned before that member layout, member dimensions and others elements like plate and stiffened plates are constant for three models. According to Figures 2 and 3 for Model-2 and Model-3 that the modification performed at the deck side corner shape. Model-2 like Model-1 (existing condition) where the inner hull at deck corner having the slope to the perpendicular direction, however the shear strake has not circle as shown in Figure 2. For the case of Model-3, the inner hull at the shear strake directly perpendicular to the end of main deck, while outer hull has circle like bilge at the bottom part. The modifications both two models are indicated by the marked with dotted circle. The modification of the deck side corner shape of the ship are symmetrically modelled on the right and left side. The material properties, such as density, Young's modulus, yield strength, and Poisson's ratio, are assumed to be constant (refer to Table 1). Other element construction like longitudinal bulkhead at the center line including their stiffened plates elements, center and side girders are similar and symmetrical placed at right and left at the bottom part as shown in Figures 1-3, respectively. Type and number of the stiffened plates attached on the cross section of the ship are also constant, but the dimensions are different like flange and web thickness.

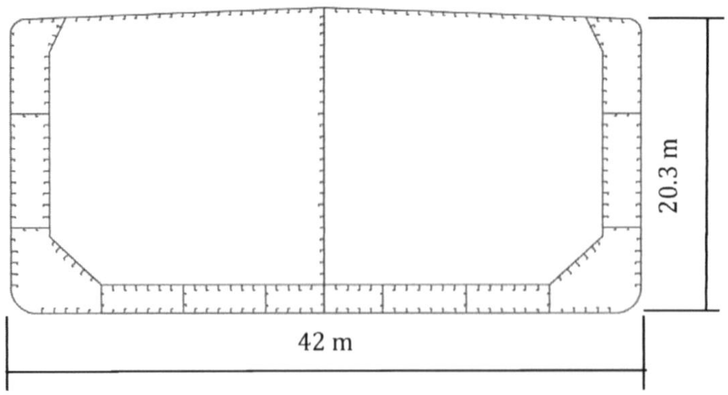

Figure 1. Model-1 (Existing condition)

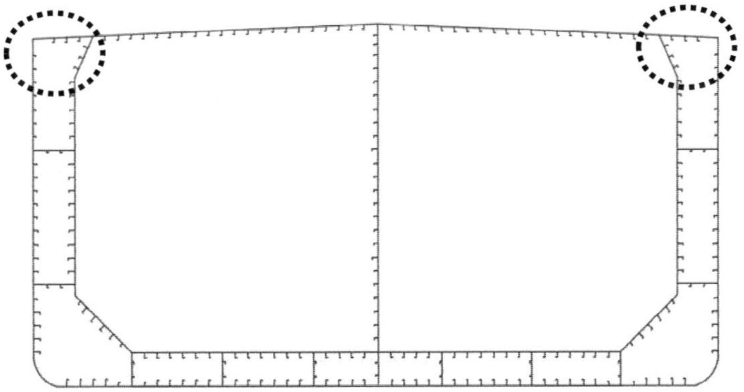

Figure 2. Model-2 (Modified of deck side corner shape)

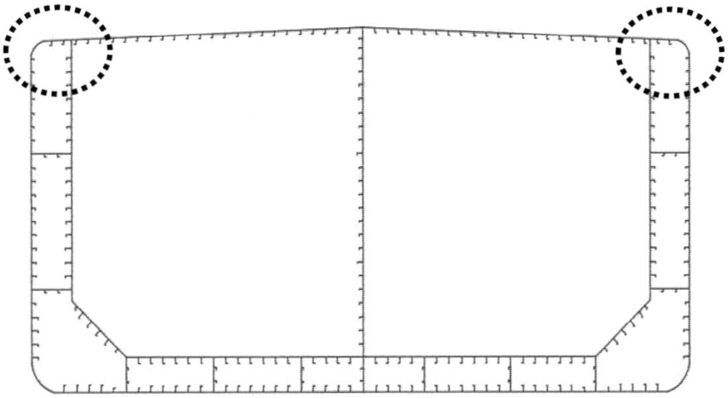

Figure 3. Model-3 (Modified of deck side corner shape)

Table 1. Properties of the material

Items	Value	Unit
Mass per unit volume	7850	kg/m^3
Elastic modulus	210000	N/mm^2
Poisson Ratio	0.3	-
Yield Strength	370	N/m^2

2.2 Force and constrained condition

This study investigates the ultimate strength by considering the effect of deck side corner shape through the application of the Multiple Point Constraint (MPC). The neutral axis is established beforehand to precisely place the MPC, which is applied to one side of the cross-section, with the opposite side being rigidly fixed (Figure 4). The plate components are modeled using four-node quadrilateral shell elements in the Finite Element Method, while the stiffeners are modeled with two-node beam elements. The shell element type is consistently used across the entire cross-section of the models. The material is considered homogeneous and isotropic, with no account

IOP Conf. Series: Earth and Environmental Science **1461** (2025) 012002 doi:10.1088/1755-1315/1461/1/012002

taken for strain-hardening effects, cracks, or damage in this simplified case. A turning force (moment) is exerted on the MPC. The ultimate strength of three double-hull tanker models is analyzed in their intact state, considering both hogging and sagging conditions.

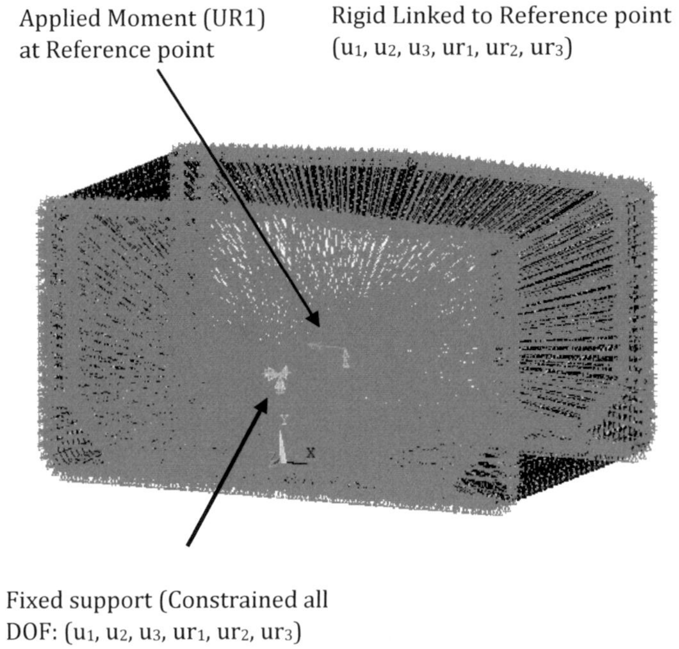

Figure 4. Force and constrained condition

3. Result and Discussion

3.1 Deformation

This study examines the ultimate strength of a double-hull oil tanker under both hogging and sagging conditions. It is well established that during hogging, the hull girder's deck is under tension, while the bottom undergoes compression.

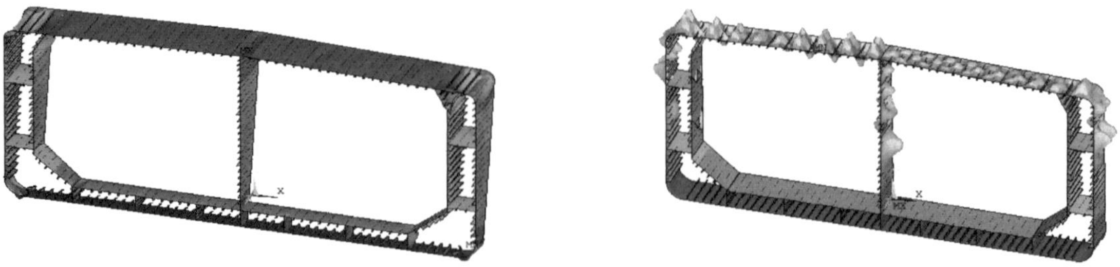

(a). Hogging **(b).** Sagging

Figure 5. Deformation pattern of Model-1 at ultimate strength

IOP Conf. Series: Earth and Environmental Science **1461** (2025) 012002 doi:10.1088/1755-1315/1461/1/012002

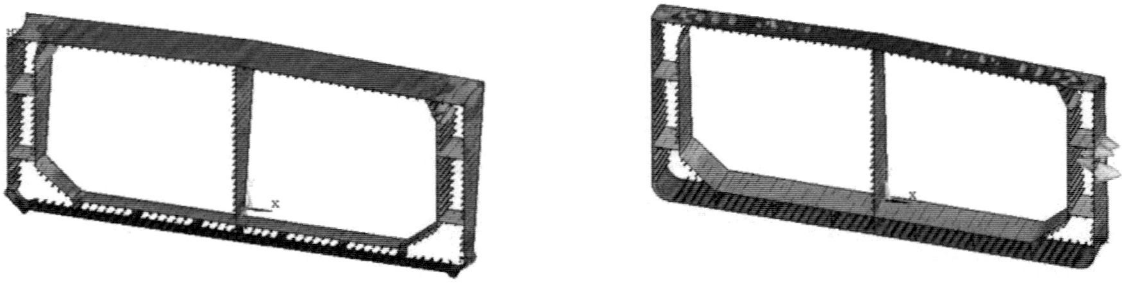

(a). Hogging **(b).** Sagging

Figure 6. Deformation pattern of Model-2 at ultimate strength.

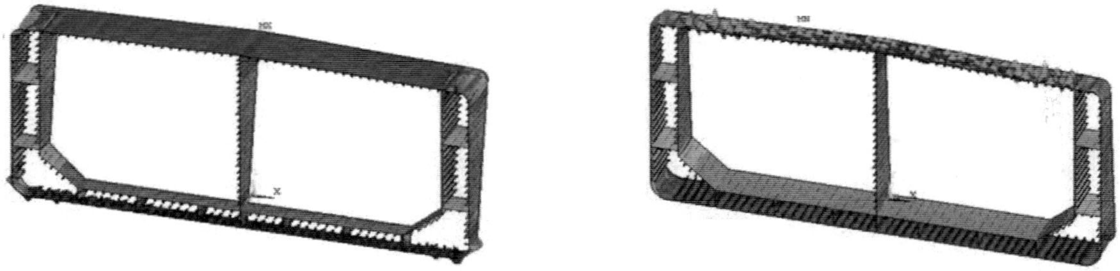

(a). Hogging **(b).** Sagging

Figure 7. Deformation pattern of Model-3 at ultimate strength.

Figures 5-7 display the deformation at the ultimate strength stage for three double-hull tanker models under both hogging and sagging conditions. The blue and red colors represent that yielding and buckling occur at the deck and bottom part or vice versa. In addition, Model-1 and Model-3 are almost identic for the deformation in the deck area because those two models have a circle radius on the side corner shape. On the other hand, Model-2 doesn't have a circle; it is a straight line vertical on the side corner shape. Therefore, the deformation shapes of the three models are completely different from one another.

3.2 Limit State

This study examines the ultimate strength and demonstrates it through the moment-curvature relationship for three different models of double-hull tankers. It compares the effect of the deck corner side shape across these three models to understand its influence. Furthermore, the ultimate strength, depicted through the moment-curvature relationship, is evaluated and compared across the three double-hull tanker models. The solid line represents Model 1, while the dashed and dotted lines are used for Model 2 and Model 3, respectively. As illustrated in Figure 8, Model-1 demonstrates the greatest ultimate strength, with Model-2 (Figure 9) and Model-3 (Figure 10) following in order, In both hogging and sagging states. The influence of the deck side

corner shape for the three models gives significant differences to the ultimate strength. The effects, such as the circle radius and inner hull on the deck side, also contribute to the differences in deformation and ultimate strength. The simulation of the deck side corner shape to the three models indicates that it should be considered in the design and construction of the ship. The contribution of this study can be a basis for ship design, construction, and regulation.
Moreover, the hogging condition is where there is no change in the double bottom construction, so the deformation is not large compared to the sagging condition. These three models provide information on the failure modes caused by deck corner shape construction changes. The shape of the corner plate, whether circular or straight, affects the ultimate strength.

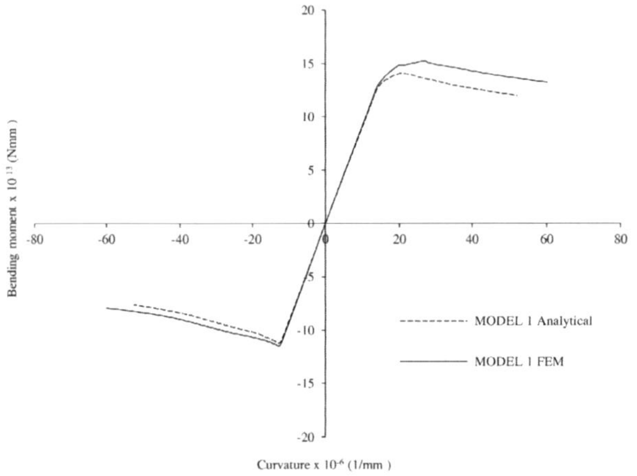

Figure 8. The comparison of the moment-curvature behavior of Model-1

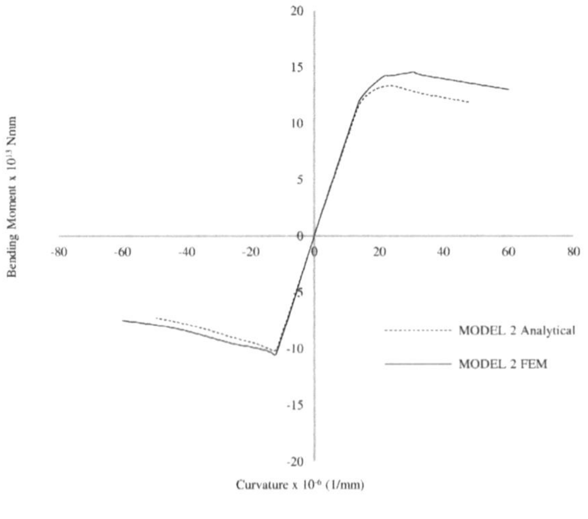

Figure 9. The comparison of the moment-curvature behavior of Model-2

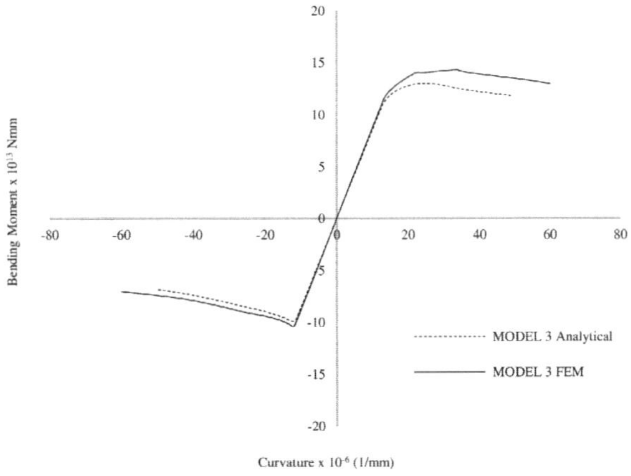

Figure 10. The comparison of the moment-curvature behavior of Model-3

Figures 8, 9, and 10 present a comparison of the bending moment and curvature between the analytical results and the FEM results for Model-1, Model-2, and Model-3. According to those figures, the ultimate strength is larger in hogging conditions compared to sagging for the three models. It is found that the ultimate strength is greater using the FEM method compared to the analytical method under hogging conditions for the three models. While in sagging conditions, the difference is very small, the ultimate strength using the FEM method is still greater than the analytical method.

Table 2. Ultimate Strength

Items	$M_{UH} \times 10^{13}$ (Nmm)	$M_{US} \times 10^{13}$ (Nmm)
Model-1	1.39	-1.16
Model-2	1.28	-0.87
Model-3	1.24	-0.79

Table 2 shows the ultimate bending moment capacity of three double-hull tanker models under both hogging and sagging conditions, as calculated using numerical methods. It is noted that Model-1 (the current condition) exhibits the greatest ultimate bending moment capacity, with Model-2 and Model-3 following in sequence. Despite only minor modifications to the deck side corner shape, these changes significantly impacted the ultimate strength. Therefore, the findings can serve as valuable recommendations for ship construction.

4. Concluding Remarks

The ultimate strength of double-hull ships for three models obtained by the numerical method has been investigated in this study. Double hull tanker is the target, which is simulated with the three models, namely Model-1, Model-2, and Model-3. It is found that Model-1 is the highest of the ultimate strength, followed by Model-2 and Model-3. The influence of the double hull Tanker's deck side corner shape significantly impacts the ultimate strength under longitudinal bending.

The contribution of this study addresses the regulations for the recommendation of ship hull design and construction.

5. Acknowledgment

The authors wish to express their gratitude for the support given by the Engineering Faculty at Hasanuddin University through the Internal Research Grant of Laboratory-Based Education (LBE), under contract number 12684/UN4.7.2/PM.01.01/2024.

References

[1] Woloszyk K, Goerlandt F and Montewka J 2024 A methodology for ultimate strength assessment of ship hull girder accounting for enhanced corrosion degradation modelling *Mar. Struct.* **93** 103530

[2] Li S and Kim D K 2022 A comparison of numerical methods for damage index based residual ultimate limit state assessment of grounded ship hulls *Thin-Walled Struct.* **172** 108854

[3] Li S, Kim D K and Benson S 2021 A probabilistic approach to assess the computational uncertainty of ultimate strength of hull girders *Reliab. Eng. Syst. Saf.* **213** 107688

[4] Liu B, Wang G, Chen B and Guedes Soares C 2021 Experimental numerical and analytical analysis of the penetration of a scaled double-hull tanker side structure *Mar. Struct.* **78** 103018

[5] Shi G jie and Gao D wei 2021 Model experiment of large superstructures' influence on hull girder ultimate strength for cruise ships *Ocean Eng.* **222** 108626

[6] Babazadeh A and Khedmati M R 2021 Progressive collapse analysis of a bulk carrier hull girder under longitudinal vertical bending moment considering cracking damage *Ocean Eng.* **242** 110140

[7] Muis Alie, M. Z, Sitepu G., Wahyuddin J, Nugraha A M and Alamsyah A. 2016 The influence of Superstructure on the longitudinal ultimate strength of an RO-RO ship *Proc. Int. Offshore Polar Eng. Conf.* **2016-January** 1022–9

[8] Liu B and Guedes Soares C 2020 Ultimate strength assessment of ship hull structures subjected to cyclic bending moments *Ocean Eng.* **215** 107685

[9] Wang Q and Wang D 2020 Scaling characteristics of hull girder's ultimate strength and failure behaviors: An empirically modified scaling criterion *Ocean Eng.* **212** 107595

[10] Muis Alie, M. Z, Sitepu G, and Latumahin S I 2018 The Assessment of the Ultimate Hull Girder Strength of RO-RO Ship after Damages *IOP Conf. Ser. Earth Environ. Sci.* **135** 880653

[11] Muis Alie, M. Z, and Latumahina S I 2019 Progressive collapse analysis of the local elements and ultimate strength of a Ro-Ro Ship *Int. J. Technol.* **10** 1065–74

[12] Wang Q and Wang D 2020 Ultimate strength envelope of a 10,000TEU large container ship subjected to combined loads: From compartment model to global hull girder *Ocean Eng.* **213** 107767

[13] Xia T, Yang P, Li C and Hu K 2019 Numerical research on residual ultimate strength of ship hull plates under uniaxial cyclic loads *Ocean Eng.* **172** 385–95

[14] Zhang Y, Guo J, Xu J, Li S and Yang J 2021 Study on the unequivalence between stiffness loss and strength loss of damaged hull girder *Ocean Eng.* **229** 108986

[15] Wang Q and Wang D 2020 Warping and shear behaviors involved in hull girder's torsional collapse process *Thin-Walled Struct.* **157** 107123

[16] Ma H, Wang Q and Wang D 2022 Scaling characteristics of the hull girder's ultimate strength subjected to the combined hogging moment and bottom lateral pressure: An empirically modified scaling criterion *Ocean Eng.* **257** 111520

IOP Conf. Series: Earth and Environmental Science **1461** (2025) 012003 doi:10.1088/1755-1315/1461/1/012003

Impact of Fillet Geometry on Flow Patterns in Seawater Cooling System T-Type Tee Junctions of Multi-Purpose Support Vessel

Roslin Ramli[1*], Muhammad Aysar Roslan[1], Iwan Zamil Mustaffa Kamal[1], Noorlee Boonadir[2] and Mohd Hisbany Mohd Hashim[3]

[1] Maritime Engineering Technology, Universiti Kuala Lumpur Malaysian Institute of Marine Engineering Technology, Lumut, Perak, Malaysia
[2] Maritime Management, Universiti Kuala Lumpur Malaysian Institute of Marine Engineering Technology, Lumut, Perak, Malaysia
[3] School of Civil Engineering, Universiti Teknologi MARA, Selangor, Malaysia

*roslin@unikl.edu.my

Abstract. This research investigates the effect of fillet geometry on the flow pattern in a T-type tee junction of a seawater cooling system. Emphasising the rarity of research on converging flow in seawater cooling systems, the study addresses cavitation corrosion tendencies and damaging effects. This research evaluates velocity, pressure, and turbulence energy parameters using Pipe Flow simulation software and SolidWorks software. Simulations in three different fillet radii (0 mm, 20 mm, and 30 mm) were tested to simulate the flow behaviour in T-type tees commonly used in multipurpose support vessels (MPSVs). The results show fillet geometry significantly affects flow separation, recirculation zone, and pressure drop characteristics in T-type tee junctions. Larger fillet radii were found to reduce flow behaviour, pressure loss, and turbulent energy, leading to more efficient seawater cooling system performance.

1. Introduction

In pipeline systems, hydrodynamic energy losses are an inherent challenge, primarily arising from friction along the pipe walls and the geometrical complexities of local resistance. Research indicates that hydraulic losses in transmission and distribution systems are predominantly linked to resistance at junctions, particularly in T-type tees. Which are critical components in seawater cooling systems [1]. Historically, the design of these piping systems has concentrated on fluid transport, often overlooking the significance of tee-fitting geometry.

However, with advancements in computational fluid dynamics (CFD) and engineering design methodologies, the impact of tee geometry on system performance has gained increasing attention. The overall coefficients can be determined through experimentation or by conducting fluid dynamics (CFD) simulation [2].

The primary objective of this research is to gain a thorough understanding of different arrangements of ties impact on the flow fluid, distribution of pressure, and energy dissipation in the specific setting of seater cooling. Through the utilisation of advanced simulation techniques, such as flow analysis utilising Pipe Flow software and the SOLIDWORK software method, which is an intuitive computational fluid dynamics (CFD) solution embedded within the SOLIDWORKS 3D CAD method [3].

1.1 Existing T-type Tee Junctions

There are three parts to existing methods, such as curvature distribution methods, measurement techniques, and flow simulation models. For the curvature distribution techniques, the traditional often rely on curvature distribution to analyse flow patterns around the junction. However, these approaches can be sensitive to noise in the data, leading to inaccuracies in detecting features like fillets[4].

The measurement technique, which used common measurement tools, such as coordinate measuring machines (CMMs), faces challenges when capturing the intricate geometries of filled edges [5]. These instruments typically operate through point contact, which can introduce errors, particularly in complex shapes. This limitation can hinder the assessment of flow characteristics and lead to suboptimal design evaluations.

Hence, the flow simulation models with many existing computational fluid dynamics (CFD) models may not adequately account for the effects of fillet geometry on flow patterns [6], [7]. These models often simplify the geometrical representation, which can overlook critical interactions between fluid flow and the junction's physical features. Consequently, this can lead to miscalculations in parameters such as pressure drop and turbulence intensity.

Incorporating fillet geometry can address these weaknesses by improving the curvature distribution; enhancing the capability enables precise assessment of flow characteristics around the filled edges. This is the quantitative benefit of fillet geometry, as it reduces pressure losses and enhances flow uniformity. Accordingly, the added fillet geometry not only improves flow patterns but also contributes to more reliable performance metrics in seawater cooling systems.

1.2 System Component Loss

Strategic arrangements of valves, pumps and turbines are vital for mitigating pressure losses throughout the network, underscoring the necessity of this research. This research aims to address the knowledge gap surrounding the optimisation of the tee geometry in seawater cooling piping systems. By emphasising the optimisation of fit geometry in T-type tees, thus, it reduces the flow behaviour, pressure loss, and turbulent energy. This will lead to more efficient seawater cooling system performance.

Besides, studies on optimisation of the tee geometry in seawater helps developing a comprehensive understanding of how various tie configurations influence fluid dynamics, pressure distribution and energy losses in seawater cooling contexts. Despite the critical nature of energy dissipation within the tees, this area remains largely uncharted due to the multitude of variables involved, including tee geometry, seawater flow patterns, wall roughness, and operational conditions[8].

The anticipated outcomes of this research are expected to make a substantial contribution to the field of marine engineering, providing insights that can inform the design and operation of seawater cooling system[9]. Additionally, the finding will enhance understanding of the specific resistance characteristics exhibited by T-type tee, which could positively influence the design and operational efficiency of the liquid network across various application such as the efficiency of liquid networks across various applications, such as drainage, irrigation, and building water supply.

IOP Conf. Series: Earth and Environmental Science **1461** (2025) 012003 doi:10.1088/1755-1315/1461/1/012003

In addressing contemporary engineering challenges, it is important to highlight the approach to tee geometry optimisation by integrating the development of robust piping systems. Consequently, this study advances the field towards more efficient and sustainable solutions for the challenges.

2. Methodology

This methodology section details the systematic approach adopted for the development of the seawater piping system. It encompasses critical aspects such as system design, material selection, pipe routing, fitting and valve selection and pipe analysis. By adhering to a rigorous methodology, this project aims to create an efficient and reliable piping system that fulfils the stringent requirements of the maritime industry.

2.1 Critical Area

The design of the existing seawater cooling system involves the use of pipe flow software to identify critical areas. The process of identifying critical areas in a seawater piping system entails the detection of locations where is a notable decrease in pressure or the potential for disruption in flow. The examination begins by conducting a thorough examination of the comprehensive piping layout, precisely identifying all bends, fittings, tees, valves, and other associated components that have the potential to cause pressure loss or disrupt the smooth flow of fluids as shown in Figure 1. Each of the components in the system is assigned a pressure loss coefficient (K value) or equivalent length, which affects the overall head loss of the system. The critical part considered for analysis is the discharge from tee fittings, specifically focusing on velocities exceeding 3 m/s as shown in Figure 2.

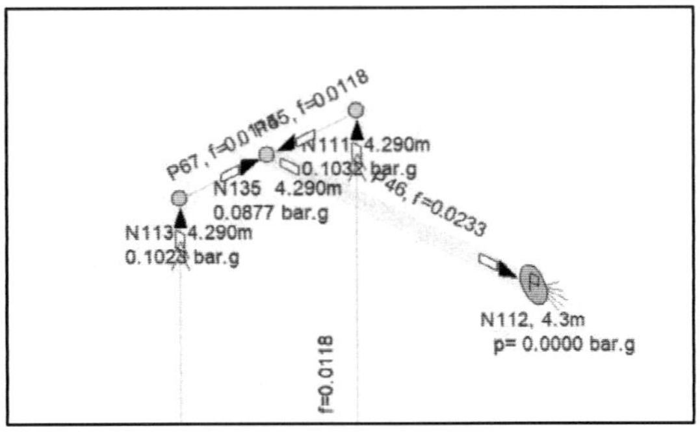

Figure 1. Critical area from Pipe Flow Software

IOP Conf. Series: Earth and Environmental Science **1461** (2025) 012003 doi:10.1088/1755-1315/1461/1/012003

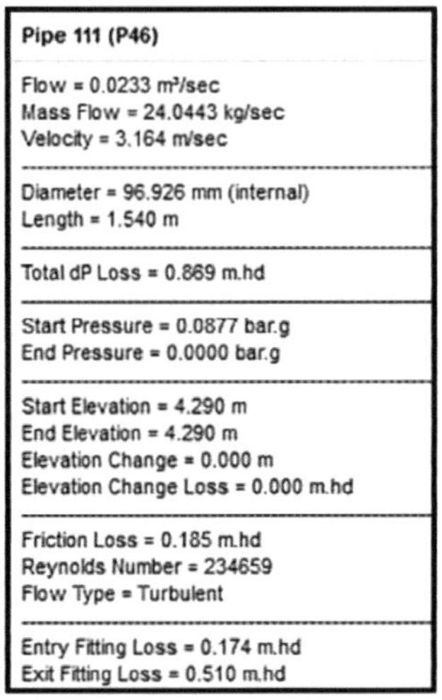
Pipe 111 (P46)

Flow = 0.0233 m^3/sec
Mass Flow = 24.0443 kg/sec
Velocity = 3.164 m/sec

Diameter = 96.926 mm (internal)
Length = 1.540 m

Total dP Loss = 0.869 m.hd

Start Pressure = 0.0877 bar.g
End Pressure = 0.0000 bar.g

Start Elevation = 4.290 m
End Elevation = 4.290 m
Elevation Change = 0.000 m
Elevation Change Loss = 0.000 m.hd

Friction Loss = 0.185 m.hd
Reynolds Number = 234659
Flow Type = Turbulent

Entry Fitting Loss = 0.174 m.hd
Exit Fitting Loss = 0.510 m.hd

Figure 2. Critical area Pipe 111 (P46)

2.2 Design T-type Tee

In SolidWorks, designing the T-type tee geometry entails creating a detailed three-dimensional model of a tee component, commonly used in piping systems, plumbing, and structural designs. SolidWorks, a powerful computer-aided (CAD) software, provides comprehenesive tools for designing, simulating, and manufacturing various mechanical components.

The design process begins with sketching the outline of the tee and defining the dimensions. Setting the fillet radius at 0.1 times the branch's outer diameter (0.1 douter) is a critical aspect of these designs. This parameter is essential for ensuring smooth flow transitions and reducing turbulence at the junctions as shown in Table 1.

Table 1. T-type tee parameters

Parameter	Value
Diameter of Pipe, mm	114
Length, mm	210
Fillet Radius, mm	0, 20, 30
Roughness, mm	0.15
Inlet Velocity, ms^{-1}	1:1
Material	Galvanized Steel ANSI
Outside Pressure	1:1
Reynolds Number	$(115000<Re<200000)$

Specific ratios can facilitate the design by incorporating other values, allowing for flexibility in the geometry while maintaining optimal performance. By leveraging SolidWorks' advanced features, engineers can effectively model the tee geometry, analyse flow patterns, and evaluate the impact of design choices on system efficiency. This methodology not only enhances the accuracy of the design but also contributes to the overall reliability and functionality of the piping system.

2.3 Simulation T-type tee

In SolidWork, these settings collectively determine the fineness or coarseness of the mesh in various regions of the simulation domain as shown in Figure 3. A well-defined mesh is essential for ensuring the accuracy of CFD simulations conducted in SolidWorks as a Table 2. The selection of the mesh settings should be based on the critical aspects of developing simulations. The aspects included arethe physics being simulated, the complexity of the geometry, and the computational resources available.

The wall boundary conditions, the inlet and outlet velocities are set to be equal, ensuring consistency in the simulation. The initial mesh is generated automatically after defining the refinement criteria for solid cells, allowing for an efficient and effective representation of the simulation domain. By optimising the mesh, the simulation can better capture the intricate flow characteristics and enhance the reliability of the results.

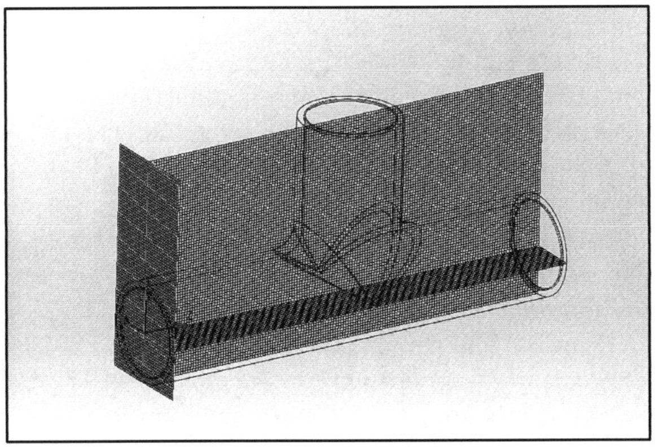

Figure 3. Design a T-type tee with a specific geometry configuration.

Table 2. T-type tee global mesh parameters

Global Mesh Setting	Setting Value
Basic Mesh	Nx=36, Ny=83, Nz=159
Refining Solid Cells	2
Refining Cells at Fluid/Solid	3
Small Solid Featured Refine	3
Tolerance Level	3
Tolerance Refinement Criterion	0.00072m

3. Results and discussion

The research focused on the geometry of T-type tee seawater cooling piping systems. Simulations included pressure distribution, turbulent kinetic energy, local resistance coefficients, and energy losses. A preliminary layout of the piping system was developed, and flow distribution patterns across different tie geometries were investigated through simulations. This included determining the local resistance coefficients and hydrodynamic pressure losses associated with various tee configurations.

The findings from this rigorous analysis reveal significant insights into the performance of the piping systems under various conditions. A comparative analysis was conducted specifically for the converging case, evaluating the effects of different inlet velocities alongside varying fillet radii of 0, 20, and 30. These results provide a deeper understanding of how geometric variations influence flow dynamics and system efficiency, offering valuable implications for the design and optimisation of seawater cooling systems.

The analysis of pressure distribution reveals significant differences between T-type tee junctions with varying fillet radii of 0, 20, and 30. The maximum pressure recorded for fillet radius 0 was 156,351 Pa, while for radii 20 and 30, the values were 137,375 Pa and 136,952 Pa, respectively as shown in Figure 4. The simulations indicate that the presence of a fillet radius results in a smoother pressure gradient around the tee's curve. This gradual change in fluid direction reduces high-pressure gradients and stress concentrations, thereby minimising the likelihood of flow separation, a common contributor to increased pressure drop and turbulence. In contrast, the configuration without a fillet exhibits sharper pressure gradients, particularly at the abrupt directional change, creating a high-pressure zone marked by intense red colouring. This sharp angle not only induces flow separation. It also leads to a low-pressure zone downstream., as indicated by the yellow to green gradient in pressure distribution.

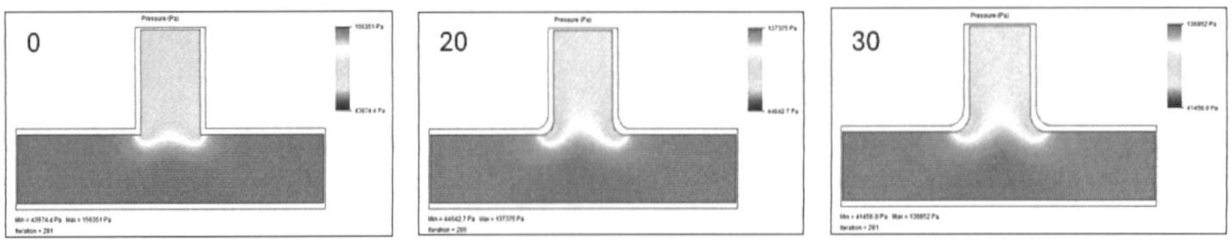

Figure 4. Pressure Distribution of radius 0, 20 and 30

The analysis of turbulent kinetic energy within the T-type tee junctions reveals significant differences between the fillet radii of 0, 20, and 30. The maximum turbulent kinetic energy recorded for fillet radius 0 was 6.89197 J/kg, while for radii 20 and 30, the values were 0.704417 J/kg and 0.549385 J/kg, respectively as shown in Figure 5. The colour contours indicate the intensity of turbulence kinetic energy within the fluid flow. Warmer colours representing higher energy levels. The simulation incorporating a fillet radius shows a more contained and lower intensity of turbulent energy, as evidenced by the predominance of cooler colours in the corresponding contour plot. The presence of a fillet radius typically promotes smoother flow transitions. Thereby, reducing the generation of turbulence. The smooth contour of the fillet helps maintain a more laminar flow. Thus, it dissipates less energy into turbulence. This substantial reduction in turbulent energy suggests a smoother flow with diminished energy losses due to turbulence when a fillet radius is present.

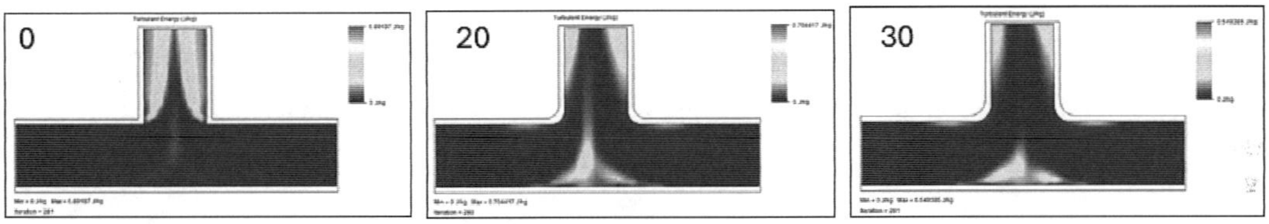

Figure 5. Turbulent Kinetic Energy of radius 0, 20 and 30

The local resistance coefficients, represented as ξ, were calculated for each T-type tee by assessing the pressure drop across the junctions at an inlet velocity of 5 m/s for fillet radii of 0, 20, and 30. These coefficients were derived by considering various factors, including the outer diameters of the tees (Dout), outlet pressures (P), and the specific designs of the junctions connecting the main line to the branches. Using these coefficients, the corresponding pressure (energy) losses, denoted as h, computed for each tee, with results also presented in Table 3. Additionally, it includes the pressure differential values at the tee outlets obtained through SolidWorks Computational Fluid Dynamics (CFD) simulations. This comprehensive evaluation highlights the relationship between tee geometry and local resistance, providing valuable data for optimising piping system designs.

Table 3. Local Coefficient and Energy Loss Data at 5m/s

Fillet	Pressure Difference	Local Coefficient	Energy Loss
0	2345	0.18	2.1051
20	1609	0.13	1.4444
30	786	0.06	0.7056

4. Conclusion

The research demonstrates that fillet geometry plays a crucial role in the flow dynamics of T-type tee junctions in seawater cooling systems. Explicitly, the results indicate that larger fillet radii (20 and 30) significantly reduce flow separation, recirculation zones, and pressure drop characteristics compared to a fillet radius of 0. This leads to smoother flow distribution, decreased turbulent kinetic energy, and improved overall efficiency. By employing computational fluid dynamics (CFD) simulations, the study established a relationship between local resistance coefficients and geometric parameters, providing a valuable equation for designers. These insights can guide engineers in optimising tee designs to minimise energy losses and enhance system performance.

References

[1] T. Liu, S. Li, C. Jiang, X. Zhang, and Z. Tan, "Local Resistance Characteristics of T-Type Tee Based on Chamfering Treatment," *Sustain.*, vol. 15, no. 19, 2023.
[2] J. Štigler, R. Klas, and O. Šperka, "Characteristics of the T-junction with the equal diameters of all branches for the variable angle of the adjacent branch," in *EPJ Web of Conferences*, 2014, vol. 67.
[3] SolidWorks, "An Introduction to Flow Analysis Applications with SolidWorks Flow Simulation," *Eng. Des. Technol. Ser.*, 2010.
[4] E. P. Application, "* EP004345673A1 *," vol. 1, no. 19, pp. 1–35, 2024.
[5] P. Saunders, A. Wilson, N. Orchard, N. Tatman, and P. Maropoulos, "An exploration into measurement consistency on coordinate measuring machines," *Procedia CIRP*, vol. 25, no. C, pp. 19–26, 2014.
[6] C. Dianita, R. Piemjaiswang, and B. Chalermsinsuwan, "CFD simulation and statistical experimental design analysis of core annular flow in T-junction and Y-junction for oil-water system," *Chem. Eng. Res. Des.*, vol. 176, 2021.
[7] O. Chinedum Vincent, "Analysis of Water Flow in Tee-junction Pipes Using CFD," 2020.
[8] Y. Doroshenko and I. Rybitskyi, "Investigation of the influence of the gas pipeline tee geometry on hydraulic energy loss of gas pipeline systems," *Eastern-European J. Enterp. Technol.*, vol. 1, no. 8–103, 2020.
[9] M. Chaal, A. Bahootoroody, S. Basnet, O. A. Valdez Banda, and F. Goerlandt, "Towards system-theoretic risk assessment for future ships: A framework for selecting Risk Control Options," *Ocean Eng.*, vol. 259, 2022.

Concept development of new type anchoring system for a tanker

Chunsik Shim[1], Deokyeon Lee[2*], Daseul Jeong [2], Kangho Kim[2], Daeseong Jeonge[3], Kyuphil Choi[3], Kwangok Kim[3], Hyeonseok Lee[4]

[1] Department of Naval Architecture and Ocean Engineering, Mokpo National University, Mokpo, South Korea.
[2] Department of Naval Architecture and Ocean Engineering, Graduate School Mokpo National University, Mokpo, South Korea.
[3] DAEHAN shipbuilding Co., Ltd, Haenam-gun, South Korea.
[4] Gwangu Jeonnam Regional Innovation Platform, South Korea.

*E-mail: dy_lee@mokpo.ac.kr

Abstract. Ships and offshore structures use anchors for safe mooring or emergency maneuvers to stop the vessel's operation in critical situations. Among the anchoring types, vessels with large block coefficients (e.g., tankers and LNG carriers) use bell mouth types made of cast iron installed between the hull exterior and the anchor. However, since these are made from cast steel, operational issues frequently arise due to discrepancies between the planned dimensions and the actual dimensions compared to steel cutting and processing. To address these problems, this study has developed a new anchoring system that modifies the vessel's bow structure and does not require a bell mouth. Before applying it to actual vessels, a mock up test was conducted to examine the anchor's motion and identify critical structural points in the developed anchoring system. Future research will focus on evaluating the structural strength of the vessel's bow equipped with the new anchoring system using Korean Register's structural analysis program, See Trust Hull Scan. The newly developed type of anchoring system is expected to resolve operational issues in advance through 3D simulations and reduce costs by doing away with the need for continuous mock up testing.

IOP Conf. Series: Earth and Environmental Science **1461** (2025) 012004 doi:10.1088/1755-1315/1461/1/012004

1. Introduction

The anchor is used for various purposes, such as safely mooring the vessel or performing emergency maneuvers to stop the ship in emergency situations, as shown in Fig. 1. It is also used to protect the vessel during operations under adverse conditions.

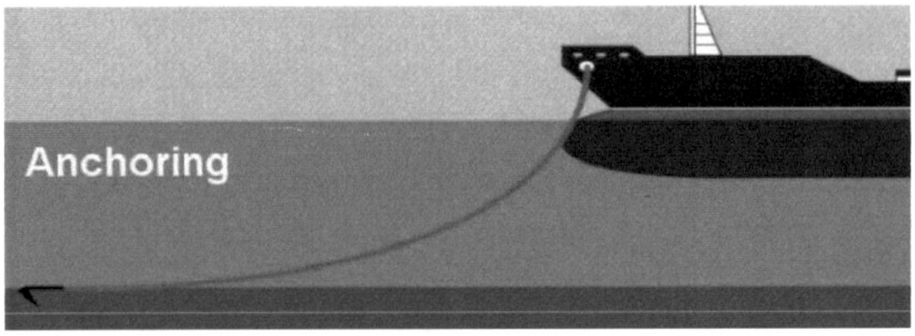

Figure 1. Use of anchor

The anchoring system is installed with one anchor on each of the port and starboard sides of the bow, as shown in Fig. 2. The system consists of the anchor, which secures the vessel at sea; the chain locker, where the anchor is stored; the hawse pipe, a passage through which the anchor chain moves up and down; the windlass, which transmits power to the anchor; the chain compressor, which guides and secures the chain; and the dog bar, which locks the anchor to the seabed and secures it to the vessel.

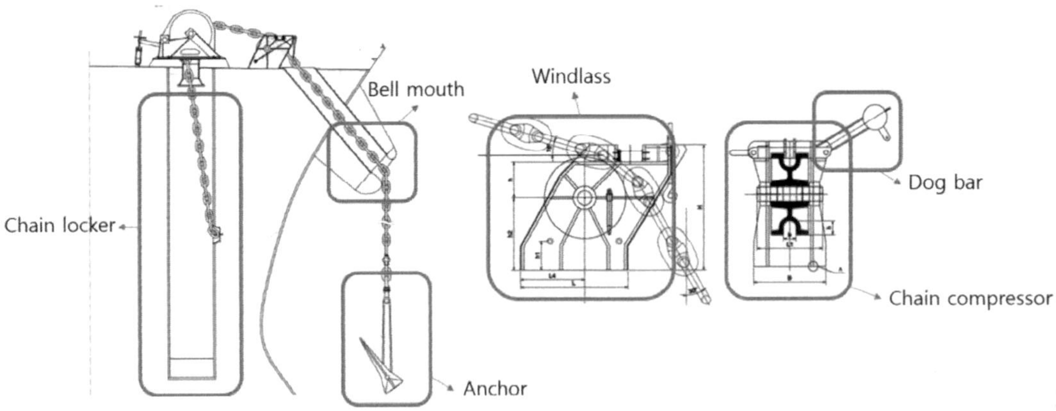

Figure 2. Anchoring system components

The type of anchoring system varies depending on the type of ship and the shape of the bow area. As shown in Fig. 3, it is classified into pocket and bell mouth anchoring types. This study focuses on the bell mouth anchoring system used in tanker vessels.

Figure 3. Ship anchoring system

The bell mouth is made of cast steel, and due to the nature of being manufactured from molds, operational issues arise from dimensional deviations compared to steel cutting and machining processes. Additionally, the complex shape of cast steel makes 3D modeling difficult, limiting the ability to accurately identify potential problems during the design and manufacturing stages. In particular, separate mock up tests are conducted to check issues such as the movement of the anchor, ensuring the anchor fluke achieves 3-point contact within the bell mouth. However, these tests require considerable time and cost.

Fig. 4 illustrates problems encountered during the use of the bell mouth. These issues include the anchor becoming stuck in the bell mouth, the fluke failing to make 3-point contact with the bell mouth, and the anchor shank failing to align with the groove in the bell mouth. To address these issues, workers must perform high-altitude tasks on the ship's outer shell, leading to delays in the process, increased labor, and reduced profitability.

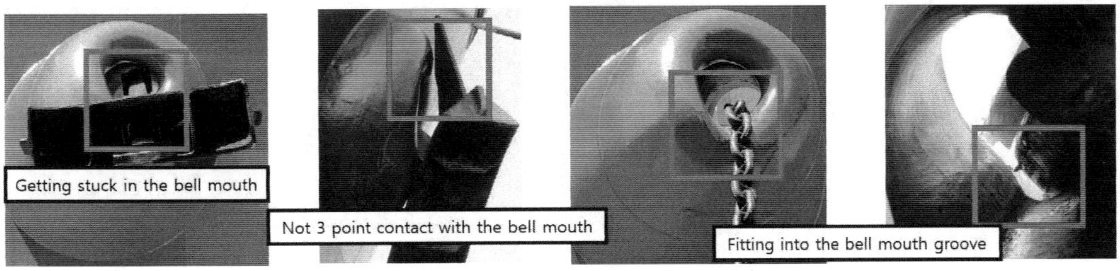

Figure 4. The problems encountered when using the Bell Mouth

Kang et al. (2013) conducted a study on the modeling and design issues of the windlass, as shown in Fig. 5, to address problems related to the active operation of the windlass that transmits power in the anchoring system. They formalized the effects of external forces that may occur on the windlass in constrained environments and validated the modeling techniques of the windlass system and the effectiveness of the designed controller through experiments and simulations.

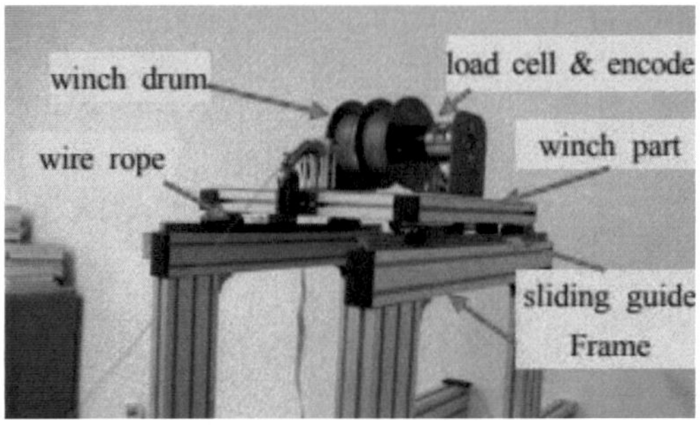

Figure 5. Single type pilot model of mooring winch system

Kang et al. (2020) argue that a guide bar should be installed to address issues related to the shape of the bell mouth in anchoring systems that store anchors. They analyzed the effects of installation conditions for the guide bar through mock up tests and later applied their findings to actual vessels, confirming that installing the guide bar can improve performance when there are issues with the bell mouth shape.

Kim et al. (2013) point out that the anchoring device has the disadvantage of the entire anchor chain being formed from a heavy steel material, which inevitably increases the size of the chain locker space, and that the load required for deploying the heavy anchor chain also increases. Accordingly, by replacing the existing anchor chain with an anchor wire and designing the structure to wind the anchor wire around the chain drum of the windlass for storage, a new anchoring device has been developed that reduces the chain load, eliminates the need for a chain locker, and dramatically decreases the storage space for the anchor wire.

W.N. Joo. (2013) stated that the cable clench functions to hold the end of the anchor chain. However, in the case of large vessels, the size and weight of the anchor chain are substantial, and consequently, the size of the cable clench must also increase, putting a significant load on the chain locker. To address this, a new cable clench has been developed that simplifies the structure while securely supporting the anchor chain.

Kim et al. (2021) developed a new anchoring system to replace the bell mouth, which is made of cast steel during anchoring. They found that using the bell mouth did not change the anchor's insertion angle, making precise mooring difficult.

Jeon (2017) stated that the phenomenon of the bell mouth getting caught between the anchor's shank and fluke requires stopping the vessel's operation to resolve the issue. To address this, an anchoring system was developed that can immediately resolve the issue when the anchor's shank and fluke get caught in the bell mouth.

In this study, a new type of anchoring system without the bell mouth was developed to address the issues of the existing anchoring system that uses the bell mouth, aiming to reduce process delays and improve profitability. Before applying it to actual vessels, mock-up tests were conducted to identify the motion of the anchor and the structural vulnerabilities of the developed anchoring system. In future research, the structural strength of the bow of the vessel with the new anchoring system applied will be evaluated according to the IACS Common Structural Rules and DNV, as shown in Fig. 6.

Chun et al. (2024) conducted a structural strength evaluation test on U-bolts used to support piping installed on ships or offshore structures to verify their safe working loads. Strain gauges were attached to the U-bolt fastening points to obtain load-displacement curves, and the validity of the experimental results was verified through numerical analysis. Based on these findings, a method for estimating the safe working load of U-bolts was proposed.

Moon et al. (2023) performed a structural analysis of a 38-foot HDPE powerboat by applying design loads specified by ISO and various classification societies. They analyzed and compared the validity of each design load.

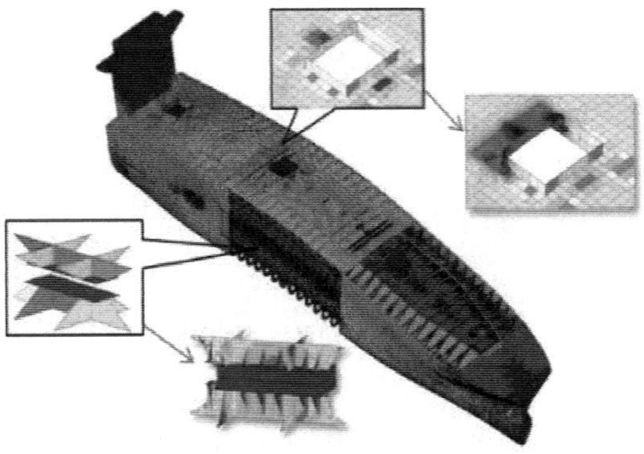

Figure 6. Example of hull structural stability assessment with the application of the new type anchoring system

Source : KR(Korean Register)

2. Anchoring system

Fig. 7 shows the flowchart of the structural strength evaluation process for the bow of a ship applying the new type anchoring system, and in this study, the process was conducted up to the ship structural model.

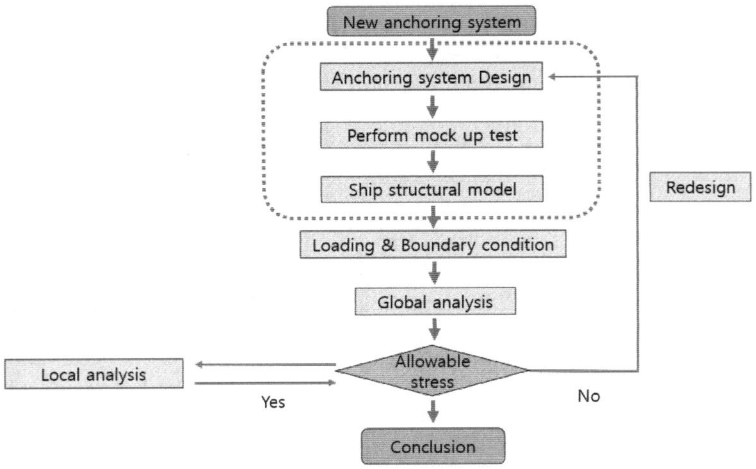

Figure 7. The procedure for the development of a new type anchoring system

The anchor allows a vessel to safely moor at any location at sea and is used for various purposes, such as mooring a ship in open waters, maneuvering the bow in narrow waterways, urgently reducing speed to prevent collisions, maintaining stability while drifting, or securing a vessel that has run aground. Fig. 8 shows the H.H.P. (High Holding Power) anchor used in this study. According to Y.K. Kim (2009), based on KS V 8877, this anchor possesses high holding power, allowing it to maintain the same holding capacity with only 75% of the weight of other types of anchors. The anchor consists of a crown, which allows it to sink; a fluke, which secures the hull once the anchor is set; a shank, which helps maintain the correct orientation on the seabed for the fluke to engage properly; and a swivel on the anchor chain, which allows the anchor to be stored within the hull when raised.

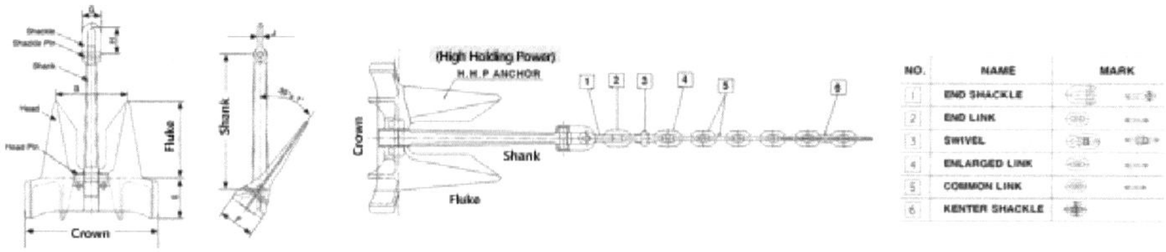

Figure 8. Components of the anchor

The anchoring system is classified into two types, as shown in Fig. 9: Bell mouth type and Pocket type.

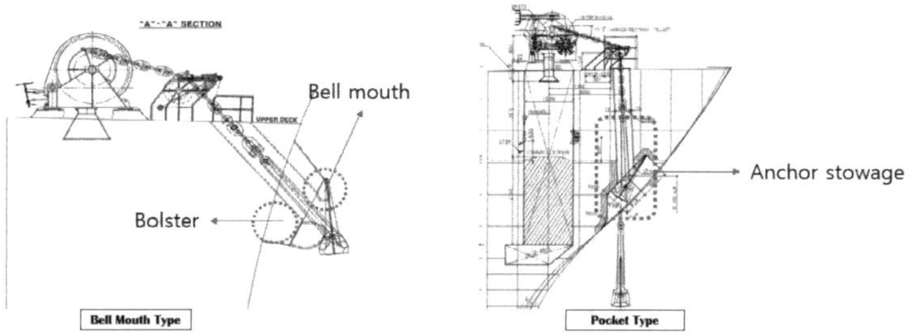

Figure 9. Types of anchoring systems

The Bell mouth type consists of cast steel components, including the bolster and the bell mouth. The bell mouth is primarily used for tankers, and due to the motion of the vessel, anchor swing occurs, necessitating a minimum distance of 0.7 meters between the hull and the anchor. Additionally, the anchor chain must include a swivel; without it, the fluke may not achieve 3-point contact with the bell mouth when the anchor is raised, which is crucial. There is also a concern that pirates may attach a rope to the bell mouth to penetrate the cabin during an attack.
According to Ha et al. (2012), the pocket type has a thick structure, which poses a significant challenge during operation if the position of the connection points with the surrounding structures and the hull's curved plating cannot be accurately aligned. This issue makes subsequent modification work very difficult, resulting in excessive time consumption and the need for rework due to welding deformation, which always requires excessive labor and time during production. In contrast, unlike the bell mouth, the pocket type has a space for stowing the anchor within the hull, optimizing the internal space of the vessel. Additionally, the anchor chain does not have a swivel, and 3D simulation is possible.For the bell mouth type, since 3D simulation is not possible, a separate mock up test is conducted to select the shape. The mock up is made at a 1/10 scale of the actual vessel, and it is essential that the anchor fluke achieves 3-point contact with the bell mouth, as shown in Fig. 10, which necessitates a swivel on the anchor chain.

Figure 10. Anchor swivel of the anchor chain

Fig. 11(a) shows that the anchor must rotate and make three-point contact with the bell mouth. Otherwise, noise will occur during operation, and continuous impact will be applied to the bell mouth. If the anchor swivel is not used, as shown in Fig. 11(b), the anchor fluke will catch on the bell mouth, preventing the dog bar of the chain compressor from closing, making it impossible to secure the anchor to the hull.

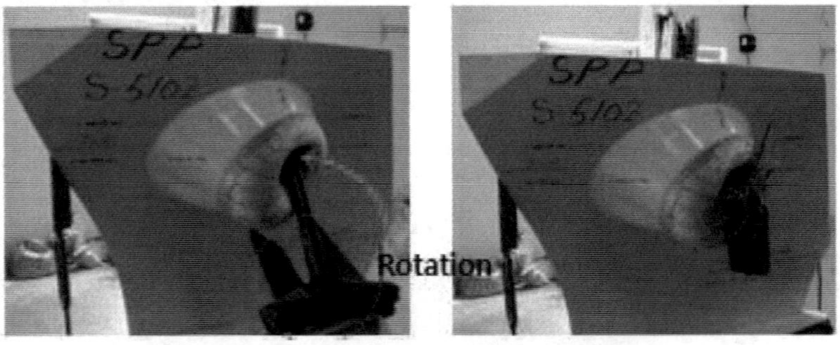

(a) Three-point contact of the anchor fluke with the bell mouth

(b) The situation of the anchor flukes getting caught on the bell mouth
Figure 11. Results of the bell mouth type mock up test

IOP Conf. Series: Earth and Environmental Science **1461** (2025) 012004 doi:10.1088/1755-1315/1461/1/012004

According to WOOHYUN SHIP DESIGN TECHNOLOGY, the production period for the bell mouth takes a total of 10 to 15 days, and the production process is shown in Fig. 12(a). After that, the

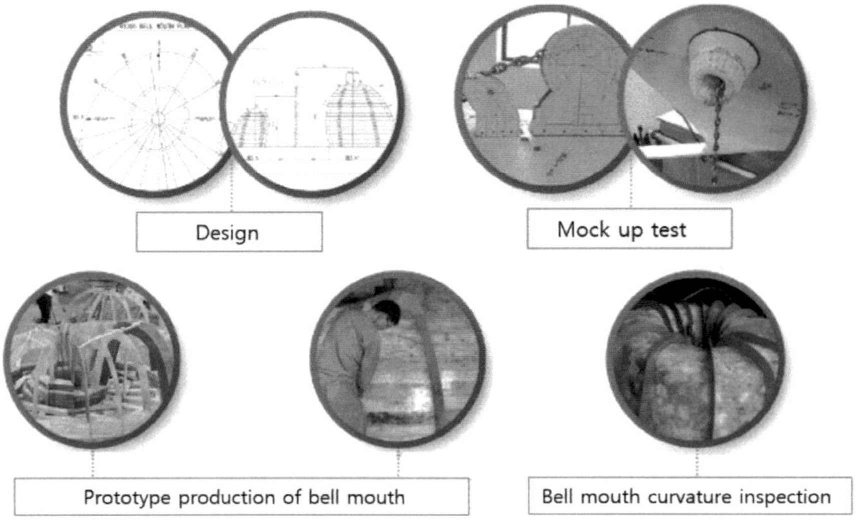

(a) Flowchart of the bell mouth manufacturing process

curvature of the final bell mouth is inspected, and it is used for ship anchoring as shown in Fig. 12(b).

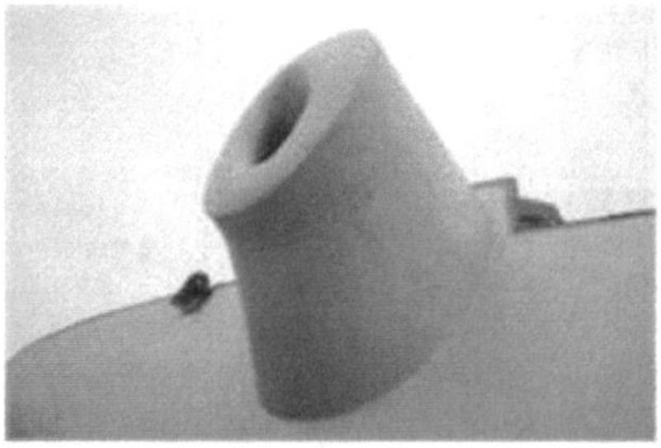

(b) Application of the Bell mouth type to the vessel

Figure 12. Manufacturing process of the bell mouth type

IOP Conf. Series: Earth and Environmental Science **1461** (2025) 012004 doi:10.1088/1755-1315/1461/1/012004

3. New type anchoring system

In the case of the bell mouth, a mock up test must be conducted each time before application to the vessel. If issues arise during operation, it may be necessary to remake or repair the bell mouth, which weighs over 3 tons. To address operational problems and reduce process delays and costs, a new type of anchoring system was developed by combining the anchor stowage structure of the pocket type with the simple shape of the bell mouth type, as shown in Fig. 13.
Park. (2011) proposed a design method for ship bolsters that contributes to optimizing their shape, improving productivity, and enhancing quality.

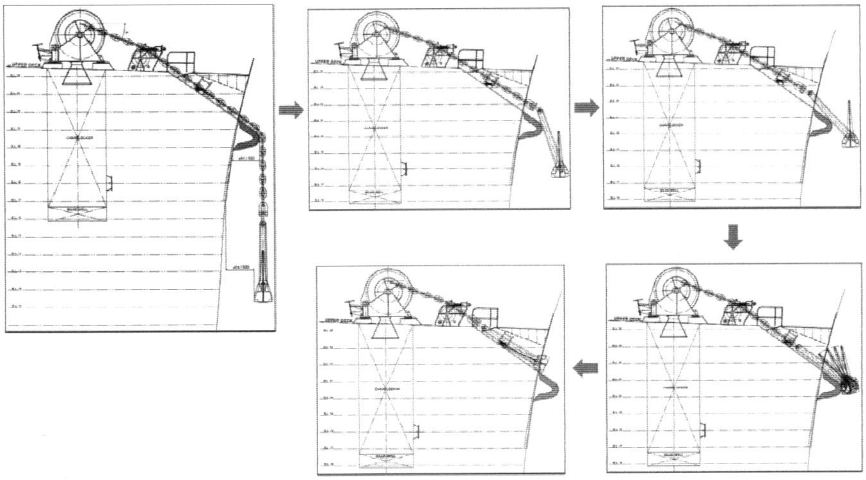

Figure 13. Semi pocket type anchoring system drawing

Fig. 14 shows the profile view of the new anchoring system. The hawse pipe is shorter than those of the existing bell mouth type and pocket type. The bolster is designed to increase strength by deforming the hull's outer plating and using stiffeners.

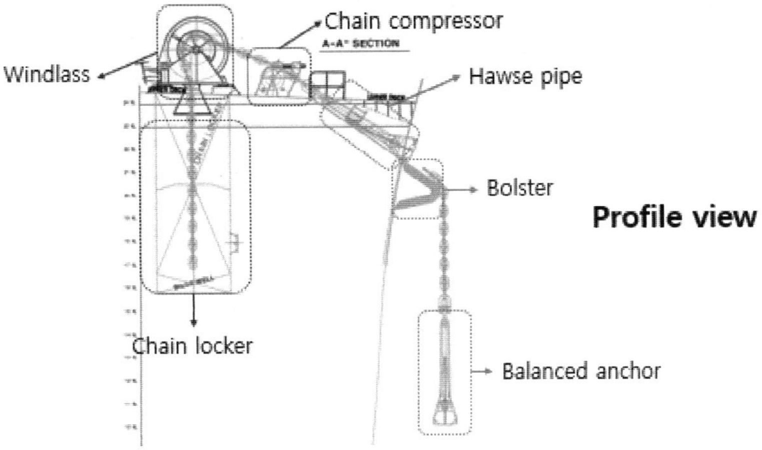

Figure 14. Semi pocket type anchoring system components (Profile view)

IOP Conf. Series: Earth and Environmental Science **1461** (2025) 012004 doi:10.1088/1755-1315/1461/1/012004

Fig. 15 shows the plan view of the new anchoring system, which includes a stopper that secures the anchor fluke when the anchor is raised and a chock that guides the chain without impacting the hull.

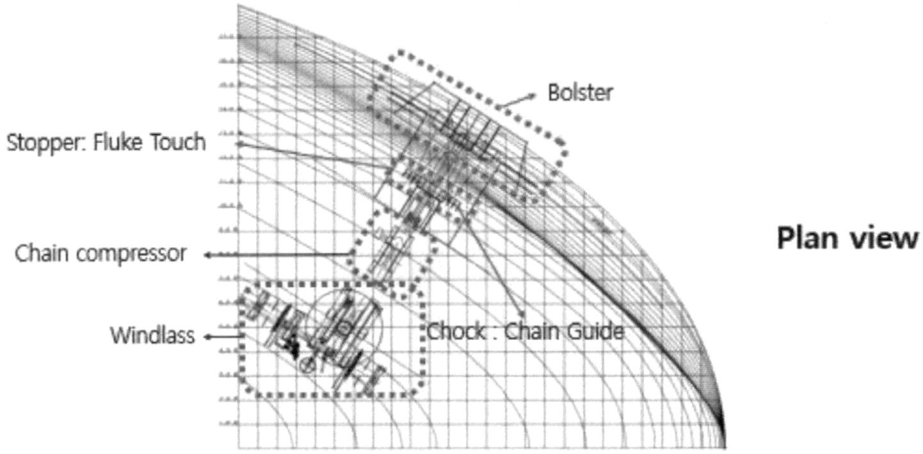

Figure 15. Semi pocket type anchoring system components (Plan view)

Mock up tests were conducted to verify the validity of the motion of the developed New anchoring system and the critical points during anchor use. First, Fig. 16(a) shows the balanced anchor, Fig. 16(b) represents the mock up of the Windlass and Chain compressor, and Fig. 16(c) illustrates the mock-up of the New anchoring system.

(a) Balanced anchor **(b)** Windlass and Chain compressor **(c)** Semi pocket type

Figure 16. Production of a mold for mock up testing

Subsequently, a mock up test was conducted as shown in Fig. 17.

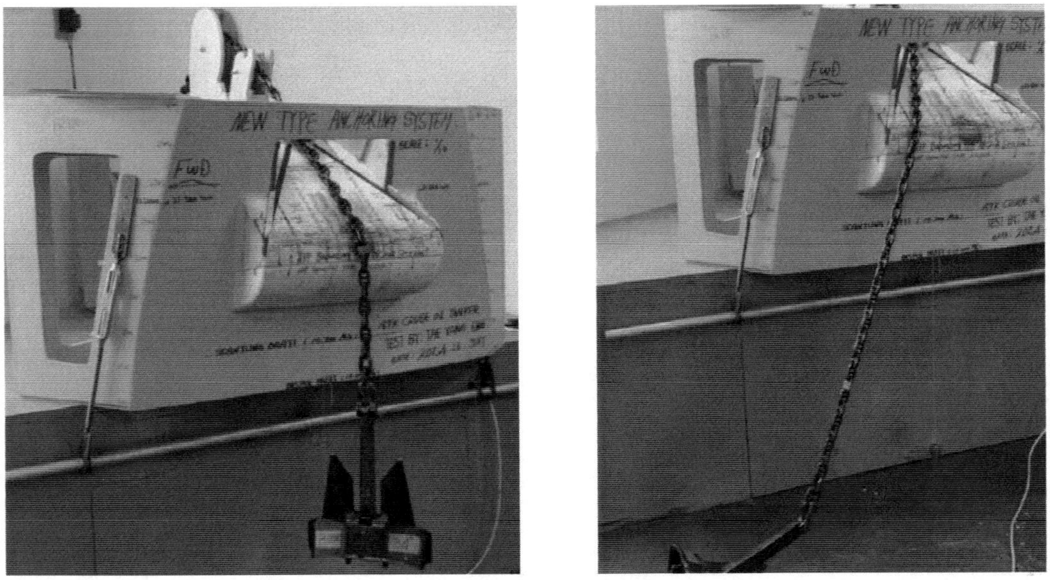

Figure 17. Results of the new anchoring system type

It is composed of a round bar and the guide bar of the bolster to prevent the anchor from contacting the ship's hull, and as shown in Fig. 18, damage was observed during the lowering and raising of the anchor.

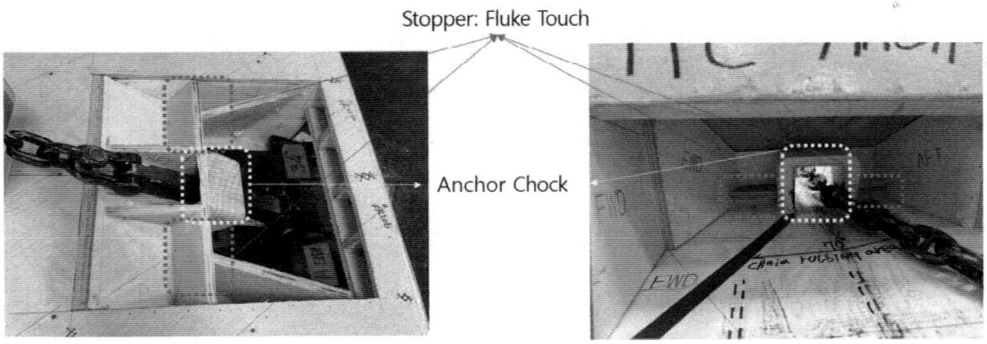

Figure 18. Critical structural point of the new anchoring system

IOP Conf. Series: Earth and Environmental Science **1461** (2025) 012004 doi:10.1088/1755-1315/1461/1/012004

Fig. 19 shows a mock up test photo when using an anchor chain with a swivel.
When using an anchor chain with a swivel, it was observed that the anchor fluke gets caught on the bolster, causing the anchor chain to misalign with the chain compressor, as shown in Fig. 20.

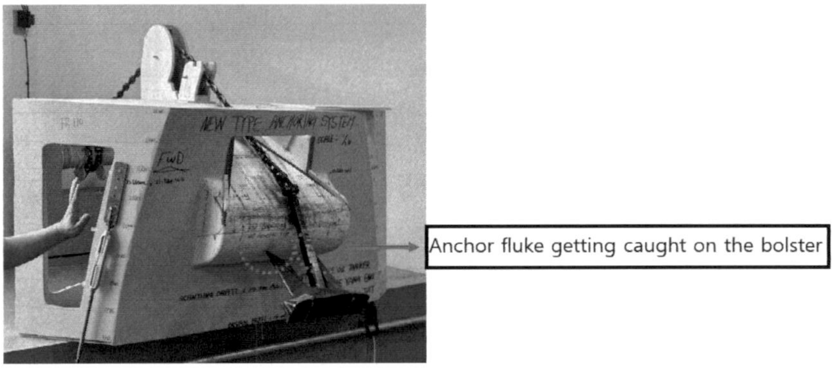

Figure 19. Problem arising in the new type anchoring system when using the anchor swivel

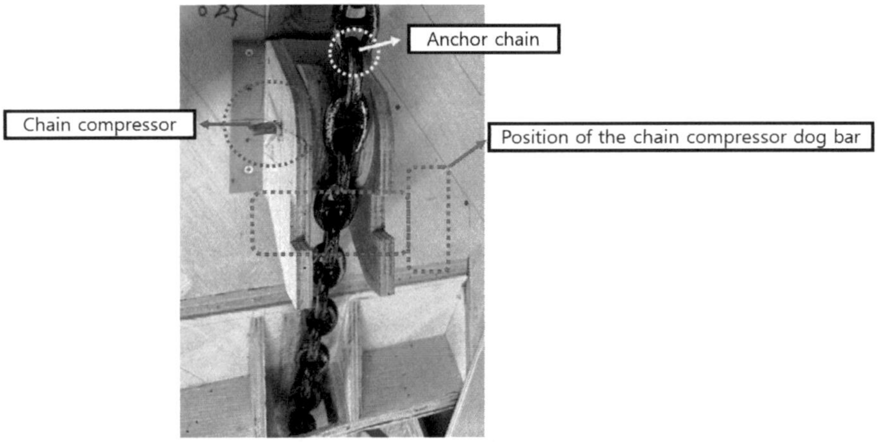

Figure 20. The misalignment of the anchor chain with the compressor

4. Conclusion

(1) As shown in Figure 21, a comparison of the bell mouth type, pocket type, and new anchoring system has been made, and the information is listed in Table 1. In this study, we developed a new anchoring system by combining various advantages, such as the anchor stowage structure of the pocket type and the simple design of the bell mouth type.

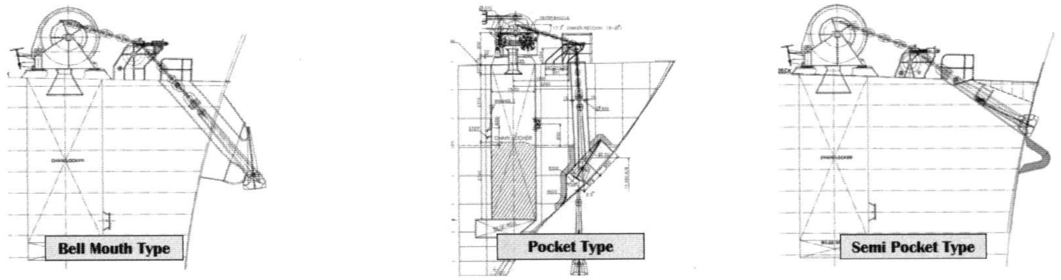

Figure 21. Comparison of anchor systems

Table 1. Comparison of anchor systems information

Type	Bell mouth type	Pocket type	Semi pocket type
Components	Bell Mouth, Original Bolster Hawse Pipe	No casting	Horizontal bolster Storage pocket, Chock, Fluke Stopper
Hull Interference	Anchor Swing and hull interference (Minimum separation of 0.7 meters)	No swivel (No swing)	1.5 meters hull separation
Dimension/Degree management items	Fabrication/Assembly: Bell Mouth, Bolster, Hawse Pipe Anchor Fabrication: Fluke dimensions/ angles, Shank dimensions	All items except materials can be designed by the company	Fabrication/Assembly : Horizontal management of the bolster Chain : Short link length
Chain Compressor	Large chain bend angle	Chain compressor alignment is possible	Chain bend angle is small (reduced load)
Major Issues	Modification of Bell mouth (Anchor 3-point touch, turning, interference)	Bow impact analysis required (risk of displacement due to movement)	Anchor storage space is small

(2) We identified the critical points of the key components in the new type anchoring system according to anchor motion. In future research, we plan to select the loads occurring on the bolster, stopper, and anchor chock due to vessel movement when the anchor is fixed to the seabed. Additionally, we aim to evaluate the structural strength of the bow of the vessel equipped with the new type anchoring system by determining the loads generated during anchoring operations.

(3) The newly developed anchoring system in this study enables 3D simulations, allowing potential operational issues to be addressed in advance. Additionally, by not using a carbon fiber bell mouth, it helps reduce environmental pollution during production and disposal.

(4) The new anchoring system is designed by modifying the hull plating instead of using a bell mouth weighing over 3 tons and can be repaired and reused in the event of an issue. Furthermore, the elimination of continuous mock-up testing is expected to result in cost savings.

Acknowledgement

This research was supported by "Regional Innovation Strategy (RIS)" through the National Research Foundation of Korea(NRF) funded by the Ministry of Education(MOE)(2021RIS-002), and this work was supported by korea institute for Advancement of Technology (KIAT) grant funded by the Korea Government (MOTIE) (P0017006, HRD Program for Industrial Innovation,2024).

References

[1] M.S. Chun, J.Y. Kim, K.H. Kim, D.S. Jeong, D.Y. Lee, S.K. Wi, B.H. Kim, C.H. Kim, C.S. Shim, 2024. A fundamental study on structural strength assessment of U-bolts for expanded application to shipbuilding and offshore piping systems. International Journal of Naval Architecture and Ocean Engineering, Volume 16, 100561.
[2] Common Structural Rules for Bulk Carrier and Oil Tankers, 2024. International Association of Classification Societies(IACS).
[3] S.K. Ha, S.B. Soo, K.W. Kang and J.H. Bae, 2012. Anchor pocket structure for Vessels and Assembly Method Thereof. [Online] (Updated 02 Septemberr 2013) Available at: http://www.kipris.or.kr/khome/main.jsp [Accessed 15 September 2024].
[4] Hull equipment, supporting structure and appendages, 2024. Det Norske Veritas(DNV).
[5] W.N. Joo, 2013. Cable clench structure equipped to the chain locker of anchoring system for ship. [Online] (Updated 19 November 2012) Available at: http://www.kipris.or.kr/khome/main.jsp [Accessed 10 September 2024].
[6] M.J. Jeon, 2015. Anchoring apparatus and floating offshore structure having the same. [Online] (Updated 02 July 2017) Available at: http://www.kipris.or.kr/khome/main.jsp [Accessed 1 September 2024].
[7] D.J. Kang, D.H. Cho, S.W. Song, and J.G. Kim, 2020. A Study on the Effects of Guide bar Installation Conditions on the Ship's Anchor Containment. Journal of the Korean Society of Mechanical Technology, Vol. 22, pp. 769-773.
[8] C.N. Kang, J.H. Jeong, Y.B. Kim, (2013). A Study on Control System Design for Ship Mooring Winch System. Journal of the Korean Society for Power System Engineering, Vol. 17, pp. 89-98.
[9] J.K.Kim, S.H.Lee, J.R.Bae, and S.S.Kim, 2013. Anchoring Appartus of ship. [Online] (Updated 03 September 2015) Available at: http://www.kipris.or.kr/khome/main.jsp [Accessed 4 September 2024].
[10] Y.J.Kim, 2009. A Study on Shape of Bell Mouth for Optimal Anchoring Condition. Ph.M. Cambridge: University of Ulsan.
[11] T.W. Kim, J.S. Park, 2020. Anchoring Appartus. [Online] (Updated 23 November 2021) Available at: http://www.kipris.or.kr/khome/main.jsp [Accessed 1 September 2024].
[12] Korean Standards & Certification, 2023. KS V 8877, Korean Agency for Technology and Standards
[13] B.Y. Moon, H.J. Hong, D.H. Kim, W.M. Lee, S.M. Lee, 2023. Comparative Study on Load Criteria by Class Based on Structural Analysis of 38ft HDPE Power Boat. Journal of the Society of Naval Architects of Korea, Vol. 60, pp. 38-47.
[14] S.J. Park, 2018. http://whship.co.kr [Accessed 8 September 2024].

[15] K.W. Park, 2011.Method for Designing Bolster of Vessels. [Online] (Updated 17 February 2011) Available at: http://www.kipris.or.kr/khome/main.jsp [Accessed 19 July2024].

Study of polyurethane elastomer cores interfacial fracture resistance of the sandwich materials for ship structures

H Siswanti[1,4], A Zubaydi[2*], B Piscesa[3] H Syahab[2], R C Ariesta[2]

[1] Department of Ocean Engineering, Faculty of Marine Technology, Institut Teknologi Sepuluh Nopember, Surabaya, Indonesia
[2] Department of Naval Architecture, Faculty of Marine Technology, Institut Teknologi Sepuluh Nopember, Surabaya, Indonesia
[3] Department of Civil Engineering, Faculty of Civil, Planning, and Geo-Engineering, Institut Teknologi Sepuluh Nopember, 60111, Surabaya, Indonesia
[4] Department of Marine Technology, Politeknik Negeri Madura, 69281, Sampang, Indonesia

*E-mail: zubaydi@its.ac.id

Abstract. The use of sandwich materials continues to grow in several engineering applications. In line with this trend, the marine industry is also adopting steel-based sandwich materials with polyurethane elastomer cores as primary materials to substitute the steel ship structures. Using lightweight and thick core material gives added stiffness without significantly increasing the structure's weight. However, it poses a challenge when dealing with damage vulnerability due to the differences in faceplates and core material properties, especially at the interface of two layers. Therefore, the resistance to fracture is an important design parameter to be determined. This study approaches measuring the interfacial fracture resistance of the steel faceplate and polyurethane elastomer core. The interfacial fracture resistance is quantified as the Energy Release Rate (ERR) estimated due to mode-I and mode-II loading. The interfacial fracture resistance employs experimental and numerical methods, and the result under mode-I loading shows valid agreement. However, more accurate numerical methods are necessary to represent the experimental conditions for mode-II loading.

1. Introduction

Sandwich materials are increasingly adopted in various engineering applications due to their high strength-to-weight ratio, corrosion resistance, and versatile performance characteristics [1, 2]. In particular, the marine industry has adopted sandwich structures to replace traditional steel ship structures. Applying these sandwich structures promises about 4-10% structure weight reduction. Besides that, using these sandwich structures offers significant advantages in impact resistance, fatigue performance, and overall durability [3, 4, 5, 6].

Sandwich materials are characterized by their layered construction of strong face sheets and a lightweight core. One such material is the steel-based hybrid sandwich, which uses steel for the face sheet layer and polyurethane elastomer for the core layer. The polyurethane elastomer core and steel faceplates form a robust structural system that efficiently manages the diverse and demanding load conditions encountered in naval and commercial vessels [7, 8]. However, one of the critical challenges in ensuring the long-term performance of these materials is the phenomenon of face-core debonding. Face-core debonding occurs when the faceplates and the polyurethane core bond fail. This condition leads to a separation of each layer that can severely compromise the sandwich plate's structural integrity. Various factors, including mechanical loading, environmental degradation, and manufacturing defects, can trigger this face-core debonding. Since the faceplate and core are bonded with the adhesive, the face-core debonding can also cause by the adhesive-bonded failure. In steel sandwich plates with polyurethane cores, debonding is prone to occur at the interface between the faceplate and core layer, known as the interfacial debonding phenomenon [5, 9, 10, 11].

Understanding the mechanisms of face-core debonding is essential for designing and maintaining sandwich structures, which must possess sufficient toughness to prevent debonding under various loading conditions. One critical aspect of sandwich material performance that must be learned is the interfacial fracture resistance. Fracture resistance, quantified as the Strain Energy Release Rate (SERR), is a critical parameter determining how the material can withstand debonding crack propagation and maintain structural integrity [11]. Accurately assessing SERR in sandwich materials involves addressing unique challenges arising from the differing mechanical properties of the faceplate and core materials. Fracture resistance or fracture toughness of the material may be quantified as energy release rate (ERR) or stress intensity factor (K). It describes the material's ability to resist the growth of cracks. Fracture resistance plays a critical parameter for assessing the integrity and reliability of material against crack propagation. It is a measure of the energy released per unit area of crack surface as a crack propagates [12].

Sandwich plate damage behaviour is usually evaluated based on the principle of fracture mechanics, with mode-I, mode-II loading, or mix-mode I and II [13, 14, 15, 16]. These tests measure the fracture resistance characteristics of composite sandwich specimens by experimental and numerical analysis. However, more research is needed into the fracture resistance of steel polyurethane sandwich plates. Furthermore, this study investigates the interfacial fracture resistance of polyurethane elastomer cores in sandwich plates designed explicitly for ship structures. Experimental and numerical studies analyzed the interfacial fracture resistance under mode-I and mode-II loading to provide valuable insights into the failure mechanisms associated with face-core debonding. The findings will contribute to developing improved design guidelines and manufacturing practices to enhance the durability and reliability of steel-polyurethane sandwich plates in marine applications.

2. Method

2.1 Experimental Analysis

The experimental study to characterise the interfacial fracture resistance of sandwich plates involves two specific test methods: opening mode loading (mode-I) and shearing mode loading (mode-II). These tests are designed to evaluate how well the interface of faceplates and core resists crack debonding propagation. The sandwich double cantilever beam (SDCB) specimen is used for opening mode loading, and the sandwich End-notch flexure (AENF) is also used for shearing mode. The dimension of sandwich DCB and ENF specimen is 300 mm length and 40 mm width (b). The sandwich specimen thickness consist of 4 mm faceplates thickness (t_1 & t_2) and 15 mm polyurethane core thickness (t_c). The dimension of the specimen is shown in Figure 1.

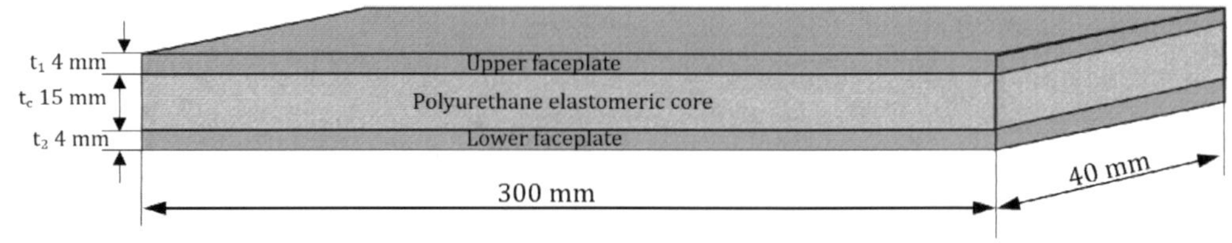

Figure 1. Sandwich plate specimen dimension

The sandwich DCB test method determines the Mode-I interfacial fracture energy (G_{IC}). A 50 mm pre-crack is introduced at the interface between the faceplates and the core as a debonding initiation site. The specimen is clamped and hinged at the ends of the faceplates, and the forces are applied perpendicular to the specimen's longitudinal direction. The mode-I loading is applied continuously at a 20 mm/minute speed, with a maximum opening displacement of 30 mm. The strain energy release rate (SERR/G_{IC}) is calculated using the recorded data, and the fracture resistance is evaluated by correlating the (G_{IC}) with the crack length. The critical energy release rate value (G_{IC}) is achieved when the GIC value reaches a steady state.

Meanwhile, the sandwich ENF test method determines the Mode-II interfacial fracture energy (G_{IIC}). The test specimen geometry is similar to the DCB specimen but is tested in a different configuration to promote shear loading. The ENF specimens are notched at the interface as the crack initiation site. During the test, the load and displacement are recorded, and the crack length is monitored. The strain energy release rate (SERR/G_{IIC}) is calculated using the applied load, displacement, and specimen dimensions. Like the DCB test, the G_{IIC} is plotted against the crack length to determine the steady-state fracture resistance.

In this study, the data reduction method to determine the strain energy release rate of mode-I loading test (G_{IC}) is derived from beam theory.

$$G_{IC} = \beta \frac{\partial U}{\partial a} = \frac{P^2 a^2}{2EIb} \tag{1}$$

Consider crack opening displacement, derived from Euler Bernoulli Beam theory:

$$\delta = \frac{Pa^3}{3EI} \tag{2}$$

Substituting we obtain G_{IC} final from

$$G_{IC} = \frac{3P\delta}{2ba} \tag{3}$$

Next compliance is obtained by dividing displacement by load, and the $|\Delta|$ must be determined using the curve shown in Figure 3 to remedy the not-perfectly built-in clamp issue.

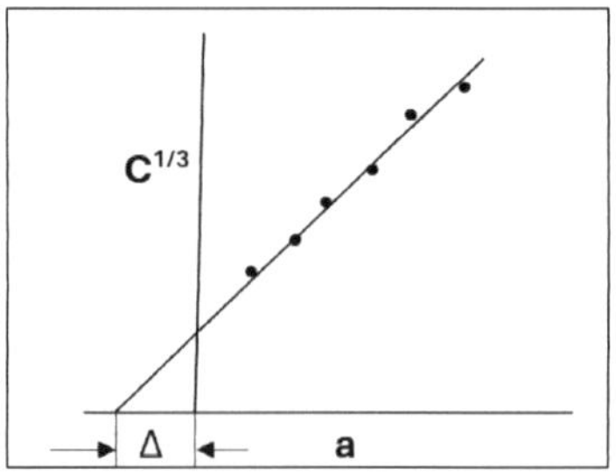

Figure 2. Curve to determine $|\Delta|$ [18]

So, the determination of the strain energy release rate of mode-I loading test (G_{IC}) is derived from the equation:

$$G_{IC} = \frac{3P\delta}{2b(a+|\Delta|} \tag{4}$$

The determination of the strain energy release rate of mode-II loading test (G_{IIC}) is derived from the equation:

$$G_{IIC} = \frac{9P^2a^2C}{2B(2L^3+3a^3)} \tag{5}$$

The sandwich plate specimens were manufactured using polyurethane core and steel faceplates; the material's properties are shown in Table 1.

Table 1. Sandwich material properties [17]

Part	Faceplates	Core
Thickness (mm)	4	15
Material	Steel	Polyurethane Elastomer
Density (Ton/m³)	7.85	1.1
Elastic Young's Modulus (GPa)	206	0.9
Poisson's Ratio	0.3	0.36

2.2 Numerical Finite Element Analysis

Failure to analyze sandwich materials is quite complex and challenging. Finite element analysis is a numerical method to determine the failure progression, such as interfacial debonding propagation at the sandwich plate. In this study, the numerical finite element analysis is conducted to simulate the interfacial debonding of sandwich plate specimens. A bilinier traction-separation law employ to captures the debonding behavior at the interface between the faceplates and the core. This method is essential for representing the complex behavior of the adhesive layer under loading. The bilinier traction defines the relationship between the traction (force per unit area) and the separation (displacement) of the interface in two distinct stages: an initial elastic region followed by a softening region. The bilinier traction curve shows in Figure 3. The bonded region expressed initial elastic condition, the debonding and debonded region expressed the softening due to material degradation and the complete material failure as well. The application of this method within FEA framework effectivelly predict the initiation and propagation of face/core debonding in this sandwich plate models.

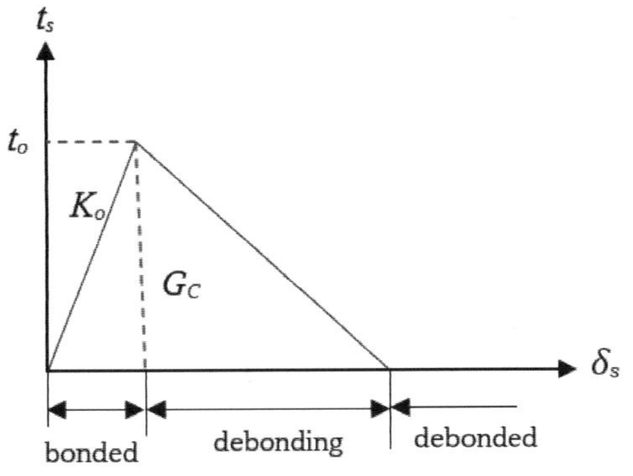

Figure 3. The bilinier traction separation curve

IOP Conf. Series: Earth and Environmental Science **1461** (2025) 012005 doi:10.1088/1755-1315/1461/1/012005

The numerical simulation model has the same dimensions and material properties as the experiment specimen. Numerical analysis using the finite element method involves a meshing process, which divides a complex structure into a finite number of smaller and simpler shapes called elements. The mesh size of the model can sometimes vary in type and size according to the requirements. Areas requiring high precision are divided into smaller meshes, while less critical areas use larger mesh sizes, as illustrated in Figure 4

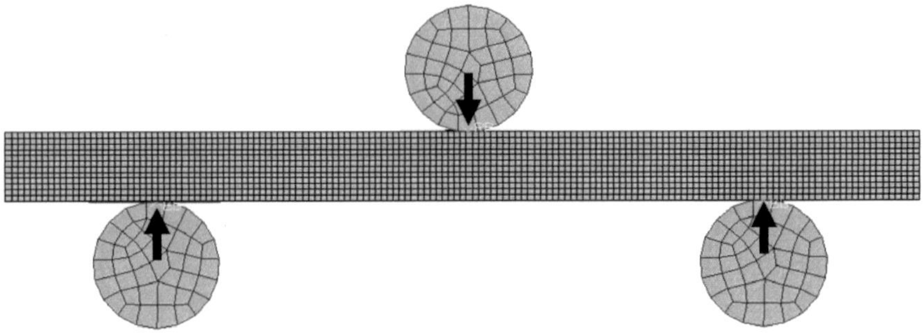

Figure 4. The different mesh size of the model

This study used an eight-node plane-strain solid element to model the faceplate and core layers. In contrast, a six-node zero-thickness cohesive element is utilized for the interface layer between the faceplate and core, and a numerical inverse procedure aims to reproduce the experimental load-displacement curve. In the numerical simulation with mode-I loading, crack propagation was observed by tracking the crack tip movement at the interface between the faceplate and core, as shown in Figure 5. A zero-thickness cohesive element is utilised for the faceplate interface and core element.

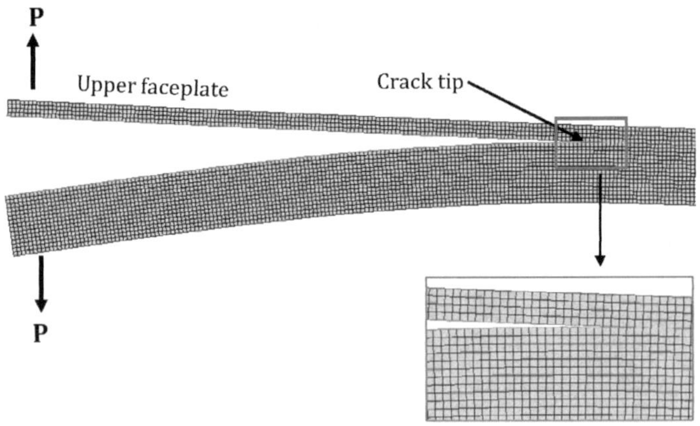

Figure 5. The different mesh size of the model

3. Result and Discussion

3.1 Mode-I loading test results

The mode-I loading test result is shown in a load-displacement curve in Figure 4. The experimental peak load or critical load (***P_C***) for mode-I is 1684.45 N. Meanwhile, the numerical peak load result is 1762 N. The critical load (***P_C***) values from the numerical simulation are higher than the experimental results. This difference may be caused by several factors. In computational simulations, the specimen material properties are usually assumed to be uniform and based on nominal values from literature or prior sample tests. However, in experiments, actual material properties can vary due to imperfect manufacturing processes or environmental conditions. Since the specimens for experimental experiments were manually fabricated, there is the possibility of uniform adhesive or inaccurate dimensions. Whereas in numerical simulations usually rely on idealized conditions and assumptions, such as perfect material properties, boundary conditions, and geometry. The mesh size and discretization in FEM numerical simulation are also sensitive to increase the stiffness of the model, leading to higher predicted loads. Measurement errors, noise, and limitations of the testing equipment also influenced the experimental results. These factors can lead to underestimations of peak load and ERR in experimental measurements. The load-displacement curve in Figure 6 shows that the experiment results are more scattered data. However, the trend of the load-displacement curve resulting from experiments and numerical simulations shows quite valid agreement. The results of this numerical simulation can be considered quite representative of experimental conditions.

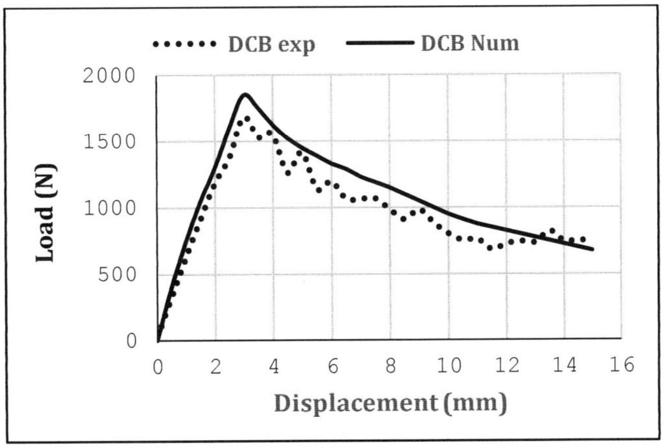

Figure 6. Load-displacement curve for mode-I loading specimens (DCB)

The interfacial fracture resistance is expressed by the energy release rate (ERR), which represents the amount of potential energy converted into crack energy during the propagation of interfacial debonding cracks. The ERR under mode-I loading is usually notated as GI and displayed as a curve correlated with the crack length, as shown in Figure 7. The shape of the resistance curve (R-curve) from the damage testing of the steel polyurethane sandwich specimen produces a rising R-curve. This curve shows an initial linear state, followed by a plateau region. At the linear region, the resistance to crack growth increases steadily as the load increases. This initial linear part represents the early stages of crack growth, where the material is still relatively intact, and energy release is low. The plateau indicates a critical point where the material's resistance to crack growth stabilizes. The plateau is crucial because it represents the critical strain energy release

IOP Conf. Series: Earth and Environmental Science **1461** (2025) 012005
doi:10.1088/1755-1315/1461/1/012005

rate ($\boldsymbol{G_{IC}}$), which is the point where the crack begins to propagate more freely, and the material can no longer increase its resistance to further crack growth. The resistance curve in Figure 7 shows that the critical ERR ($\boldsymbol{G_{IC}}$) from the mode-I loading experiment is 2.4 J/mm^2, and for the numerical result, it is 2.5 J/mm^2. The difference in peak load has implications for differences in G_I and G_{IC}. The resistance curve in Figure 7 shows that the experiment results are more scattered data as well. However, this numerical simulation can be considered quite representative of experimental conditions.

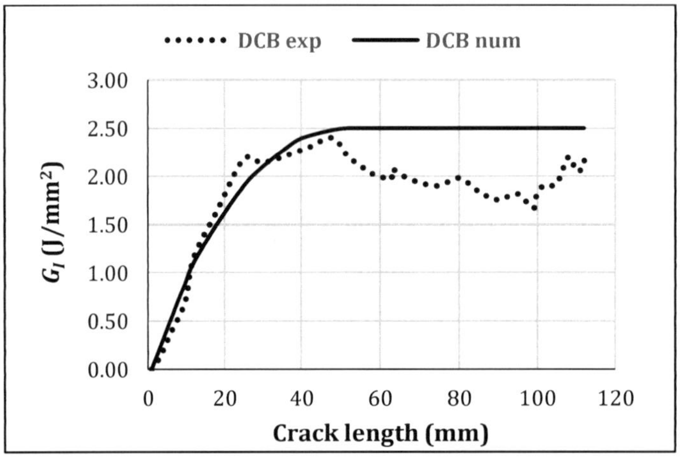

Figure 7. Resistance curve mode-I loading specimens (DCB)

3.2 Mode-II loading test results

The mode-II loading test result is shown in a load-displacement curve in Figure 8. This curve plots the applied load against the corresponding displacement during the test.

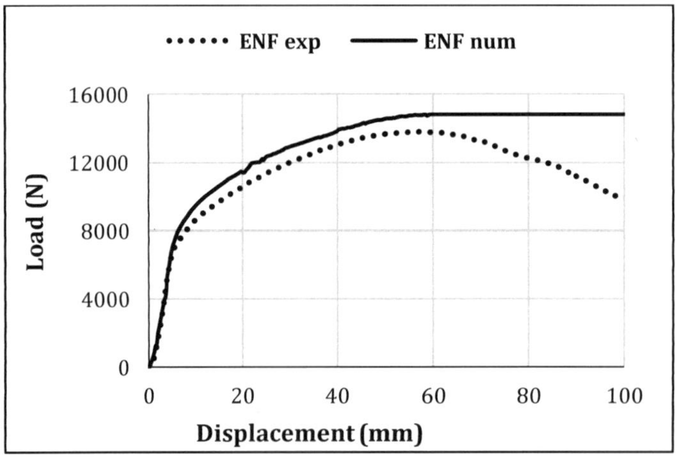

Figure 8. Load-displacement curve for mode-II loading specimens (ENF)

The experimental result show that at the beginning of the curve, there is a linear region where the load increases proportionally with displacement. Follow with the non-linear region and reach a point of yield at the peak point. The peak point load represents the maximum load the material can withstand before significant failure. After reaching the peak load, the curve shows post-peak softening region, indicates the decrease of load-carrying capacity. A gradual decline suggests controlled damage and stable crack propagation. Meanwhile, the numerical simulation result shows relativelly different trend. After reach peak load the curve not show softening but tend to plateau region. The peak load for mode-II loading from experimental result is 13791 N, and the numerical result is 14800 N. The difference between the experiment and the numerical simulation results may caused by the assumption that the material is linear elastic, and does not include the plasticity properties of the material at the numerical simulation.

The resistance curve in Figure 9 shows the ERR values from experimental and numerical simulation results due to mode-II loading (G_{II}). The resistance curves from both analysis methods display a rising R-curve. Initially, the curve has a gradual slope, then rises as the crack length increases. The critical ERR for mode-II loading (G_{IIC}) from experimental result is 249.5 J/mm², and the numerical result is 267.6 J/mm². The G_{II} curve estimated from the numerical simulation shows a more significant increase than the experimental results. This implies the difference in load values obtained from the two analysis methods. At the last stage, the G_{II} from the numerical simulation results is higher than the experimental results. This overestimation occurs because the numerical simulation has yet to include the plasticity properties of the core material, so it cannot capture the material degradation due to core shear.

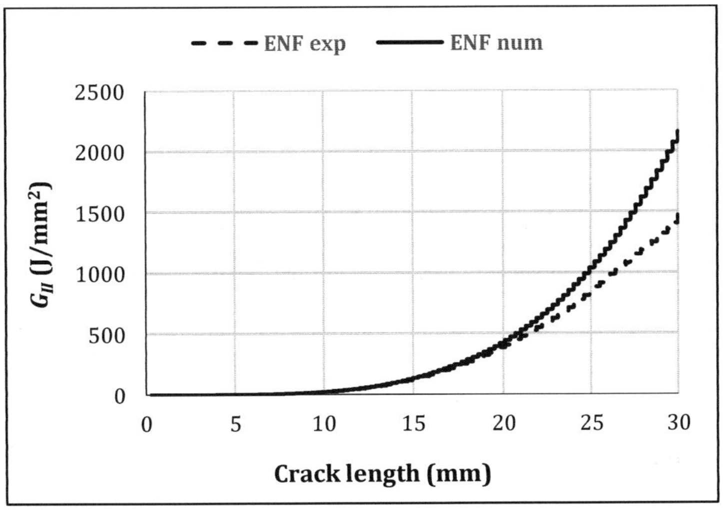

Figure 9. Resistance curve for mode-II loading specimens (ENF)

4. Future Work

The study of the sandwich plate's interfacial fracture resistance is an initial part of the damage tolerance analysis of the sandwich structure. The findings of this study offer significant insight into the investigation of the damage characteristics of sandwich plates consisting of steel faceplates and polyurethane elastomer cores. The experimental data contributed to further damage studies of sandwich structures. However, it may be necessary to apply other more accurate numerical methods that can better represent the experimental conditions.

5. Conclusion

The interfacial fracture resistance of sandwich plates has been investigated. The damage characteristic due to Mode-I loading is known to have a lower critical strain-energy release rate than mode-II loading. The numerical simulation result of the mode-I loading test is higher than the experimental result. However, the trend of the curve shows quite valid agreement. Meanwhile, for the mode-II loading test, the numerical simulation result shows a relatively different trend in a curve; it is necessary to use other accurate numerical methods to represent the experimental conditions better. At least the results of this study provide valuable insights for characterizing the interfacial fracture resistance of steel polyurethane sandwich plates for ship structure applications that have yet to be widely studied. The interfacial fracture resistance from this study helps analyse the damage behaviour of sandwich plates. Estimating the interfacial fracture resistance of sandwich plates in ship sandwich structures is crucial for optimizing design and ensuring compliance of sandwich structures.

References

[1] P. Kujala and A. Klanac, "Steel Sandwich Panel in Marine Applications," *BRODOGRADNJA,* vol. 56, no. 4, pp. 305-314, 2005.

[2] SANDCORe, "Best Practice Guide for Sandwich Structures in Marine Applications," University of Newcastle upon Tyne, New Castle, 2013.

[3] Momcilovic and M. Motok, "Estimation of ship lightweight reduction by means of application of sandwich plate system," *FME Trans vol. 37, no. 3,* p. 123–128, 2009.

[4] S. Sujiatanti, A. Zubaydi and A. Budipriyanto, "Finite element analysis of ship deck sandwich panel," *Applied Mechanics and Materials 874,* pp. 134-139, 2018.

[5] T. J. Grafton and J. R. Weitzenbo€ck, "Steel-concrete-steel sandwich structures in ship and offshore engineering," in *Advance Marine Structure Proceedings of the 3rd International Conference on Marine Structures*, 2011.

[6] A. Ismail, A. Zubaydi, B. Piscesa, Tuswan and R. C. Ariesta, "Study of Sandwich Panel Application On Side Hull Of Crude Oil Tanker," *Journal of Applied Engineering Science,* 2021.

[7] A. Ismail, A. Zubaydi, B. Piscesa and E. Panangian, "A comparative study of conventional and sandwich plate side-shell using finite element method," in *2nd International Conference on Mechanical Engineering Research and Application (iCOMERA 2020)*, 2020.

[8] A. Ismail, A. Zubaydi, B. Piscesa and Tuswan, "A novel fiberglass-reinforced polyurethane elastomer as the core sandwich material of the ship–plate system," *Journal of the Mech Behavior of Mat,* vol. 32, 2023.

[9] G. Irven, A. Whitehouse, D. Carolan, A. Fergusson and J. P. Dear, "Toughening of face-sheet core bonds in sandwich structures," *Engineering Fracture Mechanics,* vol. 290, 2023.

[10] D. Zenkert, "Damage Tolerance of Naval Sandwich Panels," Department of Aeronautical and Vehicle Engineering, Stockholm, 2009.

[11] E. Oterkus, C. Diyaroglu, D. De Meo and G. Allegri, "Fracture modes, damage tolerance and failure mitigation in marine composites," in *Marine Applications of Advanced Fibre-Reinforced Composites*, Cambridge, Woodhead Publishing, 2016, pp. 79-102.

[12] L. Anderson, Fracture Mechanics Fundamentals and Applications, 3rd ed., Taylor & Francis, 2005.

[13] M. H. Rarani, M. M. Shokrieh and P. Camanho, "Finite element modelling of mode I delamination growth in laminated DCB specimens with R-curve effects," vol. 45, 2013.

[14] A. Ramji, Y. Xu, M. Yasaee and M. Grasso, "Delamination migration in CFRP laminates under mode I loading," vol. 190, 2020.

[15] C. D. Chen and W. L. Lai, "The analysis of mode II strain energy release rate in a cracked sandwich beam based on the refined zigzag theory," *Theoretical and Applied Fracture Mechanics,* vol. 107, 2020.

[16] A M. M. Shokrieh, A. Zeinedini and S. M. Ghoreishi, "On the mixed mode I/II delamination R-curve of E-glass/epoxy laminated composites," vol. 171, 2017.

[17] A. Ismail, A. Zubaydi, B. Piscesa and Tuswan, "Study of Sandwich Panel Application On Side Hull Of Crude Oil Tanker," *Journal of Applied Engineering Science,* 2021.

[18] *ASTM D5528-13 Standard Test Method for Mode I Interlaminar Fracture Toughness of Unidirectional Fiber-Reinforced Polymer Matrix Composites,* vol. 13, ASTM, 2013.

Comparative study of ship resistance calculated from CFD and obtained from towing tests: Case study of LCT vessel

D Suryawan[1], I K Suastika[1], and B Ali[2]

[1]Department of Naval Architecture, Institut Teknologi Sepuluh Nopember (ITS), Surabaya, Indonesia
[2]Indonesian Hydrodynamic Laboratory, BRIN, Surabaya, Indonesia

E-mail: suastika@its.ac.id

Abstract. A comparative study of the resistance of a landing craft tank (LCT) is reported, which is obtained from experiments (EFD) and computational fluid dynamics simulations (CFD). LCTs are naval vessels designed for amphibious warfare, specifically to transport tanks, vehicles, and troops to shore without the need for docks or piers. They are playing an important role in naval and amphibious operations, for example, in transporting heavy vehicles, equipment, and troops to unprepared shores. The LCT considered in this study was designed and constructed by the Indonesian Hydrodynamics Laboratory (IHL), BRIN, Surabaya, Indonesia, in 2021. The towing tests were conducted by the IHL BRIN to measure the vessel's resistance at various speeds, and CFD calculations were carried out to verify the experimental results. Results of the study show that the vessel's total resistance calculated from the CFD agrees well with the experimental data with an averaged discrepancy of 2.98%. This result is supported by the agreement in wave patterns obtained from the CFD and EFD.

1 Introduction

Landing craft tanks (LCTs) play an important role in naval and amphibious operations. Understanding their hydrodynamic characteristics is crucial for optimizing their performance. The LCT model considered in this study was designed and manufactured by the Indonesian Hydrodynamics Laboratory (IHL), BRIN, Surabaya, Indonesia, in 2021. This test model facilitates the experimental investigations on its hydrodynamic characteristics. All model tests were conducted by the IHL BRIN. The vessel's geometry and design parameters form a basis for the subsequent computational fluid dynamics (CFD) simulations. The CFD method can serve as a predictive model for future LCT designs. CFD has become a powerful tool for fluid flow simulations, which complements the theoretical and experimental studies of fluid dynamics [1, 2, 3].

The field of naval architecture is inherently linked to the optimization of maritime vessel design, with a particular emphasis on enhancing efficiency, safety, and performance. The resistance of a vessel significantly influences its overall performance [4]. LCTs have distinct hull forms and operational requirements. For LCTs, which are specifically designed for the efficient transport of military personnel, vehicles, and equipment from ship to shore, minimizing resistance is crucial for achieving operational success. Further, LCTs have a flat side inner (FSI) bottom hull, causing large waves [5].

The accurate prediction and understanding of resistance not only contribute to improved fuel efficiency but also impact the vessel's speed, maneuverability, and seakeeping abilities. This study aims at understanding the vessel's resistance through comparative study of the experimental results (EFD) with results from CFD simulations. The objectives of this study are to compare the LCT total resistance and the wave patterns obtained from the experiments (EFD) performed by IHL BRIN with the results of CFD simulations performed in this study.

2 Method

2.1 Governing equations

The RANS model is utilized to simulate the turbulent flow passsed the vessel's hull [3, 6, 7]. For an incompressible flow without heat transfer, the RANS equations for the mean flow consist of continuity and momentum equations as follows:

a. Continuity equation:

$$\nabla \cdot [\rho \mathbf{U}] = 0 \tag{1}$$

b. Momentum equation:

$$\frac{\partial [\rho \mathbf{U}]}{\partial t} + \nabla \cdot \{\rho \mathbf{U}\mathbf{U}\} = -\nabla P + \left[\nabla \cdot \{\overline{\tau}_{ij} - \rho\overline{\mathbf{u}'\mathbf{u}'}\}\right] + \rho \mathbf{g} \tag{2}$$

In Eqs. (1) and (2), $\mathbf{U}$ is the mean velocity, P is the mean pressure, $\overline{\tau}_{ij}$ is the mean viscous stress tensor, ρ is the fluid density, $\mathbf{g}$ is the gravitational acceleration, and t is the time. The term $-\rho\overline{\mathbf{u}'\mathbf{u}'}$ in Eq. (2) is the Reynolds stress tensor, denoted as τ^R.

Utilizing Boussinesq' hypothesis [8, 9, 10], the Reynolds stress tensor is represented as

$$\tau^R = -\rho\overline{\mathbf{u}'\mathbf{u}'} = \mu_t \left\{\nabla \mathbf{U} + (\nabla \mathbf{U})^T\right\} - \frac{2}{3}(\rho k)\mathbf{I} \tag{3}$$

where μ_t is the turbulence viscosity, k is the turbulence kinetic energy, and $\mathbf{I}$ is an identity tensor. The turbulence model utilized in this study to model the Reynolds stress tensor τ^R is the $k-\omega$ SST (shear stress transport) turbulence model [11]. In this model, the turbulence viscosity is calculated as

$$\mu_t = \rho\frac{k}{\omega} \tag{4}$$

where ω is the dissipation rate of the turbulence kinetic energy.

The equations for the turbulence kinetic energy k and the turbulence dissipation rate ω are as follows.

a. Equation for the turbulence kinetic energy k:

$$\frac{\partial}{\partial t}(\rho k) + \nabla \cdot [\rho k \mathbf{U}] = \nabla \cdot [\mu_{eff,k}\nabla k] + \underbrace{P_k - \beta^* \rho k \omega}_{S_k} \tag{5}$$

b. Equation for the turbulence dissipation rate ω:

$$\frac{\partial}{\partial t}(\rho \omega) + \nabla \cdot [\rho \omega \mathbf{U}] = \nabla \cdot [\mu_{eff,\omega}\nabla \omega] + \underbrace{C_{\alpha 1}\frac{\omega}{k}P_k - C_{\beta 1}\rho\omega^2}_{S_\omega} \tag{6}$$

For further details on the $k-\omega$ SST turbulence model, see, e.g., references [11, 12].

2.2 Computational domain and boundary conditions

Figure 1 shows the computational domain with inlet, outlet, symmetry plane, side wall, bottom and upper walls [6, 13]. The boundary conditions are summarized in Table 1. Two-phase fluid flow simulations were conducted, i.e., water below the free surface and air above the free surface. Generation of surface waves is modelled by using the volume of fluid (VoF) method [14]. The transport equation for the volume fraction α is given as follows:

$$\frac{\partial \alpha}{\partial t} + \nabla \cdot [\alpha \mathbf{U}] = 0 \tag{7}$$

The vessel speeds simulated in the CFD calculations are the same as those measured in the towing tests, which are eight speeds as follows: 0.762, 0.918, 1.066, 1.218, 1.367, 1.527, 1.677 and 1.827 m/s.

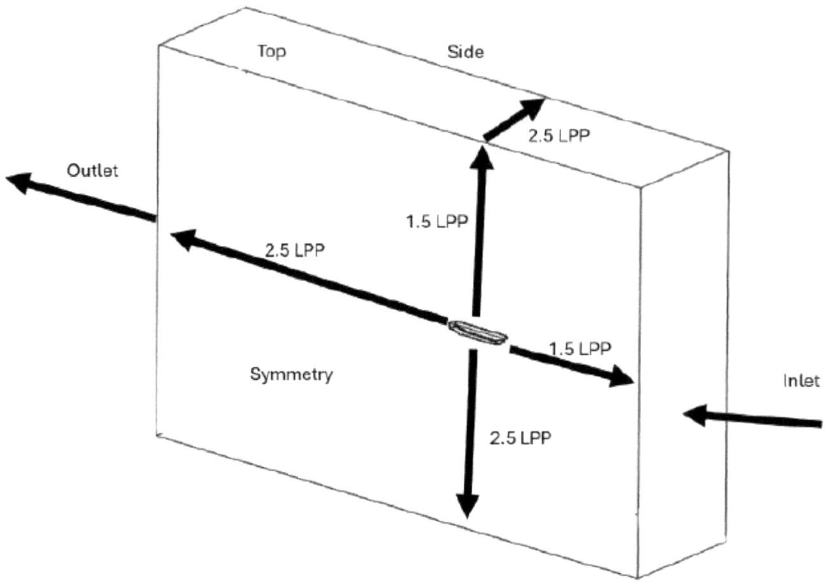

Figure 1: Computational domain.

Table 1: Boundary conditions.

Boundary	Condition
Inlet	Prescribed velocity
Outlet	Pressure
Top wall	Symmetry
Bottom wall	Symmetry
Side wall	Symmetry
Symmetry plane	Symmetry
Vessel's hull	No slip

2.3 Geometrical modelling and meshing

Figure 2 shows result of the three-dimensional (3-D) geometrical modelling of the LCT vessel. This 3-D geometrical model is required for the CFD simulations. Table 2 shows a comparison of parameters of the experimental model utilized in the towing tests (EFD) and the numerical model utilized in the CFD simulations. Table 2 shows that the discrepancies between the EFD and CFD models are less than 4% for all parameters considered, indicating an accurate numerical modelling result.

In the meshing process, the first step is to make a mesh and focus on the shape of the vessel with the aim of getting accurate calculation results. The smaller the mesh size, the more accurate the results will be. The body sizing process is done by adding geometry that is unified in the domain and geometry of the vessel with the purpose to get a good sizing mesh and accurate geometrical modelling around the vessel's hull. The result of the body sizing process is shown in Fig. 3. It can be seen in Fig. 3 that the mesh is fairly densed near the vessel's hull, from the baseline area to the water line. The body sizing area is as wide as the domain that has been created, and body sizing is also useful for adjusting the mesh size near the vessels's hull.

The resulting mesh without body sizing differs from that using a body sizing. Body sizing only focuses on the simulated area in order to get accurate results. In addition to the body sizing process is the face sizing process, which is useful for adjusting the cell size on the face of the geometry. Both meshing processes aim at getting accurate simulation results. The mesh quality is checked in the grid independence tests and grid convergence index, discussed in the following sub-section.

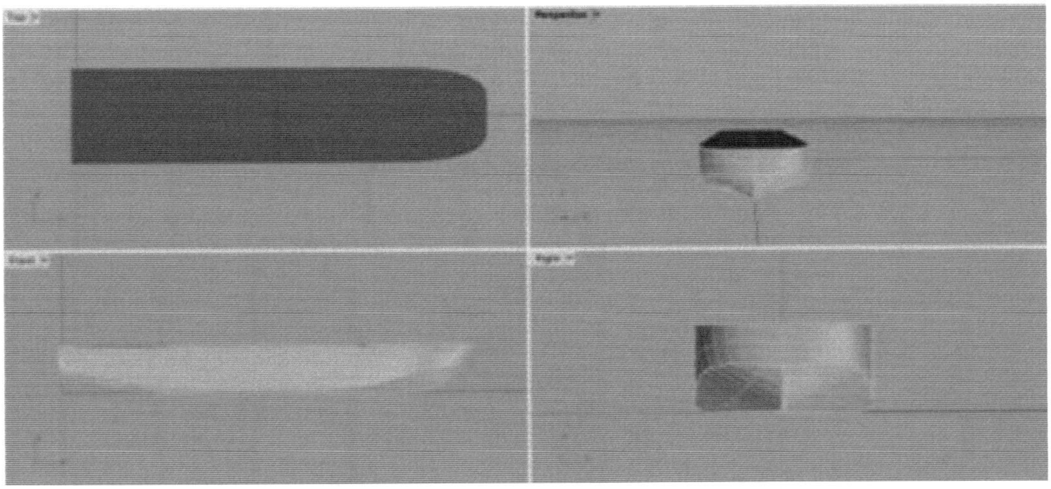

Figure 2: Result of 3-D geometrical modelling of the LCT vessel.

Table 2: Comparison of geometrical parameters between experimental and numerical models.

Parameter	Experimental model	Numerical model	Discrepancy [%]
Displacement [m^3]	0.76	0.78	2.63
Length overall, LOA [mm]	4308.2	4308.2	0
Length between perpendiculars, LPP [mm]	4000	4000	0
Length of water line, LWL [mm]	4082.9	4082.9	0
Beam, B [mm]	997.5	997.5	0
Height, H [mm]	306.3	306.3	0
Draft, T [mm]	218.8	218.8	0
Wetted surface area, WSA [m^2]	5.54	5.56	3.61
C_B [-]	0.41	0.42	2.43
C_{WP} [-]	0.63	0.65	3.17

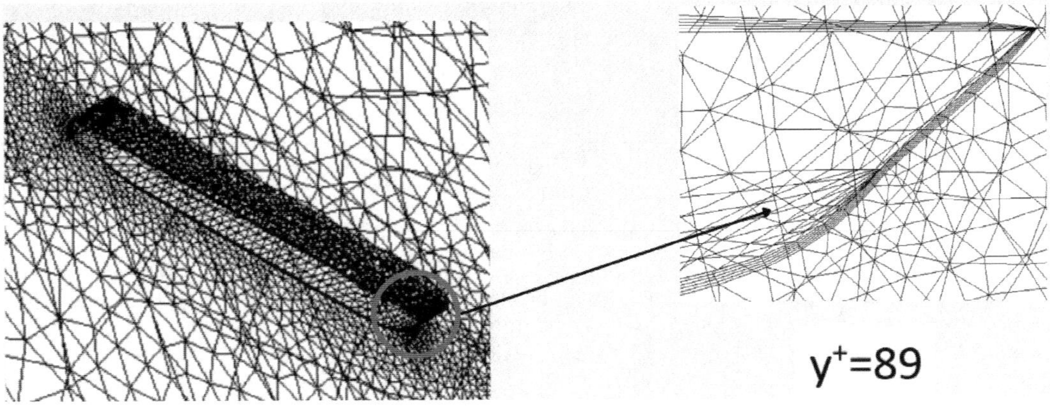

Figure 3: Fine mesh near the hull surface.

2.4 Grid independence tests and grid convergence index

In the grid independence tests the total vessel resistance is calculated by using an increasing number of cells in the CFD simulations. The number of cells in the current run is approximately twice that in the previous run. Table 3 summarizes results of the grid independence tests. The percent relative difference between run no. 4 and 5 is 1.78%, which is less than 2% used as the criterion for grid independence [1]. Based on the results tabulated in Table 3, run no. 4 with number of cells of 2 137 090 is considered as the most optimum number of cells.

The grid convergence index (GCI) is a parameter to quantify whether the grid has been sufficiently refined to provide accurate and stable results [15]. Results of calculations to determine the GCI are summarized in Table 4. Convergence refers to how closely the numerical solution approaches the exact solution as the grid size is reduced. The order of accuracy measures how quickly errors decrease as the grid is refined, while the extrapolated relative error estimates the relative error between the numerical and exact solutions. As shown in Table 4, the approximate relative error is in the range 0.018–0.025%, while the extrapolated relative error is in the range 0.045-0.063%. The GCI is calculated with a safety factor of 1.25 [16], which is in the range 0.053–0.074%. A GCI value of less than 1% indicates that discretization errors are small, demonstrating that the solution is sufficiently accurate.

Table 3: Grid independence study.

Run no.	Number of cells	Total resistance [kN]	Discrepancy [%]
1	197 592	7.24	-
2	573 202	7.47	3.10
3	1 051 488	7.03	6.32
4	2 137 090	6.86	2.51
5	4 274 180	6.74	1.78

Table 4: Grid convergence index.

Parameter	Equation	Value
Difference of estimation	$\epsilon_{21} = C_{T2} - C_{T1}$	0.074×10^{-3}
	$\epsilon_{32} = C_{T3} - C_{T2}$	0.10×10^{-3}
Convergence ratio	$R_i = \epsilon_{21}/\epsilon_{32}$	0.71
Refinement ratio	$r_i = r_{21} = (NG_2/NG_1)^{1/3} = r_{32} = (NG_3/NG_2)^{1/3}$	1.3
Order of accuracy	$p = \ln(\epsilon_{32}/\epsilon_{21})/\ln(r_i)$	1.328
Extrapolated value	$C_{T,ext-21} = (r_{21}^p C_{T1} - C_{T2})/(r_{21}^p - 1)$	0.044
	$C_{T,ext-32} = (r_{32}^p C_{T2} - C_{T3})/(r_{32}^p - 1)$	0.044
Approximate relative error [%]	$\epsilon_{a-21} = [(C_{T1} - C_{T2})/C_{T1}] \times 100\%$	0.018
	$\epsilon_{a-32} = [(C_{T2} - C_{T3})/C_{T3}] \times 100\%$	0.025
Extrapolated relative error [%]	$\epsilon_{ext-21} = [C_{T,ext-21} - C_{T1}/C_{T,ext-21}] \times 100\%$	0.045
	$\epsilon_{ext-32} = [C_{T,ext-32} - C_{T2}/C_{T,ext-32}] \times 100\%$	0.063
Grid convergence index (GCI) [%]	$GCI_{21} = 1.25\epsilon_{21}/(r_{21}^p - 1)$	0.053
	$GCI_{32} = 1.25\epsilon_{32}/(r_{32}^p - 1)$	0.074

3 Results and discussion

3.1 Vessel's total resistance

Results of the CFD calculations for the vessel's total resistance are compared with those results obtained from the towing tests (EFD). The comparisons are summarized in Table 5. In addition, the total resistance coefficients C_T are also calculated and summarized in Table 5. The total resistance coefficient C_T is calculated as follows:

$$C_T = \frac{R_T}{(1/2)\rho V^2 S} \tag{8}$$

Table 5: Total resistance and coefficient of total resistance.

Vessel's speed [m/s]	Fr	R_T [kN]		$C_T \times 10^3$		Discrepancy [%]
		EFD	CFD	EFD	CFD	
0.762	0.12	6.72	6.86	4.16	4.25	2.1
0.918	0.15	9.16	9.60	3.91	4.10	4.8
1.066	0.17	12.90	12.80	4.08	4.05	0.8
1.218	0.19	17.52	19.00	4.25	4.61	8.4
1.367	0.22	23.42	23.20	4.51	4.47	0.9
1.527	0.24	33.47	33.56	5.16	5.18	0.3
1.677	0.27	43.66	45.20	5.58	5.78	3.5
1.827	0.29	62.64	62.80	6.75	6.77	0.3

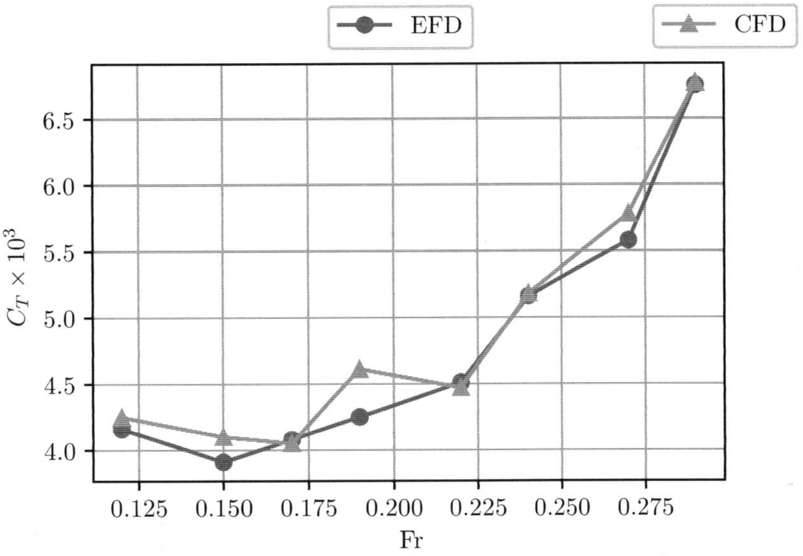

Figure 4: Coefficient of total resistance C_T as function of Froude number Fr.

where R_T is the total resistance, ρ is the water density, V is the vessel's speed and S is the wetted surface area.

Table 5 shows that the resistance obtained from the experiment is close to the CFD results, where the discrepancy is below 5%, except at Fr = 0.19. The average discrepancy from all the data is 2.98%, indicating that the CFD simulation results are consistent with the experimental data (EFD).

Figure 4 visualizes the variation of the total resistance coefficient C_T as function of the Froude number Fr. Figure 4 shows that C_T initially decreases and then tends to increase with increasing Fr.

3.2 Wave pattern

A visualization of the wave pattern is important to understand the differences in wave patterns at various speeds, particularly the bow and stern waves. Further, the simulated wave patterns are compared with those wave patterns observed in the towing tank experiments.

Figure 5 shows the wave patterns at Fr = 0.22, 0.24, 0.27 and 0.29 obtained from the CFD simulations. Differences in wave patterns and wakes generated at different Fr can be observed. At Fr = 0.22 the wave patterns are visible at the bow and stern of the LCT hull, with a rather weak wake. As the speed increases, both the bow waves and wakes become more pronounced. At Fr = 0.24 waves begin forming at the bow and stern of the LCT vessel, which become more pronounced at Fr = 0.27 and 0.29.

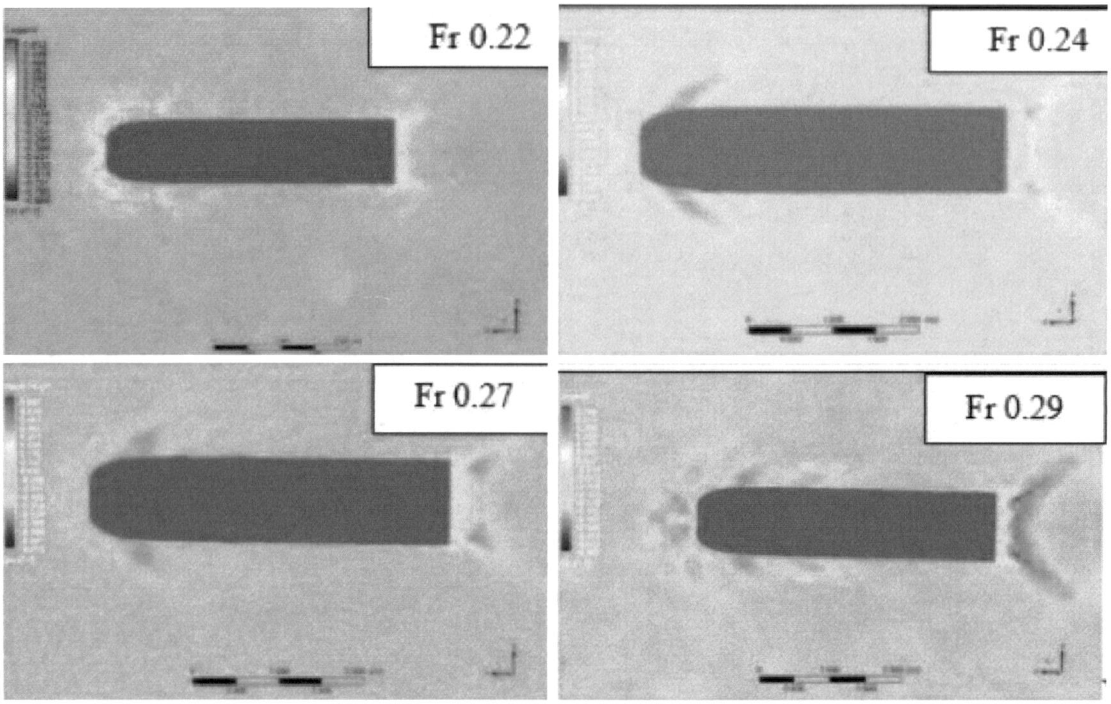

Figure 5: Wave patterns at Froude number Fr = 0.22, 0.24, 0.27 and 0.29.

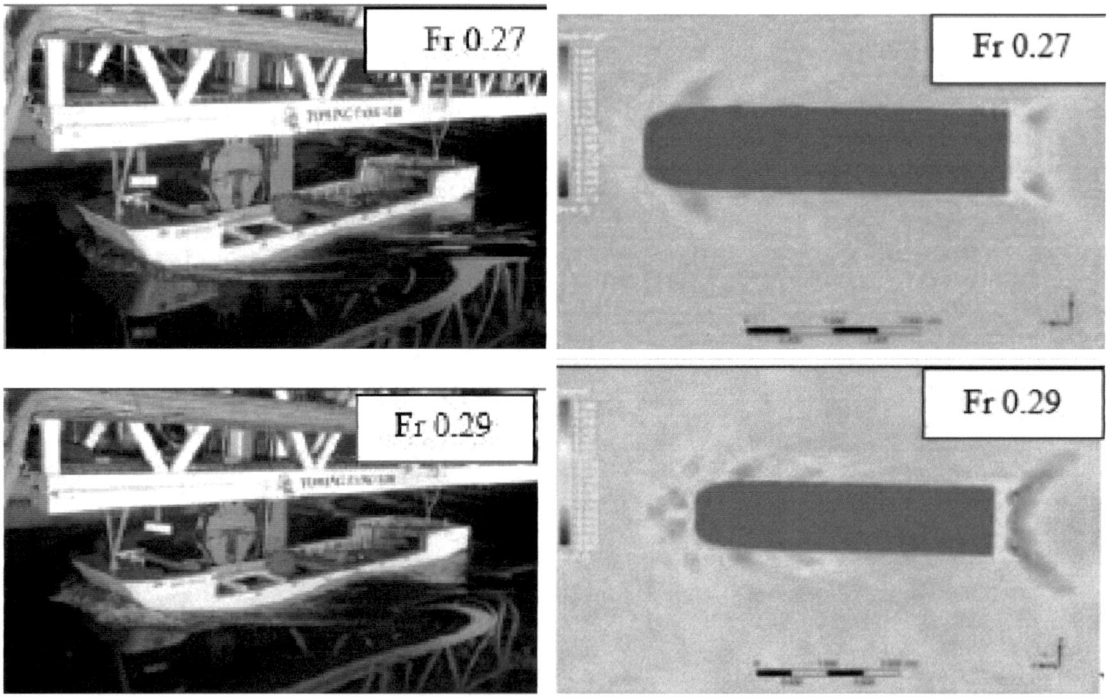

Figure 6: Comparison between wave patterns obtained from EFD (left column) and CFD (right column) at Froude number Fr = 0.27 and 0.29.

A comparison between the wave patterns obtained from CFD simulations and towing tank experiments (EFD) is shown in Fig. 6 at Fr = 0.27 and 0.29. Figure 6 shows similar wave patterns obtained from the CFD and EFD at the bow, middle body, and stern of the LCT vessel.

4 Conclusions

A comparative study of the resistance of an LCT vessel is reported in this study where results of CFD simulations are compared with data obtained from the towing tests (EFD). The CFD results agree well with the experimental data with an averaged discrepancy of 2.98%. The total resistance coefficient C_T initially decreases and then tends to increase with increasing Fr. The agreement in vessel's total resistance is supported by the agreement in wave patterns obtained from the CFD and EFD.

Acknowledgement

The authors gratefully acknowledge the financial support for this work from Institut Teknologi Sepuluh Nopember (ITS), Surabaya, Indonesia, under the project scheme Publication Writing and IPR Incentive Program (PPHKI) 2024.

References

[1] J. D. Anderson. *Computational Fluid Dynamics: The Basics with Applications.* McGraw-Hill, Inc., 1995.

[2] J. H. Ferziger and M. Peric. *Computational Methods for Fluid Dynamics.* Springer, 2002.

[3] H.K. Versteeg and W. Malalasekera. *An Introduction to Computational Fluid Dynamics: The Finite Volume Method.* Longman Scientific, Harlow, UK, 2007.

[4] A. F. Molland, S. R. Turnock, and D. A. Hudson. *Ship Resistance and Propulsion.* Cambridge University Press, New York, USA, 2011.

[5] O. B. Yaakob, A. Nasirudin, M. P. Abdul Ghani, T. Mat Lazim, M. A. Abd Mukti, and Y. M. Ahmed. Parametric study of a low wake-wash inland waterways catamaran. *Scientia Iranica*, 19(3):463–471, 2012.

[6] ITTC. Recommended procedures and guidelines. practical guidelines for ship cfd applications. Technical report, International Towing Tank Conference, 2011.

[7] F. Moukalled, L. Mangani, and M. Darwish. *The Finite Volume Method in Computational Fluid Dynamics.* Springer, Switzerland, 2016.

[8] J. Boussinesq. Essai sur la théorie des eaux courantes. *Mémoires présentés par divers savants à l'Academie des Sciences*, 23(1):1–680, 1877.

[9] H Schlichting. *Boundary-layer Theory.* McGraw Hill, 1968.

[10] F. G. Schmitt. About boussinesq's turbulent viscosity hypothesis: historical remarks and a direct evaluation of its validity. *Comptes Rendus Mécanique*, 335(9 and 10):617–627, 2007.

[11] F. R. Menter. Two-equation eddy-viscosity turbulence models for engineering applications. *AIAA J.*, 32(8):1598–1605, 1994.

[12] D. C. Wilcox. *Turbulence Modeling for CFD.* DCW Industries, USA, 2nd edition, 1998.

[13] K. S. Min and S. H. Kang. Study on the form factor and full-scale ship resistance prediction method. *J. Marine Sci. Tech.*, 15(2):108–118, 2010.

[14] C. W. Hirt and B. D. Nichols. Volume of fluid (vof) method for the dynamics of free boundaries. *J. Comput. Phys.*, 39:201–225, 1981.

[15] ITTC. Recommended procedures and guidelines. uncertainty analysis in cfd verification and validation, methodology and procedures. Technical report, 2021.

[16] P. J. Roache. Uncertainties and cfd code validation. *J. Fluids Eng.*, 120(3):635, 1998.

Numerical simulation of sloshing with floating baffles using smoothed particle hydrodynamics

Rozaq Abdullah[1], Andi Trimulyono[1]*, Samuel[1], Berlian Arswendo Adietya[1]

[1] Department of Naval Architecture, Faculty of Engineering, Universitas Diponegoro, Semarang, Indonesia

*E-mail: anditrimulyono@live.undip.ac.id

Abstract. The rising demand for liquefied natural gas (LNG) has made membrane-type LNG tanks popular due to their effectiveness in reducing sloshing effects. However, incidents involving free surface effects during voyages still occur. This study examines the impact of floating baffles on sloshing waves under force oscillation conditions, including coupled motions (sway and rolling) to represent actual ship tank scenarios. Tanks were filled to 25% ratios to assess the free surface effects. The baffle height was adjusted based on previous research, targeting a ratio of ≥ 0.75 between the baffle's vertical distance and the water surface height. Simulations were performed using mesh-free CFD, i.e., Smoothed Particle Hydrodynamics (SPH), a mesh-free method for handling large fluid deformations and violent flows. The results show that floating baffles can mitigate and worsen sloshing effects, depending on conditions, as evidenced by wave height, hydrostatic pressure, and dynamic pressure measurements within the tank.

1. Introduction

Demand for liquefied natural gas (LNG) has increased significantly in recent years. The use of membrane-type LNG tanks has become the main choice because of their superior ability to overcome sloshing effects. Membrane-type prismatic tanks, in particular, are often used in the liquid cargo tanker industry due to their large capacity. Sloshing, defined as the free movement of fluid in tanks, is a serious problem in the distribution of liquid cargo on ships, ranging from reducing the stability of the ship to the potential for fire that can cause the sinking of the ship. As a complex nonlinear phenomenon in fluid dynamics, sloshing is very difficult to overcome. One effective way to mitigate sloshing is to use baffles in the tank. Various analyses, both experimental and numerical, have been carried out and continue to develop along with technological advances to increase analytical efficiency. This research uses Smoothed Particle Hydrodynamics (SPH), a fully Lagrangian-based numerical approach that does not rely on mesh and has the advantage of overcoming non-linear phenomena.

SPH was originally for astrophysical problems, however, Monaghan expanded the use of SPH to free surface flows, including the propagation of water waves and dam breaks [1]. SPH is the preferred option when it comes to sloshing because of its exceptional capacity to manage large fluid deformations [2,3]. It has been demonstrated that SPH works well at simulating the fluid resonance in the tank that is caused by outside vibrations, closely resembling the features of the sloshing phenomena that occur in liquid transport vessels. The applications of SPH have widened not only for sloshing but for water wave propagation [4,5], which shows SPH has a promising method for free surface problems in marine engineering.

The implementation of Smoothed Particle Hydrodynamics (SPH) has been proven effective in simulating sloshing behavior, showing that different tank designs can significantly influence sloshing dynamics [6,7]. In a study on the sloshing effects on LNG ships, simulated with two oscillatory motions, the SPH method has been shown to have advantages in simulating complex internal and external flows [8]. An analytical study using single and two-phase SPH on a cylindrical tank with a low-pass filter technique successfully dampened pressure oscillations caused by WCSPH [9]. Optimal placement of vertical baffles has been shown to be effective in attenuating sloshing effects in cylindrical tanks [9]. A study conducted by Trimulyono et al [10] shows that the use of vertical or T-shaped partitions has a significant impact on reducing wave height. Another study explores the use of floating baffles as an alternative in handling sloshing effects, where the baffle height can be adjusted according to the filling ratio and has been proven effective in reducing peak fluid pressure at lower fill levels [11].

This research investigates the sloshing effect in a prismatic tank by varying the number of baffles on the floating baffle. The baffle height is based on previous research [12]. The baffle height is based on previous research, which states that the optimal baffle height ratio is $0.75 \leq hb/hw \leq 1$, with hb being the baffle height and hw being the water surface height in the tank. Pressure sensors are used to validate simulation results with experimental data conducted by Trimulyono et al [13]. In addition, a water level with a ratio of 25% was used to observe the deformation effect of the free surface in the tank. Most research on sloshing usually uses only one oscillatory movement to study free surface effects in the tank. However, to approach actual shipping conditions, this research combines two oscillatory movements, namely rolling and swaying, simultaneously in one simulation. The movements used are rolling and swaying combined simultaneously. In this study, DualSPHysics code [14] was employed and developed to reduce high computational costs, by implementing GPU strategies that significantly improve performance, including reduced global memory accesses and increased occupancy for better processing efficiency, in addition, advanced visualization was also carried out using Blender version 2.92 [15].

2. Numerical setup

2.1 Numerical settings of sloshing simulation

Smoothed Particle Hydrodynamics (SPH) is a completely Lagrangian approach without a mesh that uses smooth particles to identify mass, velocity, and position. This method allows for estimating physical values and derivatives of continuous fields using discrete evaluation points. These values are obtained from adjacent particles within the rarefaction length range (h), which aims to reduce the contribution from particles located at a greater distance. The main features of SPH, which are based on integral interpolation, are described in detail in refs [14].

Tanks with the same dimensions as those used in experimental trials by Trimulyono et al. [13] were used in simulations using SPH to observe the sloshing phenomenon that occurs after adding floating baffle variations. Figure 1 shows the sloshing experimental setup in a prismatic tank carried out by Trimulyono et al [13], whose data is used as validation in this research. In this research, three pressure sensors were used to obtain pressure data at certain points. However, in the SPH simulation validation, only one sensor was used, at a fill ratio of 25% (Figure 2). The pressure sensor used is a sensor located at the bottom of the tank because, at a filling ratio of 25%, this sensor is close to the free surface. More detailed information regarding this experimental sloshing experiment can be found in refs [13].

Figure 2 shows the prismatic tank geometry used in SPH calculations, where h, *l*, *L*, and *D* represent the height, width, length of the tank, and water depth, respectively. Table 1 presents the main dimensions of the prismatic tanks used. The amplitude and frequency of the oscillatory movements used are 6.52 mm/degree and 1.08 Hz for swaying movements, and 8.66 mm/degree and 1.04 Hz for rolling movements. Previous studies have shown that this excitation frequency is close to the natural frequency of prismatic tanks.

(a) (b)

Figure 1. (a) Overview of the oscillating machine for the sloshing test (b) Photo of the tank and 3 pressure sensors

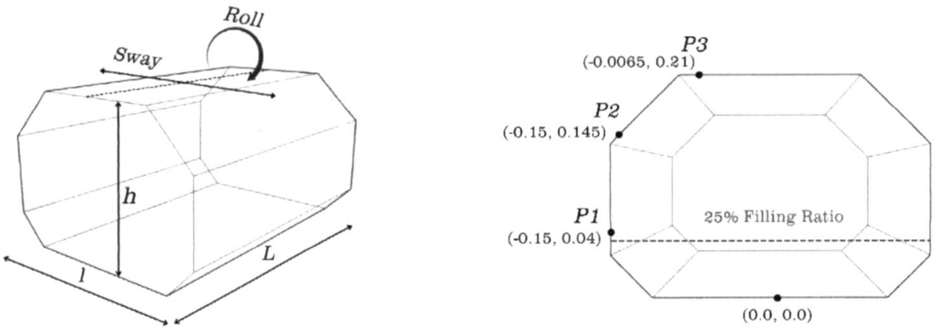

(a) (b)

Figure 2. (a) Tank geometry (b) Fill ratio and 3 pressure point coordinates

Table 1. Principal dimensions of the prismatic tank

Dimension	(m)
H	0.21
l	0.30
L	0.38
D	0.0525 (25%)

External oscillations are applied based on movements determined from the displacement time history, which comes from experiments carried out by Trimulyono et al [13]. In the previous simulation, 5 seconds without external movement was added to allow the fluid to reach a steady state before oscillations began. Table 2 presents the numerical parameter data used during the SPH simulation. The Wenland kernel function is used in this simulation. The use of Wenland kernels helps reduce numerical errors and improve simulation accuracy, especially in thin structures with high aspect ratios [16]. In this scenario, since the simulation is done in 3D, the value of "Coefh" can be defined as $Coefh = \frac{h}{dp\sqrt{3}}$, where dp is the distance between the particles in the simulation. Artificial viscosity coefficients (α) need to be included to achieve proper dissipation [8]. In the previous simulation, the α value used was 0.01; In this study, the same artificial viscosity coefficient values were used. According to Trimulyono et al [13], The speed of sound has a significant influence on the pressure field based on the equation of state. The speed of sound is defined by $Coefsound = \frac{Cs}{\sqrt{gdswl}}$, where Cs and dswl are the speed of sound and the height of the water when it is in a calm position. The Coefsound value used is 60, which was obtained from the results of previous numerical tests. The CFL number is used to calculate the time steps in the simulation to satisfy the Courant-Friedrichs-Lewy (CFL) condition. CFL coefficients play an important role in ensuring the numerical stability of fluid flow simulations. If the CFL conditions are not met, the simulation results may become unstable, resulting in errors in the prediction of flow behavior [17]. The delta-SPH coefficient value was used to reduce density fluctuations, as recommended by Holteni and Colagrossi [18], with a value of 0.1 used in the simulation. Due to limited equipment for running simulations, an interparticle distance of 1.7 mm was chosen, because this value is close to the optimal particle distance that has been studied by Trimulyono et al [13], and the simulation time is 28 seconds.

Table 2. Computational setting in DualSPHysics

Kernel Function	**Wendland**
Time step algorithm	Symplectic
Artificial Viscosity Coefficient (α)	0.01
Coefsound	60
Particle Spacing (mm)	1.7
Koefh	1.2
CFL	0.2
Delt-SPH ($\delta\varphi$)	0.1
Simulation Time (second)	28

This research uses two variations of floating baffles and tanks. Floating baffle variations involve varying the number and distance of baffles and tank variations involve the use of tanks with and without rails to limit the movement of floating baffles in the tank. Figure 3 shows the main geometry and position of the baffle in the tank, where *b, hb, Lb*, and *lb* (see Table 3)refer to the thickness, height, length, and width of the baffle, respectively. The position of the floating baffle is adjusted to the filling ratio in the tank, where half of the baffle body is submerged in water (Figure 3). To achieve buoyancy according to the plan, the floating baffle is made of aluminum with a cavity inside (see Table 4). Specification data for floating baffle variations can be seen in Table 3, while the model geometry can be seen in Figure 4. The second tank variation involves adding rails that are adapted to the shape of the baffle and mounted on the inner wall of the tank

(Figure 5). The rail installation functions to limit the movement of the floating baffle during the simulation. The rail geometry can be seen in Figure 5.

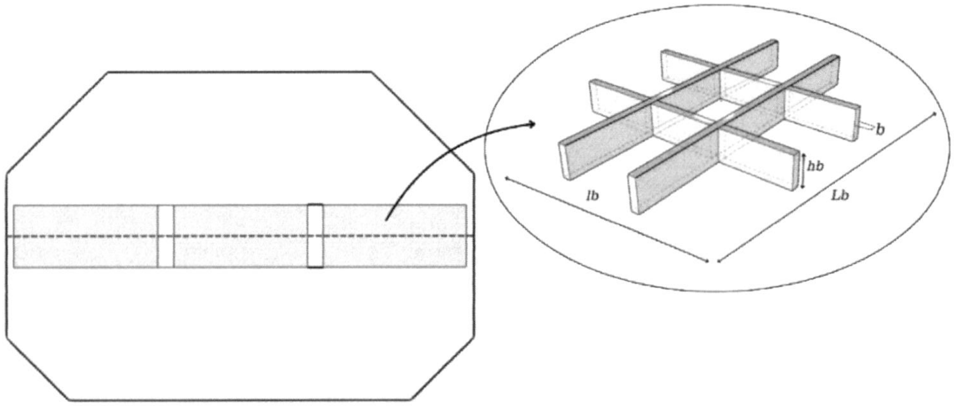

Figure 3. Floating Baffle Position and Geometry

Table 3. Principal dimension of Floating Baffle

Dimension	(m)
b	0.01
hb	0.04
Lb	0.37
lb	0.29

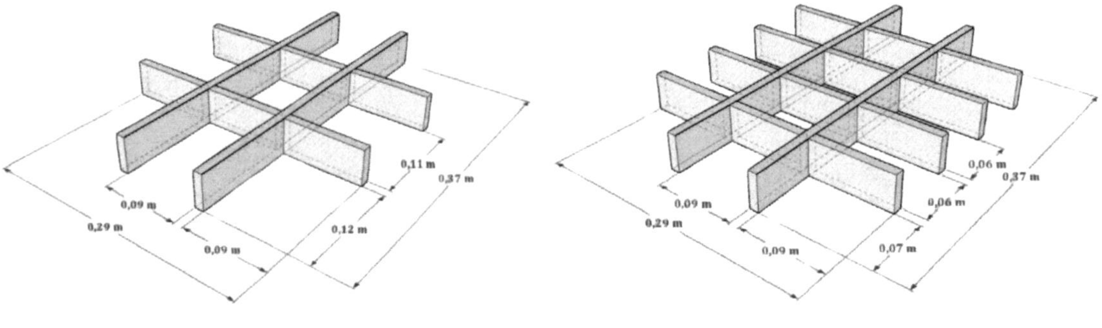

Figure 4. Geometry Floating Baffle

Table 4. Specifications for Floating Baffle Variations

Variation	Mass (Kg)	Vol. (m^3)	Properties	Density (kg/m^3)
Floating Baffle 1	0.256	0.000512	Aluminum	2700
Floating Baffle 2	0.364	0.000728	Aluminum	2700

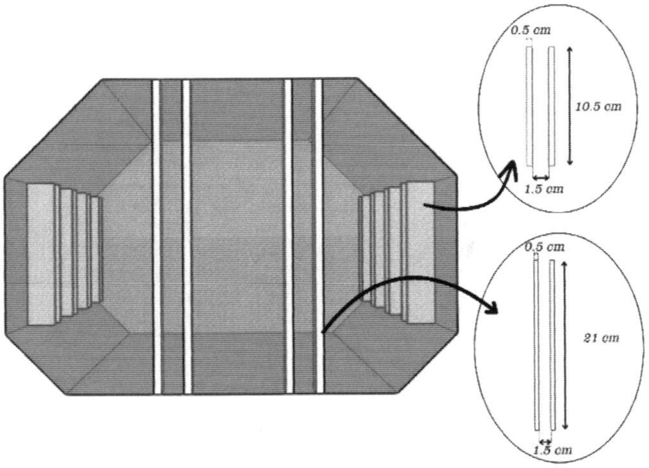

Figure 5. Geometry Rails

3. Results and discussion

3.1 Hydrostatic and Hydrodynamic Pressure

Figure 6 shows a comparison of hydrostatic pressure at a filling ratio of 25%, both with and without floating baffles. The hydrostatic pressure was reproduced well by the SPH method, with a difference of 3% compared to the analytical results [13]. From Figure 6, it can be seen that the hydrostatic pressure pattern corresponds to the analytical solution, where the highest pressure is at the bottom of the tank and the lowest near the free surface. Figure 7 displays the dynamic pressure due to sloshing in the baffle configuration at t=19.0 s. The free surface experiences more significant deformation in tanks without floating baffles while floating baffles with rails are proven to be more effective in reducing free surface deformation compared to those without rails.

Figure 8 presents a comparison graph of the dynamic pressure detected by the sensor at the bottom of the tank (see Figure 2). Dynamic pressure is obtained by subtracting static pressure from total pressure. In Figure 8(a), the impact of hydrodynamic pressure at a filling ratio of 25% can be seen, where the red and yellow lines represent the floating baffle with rails for variations one and two, while the blue line shows the test results without the floating baffle. Figure 8(b) shows the impact of hydrodynamic pressure at a filling ratio of 25%, with the orange and purple lines depicting the floating baffle without rails for variations one and two, while the blue line still represents the results without the floating baffle. The average difference in peak pressure between tanks without baffles and floating baffles with rails for variations one and two is 9.3% and 8.7% respectively. In contrast, in the railless configuration, the floating baffle negatively impacts the pressure measured at the bottom of the tank (Figure 2), with dynamic pressure increases of 9.9% and 8.8% for variations one and two.

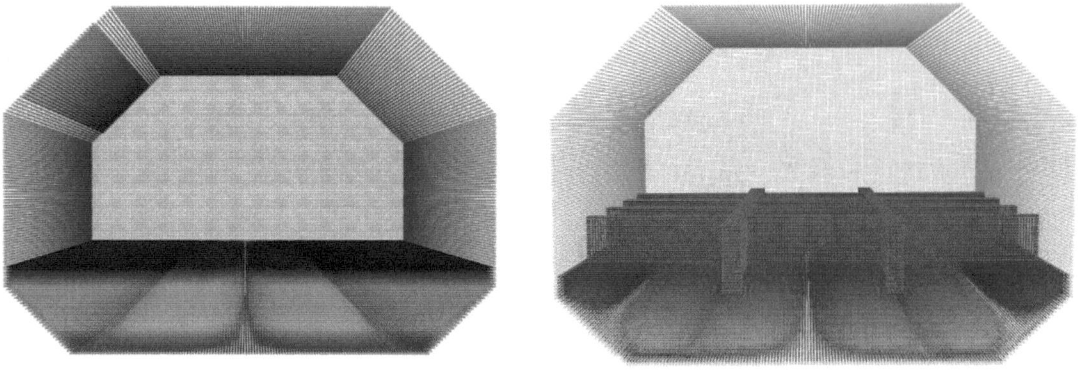

Figure 6. Hydrostatic pressure at 25% filling ratio without and with floating baffle

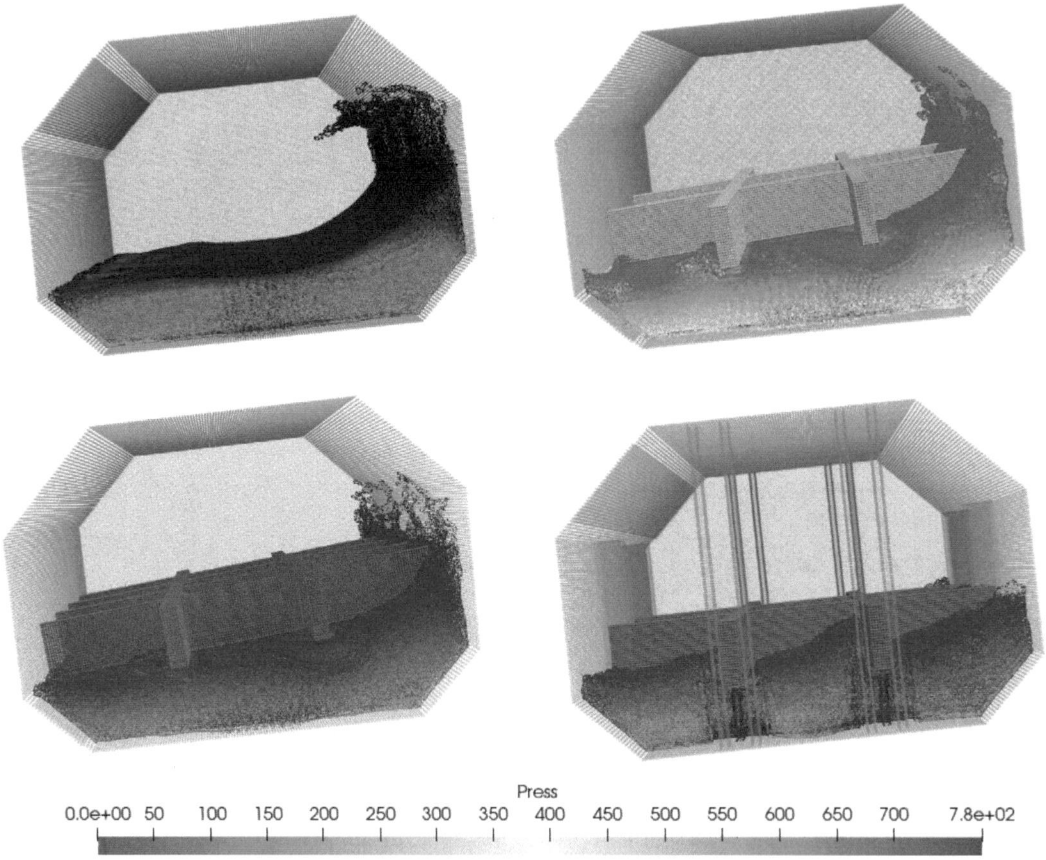

Figure 7. Dynamic pressure comparison at t = 19.0 sec

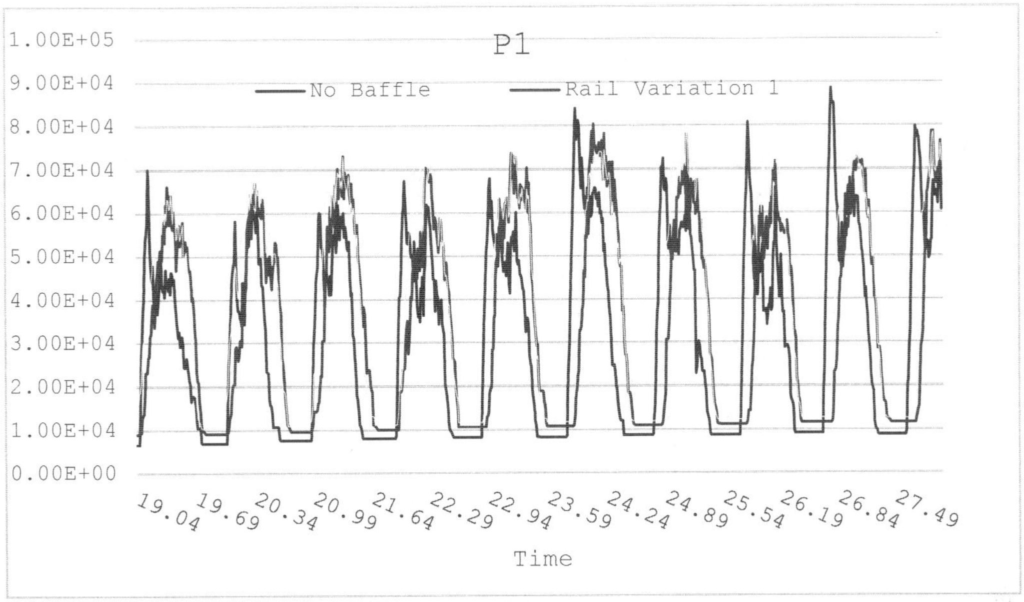

(a)

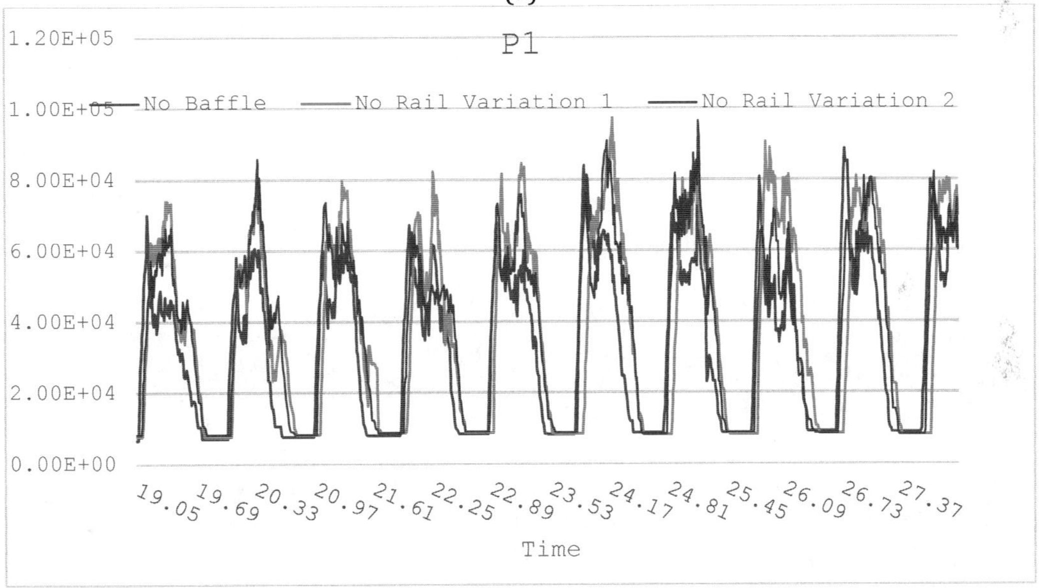

(b)

Figure 8. (a) Comparison of the dynamic pressure of a rail tank with a floating baffle (b) Comparison of the dynamic pressure of a tank without a rail with a floating baffle.

3.2 Free Surface Deformation

The advanced visualization process was carried out using Blender version 2.92, an open-source software that can be downloaded for free via https://www.blender.org/. With the latest visualization techniques, SPH post-processing becomes more realistic and closer to the reality condition. Figure 9 shows a more detailed comparison of surface texture visualization using Blender, which shows a fluid representation that is more similar to real conditions and more aesthetically pleasing compared to particle-based visualization.

The free surface deformation in the tank is influenced by the excitation force. The sloshing energy in fluid motion is often complex and irregular, especially when the frequency of the excitation force is close to the natural frequency of the tank. To effectively reduce fluid movement, the use of baffles is highly recommended. Figure 10 shows that without baffle installation, fluid run-up reaches the top of the tank. In contrast, after the installation of the baffles, the deformation of the fluid surface is significantly reduced.

The use of floating baffles with rails can reduce the maximum wave height by 8.48% and 4.33% for variations one and two respectively, as well as slightly reducing dynamic pressure, as shown in Figure 8(a). Meanwhile, floating baffles without rails are not effective in reducing free surface effects when the tank is subjected to a couple of oscillatory motions. In fact, in the floating baffle variation without rails in variation 1, there is a small increase in the height of the fluid-free surface of 1.44% at t = 17s – 18s. On the other hand, the floating baffle without rails in variation 3 succeeded in suppressing the maximum surface waves by 4.33%. Vertical Baffles with a height ratio of 0.75 to 1.0 to the fluid surface height have proven to be effective in overcoming the sloshing phenomenon. The energy produced by fluid movement can be dissipated well at the surface and bottom of the flow through the use of vertical baffles, which play a significant role in reducing the impact of sloshing due to rolling movements. The flow at the bottom of the fluid has an effect by interacting with the tank walls. When the sloshing wave moves, the flow at the bottom can influence the surface effect of the fluid which can cause an increase in pressure at certain points.

Based on visual observations, floating baffles with rails are proven to be more effective than those without rails. This is caused by the rail which functions to stabilize the movement of the floating baffle so that it is not too affected by the effects of free fluid spills in the tank. Thus, floating baffles with rails can reduce fluid-free surface deformation better than those without rails. However, the drawback of this floating baffle design is that there is a transverse partition that is locked on the bottom wall of the tank, which allows fluid to escape through the gap under the floating baffle, especially at low filling ratios and under certain conditions (Figure 12).

(a) (b) (c)

Figure 9. (a) Particle Visualization (b) Isosurface (c) Surface texture

Figure 10. Comparison of free surface deformation in tanks with and without floating baffles (a) Without Baffles (b) Floating baffles without rails (c) Floating baffles with rails.

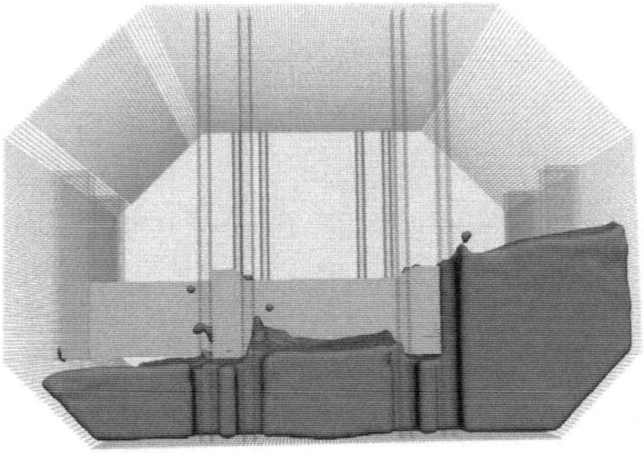

Figure 11. Fluid flow that passes under the Floating Baffle

3.3 Hydrodynamic force

Sloshing experiments were carried out using a forced oscillation machine at four degrees of freedom. Oscillatory movements such as heave and sway are often used in sloshing tests, both numerically and experimentally. Faraday movement is applied in testing the effects of liquid spills using a floating baffle with rails [19]. Floating baffles have proven effective in dealing with the effects of fluid spills due to regular sinusoidal movements. This research uses combined motion (coupled motion) to evaluate the effectiveness of floating baffles in overcoming the sloshing phenomenon. Floating baffles are rarely applied in the analysis of liquid surface spills involving rotational forces, while previous research has focused more on the use of floating baffles to dampen translational movements.

Hydrodynamic forces occur when the fluid in the tank is forced to move by an oscillating machine. Figure 12 shows the impact of hydrodynamic forces at a filling ratio of 25%. The blue line represents the hydrodynamic force without baffles, while the red, yellow, orange, and purple lines represent the hydrodynamic force for floating baffles without rails in the first and second variations and floating baffles with rails in the first and second variations. In floating baffles without rails, the maximum hydrodynamic pressure value is reduced by 1.15% and 5.36% for variations one and two respectively. However, in the floating baffle with rails, the hydrodynamic force value increased significantly in the second variation by 39.21% and increased by 2.88% in the first variation. This increase is caused by the addition of hydrodynamic force and weight on the floating baffle, which results in a significant increase in the hydrodynamic force value.

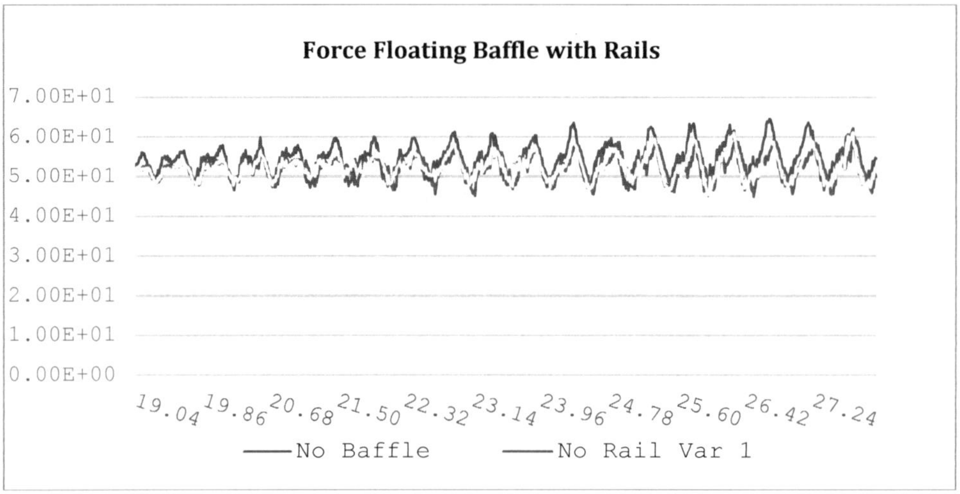

(a)

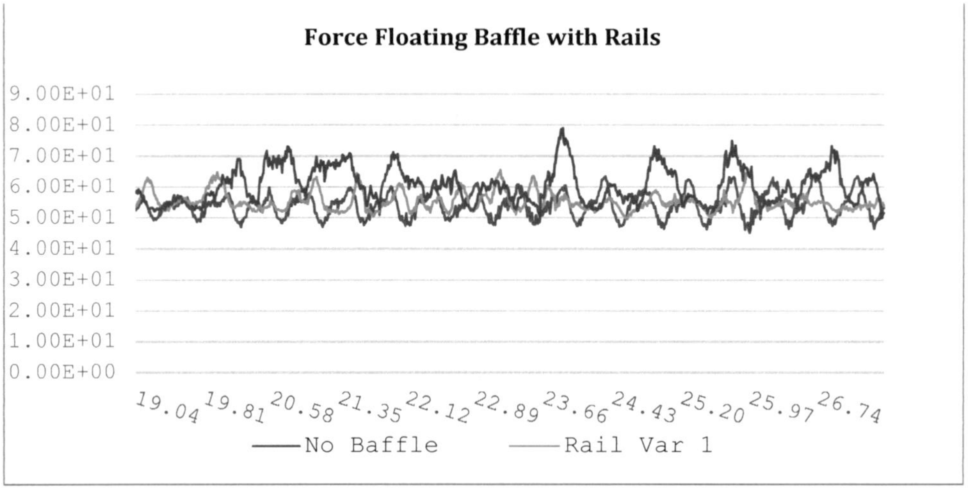

(b)

Figure 12. (a) Hydrodynamic force of floating baffle without rails (b) With rails

4. Conclusions

The sloshing simulation in a prismatic tank with a filling ratio of 25% was successfully carried out using the Smoothed Particle Hydrodynamics (SPH) method, which is very effective in handling fluids with large deformations. Simulation results reveal that floating baffles, under certain conditions, are able to reduce or even worsen the sloshing effect in the tank. Floating baffles have proven to be quite efficient in reducing sloshing waves due to sway movements, but are less than optimal for reducing rolling movements because fluid flow at the bottom of the tank can easily pass through the bottom of the baffle. The first variant of the floating baffle with rails shows the best performance in reducing free surface effects in the tank. Overall, the use of floating baffles with rails is effective in reducing dynamic pressure and free surface effects, although it is less efficient in handling hydrodynamic forces generated by fluids. On the other hand, floating baffles without rails are able to reduce hydrodynamic forces but are not as effective in resisting dynamic pressure and controlling fluid-free surface effects. Future works of sloshing two-phase with floating are undergoing research to know the effect of compressibility of air in the tank [20].

Acknowledgment

The author would like to thank the ship hydrodynamics laboratory of the Naval Architecture department at Diponegoro University for the conference registration fees and computer facilities provided.

References

[1] R. A. Gingold and J. J. Monaghan, "Smoothed particle hydrodynamics: theory and application to non-spherical stars," *Comput. Phys.*, no. 181, pp. 375–389, 1994. doi: https://doi.org/10.1093/mnras/181.3.375.

[2] C. Ma, C. Xiong, and G. Ma, "Numerical study on suppressing violent transient sloshing with single and double vertical baffles," *Ocean Eng.*, vol. 223, 2021. doi: https://doi.org/10.1016/j.oceaneng.2020.108557

[3] A. Trimulyono, S. Utami, D. Chrismianto, and P. Manik, "Investigation of Sloshing in Different Tank Shapes using Smoothed Particle Hydrodynamics," *CFD Lett.*, vol. 15, no. 12, 2023. doi: https://doi.org/10.37934/cfdl.15.12.1933

[4] A. Trimulyono, and H. Hashimoto, "Experimental validation of smoothed particle hydrodynamics on generation and propagation of waterwaves," *J. Mar. Sci. Eng.,* vol. 7, no. 8, 2019. doi: https://doi.org/10.3390/jmse7010017.

[5] M. A. Salvikran, A. Trimulyono, E. S. Hadi, "Validation of multiphase water wave propagation using smoothed particle hydrodynamics," *IOP Conf. Ser. Earth Environ. Sci.,* vol. 1166, 2023. doi: :10.1088/1755-1315/1166/1/012032

[6] A. Trimulyono, B. A. Adietya, J. I. Alvianto, D. Chrismianto and H. Syahab, "Numerical simulation of sloshing in different tank shape with vertical and horizontal baffle using smoothed particle hydrodynamics," *Kapal: J. Ilmu Pengetah. dan Teknol. Kelaut.*, vol. 21, no. 2, pp. 50–57, 2024. doi: https://doi.org/10.14710/kapal.v21i2.64288.

[7] Syamsuri, M. J. Chern, and N. Vaziri, "SPH model for interaction of sloshing wave with obstacle in shallow water tank," J. King Saud Univ. - Eng. Sci., vol. 34, no. 2, pp. 126–138, 2022. doi: 10.1016/j.jksues.2020.07.009.

[8] J. Jiao, M. Zhao, G. Jia, and S. Ding, "SPH simulation of two side-by-side LNG ships' motions coupled with tank sloshing in regular waves," *Ocean Eng.*, vol. 297, 2024. doi: https://doi.org/10.1016/j.oceaneng.2024.117022

[9] A. Trimulyono, S. Samuel, and M. Iqbal, "Sloshing simulation of single-phase and two-phase SPH using DualSPHysics," *Kapal: J. Ilmu Pengetah. dan Teknol. Kelaut.*, vol. 17, no. 2, pp. 50–57, 2020. doi: 10.14710/kapal.v17i2.27892.

[10] A. Trimulyono, H. Atthariq, D. Chrismianto, and S. Samuel, "Investigation of sloshing in the prismatic tank with vertical and T-shape baffles," *Brodogradnja*, vol. 73, no. 2, pp. 43–58, 2022. doi: 10.21278/brod73203.

[11] U. G. Md. Arif *et al.*, "Suppression of hydrodynamic sloshing in liquefied natural gas tank with floating baffle: Experimental and numerical studies," *IOP Conf. Ser. Earth Environ. Sci.*, vol. 463, no. 1, 2020. doi: 10.1088/1755-1315/463/1/012111.

[12] C. Ma, C. Xiong, and G. Ma, "Numerical study on suppressing violent transient sloshing with single and double vertical baffles," *Ocean Eng.*, vol. 223, 2021. doi: https://doi.org/10.1016/j.oceaneng.2020.108557.

[13] A. Trimulyono, H. Hashimoto, and A. Matsuda, "Experimental validation of single- and two-phase smoothed particle hydrodynamics on sloshing in a prismatic tank," *J. Mar. Sci. Eng.*, vol. 7, no. 8, 2019. doi: 10.3390/jmse7080247.

[14] J. M. Domínguez *et al.*, "DualSPHysics: from fluid dynamics to multiphysics problems," *Comput Part Mech*, vol. 9, no. 5, pp. 867–895, 2022. doi: https://doi.org/10.1007/s40571-021-00404-2

[15] O. García-Feal, A. J. C. Crespo, and M. Gómez-Gesteira, "VisualSPHysics: advanced fluid visualization for SPH models," *Comput Part Mech*, vol. 9, no. 5, pp. 897–910, 2022. doi: https://doi.org/10.1007/s40571-020-00386-7.

[16] X. Tang, D. Wu, Z. Wang, O. Haidn, and X. Hu, "Simulating plate and shell structures with anisotropic resolution using adaptive smoothed particle hydrodynamics," *Eng. Anal. Bound. Elem.*, vol. 167, 2024. doi: 10.1016/j.enganabound.2024.105886.

[17] Z. Kowalczuk and M. S. Tatara, "Analytical 'steady-state'-based derivation and clarification of the courant-friedrichs-lewy condition for pipe flow," *J. Nat. Gas Sci. Eng.*, vol. 91, 2021. doi: 10.1016/j.jngse.2021.103953.

[18] D. Molteni and A. Colagrossi, "A simple procedure to improve the pressure evaluation in hydrodynamic context

using the SPH," *Comput. Phys. Commun.*, vol. 180, no. 6, pp. 861–872, 2009, doi: 10.1016/j.cpc.2008.12.004.

[19] J. Lu, H. Jin, M. A. Xue, J. Shen, and S. Li, "Experimental investigation of the effect of the floating damping baffle on liquid tank sloshing under vertical excitation," *Appl. Ocean Res.*, vol. 147, 2024, doi: 10.1016/j.apor.2024.103976.

[20] A. Trimulyono, D. Chrismianto, S. Samuel, and M. H. Aslami., "Single-Phase and Two-Phase Smoothed Particle Hydrodynamics for Sloshing in the Low Filling Ratio of the Prismatic Tank," *International Journal of Engineering, Transactions B: Applications,* vol. 34, no. 25, 2021. doi: https://doi.org/10.5829/ije.2021.34.05b.30.

An Evaluation of Stability Parameter of Small Purse Seiner

Wolter R Hetharia[1*], Eliza R de Fretes[2], Fella Gaspersz[3] and Meily A Feninlambir[4]

[1,2,3,4]Department of Naval Architecture Faculty of Engineering, Pattimura University, Ambon, Indonesia
*E-mail: wolter.hetharia@lecturer.unpatti.ac.id

Abstract. Purse Seiner is a kind of fishing boat which is operated to catch small pelagic fishes with circle net at night time. This boat is constructed by local boat builders without considering any design parameters which impacts on limitation in boat's operation particularly stability parameters. In order to evaluate stability parameter inclining test was executed to obtain basic stability parameters of metacentre height GM and vertical centre of gravity KG. A case study was conducted with an observed boat of "*Purse Seiner Yana*" at Hitu Village in Ambon Island. The study began with providing related documents and executing inclining test. The value KG of 0.874 meter and nondimensional ratio KG/H of 0.740 were found at empty ship (departure) condition, where H is deck height. These values were used further to evaluate stability parameter at large angles of inclination at any loading conditions (arrival). A scenario of variation of placement cargo (fishes) on deck and at fish hold was executed. Placement of cargo on deck decreased stability value followed by combination of placement cargo on deck and fish hold. In fact, all load variations were result in the boat is still stable particularly cargo 3 tons on deck and 3 tons in fish hold. The results of cargo variation are presented in value of KG and ratio of KG/H according to loading positions and quantities. In addition, regression equations are provided to assist designers or boat operators to find any values of loading conditions, particularly at early design phase. The results may be applied as justification for the application of Mobile Phone to evaluate boat's stability during operation.
Keywords: stability fishing vessel, loading conditions, Inclining test, load variation

1. Introduction

Fishing boats *purse seiner* have great contribution in catching small pelagic fishes for Indonesian domestic consumption and international export. Those boats use circle net in catching operation in the night time when the fishes have existed under *rumpon (a device to collect pelagic fishes)*. Those boats operate using circle net and were found mostly at Indonesian waters. In fact, there are problems faced by those boats such as safety and other technical and economic aspects. One important parameter which should be considered is stability parameter. Previous study conducted by the authors recommended to evaluate stability parameter of those boats [1]. Stability parameter is imposed to all kinds of ships to fulfil standard criteria of International Maritime Organization (IMO). Fishing vessels stability parameter should meet The International Conference for Safety of Fishing Vessels 1977 (IMO 1977) at all loading conditions [2, 3]. Autrhors' previous study [4] also found that such fishing boats had not any documents particularly stability documents. This condition is true due to the fact that such fishing boats were built based on traditional experiences without considering any design aspects.

2. Review of stability fishing vessels

Pinsker [5] explained that stability of fishing vessels should be evaluated for all loading conditions which are departure and return with full loads. Meanwhile, Fyson [6] stated that fishing vessels above 24 m in length should satisfy regulation of IMO 1977 with stability criteria at large angle of inclination. To achieve the evaluation of stability parameter several related documents should be prepared before executing inclining test. Those documents had been prepared prior to the research study by the authors [4]. Parameter height of metacentre above baseline (*KM*) have been provided while other parameter of vertical centre of gravity (*KG*) should be determined during inclining test then all stability parameters of at any loading conditions may be obtained.

Safety aspects of fishing vessels are regulated in Torremolinos Protocol and The Torremolinos International Convention for the Safety of Fishing Vessels, consolidated edition, 1995, adopted in Torremolinos, Spain 2 April 1993. This protocol stipulated standard requirements of fishing vessels' stability [5]. Insight report Lloyd's Register Foundation [6] clarified that the reason of fishing vessels' accidents caused by the lack of regulation and weak implementation of regulation. Moreover, safety code issued by Implementation Guidelines on Part B of the Code, the Voluntary Guidelines and the Safety Recommendations (FAO UN, 2014) [8] stated one practical code of ship safety is maintaining watertight integrity and ship stability. Regulation of stability for ships above 24 meter in length firmly stated in (International Conference on Safety of Fishing Vessels, 1977, IMO, London, 1977) which is the complete text from Torrelinos International Convention for the Safety of Fishing Vessels, 1977 and The International Conference on Safety of Fishing Vessels, IMO, London, 1977 [5, 9]. The concept of ship stability component mostly in order to find component of ship stability [10, 11, 12, 13]. The scenario of inclining test is presented at Figure 1.

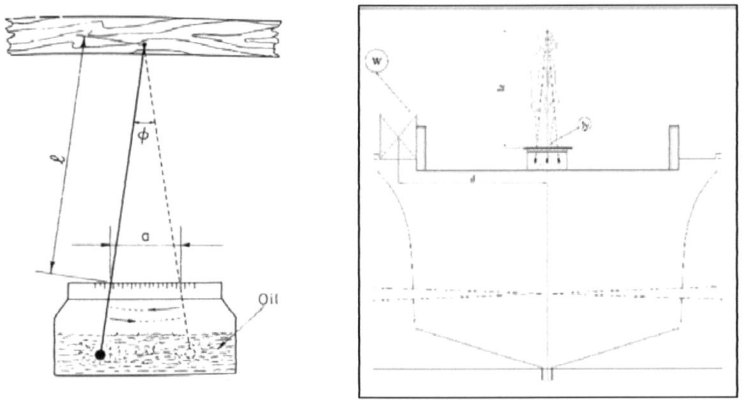

Figure 1. Scenario of Inclining Test

The value of inclination angle ϕ is obtained from:

$$\phi = a/l \text{ (in radians) and } \phi = 57.3 \times (a/l) \text{ (in degrees)} \quad (1)$$

IOP Conf. Series: Earth and Environmental Science **1461** (2025) 012008 doi:10.1088/1755-1315/1461/1/012008

The criteria are contained in Regulation 28 of Chapter III (International Conference on Safety of Fishing Vessels, 1977, IMO, London, 1977) [1,5]. At present, there are no such criteria for vessels below 24 m in length, but in Reference [14], it is presently recommended to use the same criteria also below 24 m for decked vessels. There are not yet any internationally recognized stability regulations for vessels with length L < 24 m. The regulations issued by The Maritime Coastguard Agency [15] for stability criteria of fishing vessels less than 15 metres. Fishing vessels should be so designed, constructed and operated that the minimum stability criteria established by the competent authority will be met in all operation conditions. The following minimum stability criteria are recommended for decked fishing vessels less than 24 m [16, 17] which is based on the FAO/ILO/IMO Voluntary Guidelines for the Design, Construction, and Equipment of Small Fishing Vessels, 2005.

Vertical centre of buoyancy (KB) and radius of metacentre (BM) or KM (vertical distance from base line to metacentre point) are obtained from hydrostatics properties. Furthermore, these relationships are required for inclining test.

$$GM_T = (w \cdot d \cos \phi) / (W \sin \phi) = (w \cdot d) / (W \tan \phi) \quad (2)$$

$$KG = KM_T - GM_T \quad (3)$$

$$KM_T = KB + BM_T \quad (4)$$

Where:

KB = vertical centre of buoyancy

BM_T = transverse radius of metacentre

KM_T = vertical distance from base line to metacentre point

W = displacement of boat according to boat draft

w = moving weight at the distance d from boat centreline

ϕ = angle of inclination

The value of KB and BMT are obtained from hydrostatics properties of the boat according to draft of the boat. Those values were obtained from previous work from the authors [1]. Once the value of GMT is obtained then vertical centre of gravity of the boat will be obtained.

3. Methodology

3.1. Preparation of boat's documents

In executing inclining test some detail of boat should be prepared before. Previous research was executed for the task of providing vessel's documents such as general arrangement and lines plan and hydrostatics properties [1]. Stability parameters such as vertical centre of buoyancy (KB) and radius of metacentre (BM) or KM (vertical distance from base line to metacentre point) were obtained from hydrostatics properties. Dimensions of observed purse seiner for inclining test are presented at Table 1.

Table 1. Principle dimension of purse seiner 'Yana'

No	Boat parameter	Value	Unit
1	Length overall, Loa	20.90	m
2	Length of waterline, Lwl	18,336	m
3	Beam maximum, B	3.07	m
4	Beam of waterline, Bwl	2.975	m
5	Max draft, T	0.585	m
6	Deck height, H	1.182	m
7	Weight displacement, ▲	16.190	ton
8	Speed (empty load), Vs	6 to 7.5	knot

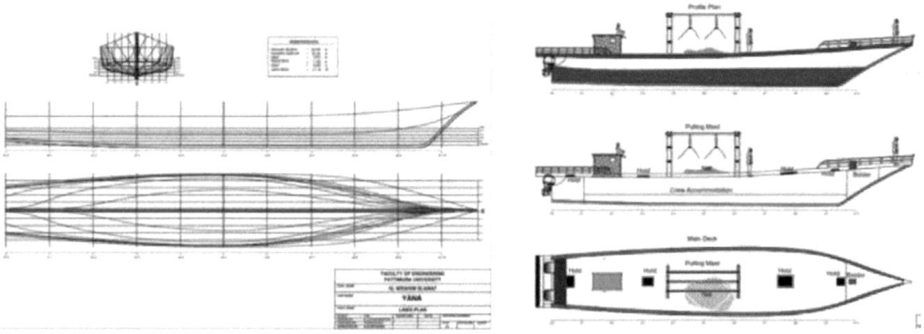

Figure 2. Lines plan and general arrangement of purse seiner 'Yana'

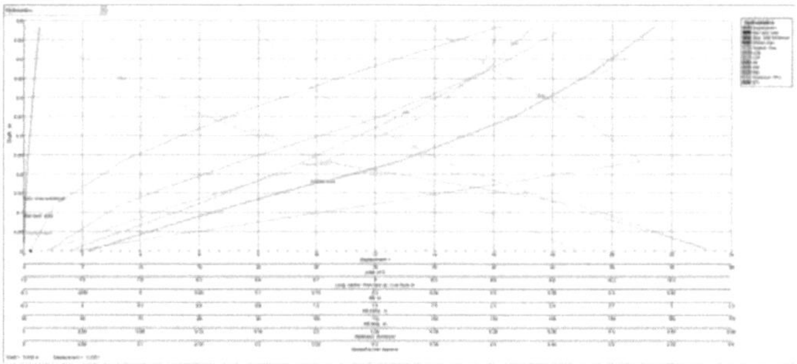

Figure 3. Hydrostatic curve of purse seiner "Yana"

3.2. Preparation and execute inclining test

The steps of inclining test include:

- Preparation of boat and measurement devices: this work was executed to ensure all measurements on board should fit standard inclining test. Based on author's experience, and related references [19, 20] some special attentions should be kept during the test which are: mooring ropes are in free condition without tension, the boat is protected from current (drift), wind, waves and other external effects.
- Measurement angle of inclination ϕ, GMt and KG of the boat
    - Measurement for scenario 1: few crew and additional loads (water)
    - Measurement for scenario 2: all crew on board without additional load
- Gain results of measurement and evaluate the value of KG

Figure 4. Preparation of boat and devices

Figure 5. Measuring and developing result of inclining test

3.3. Scenario of loading conditions

Once the value of KG for empty boat is obtained then several loading conditions are executed in this study. The purpose of developed several loading conditions is to evaluate boat stability at several loading conditions which are departure to fishing ground (empty condition) and return to home base (full condition with captured fishes). Meanwhile additional scenario was developed for variation of loads (captured fishes) which are:

- Variation of loads on deck
- Variation of loads in fish hold
- Variation of loads on deck and fish hold

The habit of fishermen in placing captured fishes on boat deck will decrease the level of boat stability. In addition, there are several fish holds under main deck which are rarely used by local fishermen due to one night fishing with single catching operation. However, this scenario should be provided in order to increase catching operation several times at night. With this scenario the captured fished may be placed at holds and deck. Therefore, evaluation of stability of all possible loading conditions should be provided in this study. The amount of maximum catching capacity of 3 tons is provided for this fishing boat due to available space on deck.

4. Results and Discussion

41. Result of the inclining test

Result of inclining test is presented at Table 2. In this test, scenario 1 was arranged for few crews, additional loads (water) and all equipment on board while scenario 2 was arranged for loads with full crew and all equipment. In addition, Scenario E or F was arranged for repeating measurement to achieve an average value of KG, this inclining test was set for empty boat at departure condition (without captured fishes). All items such as equipment, fuel and crew were included at departure condition.

Table 2. Results of inclining test scenario E & F *'Purse seiner Yana'*

No	Boat parameter	Unit	Value			
			Scenario 1E	Scenario 2E	Scenario1F	Scenario. 2F
1	Additional load (water)	ton	0.667	0.000	0.667	0.000
2	Personal load (crew)	ton	0.900	1.725	0.900	1.725
3	Total load	ton	1.567	1,725	1.567	1.725
4	Average draft midship	m	0.514	0.522	0.514	0.522
5	Trim (by stern)	m	0.202	0.200	0.202	0.200
6	Volume displacement	m^3	13.502	13.815	13.502	13.815
7	Weight displacement	tons	13.840	14.160	13.840	14.160
8	Metacenter height KM	m	1.915	1.893	1.915	1.893
9	Distance weight from CL	m	1.171	1.200	0.729	0.650
10	Tan ϕ	rad	0.125	0.143	0.077	0.077
11	Metacenter height GMt	m	1.062	1.020	1.066	1.024
12	Vertical center of gravity KG	m	0.853	0.873	0.849	0.876
13	Ratio KG/H	-	0.723	0.740	0.722	0.737
14	Inclination angle	deg.	7.171	8.153	4.424	4.424
	Resume		Scenario 1	Scenario 2		
1	Average KG	m	***0.854***	***0.874***		
2	Ratio KG/H	-	***0.723***	***0.740***		

Evaluation of stability parameters was executed for loading conditions: empty boat and cargo (captured fishes) on deck with the capacity of 1, 2 and 3 tons. The result of evaluation is presented at Table 3.

Table 3. Results of evaluation of stability parameter (cargo on deck)

Code: A.749(18) Ch.3 Applicable to all ships and Reg.28.Ch III.1977 for fishing vessels								
No	Criteria	Value	Unit	Actual	(cargo on deck)			Status
				Empty	1 ton	2 tons	3 tons	
A.	**Empty condition (departure)**							
1	Area 0 to 30. >=	0.055	m.rad	0.116	0.102	0.101	0.094	pass
2	Area 0 to 40. >=	0.090	m.rad	0.180	0.156	0.155	0.144	pass
3	Area 30 to 40. >=	0.030	m.rad	0.064	0.056	0.054	0.049	pass
4	Max GZ at 30 or >. >=	0.20	m	0.368	0.334	0.312	0.295	pass
5	Angle of Max GZ. >=	25	deg	33.6	31.8	31.5	30.9	pass
6	Initial GMT. >=	*0.350*	*m*	1.020	0.887	0.833	0.778	pass
7	Crew crowding. <=	10	deg	8.3	8.55	8.70	8.90	pass
8	Turning angle. <=	10	deg	1.0	1.2	1.26	1.30	pass
9	Severe wind. <=	16	deg	1.3	1.30	1.30	1.30	pass

Similarly, the evaluation of stability parameters was executed for loading conditions: empty boat and cargo (captured fishes) in fish hold with the capacity of 1, 2 and 3 tons. The result of evaluation is presented at Table 4.

Table 4. Results of evaluation of stability parameter (cargo on fish hold)

Code: A.749(18) Ch.3 Applicable to all ships and Reg.28.Ch III.1977 for fishing vessels

No	Criteria	Value	Unit	Actual (cargo on fish hold) Empty	1 ton	2 tons	3 tons	Status
A.	**Empty condition (departure)**							
1	Area 0 to 30. >=	0.055	m.rad	0.116	0.117	0.115	0.115	pass
2	Area 0 to 40. >=	0.090	m.rad	0.180	0.182	0.180	0.181	pass
3	Area 30 to 40. >=	0.030	m.rad	0.064	0.065	0.065	0.066	pass
4	Max GZ at 30 or >. >=	0.20	m	0.368	0.377	0.376	0.379	pass
5	Angle of Max GZ. >=	25	deg	33.6	33.60	33.60	33.6	pass
6	Initial GMT. >=	*0.350*	*m*	1.020	0.979	0.948	0.914	pass
7	Crew crowding. <=	10	deg	8.3	7.800	7.400	7.00	pass
8	Turning angle. <=	10	deg	1.0	0.900	0.80	0.80	pass
9	Severe wind. <=	16	deg	1.3	1.20	1.00	1.00	pass

The evaluation of stability parameters was executed also for combination of loading conditions: cargo on deck with capacity 1, 2 and 3 tons and cargo in fish hold with the capacity of 3 tons. The result of evaluation is presented at Table 5. In this evaluation, stability criteria used to evaluate stability parameters based on *Code: A.749(18) Ch.3 Applicable to all ships and Reg.28.Ch III.1977 for fishing vessels.*

In this evaluation, the criteria cover: Area 0 to 30, Area 0 to 40, Area 30 to 40, Maximum GZ at 30 degree, Angle of GZ maximum

Table 5. Results of evaluation of stability parameter (cargo on fish hold)

Code: A.749(18) Ch.3 Applicable to all ships and Reg.28.Ch III.1977 for fishing vessels

No	Criteria	Value	Unit	Actual (cargo 3 t on hold + deck) Empty	1 ton	2 tons	3 tons	Status
A.	**Empty condition (departure)**							
1	Area 0 to 30. >=	0.055	m.rad	0.116	0.104	0.094	0.082	pass
2	Area 0 to 40. >=	0.090	m.rad	0.180	0.162	0.144	0.122	pass
3	Area 30 to 40. >=	0.030	m.rad	0.064	0.058	0.050	0.041	pass
4	Max GZ at 30 or >. >=	0.20	m	0.368	0.335	0.294	0.246	pass
5	Angle of Max GZ. >=	25	deg	33.6	31.80	30.00	28.20	pass
6	Initial GMT. >=	0.350	m	1.020	0.819	0.734	0.644	pass
7	Crew crowding. <=	10	deg	8.3	7.300	7.600	8.100	pass
8	Turning angle. <=	10	deg	1.0	0.900	1.100	1.400	pass
9	Severe wind. <=	16	deg	1.3	1.00	0.900	0.900	pass

IOP Conf. Series: Earth and Environmental Science **1461** (2025) 012008 doi:10.1088/1755-1315/1461/1/012008

Among all stability criteria, two important criteria which are transverse metacentre height (GM_T) and lever arm (GZ) are presented to be evaluated (Figure 6). These criteria may be easily seen as the cargo quantity and position are changed.

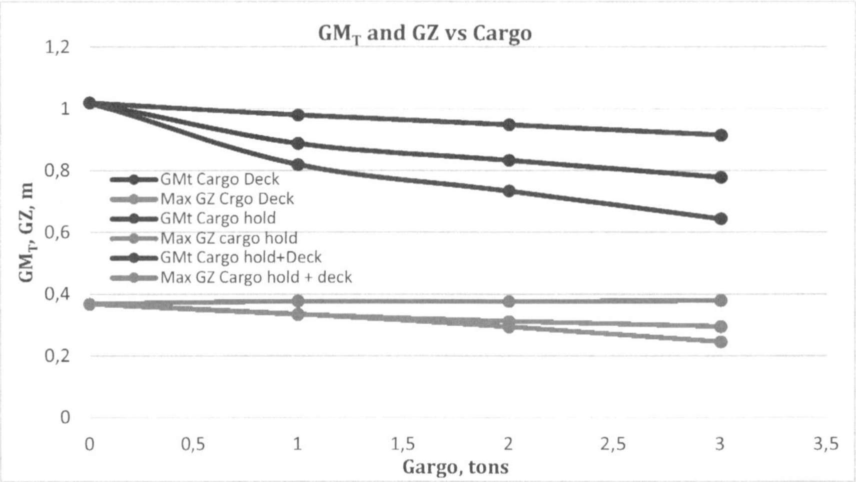

Figure 6. The changing of GM_T and GZ due to position of cargo on board

Finally, the purpose of study was achieved by providing the value of KG ratio of KG/H as seen at Table 6 and Figure 6. Those values KG and ratio KG/H are presented for loading conditions of empty boat, cargo on deck, cargo in fish hold and combination of cargo on deck and in fish hold.

Table 6. Ratio of KG Vs Load for Several Loading Conditions

No	Configuration of Cargo	Vertical Centre of Gravity, KG (m)	Deck Height H, (m)	Ratio KG/H
1	Empty boat	0.874	1.182	0.740
2	1 ton on deck	0.925	1.182	0.783
3	2 tons on deck	0.972	1.182	0.822
4	3 tons on deck	1.038	1.182	0.878
5	Empty boat	0.874	1.182	0.740
6	1 ton in fish hold	0.829	1.182	0.701
7	2 tons in fish hold	0.795	1.182	0.673
8	3 tons in fish hold	0.764	1.182	0.646
9	1 ton on deck plus 3 tons in fish hold	0.813	1,182	0.688
10	2 tons on deck plus 3 tons in fish hold	0.858	1.182	0.726
11	3 tons on deck plus 3 tons in fish hold	0.919	1.182	0.777

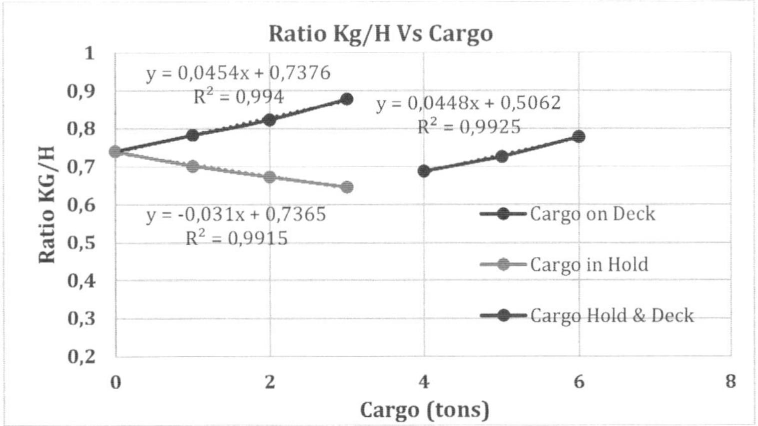

Figure 7. Ratio of KG/H for several loading conditions

Regression lines of ratio KG/H as function of cargo position and capacity are presented also in case a boat designer will have other values of cargo.

- Cargo on deck: Y = 0.045 X + 0.737 with R^2 = 0.994 (5)
- Cargo in fish hold: Y = - 0.031 X + 0.736 with R^2 = 0.991 (6)
- Cargo on deck & fish hold: Y = - 0.031 X + 0.736 with R^2 = 0.991 (7)

Where: X = the amount of cargo (captured fishes)
Y = ratio of KG/H

This value is important in order to evaluate ship stability at early design phase for all ships [18] particularly for purse seiner fishing boat observed in this study could be a reference for designer

4.2. Discussion

Vertical centre of gravity KG of observed purse seiner at empty loading condition (departure) gives a contribution value in evaluating stability parameter of boat operation. The value KG obtained from inclining test which is 0.874 meter was used further to determine stability parameter of the boat at large angle of inclination. In addition, a nondimensional value of ratio KG/H of 0.740 was found during this study (Table 2). This value was obtained as an average values of loading condition and position of loads. This value may be used as reference at initial design phase of similar fishing boats. Furthermore, this value is used more for evaluation of stability parameter of the boat at any loading conditions.

Evaluation of stability parameters at large angle of inclination was done by combination load case at empty condition (departure) and variation of loads (arrival condition). Variation of loads were executed by assessment load capacity and position of centre of gravity (VCG or KG and LCG). Loading position are on deck, fish hold and combination of both positions. Furthermore, stability parameters at large angle of inclination were computed and presented at Tables 3, 4, and 5. Evaluation of stability parameters with cargo (captured fishes) on deck as presented at Table 3 found that the boat is in stable conditions. As the load increase the stability parameters decrease but the boat is still in stable condition. Similarly for evaluation of stability parameters with cargo (captured fishes) in fish hold as presented at Table 4 found that the boat is in stable conditions. As the load increase the stability parameters increase at certain level with small changing and the boat is still in stable condition. Table 5 represent evaluation of stability parameters with combination of cargo on deck and in fish hold. As the load increase the stability parameters decrease but the boat is still in stable condition.

Stability parameters such as metacentre height (GM) and lever arm (GZ) is presented at Figure 6. These values are affected greatly by position and amount of cargo. Placing cargoes on deck affect the decrease of GM and GZ significantly followed by combination position on deck and in fish hold. A safe way is the position of cargo in fish hold. In fact, the habit of fishermen to place cargo on deck for the reason of easy loading and unloading. However, from safety perspective this position of cargo is not safe during operation. In general, stability parameters of the boat that imposed the maximum cargoes (3 tons on deck and 3 tons in fish hold) that makes the boat is still stable (satisfy criteria of IMO and fishing vessels). In addition, placement fish on cargo holds as well as deck area will give opportunity for the fishermen to make several catchments during one night catching operation.

In general, providing the value of KG and ratio KG/H will have benefit for designer and boat owner/operator. The results of those values as shown at Table 6 and Figure 6 may be used by designer as reference in other design process. Meanwhile, boat owner/operator would have understanding of safety of boat and increasing catching operation. Regression equations are provided in equations (5), (6) and (7) that help designer and owner may find other cargo capacity rather than these presented in this study.

5. Conclusion and Recommendation

5.1 Conclusion

Some points may be withdrawn from this study including:

1. The real value KG obtained from inclining test (empty condition) which is 0.874 meters and nondimensional ratio of KG/H is 0.740. These values may be used as reference for new initial design phase or other operational purposes such as the application of MobileApp to evaluate stability parameter of the boat.
2. Generally, loading condition of placing cargo on deck or combination placement in fish holds and deck will decrease stability quality at significant but in the case of observed purse seiner the boat is still in stable condition.
3. Catching proses at one-night fishing may be doubled where the first catchment may be placed in fish holds and the second catchment may be placed on deck. With the maximum catchment of 3 tons each then the boat is still stable condition. However, with total catchment of 6 tons the stability parameter is decreasing at significant level.

5.2. Recommendation

1. It is recommended to use the nondimensional ratio of KG/H as reference at initial design stage or other operational purposes.
2. It is also to recommend these results as reference to justify the application of MobileApp during calibration and operation process.

Acknowledgement

Special appreciation is delivered to Dr Andrea Grech La Rosa and the Institution of University College London (UCL), UK for the collaboration and contribution in this research with the contract SG6\100039 dated 1 June 2023.

References

[1] Hetharia, W. R, 2022, Report of Research Part A entitled *"Ensuring the Safety of Seafarers and Fsihers in the Era of Covid-19 and Beyond,* the research was continued to investigate stability parameters of fishing vessels

[2] IMO (1977), International Convention for Safety of Fishing Vessels, Res. A.16, See Coast Guard, 1976

[3] Lewis E. V, 1988, Principles of Naval Architecture, Volume 1 Stability and Strength, The Society of Naval Architects and Marine Engineers , 601 Pavonia Avenue, Jersey City, NJ. pp. 106-108

[4] Hetharia, et. al. 2019, Laporan hasil survey perikanan di Perairan Maluku dan Maluku Utara Periode 2017-2019.

[5] Pinkster, J 'Fishing Vessels', 2004, Ship Design and Costruction, Written by an International Group of Authorities, Thomas Lamb, Editor. The society of Naval Architects and Marine Engineers, 601 Pavonia Avenua, Jersey City, NJ. 07306. Pp. 41.4 – 41.57

[6] John Fyson – Design of Small Fishing Vessels, Published by arrangement with the Food and Agriculture Organization of the United Nations. By Fishing News Book Ltd, Farnham – Surrey – England, FAO 1985, ISBN: 0 85238 133 6, pp. 155 s/d 172

[7] Lewis E. V, 1988, Principles of Naval Architecture, Volume 1 Stability and Strength, The Society of Naval Architects and Marine Engineers, 601 Pavonia Avenue, Jersey City, NJ, pp. 106-108

[8] Insight Report Lloyd's Register Foundation Report Series: No. 2018.3) Lloyd's Register Foundation Report Series: No. 2018.3 (Insight report on safety in the fishing industry - A global safety challenge), June 2018 –

[9] Implementation Guidelines on Part B of the Code, the Voluntary Guidelines and the Safety Recommendations FOOD AND AGRICULTURE ORGANIZATION OF THE UNITED NATIONS - INTERNATIONAL LABOUR ORGANIZATION INTERNATIONAL MARITIME ORGANIZATION, Rome, 2014, ISBN 978-92-5-108213-3 (print), E-ISBN 978-92-5-108214-0 (PDF), © FAO/ILO/IMO, 2014

[10] Rawson, K. J. and Tupper, E. C., 1984, *Basic Ship Theory*, Vol. 1, 3rd Edition, Longman, New York.

[11] Biran A. B, 2003, Ship Hydrostatics and Stability, Butterworth-Heinemann An Imprint of Elsevier, Linacre House, Jordan Hill, Oxford, OX2 8DP, pp 166 – 170.

[12] Barrass B and Derrett D. R, Ship Stability for Masters and Mates, 2006, Butterworth-Heinemann An Imprint of Elsevier, Linacre House, Jordan Hill, Oxford, OX2 8DP, pp 286 – 289.

[13] Moore, C S, 2010, Intact Stability - The Principles of Naval Architecture Series, The Society of Naval Architects and Marine Engineers, 601 Pavonia Avenue, Jersey City, New Jersey, 07306, pp. 59-65

[14] Van Dokkum, K, 2003, Ship Knowledge - A Moderen Encyclopedia, DOKMAR, Enkhuizen, The Netherland.

[15] Malmin, K, Ø., Design of a small fishing vessel in alternative building materials, B.Sc Thesis, NTH, 1978

[16] Maritime Coastguard Agency, The Code of Practice for The Safety of Fishing Vessels of Less Than 15 metres Length Overall. The Maritime and Coastguard Agency – Spring Place 105 Commercial Road, Southhampton SO15 1EG.

[17] Gudmundsson, A, Safety Practices Related to Small Fishing Vessel Stability – Fishery Industry Officer (Vessels) – Fishing Technology Service Fish Products and Industry Division FAO, Rome. FAO Fisheries and Aquaculture Technical Paper No. 517 Food and Agriculture Organization of The United Nations, Rome, 2009. ISBN 978-92-5-106202-9

[18] Parsons, M. G, 2003, Parametric Design – Chapter 11 in Ship Design and Construction – Editor Thomas Lamb, Volume 1, SNAME Publication, ISBN 0-939773-40-6, United Stated of America, pp 11-26 to 11-27.

References from the website:

[19] Marine Insight: https://learn.marineinsight.com/eBooks/oto274580297dkkmght/

[20] Cultofsea – Maritime Knowledge Base: https://www.cultofsea.com/ship-stability/the-inclining-experiment/

IOP Conf. Series: Earth and Environmental Science **1461** (2025) 012009 doi:10.1088/1755-1315/1461/1/012009

Enhancing Engineering Intuition for Shipyard Investment Cost Estimation through a Computer-Aided Training Framework

Triwilaswandio Wuruk Pribadi[1], Sufian Imam Wahidi[1,2], Athallah Agsa Parahita[1], Imam Baihaqi[1*], and Heri Supomo[1]

[1] Department of Naval Architecture, Institut Teknologi Sepuluh Nopember, Surabaya, 60111, Indonesia
[2] Department of Naval Architecture, Ocean and Marine Engineering, University of Strathclyde, Glasgow, G4 0LZ, United Kingdom

*E-mail: imam.baihaqi@its.ac.id

Abstract. Estimating investment costs for a shipyard requires a high level of expertise, which typically takes time and experience to develop. This research presents a computer-aided training framework to train intuitive skills in estimating investment costs for new shipyards. The study begins with a review of the parametric method, a relevant approach for estimating shipyard investment costs. Subsequently, the recognition-triggered decision method, a technique based on intuitive decision-making, is integrated into a predetermined estimation method. Program and application algorithms are developed accordingly. A prototype of the application is tested on ten participants using the "Two Mean Difference Test." Prospective estimators undergo pre-test and post-test assessments to evaluate the improvement in their intuitive skills. The results demonstrate that the application effectively enhances the intuitive skills of prospective estimators in estimating the investment costs of new shipyards. This conclusion is supported by the values of "t" and "t of the table," where 11.53 > 2.262, within a critical range of 5%.

1. Introduction

Intuition plays a pivotal role in engineering, serving as an essential skill that enhances problem-solving and decision-making capabilities [1]. Defined as the ability to understand and address complex issues without relying solely on analytical reasoning, engineering intuition is developed through extensive experience and education [2]. This skill enables engineers to recognize patterns, predict outcomes, and make swift decisions even when faced with incomplete information. As such, it differentiates exceptional engineers from their peers, allowing them to navigate challenges effectively in various fields [3].

In the global context, intuition is increasingly recognized as a crucial soft skill across engineering disciplines [4]. It facilitates creative problem-solving and quick identification of potential issues, thereby improving project efficiency and outcomes. Engineers who harness their intuition can adapt to rapidly changing environments and make informed choices that align with both technical requirements and broader project goals. This adaptability is particularly vital in today's fast-paced technological landscape, where the ability to synthesize knowledge from diverse experiences can lead to innovative solutions. The paper "Situating Intuition in Engineering Practice" [5] explores how intuition, developed through experience, complements logical analysis in engineering decision-making. The authors define intuition as a subconscious process that allows experienced engineers to assess situations or predict outcomes efficiently, especially when under constraints like time or incomplete data. Through the Leveraging Intuition Toward Engineering Solutions (LITES) framework, they illustrate how intuition integrates with problem-solving by blending instinctive responses with holistic and data-based approaches. Intuition, linked to expertise and System 1 thinking, helps engineers make quick, informed decisions, particularly when traditional analytical methods are insufficient. This highlights the importance of intuition as a critical skill for navigating complex and ambiguous engineering challenges [6].

Specifically in shipbuilding engineering, intuition becomes even more significant due to the intricate nature of maritime projects. Shipbuilding involves complex systems that must operate reliably under various conditions, necessitating a deep understanding of both engineering principles and practical applications. Experienced shipbuilders often rely on their intuitive insights to foresee potential design flaws or operational challenges before they arise. This foresight not only enhances safety [7] but also optimizes resource allocation and project timelines [8].

An attempt has been proposed by Pradisa [9], in utilising the knowledge of intuition to predict the new shipbuilding cost by using computer applications. The framework proposed that through the computer applications, the parameter used in estimating the new shipbuilding price can be trained to the human. The learning process through this approach can be used to initiate the intuition knowledge in engineering, in this case, estimating the new shipbuilding price.

In this article, through the similar approach, it is applied to predict the shipyard investment cost. The breakdown investment cost in shipyard is defined as several item contribute to the investment cost. Each cost components are then ranged into several variable sized to accommodate the different size of the shipyard.

2. Literature Study

2.1 Cost Estimation Method

Estimating investment costs in shipyards is a multifaceted process that requires a systematic approach to ensure accuracy and reliability. One of the primary methodologies used is Cost Estimating Relationships (CERs), which express the relationship between various performance parameters and total costs. This method allows estimators to predict how changes in design specifications, such as a 5% increase in engine power, will impact overall costs. By utilizing historical cost data from previously built vessels, engineers can identify patterns and develop regression models that articulate unit costs per cost-driving parameter for different vessel types. This parametric top-down estimation approach not only enhances precision but also helps inexperienced personnel gain insights into the complexities of shipbuilding costing [10].

Another effective methodology employed in shipyard cost estimation is Activity-Based Costing (ABC). This approach allocates costs based on the actual activities involved in ship construction, providing a clearer picture of both direct and indirect costs [11]. By developing a Work Breakdown Structure (WBS), shipyards can break down all necessary activities for construction, enabling more accurate cost allocation. This method also helps identify inefficiencies and hidden costs that may not be apparent through traditional costing methods [12]. ABC is particularly beneficial for shipyards transitioning from military to commercial shipbuilding, as it allows for a detailed analysis of operational costs and resource utilization.

Preliminary cost estimation models are also crucial for providing early assessments of investment needs based on high-level design specifications. These models utilize initial design parameters to generate rough estimates of acquisition and lifecycle costs. The iterative nature of this process allows for continuous refinement as more detailed design information becomes available. By focusing on the consequences of changes in high-level performance requirements, shipyards can respond more quickly to customer inquiries and improve their competitive advantage in the market [13].

Advanced methodologies such as simulation and predictive modelling are increasingly being utilized to estimate investment costs in shipyards. These techniques involve scenario analysis to forecast potential cost outcomes based on various design choices and operational conditions. By employing data-driven approaches, engineers can anticipate future costs more accurately, taking into account factors such as market demand, material availability, and technological integration [14]. Overall, these methodologies collectively enhance the accuracy and efficiency of investment cost estimations in shipbuilding, ultimately contributing to better financial planning and resource management within the industry.

Shipyard is a place designed to build and / or repair and maintain ships. Generally, shipyards are built on land directly adjacent to the water or sea level [15]. In estimating the cost of a new building shipyard investment project, there are many methods that can be used. One of the methods that can be used is the parametric method. The parametric method is a method that estimates the cost of investment components of new building shipyards by considering parameters that affect the cost of these components.

Cost estimation methods have been used in research on designing applications to train intuition in estimating the price of new ships [9]. According to the research, this method is suitable to be applied to intuition training because the information that will be provided for estimating is very minimal, so that prospective estimators must analyze and explore for themselves the information obtained as material for making intuitive estimation decisions. Therefore, this method can also be used in this study because the application designed is also related to practicing intuition. The use of this parametric method will adjust for the parameters that affect the cost of the investment in a new building shipyard. The parameters used in the application are:

Geography and Topography of The Land. Geographical conditions are the conditions or circumstances of a region that are seen from its state in relation to geographical aspects. Those aspects include the location, interrelationships, area, shape, and coordinate position of the map. While topography is the surface of a land that is grouped or determined according to the height of the land from the earth's surface. The topography of the land will determine the magnitude of the land price. The price of land will be higher if the location of the land is more ideal for a development. In addition to affecting land prices, soil topography also affects the cost of earthwork, which includes land clearing, soil management, and soil compaction.

Bathymetry. Bathymetry is the mapping of the depths of the waters, also known as sea topography. Bathymetry will lead to cost-related expenditures in the form of land dredging. The dredging of the soil is caused not only by the depth of the waters, but also by the height of ship`s.

IOP Conf. Series: Earth and Environmental Science **1461** (2025) 012009 doi:10.1088/1755-1315/1461/1/012009

If the height of ships is greater than the depth of the waters at the site, then dredging in the water needs to be carried out.

Ship Size. The size of the ship that is planned to be built in the shipyard greatly affects the investment costs that will be incurred. From the size of the ship or the DWT of the planned ship, the cost of meeting the ship's production process at the shipyard can be determined. These costs include production equipment, buildings at the shipyard, and other costs in the production process [16].

Technology. In general, the application of technology management in the shipbuilding industry, especially in shipyards in Indonesia, is still developing. Several developed tools are integrated with digital technologies, including wearable devices for identifying welders [17], radio-frequency identification (RFID) [18], quick-response code application [8], indoor positioning technologies [19], and robotic welding [20], all of which contribute to improving the efficiency of shipyard operations [21].

2.2 Intuitive Proficiency Improvement Method

Intuition is a personal cognitive resource or a person cultural capital [22]. In Figure 1, you can see the flow of the intuitive proficiency improvement method that will be used in the application, namely the recognition-triggered decision method [3]. The flow of this method will be implemented in the application as a way to improve intuitive proficiency in estimation. Studies point towards that work-related flow (work motivation, enjoyment and absorption) increase personal and organizational resources, a positive mood, and that flow is connected to situational characteristics [6].

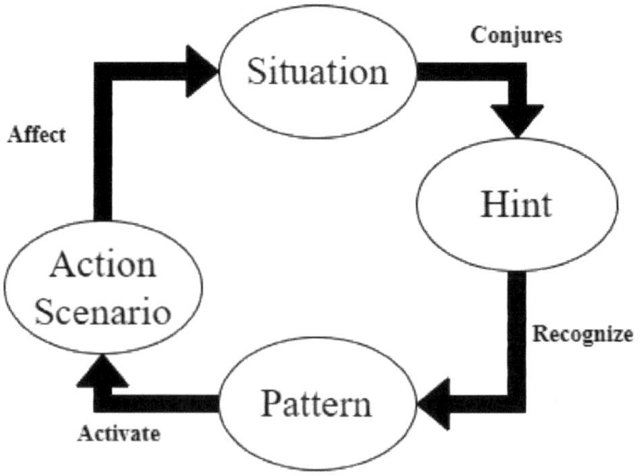

Figure 1. KDP Method (Source: The Power of intuition Book)

First, an understanding of the situation. The situation that the prospective estimator will understand is the relationship between the cost influence parameter and the cost-forming component. The appearance of this relationship is a parametric flowchart. Understanding this situation is also helped by the introduction of the components that make up investment costs and examples of case studies of estimates that have been given previously. In this situation understanding session, the prospective estimator will also enter inputs in the form of influence parameter values and produce outputs in the form of the cost of forming components. The prospective estimator can then analyze the relationship for himself. Secondly, the understood situation gives rise to clues.

The clue obtained after understanding the situation is an empirical graph of the investment cost of each component. This empirical graph will also aid aspiring estimators in pattern recognition. Third, from pattern recognition and clues in the form of empirical graphs, prospective estimators will recognize patterns in estimating the investment of each component. Fourth, perform actions based on patterns. The familiar pattern will help potential estimators take action in the simulation session of the estimated total investment. This simulation will be carried out repeatedly under different conditions. The more simulations you run, the more patterns you will recognize.

2.3 Intuitive Decision Making

A key issue in decision research has been to understand how individual dispositions are related to decision making [23]. A person who has repeated knowledge and experience of something will find it easier and more precise to make a decision related to it [24]. It is an intuition formed in the mind because of the knowledge and experience that have been gained by that person. Intuitive judgments are automatic and effortless. It become particularly salient in the presence of risk and ambiguity [25]. From the flow of intuitive proficiency improvement methods to be carried out, prospective estimators can make informed decisions related to the estimated investment costs of new building shipyards. This decision is taken based on the pattern that has been obtained after conducting training.

3. Methodology

Problem Identification. The background of this research is the importance of estimating before starting a construction project, especially the construction of a new building shipyard. However, to estimate something requires knowledge and experience about the field or thing you want to estimate. It takes a long time for an estimator to gain such knowledge and experience. Therefore, a means is needed in this study in the form of a web-based application that can train a person's intuitive skills in estimating the investment costs of a new building shipyard.

Literature Studies. Literature study is necessary to understand the theories that will be used in working on this research. Literature studies can be supplemented with references from relevant scientific journals, books, articles, previous final projects, and the internet.

Analysis of the Implementation of Intuition Training in the Investment Estimation of New Building Shipyard. An analysis of how to implement intuition in the estimated investment cost of a new building shipyard in the form of determining how to estimate the investment cost of a new building shipyard, determining the method of increasing intuition, and how to implement both in the application.

Application Design. The computer application will be designed based on the analysis that has been carried out so that the purpose of this research can be fulfilled. The goal is to create a tool that can train user intuition in estimating the investment cost of a new building shipyard.
In the design of this application, there is also a validity test for the cyst. The data calculated in Excel calculations using the flow system is the same as the flow of the program or application. The value to be validated is the value of the investment cost of the new building shipyard for each component.

Application Analysis and Discussion. Computer applications that have been designed will be tested on users with regard to the accuracy and effectiveness of the purpose and function of making applications according to the hypothesis. The hypothesis is to train intuition in the estimation of new building shipyard investments. The correctness of this hypothesis will be tested using one of the parametric statistical tests, namely the "t" test.

Conclusion. Conclusions will be reached after all stages are completed. This conclusion will address whether the study's findings are hypothesized or even contradictory.

4. Development of the Computer Aided-Training Framework

4.1 Determination of the Components that Make Up the Investment Cost of a New Building Shipyard

In Figure 2, you can see the flowchart of the components that make up the investment cost of the new building shipyard. The investment cost of the shipyard is the total of the estimated costs of all forming components.

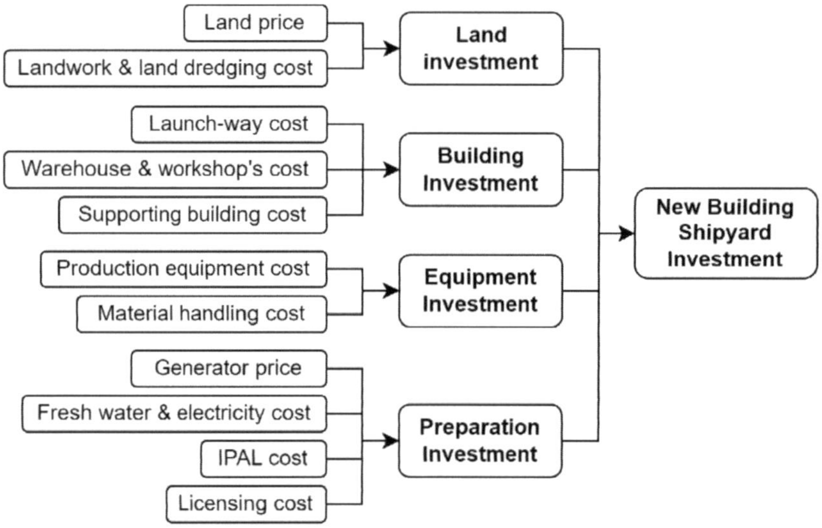

Figure 2. Component Forming Investment Cost.

These forming components need to be detailed more specifically so that it is known what sub-components are contained in each component. Here is a breakdown of the components that make up the investment cost of the new building shipyard. will be broken down into smaller components as follows:

1. Land investment can be broken down into land prices, land clearing, soil decomposition, soil compaction, and dredging in water.
2. Building investment can be broken down into launch facilities, storage warehouses, preparatory workshops, fabrication workshops, sub assembly workshops, assembly workshops, blasting and painting workshops, security posts, parking areas, offices, prayer rooms, toilets, and electrical rooms.

3. Equipment investment can be detailed into rack, straightening machine, shot blasting machine, CNC plasma cutting, flame planner, bending machine, SAW welding machine, FCAW welding machine, forklift, overhead crane, mobile crane, and transporter
4. Preparatory investments can be broken down into generators, clean water and electricity installations, IPAL, and licensing.

4.2 Relationship of Cost Effect Parameters with Cost-Forming Components of New Building Shipyards.

In Table 1, it can be seen the relationship between the parameters and components of the investment cost of the new building shipyard. The cost influence parameter is the condition that causes the large investment value of the forming component. This relationship needs to be known so the investment cost of each component of the new building shipyard can be estimated. Components that have no relationship with the influence parameter are assumed to be equally costly under each known condition of the application. This is because the size of the ship that will be used as a reference in intuition training does not have a significant size difference, which is 5000–9999 DWT. This component consists of supporting buildings, material handling prices, and preparation investment costs.

Table 1. Parameters Affecting Cost of the Components

Component	Parameters
Land Price	Location and Land Area
Landwork Cost	Rock Structures, Slope of the Area, and Land Area
Dredging Cost	Depth of Water, Size of Ship, and Land Area
Launchway Cost	Size of Ship and Type of Launch-way
Warehouse and Workshops Cost	Size of Ship
Supporting Building Cost	Assumed to be Same
Production Equipment Cost	Size of Ship
Material Handling Cost	Assumed to be Same
Generator Price	Assumed to be Same
Fresh Water and Electricity Installation Cost	Assumed to be Same
IPAL Cost	Assumed to be Same
Licensing Cost	Assumed to be Same

4.3 Estimated Investment Cost of a New Building Shipyard Based on Component needs

The next thing to do after knowing the relationship between the parameters that affect costs and the components that make up the investment costs of the new building shipyard is to start estimating the cost of each component according to known conditions or parameters [11]. Especially with the use of slipway launching facilities and airbag systems, a winch is needed to hold the ship when launching. Therefore, it is necessary to calculate the winch needs on each DWT ship. Table 2 shows the formula for estimating the investment cost of each component based on its needs.

Table 2. Component Investment Cost Estimation Formula

Component	Estimation Formula
Land Price	Land area x Land area cost

Landwork Cost	Land area x Landwork Cost
Dredging Cost	Dredging volume x Dredging Cost
Launchway Cost	Size or amount of the launchway x Launchway Cost
Warehouse and Workshops Cost	Building Area x Warehouse and Workshops Cost
Production Equipment Cost	Amount of machine x Production Equipment Cost

4.4 Implementation of Intuition Training in Estimating the Investment Cost of New Building Shipyards

The value or cost of investing in components that have been obtained in the estimation calculation will be used as a database in this intuition training application. Here is the flow of the intuition training method implemented into the investment cost estimation.

In Figure 3, you can see the flow of intuition training in the estimated investment cost that was applied to the application. In each training session, prospective estimators will be given an empirical graph related to the cost of the components being trained. Here is an empirical graph of each component obtained using the calculation of the estimation formula in Table 2.

Land & land-working cost component. In Figure 4a, you can see a pattern in the form of an empirical graph of the investment costs of the land price component in Indonesian Rupiah (Rp). This graph displays the investment cost of the land price component against the land area. The cost of the land price component is divided into three places, namely Lamongan, Gresik, and Batam. In Figure 4b, a pattern can be seen in the form of an empirical graph of the investment costs of the landwork component relative to the land area. This graph displays three types of earthwork costs, namely land clearing costs, soil compaction costs, and soil compaction costs.

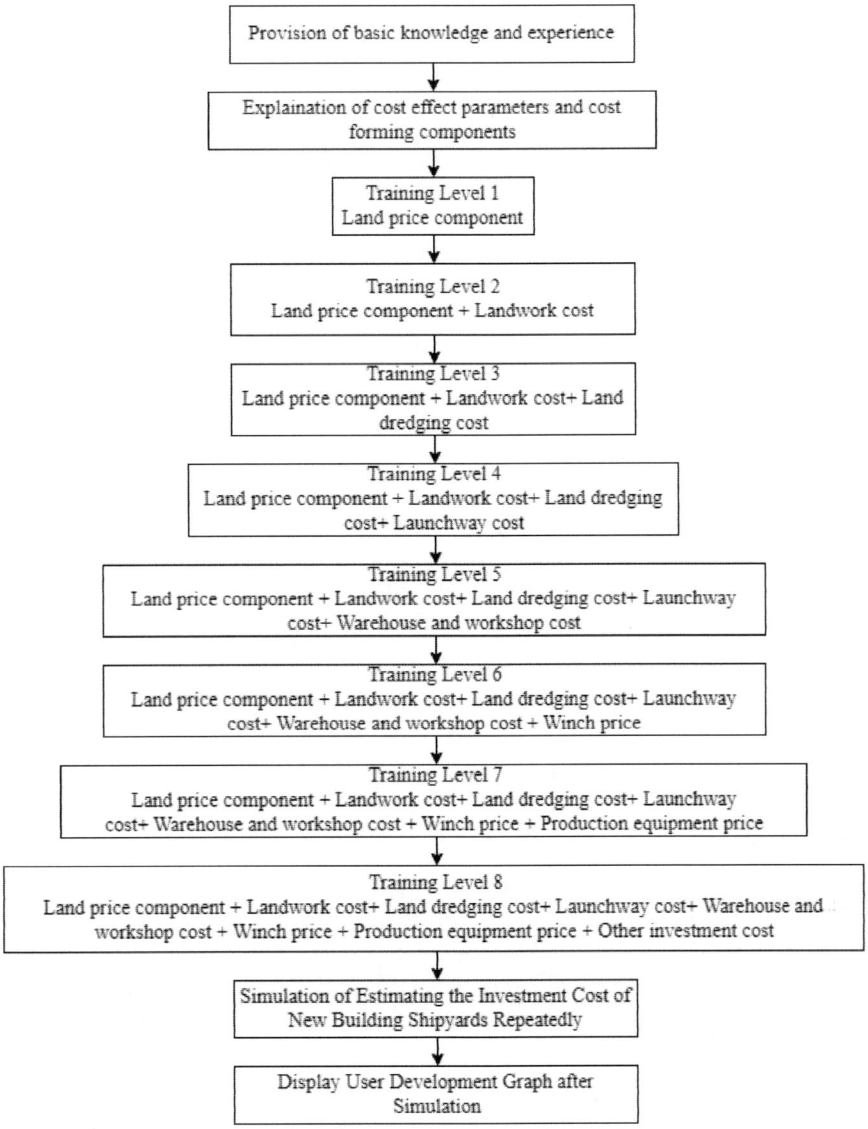

Figure 3. Training Flow

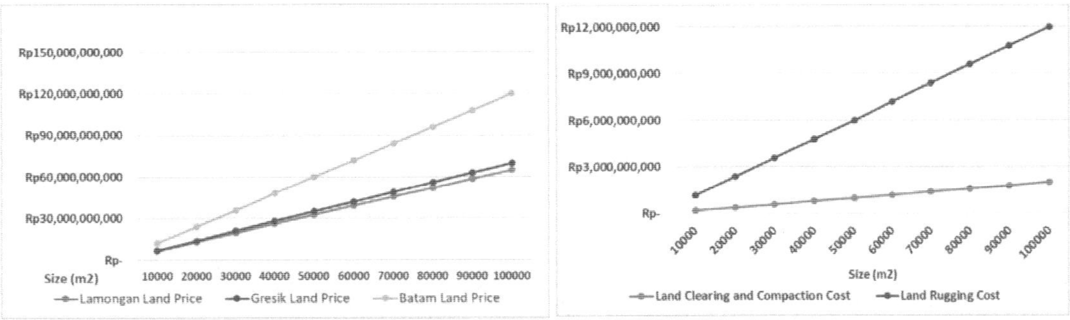

a b

Figure 4. Empirical Graph of Investment Cost of Land Price & Land-working Components

Land Dredging Cost & Cost of Launch Facilities. In Figure 5a, you can see one of the patterns in the form of an empirical graph of the investment cost of the soil dredging component with a ship size of 5000 DWT. The cost component of dredging land has three variables in estimating its cost: location, ship size, and land area. Therefore, the cost of this component cannot be displayed in a single graph. On the application, an empirical graph of the cost of the soil dredging components will be displayed based on the vessel sizes of 5000 DWT and 9999 DWT. In Figure 5b, you can see a pattern in the form of an empirical graph of the investment cost of one of the launching, namely the graving dock. In the application, the launch facilities that will be used are graving docks, slipways, and airbag systems.

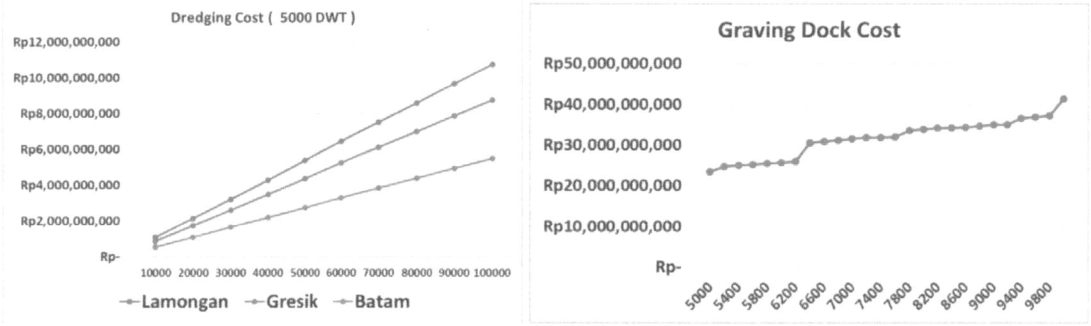

a b

Figure 5. Empirical Graph of Investment Cost of Soil Dredging Components with Vessel Size 5000 DWT & Cost of Graving Dock Components

Cost of Building Warehouse and Workshops & cost of Winch. In Figure 6a, you can see a pattern in the form of an empirical graph of the investment costs of warehouse and workshop components. A warehouse is a temporary storage place for raw materials, while a workshop is a place to produce a ship. In Figure 6b, you can see a pattern in the form of an empirical graph of the investment cost of the winch price component. There are three winch capacities to be used in the application, namely 250 KN, 320 KN, and 400 KN.

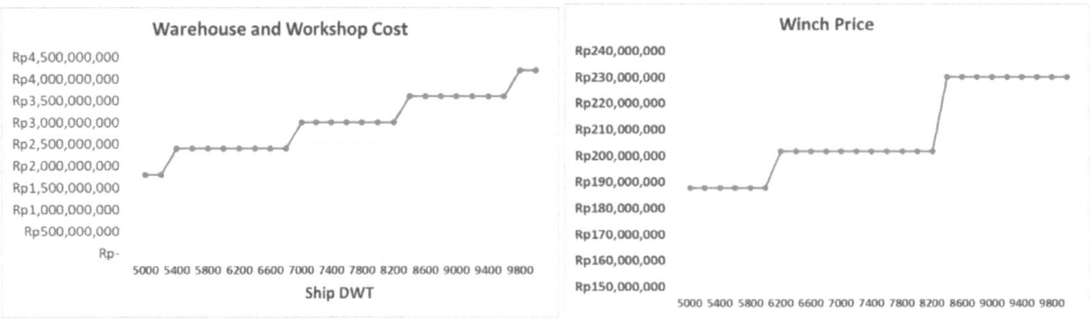

a b

Figure 6. Empirical Graph of Investment Costs of Warehouse and Workshop Components & winch cost.

Production Equipment Price & Other Component Cost. In Figure 7, you can see a pattern in the form of an empirical graph of the investment costs of production equipment components. The specifics of the production equipment used are the same. Therefore, the variable that changes with each size of the ship is the amount of production equipment.

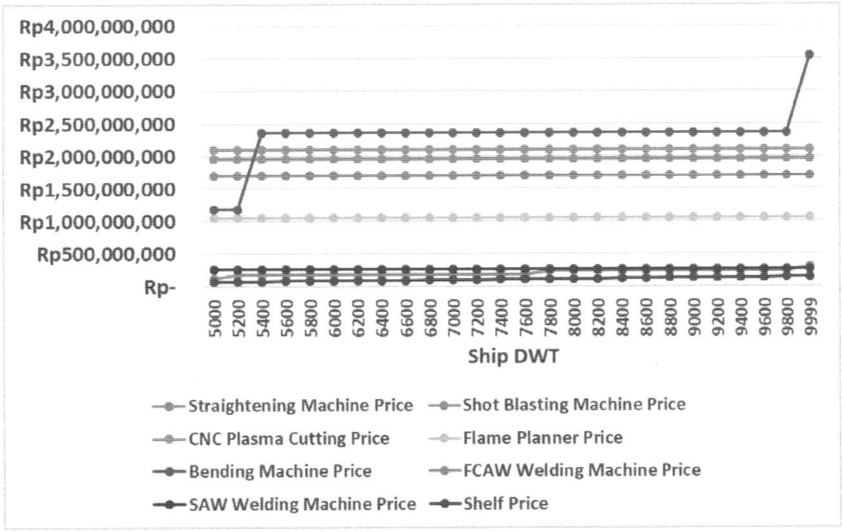

Figure 7. Empirical Graph of Investment Cost of Production Equipment Components

The investment cost of the other component is a component that has no relationship with the cost influence parameter. The investment cost of these components will be equal under each condition. A list of investment costs for other components can be seen in Table 3.

Table 3. Investment Costs of Other Components

Cost of Supporting Building (Rp)	
Security	200,000,000
Parking Area	800,00,000
Office	3,750,000,000
Prayer room	500,000,000
Toilet	75,000,000
Electricity Room	250,000,000
Cost of Material Handlings (Rp)	
Forklift	322,000,000
Overhead Crane 5 ton	435,000,000
Mobile Crane 25 ton	2,175,000,000
Mobile Crane 50 ton	3,088,500,000
Transporter	4,350,000,000
Cost of Preparatory Installation (Rp)	
Generator (100 KVA)	1,800,000,000
Generator (80 KVA)	650,000,000
Generator (60 KVA)	430,000,000
Clean Water and Electricity Installation	50,000,000
IPAL	40,000,000
Licensing	10,000,000

The implementation of this intuition training will be carried out in stages, as can be seen in Figure 3. Level one intuition training consists of the components of land price and the parameters that affect them; level two intuition training consists of components of land price and earthworking costs as well as parameters that affect both. And so on until the last level, level eight, which is intuition training in estimating the investment cost of components as a whole and each parameter that affects each component.

At this level of training, the first thing to train is to understand the situation. The situation that will be understood by the user is the relationship between the cost influence parameter and the cost-forming component. When understanding the situation, the user will enter input in the form of parameter values that affect the cost of the new component being trained. By entering inputs in the form of influential parameter values, users will understand what situations affect the investment size of each component.

After understanding the situation, the user will get a hint in the form of component investment costs. When the user enters input in the form of a parameter size, the cost of investing in the component will change depending on the value that has been entered. This is what will be the clues that users will get when they are doing training. After getting the clues, the user can figure out the pattern based on the clues that have been obtained. The pattern here is a graph of the cost of components for each condition.

The final session of this intuition training per level is an evaluation of the user's understanding by answering the questions to be given. Users must answer all questions correctly before proceeding to the next training. After the per-level intuition training is completed, users will be faced with a simulation session of estimating the investment cost of the new building shipyard. This simulation is repeated with different conditions in order for the user's intuitive skills to appear.As already explained, it takes knowledge and experience to develop intuition. The more knowledge and experience a person has, the more intelligent he will be in making decisions related to what he is familiar with. After each evaluation, the user will receive feedback in the form of errors he made.From this feedback, intuitive proficiency will emerge in the user regarding the estimated investment cost of the new building shipyard.

4.5 Application Design

Based on the implementation of intuition training in the investment estimates of new building shipyards that have been delivered, the flow of intuition training per level can be seen in Figure 3. Before conducting the training, users will be provided with basic knowledge and engineering experience. Basic knowledge in the form of component introduction and experience engineering in the form of example estimation case studies. After doing the simulation, there is a score graph for users to see their progress. Here is the arrangement of the application interface.

1. *Interface* 1. Interface 1 in this application displays the "login" feature. Every user who wants to run the application must first create an account by registering on this interface
2. *Interface* 2. After the user registers and enters the application, the user will be faced with Interface 2, which displays the operational systematics of the application in the form of functions, objectives, application limitations, and application operational methods.
3. *Interface* 3. The basic knowledge that will be presented to users is the component that forms the investment cost of a new building shipyard as a form of representation of user understanding related to the components that make up the investment cost of a new building shipyard. In addition to the introduction of components, interface 3 also provides a case study of the estimated investment cost of the new building shipyard, which can be seen in Table 4 as a material in the creation of an engineering experience for the user.

Table 4. Example of Case Study Estimation

Parameter	Value
Shipyard Location	Socah Village, Bangkalan City, Madura, Indonesia
Size of Land	160 x 110 m^2
Depth of Water	5 m
Land Price	Rp 600,000/m^2
Ship Size (Tanker Ship 6400 DWT)	
Length (L)	98.45 m
Breadth (B)	17.09 m
Height (H)	8.76 m
Actual Draft (Tactual)	6.24 m
Ship Steel Weight (Wst)	1755.96 Ton

4. *Interface 4.* Interface 4 is an introduction to the pattern of relationship between cost-building parameters (known conditions) and cost-forming components. By knowing the relationship between the two, it will be easier for users to estimate the investment cost of a new building shipyard in a simulation session based on the clues and patterns obtained.
5. *Interface 5.* Interface 5 is the initial stage of the user conducting intuition training. In this interface, users will be faced with level 1 intuition training, namely intuition training in estimating the cost of the land price component. First, the application will display the limitations on training to make it clear what components will be trained in this application. Furthermore, there are parameters in the form of location and land area that must be filled in by the user after understanding the situation. After the user inputs the value of the parameter, the cost of investing in the land price will appear based on the value of the parameter that has been input. This will be a clue to the user about what parameters affect the price of land. After that, users will get a pattern in the form of a graph of land prices based on each condition. Finally, users will be faced with an evaluation to answer the questions already provided. The user's answer must be all correct if you want to continue to the next level of training.

Table 5. Intuition Training Level 1

Parameter	Value
Location	Batam
Land Area	45000 m^2
Component	
Land Price	Rp 54,000,000,000
Total Investment	Rp 54,000,000,000

As shown in Table 5, the user will input the parameters for location and land area and will receive an output in the form of component investment costs that are affected by these parameters. From here, the user will analyze the relationship between the cost influence parameter and the cost-forming component.

6. *Interface 6.* Interface 6 is an intuition training in estimating the cost of land price components and earthwork costs. The inputted parameters will increase according to the addition of

components. In this interface, the parameters to be added are the rock structure and the slope of the soil. By entering the value of this parameter, the components of the cost of soil work in the form of land clearing costs, soil management costs, and soil compaction costs will change based on the inputted values. After that, a graph of the cost of earthwork is shown, and then there will be an evaluation.

7. *Interface 7.* Interface 7 is an intuition training in estimating the cost of the land price component, the cost of earthworks, and the cost of dredging the land. The inputted parameters will increase according to the addition of components. In this interface, the parameters to be added are the DWT of the ship and the depth of the waters. By entering the value of this parameter, the cost component of dredging the soil will change based on the inputted value. After that, a graph of the cost of dredging the land is shown, and then there will be an evaluation.
8. *Interface 8.* Interface 8 is an intuition training program for estimating the cost of land investment and launch facilities. By entering the value of the ship's DWT parameter and the type of launch means, the cost component of the launch means will change based on the inputted value. After that, a graph of the cost of the means of launch is shown, and then there will be an evaluation.
9. *Interface 9.* Interface 9 is an intuition training in estimating the cost of land investment, the cost of launching facilities, and the cost of building warehouses and workshops. By entering the value of the ship's DWT parameters, the cost component of building warehouses and workshops will change based on the inputted values. After that, a graph of the cost of building warehouses and workshops is shown, and then there will be an evaluation.
10. *Interface 10.* Interface 10 is an intuition training in estimating the cost of land investment, building investment, and winch. By entering the value of the ship's DWT parameters, the cost component of the winch needs will change based on the inputted value. After that, a graph of the cost of winch needs is shown, then there will be an evaluation.
11. *Interface 11.* Interface 11 is a training in intuition for estimating the cost of investing in land, building investments, winches, and manufacturing equipment. By entering the value of the ship's DWT parameter, the cost component of the production equipment will change based on the inputted value. After that, a graph of the cost of production equipment is shown, and then there will be an evaluation.
12. *Interface 12.* Interface 12 is an intuition training in estimating the total cost of investing in a new building shipyard. In this interface, other components will be added that have no relationship with parameters, so that the investment cost of these components will be equal in each condition. After that, there will be an evaluation.
13. *Interface 13.* Interface 13 user proficiency evaluation page after gaining knowledge about the estimated cost of investing in a new building shipyard. In this interface, a simulation will be carried out to estimate the investment cost of the new building shipyard based on the clues and patterns that have been obtained by the user. Every time they make one evaluation, users will be given feedback as a consideration for doing the next simulation. This simulation is carried out at least three times until the user feels that he has been trained in his intuitive skills in estimating the investment cost of the new building shipyard. A simulation of the

estimate can be seen in Table 6. Prospective estimators are required to fill in the empty answer fields so that the value of each simulation can be known.

Table 6. Investment Cost Estimation Simulation

Parameter	Value
Location	Lamongan
Depth of Water	2.5 m
Land Area	235 x 100 m^2
Ship DWT	5800 DWT
Length (L)	95.54 m
Breadth (B)	16.62 m
Height (H)	8.35 m
Actual Draft (Tactual)	6.03 m
Ship Steel Weight (Wst)	1580.94 ton
Launchway	Graving Dock

14. *Interface 14.* Interface 14 is the last page of this web-based application program. This interface displays a diagram of the development of user proficiency in estimating the investment costs of a new building shipyard. The progress graph will show the score obtained by the user after performing an estimated simulation. The purpose of this graph is to provide a result of evaluation that can be used as a representation of the development or progress of the user.

4.6 Application Validation Test

In this sub-chapter, we will discuss the validation of systems that have been made with manual calculations using Excel. The calculated data in Excel calculations uses the same system flow as the program or application flow that is built and is based on the calculation reference as calculated in the previous chapters. The value to be validated is the value of the investment cost of the new building shipyard for each component.

5. Analysis and discussion

5.1 Application Implementation Environment

The implementation environment is a specific area or condition where this training application can be operated. The result of this design is a web-based application using a MySQL database.

5.2 Application Verification Test

The prototype verification test of this application will use one of the parametric statistical tests, namely the "t" test. The t-test is used in testing to determine whether there is a significant difference between the two variables to be compared. The t test that will be used to test this application is the paired sample t test, which is a test that involves two measurements of the same subject against a certain influence. These two variables are the pre test given to the user before training using the application and the post test given to the user after training using the application. Table 7 is a recapitulation of respondents' values when filling out the pre-test and post test.

Table 7. Recapitulation of Respondent's Grade

Respondent	*Pre Test*	*Post Test*	d (Post – Pre)	d2
Respondent 1	0	6	6	36
Respondent 2	0	4	4	16
Respondent 3	2	7	5	25
Respondent 4	0	4	4	16
Respondent 5	0	6	6	36
Respondent 6	1	5	4	16
Respondent 7	0	6	6	36
Respondent 8	0	5	5	25
Respondent 9	1	7	6	36
Respondent 10	0	5	5	25

The formula used in this statistical test is:

$$t = \frac{\sum d}{\sqrt{\frac{N \sum d^2 - \left(\sum d\right)^2}{N-1}}}$$

The result obtained after testing is that the calculated t value is greater than the statistical table value t with a critical range of 5%, which is 11.53 > 2.262. This proves that this application can increase users' intuitive skills in estimating the investment costs of new building shipyards.

6. Conclusion

The following results were obtained after conducting experiments and analysis in this study:

1. A parametric method is the proper estimation method as a representation of performing an investment cost estimation of a new building intuitively. The parametric method is a method that estimates the cost of investment components of new building shipyards by considering parameters that affect the cost of these components.
2. The method of building the right intuition pattern to be implemented into the decision-making process of the investment cost estimation of the new building shipyard is the triggered decision-making (KDP) method. This KDP method consists of 4 sessions, namely situations, hints, patterns, and action scenarios. These four sessions will be conducted repeatedly so that intuitive skills can be obtained.
3. By dividing the application into two stages, it is possible to design an application program that can provide intuition training in estimating the investment costs of a new building shipyard. First, intuition training will be provided to users for estimating the cost of the components that make up the investment cost of a new building shipyard by considering the parameters of the cost effect. Second, after the user can estimate the cost of each component, a simulation of the total cost of investing in a new building shipyard is held on a recurring basis. Every time you finish estimating one condition, feedback is given so that you can find out the estimate of components that are not right

4. The prototype application created can train intuitive skills in estimating the investment costs of new building for small tankers of sizes 5000–9,999 DWT, among other limitations. This is based on the results of the verification test shown by taking response data from 10 respondents and using the "Two-Mean Difference Test" or "Paired T-Test (Pre-Post)" method, which produces a number of "t" and "t table" values of 11.53 > 2,262.

References

[1] H. Wu, "Intuition in Investment Decision-Making Across Cultures," *J. Behav. Financ.*, vol. 23, no. 1, pp. 106–122, 2022, doi: 10.1080/15427560.2020.1848839.

[2] V. Dörfler and F. Ackermann, "Understanding intuition: The case for two forms of intuition," *Manag. Learn.*, vol. 43, no. 5, pp. 545–564, 2012, doi: 10.1177/1350507611434686.

[3] G. A. Klein, "The Power of Intuition," in *Sources of Power*, The MIT Press, 2018. doi: 10.7551/mitpress/11307.003.0010.

[4] C. Akinci and E. Sadler-Smith, "Intuition in Management Research: A Historical Review," *Int. J. Manag. Rev.*, vol. 14, no. 1, pp. 104–122, 2012, doi: 10.1111/j.1468-2370.2011.00313.x.

[5] E. E. Miskioğlu *et al.*, "Situating intuition in engineering practice," *J. Eng. Educ.*, vol. 112, no. 2, pp. 418–444, 2023.

[6] S. Mattsson and Fast-Berglund, "How to Support Intuition in Complex Assembly?," *Procedia CIRP*, vol. 50, pp. 624–628, 2016, doi: 10.1016/j.procir.2016.05.014.

[7] S. I. Wahidi, T. W. Pribadi, W. S. Rajasa, and M. S. Arif, "Virtual Reality Based Application for Safety Training at Shipyards," *IOP Conf. Ser. Earth Environ. Sci.*, vol. 972, no. 1, 2022, doi: 10.1088/1755-1315/972/1/012025.

[8] M. S. Arif, H. Supomo, W. U. Alifia, and S. I. Wahidi, "Ship production process monitoring application using QR-code technology," *IOP Conf. Ser. Earth Environ. Sci.*, vol. 972, no. 1, p. 012016, Jan. 2022, doi: 10.1088/1755-1315/972/1/012016.

[9] G. R. Pradisa and T. W. Pribadi, "Computer-Based Application Design to Improve Intuition in New Ship Price Estimation," *J. Tek. ITS*, vol. 4, no. 1, 2016.

[10] C. K. Lin and H. J. Shaw, "Feature-based estimation of preliminary costs in shipbuilding," *Ocean Eng.*, vol. 144, no. October 2016, pp. 305–319, 2017, doi: 10.1016/j.oceaneng.2016.11.040.

[11] S. I. Wahidi, V. M. Virmansyah, and T. W. Pribadi, "Study on Implementation of Activity-Based Costing (ABC) System on Determination of Indirect Costs in Ship Production," *Kapal J. Ilmu Pengetah. dan Teknol. Kelaut.*, vol. 18, no. 1, pp. 1–7, 2021, doi: 10.14710/kapal.v18i1.33000.

[12] S. I. Wahidi, T. W. Pribadi, S. R. W. Pribadi, and S. Megawati, "Implementation Study of Activity Based Costing System to Define Indirect Costs on Ship Repair Industries," *IOP Conf. Ser. Mater. Sci. Eng.*, vol. 1052, no. 1, p. 012049, 2021, doi: 10.1088/1757-899x/1052/1/012049.

[13] I. Baihaqi, I. Lazakis, and H. Supomo, "Integrated Value Engineering and Risk Assessment Performance Measurement Framework in Ship-Manufacturing Industry towards Net Zero Emissions Using Fuzzy DEMATEL-AHP," *Machines*, vol. 11, no. 8. 2023. doi: 10.3390/machines11080799.

[14] I. Baihaqi, I. Lazakis, and H. Supomo, "A novel shipyard performance measurement approach through an integrated Value Engineering and Risk Assessment (VENRA) framework using a hybrid MCDM tool," *Proc. Inst. Mech. Eng. Part M J. Eng. Marit. Environ.*, vol. 238, no. 4, pp. 911–932, 2024, doi: 10.1177/14750902231219533.

[15] T. W. Pribadi, R. N. Adam, and S. I. Wahidi, "Technical and Economical Analysis of Shipyard Re-Layout for Product-Oriented Work Breakdown Structure Implementation," *IOP Conf. Ser. Mater. Sci. Eng.*, vol. 1052, no. 1, p. 012046, 2021, doi: 10.1088/1757-899x/1052/1/012046.

[16] S. I. Wahidi, S. Oterkus, and E. Oterkus, "Simulation of a Ship's Block Panel Assembly Process: Optimizing Production Processes and Costs through Welding Robots," *J. Mar. Sci. Eng.*, vol. 11, no. 8, p. 1506, Jul. 2023, doi: 10.3390/JMSE11081506.

[17] T. W. Pribadi and T. Shinoda, "Hand Motion Analysis for Recognition of Qualified and Unqualified Welders using 9-DOF IMU Sensors and Support Vector Machine (SVM) Approach," *Int. J. Technol.*, vol. 13, no. 1, pp. 38–47, 2022, doi: 10.14716/ijtech.v13i1.4813.

[18] I. Salo, T. W. Pribadi, and S. I. Wahidi, "Analysis of Access and Monitoring Systems Implementation on Human Resources to Improve the Shipbuilding Industry Performance," no. Senta 2018, pp. 79–86, 2020, doi: 10.5220/0008542600790086.

[19] I. CIL, F. ARISOY, A. Y. CIL, and H. Kilinc, "a Comparative Analysis of Indoor Positioning Technologies in Shipyard Digitalization Context," *J. Mar. Technol. Environ.*, vol. 2022, no. 1, pp. 15–25, 2022, doi: 10.53464/jmte.01.2022.03.

[20] S. I. Wahidi, S. Oterkus, and E. Oterkus, "Robotic welding techniques in marine structures and production processes: A systematic literature review," *Mar. Struct.*, vol. 95, no. May 2024, p. 103608, 2024, doi: 10.1016/j.marstruc.2024.103608.

[21] P. Fraga-Lamas, D. Noceda-Davila, T. M. Fernández-Caramés, M. A. Díaz-Bouza, and M. Vilar-Montesinos, "Smart Pipe System for a Shipyard 4.0," *Sensors*, vol. 16, no. 12, 2016, doi: 10.3390/s16122186.

[22] P. L. Nuthall and K. M. Old, "Intuition, the farmers' primary decision process. A review and analysis," *J. Rural Stud.*, vol. 58, no. April 2017, pp. 28–38, 2018, doi: 10.1016/j.jrurstud.2017.12.012.

[23] T. Pachur and M. Spaar, "Domain-specific preferences for intuition and deliberation in decision making," *J. Appl. Res. Mem. Cogn.*, vol. 4, no. 3, pp. 303–311, 2015, doi: 10.1016/j.jarmac.2015.07.006.

[24] J. E. Pretz *et al.*, "Development and validation of a new measure of intuition: The types of intuition scale," *J. Behav. Decis. Mak.*, vol. 27, no. 5, pp. 454–467, 2014, doi: 10.1002/bdm.1820.

[25] M. M. Roghanizad and D. J. Neufeld, "Intuition, risk, and the formation of online trust," *Comput. Human Behav.*, vol. 50, pp. 489–498, 2015, doi: 10.1016/j.chb.2015.04.025.

The Investigation of the Initial Cracks Effect on Ship Steering Gear Towards Environmental Safety

Septia Hardy Sujiatanti[1*], Lathif Jaya Nugraha[1], Muhammad Nurul Misbah[1], Rizky Chandra Ariesta[1], Roslin Ramli[2], Yuli Panca Asmara[3]

[1]Department of Naval Architecture, Institut Teknologi Sepuluh Nopember, Surabaya
[2]Maritime Engineering Technology, Universiti Kuala Lumpur Malaysian Institute of Marine Engineering Technology, Malaysia
[3]Faculty of Engineering and Quantity Surveying, INTI International University, Malaysia

E-mail: septia@its.ac.id

Abstract. This research aims to analyze the effect of initial cracks on the damage of ship Steering gear using the finite element method software. The study focuses on the hydraulic cylinder of the ship's Steering gear and employs two material variations: stainless steel 316 and AISI 1045. In this research, the initial cracks vary on the outer surface of the hydraulic cylinder of the steering gear with a semi-elliptical crack shape. The simulation results show that stainless steel 316 and AISI 1045 materials exceed their ultimate tensile strength at a crack length of 12 mm. For stainless steel 316, the ultimate tensile strength is 580 MPa, while for AISI 1045, it is 585 MPa. Crack propagation analysis reveals that the crack propagation rate in the hydraulic cylinder of the Steering gear with stainless steel 316 material is faster than that of the hydraulic cylinder with AISI 1045 material. Additionally, when comparing crack sizes, the crack propagation occurs more rapidly in the cylinder model with a deeper initial crack compared to the hydraulic cylinder with a shallower initial crack.

Keywords: Steering Gear, Crack Propagation, Initial Crack, Stainless Steel 316, AISI 1045, Finite Element Analysis

1. Introduction

The steering gear system plays a critical role in a ship, serving as the primary mechanism for determining the vessel's direction. Its function extends beyond merely enabling the ship to turn to the left (port side) or right (starboard side), as it forms the core of the entire navigation system. In complex maritime environments with varying weather conditions, precise and responsive control of the steering gear is essential for maintaining the ship's stability and ensuring safe passage. The steering gear system consists of several key components, including the rudder, rudder stock, hydraulic rams, hydraulic pumps, and reservoirs, all of which work together with control valves, piping, and hoses. The hydraulic system in the steering gear is a significant technological achievement, allowing for rapid and accurate control of the vessel's direction by manipulating fluid flow. This capability enables ships to navigate complex routes and adapt swiftly to changing conditions at sea [3].

Failures in the hydraulic steering gear system can occur due to cracks in the hydraulic cylinder. These cracks may develop from repeated loading, high pressure, impacts, or manufacturing defects, gradually reducing the cylinder's structural integrity. As cracks propagate, they can cause fluid leakage, lower system efficiency, and ultimately lead to total system failure. When this happens, the ship's maneuverability may be compromised, posing potential risks to the safety of the voyage. Therefore, analyzing crack propagation in the hydraulic cylinder is crucial to understanding the impact of such damage on the system. Any failure in these components can affect the overall safety and performance of the ship. Through crack propagation analysis, it is possible to detect structural damage early and prevent more severe failures over time. This knowledge allows for the planning of effective preventive or corrective measures, ensuring the integrity and reliability of the steering gear system.

2. Theoretical Based

2.1 Hydraulic system
A hydraulic system uses pressurized fluid to transmit motion and force, based on Pascal's law, which ensures uniform pressure in all directions. It employs a prime mover like an electric motor and can create movement with attachments such as jacks and gears. Its key advantage is producing large forces with minimal input [1].

2.2 crack
A crack is a factor that causes a structure to fail[6]. Failure occurs when the crack is subjected to cyclic loads, leading to a change in size until it reaches a certain point that causes fracture. The fracture due to the crack happens in several stages, including initial crack, crack propagation, and fracture[5][9].

2.3 Fracture Mechanics
Fracture mechanics is a method of analyzing fracture resolution based on stress analysis around cracks or notches. The method is divided into two categories: linear-elastic (LEFM) and elastic-plastic (EPFM) [2]. LEFM is an approach that describes the relationship between the stress pattern around the crack tip and factors such as the size, shape, crack orientation, and material properties affected by nominal stress in the structure. This method utilizes the parameter K, or SIF (Stress Intensity Factor), to identify the characteristics of the formed stress pattern. On the other hand, EPFM is more commonly used for ductile materials, where elastic-plastic phenomena occur due to the load applied to the material. EPFM is a development of LEFM, focusing on load analysis that can show the plastic deformation of the material.

2.4 Stress Intensity Factor
The Stress Intensity Factor (SIF) is one of the most fundamental and significant parameters in fracture mechanics. The Stress Intensity Factor describes the stress distribution at the crack tip[8]. This stress distribution is influenced by external loads, geometric shape, as well as the form and size of the existing crack. It is related to the crack growth rate and is used to establish failure criteria caused by fracture [2].

2.5 Crack Propagation Rate
The rate of crack propagation in a structure can be influenced by several factors, such as the thickness of the structure, the load applied to it, the size of the crack, and the type of material. The thickness of the material structure can affect the rate of crack propagation as it influences the

IOP Conf. Series: Earth and Environmental Science **1461** (2025) 012010 doi:10.1088/1755-1315/1461/1/012010

calculations. The load on the structure also plays a role, as crack propagation can only occur when the applied load is sufficient to cause the Stress Intensity Factor (SIF) to exceed the fatigue threshold region. Additionally, the initial crack size impacts the propagation because the stress intensity factor varies with different crack sizes. The type of material is another important factor since each material has its own specifications and toughness. Crack propagation can begin once the crack passes region 1, also known as the fatigue threshold region (ΔK), and can only occur when the stress intensity factor exceeds the threshold value, K_th, entering region 2 where the crack propagation accelerates.

2.6 Finite Element Method
The Finite Element Method (FEM) is a numerical technique used to solve complex engineering and physics problems. Its main advantage is its ability to analyze intricate geometries with results that closely match real-world conditions. FEM works by dividing a structure into smaller elements, a process called discretization. This allows for detailed analysis of stress and displacement using polynomials defined over these small elements and their nodal values [7][4].

3. Metodology

The ship's steering gear data used in this research is from the DDM 5500 type steering gear, which already includes detailed specifications regarding material, dimensions, and the pressure acting on the steering gear cylinder. This is followed by the variation determination stage, where the steering gear is given variations in the initial crack length. The required steering gear cylinder dimensions for modeling include the cylinder size and the pressure generated. The hydraulic steering gear data can be seen in Table 1.

Table 1. Detail specifications of the ship's steering gear.

pressure	122 bar
Rod	Ø125 mm
Piston	Ø220 mm
Strok	987 mm
Total cylinder displacement	62900 cm^3
Outer Diameter of the Cylinder	Ø245 mm
Inner Diameter of the Cylinder	Ø220 mm
Wall Thickness of the Steering Gear Cylinder	12,5 mm
Length of the Cylinder	1000 mm

The hydraulic cylinder steering gear model in SAT format is needed to perform static structural crack analysis using finite element analysis software. The hydraulic cylinder model was created using Autodesk Fusion 3D software; the hydraulic cylinder model in SAT format. can be seen in Figure 1.

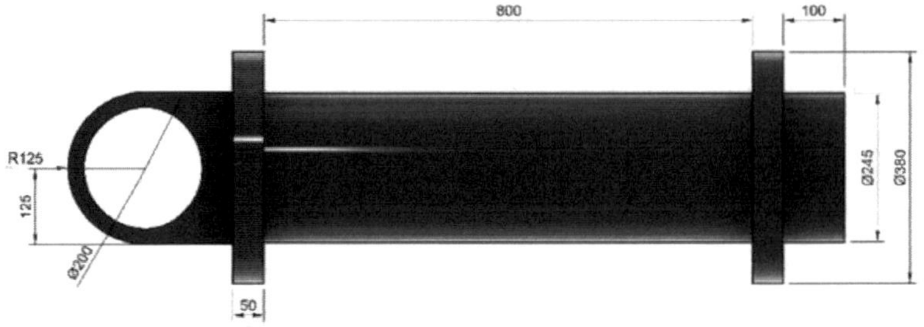

bb

Figure 1. Hydraulic Cylinder Model in SAT Format.

In this analysis, it was found that two types of materials are used: AISI 1045 and stainless steel 316. Therefore, detailed material properties for both materials are required. After that, boundary conditions for the hydraulic cylinder steering gear model are determined, including the definition of the loads applied to the model. The boundary conditions are adjusted according to the constraints of the ongoing research.

This analysis considers the load due to hydraulic fluid inside the hydraulic cylinder steering gear. The load applied by the hydraulic fluid must be considered by accounting for the operational pressure generated by the system. The loading from the hydraulic fluid in the ship's steering gear is assumed to be a pressure distributed evenly across the inner walls of the hydraulic cylinder. According to the steering gear data, the nominal working pressure is 122 bar, so the load is depicted as a pressure uniformly distributed inside the hydraulic cylinder steering gear. After that, convergence is performed using the finite element analysis application. Convergence is achieved by comparing stress results from different numbers of elements, with the goal of reaching stable results. In this analysis, the difference between the results of the two should be less than 5% of the first result in order to be considered convergent.

The stress intensity factor describes the distribution of stress at the crack tip. The stress distribution is influenced by external loads, geometric shape, as well as the shape and size of the crack. The calculation of the SIF (stress intensity factor) for semi-elliptical cracks can be analyzed using finite element analysis software. This application provides powerful tools for visualizing the stress distribution around cracks and accurately calculating the SIF. After that, crack propagation analysis is performed. Crack propagation can be predicted using the Paris-Erdogan equation. To apply this formula, the stress intensity factor (SIF) values are required, which can be obtained through analysis using finite element analysis software. This method allows for accurate simulations to understand and predict how cracks will develop in a structure.

4. Results and Discussion

4.1 *Analysis of Cylinder Strength Due to Cracks*

To analyze the strength of the hydraulic cylinder in this analysis, three parameters can be used. The material can be considered damaged if it has exceeded one of the failure criteria. The parameters or failure criteria are as follows:

4.1.1 Comparing Equivalent Stress with Ultimate Tensile Strength

In this analysis, the failure criterion that is reached first is the comparison of the equivalent stress with the ultimate tensile strength. The equivalent stress values of the two materials after analysis at each crack increment can be seen in Table 2.

Table 2. Results of Equivalent Stress Values on the Cylinder Model.

Crack length	Equivalent stress maximum (AISI 1045)	Equivalent stress maximum (Stainless steel 316)
mm	MPa	MPa
5	389.428	385.401
6	418.964	414.267
7	447.645	442.310
8	475.976	470.026
9	504.472	497.929
10	533.348	526.222
11	562.875	555.179
12	593.210	584.943
13	624.336	615.499
14	656.492	647.084
15	689.668	679.684
16	723.874	713.308
17	758.970	747.806
18	794.854	783.091
19	831.527	819.157
20	868.385	855.384
21	905.570	891.947
22	942.944	928.697

Based on the data obtained from the simulation, it is known that for both types of materials, namely stainless steel 316 and AISI 1045, at a crack length of 12 mm, the materials have exceeded their ultimate tensile strength. For stainless steel 316, the ultimate tensile strength obtained is 580 MPa. For AISI 1045, the recorded ultimate tensile strength is 585 MPa. This indicates that at a crack length of 12 mm, both materials experience a condition that exceeds the maximum tensile strength they can withstand, which potentially indicates a risk of structural failure in real-world applications.

4.1.2 Comparing SIF Values with Fracture Toughness

In this analysis, the failure criterion of comparing SIF values with fracture toughness is not met. The stress intensity factor values for each crack length analyzed can be seen in Table 3.

Table 3. SIF Values of Both Materials at Each Crack Length.

Crack length	AISI 1045	Stainless Steel 316
	SIFS (K1) Maximum	SIFS (K1) Maximum
mm	MPa mm^0.5	MPa mm^0.5
5	219,699	220,014
6	242,384	242,743
7	264,172	264,582
8	285,455	285,927
9	306,572	307,113
10	327,739	328,361
11	349,147	349,859
12	370,906	371,717
13	393,122	394,045
14	415,827	416,868
15	439,093	440,262
16	462,925	464,231
17	487,362	488,811
18	512,182	513,776
19	537,501	539,248
20	563,079	564,976
21	588,889	590,939
22	614,655	616,856

The SIF values for both materials do not exceed the fracture toughness values of each material. The fracture toughness for AISI 1045 is 1.296 MPa √mm, while the SIF value for AISI 1045 with the longest crack length of 22 mm is only 614.655 MPa √mm. For stainless steel 316, the fracture toughness is 1581 MPa √mm, and the SIF value for the longest crack length of 22 mm is 616.856 MPa √mm.

After analyzing the two steering gears with different materials, it was found that the steering gear designed with stable material will not reach the fracture toughness (KIC) value, even if the material has experienced cracks that nearly penetrated the hydraulic cylinder of the steering gear.

4.1.3 Crack Propagation Rate Analysis

Crack propagation analysis can be conducted after obtaining the Stress Intensity Factor (SIF) values. These SIF values are obtained through simulations performed on two models with different materials and various crack lengths. The complete data from the simulations with different crack lengths can be seen in Table 4.

Table 4. SIF Values of Both Materials at Each Crack Length.

Crack length	SIFS (K1) Maximum (Stainless steel 316)	SIFS (K1) Maximum (AISI 1045)
mm	MPa mm^0,5	MPa mm^0,5
5	220,01	219,70
6	242,74	242,38
7	264,58	264,17
8	285,93	285,46
9	307,11	306,57
10	328,36	327,74
11	349,86	349,15
12	371,72	370,91
13	394,04	393,12
14	416,87	415,83
15	440,26	439,09

After obtaining the Stress Intensity Factor (SIF) values, crack propagation can be calculated using the Paris-Erdogan formula to obtain the crack propagation rate (da/dN). The value of K is needed to calculate the crack propagation rate. To find the value of ΔK, the following steps are taken:

$\Delta K = K_{max} - K_{min}$

The minimum stress in this analysis is 0; therefore, the SIF value at the minimum stress (K_min) is also 0. Hence, the value of ΔK can be seen in Table 5, along with the material constants. The values of the material constants c and m for the two analyzed materials can be found in Table 6.

Table 5. Table of ΔK Values for Both Materials.

Crack length (mm)	AISI 1045	Stainless Steel 316
	ΔK (Ksi.in^0.5)	ΔK (Ksi.in^0.5)
5	6,323	6,332
6	6,975	6,986
7	7,602	7,614
8	8,215	8,228
9	8,823	8,838
10	9,432	9,450
11	10,048	10,068
12	10,674	10,697
13	11,313	11,340
14	11,967	11,997
15	12,636	12,670

Table 6. Material Constant

material	Stainless steel 316	AISI 1045
c	3 x 10^-10	3,6 x 10^-10
m	3,25	3

IOP Conf. Series: Earth and Environmental Science **1461** (2025) 012010 doi:10.1088/1755-1315/1461/1/012010

Based on the obtained data, the crack propagation calculation can now be performed. The crack propagation calculation uses the Paris-Erdogan formula as shown below:

$$da/dN = C(\Delta K)^{m}$$

$$= 3x10^{-10} x\ (6{,}323)^{3{,}25}$$

$$= 0{,}00000012\ in/cycle\ (0{,}00000307\ mm/cycle)$$

The calculation above is an example of the crack propagation rate for Stainless Steel 316 with an initial crack length of 5 mm. The results of the crack propagation rate calculation using this formula for Stainless Steel 316 can be seen in Table 7, and for AISI 1045 in Table 8.

Table 7. Initial Crack 5mm Propagation Rate (Stainless Steel 316)

Crack length (mm)	da/dN (in/cycle)	da/dN (mm/cycle)
5	0,00000012	0,00000307
6	0,00000017	0,00000422
7	0,00000022	0,00000559
8	0,00000028	0,00000719
9	0,00000036	0,00000907
10	0,00000044	0,00001127
11	0,00000055	0,00001385
12	0,00000066	0,00001687
13	0,00000080	0,00002039
14	0,00000096	0,00002449
15	0,00000115	0,00002924

Table 8. Initial Crack 5mm Propagation Rate (AISI 1045)

Crack length (mm)	da/dN (in/cycle)	da/dN (mm/cycle)
5	0,00000009	0,00000231
6	0,00000012	0,00000310
7	0,00000016	0,00000402
8	0,00000020	0,00000507
9	0,00000025	0,00000628
10	0,00000030	0,00000767
11	0,00000037	0,00000928
12	0,00000044	0,00001112
13	0,00000052	0,00001324
14	0,00000062	0,00001567
15	0,00000073	0,00001845

After obtaining da/dN, the value of N is calculated to determine how many cycles are needed until the final crack length. The data for N can be found using the equation $N = \frac{af-a0}{da/dN}$. The results of the N calculation can be seen in Table 9, and the graph of N versus crack length is shown in Figure 2.

Table 9. Calculation of N Values for Both Materials

Initial crack length (mm)	Final crack length (mm)	Stainless steel 316		AISI 1045	
		ΔN (Cycles)	ΣN (Cycles)	ΔN (Cycles)	ΣN (Cycles)
5	6	325928	325928	432700	432700
6	7	236789	562717	322226	754926
7	8	178963	741680	248893	1003818
8	9	139078	880758	197268	1201086
9	10	110247	991005	159249	1360335
10	11	88704	1079709	130342	1490677
11	12	72183	1151892	107807	1598484
12	13	59278	1211170	89925	1688409
13	14	49041	1260212	75525	1763934
14	15	40840	1301052	63817	1827750

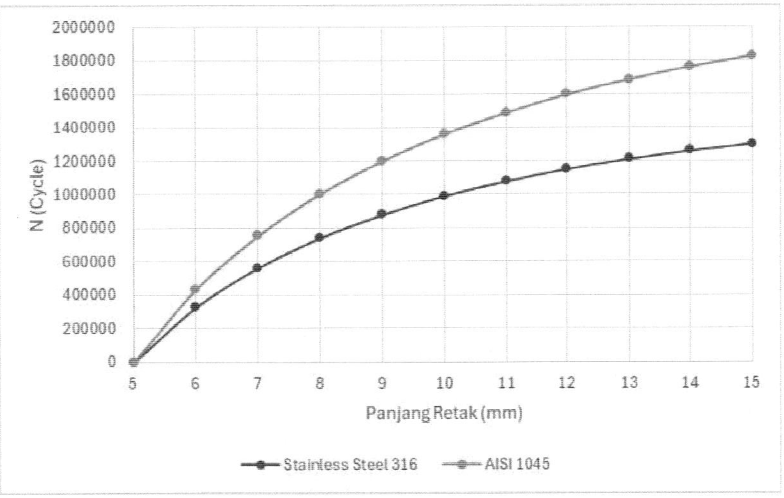

Figure 2. Comparison of Crack Length to Number of Cycles.

From the obtained data, for crack lengths ranging from 5 mm to 15 mm, the number of cycles for Stainless Steel 316 is 1,301,052 cycles, while for AISI 1045 it is 1,827,750 cycles.

5. Conclusion

After conducting experiments and research on various hydraulic cylinder models, the conclusions can be summarized as follows. It was found that the hydraulic cylinder model can no longer support loads when the crack length reaches 12 mm. Both materials tested-stainless steel 316 and AISI 1045 showed similar results in that the cylinder could not maintain its strength. The equivalent stress values for both materials exceeded the ultimate tensile strength, indicating that the hydraulic cylinder cannot operate safely or effectively under the given load conditions at this crack length. Additionally, the analysis revealed that crack propagation in the hydraulic steering gear cylinder with stainless steel 316 occurs more rapidly than in the cylinder using AISI 1045 steel. Cylinders with deeper initial cracks experience faster propagation compared to those with shallower initial cracks.

IOP Conf. Series: Earth and Environmental Science **1461** (2025) 012010 doi:10.1088/1755-1315/1461/1/012010

References

[1] Elmardi, O. M., & Khayal, S. (2017). Introduction to hydraulic systems Module 5: Hydraulic Systems Lecture Introduction. https://www.researchgate.net/publication/318672647

[2] Naess. (1985). A. Fatigue handbook: Offshore steel structures.

[3] Wang, F., Wei, Z., Li, P., Yu, L., & Huang, W. (2019). Initial crack propagation and the influence factors of aircraft pipe pressure. Materials, 12(19). https://doi.org/10.3390/ma12193098

[4] Whiteley, J. (2017). Mathematical Engineering Finite Element Methods A Practical Guide. http://www.springer.com/series/8445

[5] Barsom, J.M. and Rolfe, S.T. (1999) Fracture and Fatigue Control in Structures: Application of Fracture Mechanics. 3rd Edition, ASTM International, West Conshohocken, PA.

[6] Gunawan, M.M.R.A., Zubaydi, A., Sujiatanti, S.H., Ariesta, R.C. (2023). Analisis Crack pada Transverse Corrugated Bulkhead Kapal Tanker Menggunakan Metode Elemen Hingga.

[7] Ardianus, Sujiatanti, S.H., Setyawan. D. (2017). Analisis Kekuatan Konstruksi Sekat Melintang Kapal Tanker dengan Metode Elemen Hingga.

[8] Budiman, M.A., Misbah, M.N., Yulianto, T. (2019). Studi Kepecahan Moda Kegagalan Material A36 dengan Variasi Panjang Retak Terpusat.

[9] Pitaloka, S.D., Misbah, M.N. (2021). Studi Pengaruh Variasi Pengelasan Bracket terhadap Fatigue Life Struktur Kapal Patroli Cepat 40M Akibat Beban Gelombang.

Structural Reliability Investigation of a Passenger-Cargo-Container Ship Under Varying Loading Conditions and Wave Heights Towards Safety Management

Septia Hardy Sujiatanti[1]*, Fathur Rahman[1], Tri Achmadi[2], Aries Sulisetyono[1], Adin Sugiarso[3], Yuli Panca Asmara[4]

[1] Department of Naval Architecture, Institut Teknologi Sepuluh Nopember, Surabaya, Indonesia
[2] Department of Marine Transportation, Institut Teknologi Sepuluh Nopember, Surabaya, Indonesia
[3] National Ship Design and Engineering Center, Surabaya
[4] Faculty of Engineering and Quantity Surveying, INTI International University, Malaysia

*E-mail: septia@its.ac.id

Abstract. This study analyzes the reliability of a passenger-cargo-container ship operating in Indonesian waters. The primary focus is to evaluate the impact of variations in loading conditions and wave height on the structural integrity of the vessel. The research methodology involves calculating the longitudinal strength of the ship under various wave conditions in its operational area, and integrating Monte Carlo simulations to analyze the structural reliability under different operational scenarios. The yield strength of marine grade A material is used as the failure definition. The results show that structural reliability decreases with increasing wave height, with the highest probability of failure reaching 0.60% at a wave height of 3 meters. Loading condition analysis reveals that the scenario with passenger-only loading shows a significant decrease in reliability, reaching 99.30%, compared to other loading conditions that maintain reliability above 99.8%. Monte Carlo simulations yield structural reliability values of 99.72% for the ship's deck and 99.81% for the ship's bottom, with failure probabilities of 0.28% and 0.19%, respectively. These findings indicate that the deck structure is more prone to failure compared to the bottom structure.

Keywords: Reliability, Safety Management, Ship Structure, Monte Carlo Simulation

1. Introduction

The passenger-cargo-container vessel presents an innovative solution to optimize maritime transport efficiency, especially in archipelagic nations like Indonesia [1]. Its multifunctional design enables the simultaneous transportation of passengers, cargo, and containers, enhancing operational flexibility and economic efficiency [2]. However, this design complexity introduces new challenges in terms of structural reliability. Structural reliability is crucial for ensuring safe operations, especially considering the significant variations in loading conditions on passenger-cargo-container vessels [3]. Dynamic shifts in load distribution resulting from the combination of passengers, cargo, and containers can compromise the overall structural integrity of the vessel [4]. Furthermore, Indonesia's diverse wave conditions add complexity to structural reliability analysis [5]. Despite existing research on the structural analysis of multi-purpose vessels [6], a comprehensive understanding of how varying loading conditions and wave heights impact the structural reliability of passenger-cargo-container vessels, particularly in Indonesian waters, remains elusive. Therefore, this study aims to assess the structural reliability of passenger-cargo-container vessels by considering variations in loading conditions and wave height effects. The study will identify critical structural areas susceptible to failure under different operational scenarios. A passenger-cargo-container vessel operating in Indonesia was used for this research. The findings are expected to provide valuable insights for naval architects, operators, and regulators in enhancing the safety and operational efficiency of Indonesian waters. Additionally, these findings can contribute to the development of more comprehensive safety standards for multipurpose vessels in the future [7].

2. Literature Review

2.1 Passenger-Cargo-Container Vessel

A passenger-cargo-container vessel is an innovative ship design that combines the functions of transporting passengers, general cargo, and containers on a single platform. This concept has attracted the attention of the maritime industry due to its potential to increase operational efficiency and service flexibility, especially for routes with high demand variations [2]. According to Papanikolaou [8], the design of the vessel requires a holistic approach that considers various aspects, including:

a. Optimal space layout to accommodate passengers, cargo, and containers.
b. Efficient cargo handling systems for various types of cargo.
c. Ship stability under various loading conditions.
d. Passenger and cargo safety.

Lee et al. [9] proposed a modular design that allows for the reconfiguration of cargo space according to demand, increasing operational flexibility by up to 40% compared to conventional ships. Zhang and Moan [10] showed that the ships can increase the average load factor to 75%, compared to 60% for passenger or cargo-only ships. This results in a 15-20% increase in fuel efficiency per ton-mile of cargo transported.

The structural design of passenger-cargo-container ships faces unique challenges due to significant load variations. Shin and Lee [11] identified that the bending moment at the midship section can vary by up to 30% depending on the cargo configuration, requiring a stronger and more flexible structural design. Safety is a primary concern in the design of the ships. Vassalos et al. [12] emphasized the importance of integrated safety analysis that considers the interaction between passengers and cargo operations. They proposed a "design for safety" approach that can reduce accident risks by up to 50% compared to traditional approaches.

The International Maritime Organization (IMO) does not have specific regulations for the ships. However, Karahalios [13] suggests that a combination of the SOLAS (Safety of Life at Sea) for passenger ships and the IMDG Code (International Maritime Dangerous Goods Code) for hazardous cargo can be applied with certain modifications. In Indonesia, the concept of passenger-cargo-container ships has started to be adopted to support inter-island connectivity. Research by Syafiul et al. [1] shows that implementing passenger-cargo-container ships on strategic routes can reduce logistics costs by up to 25% and increase service frequency by up to 40%.

2.2 Longitudinal Strength of Ships
In designing a ship's structure, it is crucial to consider the loads that the structure will experience throughout its operational life. These loads include shear force and bending moment along the ship's length. The hull structure of a ship operating at sea will experience external loads primarily from fluid pressure on the hull surface, gravitational forces, and inertial forces. Fluid pressure on the hull surface will generate hydrodynamic forces. These forces cause shear forces and bending moments to arise [14].

The pattern and intensity distribution of shear forces and bending moments in the longitudinal direction of a warship at each station are calculated from the difference between the distribution of the ship's weight and buoyancy. This assumes the ship's condition as a static beam

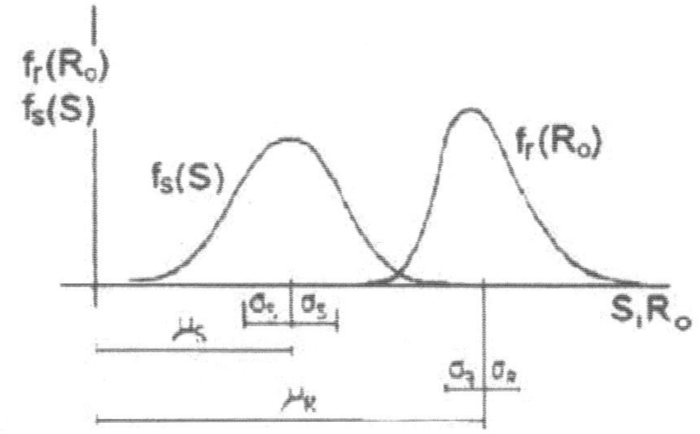

Figure 1. Joint Probability Function variable R and S

supported at the middle or at both ends.

2.3 Reliability Analysis Concept
Structural Reliability Theory focuses on assessing uncertainty and how to improve and streamline the design process. This means that reliability analysis must clearly define the amount of load capacity and the uncertainty of the amount of load, as well as the probability of events so that the accuracy of the safety margin can be relied upon. Another definition of reliability is the probability of the ability of a structural system to maintain its performance throughout its service life, which shows the probability density function (PDF) of the structural capacity (R) and variable load (S).

2.4 Monte Carlo Simulation Method
The main element of Monte Carlo simulation is the use of a random number generator (RNG), which is done by sampling a set of random variables based on the probability distribution of those random variables. Basically, the sample taken based on the RNG is taken as input to the performance function FK(x) and then the value of FK(x) is calculated. For example, for a certain performance function, a system or component is considered to have failed if FK(x) > 0. If the number of samples is N, then the number of occurrences of FK(x) > 0 is recorded n times. Thus, the probability of failure of the system/component under consideration is the ratio between the number of failed events and the number of completed events.

IOP Conf. Series: Earth and Environmental Science **1461** (2025) 012011 doi:10.1088/1755-1315/1461/1/012011

3. Methodology

3.1 Ship Data

This study involves a passenger-cargo-container ship operating in the waters of Central Indonesia. The ship is designed to carry passengers, cargo, and containers during its voyage. Table 1 presents the main dimensions of the ship, while the hull modeling used in the longitudinal strength analysis is shown in Figure 3. The ship's Midship Section is displayed in Figure 2.

Table 1. Ship Particular

Name	Dimension	Unit
LOA	146.5	meter
B	23.4	meter
H	8.2	meter
T	5.9	meter
LWT	6863.52	ton

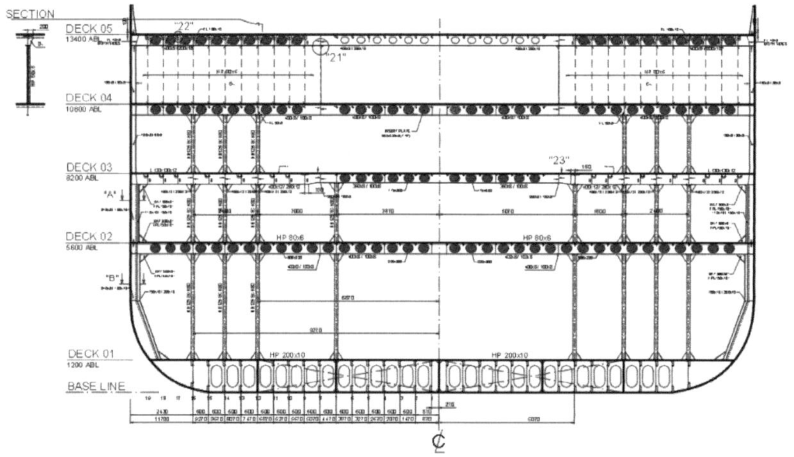

Figure 2. Midship Section Plan of the Ship

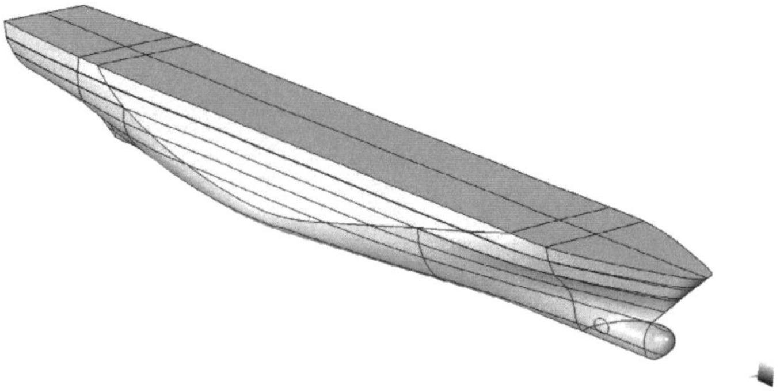

Figure 3. Hull Modelling

IOP Conf. Series: Earth and Environmental Science **1461** (2025) 012011 doi:10.1088/1755-1315/1461/1/012011

To understand the distribution of the ship's longitudinal strength, the location of the loads and tanks on the ship was also modeled, using the load distribution data and the ship's general arrangement. Figure 4 shows the modeling of the load positions and the consumable tanks.

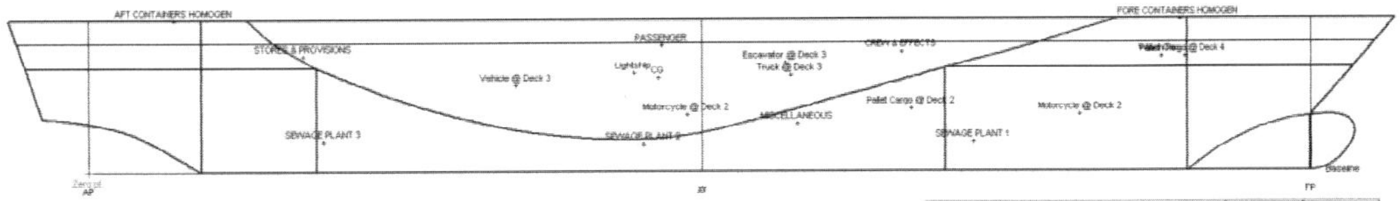

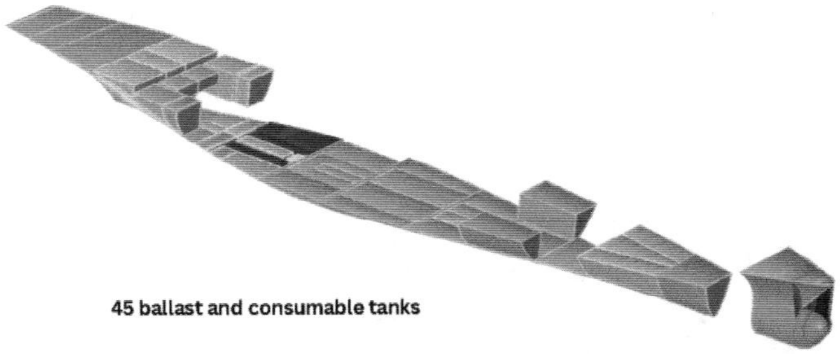

	Item Name	Quantity	Unit Mass tonne	Total Mass tonne
1	Lightship	1	6864,520	6864,520
2	FORE CONTAINERS HOM	1	345,000	345,000
3	AFT CONTAINERS HOMO	1	800,000	800,000
4	Vehicle @ Deck 3	39	2,000	78,000
5	Truck @ Deck 3	9	24,000	210,000
6	Escavator @ Deck 3	4	40,000	160,000
7	Pallet Cargo @ Deck 2	153	1,000	153,000
8	Motorcycle @ Deck 2	96	0,100	9,600
9	Motorcycle @ Deck 2	48	0,100	4,800
10	Pallet Cargo @ Deck 4	48	1,000	48,000
11	Winch Pos	1	2,000	2,000
12	PASSENGER	1	146,165	146,165
13	CREW & EFFECTS	146	0,150	21,900
14	STORES & PROVISIONS	1	100,000	100,000
15	MISCELLANEOUS	1	160,000	160,000
16	SEWAGE PLANT 1	1	25,000	25,000
17	SEWAGE PLANT 2	1	25,000	25,000
18	SEWAGE PLANT 3	1	25,000	25,000

Figure 4. Load and Consumable Tank Modelling

This ship uses marine-grade A steel material. Table 2 below presents the yield strength distribution data of the steel used, based on data from SSC [14].

Table 2. Yield Strength Distribution of Grade A Steel
Source: SSC-375 UNCERTAINTY INSTRENGTH MODELS FOR MARINE STRUCTURES, Tabel 6-13

Variations	Ultimate Yield Stress (MPa)	μs (Mean)	σs (Stdev)
1	256		
2	187		
3	132		
4	196		
5	185		
6	130		
7	179		
8	162		
9	121		
10	114	151,21	39,16
11	190		
12	134		
13	127		
14	153		
15	145		
16	117		
17	128		
18	122		
19	94		

IOP Conf. Series: Earth and Environmental Science **1461** (2025) 012011 doi:10.1088/1755-1315/1461/1/012011

3.2 Environmental Data

To analyze the ship's reliability, wave scatter data from the ship's operational area must be included. The ship sails from Makassar, Parepare, Balikpapan, Tarakan, Nunukan, Maumere, Lewoleba, and Kupang, as shown in Figure 5. With the assistance of the website www.metoceanview/hindcast, wave scatter data was obtained for the period from 1979 to 2021 (Table 3). This data provides information about the sea wave characteristics frequently encountered during the ship's journey. In the reliability analysis using wave scatter data, the primary focus is on the likelihood of encountering waves with a probability above 1.5% of the collected data range. Waves with a probability above this threshold can be considered extreme conditions that need to be taken into account in the ship's load calculations. Therefore, wave data with a probability above 1.5% will be used as input in the ship load calculations.

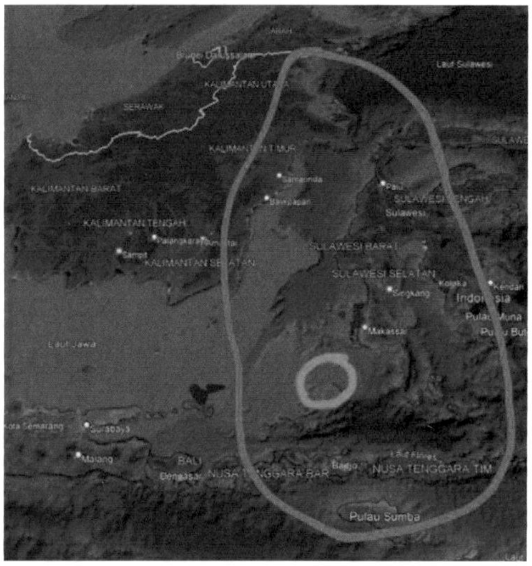

Figure 5. Operational Area of the Vessel

Table 3. Wave Scatter Data on the Vessel's Operational Route

Relative Peak Period (Smooth) s; rows: Significant wave height m

Significant wave height m	9-10	10-11	11-12	12-13	13-14	14-15	15-16	16-17	17-18	18-19	19-20
0 - 0.5	0%	0%	0%	0%	0%	0%	0%	0%	0%	0%	0%
0.5 - 1	<0.1%	0.20%	0.80%	**1%**	0.80%	0.30%	0.10%	0.20%	<0.1%	<0.1%	<0.1%
1 - 1.5	0.20%	0.50%	2.40%	7.90%	**11.10%**	6.20%	2.20%	2.70%	0.60%	0.80%	0.20%
1.5 - 2	0.20%	0.30%	0.60%	3.10%	9.50%	**11%**	5.20%	4.80%	1.20%	1.10%	0.50%
2 - 2.5	0.20%	0.20%	<0.1%	0.20%	1.30%	3.70%	**3.80%**	3.80%	1.20%	0.70%	0.30%
2.5 - 3	0.10%	<0.1%	<0.1%	<0.1%	<0.1%	0.20%	0.70%	**1.60%**	0.70%	0.30%	0.10%
3 - 3.5	<0.1%	<0.1%	<0.1%	<0.1%	<0.1%	<0.1%	<0.1%	0.10%	0.20%	0.10%	<0.1%
3.5 - 4	<0.1%	<0.1%	<0.1%	<0.1%	0%	0%	<0.1%	<0.1%	<0.1%	<0.1%	<0.1%
4 - 4.5	<0.1%	<0.1%	<0.1%	<0.1%	<0.1%	0%	0%	0%	0%	0%	0%
4.5 - 5	<0.1%	<0.1%	<0.1%	0%	<0.1%	0%	0%	0%	0%	0%	0%
5 - 5.5	0%	<0.1%	<0.1%	0%	<0.1%	0%	0%	0%	0%	0%	0%
5.5 - 6	0%	<0.1%	<0.1%	<0.1%	0%	<0.1%	0%	0%	0%	0%	0%
6 - 6.5	0%	0%	<0.1%	0%	0%	<0.1%	<0.1%	0%	0%	0%	0%
6.5 - 7	0%	0%	0%	0%	0%	0%	<0.1%	0%	0%	0%	0%

IOP Conf. Series: Earth and Environmental Science **1461** (2025) 012011 doi:10.1088/1755-1315/1461/1/012011

3.3 Ship Loading Scenarios

In this reliability analysis, the combination of longitudinal strength loading from 5 wave heights is based on the highest wave height probabilities from the wave scatter data presented in Table 3. Additionally, a calm water condition is included, with a wave height of 0 m. In these 6 wave conditions, 8 ship loading scenarios are modeled as shown in Table 4. Each loading scenario represents different operational conditions. C1 and C2 represent homogeneous container loading conditions (67 TEU @ 15t) during departure and arrival. C3 and C4 represent non-homogeneous container loading conditions (67 TEU with loads below 15t). C5 and C6 represent loading conditions with only passengers, and C7 and C8 represent the heaviest loading conditions where homogeneous containers are combined with crane loads during container loading.

Table 4. Loading Conditions

Loading Condition	**LWT**	**C1**	**C2**	**C3**	**C4**	**C5**	**C6**	**C7**	**C8**
Container Homogen	-	100%	100%	-	-	-	-	100%	100%
Container Non-Homogen	-	-	-	100%	100%	-	-	-	-
Container Loading	-	-	-	-	-	-	-	100%	100%
Vehicle	-	100%	100%	100%	100%	-	-	100%	100%
Pallet Cargo	-	100%	100%	100%	100%	-	-	100%	100%
Passanger	-	100%	100%	100%	100%	100%	100%	100%	100%
Consumables	-	100%	10%	100%	10%	100%	10%	100%	10%
Provisions	-	100%	10%	100%	10%	100%	10%	100%	10%

From the wave height data and ship loading conditions, a combination of 48 ship loading scenarios was derived for analysis in this study. These scenarios are presented in Table 5. Each condition reflects a different combination of the ship's carried load and the sea wave characteristics encountered during the voyage. These 48 scenarios will be used as input in the calculation of the ship's longitudinal strength.

3.4 Bending Moment Calculation

This study only considers vertical wave loads, as the dominant motion of small ships when underway is vertical motion, namely heave and pitch. The load that most significantly affects the longitudinal strength of the ship's structure is the vertical bending moment in the midship area. The vertical bending moment (MB) is calculated under two free surface conditions: calm water (MS) and wave conditions (MW), as explained in equations (1) and (2).

$$MB = MS + MW \tag{1}$$
$$MW = M\zeta \tag{2}$$

The wave bending moment (Mw) is the bending moment caused by waves (Mζ). In this study, the bending moment from waves is calculated with the aid of Maxsurf Stability software. The inputs required for Maxsurf Stability include wave length, wave height, and the position of the wave crest relative to the ship's length. Since a static analysis method is used in this study, the position of the wave crest chosen is the one that results in the highest bending moment. Figure 6 shows the wave modeling used to obtain the bending moment.

Table 5. Longitudinal Strength Loading Scenario of the Vessel

	Container Homogen	Container Non-Homogen	Container Loading	Vehicle	Pallet Cargo	Passanger	Consumables	Provisions
Case	**Still Water Condition / T = 0 s ; H = 0 m**							
1	100%	-	-	100%	100%	100%	100%	100%
2	100%	-	-	100%	100%	100%	10%	10%
3	-	100%	-	100%	100%	100%	100%	100%
4	-	100%	-	100%	100%	100%	10%	10%
5	-	-	-	-	-	100%	100%	100%
6	-	-	-	-	-	100%	10%	10%
7	100%	-	100%	100%	100%	100%	100%	100%
8	100%	-	100%	100%	100%	100%	10%	10%
Case	**Wave Condition / T = 13 s ; H = 1 m**							
9	100%	-	-	100%	100%	100%	100%	100%
10	100%	-	-	100%	100%	100%	10%	10%
11	-	100%	-	100%	100%	100%	100%	100%
12	-	100%	-	100%	100%	100%	10%	10%
13	-	-	-	-	-	100%	100%	100%
14	-	-	-	-	-	100%	10%	10%
15	100%	-	100%	100%	100%	100%	100%	100%
16	100%	-	100%	100%	100%	100%	10%	10%
Case	**Wave Condition / T = 14 s ; H = 1,5 m**							
17	100%	-	-	100%	100%	100%	100%	100%
18	100%	-	-	100%	100%	100%	10%	10%
19	-	100%	-	100%	100%	100%	100%	100%
20	-	100%	-	100%	100%	100%	10%	10%
21	-	-	-	-	-	100%	100%	100%
22	-	-	-	-	-	100%	10%	10%
23	100%	-	100%	100%	100%	100%	100%	100%
24	100%	-	100%	100%	100%	100%	10%	10%
Case	**Wave Condition / T = 15 s ; H = 2 m**							
25	100%	-	-	100%	100%	100%	100%	100%
26	100%	-	-	100%	100%	100%	10%	10%
27	-	100%	-	100%	100%	100%	100%	100%
28	-	100%	-	100%	100%	100%	10%	10%
29	-	-	-	-	-	100%	100%	100%
30	-	-	-	-	-	100%	10%	10%
31	100%	-	100%	100%	100%	100%	100%	100%
32	100%	-	100%	100%	100%	100%	10%	10%
Case	**Wave Condition / T = 16 s ; H = 2.5 m**							
33	100%	-	-	100%	100%	100%	100%	100%
34	100%	-	-	100%	100%	100%	10%	10%
35	-	100%	-	100%	100%	100%	100%	100%
36	-	100%	-	100%	100%	100%	10%	10%
37	-	-	-	-	-	100%	100%	100%
38	-	-	-	-	-	100%	10%	10%
39	100%	-	100%	100%	100%	100%	100%	100%
40	100%	-	100%	100%	100%	100%	10%	10%
Case	**Wave Condition / T = 17 s ; H = 3 m**							
41	100%	-	-	100%	100%	100%	100%	100%
42	100%	-	-	100%	100%	100%	10%	10%
43	-	100%	-	100%	100%	100%	100%	100%
44	-	100%	-	100%	100%	100%	10%	10%
45	-	-	-	-	-	100%	100%	100%
46	-	-	-	-	-	100%	10%	10%
47	100%	-	100%	100%	100%	100%	100%	100%
48	100%	-	100%	100%	100%	100%	10%	10%

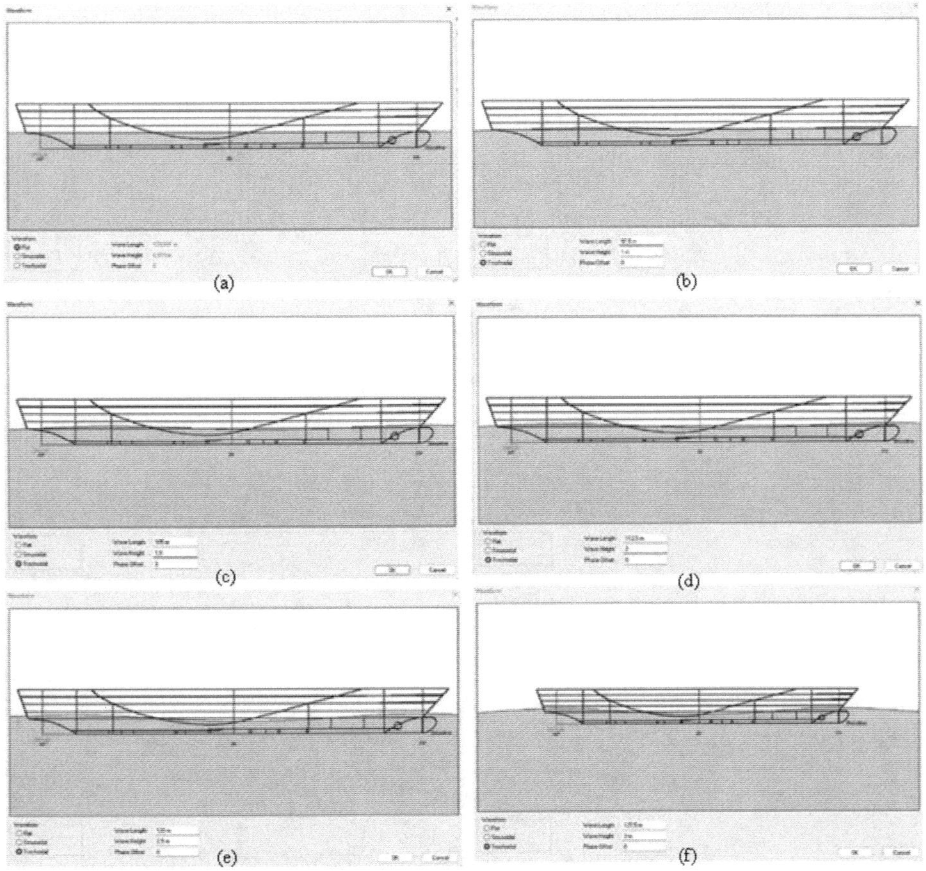

Figure 6. *Wave Modeling in Longitudinal Strength Calculation: (a) H=0 m & λ=0 m, (b)H=1 m &λ=97.5 m, (c)H=1.5 m & λ=105 m, (d) H=2 m & λ=112.5 m, (e)H=2.5 m & λ=120 m, (f)H=3 m & λ=127.5 m*

The bending moment calculation is performed for 48 loading scenarios that could occur on the ship. As shown in Figure 7, the bending moments from these 48 loading scenarios vary depending on the ship's load distribution and the characteristics of the waves affecting it.

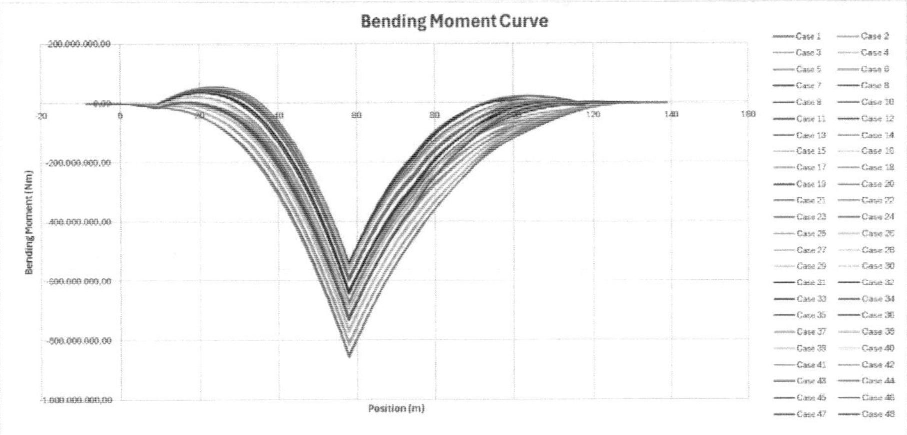

Figure 7. *Bending Moment Curve in 48 Loading Scenarios*

3.5. Load Stress Calculation

The maximum bending moment response is used to calculate the maximum stress occurring in the structural design. The maximum stress (σ) typically occurs at the bottom and deck sections of the ship's midship construction. It is calculated by multiplying the vertical bending moment by the distance between the center of gravity and the construction section of the ship (Z), and then dividing the result by the moment of inertia of the midship construction area (Iy), as shown in equation (3). The deck used for this calculation is Deck 5, which has the highest position. The results of the moment of inertia calculations and the distance to the center of gravity for the two points are shown in Figure 8.

$$\sigma = (M \times Z) / Iy \tag{3}$$

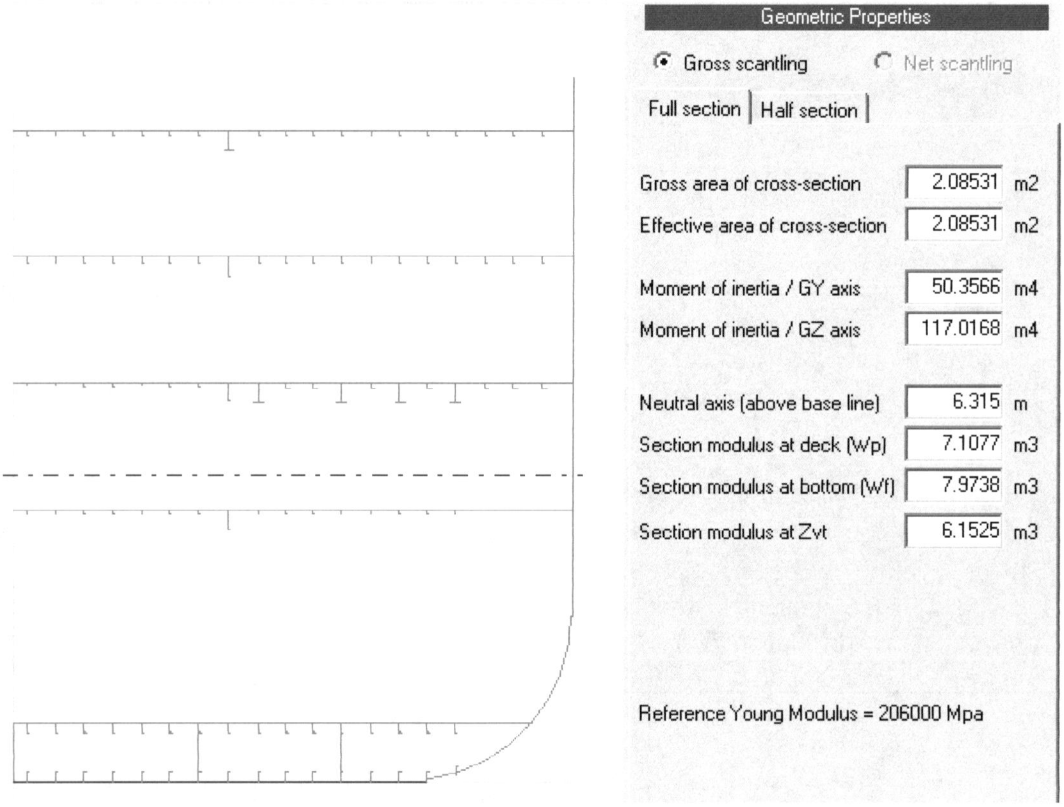

Figure 8. *Calculation of Geometric Properties of the Ship's Midship Section*

In this study, the stresses considered are in the bottom and deck sections. Therefore, in calculating the stress values, it is important to consider the value of (Z), which represents the distance from the construction section to the neutral axis of the ship's cross-sectional area. Stress is calculated along the deck and bottom of the ship. Figures 9 and 10 show the distribution of stress along the ship. It can be seen that the maximum stress occurs in the midship area, with the maximum stress on the deck being higher than that on the bottom.

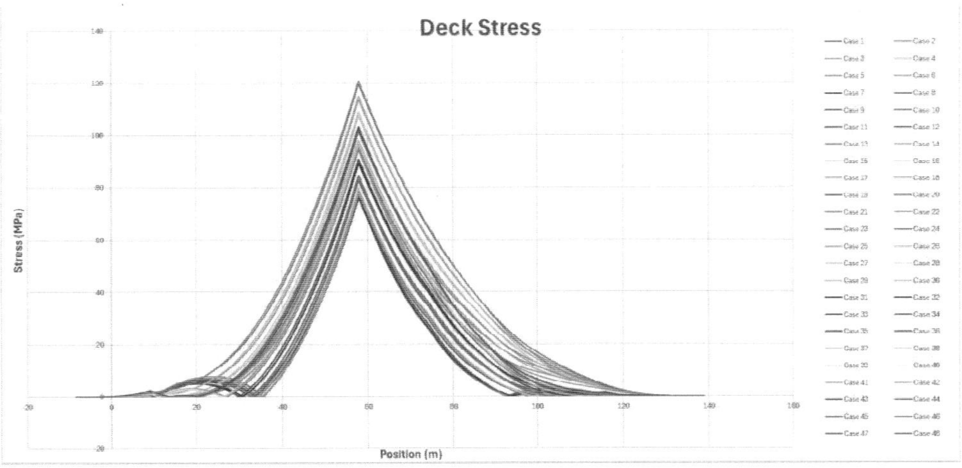

Figure 9. *Stress on the Ship's Deck*

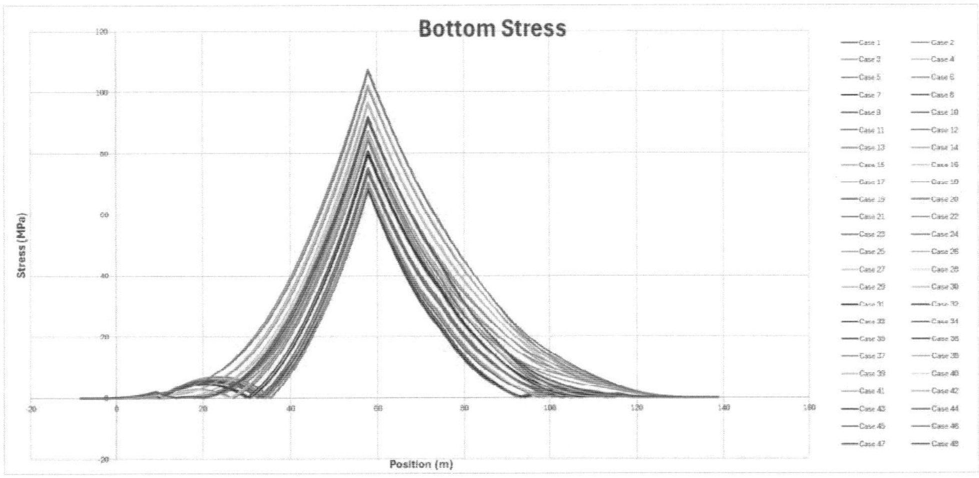

Figure 10. *Stress on the Ship's Bottom*

3.6. Structural Reliability Analysis with Monte Carlo Simulation

The joint probability function is a concept in structural reliability analysis used to model and estimate the joint failure probabilities of multiple components or structural elements. In this context, structural reliability refers to the likelihood that a structure or system will perform adequately or not fail under specified conditions. To calculate structural reliability, failure modes and specified loading data are required to determine the stress values under each loading condition.

Once the stress values are obtained, the mean and standard deviation can be calculated. These mean and standard deviation values are used as input to create a Probability Density Function (PDF) graph. The mean and standard deviation can be determined using data processing software that automatically calculates these values, resulting in a PDF graph. The PDF graphs compared are those of the load data from stress and the strength data from material yield strength. The Monte Carlo simulation generates random values from the PDF of load and strength and then calculates the probability of failure.

4. Results and Discussion

4.1 Probability Density Function

Based on the stress distribution shown in Figures 9 and 10, the data was processed to obtain the probability density function (PDF) of the load. Table 6 provides a summary of the PDF of the load for each loading condition at the deck and bottom positions using a normal distribution. According to the PDF results, the average stress ranges from 11.67 to 35.55 MPa, with a standard deviation between 17.09 and 35.06 MPa. These values indicate that the stress experienced by the ship due to static loads is relatively low compared to the average and standard deviation of the material yield strength shown in Table 2, which are 151.21 MPa and 39.16 MPa, respectively.

Table 6. Probability Density Function Load

	Case	Stress on Deck (Mpa)		Stress on Bottom (Mpa)			Case	Stress on Deck (Mpa)		Stress on Bottom (Mpa)	
		μ_s (Mean)	σ_s (Stdev)	μ_s (Mean)	σ_s (Stdev)			μ_s (Mean)	σ_s (Stdev)	μ_s (Mean)	σ_s (Stdev)
Still Water Condition / T = 0 s ; H = 0 m	1	13,09	19,74	11,67	17,60	Wave Condition / T = 15 s ; H = 2 m	25	18,33	24,42	16,34	21,77
	2	13,09	19,18	11,67	17,09		26	17,40	23,82	15,51	21,23
	3	13,23	19,85	11,79	17,70		27	18,58	24,51	16,56	21,85
	4	13,19	19,32	11,76	17,22		28	17,56	23,95	15,65	21,35
	5	23,05	26,44	20,54	23,57		29	29,11	31,03	25,95	27,66
	6	23,66	26,94	21,09	24,02		30	29,50	31,42	26,30	28,01
	7	13,29	19,92	11,84	17,76		31	18,65	24,57	16,62	21,90
	8	13,24	19,39	11,80	17,29		32	17,63	24,01	15,71	21,41
Wave Condition / T = 13 s ; H = 1 m	9	14,09	20,43	12,56	18,21	Wave Condition / T = 16 s ; H = 2,5 m	33	20,96	26,71	18,68	23,80
	10	13,82	19,79	12,32	17,64		34	19,95	26,05	17,78	23,22
	11	14,29	20,52	12,73	18,29		35	21,22	26,79	18,91	23,88
	12	13,93	19,93	12,42	17,77		36	20,13	26,18	17,94	23,33
	13	24,11	27,29	21,49	24,33		37	32,14	33,01	28,65	29,42
	14	24,64	27,82	21,97	24,80		38	32,53	33,33	29,00	29,71
	15	14,35	20,58	12,79	18,35		39	21,30	26,84	18,98	23,93
	16	13,98	20,01	12,47	17,83		40	20,21	26,23	18,01	23,38
Wave Condition / T = 14 s ; H = 1,5 m	17	16,00	22,20	14,26	19,79	Wave Condition / T = 17 s ; H = 3 m	41	23,70	28,68	21,13	25,56
	18	15,36	21,57	13,69	19,23		42	22,72	27,99	20,25	24,95
	19	16,23	22,28	14,46	19,86		43	23,98	28,75	21,37	25,63
	20	15,49	21,71	13,81	19,36		44	22,91	28,10	20,42	25,05
	21	26,33	29,03	23,47	25,87		45	35,13	34,79	31,31	31,01
	22	26,77	29,49	23,86	26,29		46	35,55	35,06	31,69	31,01
	23	16,30	22,34	14,53	19,92		47	24,06	28,80	21,45	25,67
	24	15,55	21,78	13,86	19,42		48	23,00	28,15	20,50	25,10

4..1.1. PDF Based on Wave Conditions

Before calculating structural reliability, to assess the impact of wave conditions on reliability, the Probability Density Function (PDF) was created for each wave condition from the existing data. Data from different loading conditions but with the same wave height were grouped and the PDF was calculated. Table 7 presents the PDF data for each wave condition, showing that the average value and standard deviation of the load on the deck and bottom increase with higher wave heights and periods.

Table 7. PDF Load Based on Wave Conditions

Wave Conditions	H (m)	Deck Stress (MPa)		Bottom Stress (MPa)	
		μ_s (Mean)	σ_s (Stdev)	μ_s (Mean)	σ_s (Stdev)
Still Water Condition / T = 0 s ; H = 0 m	0	22,18	17,70	19,55	16,08
Wave Condition / T = 13 s ; H = 1 m	1	23,21	18,31	20,47	16,63
Wave Condition / T = 14 s ; H = 1,5 m	1,5	25,54	19,69	22,55	17,89
Wave Condition / T = 15 s ; H = 2 m	2	28,45	21,40	25,15	19,45
Wave Condition / T = 16 s ; H = 2.5 m	2,5	31,60	23,17	27,96	21,06
Wave Condition / T = 17 s ; H = 3 m	3	34,63	24,81	30,67	22,55

4..1.2. PDF Based on Laoding Conditions

The impact of loading conditions on stress values is examined in this section. Scenarios with the same loading are combined to observe their effect on the average stress and standard deviation. The results of the PDF can be seen in Table 8. It is observed that the average value and standard deviation are generally similar across all loading conditions. However, for scenarios C5 and C6, the stress values increase significantly to 36.53 and 25.46 MPa, respectively. This indicates that loading conditions with only passengers represent the most critical operational condition for the ship.

Table 8. PDF Load Based on Loading Conditions

Case	Deck Stress (MPa)		Bottom Stress (MPa)	
	μ_s (Mean)	σ_s (Stdev)	μ_s (Mean)	σ_s (Stdev)
C1	24,91	19,30	21,99	17,54
C2	24,07	18,74	21,24	17,03
C3	25,11	19,40	22,17	17,62
C4	24,26	18,85	21,41	17,13
C5	35,96	25,15	31,86	22,88
C6	36,53	25,46	32,37	23,16
C7	25,20	19,45	22,24	17,67
C8	24,34	18,90	21,48	17,17

4..1.3. Overall Ship PDF

By combining all the available data, the PDF of the stress values for the ship as a whole is obtained. For the deck position, the average stress value is 27.63 MPa with a standard deviation of 21.49 MPa. At the bottom position, the average stress value is lower at 23.42 MPa with a standard deviation of 19.50 MPa. These results are consistent with the difference in the distance from the centroid of the cross-sectional area of the ship to the deck and bottom, where the distance to the deck is greater. Therefore, the stress on the deck due to vertical bending moments will be higher compared to the stress on the bottom, which has a shorter distance.

4.2 Structural Reliability Calculation

After determining the Probability Density Function of load and strength, the structural reliability can be calculated using Monte Carlo simulation, with the input being the average values and standard deviations of the loads and material strength. The results of the failure probability calculations for the ship are summarized in the following subsections.

4..2.1. Reliability Based on Wave Conditions

Using the Joint Probability Function of Load and Strength, the probability of structural failure for the ship is obtained. Figure 11 shows the Joint PDF curves for each wave condition at the deck and bottom positions. The reliability summary is presented in Table 9. As shown in Figure 12, wave height significantly affects the structural reliability of the ship under study. In calm water conditions, the reliability is 99.87% and 99.91%, whereas at a wave height of 3 meters, the reliability decreases to 99.40% and 99.62%. It can be concluded that the structural reliability decreases as wave height increases at both the deck and bottom positions.

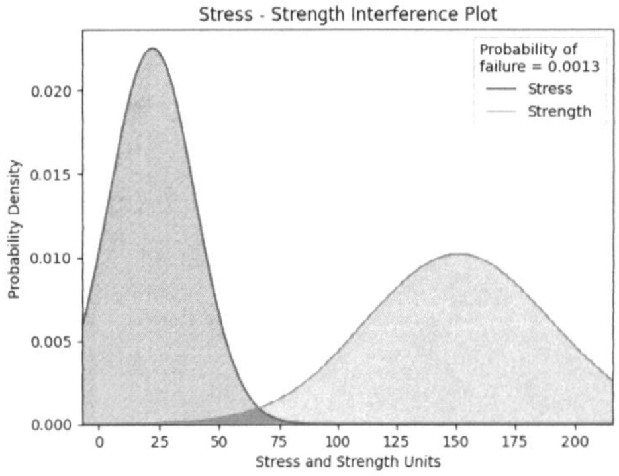

(a) T = 0 s ; H = 0 m, Deck

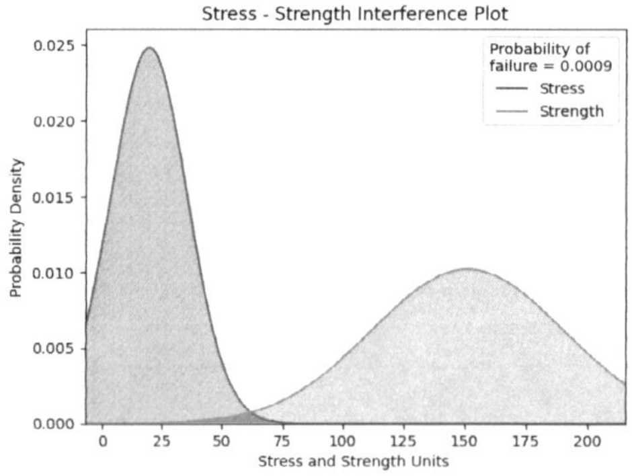

(b) T = 0 s ; H = 0 m, Bottom

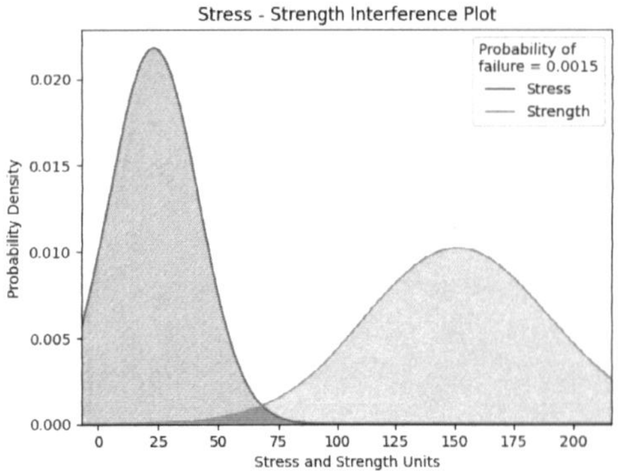

(c) T = 13 s ; H = 1 m, Deck

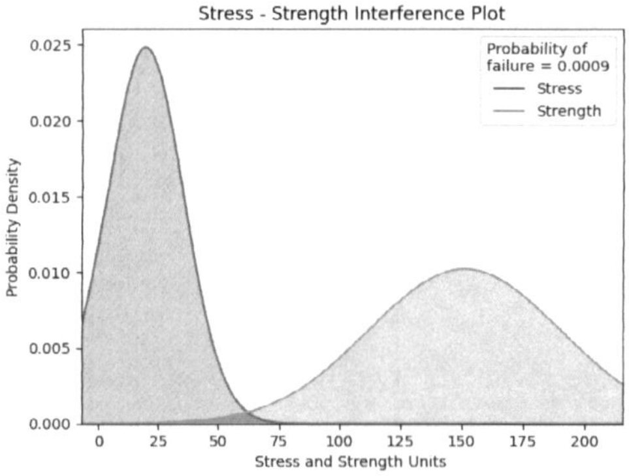

(d) T = 13 s ; H = 1 m, Bottom

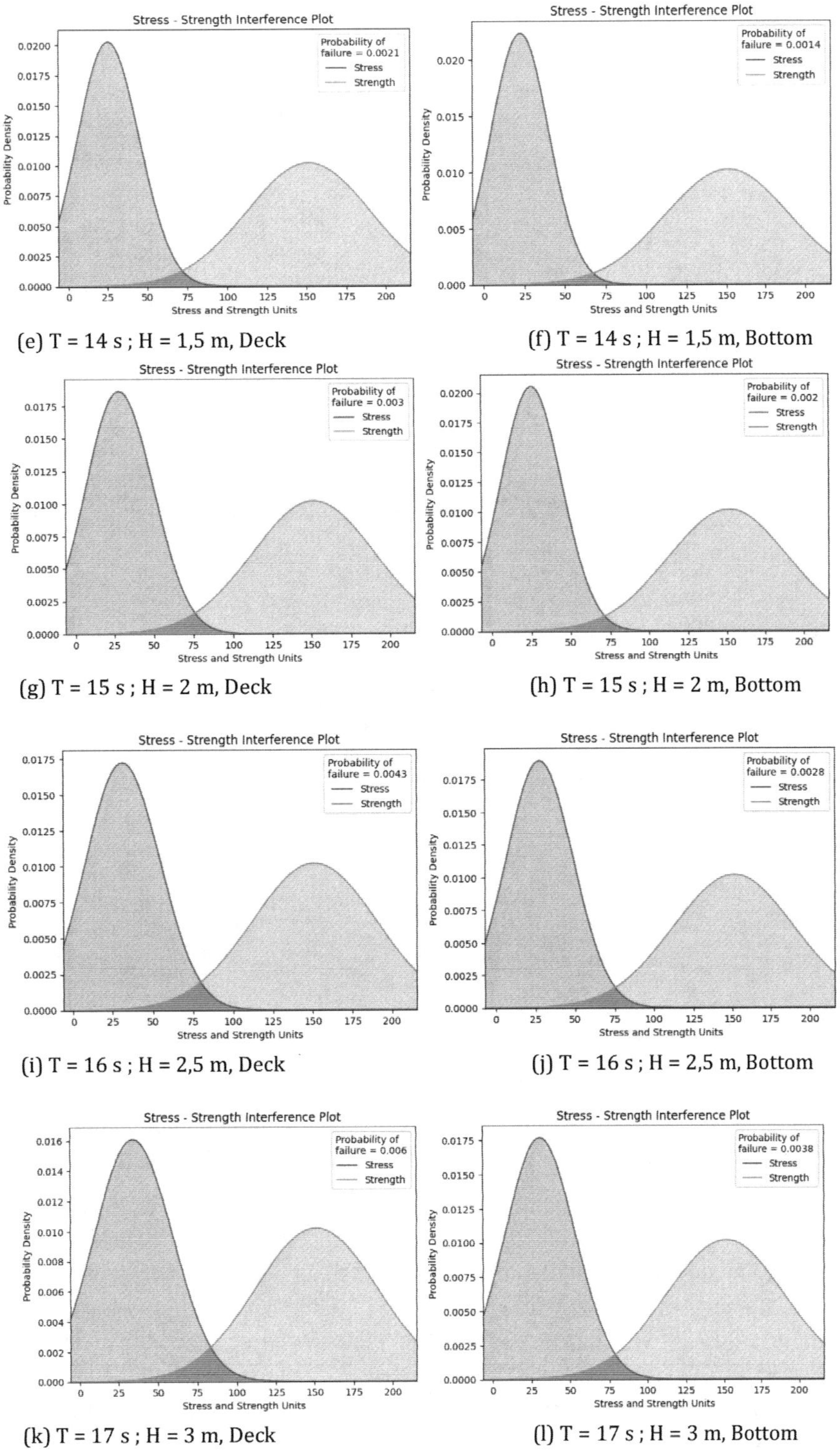

(e) T = 14 s ; H = 1,5 m, Deck (f) T = 14 s ; H = 1,5 m, Bottom

(g) T = 15 s ; H = 2 m, Deck (h) T = 15 s ; H = 2 m, Bottom

(i) T = 16 s ; H = 2,5 m, Deck (j) T = 16 s ; H = 2,5 m, Bottom

(k) T = 17 s ; H = 3 m, Deck (l) T = 17 s ; H = 3 m, Bottom

Figure 11. *Joint PDF Based on Wave Conditions*

Table 9. PDF Reliability Summary for Wave Conditions

Wave Conditions	H (m)	Deck		Bottom	
		Failure	Reliability	Failure	Reliability
Still Water Condition / T = 0 s ; H = 0 m	0	0.13%	99.87%	0.09%	99.91%
Wave Condition / T = 13 s ; H = 1 m	1	0.15%	99.85%	0.11%	99.89%
Wave Condition / T = 14 s ; H = 1,5 m	1.5	0.21%	99.79%	0.14%	99.86%
Wave Condition / T = 15 s ; H = 2 m	2	0.30%	99.70%	0.20%	99.80%
Wave Condition / T = 16 s ; H = 2.5 m	2.5	0.43%	99.57%	0.28%	99.72%
Wave Condition / T = 17 s ; H = 3 m	3	0.60%	99.40%	0.38%	99.62%

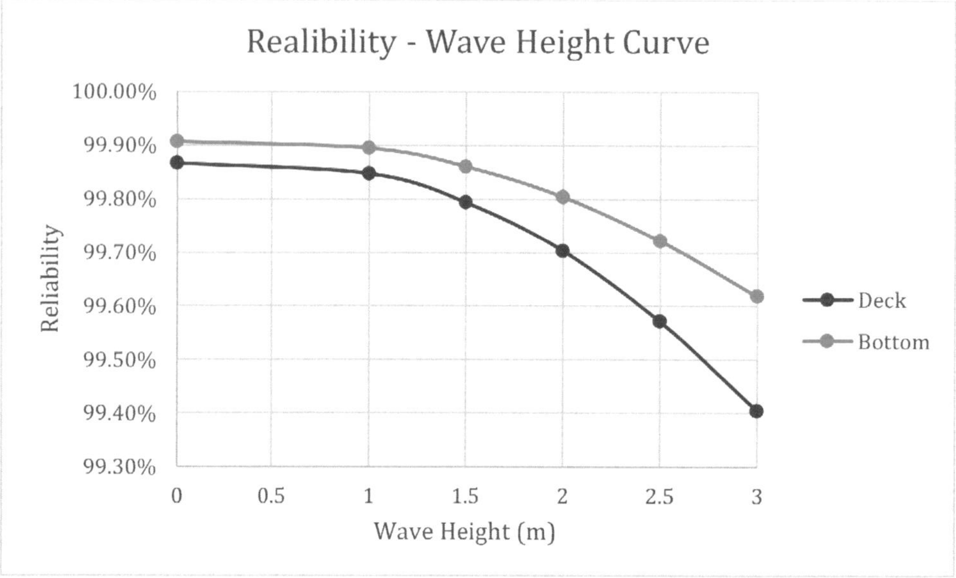

Figure 12. *Impact of Wave Height on Reliability*

4..2.2. Reliability Based on Wave Conditions

Based on the 8 loading conditions analyzed, the joint PDF is as shown in Figure 13. The reliability summary for the loading conditions is presented in Table 10, and the effect of loading conditions on reliability can be seen in Figure 14. It is observed that the highest reliability is achieved in the bottom section for condition C2, with a reliability of 99.8831%, while the lowest reliability occurs in conditions C5 and C6, with values significantly lower than other conditions. In condition C6, the lowest reliability is observed at 99.30% at the deck and 99.55% at the bottom. This low reliability is likely due to the uneven weight distribution of the ship in loading conditions C5 and C6, resulting from the absence of a uniform load from cargo and containers.

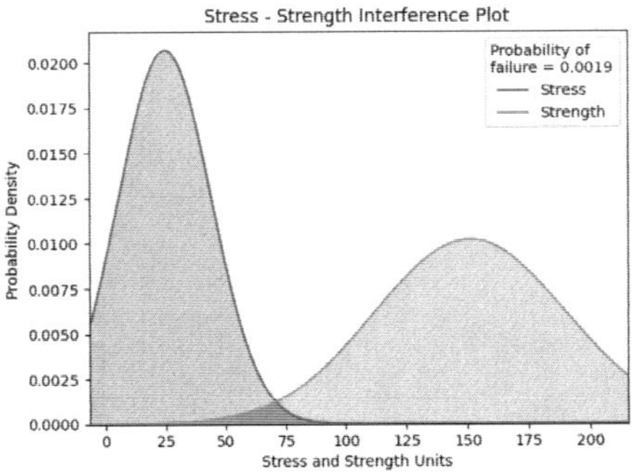

(a) C1, Deck

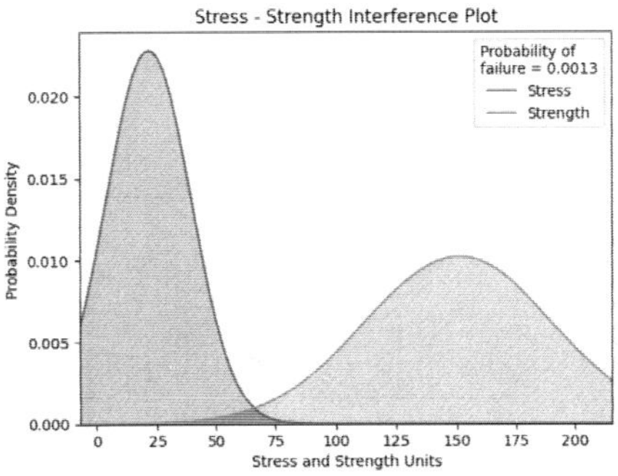

(b) C1, Bottom

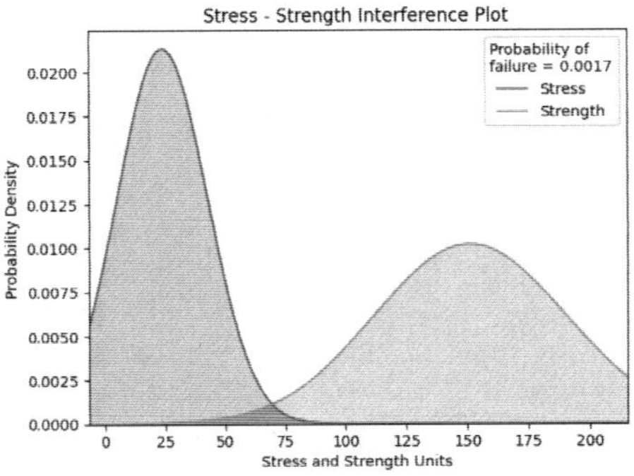

(c) C2, Deck

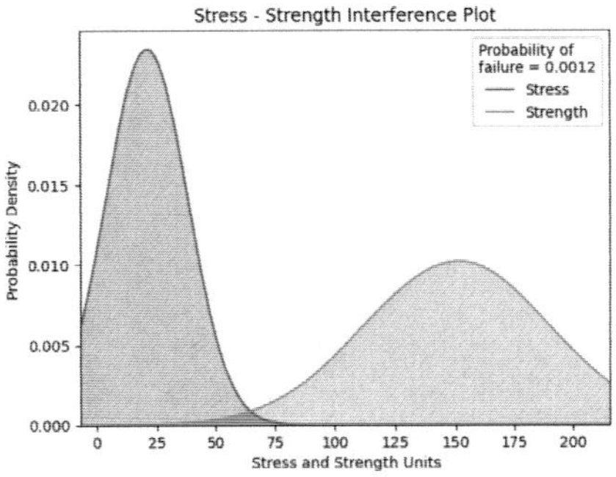

(d) C2, Bottom

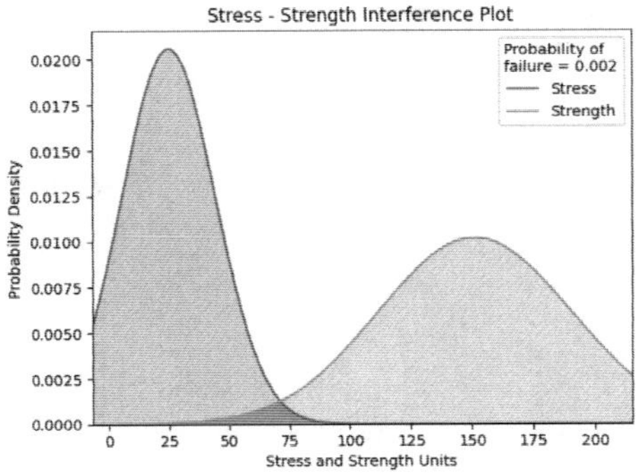

(e) C3, Deck

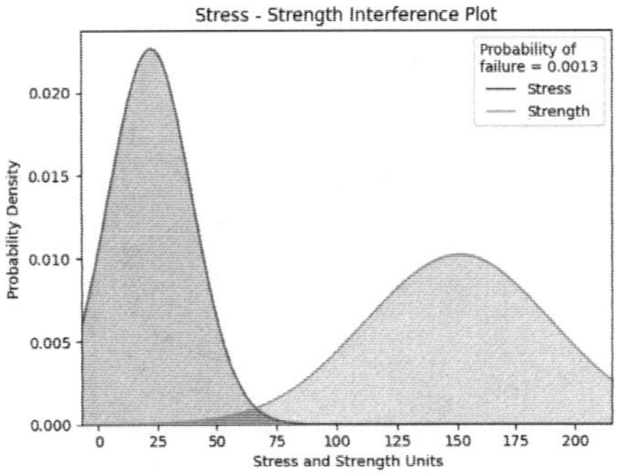

(f) C3, Bottom

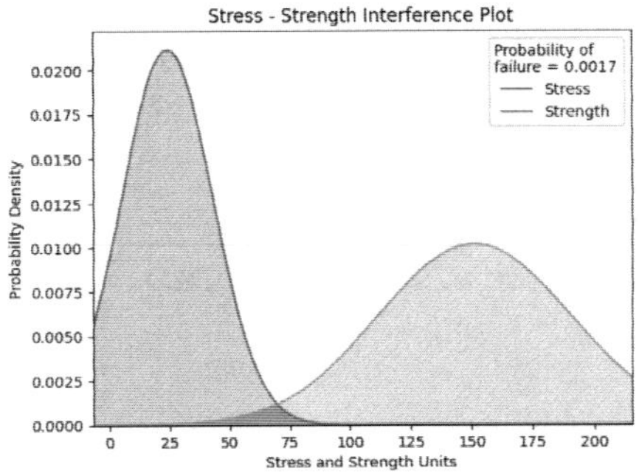

(g) C4, Deck

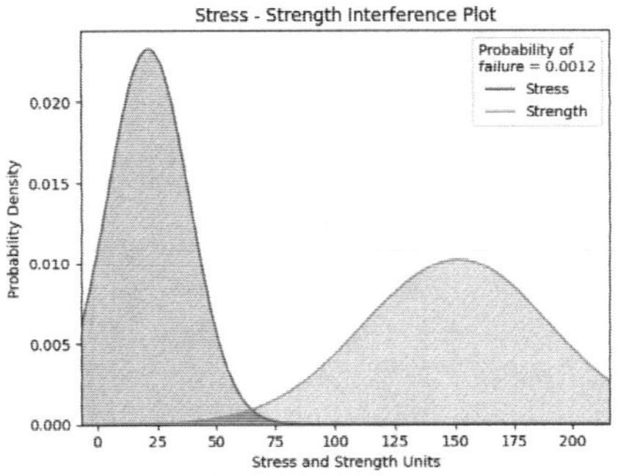

(h) C4, Bottom

IOP Conf. Series: Earth and Environmental Science **1461** (2025) 012011 doi:10.1088/1755-1315/1461/1/012011

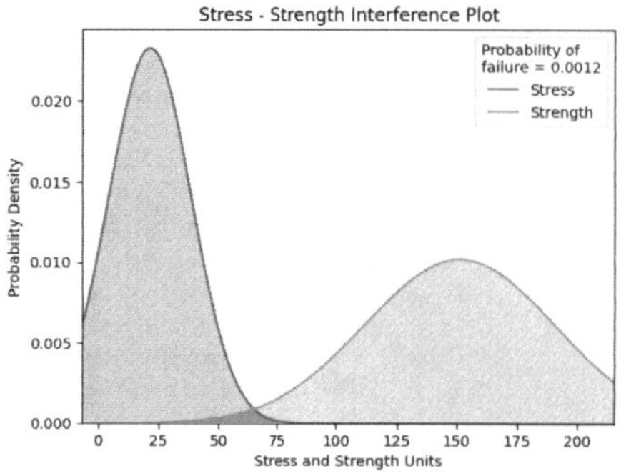

(i) C5, Deck

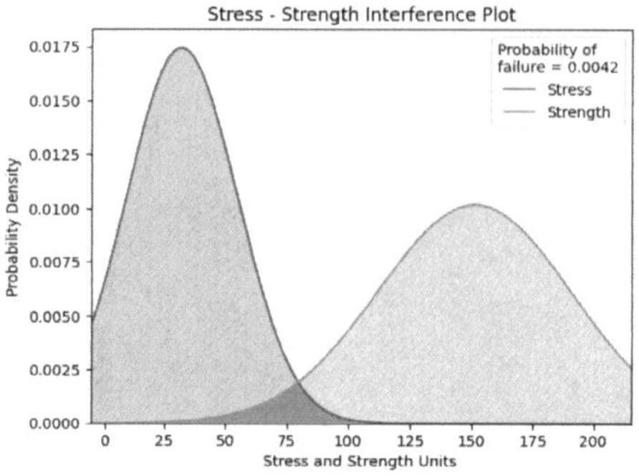

(j) C5, Bottom

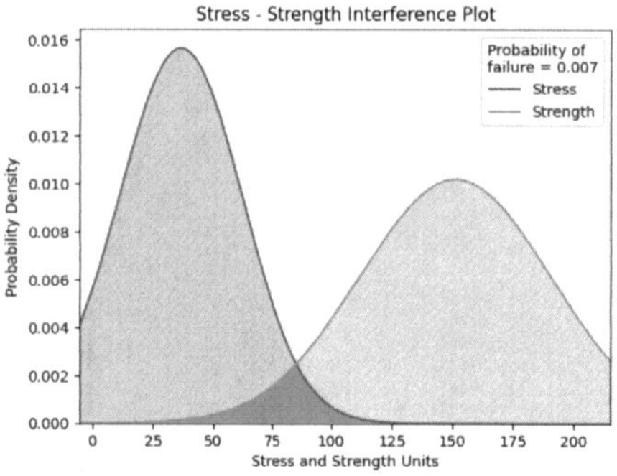

(k) C6, Deck

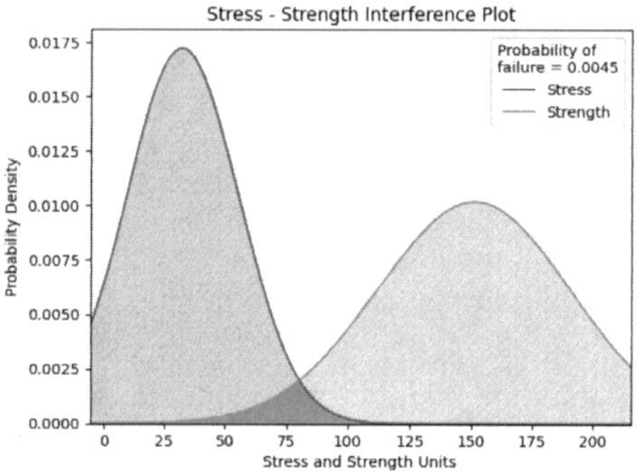

(l) C6, Bottom

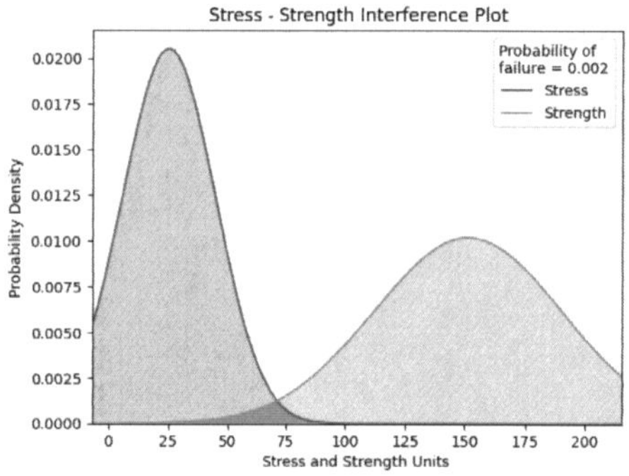

(m) C7, Deck

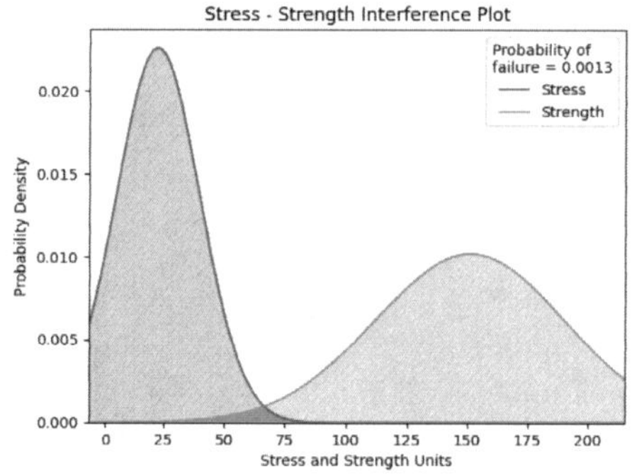

(n) C7, Bottom

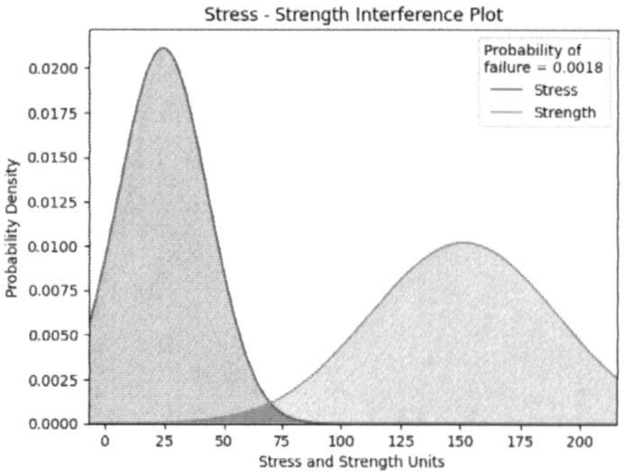

(o) C8, Deck

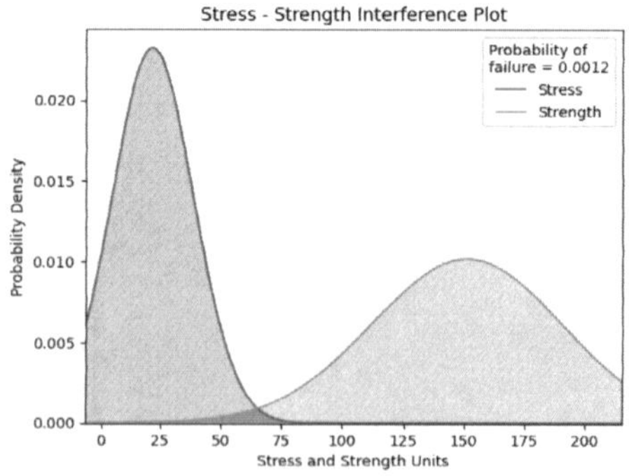

(p) C8, Bottom

Figure 13. *Joint PDF Based on Loading Conditions*

Table 10. Realibility Summary for Loading Conditions

Case	Deck		Bottom	
	Failure	Reliability	Failure	Reliability
C1	0.19%	99.81%	0.13%	99.87%
C2	0.17%	99.83%	0.12%	99.88%
C3	0.20%	99.80%	0.13%	99.86%
C4	0.17%	99.83%	0.12%	99.88%
C5	0.66%	99.34%	0.42%	99.57%
C6	0.70%	99.30%	0.45%	99.55%
C7	0.20%	99.80%	0.13%	99.86%
C8	0.18%	99.82%	0.12%	99.88%

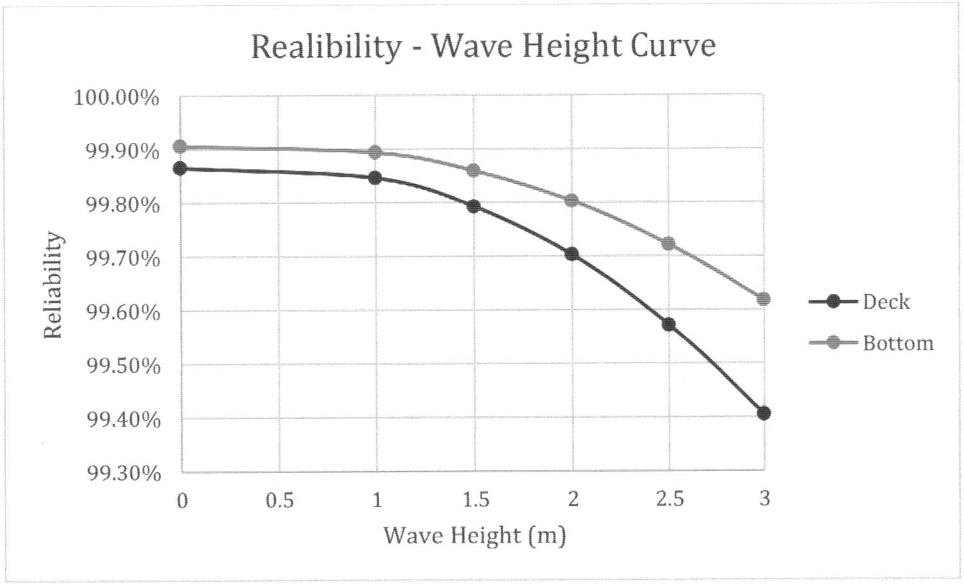

Figure 14. *Impact of Load Case on Reliability*

4.2.3. Overall Ship Reliability

By combining the Probability Density Function (PDF) of all 48 load data and the PDF of the material strength data for Marine Grade A, the ship's reliability is calculated using Monte Carlo simulation as follows. The probability of failure is found to be 0.2832% at the deck and 0.18757% at the bottom. The overall structural reliability is 99.72% at the deck and 99.81% at the bottom. From these figures, it can be concluded that the probability of structural failure due to static loads under various wave and loading conditions is relatively very low, below 1%, indicating that the ship's structure can be considered to have sufficient reliability to meet its operational demands.

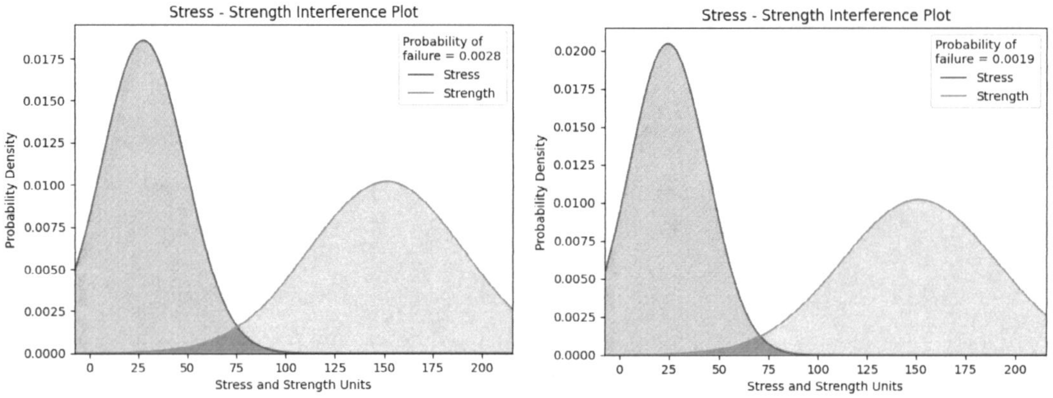

(a) Deck (b) Bottom

Figure 15. *Joint PDF of Passenger-Cargo-Container Vessel*

5. Conclusion

This study has analyzed the structural reliability of the ship using Marine Grade A material under static loads with varying loading conditions and wave heights. The conclusions drawn from the analysis are as follows:

1. The overall reliability of the ship, designed with Grade A material, is 99.72% at the deck and 99.81% at the bottom. These values represent a very high reliability, with a failure probability below 0.5%.
2. Reliability values show a decreasing trend with increasing wave height, while the probability of failure increases with higher wave heights. The smallest failure probability occurs most frequently at a wave height of 0 meters, with 0.09% at the bottom and a reliability of 99.91%, whereas the highest failure probability is observed at a wave height of 3 meters, with 0.60% at the deck and a reliability of 99.40%.
3. Reliability values for each load case are generally similar but decrease in loading conditions C5 and C6, which involve only passenger loads. In these conditions, reliability decreases to 99.30% at the deck and 99.55% at the bottom, compared to other cases with reliability above 99.8%. This indicates that the condition with only passenger loads presents the highest risk of failure.

References

[1] A. Syafiul, et al., "The development of multi-purpose vessels for Indonesian archipelagic sea transportation," Marine Technology, vol. 55, no. 4, pp. 209-222, 2018.

[2] J. R. Shin and J. Lee, "A study on optimal structural design of multi-purpose cargo ships," International Journal of Naval Architecture and Ocean Engineering, vol. 10, no. 5, pp. 652-666, 2018.

[3] D. A. Hudson, T. Moan, and R. A. Shenoi, "Reliability-based design of ship structures," Progress in Structural Engineering and Materials, vol. 6, no. 4, pp. 210-218, 2004.

[4] Y. Bai, Marine Structural Design. Amsterdam: Elsevier, 2003.

[5] M. A. Djatmiko and I. K. A. P. Utama, "Analysis of wave characteristics in Indonesian waters for ship design," IOP Conference Series: Earth and Environmental Science, vol. 135, no. 1, p. 012019, 2018.

[6] L. Zhu and P. Moan, "Reliability analysis of ship structures under combined loads," Proceedings of the Institution of Mechanical Engineers, Part M: Journal of Engineering for the Maritime Environment, vol. 225, no. 2, pp. 135-145, 2011.
[7] International Maritime Organization, Goal-based ship construction standards for bulk carriers and oil tankers, MSC.287(87), London: IMO, 2010.
[8] A. Papanikolaou, Ship Design: Methodologies of Preliminary Design. Cham: Springer, 2014.
[9] J. Lee, J. R. Shin, and J. Lee, "Modular design approach for flexible cargo space in multi-purpose vessels," Ocean Engineering, vol. 121, pp. 181-195, 2016.
[10] X. Zhang and T. Moan, "Effect of cargo distribution on ship responses in waves," Marine Structures, vol. 59, pp. 1-24, 2018.
[11] J. R. Shin and J. Lee, "Fatigue damage assessment of ship structural details in multi-purpose cargo ships," Marine Structures, vol. 56, pp. 143-156, 2017.
[12] D. Vassalos, et al., "A risk-based design methodology for passenger ships," Journal of Ship Research, vol. 62, no. 4, pp. 240-258, 2018.
[13] H. Karahalios, "The application of the safety regulations for cargo ships," Safety Science, vol. 103, pp. 288-296, 2018.
[14] R. Bhattacharya, Dynamics of Marine Vehicles. New York: Wiley, 1978.
[15] O. Hughes, E. Nikolaidis, and B. M. Ayyub, "Uncertainty in Strength Models for Marine Structures," 1994.

Safety and Reliability Analysis of the Ship Structure due to Sagging and Hogging Loads for Infrastructure Sustainability

Septia Hardy Sujiatanti[1], Akhdan Baihaqi [1], Aries Sulisetyono[1,] Rutma Pujiwat[2], Satriyo Rahmadianto[3], Roslin Ramli[4]

[1] Department of Naval Architecture, Institut Teknologi Sepuluh Nopember, Surabaya, Indonesia
[2] Research Center for Transportation Technology, National Research and Innovation Agency, Serpong, Indonesia
[3] PT Surabaya Maritim Konsultan (SMARTKONS), Surabaya, Indonesia
[4] Maritime Engineering Technology, Universiti Kuala Lumpur Malaysian Institute of Marine Engineering Technology, Malaysia

*E-mail: septia@its.ac.id

Abstract. In this study, an analysis of the reliability of the structure of the 5500 DWT tanker was carried out under the influence of hogging and sagging loads with different wave heights. Reliability analysis was performed by the Monte Carlo method where the pressure experienced by the ship due to the condition of hogging and sagging by using the finite element method. The computation results show that an increase in wave height had a significant effect on increasing pressure on the ship structure. The lowest structural reliability occurs in hogging conditions with a wave height of 6 meters, where the level of reliability only reaches 90.25%.

Keywords: Reliability Analysis, Sustainable Infrastructure, Ship Structure, Monte Carlo Simulation

1. Research Background

Ship accidents have frequently been a major concern in the maritime world, especially when they involve significant structural damage. One notable incident that drew attention was the breaking of a cargo ship in Japanese waters, as reported by Tempo in an article titled "Kapal Kargo Patah di Perairan Jepang, Ini yang Terjadi" ("Cargo Ship Breaks in Japanese Waters, Here's What Happened"). This event highlights the importance of understanding and addressing issues related to structural failures in ships.

Figure 1.1 Panama Cargo Ship Slitting inti Two Parts

According to the report, the Panama-flagged cargo ship, Crimson Polaris, broke in half while sailing in Japanese waters on August 11, 2021. This incident resulted in an oil spill and the ship splitting into two parts, which eventually stranded on the coast. Preliminary investigations indicated that bad weather and strong winds were the primary factors contributing to this damage. These extreme weather conditions created substantial lateral loads on the ship's structure, ultimately causing a stability failure and buckling of the ship's deck [1].

This incident underscores the critical importance of understanding the structural response of ships to extreme loads. Structural failure in a ship not only impacts the vessel's structural integrity but also poses significant threats to environmental and economic safety. Considering that more than 80% of the world's trade volume is transported by ships, and around 70% of the trade value involves the use of cargo vessels such as bulk carriers, container ships, and tankers [2], the role of ships is vital. Given this significance, ships are required to have a high level of safety to ensure the security of both the crew and the cargo. The safety of a ship can be enhanced by optimizing various aspects, particularly the reliability of its construction. The stronger the ship's construction, the safer the vessel will be during its operations.

2. THEORETICAL BACKGROUND

2. 1. Ship Construction

Ship Construction involves the process of building a ship, including design, fabrication, assembly, testing, and certification. A ship is required to have a robust construction to ensure the safety of its crew, cargo, and the vessel itself. One critical aspect that must be considered in ship construction is the design of the structure. In the design phase, the structure's layout and dimensions must be able to withstand the loads the ship will encounter while minimizing stress and strain as much as possible. Even though a ship's construction is optimized, failures such as buckling and deformation can still occur. The construction system of a ship is categorized into three types: longitudinal, transverse, and mixed systems. These different construction systems, particularly the longitudinal and transverse systems, have distinct characteristics. In the transverse construction system, all components, including deck plates and their supporting beams such as deck beams, strong beams, and deck girders, are present. On the other hand, the longitudinal construction system includes deck longitudinal, deck transverse, and deck girders[3].

IOP Conf. Series: Earth and Environmental Science **1461** (2025) 012012 doi:10.1088/1755-1315/1461/1/012012

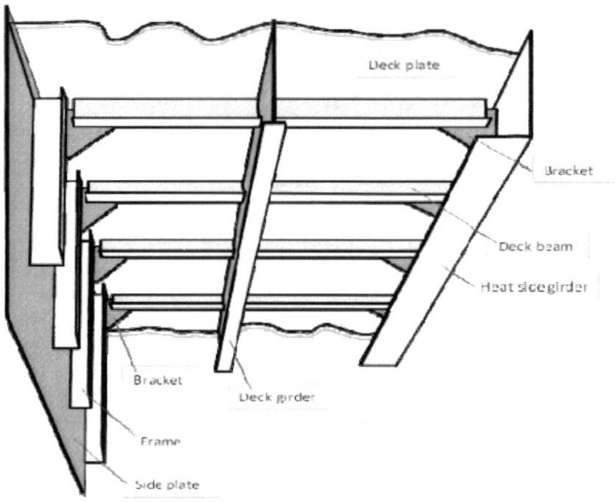

Figure 2.1. Transverse Construction System of the Ship's Deck Structure

2. 2. Bending Moment

Bending moment is a reaction that occurs in a structural element when external forces or moments are applied, causing the element to bend. The occurrence of a bending moment is closely related to the effect of shear force on the structural element, where shear force is a force acting perpendicular to the structural element. In ships, this force arises as a result of water pressure against the hull and the weight of the ship itself, both of which play a crucial role in determining the ship's stability.

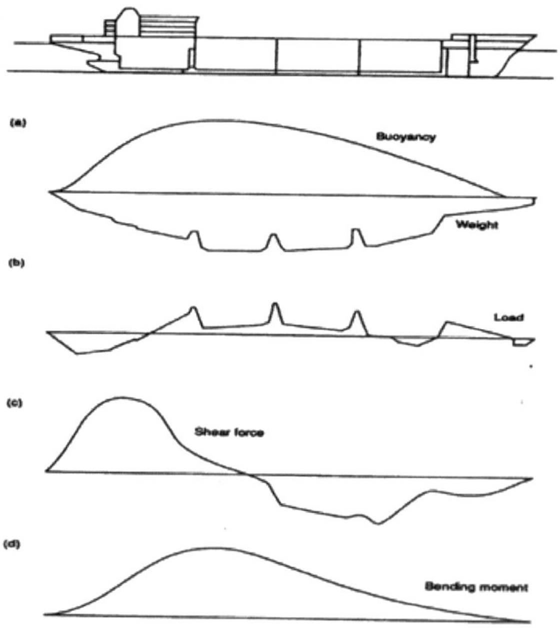

Figure 2.2 Ships Condition due to Bending Moment

2. 3. Longitudinal Strength

In designing ship structures, it is crucial to consider the loads that will be experienced throughout the structure's operational lifespan. These loads include shear forces and bending moments along the length of the ship. The hull structure of a ship operating at sea will be subjected to external loads primarily from fluid pressure against the hull surface, gravitational forces, and inertial forces. Fluid pressure on the hull surface generates hydrodynamic forces. These forces result in shear forces and bending moments. The distribution pattern and intensity of shear forces and bending moments along the length of the ship at each station are calculated from the difference between the weight distribution and the buoyant forces of the ship. This calculation is based on Basic Ship Theory, which assumes the ship as a static beam supported at its midpoint or at its ends [4].

2. 4. Stress

Stress is defined as force per unit area and is denoted by the Greek letter sigma (σ). Generally, the stress acting on a surface may be uniform across the entire area or may vary in intensity from one point to another. It is assumed that the stress acting on a cross-section is uniformly distributed across that area. Consequently, the resultant force from the stress must be equal to the stress multiplied by the cross-sectional area, A , of the bar. This simplification allows stress to be represented as a vector quantity, which is crucial for many engineering calculations and for determining material properties. The term "vector" typically refers to a quantity that has both magnitude and direction. For instance, the stress on a bar subjected to axial loading is equal to the applied force divided by the cross-sectional area [5], [6].

$$\sigma = \frac{P}{A} \tag{2.1}$$

Where:
P : Force (N)
A : Sectional area (m^2)
σ : Stress (N/mm^2 atau Pascal)

When a force causes an increase in the length of an object, it is referred to as tensile stress. Conversely, if the force results in a decrease in the length of the object, it is known as compressive stress. There are various types of stress, including normal stress, torsional stress, tensile stress, compressive stress, and bending stress [7].

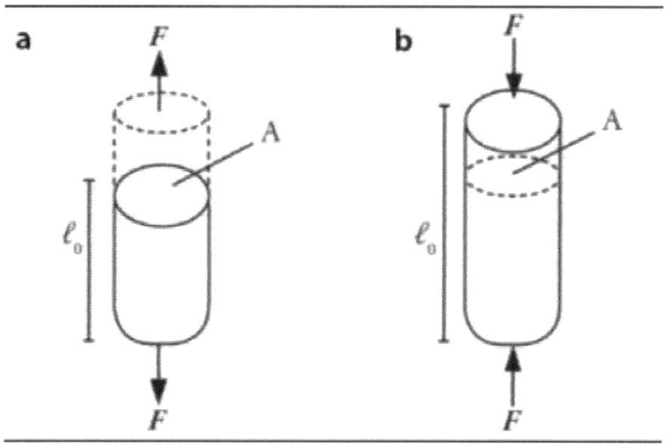

Figure 2.3. Normal Stress

2. 5. *Finite Element Method*

The Finite Element Method (FEM) is a numerical technique used to solve engineering and mathematical problems. FEM is particularly effective for addressing complex engineering problems, such as fluid mechanics analysis. The method simplifies the solution of problems with complex shapes by modeling an object and subdividing it into smaller elements, each connected by nodes. The analysis is based on representing the structural system as a network of small elements, with these elements forming a mesh through connections at nodal points . FEM divides the geometry into smaller parts called finite elements. This process, known as discretization, involves breaking down a continuous domain into discrete elements. Finite elements use polynomial approximations defined over small regions (elements) and are expressed in terms of the nodal values of the function [8-11].

In FEM, the element stiffness matrix [k] and the global stiffness matrix [K] play crucial roles in formulating the finite element equations. The stiffness matrix [k] for an individual element is calculated by integrating the displacement functions and material properties, while the global stiffness matrix [K] is constructed by assembling the stiffness matrices of all the elements within the system.

$$[K] = \sum_{e=1}^{n} [ke] \tag{2.2}$$

Where [K] is the global stiffness matrix, n n is the number of elements, and [ke] is the stiffness matrix of element e. Using matrix notation, the nodal displacements {d} and nodal forces $\{F\}$ can be expressed by the following equation:

$$[K]\{d\} = \{F\} \tag{2.3}$$

The use of the Finite Element Method requires a thorough understanding of these fundamental concepts, as well as the application of numerical techniques to ensure accurate and efficient analysis results [12], [13].

IOP Conf. Series: Earth and Environmental Science **1461** (2025) 012012 doi:10.1088/1755-1315/1461/1/012012

3. METHODOLOGY

3. 1. Ship Data

At this stage, data collection is carried out to gather the necessary information related to this project research. The data collection is conducted to obtain crucial data that supports the success of the research. The data used in this study pertains to the 5500 DWT tanker ship, including its main dimensions, midship section, and construction profile, as described below.

Table 3. 1 Main Dimension

MAIN DIMENSION	
SHIP TYPE	OIL TANKER
LENGTH OVERALL (Loa)	102.9 m
LENGTH PERPENDICULAR (Lpp)	98.4 m
BREADTH (B)	15.6 m
HEIGHT (H)	8.4 m
DRAFT (T)	6.6 m
SERVICE SPEED (Vs)	12.5 knot

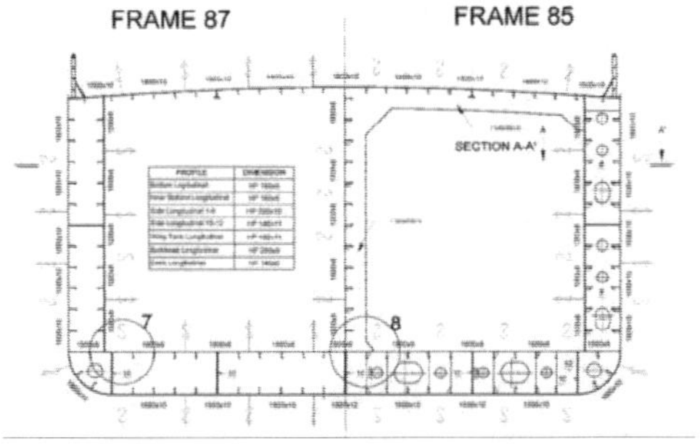

Figure 3. 1 Midship Section

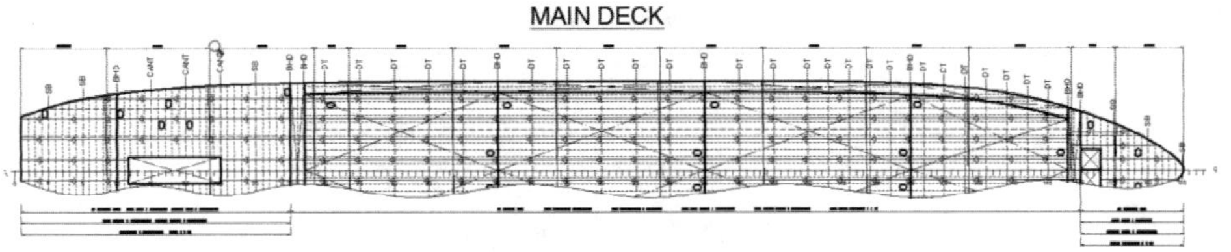

Figure 3. 2 Construction Profile

3. 2. Calculation of the Center of Gravity and Weight Distribution of the Ship

The calculation of the center of gravity is performed by determining the Light Weight Tonnage (LWT), including the ship's structural weight and Deadweight Tonnage (DWT). The ship's weight is adjusted to ensure that it can operate in an even keel condition in calm water when fully loaded. To ensure the ship is in an even keel condition, an equilibrium analysis is conducted using Maxsurf Stability. This involves inputting the calculated weight and center of gravity into the design load within the Maxsurf Stability software.

IOP Conf. Series: Earth and Environmental Science **1461** (2025) 012012 doi:10.1088/1755-1315/1461/1/012012

Table 3. 2 Load Case for Full Load Condition

	Item Name	Quantity	Unit Mass tonne	Total Mass tonne	Unit Volume m^3	Total Volume m^3	Long. Arm m	Aft. Limit m	Fwd. Limit m	Trans. Arm m	Vert. Arm m
1	Lightship	1	1211.470	1211.470			38.334	-2.460	100.689	0.000	0.000
2	Propeller and shaft	1	18.390	18.390			0.925	0.000	2.000	0.000	0.000
3	Main Engine	1	50.000	50.000			10.000	8.500	12.500	0.000	0.000
4	Electrical Unit	1	26.630	26.630			8.600	-2.460	21.000	0.000	0.000
5	Cargo Tank 1	**96%**	1460.772	1402.342	1644.459	1578.680	30.603			0.000	4.657
6	Cargo Tank 2	**93%**	1565.682	1456.084	1762.560	1639.181	48.000			0.000	4.548
7	Cargo Tank 4	**90%**	1080.347	972.312	1216.196	1094.576	81.444			0.000	4.535
8	Cargo Tank 3	**90%**	1565.682	1409.113	1762.559	1586.303	66.000			0.000	4.440
9	Forecastle	1	10.840	10.840			90.216	86.000	94.689	0.000	0.000
10	Poop Deck	1	56.400	56.400			7.750	-2.460	21.000	0.000	0.000
11	Boat Deck	1	42.830	42.830			12.909	4.200	21.000	0.000	0.000
12	Bridge Deck	1	31.100	31.100			14.131	6.269	21.000	0.000	0.000
13	Navigation Deck	1	22.840	22.840			15.600	10.200	21.000	0.000	0.000
14	Top Deck	1	15.330	15.330			17.322	12.600	21.000	0.000	0.000
15	Equipment & Outfitting	1	11.340	11.340			31.500	31.500	31.500	0.000	0.000
16	**Total Loadcase**			**6737.021**	**6385.774**	**5898.741**	**49.810**			**0.000**	**3.536**

3. 3. Equilibrium Analysis

After determining the location of the center of gravity and the weight distribution of the ship, an equilibrium analysis is performed to ensure that the ship operates in an even keel condition when fully loaded. The equilibrium analysis is conducted using Maxsurf Stability software to obtain results that verify whether the ship is in an even keel condition.

3. 4. Ship Load Calculation

This study focuses on the reliability analysis of the tanker ship structure, and therefore, the calculation of the ship's bending moment is performed under the most extreme conditions that the ship may encounter, namely hogging and sagging conditions. In the sagging condition, the ship's bottom structure experiences maximum tensile stress, while the deck structure undergoes maximum compressive stress. Conversely, in the hogging condition, the deck structure experiences maximum tensile stress, and the bottom structure undergoes maximum compressive stress.

3. 5. Compartment Modelling of the Cargo Hold

Compartment modeling is performed using Ansys SpaceClaim, where the model represents the cargo hold compartment, with dimensions of 36 meters in length and 15.6 meters in width, based on the midship section and construction profile data obtained. The modeling utilizes surface elements to represent plates and transverse structures. Beam elements are used for the profile to simplify the model and facilitate faster analysis.

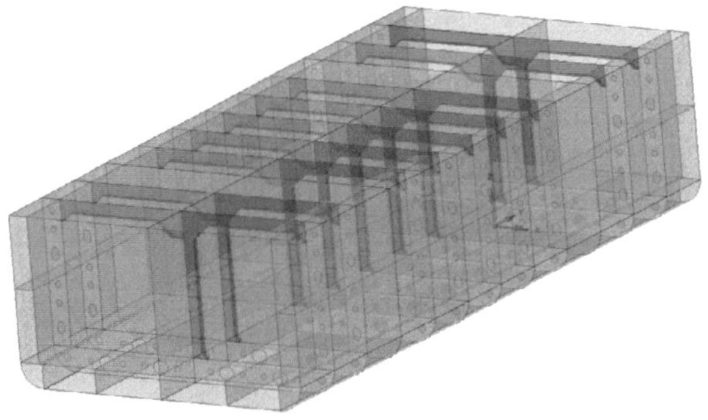

Figure 3. 3 Cargo Hold Compartment Model

3. 6. Meshing

To conduct finite element analysis, a meshing process is essential to discretize the model into smaller elements. The accuracy of the analysis is influenced by the size and aspect ratio of these elements. Typically, more accurate results are obtained with a mesh that is compact and uniformly shaped, where the elements have a low aspect ratio, ideally close to 1. Consequently, in this study, an element size of 600 mm was chosen to match the frame spacing of the ship, ensuring that most elements maintain an aspect ratio of approximately 1[12].

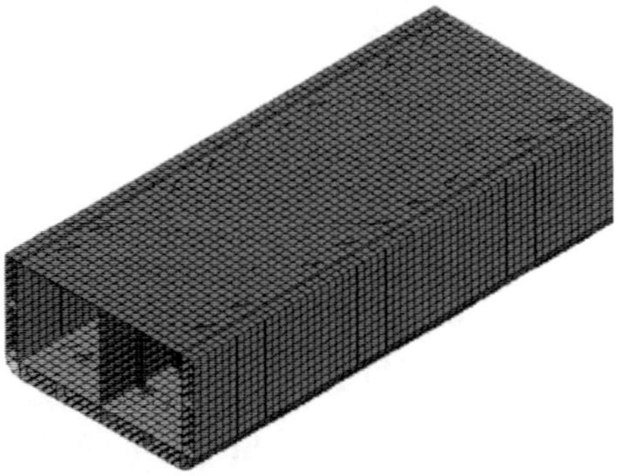

Figure 3. 4 Meshing on Cargo Hold Compartment Model

3. 7. Stress Analysis

To determine the structural response and the maximum stress occurring in the tanker ship structure, an analysis of normal stress is performed by inputting the loads and boundary conditions into the model. The loads applied to the model are those calculated from the longitudinal strength analysis, which includes bending moments and shear forces at the cross-section of the cargo hold construction model, as well as other loads experienced by the cargo hold of the ship.

In finite element analysis, it is crucial to apply boundary conditions to accurately represent the model's actual conditions and to simulate the interaction between the model and the unmodeled parts. Boundary conditions are imposed on the fore-end and after-end regions of the cargo space model. In these areas, a remote point is created to represent that all edges of the structure's cross-section are connected to an independent point located on the neutral axis. The presence of this remote point allows the application of boundary conditions and loading at the model's ends.

Boundary conditions need to be applied to the model. These conditions are imposed to restrict the model's movement patterns, both in terms of translation and rotation. The boundary conditions are applied to the previously defined remote point. The values for these boundary conditions are as follows:

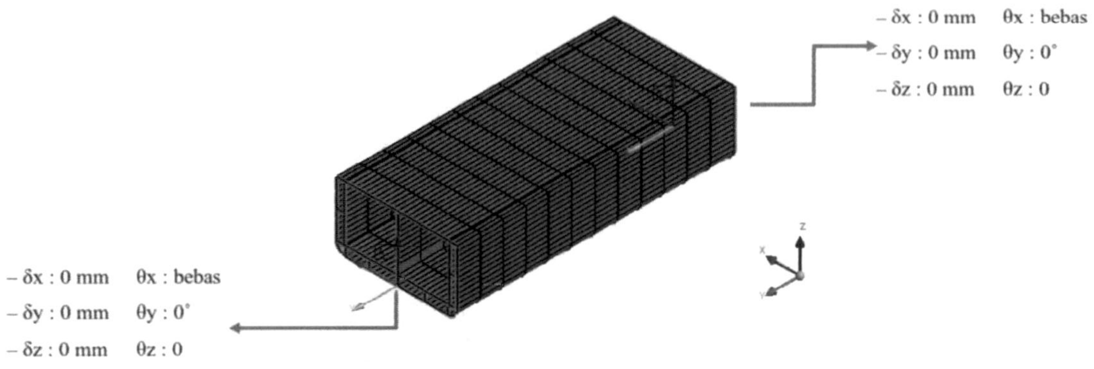

Figure 3. 5 Boundary Condition Model

With the boundary conditions in place to restrict the model's movement patterns, the model load can be applied. The bending moment is applied to both of the previously created remote points. The direction of the moment is carefully set to replicate the conditions experienced by the ship during hogging and sagging. The positive direction indicates the orientation that causes the ship to sag, while the negative direction corresponds to the opposite condition.

3. 8. Reliability Analysis

Reliability analysis was conducted by creating a normal distribution from the material strength results of grade A steel and the stresses obtained at each station of the cargo hold. The material strength data were then distributed according to a normal distribution to obtain the probability density function (PDF) values. By using a normal distribution, the spread of the material strength data can be understood, and the probability of achieving specific strength levels by the ship's material can be estimated. The strength distribution of the ship is presented in the following figure, which was generated using the mean and standard deviation values as inputs to determine the corresponding probability density function [14].

Table 3. 3 Test Results of Marine Grade A Steel Material

No Spesimen	Yield Stress [Mpa]
1	255,71
2	186,92
3	131,54
4	196,11
5	185,25
6	129,78
7	179,00
8	162,36
9	120,54
10	114,24
11	190,24
12	134,28
13	126,50
14	152,85
15	145,25
16	117,39
17	128,24
18	122,26
19	94,46
Mean	151,21
Standard Deviasi	39,16

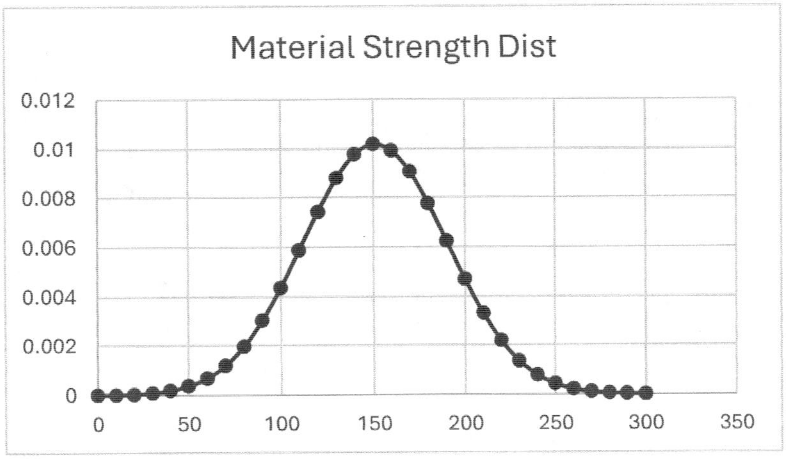

The ship's strength distribution is then used as input in the calculation of the Joint Probability Density Function (PDF) to combine the probability distribution of ship structural failure with the probability distribution of ship loads. Statistical data on ship structural failures and ship loads are used as inputs in the Joint PDF calculation. The structural failure statistics include information about the frequency and intensity of damage, while the ship load statistics encompass information on load magnitude and duration. The results from the Joint PDF calculation are used to determine the likelihood of ship structural failure based on statistical data and probabilistic analysis. These results can serve as a reference in designing and monitoring ship structures to ensure greater reliability and durability.

Thus, the Joint PDF methodology can be employed to calculate the reliability of ship structures more accurately and effectively, helping to improve the vessel's operational safety and efficiency. Interestingly, the Monte Carlo method has rarely been applied in structural reliability analysis of ships, despite other methods being used to estimate structural reliability. The Monte Carlo method could serve as a validation tool for these other methods, as it is considered to offer greater precision (Dąbrowska, 2020).

4. ANALYSIS RESULT

4. 1. Bending Moment Analysis

The bending moment was calculated using Maxsurf Stability under several different conditions. The resulting bending moment data for the 5500 DWT tanker vessel are as follows.

Table 4. 1 Bending Moment Hogging 4m

Name	Long. Pos.	Mass	Buoyancy	Net Load	Shear	Shear Force	Moment	Moment
	m	t/m	t/m	t/m	x10^3 tonne	Newton	x10^3 tonne.m	Newton.m
fr 50	29.994	97.891	-67.972	29.904	-0.623	-6111630	-13.796	-122735327.6
fr 51	30.594	97.893	-67.413	30.48	-0.605	-5935050	-14.164	-126009218.7
fr 52	31.194	97.895	-66.827	31.083	-0.586	-5748660	-14.521	-129185248.8
fr 53	31.794	97.898	-66.24	31.687	-0.556	-5454360	-14.864	-132236728.8
fr 107	64.194	87.611	-70.63	16.936	0.479	4698990	-14.832	-131952042.6
fr 108	64.794	87.613	-71.517	16.066	0.489	4797090	-14.542	-129372074.1
fr 109	65.394	87.615	-72.44	15.161	0.498	4885380	-14.246	-126738727
fr 110	65.994	87.618	-73.345	14.273	0.507	4973670	-13.944	-124052001.2

Table 4. 2 Bending Moment Hogging 5m

Name	Long. Pos.	Mass	Buoyancy	Net Load	Shear	Shear Force	Moment	Moment
	m	t/m	t/m	t/m	x10^3 tonne	Newton	x10^3 tonne.m	Newton.m
fr 50	29.994	97.845	-73.881	23.963	-0.48	-4708800	-9.467	-84222625.88
fr 51	30.594	97.833	-73.585	24.248	-0.466	-4571460	-9.751	-86749215.69
fr 52	31.194	97.822	-73.255	24.567	-0.451	-4424310	-10.026	-89195737.52
fr 53	31.794	97.811	-72.91	24.901	-0.425	-4169250	-10.288	-91526605.58
fr 107	64.194	87.651	-76.276	11.375	0.336	3296160	-10.556	-93910852.31
fr 108	64.794	87.64	-76.889	10.751	0.343	3364830	-10.353	-92104874.38
fr 109	65.394	87.629	-77.52	10.108	0.349	3423690	-10.145	-90254414.24
fr 110	65.994	87.617	-78.131	9.486	0.355	3482550	-9.934	-88377264.76

Table 4. 3 Bending Moment Hogging 6m

Name	Long. Pos.	Mass	Buoyancy	Net Load	Shear	Shear Force	Moment	Moment
	m	t/m	t/m	t/m	x10^3 tonne	Newton	x10^3 tonne.m	Newton.m
fr 50	29.994	97.845	-73.881	23.963	-0.48	-4708800	-9.467	-84222625.88
fr 51	30.594	97.833	-73.585	24.248	-0.466	-4571460	-9.751	-86749215.69
fr 52	31.194	97.822	-73.255	24.567	-0.451	-4424310	-10.026	-89195737.52
fr 53	31.794	97.811	-72.91	24.901	-0.425	-4169250	-10.288	-91526605.58
fr 107	64.194	87.651	-76.276	11.375	0.336	3296160	-10.556	-93910852.31
fr 108	64.794	87.64	-76.889	10.751	0.343	3364830	-10.353	-92104874.38
fr 109	65.394	87.629	-77.52	10.108	0.349	3423690	-10.145	-90254414.24
fr 110	65.994	87.617	-78.131	9.486	0.355	3482550	-9.934	-88377264.76

Table 4. 4 Bending Moment Sagging 4m

Name	Long. Pos.	Mass	Buoyancy	Net Load	Shear	Shear Force	Moment	Moment
	m	t/m	t/m	t/m	x10^3 tonne	Newton	x10^3 tonne.m	Newton.m
fr 50	29.994	97.845	-93.069	23.963	0.137	1343970	4.981	44313182.58
fr 51	30.594	97.833	-94.22	24.248	0.139	1363590	5.064	45051587.35
fr 52	31.194	97.822	-95.299	24.567	0.141	1383210	5.148	45798888.56
fr 53	31.794	97.811	-96.295	24.901	0.154	1510740	5.237	46590671.99
fr 107	64.194	87.651	-100.54	11.375	-0.233	-2285730	5.168	45976817.42
fr 108	64.794	87.64	-99.871	10.751	-0.24	-2354400	5.026	44713522.52
fr 109	65.394	87.629	-99.133	10.108	-0.247	-2423070	4.88	43414641.84
fr 110	65.994	87.617	-98.363	9.486	-0.254	-2491740	4.729	42071278.95

Table 4. 5 Bending Moment Sagging 5m

Name	Long. Pos.	Mass	Buoyancy	Net Load	Shear	Shear Force	Moment	Moment
	m	t/m	t/m	t/m	x10^3 tonne	Newton	x10^3 tonne.m	Newton.m
fr 50	29.994	97.891	-94.045	3.919	0.19	1863900	5.904	52524599.47
fr 51	30.594	97.893	-95.35	2.544	0.192	1883520	6.018	53538793.97
fr 52	31.194	97.895	-96.58	1.243	0.193	1893330	6.134	54570781.36
fr 53	31.794	97.898	-97.72	0.033	0.205	2011050	6.253	55629458.08
fr 107	64.194	87.611	-102.7	-14.872	-0.278	-2727180	6.35	56492413.05
fr 108	64.794	87.613	-101.916	-14.157	-0.286	-2805660	6.181	54988914.18
fr 109	65.394	87.615	-101.054	-13.366	-0.295	-2893950	6.006	53432036.66
fr 110	65.994	87.618	-100.159	-12.54	-0.302	-2962620	5.827	51839573.36

Table 4. 6 Bending Moment Sagging 6m

Name	Long. Pos.	Mass	Buoyancy	Net Load	Shear	Shear Force	Moment	Moment
	m	t/m	t/m	t/m	x10^3 tonne	Newton	x10^3 tonne.m	Newton.m
fr 50	29.994	97.891	-94.976	2.985	0.242	2374020	6.8	60495812.4
fr 51	30.594	97.893	-96.435	1.458	0.243	2383830	6.945	61785796.64
fr 52	31.194	97.895	-97.817	0.01	0.244	2393640	7.091	63084677.31
fr 53	31.794	97.898	-99.101	-1.341	0.255	2501550	7.241	64419143.76
fr 107	64.194	87.611	-104.819	-17	-0.322	-3158820	7.5	66723322.5
fr 108	64.794	87.613	-103.919	-16.167	-0.332	-3256920	7.304	64979619.67
fr 109	65.394	87.615	-102.933	-15.248	-0.341	-3345210	7.103	63191434.63
fr 110	65.994	87.618	-101.913	-14.294	-0.35	-3433500	6.895	61340974.49

4. 2. Stress Analysis

After completing the preprocessing steps, including load calculation, cargo hold geometry modeling, meshing, and applying boundary conditions, the processing phase can be carried out. The processing phase is conducted using Ansys software to perform finite element analysis. The calculations are set up to determine the normal stress magnitude at each frame of the ship's cargo hold structure.

IOP Conf. Series: Earth and Environmental Science **1461** (2025) 012012 doi:10.1088/1755-1315/1461/1/012012

Table 4. 7 Stress Analysis Results at 4m Wave Height

Hogging			Sagging	
Frame number	Compression stress (Mpa)	Tensile stress (Mpa)	Compression stress (Mpa)	Tensile stress (Mpa)
Fr 50	27.91	38.649	56.291	56.385
Fr 51	30.26	35.912	60.321	54.096
Fr 52	27.257	24.196	54.936	43.761
Fr 53	27.666	25.593	54.038	45.865
Fr 54	28.232	37.704	53.823	58.331
Fr 55	29.68	40.921	55.946	62.145
Fr 56	29.349	38.019	55.297	58.607
Fr 57	28.858	27.263	54.242	47.985
Fr 58	29.274	27.332	54.728	48.544
Fr 59	29.555	38.461	54.896	58.373
Fr 60	30.415	41.143	55.815	62.02
Fr 100	30.427	40.672	55.845	61.904
Fr 101	29.66	36.438	55.12	55.673
Fr 102	29.353	27.367	54.859	48.617
Fr 103	28.88	27.091	54.297	47.787
Fr 104	29.351	37.492	55.305	58.015
Fr 105	29.619	41.498	55.814	67.362
Fr 106	28.241	36.225	53.834	55.886
Fr 107	27.625	25.478	54.014	45.48
Fr 108	27.238	24.149	54.914	43.735
Fr 109	30.258	35.975	60.313	54.023
Fr 110	27.8	38.613	56.294	56.244

IOP Conf. Series: Earth and Environmental Science **1461** (2025) 012012 doi:10.1088/1755-1315/1461/1/012012

Table 4. 8 Stress Analysis Results at 5m Wave Height

	Hogging		Sagging	
Frame number	Compression stress (Mpa)	Tensile stress (Mpa)	Compression stress (Mpa)	Tensile stress (Mpa)
Fr 50	34.279	42.662	67.834	63.62
Fr 51	37.099	40.033	72.498	61.479
Fr 52	33.492	28.647	66.204	51.686
Fr 53	33.638	30.154	64.736	54.118
Fr 54	34.054	42.397	64.518	66.686
Fr 55	35.656	45.75	66.585	70.741
Fr 56	35.252	42.702	65.807	66.947
Fr 57	34.633	31.977	64.524	56.378
Fr 58	35.061	32.157	65.042	57.136
Fr 59	35.32	42.991	65.205	66.439
Fr 60	36.193	45.893	66.107	70.477
Fr 100	36.21	45.066	66.141	71.378
Fr 101	35.452	40.814	65.469	63.464
Fr 102	35.156	32.201	65.191	57.225
Fr 103	34.662	31.799	64.593	56.171
Fr 104	35.253	42.161	65.856	66.328
Fr 105	35.578	47.382	66.425	77.839
Fr 106	33.996	40.695	64.547	63.979
Fr 107	33.629	30.028	64.704	53.755
Fr 108	33.471	28.527	66.177	52.001
Fr 109	37.095	40.081	72.488	61.357
Fr 110	34.282	42.624	72.488	63.386

Table 4. 9 Stress Analysis Results at 6m Wave Height

	Hogging		Sagging	
Frame number	Compression stress (Mpa)	Tensile stress (Mpa)	Compression stress (Mpa)	Tensile stress (Mpa)
Fr 50	40.618	46.586	80.226	71.389
Fr 51	43.787	44.072	85.572	69.405
Fr 52	39.636	32.999	78.301	60.284
Fr 53	39.513	34.658	76.221	62.98
Fr 54	39.747	46.986	76.016	75.657
Fr 55	41.499	50.471	78.008	79.971
Fr 56	41.025	47.283	77.092	75.901
Fr 57	40.28	36.587	75.563	65.39
Fr 58	40.722	36.876	76.116	66.361
Fr 59	40.958	47.421	76.294	75.099
Fr 60	41.842	50.537	77.156	79.556
Fr 100	41.864	49.362	82.367	69.081
Fr 101	41.116	45.093	80.658	69.289
Fr 102	40.83	36.929	82.046	69.094
Fr 103	40.317	36.403	82.021	68.974
Fr 104	41.024	46.727	82.367	68.994
Fr 105	41.406	53.136	80.763	68.524
Fr 106	39.674	45.067	80.658	71.389
Fr 107	39.5	34.477	82.046	69.405
Fr 108	39.62	32.874	82.021	79.971
Fr 109	43.782	44.096	82.367	75.901
Fr 110	40.621	46.546	80.763	65.39

4. 3. *Reliability Analysis*

The reliability of the ship is calculated by creating a normal distribution from the obtained stress analysis results. This distribution is used in the Joint Probability Function, a concept in structural reliability analysis that models and estimates the joint probability of failure for multiple structural components or elements. In this study, the Joint Probability Function compares the yield stress of marine grade A material with the actual stress occurring in the cargo hold structure of the 5500 DWT tanker. The resulting probability density function (PDF) is as follows:

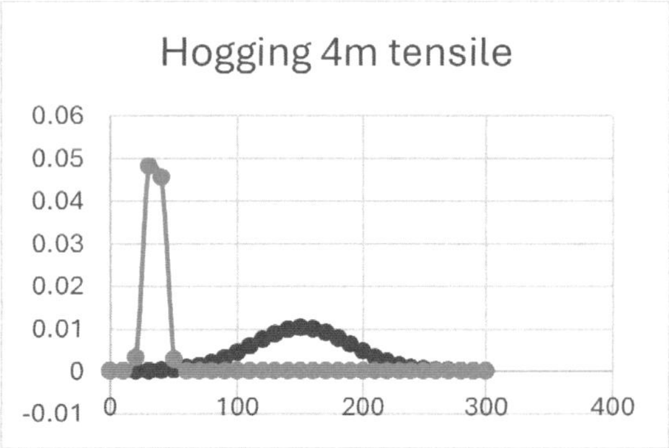

(A)

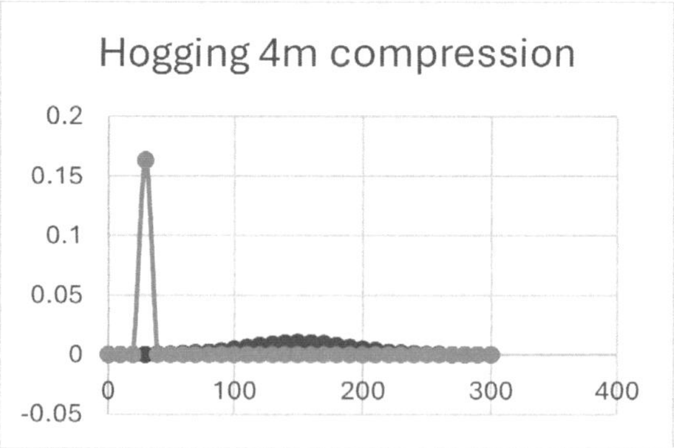

(B)

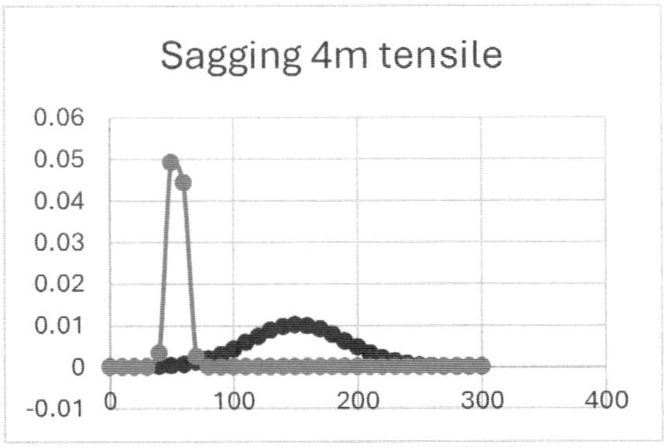
Sagging 4m tensile
0.06
0.05
0.04
0.03
0.02
0.01
0
-0.01
0
100
200
300
400

(C)

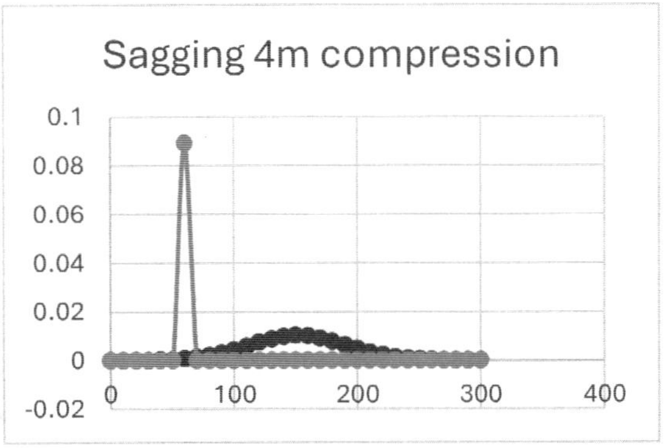
Sagging 4m compression
0.1
0.08
0.06
0.04
0.02
0
-0.02
0
100
200
300
400

(D)

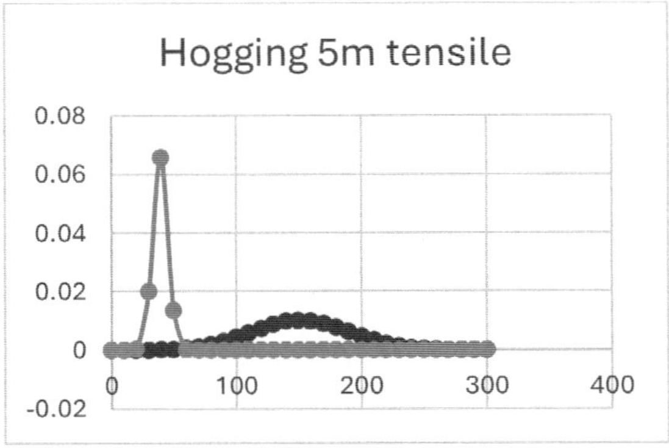
Hogging 5m tensile
0.08
0.06
0.04
0.02
0
-0.02
0
100
200
300
400

(E)

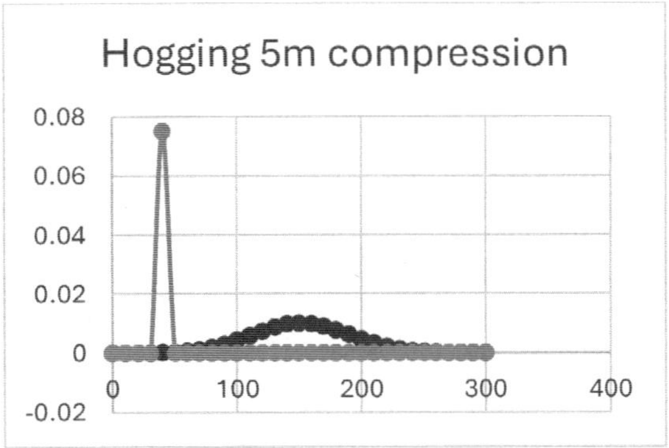
Hogging 5m compression
0.08
0.06
0.04
0.02
0
-0.02
0
100
200
300
400

(F)

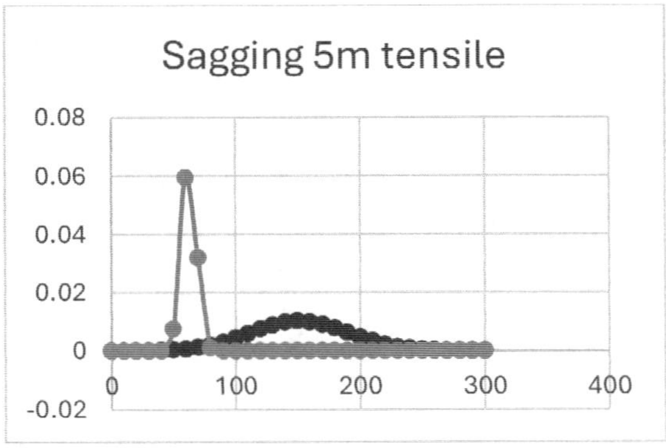
Sagging 5m tensile
0.08
0.06
0.04
0.02
0
-0.02
0
100
200
300
400

(G)

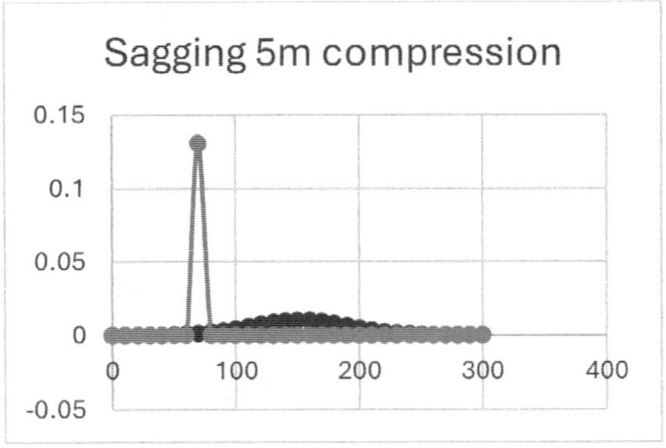
Sagging 5m compression
0.15
0.1
0.05
0
-0.05
0
100
200
300
400

(H)

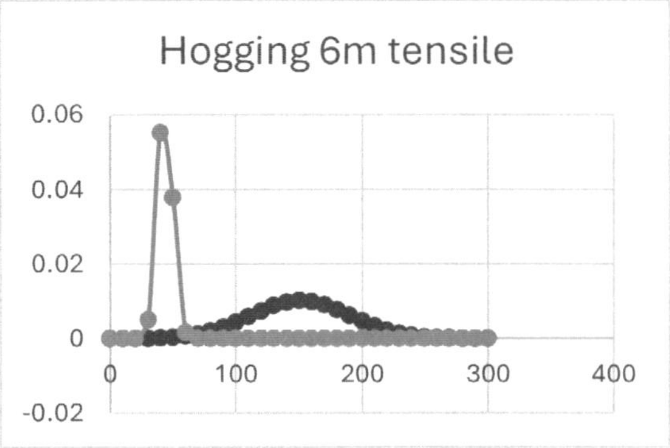

(I)

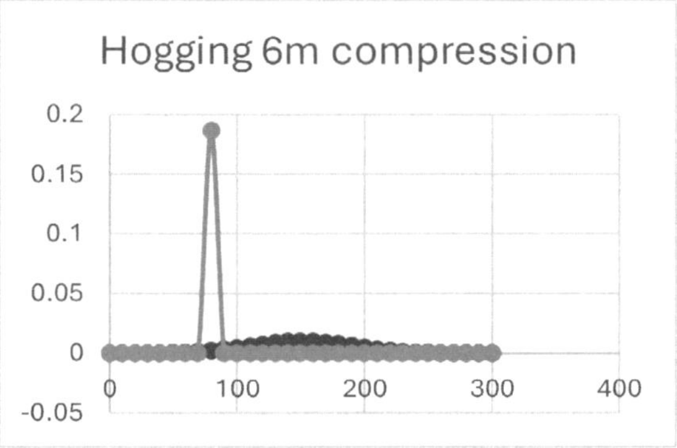

(J)

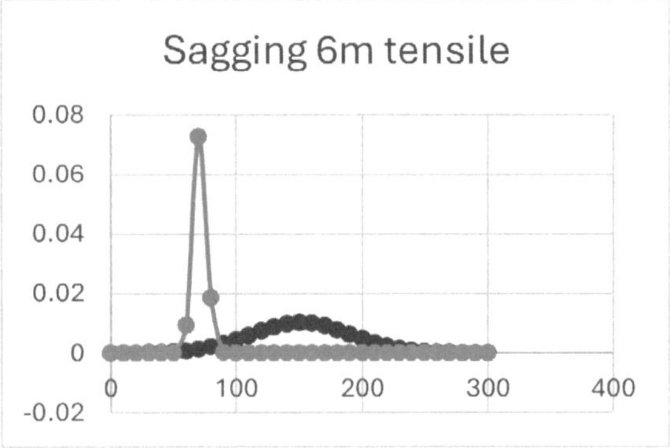

(K)

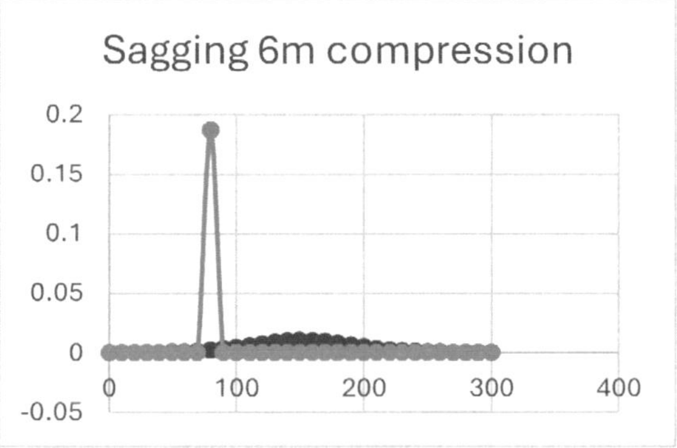

(L)

Figure 4. 1 Probability Density Function on Every Condition

In calculating structural reliability using the Monte Carlo method, data from the Joint Probability Density Function (PDF), including the mean and standard deviation, are required. The reliability determination process involves iterating 10,000, 100,000, and up to 1,000,000 times to obtain valid data and reliability values that meet the established criteria. This reliability calculation uses the Monte Carlo simulation approach, producing the following results.

Tensile stress (Hogging)			
No	wave height	PoF	Reliability
1	4	0.18%	99.82%
2	5	0.25%	99.75%
3	6	0.34%	99.66%

Compression stress (Hogging)			
No	wave height	PoF	Reliability
1	4	0.11%	99.89%
2	5	0.18%	99.82%
3	6	9.75%	90.25%

Tensile stress (Sagging)			
No	wave height	PoF	Reliability
1	4	0.75%	99.25%
2	5	1.28%	98.72%
3	6	2.15%	97.85%

Compression stress (Sagging)			
No	wave height	PoF	Reliability
1	4	0.84%	99.16%
2	5	1.74%	98.26%
3	6	3.52%	96.48%

5. CONCLUSION

Based on the analysis , the conclusions drawn from this study are as follows:

1. At wave heights of 4m, 5m, and 6m, the tensile and compressive stresses experienced by the ship's structure show significant variation. This indicates that an increase in wave height directly impacts the increase in stress experienced by the ship's deck structure.
2. The application of environment loads on the ship's such as bending moment increases the tensile and compressive stresses experienced by the structure. The increased stress with rising wave heights suggests that environment loads exacerbate the stress conditions on the ship's structure.
3. From a reliability perspective, for the 5500 DWT Oil Tanker using Grade A steel, the Monte Carlo simulation results indicate that the lowest structural reliability was found in the structure under hogging conditions with a wave height of 6m, reliability value of 90.25% and a failure probability of 9.75%. Therefore, the structure is considered less reliable in bearing the applied loads.

References

[1] Tempo, "Kapal Kargo Patah di Perairan Jepang, Ini yang Terjadi," tekno.tempo.co.
[2] Myrto Kalouptsidi, "The Role of Shipping in World Trade," ECONOFACT.
[3] American Bureau of Shipping, *Rules for Building and Classing Marine Vessel*. 2023.
[4] Rameswar Bhattacharya, *Dynamics of Marine Vehicles*. 1978.
[5] James M. Gere and Stephen P. Timoschenko, *Mekanika Bahan*, vol. 1. 1987.
[6] Owen F. Hughes and Jeom Kee Paik, *Ship Structural Analysis and Design*. 2010.
[7] Andrew Pytel and Jaan Kiusalaas, *Mechnics of Materials*. Global Engineering: Christopher M. Shortt, 2003.
[8] J. Whiteley, *Finite Element Methods*. 2017. doi: 10.1007/978-3-319-49971-0.
[9] A. Ismail, A. Zubaydi, B. Piscesa, R. C. Ariesta, and Tuswan, "Vibration-based damage identification for ship sandwich plate using finite element method," *Open engineering*, vol. 10, no. 1, pp. 744–752, 2020.
[10] G. Dhatt, E. Lefrançois, and G. Touzot, *Finite element method*. John Wiley & Sons, 2012.
[11] D. H. Norrie, "A first course in the finite element method," *Finite Elements in Analysis and Design*, vol. 3, no. 2, pp. 162–163, Jul. 1987, doi: 10.1016/0168-874X(87)90008-4.
[12] Daryl L. Logan, *A First Course in the Finite Element Method*, 4th ed., vol. 1. Chris Carson, 2006.
[13] E. Ellobody, R. Feng, and B. Young, "Linear and Nonlinear Finite Element Analyses," in *Finite Element Analysis and Design of Metal Structures*, Elsevier, 2014, pp. 56–71. doi: 10.1016/B978-0-12-416561-8.00004-4.
[14] G. J. White and B. M. Ayyub, "Reliability Methods for Ship Structure," *Naval Engineers Journal*, vol. 97, no. 4, pp. 86–96, May 1985, doi: 10.1111/j.1559-3584.1985.tb01343.x.

IOP Conf. Series: Earth and Environmental Science **1461** (2025) 012013 doi:10.1088/1755-1315/1461/1/012013

Complex Dynamics of Work-Life Balance, Employee-Organization Relationship, Flexibility, and Employee Engagement: The Role of Leadership Style and Organizational Culture in Enhancing Job Satisfaction

Riza Rosdiana*, Harmein Nasution, Nazarudin, Anizar Anizar

Industrial Engineering Science, Universitas Sumatera Utara, North Sumatra, Indonesia

*E-mail: riza.rosdiana@pelindo.co.id

Abstract. Within the framework of port management in Indonesia, this thorough study investigates the complex dynamics of work-life balance, employee-organization interactions, work-life flexibility, and employee engagement. It looks at how these elements together affect job satisfaction and looks at how organizational culture and leadership style might moderate things. Given the changing nature of employment in the post-COVID-19 era and the special difficulties experienced by the Indonesian port management industry, the research is especially relevant. Being the biggest archipelagic state in the world with more than 17,000 islands, Indonesia mostly depends on its ports for national connectivity, trade facilitation, and economic growth. Recent years have seen major changes in Indonesia's port management scene including infrastructure improvements, digitalization programs, and legislative/law reforms. These developments together with worldwide trends in work practices have produced a complex environment for port staff that calls for a better knowledge of elements influencing their job satisfaction and general well-being. This study used a mixed-method approach to gather information from 1,500 workers spread among many Indonesian port management agencies, including private operators and state-owned companies such as PT Pelabuhan Indonesia (Pelindo). To compile extensive and varied data, the study included focus groups, in-depth interviews, and a thorough quantitative survey. Key results show that, with the strength of employee-organizing interactions as a major mediator, work-life balance greatly increases job satisfaction among Indonesian port management personnel.

1. Introduction

Part of more general economic growth plans, the Indonesian government has given port infrastructure expansion and modernization top priority in recent years. Projects meant to increase port capacity, streamline logistics, and lower regional inequalities include the Sea Toll Road scheme and the National Ports Master Plan. From big international hubs like Tanjung Priok in Jakarta to smaller regional ports serving local communities, these initiatives have fundamentally changed the operational scene of ports all throughout the nation. State-owned businesses, especially PT Pelabuhan Indonesia (Pelindo), which runs significant commercial ports all throughout Indonesia, dominate the port management scene in the nation. Recent changes have, however, also provided chances for private sector involvement, therefore fostering a more varied and competitive environment. For port management staff, this changing terrain has presented new opportunities as well as problems that demand better knowledge of elements influencing their general well-being and job satisfaction. The worldwide COVID-19 epidemic has underlined the need of flexibility and resilience in the port industry as well as accelerated improvements in working methods. In the port management industry, this scenario has pushed work-life balance, flexibility, and employee well-being front stage in debates. Furthermore changing job roles and skill needs in the industry is the continuous digitalization of port operations including the application of smart port technology and automation systems.

The study aims to give a whole picture of elements impacting job satisfaction in this important industry by analyzing the intricate interaction of work-life balance, employee-organizing connections, work-life flexibility, and employee engagement. Moreover, by looking at the moderating impacts of organizational culture and leadership style, the study hopes to provide understanding of how port management organizations could design encouraging work cultures that encourage operational excellence and employee well-being.

This study contributes to the corpus of knowledge by developing a conceptual framework that ties together hitherto separately investigated aspects of employees' job lives. Furthermore more understood is the link between job happiness and work-life balance as well as the part corporate culture and leadership style play in work-life dynamics

2. Literature Review

2.1 Work-Life Balance in the Context of Port Management

Although organizational studies have conducted a lot of study on work-life balance, its application in the particular setting of port management offers special difficulties and issues. Underlined how port labor was changing from conventional dock job to more varied and specialized positions. This development affects how port environments see and control work-life balance [8]. Investigated work-life balance concerns in the marine industry in Indonesia and found that the 24/7 operational character of ports presents particular difficulties for workers in preserving a good work-life balance [11].

2.2 Employee-Organization Relationship in Port Management

In high-stakes settings like ports, the nature of employee-organization relationships is critical, emphasizing the need of perceived organizational support and psychological contracts that provide a comprehensive assessment of employee-organization relationship research [12].

2.3 Work-Life Flexibility in Port Operation

Work-life flexibility is the capacity of employees to make decisions on when, where, and for how long they participate in activities connected to their jobs [4]. Given the 24/7 operational needs of ports, applying this idea to port management offers special difficulties. Examining flexibility in European ports, the conflict between operational needs and personnel requirements. Their research offers insightful analysis on how ports may use flexible schedules to keep operational effectiveness [17].

2.4 Maritime Industry Employee Engagement

Employee engagement has been related to several favorable results for organizations [5] [10]. The marine industry and the value of participation in shipping firms, stressing their contribution in operational excellence and safety [9]. Factors including job autonomy, feedback, and skill variety greatly affected port worker engagement in Singapore [14]. These results offer a helpful relative viewpoint for the Indonesian situation.

2.5 Leadership Style in Port Management

Transformational leadership improved staff performance and job satisfaction in state-owned port companies, according to a 2018 study particular to Indonesian ports. This study offers a basis for more investigation of how leadership affects work-life balance and involvement in Indonesian ports.

2.6 Organizational culture in Port Management

Examining organizational culture in Taiwanese maritime companies about found cultural factors supporting organizational effectiveness. Their research offers an understanding of how cultural elements could affect dynamics of work-life in Asian port environments [6]. Job Satisfaction in Port Management on job satisfaction in Chinese ports, work-life balance and career advancement possibilities were the main factors influencing satisfaction, these results offer Indonesian ports a helpful basis for investigating job satisfaction [18].

2.7 The Indonesian Port Management Context

Examining the National Single Window system's application in Indonesian ports, its consequences for port operations and staff functions. This paper emphasizes the continuous technological developments influencing Indonesian port work surroundings [13].

2.8 Literature Gap

While existing literature provides valuable insights into various aspects of work-life dynamics and port management, several gaps remain, limited integration of work-life balance, employee engagement, and job satisfaction concepts in the specific context of port management and scarcity of comprehensive studies on these issues in the Indonesian port sector, particularly considering recent reforms and technological changes. This study aims to address these gaps by providing a comprehensive, contextually grounded examination of work-life dynamics in Indonesian port management.

3. Methods

3.1 Development of the Research Models and Hypotheses

This paper uses a mixed-methods approach to investigate in Indonesian port management the complicated dynamics of job satisfaction, employee engagement, and work-life balance. The mixed-methods approach makes it possible to combine qualitative and quantitative data, so offering a sophisticated knowledge of the studied phenomena [2].

Work-Life Flexibility (WLF) and Employee Engagement (EE): Work-life flexibility is an employee's capacity to juggle demands from their personal life with their job through adaptable work schedules. Studies indicate that employee engagement can improve with work flexibility.

Employee-Organizational Relationship (EOR) and Employee Engagement (EE): Employee relationships with companies mirror the nature of contacts and ties between people and their companies. Good relationships usually improve employee satisfaction.

Work-Life Balance (WLB) and Employee Engagement (EE): Work-life balance is the capacity of a person to satisfactorily meet obligations in both their personal and professional life. Good WLB improves staff involvement.

Employee Engagement (EE) and Job Satisfaction (JS): A good mental state marked by excitement, passion, and absorption in work is what employee engagement is about. Usually, engaged workers are happy with their positions.

Leadership Style (LS) as a Moderator: The relationship between employee engagement and job satisfaction can be strengthened or weakened depending on the leadership style. Good leadership can help to maximize employee involvement's benefits.

Organizational Culture (OC) as a Moderator: Job happiness can result from employee engagement depending on the organizational culture. A supportive culture helps to enhance this favorable link.

3.2 Research Design

This study combines quantitative and qualitative techniques to acquire a complete knowledge of the topic under investigation. Two primary phases comprise the research plan:

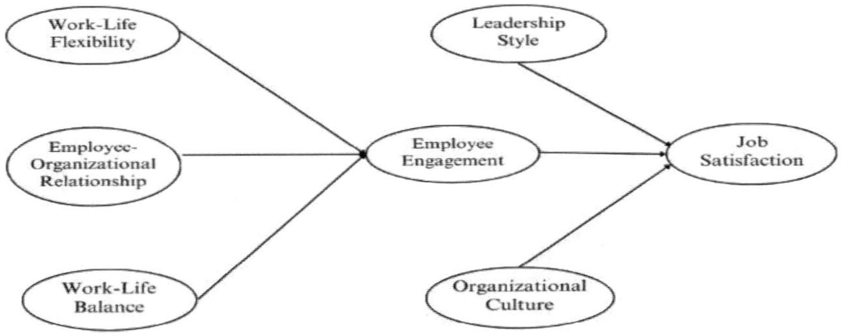

Figure 1. Research Framework source Authors (2024)

Phase 1: Quantitative Survey: A cross-sectional survey was conducted with employees from various organizations and industries. The purpose of this survey was to evaluate the main variables of the study and test the proposed hypotheses. Phase 2: Qualitative Study: In this phase, in-depth interviews and focus group discussions (FGDs) were used to gain a deeper understanding of the quantitative results and to learn about context and nuances that may not have been revealed through the survey.

3.3 Population and Sample

Employees in port management roles all throughout Indonesia make up the study's target group. This covers private port operators and state-owned companies namely PT Pelabuhan Indonesia and its affiliates. Quantitative Sample: Sample size: 1,500 employee, Sampling method: Stratified random sampling. Stratification criteria: Port location (Java, Sumatra, Kalimantan, Sulawesi, Eastern Indonesia), job level (operational, supervisory, managerial), and type of port operator (state-owned, private). Qualitative Sample: 50 in-depth interviews with employees from various levels and departments, 10 focus group discussions, each comprising 8-10 participants.

3.4 Reliability and Validity

3.4.1 Internal Validity

Method triangulation is the cross-validation of results using both qualitative and quantitative approaches. Triangulation of Data Sources.

IOP Conf. Series: Earth and Environmental Science **1461** (2025) 012013 doi:10.1088/1755-1315/1461/1/012013

3.4.2 Outsourced Validity

Many Sample: Including people from different sectors and companies to improve generalizability. Detailed Description: Addition of thorough contextual details to support findings' transferability.

3.4.3 Reliability

Procedure Standardizing::Data collecting and analysis using set procedures. Reliability between integrators: Applications of several coders for Cohen's Kappa computation and qualitative analysis. Reliability of Test-Retes: Survey was repeated two weeks apart to evaluate response stability for a subset of the sample. Audit Trail: Detailed records of every analytical choice and research technique used. Reflexivity: The principal researcher kept a reflecting notebook to be aware of and control possible prejudices.

4. Test Result

4.1 Sample Features

With 1,500 questionnaires distributed, 1,328 valid responses came back, for an 88.5%

Table 1. Sample Demographic Attributes.

Characteristic	Frequency	Percentage
Gender		
Male	712	53,60%
Female	616	46,40%
Age		
18-30	384	28,90%
31-40	496	37,40%
41-50	312	23,50%
> 50	136	10,20%

4.2 Descriptive Statistics and Interactions

Every correlation turned out to be significant and in the expected directions. Strong positive connections between job satisfaction (r =.57, p <.01) and employee-organizing relationship (r =.52, p <.01) revealed work-life balance.

Table 2. shows among the primary study variables the means, standard deviations, and correlations

Variable	Mean	SD	1	2	3	4	5	6	7
Work-Life Balance	3,42	0,89	1						
Employee-Organizational Relationship	3,56	0,95	0,52**	1					
Work-Life Flexibility	3,28	1,02	0,48**	0,41**	1				
Employee Engagement	3,71	0,88	0,45**	0,53**	0,39**	1			
Leadership Style	3,39	0.97	0,40**	0,58**	0,35**	0,49**	1		
Organizational Culture	3,45	0.92	0,43**	0,61**	0,37**	0,51**	0,55**	1	
Job Satisfaction	3,6	0,93	0,57**	0,62**	0,46**	0,59**	0,53**	0,56**	1

** p < .01

4.3 Analysis of Confounding Factors

The factor structure of the used scales was validated by means of confirmatory factor analysis (CFA.). The results displayed satisfactory match on all scales: All scales demonstrated good fit, with CFI and TLI values above .90, RMSEA values below .08, and SRMR values below .05, indicating acceptable model fit. With CFI and TLI values above.90, RMSEA values below.08, and SRMR values below.05, all scales shown strong fit.

Table 3. CFA Results.

Scale	$\chi 2/df$	CFI	TLI	RMSEA	SRMR
WLB	2,34	0,96	0,95	0,048	0,032
ERO	2,56	0,95	0,94	0,052	0,035
WLF	2,18	0,97	0,96	0,045	0,03
EE	2,72	0,94	0,93	0,055	0,038
LS	2,89	0,93	0,92	0,058	0,041
OC	2,67	0,94	0,93	0,054	0,037
JS	2,81	0,93	0,92	0,057	0,04

4.4 Modeling Structural Equation Structurally

The expected correlations were tested using structural equation modeling (SEM). With $\varphi 2/df = 2.76$, CFI =.94, TLI =.93, RMSEA =.056, SRMR =.039 the structural model suited the data rather well.

Table 4. Hypothesis Testing Results

Hypothesis	Path	Coefficient	S.E.	p-value	Result
H1	WLB → JS	0,32	0,04	<,001	Supported
H2a	WLB → ERO	0,48	0,05	<,001	Supported
H2b	ERO → JS	0,29	0,03	<,001	Supported
H3a	WLB → EE	0,41	0,04	<,001	Supported
H3b	EE → JS	0,26	0,03	<,001	Supported
H4	WLF*WLB → JS	0,15	0,02	<,001	Supported
H5	LS*WLB → JS	0,18	0,03	<,001	Supported
H6	OC*WLB → JS	0,16	0,02	<,001	Supported

4.5 Medial Analysis

The mediating effects were tested using bootstrapping analysis including 5000 sample.

Table 5. Mediation Analysis Results

Path	Indirect Effect	95% CI	p-value
WLB → ERO → JS	0,139	[0,102 ; 0,178]	<0,001
WLB → EE → JS	0,107	[0,076 ; 0.140]	<0,001

The results indicate that both employee-organization relationship (ERO) and employee engagement (EE) partially mediate the relationship between work-life balance (WLB) and job satisfaction (JS), supporting H2a and H2b. Supporting H2a and H2b, the results show that both employee-organization relationship (ERO) and employee engagement (EE) somewhat buffer the link between work-life balance (WLB) and job satisfaction (JS).Moderation Analysis: 4.1.6 Supporting H4, H5, and H6, the moderation analysis revealed that work-life flexibility (WLF), organizational culture (OC), and leadership style (LS) modify the link between job satisfaction (JS) and work-life balance (WLB).

4.6 Qualitative Study

The study of the interview and FGD data improved knowledge of the interactions among the variables by several significant themes. Participants stressed that work-life balance requires not only time but also mental commitment and energy. " Sometimes my mind is still at work even when I have time for family. That is not actual balance, remarked P17, HR Manager. Considered basic for all other aspects of the employment experience, the nature of the relationship with the company was regarded as "When I feel the organization really cares about my well-being, I'm more willing to go the extra mile," P23, Financial Analyst remarked. Flexibility, some participants noted, might make it challenging to tell personal from business life. "Flexibility lets me go to school events for my child, but it also means I often work late into the night," P08, Software Developer remarked. Participants underlined the need of balancing involvement with burnout avoidance. "Although I enjoy working extremely actively, occasionally it can be too much. One should have limits," advised P31, the project manager. It was believed that empathetic leadership forms a major factor in a successful working environment. "My management is aware that everyone of us leads life apart from our jobs. One important topic that surfaced was the twin function of technology in enabling flexibility and driving pressure to be connected. "My smartphone curses as much as it blesses. P27, the sales manager, claimed she could work from anyplace but felt she had to answer emails constantly. Generation differences in views on life goals and jobs affected work dynamics. "Younger staff members seem to value work-life balance more than my generation did," remarked Senior Director P05. Changes in remote work and flexibility brought about by the COVID-19 outbreak emerged as one of main themes that appeared. "The pandemic has changed how we view where and how work can be done," stated P50, CEO. These qualitative results enhance the quantitative results thereby facilitating a better knowledge of the complex dynamics of job satisfaction and work-life balance in the modern workplace.

5. Discussion

5.1 Job Satisfaction and Work-Life Balance

The vital relevance of work-life balance in the context of port management is confirmed by the high favorable correlation between job satisfaction (β =.32, p <.001). This result is consistent with earlier studies [3], who discovered across different cultures favorable relationships between job satisfaction and work-life balance. Nevertheless, the qualitative results expose that the idea of work-life balance in port management is more complicated than first thought. Participants underlined that balance calls not only for time but also for mental involvement and energy control. Analysis of the Indonesian maritime industry emphasize how especially complicated the port sector is because of its 24/7 operational character [11]. According to numerous participants, the difficulty of "truly disconnecting" from work implies that conventional ideas of work-life balance might have to be changed in the framework of port management. The changing character of port labor by implying that the lines separating work from personal life becoming more hazy as port roles grow more specialized and linked with worldwide supply chains [8]. Practical Implication: Port management companies should think beyond conventional work-life balance rules and take into account ways to assist in "psychological detachment" from employment. Training courses on mindfulness and stress management as well as properly defined channels of communication for off-duty hours should be part of this.

5.2 Managing Roles of Employee Engagement and Organizational Relationship

The mediation study results suggest that job satisfaction and work-life balance are partially mediated by employee engagement (indirect impact =.107, 95% CI [.076,.140]) and employee-organization relationship (indirect effect =.139, 95% CI [.102,.178]. This suggests that job happiness is influenced by work-life balance not just directly but indirectly by improving the quality of the relationship with the company and raising employee involvement levels. Particularly the theme "Critical Role of Employee-Organization Relationship," the qualitative results—which reveal employees see their relationship with the company as fundamental to other facets of their work experience—reiterate these conclusions. This is consistent with social exchange theory [1], which holds that workers usually return good treatment from their company with appropriate attitudes and actions. This result theoretically suggests the need of more integrated models considering both direct and indirect routes via which work-life balance affects job outcomes.

5.3 Workplace Flexibility: Moderating Effect

The findings reveal that job satisfaction and work-life balance have a positive moderation effect from occupational flexibility (β =.15, p <.001). This implies that employees who have more freedom in how they manage their work experience a more significant beneficial influence of work-life balance on job satisfaction. Still, the qualitative theme "Flexibility as a Double-Edged Sword" exposes significant subtleties in this result. Although most people saw it favorably, several participants pointed out that flexibility might make it difficult to distinguish personal from professional life. This is consistent with the idea of the "autonomy paradox" [7], which holds that flexibility might cause labor intensity to increase. This result suggests that companies should exercise great caution while putting flexibility policies into use. Apart from providing adaptable choices, it's crucial to create a "boundary culture" that values workers' personal time and discourages inclination toward overworking. Training managers to set clear expectations on availability and response times and to reflect good work-life boundaries could help to accomplish this.

5.4 Modulating Roles of Organizational Culture and Leadership Style

The results show that the connection between job satisfaction and work-life balance is favorably moderate by both organizational culture (β =.16, p <.001) and leadership style (β =.18, p This highlights the relevance of organizational environment in the grasp of dynamics of work-life balance. Further illumination on how these elements interact comes from the qualitative themes "Supportive Leadership" and "Organizational Culture and Unwritten Norms". While company culture defines daily expectations and actions that can either promote or impede work-life balance, sympathetic and supportive leadership is viewed as absolutely essential in establishing a pleasant work environment. Our results extend previous studies on "supportive work-family culture" by showing how leadership and culture interact with work-life balance to influence job satisfaction [15]. This result has practical relevance in terms of the need of leadership development concentrated on helping workers' work-life balance as well as methodical initiatives to change corporate culture to better promote good work-life integration.

5.5 Work-Life Limitations and Technology

The subject "Technology and Work-Life Boundaries" that surfaced from the qualitative study emphasizes still more difficulty in juggling digital era work-life balance. The dual function of technology in allowing flexibility and generating pressure to always be connected points to the need for a more complex approach to grasp and control work-life balance. This fits studies on how using a smartphone affects recovery from employment [16]. Our results imply that more study on how to control technology so as to assist rather than compromise work-life balance. Practically speaking, companies should take into account creating "right to disconnect" rules and instruction on responsible technology use to help to encourage more defined work-life boundaries.

5.6 Future Research Direction and Limitations

Although this study offers insightful analysis, many limits must be admitted: The cross-sectional design of the study restricts the capacity to create causal conclusions. Future long-term studies might assist in defining the causality direction. Cultural Context: The study was carried out in Indonesia, so generalizability to other civilizations could be limited. Comparative research across cultures helps to verify the universality of conclusions. Self-report surveys could cause common method bias depending on how you utilize them. Objective data and multi-source evaluations could be combined in the next studies. Model Complexity: Interpretation of the many-variable complex model can be difficult. Future studies could have to concentrate on subsets of factors for more detailed investigation.

6. Conclusion

The results confirm the fundamental relevance of work-life balance in determining job satisfaction and expose the intricate processes via which this effect functions. The work offers numerous important contributions: Including work-life balance, employee-organization interactions, job-life flexibility, employee engagement, leadership style, and organizational culture, it creates and validates an integrated model of job satisfaction in port management. It offers fresh understanding of the processes by which employment outcomes are influenced by employee-organization connections and employee engagement in the work-life balance and job satisfaction connection. Emphasizing the need for contextual elements in forming the work-life balance and job satisfaction link, it draws attention to the moderating roles of work-life flexibility, leadership style, and corporate culture. It reveals sector-specific difficulties and opportunities connected to work-life balance in port management, including the influence of 24/7 operations, continuous digitalization, and geographical dispersion of port sites. It offers an understanding of the generational variances and cultural subtleties driving work-life interactions in Indonesia. It has pragmatic ramifications for port management companies in terms of the need of customized work-life balance policies, leadership development concentrated on supportive actions, and careful handling of technology application.

References

[1] Blau, P. M. (1964). *Exchange and power in social life.* Wiley.

[2] Creswell, J. W., & Plano Clark, V. L. (2017). Designing and conducting mixed methods research (*3rd* ed.). Sage Publications.

[3] Haar, J. M., Russo, M., Suñe, A., & Ollier-Malaterre, A. (2014). Outcomes of work–life balance on job *satisfaction*, life satisfaction and mental health: A study across seven cultures. Journal of Vocational Behavior, 85(3), 361-373.

[4] Hill, E. J., Grzywacz, J. G., Allen, S., Blanchard, V. L., Matz-Costa, C., Shulkin, S., & Pitt-Catsouphes, M. (2008). Defining and conceptualizing workplace flexibility. Community, Work & Family, 11(2), 149-163

[5] Kahn, W. A. (1990). Psychological conditions of personal engagement and disengagement at work. Academy of Management Journal, 33(4), 692-724.

[6] Lu, C. S., & Shang, K. C. (2005). An empirical investigation of safety climate in container terminal operators. Journal of Safety Research, 36(3), 297-308.

[7] Mazmanian, M., Orlikowski, W. J., & Yates, J. (2013). The autonomy paradox: The implications of mobile email devices for knowledge professionals. Organization Science, 24(5), 1337-1357.

[8] Notteboom, T. E., & Winkelmans, W. (2002). Stakeholder relations management in ports: Dealing with the interplay of forces among stakeholders in a changing competitive environment. In Proceedings of the 2002 IAME Conference, Panama City, Panama.

[9] Progoulaki, M., & Theotokas, I. (2010). Human resource management and competitive advantage: An application of resource-based view in the shipping industry. Marine Policy, 34(3), 575-582.

[10] Schaufeli, W. B., Salanova, M., González-Romá, V., & Bakker, A. B. (2002). The measurement of engagement and burnout: A two sample confirmatory factor analytic approach. Journal of Happiness Studies, 3(1), 71-92.

[11] Siahaan, R., & Bahri, S. (2019). Pengaruh Beban Kerja dan Work-Life Balance terhadap Kinerja Karyawan PT Pelabuhan Indonesia I (Persero) Medan. Jurnal Manajemen, 5(1), 35-44.

[12] Shore, L. M., & Coyle-Shapiro, J. A.-M. (2007). The link between an employee and their organization: Where from here? Human Resource Management Review, 17(2), 166–179.

[13] Subari, A., Ricardianto, P., & Sari, D. P. (2020). Implementasi National Single Window dalam meningkatkan pelayanan di pelabuhan Indonesia. Jurnal Manajemen Transportasi & Logistik, 7(1), 85-94.

[14] Thai, V. V., Balasubramanyam, L., Yeoh, K. K. L., & Norsofiana, S. (2013). Revisiting the seafarer shortage problem: The case of Singapore. Maritime Policy & Management, 40(1), 80-94.

[15] Thompson, C. A., Beauvais, L. L., & Lyness, K. S. (1999). When work–family benefits are not enough: The influence of work–family culture on benefit utilization, organizational attachment, and work–family conflict. Journal of Vocational Behavior, 54(3), 392-415.

[16] Tims, M., Bakker, A. B., Derks, D., & van Duin, D. (2015). Work-home interference and smartphone use: Social norms' moderating effect on employee work engagement. Journal of Occupational and Organizational Psychology, 88(1), 155–177.

[17] Turnbull, P., & Wass, V. (2015). Defending Dock Workers—Globalization and Labor Relations in the World's Ports. Industrial Relations: A Journal of Economy and Society, 54(1), 72-93.

[18] Yuen, K. F., Loh, H. S., Zhou, Q., & Wong, Y. D. (2018). Determinants of job satisfaction and performance of seafarers. Transportation Research Part A: Policy and Practice, 110, 1-12.

Strength analysis of tripod jacket and skid frame structures during the loadout process

Salsabila Tsaniatur Rahma[1], Rudi Walujo Prastianto[1*], and Daniel Mohammad Rosyid[1]

[1] Ocean Engineering Department, Institut Teknologi Sepuluh Nopember, Surabaya, Indonesia

*rudiwp@oe.its.ac.id

Abstract. Offshore platforms are constructed worldwide for various needs, particularly in the oil and gas industry. Around 95% of offshore platforms used for oil and gas exploration and exploitation activities are supported by fixed jacket structures. This structure is constructed by the fabricator company and subsequently loaded out onto a barge to be transported to the installation site. The loadout process must be evaluated carefully considering the unpredictable quay side conditions that may cause the jacket structure to be overstressed. Therefore, loadout analysis of the tripod jacket structure using the skidding method was conducted in this research. Global analysis using SACS software was performed to assess the strength of the jacket structure through some loadout scenarios and ensure it satisfied the applied requirements. The results showed that the critical stress occurred at the mudmat member, which has UC (stress ratio) value of 0.91. Besides the jacket structure, a supporting loadout structure, namely the skid frame, also had been analysed. The complexity of the skid frame geometry makes it hard to model and analyse globally. Consequently, local stress analysis of skid frame structure was performed for the most critical conditions using ABAQUS software. Most local critical stress was obtained from this local analysis and had been compared to the allowable stress based on the applied requirements. It showed that there were four critical skid frame components which have von mises stress of 575.60 MPa, 522.20 MPa, 414.50 MPa, and 542.50 MPa. Those values were exceeded the allowable, even two of them exceeded the ultimate stress.

Keywords: Engineering, Infrastructure, Strength Analysis, Von Mises Stress, UC

1. Introduction

Offshore platforms are generally fabricated in an open yard nearshore to accelerate the transportation process. Before offshore platform is installed in operation site, it needs to be transported from the fabrication yard using a transportation barge or vessel. Hence, the offshore platform is not only designed to withstand harsh sea environmental conditions, but it must also have sufficient strength during the transportation process [1].

The transportation phase of an offshore platform installation process is initiated by a loadout. Loadout is a process of transferring a structure from the fabrication yard to the transportation barge or vessel [2]. This process is more complex than just simply moving the structure to a location not far from its initial location. This is because offshore platforms are typically very large and heavy, there are many supporting structures and systems that need to be prepared.

There are three ways to perform a loadout analysis, one of them is the skidding method. The skidding method performs a loadout process by sliding the structure off the quay onto the barge. In the skidding method, there is a skid way as a track for sliding the structure. In order to prevent damage to the loaded-out structure, a skid frame must be placed between the structure and the skid way. Even though the environmental condition in the fabrication yard is not as harsh as in the operation site, some nuisances might be experienced by the offshore platform during the loadout process. Thus, a loadout plan must be conducted thoroughly, and structural analysis must be performed to ensure the structural integrity of the offshore platform.

A study discussing structural analysis during loadout process had been made before for Bombay High Jacket [3]. This paper aims to ensure there is no overstress on the jacket component during loadout. This analysis was performed in SACS software to obtain the UC value for each jacket component. However, that paper was made for the analysis of loadout process using lifting method, which the process is totally different from using the skidding method. This paper used the same method as the Bombay Jacket loadout, which has used global analysis to check the strength of the jacket structure. Besides, local stress analysis used in this paper to check the strength of the skid frame, considering the complexity of the skid frame geometry. This local analysis has brought out a more accurate results [4].

Therefore, this study was conducted to analyse the strength of the loaded-out structure and the supporting structure, namely skid frame. This study aims to ensure there are no overstressed jacket and skid frame components. The structural analysis in this study will be performed in static conditions with some consideration of nuisances that might happen during the loadout process. Referring to the location of loadout analysis, which is located on the land, the environmental condition considered in this analysis is only a wind load. Additionally, the analysis will be performed using SACS for jacket analysis and performed using Abaqus for skid frame analysis.

2. Theory and Numerical Model

2.1 Loadout Scenarios

2.1.1 CoG Shift

This scenario aims to check the structural strength of the jacket and skid frame, if there is any shifting of CoG (Centre of Gravity) in the fabricated structure. The location of the CoG shifting is determined by CoG envelope, as depicted in Figure 1. The scenario was arranged according to the company specifications [5], which stated that the dimension of the CoG envelope must be 5% of the overall structure length and width.

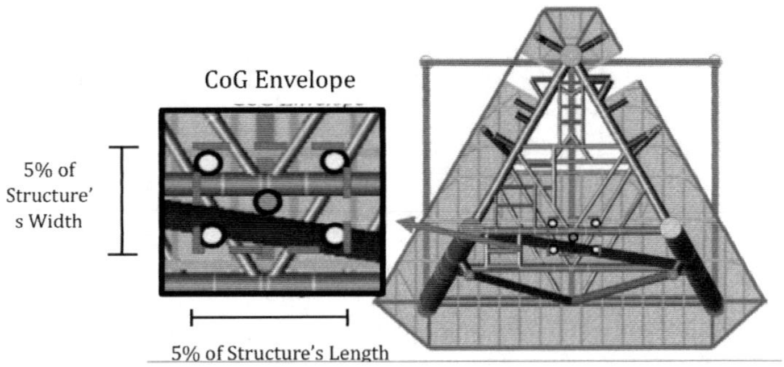

Figure 1. CoG Envelope Location.

IOP Conf. Series: Earth and Environmental Science **1461** (2025) 012014 doi:10.1088/1755-1315/1461/1/012014

The CoG shift was simulated by inserting a shift factor to move the CoG origin to the desired CoG shift. To calculate the shift factor, some dummy loads are applied to the structure. The dummy loads are applied to create the structure's moment, which are M_y and M_X. A simple mathematical equation, written in Equation 1 and Equation 2, is used in this analysis to calculate the shift factor.

$$\alpha x = \frac{|F_Z| \times (\frac{L}{2})}{M_y} \tag{1}$$

$$\alpha y = \frac{|F_Z| \times (\frac{W}{2})}{M_X} \tag{2}$$

αx is a shift factor to x direction and αy is a shift factor to y direction. F_Z is the total load applied to the structure in Z direction, while L and W are the length and the width of the CoG envelope, respectively. Lastly, M_y is the moment in y direction and M_X is the moment in x direction.

2.1.2 Ongoing Conditions

Ongoing conditions is a condition where the structure is in the state of being loaded out from the fabrication yard and continued to the barge deck. During this condition, there are some nuisances might happen which will cause the loadout process disturbed or even failed. Therefore, this condition must be considered in loadout analysis.

There are some nuisances that need to be considered, as per company specification [5], which as follows.

a. A 1.378-inch deflection in positive and negative Z direction that occurs in each skid frame support and two skid frame supports in the same row.
b. A loss of support condition that occurs in each skid frame support.

2.1.3 Initial Pulling Conditions

Initial pulling conditions is a condition where the structure is being pulled by some winch from its still condition until it moves along the loadout direction. There are two pulling scenarios used in this analysis, that are two skid support pulled forward and backward, as per company specification [6]. In performing loadout analysis for this condition, the value of pulling load must be calculated to simulate the transition from still to moving condition. According to Hooke Law II, the pulling load must have a sufficient value to encounter the structural load and the friction load, so the structure can be moving. Equation 3 can be used to calculate the load needed to pull the structure.

$$F = (m \times a) + (\mu \times m \times g) \tag{3}$$

F is the pulling load, m is the overall structural mass, a is the loadout acceleration (1 m/s^2, according to the company specification), g is the gravitation acceleration (g=9.81m/s^2), and μ is the friction coefficient.

2.2 Finite Element Method (FEM)

Finite element method (FEM) is a numerical technique commonly used in structural analysis. This method enables to simplify complex structural problems while maintaining accuracy of the results. This method works by discretize a large complex structure into smaller and simpler parts, namely finite elements [6]. There are three types of finite element, which are beams (one-dimensional), plates (two-dimensional), and solids (three-dimensional), which chosen in

IOP Conf. Series: Earth and Environmental Science **1461** (2025) 012014 doi:10.1088/1755-1315/1461/1/012014

accordance with the geometry of the structure. Subsequently, each finite element part would be analysed to obtain the overall behaviour of a structure [7].

This paper will use this method to analyse the stresses in the skid frame structure. This method is used in this analysis due to the complexity of the geometry, which makes it hard to modelled and analysed using beam-base software such as SACS software. The skid frame structure will be analysed as a solids part, instead of a beam part, so that more accurate results will be obtained [4][8]. Accordingly, more detailed skid frame structure can be modelled by using 3D FEM Software, such as Abaqus software [9][10].

2.3.1 Von Mises Stress

Von mises stress is a concept used in structural analysis to analyse the yielding and the failure state of materials under complex loading conditions [11]. This method works by providing a single equivalent stress value, which then combined to determine whether a material will yield. The single equivalent stress value consisting of the von mises stress are principal stress and shear stress in each dimension [12]. This stress concept is suitable for analysing stress in finite element analysis (FEA). The von mises stress in FEA then compared to the yield and ultimate strength of the material to determine the safety and strength of the structure.

3. Methodology

3.1 Data Collection

The data utilized in this study are secondary data obtained from an offshore construction consulting firm. There are two types of data gathered for this study, which are structural data and environmental data.

3.1.1 Structural Data

A tripod jacket structure will be used to support a wellhead platform as shown in Figure 2.a. This structure is owned by PT XYZ and will be operated in Indonesia with a water depth of 54.46ft. Besides, the skid frame structure is a framework with four legs located above the skid shoe as shown in Figure 2.b. Those structure's details can be seen in Table 1.

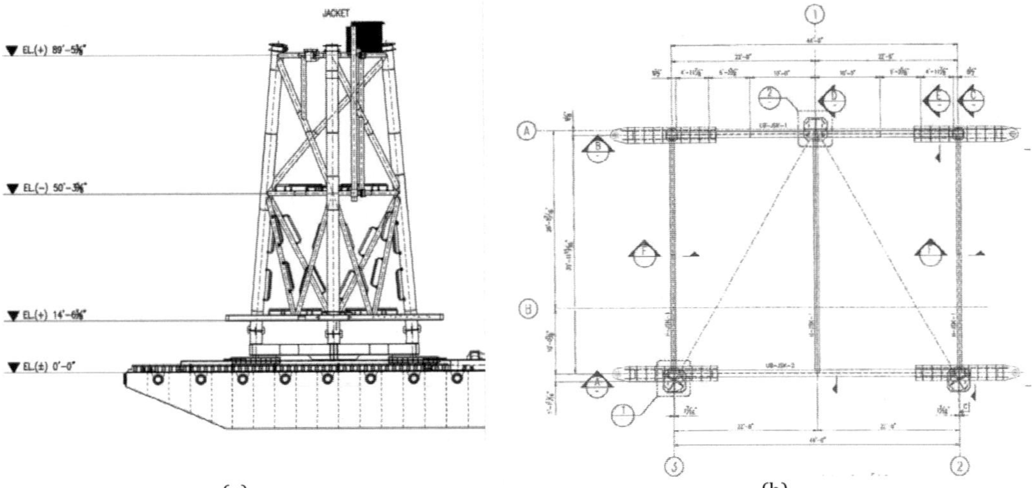

(a) (b)

Figure 2. Tripod Jacket Structural Drawing (left) and Skid Frame Drawing (right).

Table 1. Jacket and Skid Frame Structural Data

Description	**Detail**	**Description**	**Detail**
Number of Jacket Leg	3	**Number of Skid Frame Support**	4
Jacket Elevation[a]	3 (EL: (+) 89.448 ft, (+) 50.302 ft, (+) 14.531 ft	**Skid Frame Dimension**	
Leg Diameter	40" x 1.000" WT	ENS355	UB1016x305x494
Jacket Join Can[b]		ENS355	H-800x300x14x26
at EL. (+) 89.448 ft	40" x 1.500" WT	API 5L GR. X52	Ø25"x1,750"
at EL. (+) 50.302 ft	40" x 1.250" WT	API 5L GR. X52	40"x1,000"
at EL. (+) 14.531 ft	40" x 1.250" WT	ASTM A-572 Gr. 50	PL 1"
Leg Batter	1:1 (at Row A, 1); 12:7 (at Row B, 2, 3)	**Material Type**	ENS355 (F_y = 51.48 ksi); API 5L GR. X52 (F_y = 52 ksi); ASTM A-572 Gr. 50 (F_y = 50 ksi)
Material Type	API 2H GR. 50 (F_y = 50 ksi); API 5L GR. X52 (F_y = 52 ksi)		

[a] Elevation (±) 0 ft is calculated from barge deck.
[b] All jacket joint can are constructed by API 5L GR. X52 material type, except for the joint leg.

The other structural data, such as installation equipment and non-generated structures also need to be considered. To simplify the modelling process, those structures are considered as a load in this analysis. The loads and the location of those structure is referred to the drawing.

3.1.2 Environmental Data

Environmental data refers to wind speed measurements at the loadout location of the structure. The wind data is derived from a one-year return period of operational conditions, collected at a reference height of 32.8ft above mean sea level, as specified by API RP 2A WSD [13]. For this study, a wind speed of 55.77 ft/s is used, in accordance with company specifications [5]. The wind loading directions on the tripod jacket structure during loadout are based on 12 wind directions.

3.2 Model Validation

Figure 3. Structural Model of an Integrated Tripod Jacket with Skid Frame.

The model is created based on the structural drawings provided by PT XYZ. The tripod jacket structure and skid frame are modelled using SACS software. At this stage, the self-weight and the loads acting on the structure (non-generated structure load, installation equipment load, and environmental load) are also included. In the modelling of the jacket structure, the central coordinates (x=0 and y=0) are set at the support point of jacket leg A1 (see Figure 2), and the elevation of the skid frame support point is positioned according to field conditions (Z=9.57 ft from MSL). Figure 3 shows the results of the structural modelling.

After modelling, structural validation is performed to ensure the accuracy of the modelling compared to the actual structure. Validation is conducted by comparing the weight and CoG of the structure using the MAPE (Mean Absolute Percentage Error) rule. The validation results for the weight and CoG of the structure are shown in Table 2. These results have values below 10%, indicating that the modelling has a high degree of conformity with the original model.

Table 2. Structural Model Validation

Description	Net Weight (kips)	Gross Weight (kips)	Centre of Gravity (ft)		
			X	Y	Z
Jacket WCR Rev.1	471.23	518.35	-0.38	-20.6	49.2
SACS	484.97	533.47	-0.39	-21	46.97
Difference	2.92%	2.92%	2.63%	1.84%	4.53%

3.3 Global Analysis

Global analysis can be performed after the structure model has meet the MAPE validation criteria according to the WCR report. This analysis is performed to check the structural integrity of the jacket structure along with the skid frame and obtain the UC value for each member. Besides, the member deflections need to be checked according to the criteria applied. Lastly, to ensure the soil in the loadout location have a sufficient capacity to support the process, the support reaction in each skid frame support need to be checked.

There are some conditions need to be considered, which are mentioned in section 2.1. about loadout scenario. The considered conditions are generated as a load conditions, which consist of combined basic loads. However, in purpose to obtain the stability of the analysis, load conditions consist of combined CoG shift that shifting to where the loss of support occurred are not being considered.

3.4 Local Analysis

Global analysis using SACS software is used to analyse structure consist of beam elements, such as the jacket structure. However, the skid frame structure used in this paper is constructed by combining beam and plate elements, which the analysis result will be more accurate if it is modelled as a solid element. Therefore, such local analysis using finite elements method is used to analyse the skid frame structure as a solid element. This local analysis will be performed in

4. Result and Discussion

4.1 Global Analysis

The analysis was started with the CoG shift scenario. This scenario aims to check the structural strength of the jacket and skid frame, if there is a shifting of CoG in the fabricated structure. The scenario was arranged according to the company specifications, as described in point 2.1.1. The CoG envelope for this structure has a length and width of 2.96ft and 2.57 ft, respectively, as shown in Figure 4.

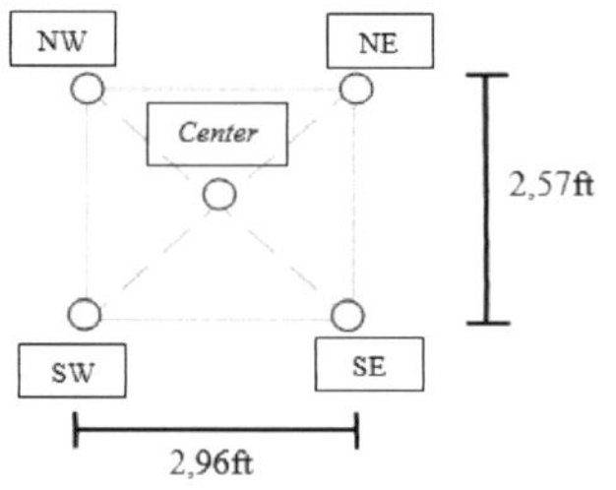

Figure 4. CoG Envelope Area for CoG Shift Scenario.

The CoG envelope then created some new CoG that shifted from the initial CoG to the corners of the CoG envelope, which its coordinates are shown in Table 3. Using Equation 1 and Equation 2, shift factor was calculated and resulting a value of αx=1.0076 and $\alpha y = 1.2304$. Subsequently, those factors are inserted as structure's load factor to simulate the CoG shift conditions in analysis.

Eventually, the CoG shift scenarios are combined with the ongoing and the initial pulling scenarios. This must be combined because the CoG shift condition has a quite significant impact in overall loadout process, which might be endangered the structure's integrity.

Table 3. CoG shift Coordinates

CoG Positions	Coordinates (ft)	
	Xc	Yc
CoG-Center	-0.3927	-20.9986
CoG-SW	-1.4008	-22.2290
CoG-SE	0.6155	-22.2290
CoG-NE	0.6155	-19.7682

The ongoing conditions scenario was arranged as described in point 2.1.2. This analysis was performed in SACS software using gap elements to simulate the loss of support conditions. Besides, deflection scenarios are simulated in this analysis by adding it in the model file. The combination of loads that used in ongoing analysis are environmental conditions, CoG shift, and loss of support or support deflection conditions.

Table 4. Highest UC Value of Structure's Members in Ongoing Condition.

No.	Description	Member Group	Member	Load Comb.	UC	Check
Jacket Structure						
1	Mudmat Primary Member	MM1	M012-201L	113B	0.91	OK
Skid Frame Structure						
2	Skid Frame SK04 Support	SK3	SK04-SK14	143B	0.90	OK

After the load combinations had been created and inserted into the model, the analysis was performed. The analysis results of members that have the highest UC are shown in Table 4.

According to the API RP 2A WSD [13], The UC value must be equal to or less than 1 as a representative of the actual stress occurring in the structure doesn't exceed the allowable stress. Since members which have the maximum UC value didn't exceeding 1, it can be concluded that the structure has a sufficient strength to perform in ongoing condition.

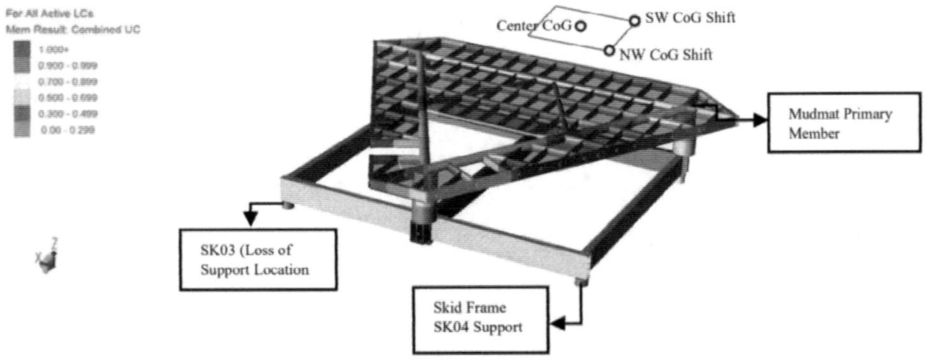

Figure 5. Critical UC Location of Structure's Members in Ongoing Condition.

Figure 5 shows the detail location of critical member, which have the maximum UC value. The maximum UC value for jacket structure has occurred at mudmat primary member during loss of support at SK03 with CoG shift to SW. This can be happened because the loss of support at SK03 combined with CoG shift to SW. This condition makes the area across it must carry more loads, which should be carried by SK03. The same principle also applies as the causes of maximum UC value at the skid frame SK04 support during loss of support at SK04 combined with CoG shift to NW condition.

Furthermore, the initial pulling conditions was arranged. The pulling loads was calculated using Equation 3 as an input for the analysis. The pulling load was simulated for two pulling directions, forward and backward which combined with CoG shift scenarios. Table 5 shows the members which have the maximum UC value. Those UC values are still satisfied the requirement applied since the values didn't exceed 1.

Table 5. Highest UC Value of Structure's Members in Initial Pulling Condition.

No.	Description	Member Group	Member	Load Comb.	UC	Check
Jacket Structure						
1	Mudmat Primary Member A	MM1	M012-201L	31XP	0.62	OK
2	Mudmat Primary Member B	MM1	202L-M015	32XN	0.62	OK
Skid Frame Structure						
1	Skid Frame SK03 Support	SK3	SK03-SK13	34XP	0.80	OK
2	Skid Frame SK04 Support	SK3	SK04-SK14	34XN	0.80	OK

It can be seen in Figure 6 the location of members which have a maximum UC value. The jacket structure has two members with the same value. The difference found at the load combination. The mudmat primary member A have the maximum UC when pulled from (+) direction combined with SW CoG shift, while the maximum UC in mudmat primary member B occurred when pulled from (-) direction combined with SE CoG shift. The skid frame structure also has two members with the same value the sam CoG shift direction. The difference found at the load combination. The skid frame SK03 support have the maximum UC when pulled from (+) direction while the skid frame SK04 support have th maximum UC when pulled from (-) direction.

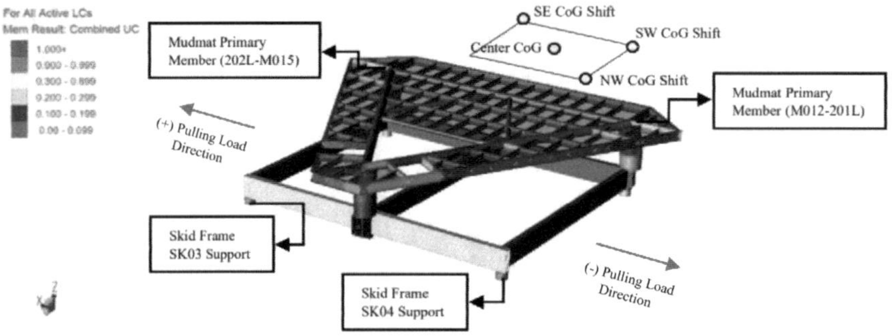

Figure 6. Critical UC Location of Structure's Members in Initial Pulling Condition.

4.2 Local Analysis

The local analysis was performed for the critical load combination for the skid frame structure according to Table 4. The skid frame was modelled as a solid structure and the jacket structure was modelled as a load obtained from the jacket legs joint reaction of previous global analysis. The loads are consisted of forces and moments, which the value can be seen in the Table 6.

Table 6. Force and Moment values Applied on Skid Frame.

Load Cond.	*Joint*	Force (N) X	Force (N) Y	Force (N) Z	Moment (N.m) X	Moment (N.m) Y	Moment (N.m) Z
	Leg 3B	114002	-26164	-574984	334456.53	24983.34	55343.95
143B	Leg 2B	31904	-108732	-401431	-84377.21	-293491.85	-55468.59
	Leg 1A	-28011	-6930	-890058	-70344.76	20977.12	2023.91

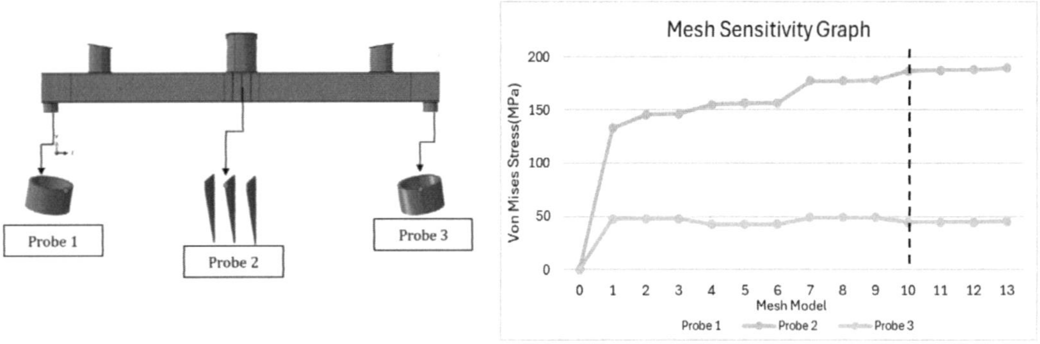

(a) (b)

Figure 7. Probe Nodes Location (left) and Mesh Sensitivity Graph (right).

IOP Conf. Series: Earth and Environmental Science **1461** (2025) 012014 doi:10.1088/1755-1315/1461/1/012014

Subsequently, a mesh sensitivity analysis was conducted by modelling various mesh configurations with progressively reduced element sizes. Concurrently, the stability of the von Mises stress was analysed at three sample points (probe nodes) for each model. It was found that the stress values began to stabilize with total 123,477 elements, which it has an element size of 4.72 inches for the coarse region and 1.15 inches for the critical region. The mesh sensitivity analysis graph and the location of probe nodes can be seen in Figure 7. The consistency of the graph expresses a more accurate results [14].

Thereafter, structural analysis of the skid frame was performed. The results of the local structural analysis of the skid frame shown the von Mises stress value located in each component of the skid frame. The analysis indicates there are four components do not meet the allowable criteria specified by the AISC 9th Edition [15], with the values detailed in Table 7 and their locations shown in Figure 8. Moreover, two of these four components exhibit stresses exceed the ultimate tensile strength of the material, which indicates failure.

Table 7. Maximum Von Misess Stress of Skid Frame's Members.

Description	Max Von Mises Stress (MPa)	Material	*Yield Strength* (MPa)	*Allowable Stress* (MPa)	*Allowable Check*	*Ultimate Strength* (MPa)	*Ultimate Check*
SK01-4	575.60	API 5L X52	360	316.01	NO	570	NO
SK05-1	522.20	EN S355	355	311.62	NO	630	OK
SK06-3	414.50	EN S355	355	311.62	NO	630	OK
Plat SK01-4	542.50	ASTM A-572	345	302.84	NO	450	NO

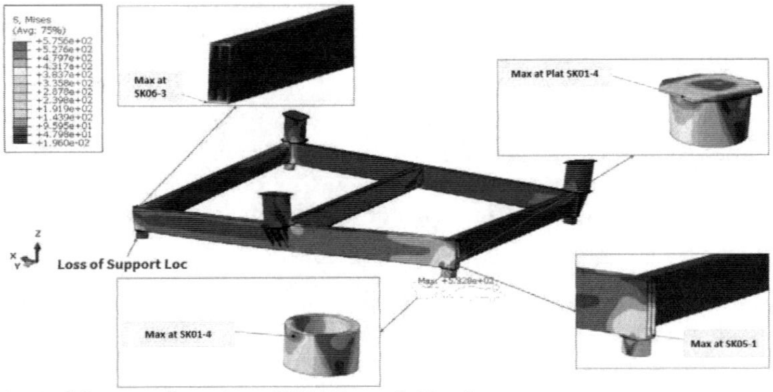

Figure 8. Critical Stresses Location of Skid Frame's Members.

5. Conclusion

According to the result of the loadout analysis, it can be concluded that the jacket structure has a sufficient strength to be loaded out with a maximum UC value of 0.91 occurred at mudmat primary member. Also, the global analysis of the skid frame structure has gave a satisfactory result, with a maximum UC value of 0.90 occurred at SK04 support. However, the skid frame structure was analysed locally due to the model limitations in global analysis. The result of local analysis has shown that there are four components exceed the allowable criteria, even two of them are exceed the ultimate criteria. Therefore, a rearrangement or redesign of the skid frame structure must be conducted.

Acknowledgement

The authors gratefully acknowledge financial support from the Institut Teknologi Sepuluh Nopember for this work, under project scheme of the Publication Writing and IPR Incentive Program (PPHKI) 2024.

References

[1] L. Y. Cheung and K. G. Foong, "Design considerations of a loadout skid frame for a 14,000-ton upper hull structure," *IES J. Part A Civ. Struct. Eng.*, vol. 1, no. 1, pp. 83–95, 2008, doi: 10.1080/19373260701620287.

[2] M. M. and H. E. H. B. Mohd Fairuz bin Mohd Idros, Johann Fruylan Infante, and Mohd Sharizal bin Mohd Nor, "Loadout of Malikai Tension Leg Platform TLP and Float-Over Topside," in *Offshore Technology Conference Asia*, Kuala Lumpur, Malaysia: Offshore Technology Conference This, 2018.

[3] S. Sabu and T. P. Somasundaran, "Analysis of Bombay High Jacket during Load out and Transportation," *Appl. Mech. Mater.*, vol. 857, pp. 41–46, 2016, doi: 10.4028/www.scientific.net/amm.857.41.

[4] R. Y. G. Kautsar, R. W. Prastianto, M. J. Rifqy, and R. F. Farhana, "Analysis of a Lower Deck Rigging Configuration and Reusable Padeye Design based on Finite Element Model for Improving Stacking Operation Efficiency," *IOP Conf. Ser. Earth Environ. Sci.*, vol. 1081, no. 1, 2022, doi: 10.1088/1755-1315/1081/1/012043.

[5] "Guidance on Load-out, Sea-fastening, Transportation, and Installation of Offshore Structures," Company Specification, 2014.

[6] H. Werkle, "Finite elements in structural analysis: Theoretical concepts and modeling procedures in statics and dynamics of structures," *Springer Tracts in Civil Engineering*, 2022, doi: 10.1007/978-3-030-49840-5.

[7] E. Oñate, Ed., "Introduction to the Finite Element Method for Structural Analysis BT - Structural Analysis with the Finite Element Method: Linear Statics," *Dordrecht: Springer Netherlands*, pp. 1–42, 2009, doi: 10.1007/978-1-4020-8733-2_1.

[8] R. W. Prastianto, Y. Hagie, S. Nahar, and Y. S. Hadiwidodo, "Local Analysis of The Padeye-Brace Clamp Structure Strength during Installation of Riser-Spool at Offshore Platform," *International Journal of Offshore and Coastal Engineering*, vol. 7, no. 2, pp. 77–83, 2023.

[9] A. Jahanpour and M. Jegarvand, "Elastic design of skid beams in offshore load-out systems," *Mar. Struct.*, vol. 84, Jul. 2022, doi: 10.1016/j.marstruc.2022.103228.

[10] T. Ribeiro, L. Bernardo, and R. Carrazedo, "Finite Element Modelling with Abaqus and Tosca for Topology Optimization of Steel," *Jounarl Build. Des. Environ.*, vol. 2, no. 1, pp. 0–0, 2023, doi: 10.37155/2811-0730-0201-3.

[11] F. Syalsabila, R. W. Prastianto, Y. S. Hadiwidodo, A. H. Adaalah, and M. R. Syarifudin, "Structural Analysis of Floating Net Cage Bracket in Current and Wave," *IOP Conf. Ser. Earth Environ. Sci.*, vol. 972, no. 1, 2022, doi: 10.1088/1755-1315/972/1/012017.

[12] I. Jong and W. Springer, "Gc 2009-267 : Teaching Von Mises Stress : From Principal Axes," *American Society for Engineering Education*, 2009.

[13] "Recommended Practice for Planning, Designing and Constructing Fixed Offshore Platforms—Working Stress Design," American Petroleum Institute, 2007.

[14] R. W. Prastianto, Y. S. Hadiwidodo, and F. I. Fasyin, "Stress concentration factor of a two-planar double KT tubular joint due to in-plane bending loading in steel offshore structures," *MATEC Web Conf.*, vol. 177, 2018, doi: 10.1051/matecconf/201817701006

[15] "ANSI/AISC 360-16 Specification for structural steel buildings." American Institute of Steel Construction, 2016.

IOP Conf. Series: Earth and Environmental Science **1461** (2025) 012015 doi:10.1088/1755-1315/1461/1/012015

Topology Optimization of Mudmat Design Based on Bottom Stability Analysis of a Three-Legs Jacket Structure

Zhahara Suci Pratiwi[1], Rudi Walujo Prastianto*[1], and Daniel Mohammad Rosyid[1]

[1] Department of Ocean Engineering, Institut Teknologi Sepuluh Nopember, Surabaya, Indonesia

*E-mail: rudiwp@oe.its.ac.id

Abstract. The initial stage in the installation of a jacket structure involves the jacket floating towards the seabed surface before being permanently anchored with piles. During this pile-free stage, the mudmat serves as a critical temporary foundation to maintain the stability of the jacket structure. On Bottom Stability (OBS) analysis is essential at this stage to ensure the structure's stability. The structural footing component integral to the OBS analysis is referred to as the mudmat. Given the mudmat's crucial role as a temporary foundation, its design and dimensions must be carefully considered in the structural design process. Consequently, topology optimization is performed to optimize the load distribution path and dimensions, thereby achieving the optimum weight. In this study, mudmat topology optimization was conducted locally by reducing the material distribution to minimize strain energy. The optimization results were then interpreted to derive a feasible design. The study yielded two optimized mudmat designs with material distributions of 30% and 20%. These designs exhibited a smaller Factor of Safety (FoS) compared to the initial design. Specifically, the FoS reduction in the OBS analysis for bearing criteria in designs 1 and 2 was 59% and 50%, respectively, overturning 43% and 21%, and sliding 67% and 59%. Nevertheless, both designs met the OBS criteria, with FoS values exceeding the thresholds of >2 and 1.5. Results show reduction structural weight up to 40% and 29%, for design 1 and 2, respectively.

1. Introduction

The jacket structure is a commonly used offshore structure for drilling and production, serving as a critical component in infrastructure projects within marine environments. During installation, environmental loads must be considered, particularly in the early stage when the jacket floats towards the seabed before pile installation. Stability is critical at this stage, with the footing component playing a key role in maintaining on-bottom stability [1]. In some installations, mudmat failure has led to instability in the jacket structure, highlighting the importance of advanced engineering solutions [2]. Soil failure occurs when the soil's bearing capacity is insufficient to support the structure's load during the OBS phase, making stability analysis essential to prevent shear failure and settlement [3].

Research on On-Bottom Stability has been conducted on various structures, including Jack-Up, Moveable Jacket, and Jacket designs, focusing on mudmats that meet stability criteria across different geographical regions by considering soil types and marine environmental conditions. These studies have resulted in stable designs for various locations [4]. However, full mudmat designs often exhibit high safety factors, well above thresholds, but are typically conservative and large to ensure safety, leading to increased manufacturing and installation costs due to added structural weight [5][6].

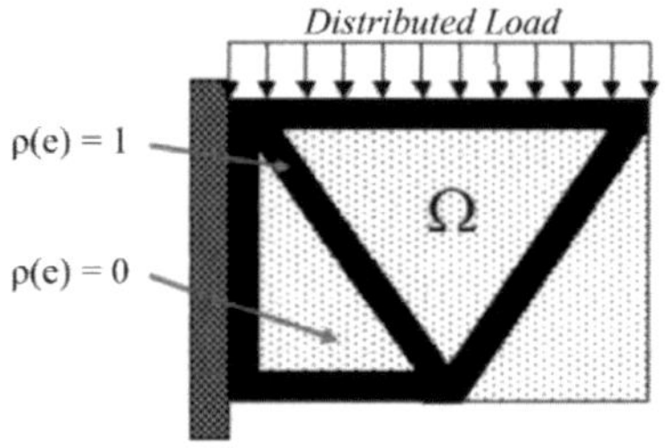

Figure 1. Optimized Material Layout Visualization [15]

Optimization processes are essential for maximizing or minimizing objective functions within specific constraints, leading to enhanced structural performance and cost reduction [7]. In this study, topology optimization was conducted using the SIMP method in ABAQUS software, as recommended in [8], to minimize the weight of the mudmat structure. Design optimization is typically divided into size, shape, and topology optimization, with topology optimization being the most widely applied method. This approach not only optimizes material distribution but also generates stress distribution and geometric details that can lead to innovative and efficient designs [9][5]. Despite its potential, topology optimization for mudmat structures in tripod jackets has not been fully explored. This study aims to optimize the mudmat structure to improve load distribution and reduce weight. The results could serve as a reference for creating more efficient designs, with expected outcomes including optimized mudmat structures, reduced maximum stress, and overall weight reduction. SACS software will be used for modeling and on-bottom stability analysis, while ABAQUS will be utilized for stress analysis and topology optimization.

2. Theoretical

2.1 On Bottom Stability

2.1.1 Bearing

This analysis is conducted to assess the soil's bearing capacity in supporting the total vertical forces against gravity and environmental loads, such as waves and currents. The soil must be able to sustain the loads from the structure placed on it without experiencing shear failure or settlement [3]. The calculation of soil bearing capacity can be done by determining the pressure at the base of the mudmat foundation. The pressure calculation at the foundation base is based on Skempton's formula.

The ultimate bearing capacity refers to the soil's ability to withstand the maximum load without failure, ensuring that the mudmat does not exert excessive pressure on the underlying soil. This capacity is calculated using Skempton's equation (Equation 1). After determining the maximum soil bearing capacity and pressure on the mudmat, the allowable bearing capacity is

IOP Conf. Series: Earth and Environmental Science **1461** (2025) 012015 doi:10.1088/1755-1315/1461/1/012015

calculated using Equation 3. The minimum Factor of Safety (FoS) for bearing capacity is 2.0 for the jacket's dead weight without environmental loads and 1.5 when environmental loads are considered [10].

$$q_u = 5.14S_u\left(1+\frac{0.2D}{B}\right)\left(1+\frac{0.2B}{L}\right) \quad (1)$$

Where,
q_u = Maximum soil bearing capacity or ultimate bearing capacity (kips/ft2)
S_u = Undrained shear strength (psf)
D = Mudmat depth (ft)
B = Mudmat width (ft)
L = Mudmat length (ft)

$$FoS = \frac{q_u}{q}; FOS \geq 2 \text{ and } 1.5 \quad (2)$$

2.1.2 *Sliding*

The analysis evaluates the lateral forces on the structure due to environmental loads resisted by the soil around the mudmat area. If these lateral forces exceed the sliding resistance, the structure may fail. Sliding stability is assessed when the structure is subjected to horizontal forces at the foundation base. The lateral force is calculated using Equation 4, and the sliding or shear resistance is determined using Equation 5.

$$H_E = \sqrt{F_x^2 + F_y^2} \quad (3)$$

H_E = Lateral force acting on the base of the foundation (kips)
F_x = Force from the x-axis acting on the base of the foundation (kips)
F_y = Force from the y-axis acting on the base of the foundation (kips)

$$H = c_u A \quad (4)$$

H = Sliding Resistance or shear force resistance (kips)
c_u = Shear force at undrained shear strength or at base level (kips/ft^2)
A = Area of mudmat foundation (ft^2)

The allowable factor of safety for sliding can be calculated using Equation 6. The minimum factor of safety (FoS) for sliding stability is 1.5 [10].

$$FoS = \frac{H}{H_E}; FOS \geq 1.5 \quad (5)$$

2.1.3 *Overturning*

Overturning analysis evaluates a structure's ability to maintain stability by ensuring the jacket's weight counteracts the overturning moment from environmental loads. For stability, the

Table 1. General Structural Criteria Data

Description		Detail
Water depth	:	54.46 Feet
Elevation[a]	:	3 EL = (+) 19.79 ft; EL = (-) 19.21 ft; EL = (-) 54.46 ft
Jacket legs Properties	:	OD 40.00" x 1.00" WT
Leg Batter	:	1 : 1 at Row A1 12 : 7 at Row B2 and B3
Jacket Material Type	:	API 2H GR. 50 (F_y=50 ksi)
Mudmat Properties	:	Plate; W10X26; W14X30; W18X50
Mudmat Material Type	:	EN10025 S355J2 and API 5L Grade B

[a]Elevation (±) 0 ft is calculated from MSL

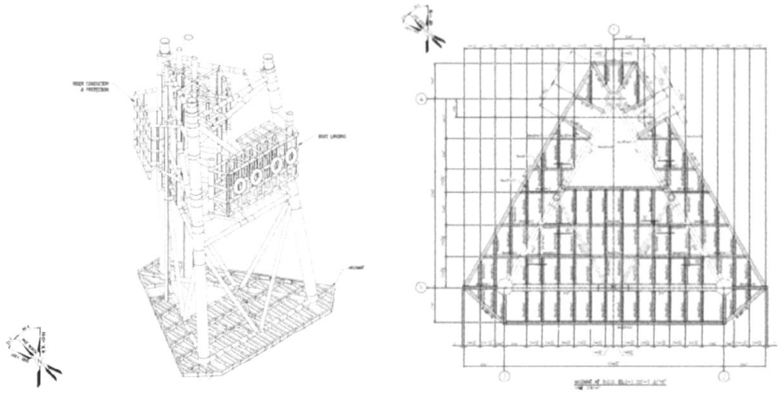

(a) (b)

Figure 2. (a) Jacket Structural Drawing; (b) Mudmat Structural Drawing

restoring moment must exceed the overturning moment during installation. The restoring moment from the payload is calculated using Equation 7, while the overturning moment from lateral loads is determined using Equation 8.

$$M_p = P \times a \tag{6}$$

Where,
M_r = Restoring moment (kips.ft)
P = Load on mudmat (kips)
a = Distance of return arm (feet)

$$M_o = F_w \times h \tag{7}$$

Where,
M_o = Overturning moment (kips.ft)
F_w = Lateral load (kips)
h = Distance from the CoG of the jacket to the mudline (ft)

The allowable factor of safety for overturning can be calculated using Equation 9. The minimum factor of safety (FoS) for overturning stability is 1.5 [10].

$$FOS = \frac{M_p}{M_o}; FOS \geq 1.5 \tag{8}$$

2.2 Topology Optimization

This optimization is applied during the design phase to predict the optimal material distribution within a structure while accounting for specifications and constraints. ABAQUS software is used for the analysis due to its effectiveness in topology optimization. The Solid Isotropic Material with Penalization (SIMP) method, used in 83% of optimization software [11], was chosen for its widespread availability in commercial software [12] and its advantages in convergence and computational efficiency compared to other methods [13].

The SIMP (Solid Isotropic Material with Penalization) method optimizes material distribution within a predefined design space by considering specific load cases, boundary conditions, manufacturing constraints, and performance requirements. As stated by Bendsøe [8], "shape optimization, in its most general form, involves determining whether material should be present or absent at each point within the design space." This approach allows for a systematic reduction in material usage while maintaining structural integrity, making it a widely used method in topology optimization.

Table 2. Environmental Data

Description	Unit	Operating 1-Year
Mean Sea Level (MSL)	(ft)	54.46
Total Tide	(ft)	0.16
Maximum Wave Height (Hmax)	(ft)	10.50
Maximum Wave Period (Tmax)	(s)	6.70
Maximum Wave Length	(ft)	237.23

In this approach, the domain is discretized into a grid of finite elements, where each element is either filled with material or left void. The material density in the design domain, ρ, is binary:

- ρ(e) = 1 where material is needed (black)
- ρ(e) = 0 where material is removed (white)

Figure 1 shows the optimized material layout for a loaded beam, where solid elements (ρ(e) = 1) are black, and void elements (ρ(e) = 0) are removed. The objective function typically maximizes stiffness by minimizing the global elasticity of the structure, as expressed in Equation 10 [14].

$$\min C(\{\rho\}) = \sum_{e=1}^{N} (\rho_e)^p [u_e]^T [K_e][u_e] \tag{9}$$

Where,
(ue) = Nodal displacement vector of element e
(Ke) = Stiffness of element e
(ρ) = Relative densities of elements

3. Methodology

3.1 Data collection

3.1.1 Structural Model

In this study, a wellhead platform-type offshore platform is used as the model. The platform is a tripod jacket structure sourced from a final project report, with the structural drawing shown

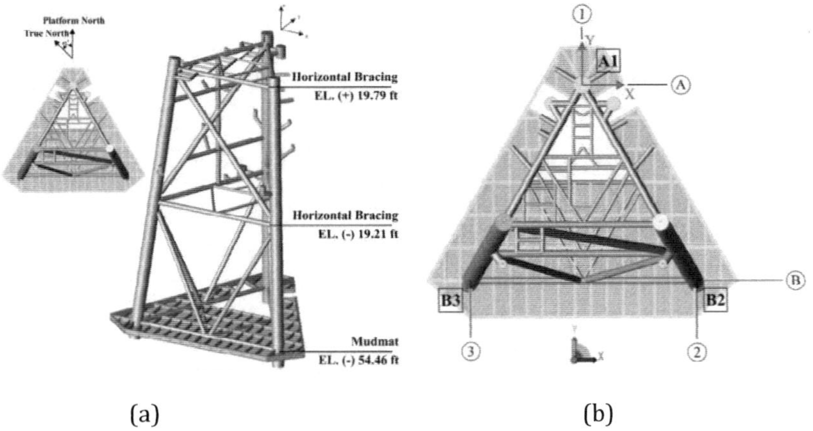

Figure 3. (a) Jacket Structural Model on SACS; (b) Global Axes System of Jacket Model

Table 3. Structure Validation

Aspect	Description
Objective Function	Minimize strain energy
Design Variable	Material distribution
	Mudmat thickness
Constraint	Mudmat size
	Support

in Figure 2(a). Additionally, the mudmat structure is also modeled based on the structural drawing, as depicted in Figure 2(b). The general structural criteria data for the tripod jacket structure are presented in Table 1. Other structural elements, such as installation equipment and non-modeled components, are considered as loads in this study for simplification purposes.

3.1.2 Environmental Data

The environmental data is based on a one-year return period of operational conditions. This includes water depth, wind, wave, and current data for the tripod jacket structure, as presented in Table 2. The loading direction for the structure, which consists of three legs, is distributed across 12 directions.

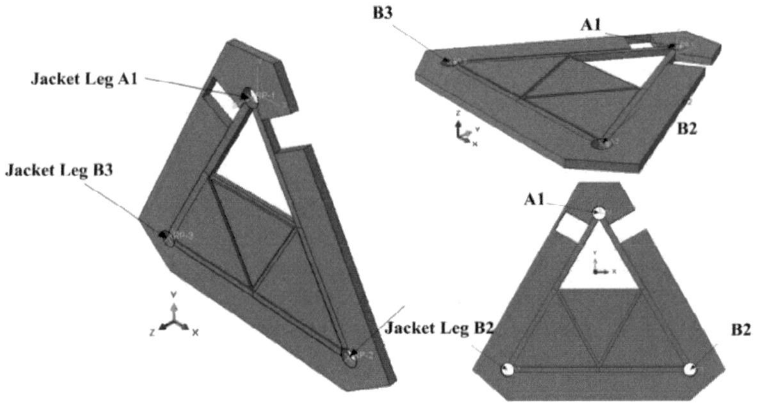

Figure 4. Mudmat Modelling on ABAQUS

Table 4. Structure Validation

Description	Structure Weight	MAPE
SACS Model	279.73 Kips	2.32%
Weight Control Report	286.36 Kips	

3.2 Validation Model

The modeled structure is a tripod jacket, excluding conductors, piles, topside, and substructure components like boat landings and riser guards (Figure 3a). Structural modeling in SACS aims to obtain reaction forces from environmental and structural loads. The global axis system is shown in Figure 3(b), with the Y-axis pointing north, the X-axis east, and the Z-axis upward.

To ensure the model's accuracy, the structural weight is validated using the Mean Absolute Percentage Error (MAPE), which compares predicted values with actual data and expresses the error as a percentage. MAPE is calculated by averaging the absolute differences between predicted and actual values, divided by the actual values, and multiplying by 100. A MAPE below 10% indicates high accuracy, though it may be less effective when actual values are very small, as this can skew the results. Overall, a lower MAPE reflects a more accurate model.

3.3 On Bottom Stability Analysis

On-bottom stability analysis is conducted using two software tools: SACS and MS Excel. In SACS, the structural model is configured with environmental loads to generate reaction forces. These reaction forces are needed for the on-bottom stability calculations performed in MS Excel. The calculation is based on the equation section 2.1. For both bearing and sliding criteria, the effective mudmat area is required, which is obtained through eccentricity calculations. Eccentricity is the distance from the center of a foundation to the point where the resultant force acts, measured parallel to the contact plane between the foundation and the soil. Once the eccentricity is determined, the effective area can be calculated using the output from the massprop command in AutoCAD. Additionally, AutoCAD can also provide the length (L') and width (B') of the effective mudmat area using the dimension command.

Table 5. Sliding Criteria Results

Load Case	Environment Load Dir. (deg)	H_E (psf)	H (psf)	FoS	Remarks
	Minimum Water Depth				
S503	60	29.16	1212.05	41.56	OK
S504	90	28.56	1196.75	41.90	OK
S505	120	28.96	1217.42	42.04	OK
	Minimum FOS			41.56	OK
	Maximum Water Depth				
S603	60	28.8	1203.89	41.81	OK
S604	90	28.33	1186.29	41.87	OK
S605	120	28.54	1209.29	42.37	OK
	Minimum FOS			41.81	OK

3.4 Topology Optimization

The optimization process involved modeling the mudmat structure in ABAQUS software, following the technical drawings. The structure was modeled as a solid element, assuming the

Table 6. Bearing criteria results

Load Case	Environment Load Dir. (deg)	q (psf)	qu (psf)	FoS	Remarks
	Minimum Water Depth				
R502	30	269.91	5584.22	20.69	OK
R503	60	260.66	5571.24	21.37	OK
R504	90	263.94	5567.03	21.09	OK
	Minimum FOS			20.69	OK
	Maximum Water Depth				
R603	60	257.39	5571.32	21.65	OK
R604	90	260.66	5567.06	21.36	OK
R605	120	256.6	5572.62	21.72	OK
	Minimum FOS			21.36	OK

maximum physical area of the component based on the SIMULIA module. As a result, specific beam details of the mudmat and jacket structure were not modeled in ABAQUS, as shown in Figure 4. The jacket structure, which was excluded from the ABAQUS model, was considered in the load and moment analysis. The loads and moments from the jacket structure were derived from the maximum reaction forces obtained from the SACS output 'member end forces'

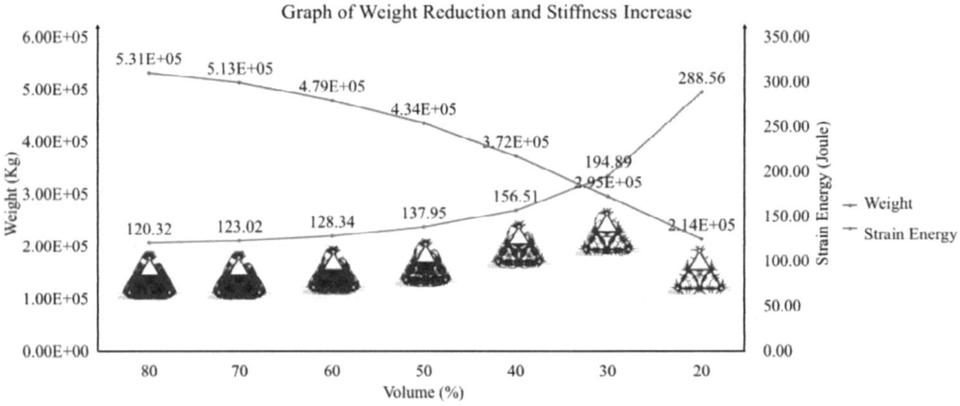

Figure 5. Graph of Weight Reduction and Stiffness Increase of Optimization Result Structure

In this study, topology optimization was conducted using the SIMP method. The penalty factor was adjusted from higher to lower values to evaluate its effect on the mudmat structure design. In SIMP topology optimization, the penalty factor (p) is crucial for controlling the final material density distribution. The objective function, constraints, and design variables used in this study are presented in Table 3.

3.5 Interpretation

The interpretation of topology optimization results is similar to traditional design processes, requiring engineering knowledge to translate optimization outcomes into feasible designs[12]. Interpretation approaches can be divided into three categories: image interpretation, density contour interpretation, and geometric reconstruction[15]. The image interpretation method uses graphical tools or computer vision to define the boundaries of black-and-white topology results. However, using the central line of primary material distribution as a baseline for arranging components may not always yield the best results due to irregular layouts and the presence of many low- to medium-density elements in the SIMP method [16]. To address this, a sizing phase simplifies mass estimation and ensures design consistency. In this study, an additional interpretation was conducted based on the thickness of the optimized topology, determined by mesh layers, to define appropriate material sizes. This approach improved the design process by refining the material distribution.

4. Result and Discussion

4.1 Validation

The structural weight data from the modeling is used for validation to ensure that the model accurately represents the actual structure, as shown in Table 4. Validation is performed by comparing the modeled structure's weight with the data using the MAPE (Mean Absolute

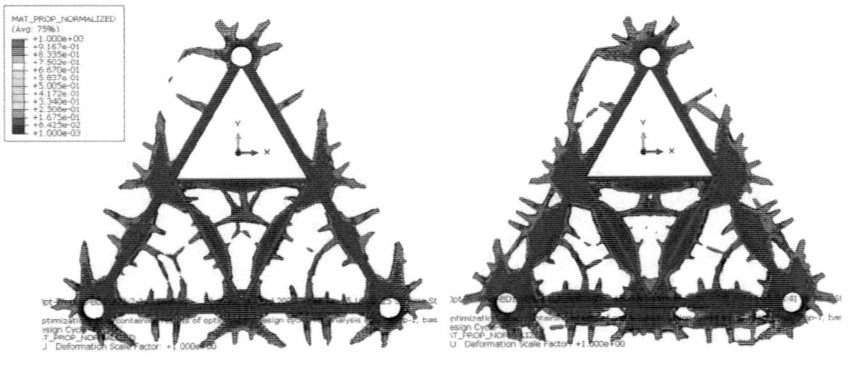

(a) (b)

Figure 6. Topology Optimization Results a) Vol. 20%; b) Vol. 30%

Percentage Error) criterion. The validation results for the tripod jacket structure model indicate a MAPE percentage of 2.32%. The percentage is below 10%, therefore the model qualifies as "highly accurate forecasting."

4.2 On Bottom Stability Result

The on-bottom stability analysis in this study followed Section 2.1. For the bearing criteria, the ultimate bearing capacity and mudmat pressure were used to calculate the Factor of Safety (FoS) based on Equation 3. If the SF exceeds 2, the structure is considered safe from bearing capacity failure. The analysis results in Table 5 show that the FoS for the tripod jacket model is greater than 1.5, meeting API RP 2A WSD standards. The lowest FoS, 20.69, occurs at minimum water depth with a 30-degree environmental load direction, indicating the structure is safe from bearing failure. The small difference in FoS at various depths is influenced by soil conditions and structural load.

After determining the sliding resistance and lateral forces, the safety factor can be calculated. The calculation of the FoS is based on Equation 6, and if the SF is greater than 1.5, the structure is considered safe from sliding failure. Based on the sliding analysis results in Table 6, the FoS for the tripod jacket model exceeds 1.5, meeting the API RP 2A WSD criteria. The lowest FoS occurs at the minimum water depth with an environmental load direction of 60 degrees. An FoS of 41.56 satisfies the criteria, indicating that the structure is safe from sliding failure.

For the overturning criteria, after obtaining the restoring moment and overturning moment, the FoS can be calculated. According to API RP 2A WSD, the FoS must be greater than 1.5. Based on the overturning analysis results shown in Table 7, the FoS for the tripod jacket model exceeds

Table 7. Overturning criteria results

Load Case	Environment Load Dir. (deg)	M_o (psf)	M_p (psf)	FoS	Remarks
	Minimum Water Depth				
V606	150	999.34	5780.9	5.78	OK
V607	180	904.74	6076.73	6.72	OK
V608	210	1042.23	6047.96	5.80	OK
	Minimum FOS			5.78	OK
	Maximum Water Depth				
V506	150	946.97	5853.72	6.18	OK
V507	180	847.29	6140.12	7.25	OK
V508	210	988.5	6111.79	6.18	OK
	Minimum FOS			6.18	OK

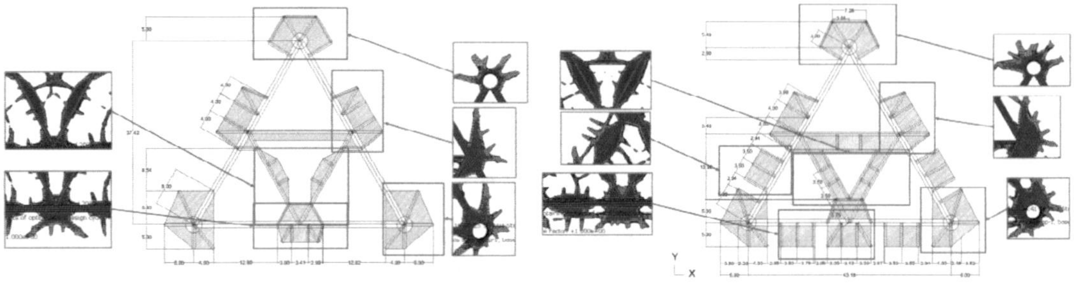

(a) (b)

Figure 7. Mudmat Design Structure Interpretation Process (a) Desain 1- Vol. 20% (b) Desain 2-Vol. 30%

1.5, meeting the API RP 2A WSD criteria. The lowest FoS occurs at the maximum water depth with an environmental load direction of 150 degrees. An FoS of 5.78 satisfies the criteria, indicating that the structure is safe from overturning failure.

4.3 Topology Result

Based on the results of the OBS analysis, the initial design structure meets the on-bottom stability parameters. However, the FoS obtained from the OBS analysis is still significantly higher than the standard specified by API RP 2A WSD, prompting an optimization process using the model shown in Figure 4. The optimization results for the mudmat structure are presented in Figure 5, which shows a reduction in the structure's weight in the ABAQUS modeling results. Meanwhile, the structural stiffness increased due to the topology optimization process. This occurs because the objective function of the optimization process was set to minimize strain energy. The topology optimization results produced two outcomes, showing material distribution and geometric details. These two results, representing optimizations with volume constraints of 20% and 30%, are presented in Figure 6.

4.4 Interpretation

In the interpretation process of the topology optimization results, most components are arranged using the centerline of the primary material distribution as a baseline. The interpretation of Design 1, with a volume optimization result of 20%, is presented in Figure 7 (a). Meanwhile, the interpretation of Design 2, with a volume optimization result of 30%, is shown in Figure 7 (b).

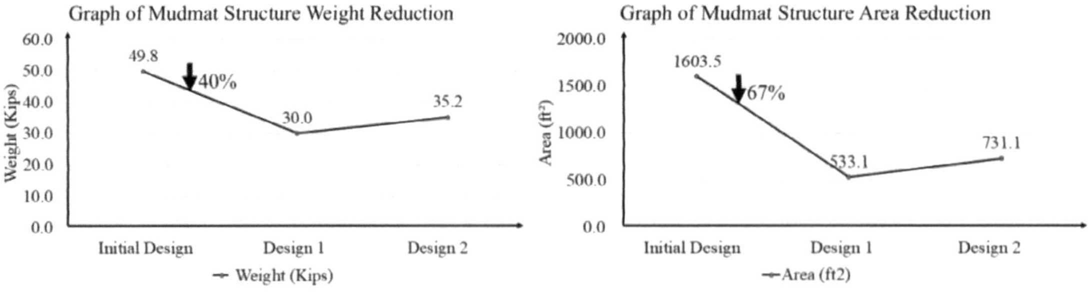

(a) (b)

Figure 8. Graph of Decrease (a) Weight of Interpreted Structure Result (b) Area of Interpreted Structure Result

4.5 On Bottom Stability on Interpretation Design

The topology optimization results revealed changes in the mudmat design, which affected the overall mudmat area. A graph showing the reduction in mudmat area is presented in Figure 8 (a). Additionally, by modeling in SACS, the structural weight after interpretation can be determined. A comparison of the structure's weight for Design 1 and Design 2 is shown in Figure 8 (b).

After interpreting and modeling the optimized mudmat structure design, a re-analysis of the OBS is necessary to ensure that the design meets the on-bottom stability parameters. The On-Bottom Stability analysis process for the optimized design was conducted in the same way as for the initial design. The results of the OBS analysis for the optimized design are presented in Figure 9. Based on this figure, it is evident that the structure still meets the on-bottom stability parameters. Additionally, a significant decrease in the FoS is observed from the initial design to Design 1.

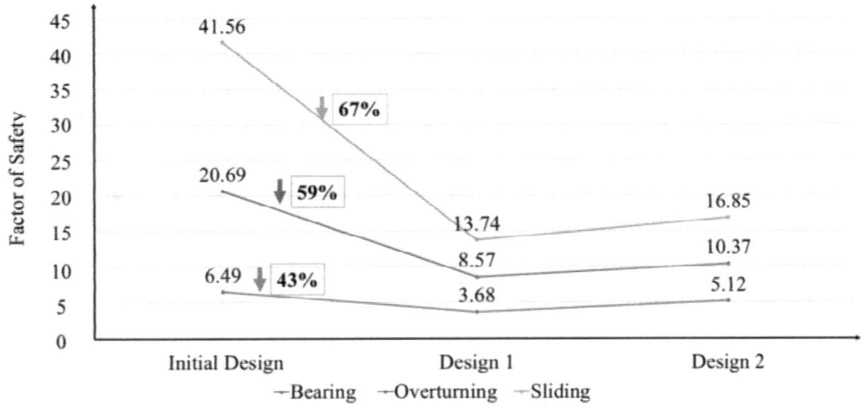

Figure 9. Graph of the Decrease in OBS FoS for the Interpreted Structure.

5. Conclusion

The research shows that the initial structure design meets on-bottom stability criteria, with a Factor of Safety (FoS) well above 2, ensuring stability before pile installation. Topology optimization produced two designs with improved material distribution and geometric details, reducing structure weight, mudmat area, and Unity Check Ratio. The optimized designs maintained an FoS greater than 2, confirming stability during the on-bottom phase.

1. OBS analysis of the initial tripod jacket design satisfies the criteria for bearing, overturning, and sliding, with the lowest FoS values being 20.69 for bearing, 5.78 for overturning, and 41.56 for sliding.
2. Topology optimization reduced material distribution by 30% and 20%, with corresponding decreases in OBS FoS—59% and 50% for bearing, 43% and 21% for overturning, and 67% and 59% for sliding—while remaining above the required thresholds.
3. Design 1 and Design 2 are recommended, achieving mass reductions of 40% and 29%, respectively, without compromising stability.

Acknowledgement

The authors gratefully acknowledge financial support from the Institut Teknologi Sepuluh Nopember for this work, under project scheme of the Publication Writing and IPR Incentive Program (PPHKI) 2024.

References

[1] T. H. Dawson, *Offshore Structural Engineering*. Michigan University: Prentice-Hall, 1983.
[2] R. Mahanta and R. K. Ghanekar, "Foundation failure and instability of an offshore jacket structure during installation—a case study," *Lect. Notes Civ. Eng.*, vol. 92, pp. 207–219, 2020, doi: 10.1007/978-981-15-6832-9_10.
[3] J. E. Bowles, *Foundation Analysis and Design*. New York: Mc-Graw-Hill, 1996.
[4] A. A. Askira, "Analisis On Bottom Stability Struktur Platform Terpancang Dengan Konsep Berpindah," Surabaya : ITS Press, 2021.
[5] X. Tian, Q. Wang, G. Liu, Y. Liu, Y. Xie, and W. Deng, "Topology optimization design for offshore platform jacket structure," *Appl. Ocean Res.*, vol. 84, no. January, pp. 38–50, 2019, doi: 10.1016/j.apor.2019.01.003.
[6] W. Musial and B. Ram, "Large-Scale Offshore Wind Power in the United States," *Risk Anal.*, no. September, 2010.
[7] T. R. Belegundu, A. D., & Chandrupatla, *Optimization Concepts and Applications in Engineering (Second Edition)*. Cambridge: Cambridge University Press, 2011. doi: https://doi.org/10.1017/9781108347976.
[8] M. P. Bendsøe and O. Sigmund, *Typology Optimization ebook*. berlin: springer, 2004. [Online]. Available: https://link.springer.com/book/10.1007/978-3-662-05086-6#toc
[9] O. M. Querin, *Topology Design Methods for Structural Optimization*. London: Elsevier, 2017.
[10] API, "Recommended Practice for Planning , Designing and Constructing Fixed Offshore Platforms — Working Stress Design," 2007 [Online]. Available: http://scholar.google.com/scholar?hl=en&btnG=Search&q=intitle:Recommended+Practice+for+planning,+designing+and+constructing+fixed+Offshore+Platforms+-+Working+stress+design#0
[11] E. Tyflopoulos and M. Steinert, "A Comparative Study of the Application of Different Commercial Software for Topology Optimization," *Appl. Sci.*, vol. 12, no. 2, 2022, doi: 10.3390/app12020611.
[12] A. Dugré, A. Vadean, and J. Chaussée, "Challenges of using topology optimization for the design of pressurized stiffened panels," *Struct. Multidiscip. Optim.*, vol. 53, no. 2, pp. 303–320, 2016, doi: 10.1007/s00158-015-1321-1.
[13] J. D. Deaton and R. V. Grandhi, "A survey of structural and multidisciplinary continuum topology optimization: Post 2000," *Struct. Multidiscip. Optim.*, vol. 49, no. 1, pp. 1–38, 2014, doi: 10.1007/s00158-013-0956-z.
[14] Dassault Systèmes, "Topology and Shape Optimization with Abaqus." [Online]. Available: https://www.scribd.com/document/253719587/Topology-and-Shape-Optimization-With-Abaqus.
[15] Y. L. Hsu, M. S. Hsu, and C. T. Chen, "Interpreting results from topology optimization using density contours," *Comput. Struct.*, vol. 79, no. 10, pp. 1049–1058, 2001, doi: 10.1016/S0045-7949(00)00194-2.
[16] Y. Yu, M. Wei, J. Yu, Y. Cui, R. Gao, Z. Dong, and X. Wang, "Reliability-based design method for marine structures combining topology, shape, and size optimization," *Ocean Eng.*, vol. 286, no. P1, p. 115490, 2023, doi: 10.1016/j.oceaneng.2023.115490.

IOP Conf. Series: Earth and Environmental Science **1461** (2025) 012016 doi:10.1088/1755-1315/1461/1/012016

Enhancing Container Terminal Productivity by Eliminating Break Times Through Optimized Crane Operator Scheduling

Prabowo Budhy Santoso[1*], Haryo Dwito Armono[2], Raja Oloan Saut Gurning[3], and Danang Cahyagi[4]

[1]Interdisciplinary School of Management Technology, Institut Teknologi Sepuluh Nopember, Surabaya Indonesia
[2] Department of Ocean Engineering, , Institut Teknologi Sepuluh Nopember, Surabaya Indonesia
[3] Department of Marine Engineering, Institut Teknologi Sepuluh Nopember, Surabaya Indonesia
[4] Department of Mechanical Engineering, Politeknik Negeri Batam, Batam, Indonesia

[*]E-mail: 7032221013@mhs.its.ac.id

Abstract. Container crane operators play a crucial role in maintaining the productivity of container terminals. The presence of well-rested and efficient crane operators is essential for maximizing terminal performance. This study highlights a strong correlation between optimized scheduling of operator work hours and increased productivity. A key innovation in this approach is the elimination of equipment downtime that is typically caused by operator breaks. This is achieved through dynamic scheduling, which ensures that operators receive adequate rest while minimizing interruptions. As a result, both operator performance and safety during loading and unloading operations improve, leading to enhanced overall terminal efficiency. In addition to boosting productivity, the proposed scheduling method takes into account the sustainability of human resources. It aligns with broader policy considerations, socioeconomics, and legal regulations, particularly labor laws. Furthermore, this study addresses challenges related to infrastructure and sustainability in container terminal operations by optimizing resource use and improving operational efficiency. Various scenarios involving different numbers of operators and workload patterns are evaluated against container handling performance, while keeping the same equipment and operational conditions. The results show that, although increasing the number of crane operators and improving workload management leads to higher operational costs, it significantly enhances both productivity and capacity at the container terminal.

1. Introduction

In optimizing container handling safety, human operators is an important role, and there are many sources of work stress that can increase the risk of serious accidents (1,2). Individual factors contributing to work fatigue in container crane operators include age and work experience (3). The study correlates mental workload strongly with work fatigue in container crane operators, while age shows the weakest correlation (4). Fatigue levels among Container Crane operators with most experiencing low fatigue. Evening shift workers and those with lack of sleep tend to report higher fatigue levels (5,6). The study analyses operator fatigue and performance in container terminal ship-to-shore crane tasks, focusing on human factors and safety at port, highlighting the impact of fatigue on operator tasks. A study on operator fatigue revealed that crane operators can experience varying levels of fatigue, ranging from low, moderate, to high levels. The study also found a correlation between fatigue levels and nutritional status, where inadequate nutritional intake can accelerate the onset of fatigue in operators (7,8).

Container terminals located in river areas with naturally shallow draft conditions and limited land face significant challenges in achieving efficiency and maintaining a competitive advantage compared to ocean terminals (9,10). Berth allocation must take into account tidal conditions and water depth, requiring advanced scheduling techniques to optimize vessel assignments (11). Many river-based terminals are constrained by shallow waters and restricted land availability, limiting their ability to expand capacity and improve handling efficiency (12). A case study highlights the need for integrated planning in ports with limited space, balancing environmental concerns with functional and sustainable development requirements (13).

Addressing the issue of increasing container terminal capacity by eliminating operator break time is complex, but can be approached through automation and optimization strategies (14). Research shows that automation can significantly boost terminal efficiency, potentially reducing the necessity for operator breaks (15). However, rather than directly focusing on eliminating breaks, other studies emphasize optimizing processes, scheduling, and layout to enhance efficiency and throughput (16). One key challenge is in developing effective operator scheduling to minimize factors like inter-crane interference, which can substantially impact crane utilization and overall productivity (17,18). A research shows Fatigue can be alleviated through adequate rest, a well-balanced, nutrient-rich diet, and appropriate exercise. Additionally, rationalizing work schedules, promoting social interactions, and providing training on the importance of fatigue awareness can help reduce excessive mental workloads (19).

This study acknowledges that meeting the regulated working hours not only improves operator performance but also contributes to enhancing safety within the terminal operations.

2. Method

2.1 Method Explanation

To model the operational performance of container terminals with an optimized crane operator shift scheduling strategy, this study employs a discrete event simulation (DES) approach. DES is a modeling technique that simulates the behavior of a dynamic system by capturing changes in state at specific points in time, driven by discrete events. Each state change is governed by a sequence of events occurring at particular moments, rather than continuously. This approach is well-suited for representing the operational performance of systems with multiple performance parameters, including equipment efficiency, ship arrivals, loading and unloading processes, and operator working hours.

The study begins by establishing basic assumptions, particularly those related to container terminal specifications and supporting parameters. The next phase involves identifying operator break times at the terminal, which are targeted for optimization. Following this, operator scheduling is mapped based on government regulations governing working hours. This scheduling is then modeled using the DES approach to determine the optimal capacity the container terminal can accommodate.

The primary challenge in this research is to develop a scheduling system that enables continuous terminal operation while ensuring compliance with labor regulations. This system must also account for operator working hours to ensure that crane operators remain in peak condition, balancing operational efficiency with the well-being of the workforce.

3.2 Related Equation

The approach used in this study to calculate container terminal capacity is based on the equation provided by Notteboom in his work on container terminals (20). To determine the optimal capacity of a container terminal, several key parameters must be considered, including the number of berths, dock operating hours, loading and unloading productivity rates, and the occupancy rate permitted by the terminal and relevant regulatory bodies. Equation (1) illustrates the container terminal capacity calculation method.

$$C_b = n\,x\,\emptyset\,x\,t_{year}\,x\,P \tag{1}$$

C_b : Berth capacity
n : Number of berth
$\emptyset$: Acceptable berth occupancy rate (%)
n : Annual average productivity of vessel (conth/h)

3. Result and Discussion

This section will present several research approaches, beginning with the basic assumptions, identification of operator break times, mapping of operator work schedules, and modelling of working hours using discrete event simulation. Additionally, it will explore the impact of these scheduling adjustments on container terminal production capacity.

3.1 Basic Assumption

This study focuses on a case study of a container terminal situated in a river area characterized by limited land availability and restricted water draft conditions. Typically, to enhance container throughput, container terminals expand their operational areas and add equipment through development. However, due to the spatial and water constraints of this terminal, traditional expansion is not feasible. Therefore, improving container terminal performance must be achieved through optimizing operational efficiency. This optimization involves considering key factors such as operator performance and the terminal's allowable occupancy rate. Table 1 presents the basic assumptions used in the simulation modeling.

This section details the methodologies employed in the research, encompassing the foundational assumptions, identification of operator breaks times, mapping of operator work schedules, modelling of working hours through discrete event simulation, and assessment of container terminal production capacity after the implementation of operator scheduling adjustments.

This section outlines the various methodologies employed in the research, including the underlying assumptions, identification of operator break times, mapping of operator work schedules, modeling of working hours using discrete event simulation, and evaluating container terminal production capacity following the mapping of operator schedules.

IOP Conf. Series: Earth and Environmental Science **1461** (2025) 012016 doi:10.1088/1755-1315/1461/1/012016

Table 1. Basic assumption of container terminal

No	Basic Assumption	Value	Unit
1	Length of Wharf	270	meter
2	Container Yard	75.000	m^2
3	Draft	-6	meter
4	Quay Crane	2	unit
5	Yard Crane	8	unit
6	BCH	35 – 40	box/crane/hr
7	Ships Size	500	box/ship

In addition to the basic assumptions concerning container terminal specifications, it is essential to establish parameters regarding the number and performance of container terminal crane operators. Table 2 presents the assumptions related to the number of crane operators. To operate two quay crane units, the terminal employs 12 operators who work in rotating shifts, each with a daily working duration of 8 hours. The crane operators' schedules comply with national labor regulations, typically restricting work to 40 hours per week and mandating a 1-hour break per shift.

Table 2. Basic assumption of container crane manpower

No	Basic Assumption	Value	Unit
1	Number of QCC	2	Crane
2	Total Number of QCC Operator	12	Personel
3	Max Working Hour / Day	8	Hr/day
4	Normal Working Hour / Week	40	Hr/week
5	Min. Operator Break Time	1	Hr/shift

3.2 Cummulative of Operator Break Time

Crane Operator break time is an essential entitlement linked to working hours. The immediate elimination of break time cannot be implemented without first evaluating alternative approaches. Removing operator breaks not only heightens the risk of fatigue but also negatively impacts the performance of container crane operations. Typically, container crane operators have a one-hour break during the middle of their shifts. In a three-shift system, this amounts to a total of three hours of break time per day for the terminal. When accumulated over a year, these breaks represent over a thousand hours during which container cranes remain idle. To optimize container terminal capacity, this study proposes an alternative solution by optimizing shift schedules for crane operators. This approach aims to maintain adequate rest periods for operators while minimizing or eliminating the downtime caused by breaks, thus enhancing overall terminal efficiency.

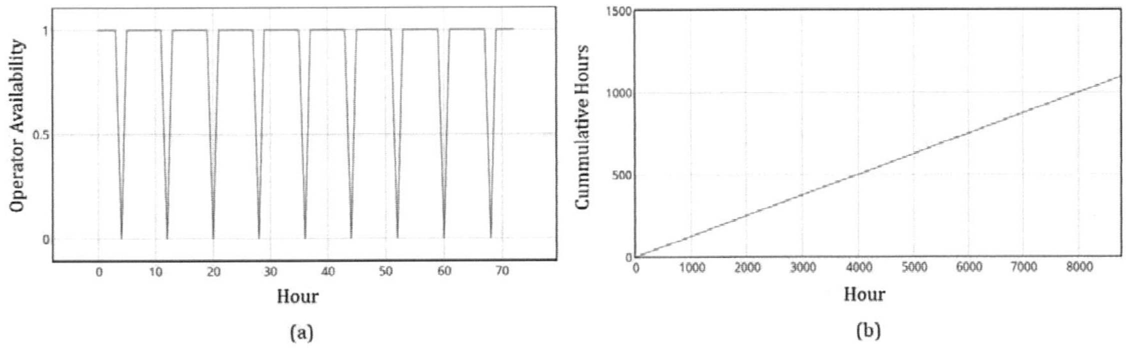

Figure 1. (a) Operator availability and break time; (b) Annual cumulative operator break time

3.3 Operator Shift Scheduling

Table 3 presents the shift scheduling framework for crane operators. The 12 operators are organized into 4 groups, each consisting of 3 operators. Each group works an 8-hour shift, with 3 shifts covering a full 24-hour operational cycle each day. In the simulation, weekly rest days are incorporated, with each group receiving two consecutive days off after 6 days of work. The scheduling also includes a dynamic shift rotation designed to minimize equipment downtime due to break periods, while still ensuring adequate rest for the operators. This approach aims to optimize both operational continuity and operator well-being.

3.4 Weekly Operator Working Hours

Figure 2 illustrates the weekly working hours for each container terminal operator. By implementing a dynamic shift scheduling system with rotating shifts, operator break times are effectively covered by other operators. This ensures continuous crane operation, even when one operator is on break. The simulation results show that the cumulative weekly working hours for operators can reach up to 48 hours, which exceeds the 40-hour limit set by government regulations. As a result, compensation in the form of overtime pay will be required for the additional hours worked each week.

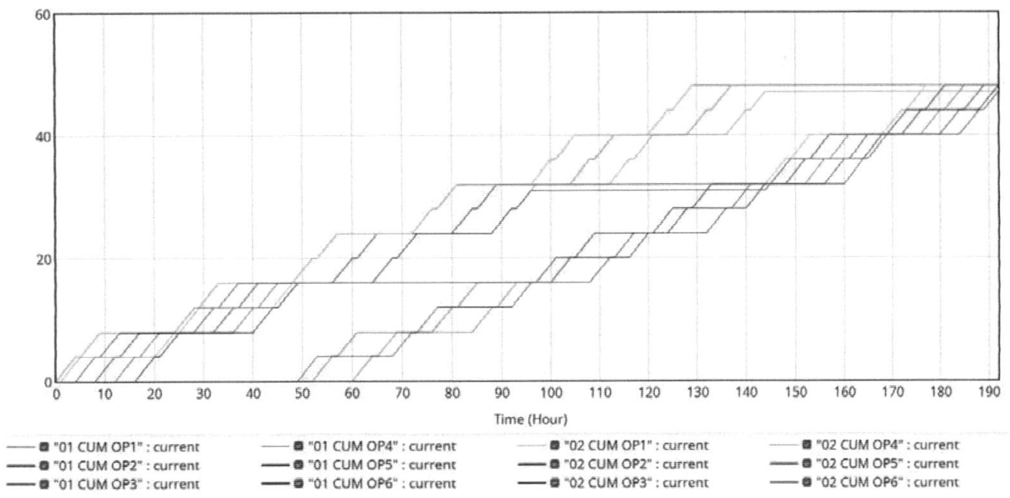

Figure 2. Weekly operator working hours

IOP Conf. Series: Earth and Environmental Science **1461** (2025) 012016 doi:10.1088/1755-1315/1461/1/012016

Table 3. Operator shift scheduling

GROUP	**Operator**	Day 1		Day 2		Day 3		Day 4		Day 5		Day 6		Day 7		Day 8	
	Operator 1	08-11/13-16	6	11-16	5	16-21	5	16-19/21-24	6	03-08	5	24-05	5	off	off	off	off
A	Operator 2	08-13	5	08-11/13-16	6	19-24	5	16-21	5	24-03/05-08	6	03-08	5	off	off	off	off
	Operator 3	11-16	5	08-13	5	16-19/21-24	6	19-24	5	24-05	5	24-03/05-08	6	off	off	off	off
	Operator 4	16-19/21-24	6	19-24	5	24-05	5	24-03/05-08	6	off	off	off	off	08-11/13-16	6	11-16	5
B	Operator 5	16-21	5	16-19/21-24	6	03-08	5	24-05	5	off	off	off	off	08-13	5	08-11/13-16	6
	Operator 6	19-24	5	16-21	5	24-03/05-08	6	03-08	5	off	off	off	off	11-16	5	08-13	5
	Operator 7	24-03/05-08	6	03-08	5	off	off	off	off	11-16	5	08-13	5	16-19/21-24	6	19-24	5
C	Operator 8	24-05	5	24-03/05-08	6	off	off	off	off	08-11/13-16	6	11-16	5	16-21	5	16-19/21-24	6
	Operator 9	03-08	5	24-05	5	off	off	off	off	08-13	5	08-11/13-16	6	19-24	5	16-21	5
	Operator 10	off	off	off	off	08-13	5	08-11/13-16	6	19-24	5	16-21	5	24-03/05-08	6	03-08	5
D	Operator 11	off	off	off	off	11-16	5	08-13	5	16-19/21-24	6	19-24	5	24-05	5	24-03/05-08	6
	Operator 12	off	off	off	off	08-11/13-16	6	11-16	5	16-21	5	16-19/21-24	6	03-08	5	24-05	5

IOP Conf. Series: Earth and Environmental Science **1461** (2025) 012016 doi:10.1088/1755-1315/1461/1/012016

3.5 Weekly Number of Operators at the Terminal

Figure 3 displays the number of operators on standby at the container terminal during working hours. The work hour mapping reveals that between 2 to 4 operators are present at the terminal at any given time. The simulation shows that both quay crane units can be operated continuously, as operators are available during every hour of operation. To ensure seamless shift transitions, the incoming crane operator is positioned on standby at the equipment just before the shift change, allowing for uninterrupted operation.

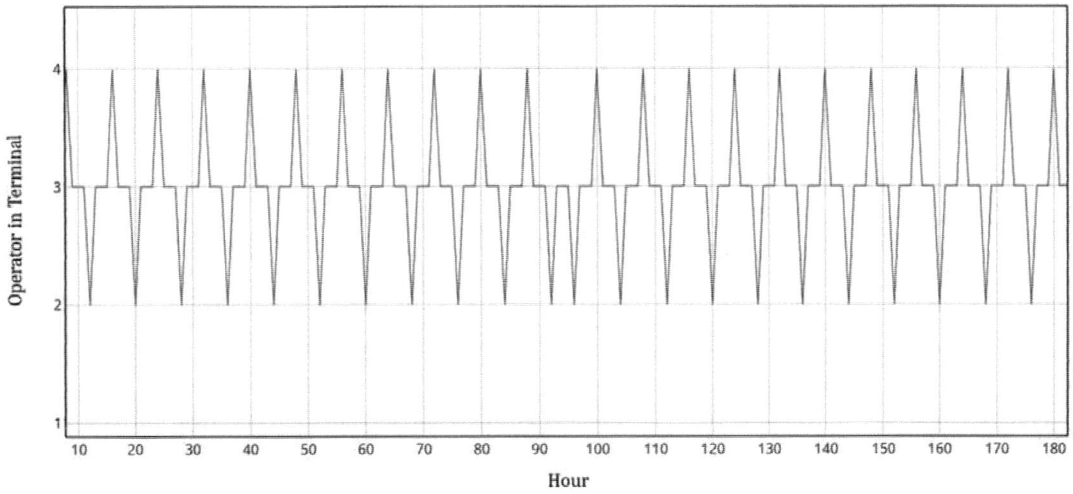

Figure 3. Weekly number operator in terminal

3.6 Container Production and Occupancy Ratio

Increasing the availability of container crane operators directly impacts the terminal's capacity. As shown in Figure 4(a), the terminal's container production has reached 329,448 containers per year. The improved operator availability has also positively affected the terminal's occupancy rate, as illustrated in Figure 4(b), with the occupancy rate reaching 51%, well within the maximum allowable threshold of 70%. This indicates that the terminal's production capacity can be further increased to its optimal level, enabling it to handle a greater volume of containers.

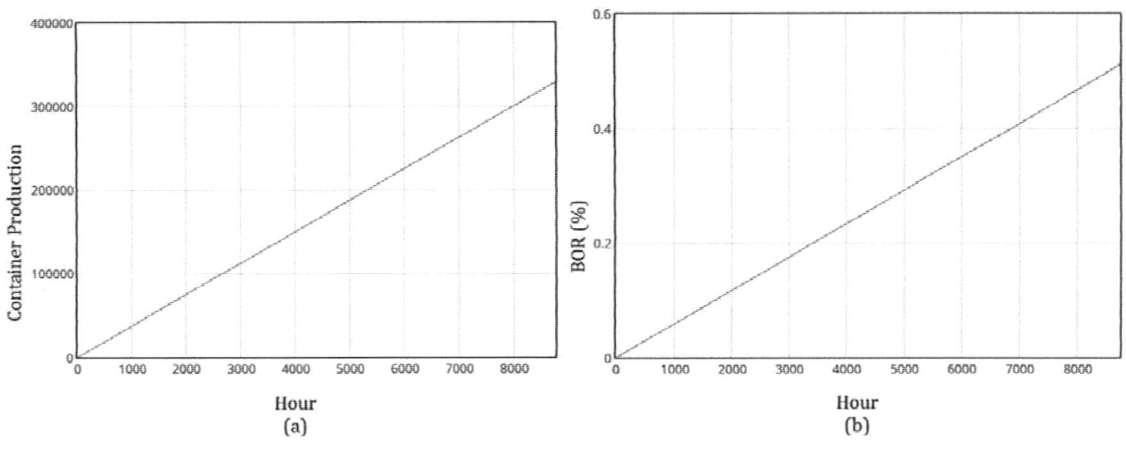

Figure 4. (a) Actual container production in terminal; (b) Terminal berth occupancy ratio

3.7 Potential for Performance Improvement

Figure 5 highlights the potential for increasing container production to reach the optimal capacity, targeting an occupancy rate of 70%. By implementing more efficient operator scheduling, the terminal could handle up to 445,480 containers annually, representing a 35% increase over the current capacity. This rise in container throughput would also lead to an increase in ship arrivals, growing from 636 to 860 arrivals per year, further enhancing the terminal's overall operational efficiency.

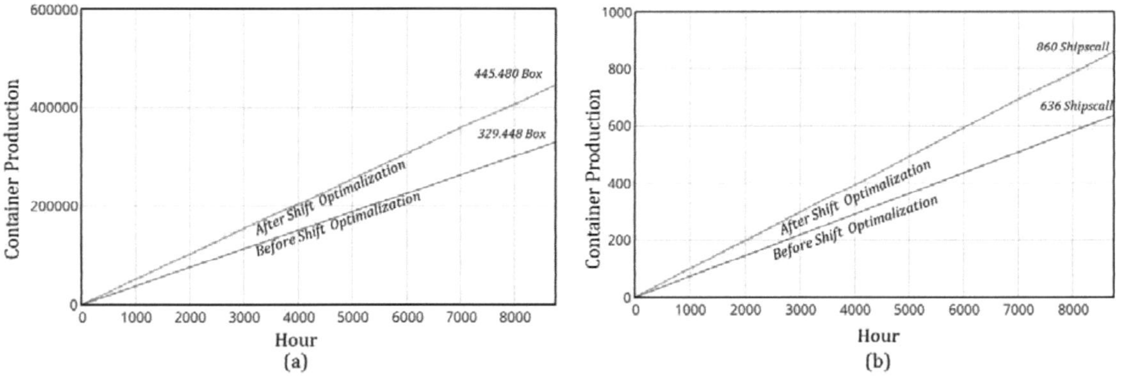

Figure 5. (a) Potential for increased container production; (b) Potential increase in ships call

3.8 Recommendations for Increasing Production Capacity Through Crane Operator Scheduling

Enhancing container terminal production capacity through crane operator scheduling requires careful consideration of multiple factors. The key is to reduce equipment downtime while ensuring operators are provided with adequate rest, comfort, and well-being. Figure 6 outlines recommendations to increase container production capacity through optimized operator work scheduling.

The drive to increase container production stems from the need for continuous operational performance, while prioritizing the safety and welfare of crane operators. Terminal operators can achieve this by implementing more efficient shift management, distributing workloads effectively, minimizing equipment downtime, and regularly evaluating and monitoring shift schedules.

To optimize operator working hours, terminal operators should develop dynamic scheduling systems, offer facilities that enhance operator performance, improve health and transportation services, ensure fair workload distribution, and actively seek feedback to refine and improve scheduling practices.

The success of these operations hinges on designing effective shift schedules, maintaining a sufficient number of operators, providing adequate rest areas, implementing comprehensive training and health programs, adjusting workloads to avoid fatigue, enabling flexible scheduling, and continuously monitoring operator performance and well-being. Regular feedback and evaluations should be used to make ongoing improvements and ensure sustainable, efficient operations.

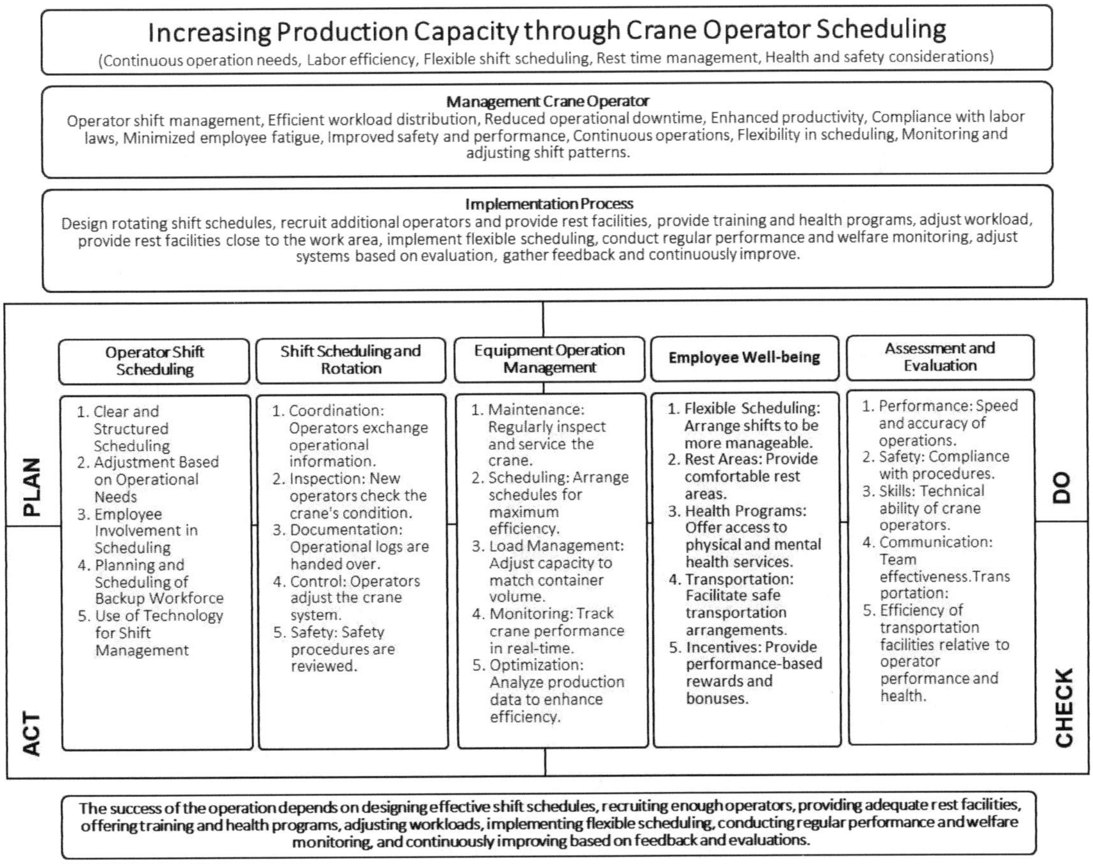

Figure 6. Recommendation of increasing production capacity trough crane operator scheduling

4. Conclusion

The availability of attentive and focused crane operators is a critical factor in enhancing container terminal performance. This study demonstrates a strong correlation between efficient operator work hour scheduling and increased terminal productivity. A key innovation of this research is the strategy to minimize equipment downtime caused by operator breaks, while still ensuring operators receive sufficient rest. By providing adequate rest periods, operator performance and safety during loading and unloading operations are significantly improved. This study proposes a dynamic scheduling model that organizes operators into groups and rotates shifts, ensuring continuous equipment operation even during operator rest periods. The optimized scheduling ensures that operators work equitable hours, meeting both daily work hour requirements and weekly rest needs. Using descriptive analysis and simulation modelling, this research evaluates container terminal production performance under various parameters. The results indicate that the terminal's current occupancy rate stands at 51%, within the optimal range of 70%. At full capacity, the terminal could handle up to 445,480 containers, equating to 860 ship arrivals annually. The overall success of terminal operations relies on effective shift scheduling, adequate staffing, provision of rest and health facilities, workload adjustments, flexible scheduling, regular performance and welfare monitoring, and continuous improvements driven by feedback and evaluation.

References

1. Fancello G, D'Errico G, Fadda P. Human factors involved in container terminal ship-to-shore crane operator tasks: operator fatigue and performance analysis at Cagliari Port. Hum Factors Assist Autom Maastricht Shak Publ. 2008;1(2):479–92.
2. Fabiano B, Currò F, Reverberi AP, Pastorino R. Port safety and the container revolution: A statistical study on human factor and occupational accidents over the long period. Saf Sci [Internet]. 2010 Oct;48(8):980–90. Available from: https://linkinghub.elsevier.com/retrieve/pii/S0925753509001635
3. Dermawan MIR, Rhomadhoni MN, Ayu F, Sunaryo M. Deskripsi Faktor Individu dan Faktor Pekerjaan terhadap Kelelahan Kerjan pada Operator Container Crane (CC) di Terminal Petikemas Nilam. JUMANTIK (Jurnal Ilm Penelit Kesehatan) [Internet]. 2022 Aug 16;7(3):301. Available from: http://jurnal.uinsu.ac.id/index.php/kesmas/article/view/12448
4. Putri AS, Dwiyanti E, Prayogi ARY, Istiqomah F. Correlation between Individual Factors and Mental Workload with Work Fatigue in Nilam Terminal Surabaya. Indones J Occup Saf Heal [Internet]. 2023 Mar 31;12(1):1–9. Available from: https://e-journal.unair.ac.id/IJOSH/article/view/26666
5. Anindya NY. Kelelahan kerja pada operator container crane di PT. X. Indones J Occup Saf Heal [Internet]. 2019 Jan 15;7(3):339. Available from: https://e-journal.unair.ac.id/IJOSH/article/view/5447
6. Fadda P, Meloni M, Fancello G, Pau M, Medda A, Pinna C, et al. Multidisciplinary Study of Biological Parameters and Fatigue Evolution in Quay Crane Operators. Procedia Manuf [Internet]. 2015;3:3301–8. Available from: https://linkinghub.elsevier.com/retrieve/pii/S2351978915004114
7. Hardianti DN, Lestari IAID, Sukraniti DP. Gambaran Status Gizi dan Kelelahan Kerja Pada Pekerja Operator Container Crane di PT. X. Gambaran Status Gizi dan Kelelahan Kerja Pada Pekerja Oper Contain Crane di PT X Overv Nutr Status Work Fatigue CC Oper Work PT X. 2023;14(1):14–21.
8. de Luca S, Cisternas LJ, Fiori C, Di Pace R. Implementation of a Methodology for the Fatigue Analysis in Quay Crane Operators Through a Multi-agent - Discrete Event Simulation: The Salerno Container Terminal Case Study. In 2022. p. 103–19. Available from: https://link.springer.com/10.1007/978-3-030-93370-8_8
9. Czermański E, Oniszczuk-Jastrząbek A, Zaucha J, Pawłowska B, Matczak M, Szydłowski Ł. Preconditions of new container terminal location in the Maritime Spatial Planning framework. A case study for the Central Port Concept in Gdańsk. Mar Policy [Internet]. 2021 Aug;130:104585. Available from: https://linkinghub.elsevier.com/retrieve/pii/S0308597X21001962
10. Notteboom T. Challenges for container river services on the Yangtze River: A case study for Chongqing. Res Transp Econ [Internet]. 2012 May;35(1):41–9. Available from: https://linkinghub.elsevier.com/retrieve/pii/S0739885911000588
11. Xu D, Li C-L, Leung JY-T. Berth allocation with time-dependent physical limitations on vessels. Eur J Oper Res [Internet]. 2012 Jan;216(1):47–56. Available from: https://linkinghub.elsevier.com/retrieve/pii/S037722171100614X
12. Metalla O, Koxhaj A, Nexhipi O. Land Area Limitation of Durres Port Containers Terminal. Glob J Manag Bus Res. 2015;15(5):29–34.
13. Momirski LA. Integration of Land and Sea in a Port Area: A Case Study of the Port of Koper. IOP Conf Ser Mater Sci Eng [Internet]. 2017 Oct;245:062044. Available from: https://iopscience.iop.org/article/10.1088/1757-899X/245/6/062044
14. Jobran Y, Kara G. Examining the Efficiency of Automation in Container Terminals. J Transp Logist [Internet]. 2022 May 31;7(1):137–55. Available from: https://iupress.istanbul.edu.tr/en/journal/jtl/article/examining-the-efficiency-of-automation-in-container-terminals
15. Moszyk K, Deja M, Dobrzynski M. Automation of the Road Gate Operations Process at the Container Terminal—A Case Study of DCT Gdańsk SA. Sustainability [Internet]. 2021 Jun 2;13(11):6291. Available from: https://www.mdpi.com/2071-1050/13/11/6291
16. Wong AKS. Optimisation of container process at multimodal container terminals. Queensland: Doctoral dissertation, Queensland University of Technology; 2008.
17. Wen C, Ekşioğlu SD, Greenwood A, Zhang S. Crane scheduling in a shipbuilding environment. Int J Prod Econ [Internet]. 2010 Mar;124(1):40–50. Available from: https://linkinghub.elsevier.com/retrieve/pii/S0925527309003600
18. Budipriyanto A, Wirjodirdjo B, Pujawan N, Gurning S. Berth Allocation Problem Under Uncertainty: A Conceptual Model using Collaborative Approach. Procedia Manuf [Internet]. 2015;4:429–37. Available from: https://linkinghub.elsevier.com/retrieve/pii/S2351978915011750
19. Pratiwi I, Oktaviara S. Job design for crane operators based on fatigue aspects and mental workload in Indonesia. Heal risk Anal [Internet]. 2023 Mar;(1):73–84. Available from: https://journal.fcrisk.ru/eng/2023/1/8
20. Notteboom T, Pallis A, Rodrigue J-P. Port Economics, Management and Policy. 2020;

Operability Analysis of Topside Platform Removal Using Dynamic Method

Murdjito [1] *, Muhammad Zakaria Al Amin[2], Eko Budi Djatmiko[1], Wahyudi[1], Fajar Rachmadiarto[3]

[1)] Lecturer, Department of Ocean Engineering, Institut Teknologi Sepuluh Nopember, Surabaya, Indonesia
[2)] Student, Department of Ocean Engineering, Institut Teknologi Sepuluh Nopember, Surabaya, Indonesia
[3)] Senior Engineer, PT ZEE Indonesia

*E-mail: murdjito@its.ac.id

Abstract. Topside removal is one of the main activities in the offshore platform decommissioning process, with one of the methods in this process being reverse installation-single lift. Corrosion is considered in this research as a parameter to reduce the strength of topside structures. The structural strength analysis is executed to check the integrity of the members and joints of the structures to avoid collapse of the structure during lifting process. The impacts of the dynamic responses to the hoisting line and sling tension are to be identified, such as due to the motions of the crane barge. The length of the hoisting line can influence significantly the members and joints tension as impacts of the changes of the stiffness, velocity and acceleration of the topside motions, thereby can increase probably the snap force of the hoisting. These conditions are considered in the determine the dynamic amplification factor (DAF) and factor of safety (FOS) of the sling. The operability analysis of the topside lifting is determined based on DnV Standards. The Cinta platform in the SES block, which is operated since the 1970s in Natuna area, is used as a research object in this study. The operability of the system is found 89.4%, at significant waves of 1.0-3.5 m and a zero up-crossing (Tz) period of 8.0 and 9.0 seconds

Keywords: *Decommissioning, Topside Removal, Wave Load, Lifting, Dynamic Analysis, DAF.*

1. Introduction

Based on data of the Directorate General of Oil and Gas of Indonesia in 2024 there are currently 113 platforms no longer in operation [3]. The decommissioning of an oil and gas platform is the process of removal of all oil and gas production facilities and restoring the site, including topside, wells, equipment, installations, and other supporting facilities after the cooperation contract ends [1]. A topside of the platform has to be transferred to a transport barge before deck removal takes place [2]. The commonly method used is heavy lifting with a crane barge in calm environmental conditions to reduce the risks that occur due to motions of the crane and the transport barge during the lifting process.

The strength of aging structures must consider corrosion factors, fatigue cracking, local dents, and other damage conditions[4]. Corrosion influences significantly material degradation [5]. In offshore structures the level of corrosion increases linearly in steel structural parts in the atmospheric and submerged zones as the structure ages [6]. Reducing the thickness of the structure results in reducing the integrity. The aging structures have problems with structural strength, especially due to corrosion [7] [8]. Although the influence of corrosion in the atmospheric zone on the global strength of the structure is not significant it is important to still evaluate the strength of the structure at the lifting point to be able to evaluate failure scenarios that probably can occur [9].

The vessel motion affects the dynamic load on offshore lifting operations. According to [10], the operating performance of a floating crane vessel is influenced by the vessel's dynamic forces on the environment and loads from the module that was appointed. A high dynamic load can influence the stress of the lifted structures significantly. The integrity of the lifted structure shall be evaluated based on von Mises stress differently in the direction of environmental load heading according to the crane barge [11].

Lifting operation is also affected by environmental loads and vessel motions that increase tension on the sling connected to the pendulum at the hook lift point [11]. Motion response crane barge is influenced by the module lifted during operation and resonance in surge and yaw motion of the lifted structure [12][13].The tension in the hoisting lines during lifting operation increase linearly corresponding to the increase of the wave height loads [10].

[10] [14] mention that the value of DAF is influenced by motions of a floating crane during lifting operation. Wave height and wave period influence significantly the structural response, linear with the RAO motion of the floating structure [15]. The value of DAF on lifting operations well in condition lift off, in air, through the splash-zone, and close to seabed is influenced significantly by roll motion of the crane barge due to the environmental heading load beam seas [16].

This research analyses impacts of lifting operations of a topside removal of an offshore platform. This research will evaluate the structural strength integrity topside during lifting. In addition, this research will examine the tension response on sling and hoisting lines, which is influenced by variations in wave loads conditions. These variations in wave loads conditions can impose significant loads and affect the tension of the lifting equipment. Finally, this research will examine the value of Dynamic Amplification Factor (DAF) and operability of the crane barge, which are also affected by changes in wave loads.

2. Literature Review

2.1. Hoisting System

The crane's hoisting system allows it to lift objects weighing more than the capacity of one line. This system usually uses multiple lines, so the connection from the crane tip to the block generally has more than one line. According to [16], the following equation (1) can be used to calculate the stiffness of the hoisting system.

$$\frac{1}{K} = \frac{1}{k_{rigging}} + \frac{1}{k_{line}} + \frac{1}{k_{soft}} + \frac{1}{k_{block}} + \frac{1}{k_{boom}} + \frac{1}{k_{other}} \quad (1)$$

Meanwhile, the stiffness of the line is calculated using equations (2) and (3) below.

$$k_{line} = \frac{EA}{L} [N/m] \quad (2)$$

$$A_{line} = \frac{\pi \cdot D^2}{4} \cdot c_f \quad (3)$$

By substituting equations (2) and (3), the stiffness value of the parallel hoisting line can be calculated using the equation (4) below.

$$k_{total} = \frac{E \cdot \pi \cdot D^2 \cdot c_f}{4 \cdot L} \cdot n \quad (4)$$

2.2. *Dynamic Amplification Factor*

The dynamic amplification factor (DAF) accounts for global dynamic effects resulting from vessel motions, boom, wire, and rigging stiffness, boom tip location and motions, crane movements, and wind loading [18]. Conventionally, DAF can be calculated using equation (5), the ratio of the largest total force on the system with the object mass in the air multiplied by the acceleration due to gravity.

$$DAF_{conv} = \frac{F_{total}}{Mg} \qquad (5)$$

2.3. *Corrosion Rate*

In the analysis, the strength of members is assumed to have uniform thickness reduction throughout the member [19]. The corrosion rate in steel structures, according to [6] is in atmospheric zone between 0.05-0.10 mm/year. In this study it is assumed to be 0.05 mm/year, totally 1.5 mm for 30-year life time.

2.4. *Hydrodynamics of Top Side*

The topside is simplified as a box as Figure 1. In the rectangular box geometry, [20] uses C_d for the force in the flow, so $C_d = C_x$. Meanwhile, for the drag properties in rotational movements, there is no standard data source. In the computational process it assumes that the contribution of the drag force df to the element area dA uses equation (6) where the value of C_d is assumed to be the same for all points on the surface.

$$df = \frac{1}{2}\rho|v|^2 C_d\, dA \qquad (6)$$

$$df = \frac{1}{2}\rho\omega z|\omega z| C_d\, x\, dz \qquad (7)$$

$$dm = \frac{1}{2}\rho\omega z|\omega z| C_d\, x\, dz\, z \qquad (8)$$

$$dm = \frac{1}{2}\rho\omega|\omega| C_d\, x\, z^3\, dz \qquad (9)$$

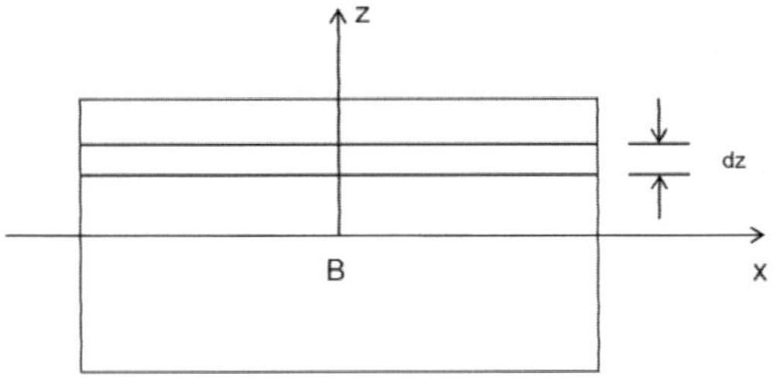

Figure 1. Integration for drag and rotation properties [23]

2.5. *Time Domain Analysis*

Lifting operations can be analysed through time-domain simulations based on the wave excitation force transfer function, which depends on the pre-generated frequency, added mass, and wave damping for the crane vessel, transport barge, and lifted object [17] Using numerical integration, time-domain analysis can be utilized to solve the floater response in the time domain.

$$\sum_{j=1}^{6}\{(M_{ij}+a_{ij}).\ddot{x}+\dot{b}.x+c.x\}=F_i \text{ for } i = 1\sim 6 \quad (10)$$

$$F = F_m + F_w + F_l \quad (11)$$

The static weight of the topside and rigging is estimated to be 696.31 MT. According to [17] this weight is categorized as a light lift because the weight of the lifted object is less than 2% of the vessel's displacement (2% of the crane barge displacement is 1010 MT). In light lift scenarios, the motion characteristics of the vessel at the crane tip are not influenced by the lifted object. The simulation is conducted for 30 minutes based on a sensitivity study[17]. The analysis results are then used to obtain extreme value statistics with a Rayleigh distribution to determine the maximum possible value during a 3-hour analysis [17].

3. Methodology

3.1. *Loading Scenarios*

The topside used in this analysis is with weight of 689 MT. The decommissioning process chosen is a single lift with the reverse installation method. The hoisting system on the crane is parallel lines with a total number of 52 lines. The specifications of the hoisting line are 60 mm diameter IWRC type with total length of 2960 m. Dynamic simulations are carried out using variation of collinear loading direction with the wind spectrum recommended by [21] is NPD spectrum. The load scenarios can be seen in the figure 2. The crane barge has principal dimensions as follows, length overall 169 m, width 42 m, draft 8.53 m and displacement of 60,500 MT.

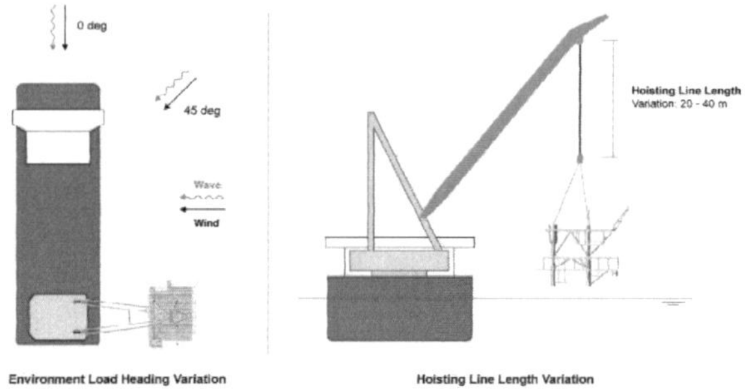

Figure 2. Illustration of load variation

3.2. *Rigging Design*

The weight of the top side is to be 613.55 MT with account corrosion of 1.5 mm during the structure's life time of 15 years. Rigging calculations are based on calculations and factors according to [18]. The design DAF value on the topside is 1.20 based on [18]. The sling size was 260mm and the total static weight of the topside, including rigging, is 696.31 MT.

3.3. *Structural Integrity*

Strength analysis of the topside justification structure is based on [22] and [23]. The lifting structure is 3-D modelled consider COG shifting by applying a force to the lift point. The structural model of the topside as seen in the figure 3. The maximum UC of the structure is 0.854 at the member in connected lift point.

IOP Conf. Series: Earth and Environmental Science **1461** (2025) 012017 doi:10.1088/1755-1315/1461/1/012017

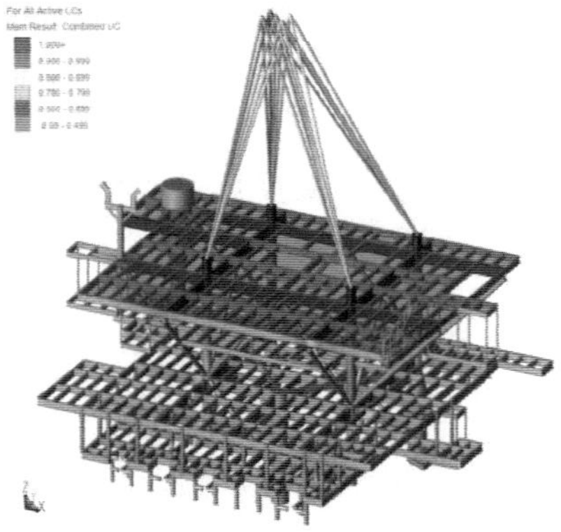

Figure 3. Lifting Structural Model of the top side

3.4. Natural Period

Decay tests are carried out to obtain the natural period of motion of the floating structure in free-floating conditions. Using the decay test it is found the natural period of heave motion is 8.78 seconds, roll motion is 10.29 seconds and pitch motion is 8.63 seconds.

3.5. Hoisting Line Stiffness

The stiffness of the hoisting line is calculated based on equation (4). From this equation, the stiffness value of the hoisting line is obtained in Figure 4. The graph shows the relationship between stiffness (in units of kN/m) and the length of the hoisting line distance between the crane tip and the crane block. The stiffness value increases exponentially with the shortening of the length of the hoisting line.

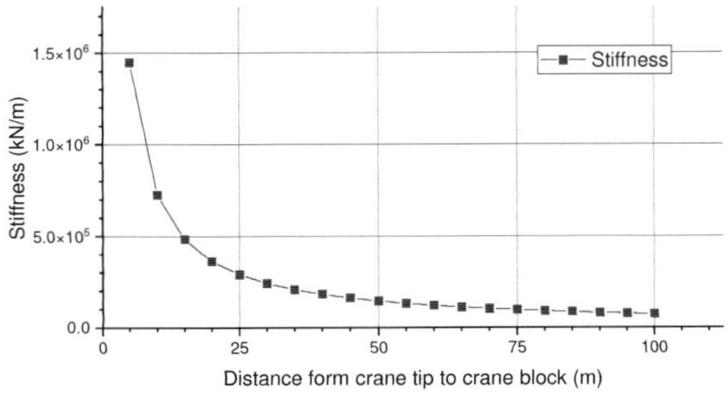

Figure 4. Hoisting Line Stiffness.

3.6. Hydrodynamic of Topside

The topside model is simplified by representing the model as a solid prism including equipment and piping systems. The property of the simplified topside is that it is a roughly rectangular prism. According to [24], the value of CD is 2.05.

4. Results and Discussion

4.1. Tension on Hoisting Line and Sling

The results of the dynamic analysis simulation showed that the maximum tension occurred in the hoisting line and sling with dynamic analysis for 30 minutes. These results were carried out using extreme value statistics with the Rayleigh distribution to obtain the most significant possible maximum tension value during the 3-hour simulation. In Figure 5, the tension on the sling is linear against the pattern of increasing tension on the hoisting line. In this case, the greater the angle at the lift point or the longer the sling, the greater the tension.

IOP Conf. Series: Earth and Environmental Science **1461** (2025) 012017 doi:10.1088/1755-1315/1461/1/012017

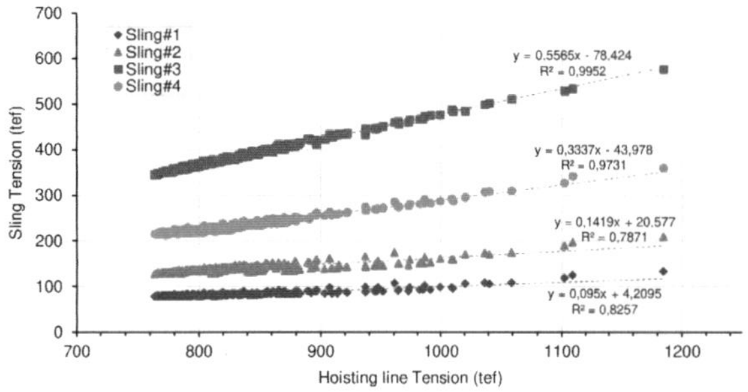

Figure 5. Correlation of tension value between hoisting line and slings.

The tension on the hoisting line and sling increases steadily, and the response pattern is similar in wave heights of 1.0 m to 3.5 m. In the wave period, the tension response also increases as the wave period increases from Tz = 8 seconds to Tz = 9 seconds. The highest tension in varying loading directions occurs in the beam seas or 90-degree direction.

The Power Spectrum Density (PSD) of the crane barge motions' responses directly influence by the environmental load heading and the hoisting line length variations. The tension of hoisting line is influenced by the crane barge's motions. The beam seas loading direction has the most significant influence on increasing the tension on the hoisting line. Using the FFT method the total energy of the PSD is calculated to determine the influences of the crane barge motions to the tension of the hoisting line. Roll motion has the most significant energy in PSD. That means has significant influence of increasing tension on the hoisting line and sling. Another influence on the tension in the hoisting line and sling is the velocity and acceleration motions of the topside. The velocity and acceleration motions cause a snap force in the lines. These can increase the tension significantly [17]. In the snap force, the shorter the hoisting line, the greater the stiffness value, and it is directly proportional to the increase in velocity and acceleration of the topside motions. Thus, the snap force that occurs becomes greater.

4.2. Dynamic Amplification Factor (DAF)

The DAF value is calculated using equation (5) by dividing the maximum tension on the hoisting line by the static weight of the lifting system. The static weight model includes the topside, crane block, and rigging, is 752.86 MT. The DAF value increases with shortening the hoisting line and increasing wave height and period. The DAF value of 1.20 is used as the limit value.

4.3. Sling Safety Factor (FOS)

The factor of safety (FOS) on the vulnerable sling decreases as the length of the hoisting line becomes shorter, followed by an increase in wave height and period. With variations in environmental loading and heading, it is found that the minimum FOS value is due to environmental loads with heading of 90 degrees or beam seas. The minimum FOS of the sling is 5.55. This FOS value is greater than 4.0 as required by sling design's standards.

4.4. Crane Capacity Safety Factor

The object's position is located on the starboard side in the simulation model. Therefore, the crane capacity is in a revolving condition. The minimum value of the crane capacity factor of safety (FOS) is found at the most extended hoisting line length (40 meters) and due to the environmental loads of 90 degrees or beam seas. The minimum FOS at crane capacity is 1.02 greater than the capacity limit of 1.0.

4.5. Wave Load Response

The Figure 6. shows that the maximum tension of the hoisting lines due to the wave loads. The tensions increase linearly with the increase of the wave height in the power regression correlation. The wave crossing up periods used (Tz) in the modelling are Tz = 8s and Tz = 9s.

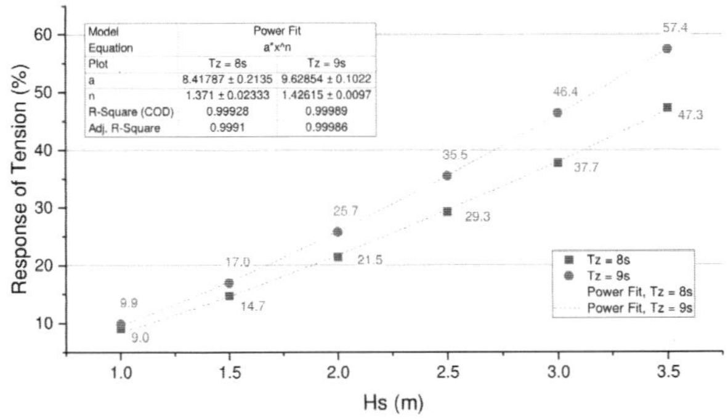

Figure 6. Maximum response of tension due to wave load

The regression parameters shown in figure above give pictures the relationship between wave height and the tensions of the hoisting line. The coefficient (a) shows the magnitude of wave height's influence to the hosting tension. The regression exponent (n) shows how the influence of the wave height to the hosting tension.

5. Probability Distribution Model

The distribution fit is carried out on the results of the DAF values to obtain a distribution model of the sample results against a particular distribution. The probability fitting is achieved by obtaining goodness of fit and index of fit values using the least squares method. The results of the distribution suitability test on the DAF value results show that the correct distribution is the 3-parameter Lognormal distribution illustrated in Figure 7igure 8 with an Anderson Darling (adj) value of 0.190 and a correlation coefficient value of 0.999.

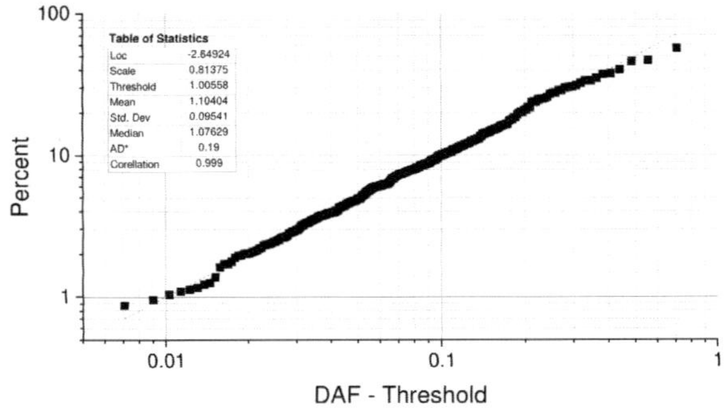

Figure 7. DAF value with 3-Parameter Lognormal with distribution fit

The study results are compared to the previous research to validate the distribution model. Similar research was carried out by [16], [25], and [26]. The 3-Parameter Lognormal distribution is the most suitable distribution model to be used as a DAF value distribution model even though the variables and data studied have many differences.

6. Operability

Table 3 below summarizes the results of the criteria analysis. From the structural strength analysis results, the UC value meets the criteria with a design DAF = 1.20 in accordance with [18]. However, in the dynamic analysis results, the DAF value shows that 34 simulations out of 300 simulations do not meet the criteria with a DAF value of more than 1.20 compared to other variables against the criteria. Therefore, DAF is the main factor in determining the operability criteria for this operating system.

Table 1. Summary of analysis results on operability criteria.

Parameter	Criteria		OK	NOT OK
Structure Strength	UC	<1.00	OK	-
DAF	DAF	<1.20	266	34
Sling	FOS	>5.55	295	5
Crane Capacity	FOS	>1.00	300	0

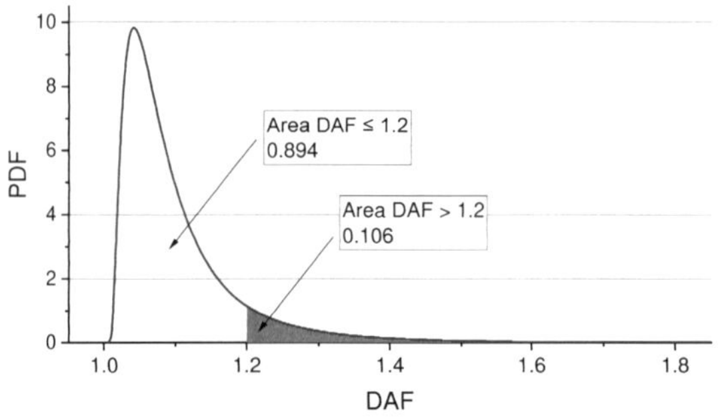

Figure 8. Probability density function of DAF

The 3-parameter Lognormal distribution as shown in Figure 8 is used for the data process to obtain a probability density function (PDF). Figure 8 shows that with the design DAF of 1.20 as the operability limit parameter, the PDF area with DAF > 1.20 is 0.106 or the PDF area with DAF ≤ 1.20 is 0.894. It means that the operability of the system in this study is 89.4%, and downtime is 10.6% at significant waves of 1.0-3.5 m and a zero up-crossing (Tz) period of 8.0 and 9.0 seconds. The operating limit with the operational sequence model in this study, is that the significant wave height at Tz = 8s is 1.87m, while at Tz = 9s, it is 1.66m.

7. Conclusion

In the decommissioning process, it must be ensured that the structure meets the strength criteria to avoid collapse conditions during the lifting process. In this study, the UC structure was found to be 0.854, less than 1.0 as required by standards.

The tension on the hoisting line and sling increases exponentially with significant increases in wave height and wave period. The tension on the hoisting line and slings is influenced by the ship's motion, with roll motion contributing to the most significant increase in tension, as well as the velocity and acceleration of topside motion. The lowest FOS at crane capacity is 1.02, which occurs on a 40m hoisting line.

The DAF value increases significantly with increases in wave height and wave period. The pattern of increasing DAF is also caused by the short hoisting line and the environmental loading direction. The 3-parameter Lognormal distribution of DAF is a suitable distribution model.

The operability of the system is 89.4%, at significant waves of 1.0-3.5 m and a zero up-crossing (Tz) period of 8.0 and 9.0 seconds. The operating limit with the operational sequence model in this study, is that the significant wave height at Tz = 8s is 1.87m, while at Tz = 9s, it is 1.66m.

References

[1] KESDM, *PERMEN ESDM Nomor 15 Tahun 2018 Tentang Kegiatan Pasca Operasi pada Kegiatan Usaha Hulu Minyak dan Gas Bumi*. 2018.

[2] ICF International, "Decommissioning Methodology and Cost Evaluation," 2016.

[3] SKK Migas, "Data Inactive Offshore Platform di Indonesia," *Unpublished*. SKK Migas, Jakarta, 2024.

[4] D. K. Kim, M. A. Zalaya, M. H. Mohd, H. S. Choi, and K. S. Park, "Safety assessment of corroded jacket platform considering decommissioning event," *INTERNATIONAL JOURNAL OF AUTOMOTIVE AND MECHANICAL ENGINEERING*, vol. 14, no. 3, pp. 4462–4485, Sep. 2017, doi: 10.15282/ijame.14.3.2017.6.0353.

[5] M. M. A. Wahab, V. J. Kurian, M. S. Liew, and D. K. Kim, "Condition Assessment Techniques for Aged Fixed-Type Offshore Platforms Considering Decommissioning: a Historical Review," *Journal of Marine Science and Application*, vol. 19, no. 4, pp. 584–614, Dec. 2020, doi: 10.1007/s11804-020-00181-z.

[6] R. E. Melchers, "Corrosion uncertainty modelling for steel structures," *J Constr Steel Res*, vol. 52, no. 1, pp. 3–19, Oct. 1999, doi: 10.1016/S0143-974X(99)00010-3.

[7] Y. Sharifi, S. Tohidi, and J. K. Paik, "Ultimate compressive strength of deteriorated steel web plate with pitting and uniform corrosion wastage," *Scientia Iranica*, vol. 23, no. 2, pp. 486–499, Apr. 2016, doi: 10.24200/sci.2016.2133.

[8] N. A. Othman, M. H. Mohd, M. A. A Rahman, M. A. Musa, and A. Fitriadhy, "Investigation of the corrosion factor to the global strength of aging offshore jacket platforms under different marine zones," *International Journal of Naval Architecture and Ocean Engineering*, vol. 15, p. 100496, 2023, doi: 10.1016/j.ijnaoe.2022.100496.

[9] A. M. A. Abdelaah, S. M. Cresswell, G. M. E. Manzocchi, and C. Thistlethwaite, "Practical Applications of Structural Analysis Support to Decommissioning," in *Day 1 Mon, May 02, 2016*, OTC, May 2016. doi: 10.4043/27128-MS.

[10] J.-H. Cha, M.-I. Roh, and K.-Y. Lee, "Dynamic response simulation of a heavy cargo suspended by a floating crane based on multibody system dynamics," *Ocean Engineering*, vol. 37, no. 14–15, pp. 1273–1291, Oct. 2010, doi: 10.1016/j.oceaneng.2010.06.008.

[11] S.-H. Ham and M.-I. Roh, "Time-domain structural analysis during block turnover and lifting using 2D flexible multibody dynamics," *Marine Structures*, vol. 75, p. 102841, Jan. 2021, doi: 10.1016/j.marstruc.2020.102841.

[12] N. Firdaus, E. B. Djatmiko, R. W. Prastianto, and Muryadin, "Experimental Study on Coupled Motion of Floating Crane Barge and Lifted Module in Irregular Waves," *IOP Conf Ser Earth Environ Sci*, vol. 972, no. 1, p. 012070, Jan. 2022, doi: 10.1088/1755-1315/972/1/012070.

[13] G. O. Tysse and O. Egeland, "Dynamic Interaction of a Heavy Crane and a Ship in Wave Motion," *Modeling, Identification and Control: A Norwegian Research Bulletin*, vol. 39, no. 2, pp. 45–60, 2018, doi: 10.4173/mic.2018.2.1.

[14] H. Cao, X. Yan, J. Xia, and D. Wei, "Effect of fluid-structural interaction on the dynamic response of stepped lifting system subjected to heave motion," *Ocean Engineering*, vol. 279, p. 113943, Jul. 2023, doi: 10.1016/j.oceaneng.2023.113943.

[15] L. Li, Z. Gao, T. Moan, and H. Ormberg, "Analysis of lifting operation of a monopile for an offshore wind turbine considering vessel shielding effects," *Marine Structures*, vol. 39, pp. 287–314, Dec. 2014, doi: 10.1016/j.marstruc.2014.07.009.

[16] S. Madland, "Dynamic Analysis for the Installation of Offshore Wind Turbine Foundations," University of Stavanger, Stavanger, 2012.

[17] DNV, "DNVGL-RP-N103 Modelling and analysis of marine operations." Det Norske Veritas, Bærum, 2017.

[18] DNV, "DNV-ST-N001: Marine Operations and Marine Warranty." Det Norske Veritas, Bærum, 2023.

[19] NORSOK, "NOSROK N-004 Design of Steel Structure." Norwegian Technology Standards Institution, Oslo, 1998.

[20] ESDU, "ESDU 71016: Fluid forces, pressures and moments on rectangular blocks.," *ESDU International*, 1978.

[21] API, *API RP 2SK: Design and Analysis of Stationkeeping Systems for Floating Structures*, 3rd ed. American Petroleum Institute, 2005.

[22] API, "API RP 2A: Recommended Practice for Planning, Designing and Constructing Fixed Offshore Platforms—Working Stress Design 21st edition," *American Petroleum Institute*. American Petroleum Institute, 2000.

[23] AISC, *Steel Construction Manual 13rd*, 13th ed. Chicago: American Institute of Steel Construction, 2005.

[24] S. F. Hoerner, *Fluid-Dynamic Drag*, 2nd ed. Washington D.C.: Hoerner, Sighard F., 1965.

[25] D. Å. Viki, "Limiting Operational Wave Criterion for Spool Installation Lift," Master's Thesis, University of Stavanger, Stavanger, 2015.

[26] S. D. Eri, "Analysis of Operability in Installing Heavy Subsea Modules," Master's Thesis, University of Stavanger, Stavenger, 2015.

The Effect of Connector Loads on The Multi-Directional Configuration of a Hexagonal Modular Floating Structure

Nur Hanani AHMAD AZLAN[1*], Nik Mohd Ridzuan SHAHARUDDIN[1], Arifah ALI[1], and Sunarsih [2]

[1] Faculty of Mechanical Engineering, Universiti Teknologi Malaysia, Johor, Malaysia
[2] Department of Marine Engineering, Institut Teknologi Sepuluh Nopember, Surabaya, Indonesia

*E-mail: *nhanani4@gmail.com*

Abstract. A hexagonal modular floating structure (HMFS) system is preliminarily designed in a linear configuration which creates network of connectors. The load of the connectors should be able to address issues related to the layout of the connector network and sustain the forces caused by wave motion. This paper presents the development of the HMFS connector network and the estimation of the connectors horizontal load due to the linear direction configurations. The design concepts of linear arrangement for HMFS configurations are proposed where analysis was carried out for regular wave in various directions of 0^0, 30^0, 45^0,60^0, 85^0 and 90^0. The effect of these various wave directions and the HMFS configurations on the connector load is analysed accordingly. The analysis disclosed that the connector load in 0^0 is generally higher than the load in other wave directions for hexagonal modular floating structure.

1. Introduction

Rising sea levels and population density in coastal areas drive the increasing demand for ocean space. Land reclamation has been widely used as a way to utilize the ocean space and expand land area. Therefore, Very Large Floating Structure (VLFS) has been a focus in ocean utilization for decades and studies have been conducted on various sizes of VLFS including floating docks, airports, fuel facilities and even cities. Various VLFS projects, such as Mega-Float in Tokyo Bay, floating oil storage bases in Kamigoto and Shirashima islands, emergency rescue bases in Yokohoma, and floating performance stage in Singapore have been conceptualised and already completed.

Thus, two types of VLFS have already been established are single module VLFS and multi-module VLFS. For multi-module VLFS, the structural integrity is directly linked to the connector's safety and also impacts the operational capability of the multi-floating structure, hence, it is highly important to develop and analyse the connector for multi-floating structures [1,2]. The interaction between floating structures is complex due to the excitation of incident waves causes the space between modules to leave wave resonance, which is influenced by relative motion. Riggs et al. (1998) and Lan et al. (2004) agreed that the longitudinal forces are in general substantially larger than the transverse and vertical forces, therefore the focus on these forces. Furthermore, increasing the number of modules can enhance the shielding effect on the central module, hence improving the stability of the modular floating structure during expansion [3,15].

As far as expansion modular floating structure is concerned, any arbitrary form such as regular polygon that can be laid out to fill the surface of the water without gaps. Traditionally, builders have primarily constructure rectangular shape, but in recent times, a growing number of designers have suggested hexagonal shapes, citing their shorter distance around the same area and increased efficiency. [9]. Dai et al. (2022) described the mostly rectangular modular unit between square, rectangular and hexagon as having the best performance in terms of the efficiency of the structural system, cost economy and design options.

2. Conceptual Design of the HMFS

The proposed HMFS is connected by five hexagonal floating structures. Each hexagonal structure is symmetrically moored to the sea floor by four tension legs in "chain-type". Fenders are installed at the bottom edge of adjacent modules to reduce possible collisions. Two types of connectors have been applied for the proposed HMFS such as rigid connectors (R) and hinged connectors (H) as shown in Figure 1. Wave kinetic energy is facing the edge of the HFMS in the 0^0 wave direction, while 90^0 wave directions are facing the vertices of the HMFS. The HFMS has proposed three and seven modules considering that an increase in the number of modules affects the horizontal connector load. The hexagonal floating structure design by Yanwei et al. (2022) is adopted and details are given in Table 1.

3.Result and Discussion

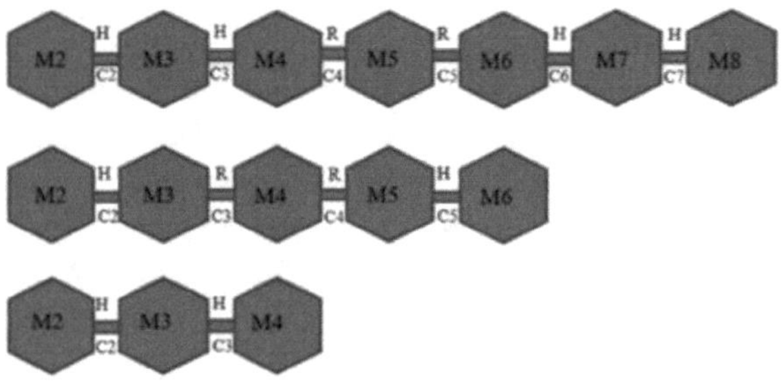

Figure 1. The arrangement of HMFS and the two connector types.

Table 1. Details of the HMFS.

Details	Value	Unit
Side length, Height	20, 12	m
Water depth, Draft	80, 10	m
Mass	6000	t
Ixx= Iyy	9.6×10^8	$kg.m^2$
Tension – leg dimension	D= 1.2; T=0.04; L = 70	m
Steel tension leg, E	2.1×10^{11}	N/m^2
Stiffness of fenders	1.0×10^7	N/m
Adjacent distance	3	m

The results of the maximum horizontal load connector (Fx) are divided into three components: (1) the effect of incident wave angle, (2) the effect of incident wave period and (3) the effect of module quantity. Incident wave angles that are included 0^0,30^0,45^0,60^0,85^0, and 90^0 wave directions in 6s wave period linked 5-module at linear arrangement. The incident wave period covered 4s to 20s. The numbers of modules quantity investigated are 3-module, 5-module and 7-module. In this research, the horizontal connector load is linked five modules in a hexagonal shape. Later, the discussion will compare the rectangular modular floating structure studied by other researchers.

3.1Effect of Incident Wave Angle

Typical regular sea conditions (H=2m, T=6s) were selected for the simulation in various waves such as 0^0, 30^0, 45^0, 60^0, 85^0 and 90^0. The main motion responses of each hexagonal module with different wave directions are compared in Figure 2.

The highest surge response and pitch response are present at the 0^0 wave direction and consistently decrease as the wave direction increases as shown in Figure 2(a,d). All modules show a similar response pattern with the highest surge and pitch at the 0^0 wave direction and the lowest at 90^0. Compared to surge and pitch, the highest sway motion response is at 90^0 and increases as the wave direction increases due to the lower module shielding effect. The heave response for all modules is relatively similar with

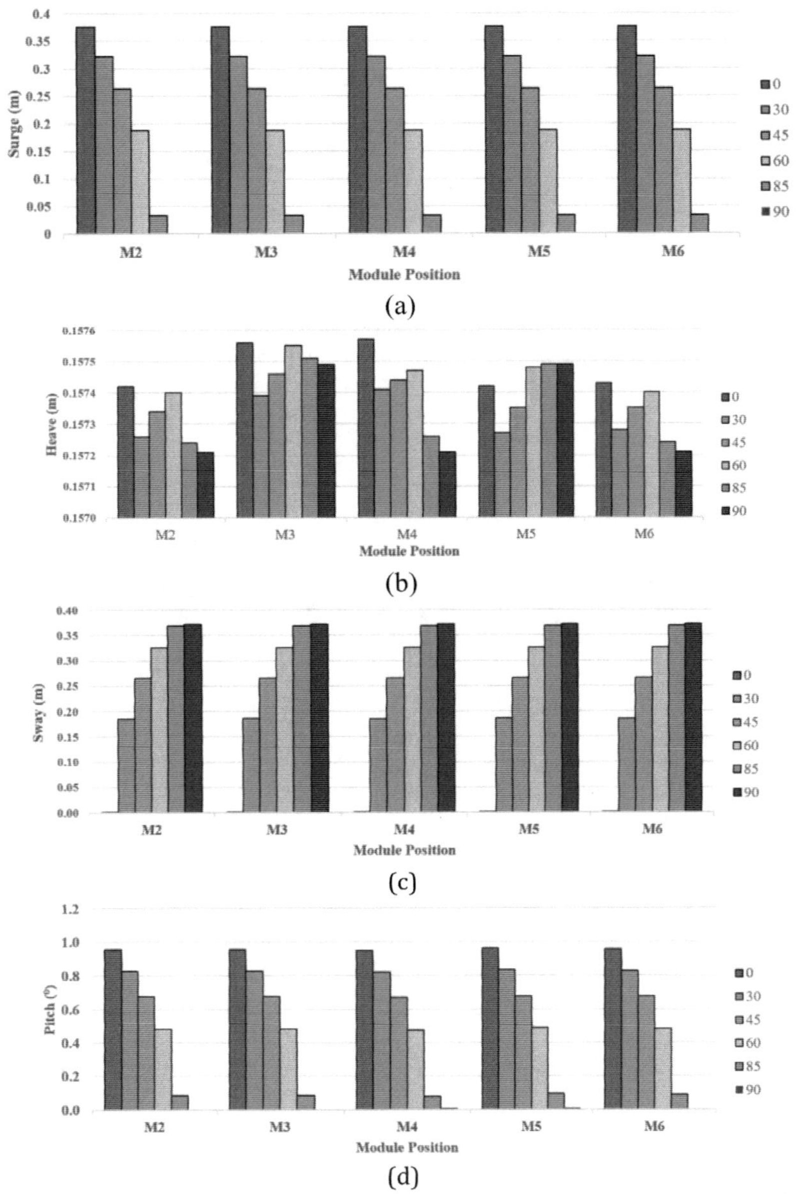

Figure 2. The main responses of each module under different incident wave angles at 6s wave period for 5 modules: (a) surge, (b) heave, (c) sway, (d) pitch.

slight variations among different wave directions in Figure 2b. For example, the wave at 30^0 has similar heave response trends to the wave at 45^0 while waves at 60^0,85^0 and 90^0 also display a similar trend. The pitch response at M3 and M4 which are placed at the back of M2 is lower due to the enhanced shielding effect of M2 near the resonance condition. (Yuan et al., 2020). According to Ziyi et al. (2023), the hexagonal modules' higher pitch motion response is in the 0^0 wave direction and their lowest pitch is 90^0, a similar tendency to the pitch motion shown in Figure 2d.

Figure 3 shows the peak maximum horizontal force of the connector for all connectors in the 0^0 wave direction. The trend indicates that as the wave direction increases, the maximum horizontal load

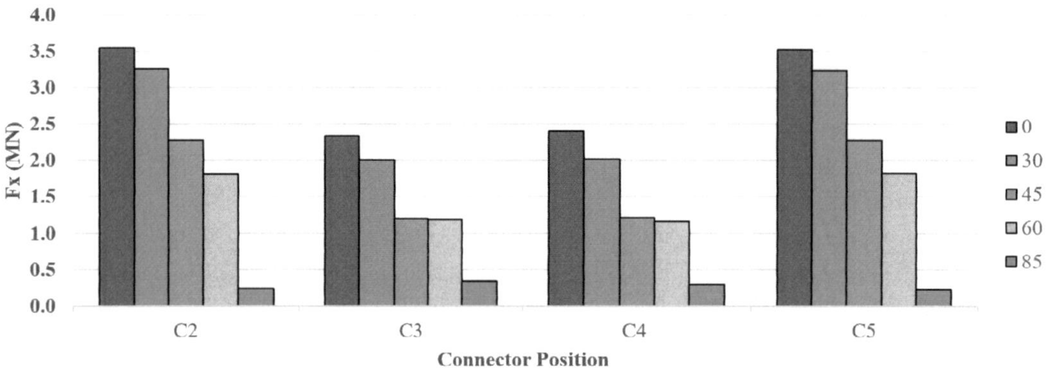

Figure 3. Maximum horizontal force of each connector on 6s wave period under different incident wave direction for 5 modules.

decreases. However, the 90^0 shows a lowest horizontal force. Similar to what Yuan et al. (2020) studied, the higher horizontal load connector at 0^0 wave direction than 45^0,85^0 and 90^0 for a hollow regular hexagon. In addition, Nianxin et al. (2022) also investigated the connector load for an oblique layout of four hexagonal multi-modules given a 0^0 wave direction given the highest horizontal connector load between 30^0,60^0 and 90^0 wave directions. The connectors at the outside module (C2 and C5) attempt to exert a higher horizontal load than the middle connector. This is due to the fact that the outside module are able to absorb the energy from the waves, which in turn reduces the stresses exerted on the central

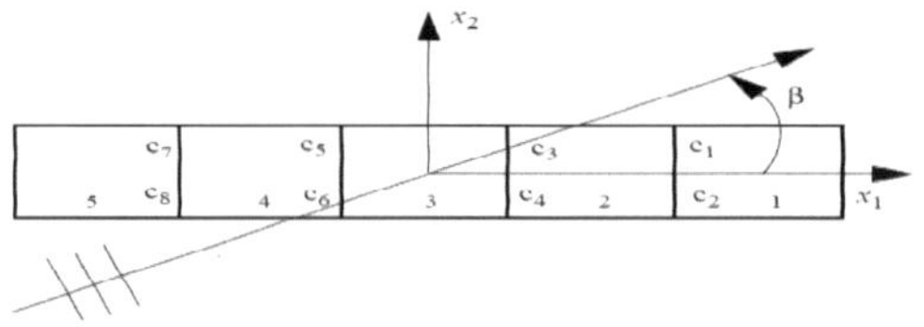

Figure 4. The studied 5-module was linked in linear arrangement by Riggs et al. (1998).

module (Park et al,2023). Furthermore, the center module has four times the inertia of the outer module.

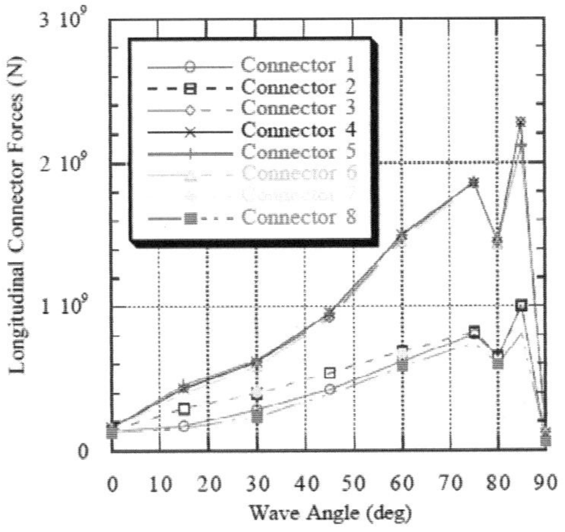

Figure 5. The longitudinal connector force studied by Riggs et al. (1998).

Riggs et al. (1998) investigated five rectangular modules linked in a linear arrangement as shown in Figure 4. The lower longitudinal connector force is in the 0^0 wave direction and the highest value is in the 85^0 wave direction as shown in Figure 5. The outer connector (C1, C2, C7 and C8) has a lower longitudinal connector force, while the inner connector (C3, C4, C5, and C6) has higher connector loads. In addition, Lan et al. (2004) also investigated five rectangular modular floating structures and discovered that the connector X-force increased dramatically at a wave angle 85^0. Additionally, Yousheng et al.(2017) explored the three rectangular modules and found a similar trend, with the highest load occurring in the 85^0 wave direction and the lowest in the 0^0 wave direction.

The largest value and lowest value of the maximum horizontal connector load at wave between 0^0 and 85^0 wave directions are reversed between the modular modules of rectangular and hexagon. The reason it happened cannot be identified. The difference between two shape modules with a linear arrangement is that the square has a straight edge while sometimes a hexagon has vertices when facing the wave direction. As in rectangular, wave energy faces the horizontal or vertical edge while in hexagonal, the wave energy is split at vertices, for instance, 90^0 wave directions. Therefore, estimation of connector force applied in different shapes floating structures should be determined according to the basic shape of function requirement in designing floating structures. Apart from that, engineers can optimize the connector design as the early estimation of connector force as allows identification of potential failure points within the structure that enhance safety.

Other than that, the different regular polygons have their own side angles thus, the distribution of connector forces in a modular floating structure approaching wave direction is also different. Besides, each connector force should be determined as the wave direction approaches different sides of the floating structure that maybe some connector bear more force than others. Haico (2022) identifies the percentage different connector force of modular rectangular floating structure for 3- modules arranged in linear arrangement and 9- modules arrangement side-by-side. Moreover, the maximum connector force of floating structure effect by various wave directions is critical to discuss due to guide the designer to create a sustainable connector design that can withstand forces from various wave angles while maintaining safety factor for extended periods. In addition, the good connector capable of handling wave angle can minimizes downtime and maintenance and establish the floating structure more eco-efficient and reliable in the long time.

3.2 Effect of Incident Wave Period

The hydrodynamic characteristics of the HFMS have been analysed with different wave periods (H=2m, $\theta = 0^0$). All modules (M2 to M6) have almost identical surge motion because they are connected as a unit structure in Figure 6a. The surge motion generally increases with the wave periods except at T=8s, which is near the natural period of the pitch motion of each module implying that the coupling effect of surge and pitch motion is similar (Yuan et al., 2020).

The heave motion is similar for all modules as shown in Figure 6b. The heave response increases with the wave period and reaches a peak period at 8s for M2, M3 and M6. After 10s, the peak period shifts to a longer wave period. The geometric nonlinearity of the connector during intense motion such as in the resonant zone, where the cable is stretched obliquely can greatly limit the movement of the modules in the vertical direction leading to a shift in the peak period of heave motion of each module in the double module floating system (Yuan et al., 2020).The pitch shows a tendency to ascend and then descend with increasing wave period, all reaching a maximum of 37^0 at a period of 8s as shown in Figure 6c the same pattern found by Ziyi et al., (2023).

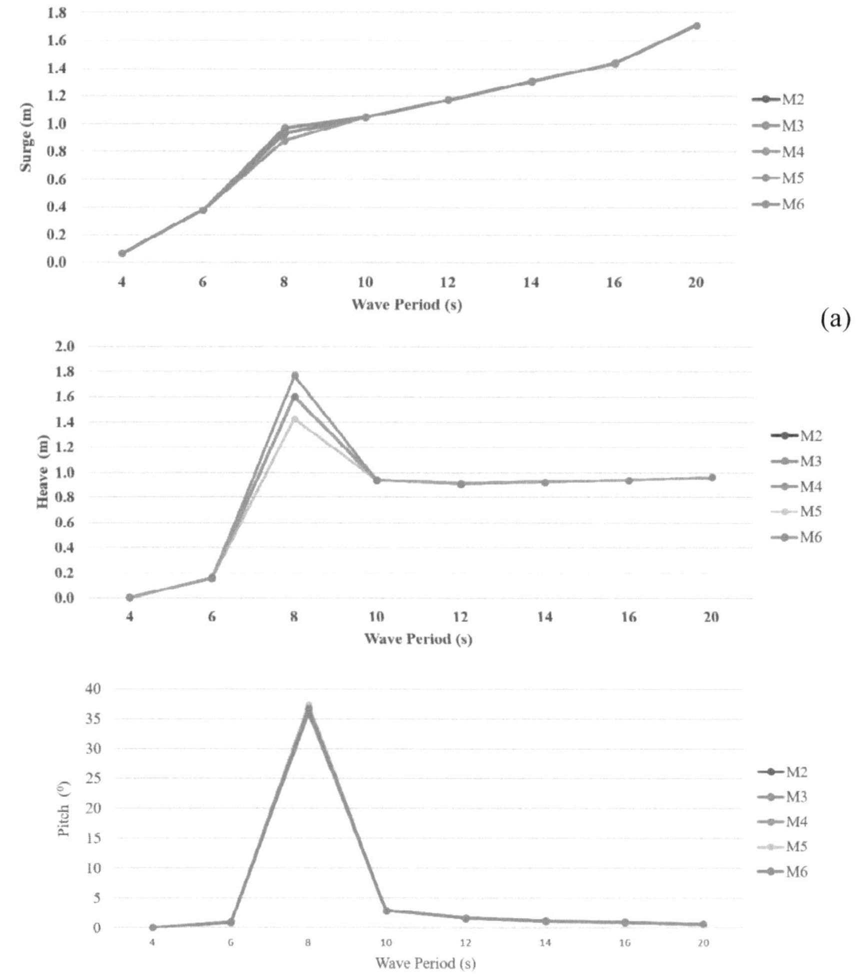

(c)

Figure 6. The main responses of each module versus wave period at 0^0 wave direction for 5 modules: (a) surge, (b) heave, (c) pitch.

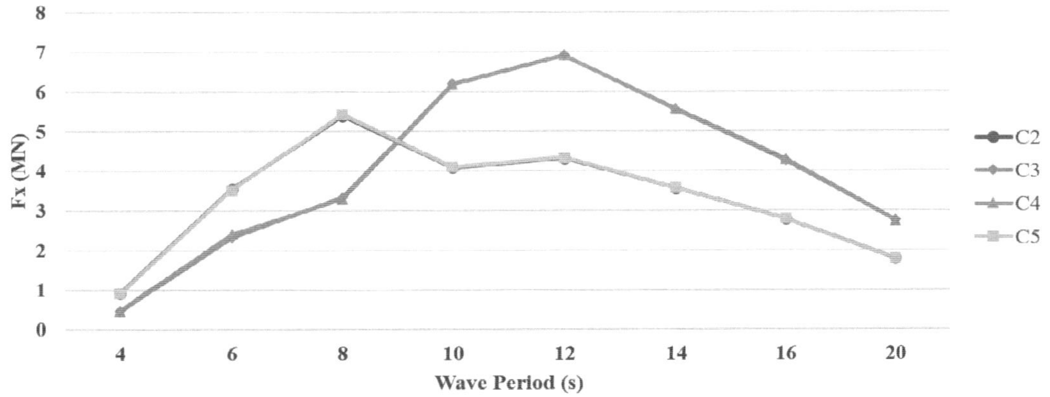

Figure 7. The maximum horizontal force of each connector versus wave period at 0^0 wave direction for 5 modules.

In general, all four connectors show an increase in the maximum horizontal load of the connector as the wave period increases from 4 seconds to 8 seconds in Figure 7. After this point, there is a general decline in Fx values at C2 and C5 located at the outer module as the wave period continues to increase up to 20 seconds.C3 and C4 exhibit a similar trend increasing from 4s to 8s, but they maintain a more consistent value from 8s to 12s before gradually decreasing. Thus, connectors C3 and C4 show the highest peak value while C2 and C5 exhibit lower and more stable values after reaching their respective peaks. The peaks of each connector occur in the wave period range of 8s and 12s, due to the coupling effect between different degrees of freedom.

The trend of maximum horizontal force of each connector versus wave periods for HMFS is similar to that of rectangular as comprised by Changqing et al. (2021). Changqing et al. (2021) states the horizontal connector load acting on connector occurred around the configuration's natural frequency were largest at first module and outer modules. Moreover, the horizontal load connector has higher value than vertical due to functional relationships of horizontal force were more complex than those of vertical forces.

The structural integrity of HMFS also considers the influence of the wave period on the connector forces, ensuring its stability and functionality even in extreme weather conditions. The wave period affects the HMFS connector, which is a key part of exploring how to make urban growth sustainable in a conceptual modular floating city planning framework. City planning includes floating housing, commercial zones and recreational zones, thus the connector of HMFS should be designed to withstand the varying wave conditions of urban waterfronts that create a stable and resilient environment. Wave period influences the connector durability studies to ensure the reliability and sustainability of floating structure infrastructures in the face of climate change, making cities more livable and adaptable to future environmental threats such as storm surges and flooding.

3.3Effect of Module Quantity

The primary connector forces of the head-wave direction have the most loaded horizontal connector load thus continuing to study the effect of the addition number of the module floating structure. The number of additional floating structures in Figure 8 has no effect on the surge motion and heave motion. However, Yanwei et al. (2022) determine that increasing the number of hexagonal modules can reduce the surge and the heave, particularly for longer wave periods in 1-module, 3-module and 5-modules. The single module exhibits the highest surge motion and heave motion.

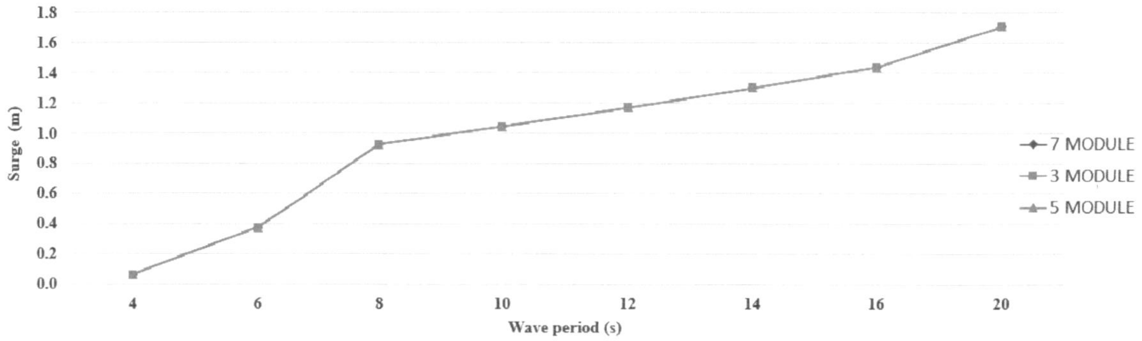

(a)

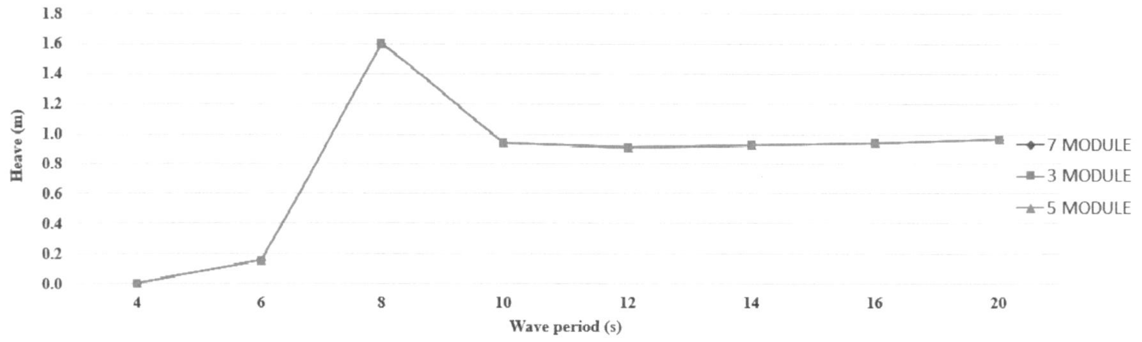

(b)

Figure 8. The main responses of each module versus wave period at 0^0 wave direction for different number of modules: (a) surge, (b) heave.

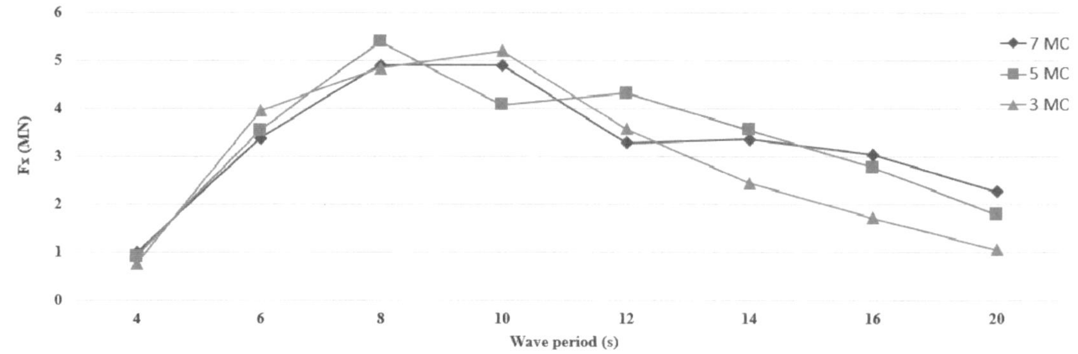

Figure 9. The maximum horizontal force of connector C2 versus wave periods at wave direction 0^0 for various number of modules.

The additional number of modules effects a consistency in load as the wave increases from 4s peaking at 8s and the load decreases gradually as show in Figure 9. The 3-module configuration bears the highest peak load until 10s while after that the 7 modules have the highest load among the 5 modules and the 3 modules. The similar pattern in the maximum force of the first connector studied by Yanwei et al. (2022) is that the increase in the number of hexagonal modules gives the highest horizontal force at 8s. In addition, the pattern indicates that within the range of 4s to 8s, fewer modules influences the maximum horizontal connector, whereas from 12s to 20s, more modules influence it.

The trend of maximum horizontal force of connector versus wave periods for HMFS is similar to that of rectangular as comprised by Lan et al. (2004) and Changqing et al (2021). Lan et al (2004) show the highest horizontal force at 6s to 8s for a 3-module rectangular floating structure. Changqing et al. (2021) investigated the horizontal force with an increase the number of modules such as 2-modules, 3-modules, 4-modules and 8-modules. The trend for all different numbers of modules given by the same pattern with the highest 8s to 10s. Thus, the trend of maximum horizontal force versus wave period with different shapes, such as rectangular and hexagonal is similar. Changing the shape of the floating structure also did not have an impact on the trend of maximum horizontal force versus wave period.

The expanded hexagonal modular floating structure with additional modules arranged in many layouts such as linear layout, tandem layout, central layout, U-shaped layout and V-shaped has considerable environmental sustainability. By adding modules, the surface area will be giving more space for integrating green technologies including solar panels, aquaculture system or floating wetlands. The hexagonal modular floating structure for renewable energy projects, like solar panels is effective at reducing land use and environmental degradation. Additionally, the maximum horizontal force of connectors has been studied to lead the development of designs that sustain the stability and durability for expanded modular floating structures to withstand dynamic marine environments without leading ecological disruption. These studied contribute to environment preservation by enhancing connector durability of modular floating structures that efficiently handle additional modules, thus reducing overall environmental impact for repairs, replacement or reconstruction of connector. These studies improve the environmental sustainability of floating structure infrastructures while also improving engineering solutions.

4.Conclusion

The present work proposes a hexagonal modular floating structure (HMFS) connected to five modules. The multi-floating structure hydrodynamic interaction effect and connector coupling effect were considered. The effect of wave direction, wave periods and module quantity on the hydrodynamic responses of the HMFS was investigated. The conclusions are summarised as follows:

(1) The 0^0 wave direction gives a higher horizontal connector load for hexagonal multi-floating and vice versa for rectangular. The lower horizontal connector loads in the 85^0 wave direction and vice versa for rectangular multi-module floating structures. Thus, the different design modules should first investigate the highest and lowest connector loads at each wave direction.

(2) Larger wave angles tend to lead to a decrease in horizontal connector force. Moreover, the absorption of wave kinetic energy tends to increase the horizontal connector forces at the outer connector.

(3) The HMFS can be more stable with expansion by providing a stronger shielding effect for the central module through the addition of more modules.

(4) The horizontal connector loads are much larger than the other connector forces, thus the horizontal connector load should focus on the connector design.

In this paper only consider the effect and trend for linear arrangement of HMFS influences by various wave directions, different wave periods and the additional of number of modules. The validation of model tests, the fatigue damage analysis of the main structure and the different layouts of floating structures should be investigated in the future. In addition, the HMFS concept is initially designed for regular waves, the future work may modify the environmental condition to explore the effective survival options for irregular wave conditions.

References

[1] Dai J, Hellan Ø, Want A and Ang KK 2022 Modular Multi-purpose Floating Structures for Space Creation *Proc. Second World Conference on Floating Solutions, Rotterdam* vol **158**
[2] Guohai D, Yanjun M, Yufei W Xiaozhou M Fang Y Xiong L and Luyao Z 2024 Experimental Study on the Hydrodynamic Performance of a Flexible Connected Double- Module Floating Structure *J. of Offshore Mechanics and Arctic Engineering* Vol. **146/ 051703** pp1-11
[3] Ikegami K, Masami M, Yuki W and Guo Z 2007 Response Characteristics of a Multi-Connected Floating Body System in Waves *Proc.26th International Conference on Offshore Mechanics and Arctic Engineering OMAE2007* pp1-10
[4] Jiang C, Moctar O and Schellin TE 2021 Hydrodynamic Sensitivity of Moored and Articulated Multibody Offshore Structures in Waves *J. Marine Science and Engineering* **9** 1028
[5] Lan Y, Run-pei L and Zhi Sh 2004 A Numerical and Experimental Study on Dynamic Responses of MOB Connectors *Proc. Fourteenth (2004) International Offshore and Polar Engineering Conference*
[6] Nianxin R, Xiang L,Yanwei L and Jinping O 2022 Hydrodynamic Responses of a Novel Modular Floating Structure System With Multi-Direction Expansion *J. Offshore Mechanics and Arctic Engineering* Vol. **145** pp1-12
[7] Park H, Jeong SK and Bo WN 2023 Numerical Analysis for Motion Response of Modular Floating Island in Waves *J. Ocean Engineering and Technology* **37(1)**, pp 8-19
[8] Riggs HR, Ertekin RC and Mills TRJ 1998.Wave-Induced Response Of A 5-Module Mobile Offshore Base *Conf OMAE '98, Lisbon Ocean Space Utilization Symposium*
[9] Yanwei L, Nianxin R, Xiang L and Jinping O 2022 Hydrodynamic Analysis of a Novel Modular Floating Structure System Integrated with Floating Artificial Reefs and Wave Energy Converters *J.l Marine Science Engineering* **10** 1091
[10] Yuan H, Chen G, Zhang W, Yin Y, Wang Y and Wang C., 2020, Design For Flexible Connector Of Multi-Floating Structure, *Proc. ASME 2020 39th International Conference on Ocean, Offshore and Arctic Engineering OMAE2020,*
[11]Yousheng W, Jun D, Zhiwei L, Xinyun N, Xiaofeng W and Chao T 2017 Hydroelastic Responses Of VLFS Deployed Near Islands And Reefs *Proc. ASME 2017 36th International Conference on Ocean, Offshore and Arctic Engineering*
[12] Changqing J, Ould el M, and Thomas ES 2021 Hydrodynamic Sensitivity of Moored and Articulated Multibody Offshore Structures in Waves *J. Marine Science Engineering* **9** 1028.
[13] Ziyi Z, Jian S, Xiaoxu H, Shaojin X, and Yufei W 2023 Optimization of a Hexagon-Shaped Floating Unit Based on the Response Surface. Available at SSRN: https://ssrn.com/abstract=4424891
[14] Haico 2022 Connector Response of a Grid Pattern Multibody Very Large Floating Structure Subject to Wave Loading *Thesis for the degree of MSc in Marine Technology in the specialisation of Ship and Offshore Structures* Delft University of Technology
[15] Yanwei L., Nianxin R., Xiang L., and Jinping O. Hydrodynamic Analysis of a Novel Modular Floating Structure System Integrated with Floating Artificial Reefs and Wave Energy Converters *Journal of Marine Science and Engineering* **10** 1091

Developing Port Sustainability Awareness through Implementing Serious Games: Recommending innovative approaches

David Rumeser[1*]

[1] Department of Strategic Initiative Monitoring, PT Pelabuhan Indonesia (Persero), Jakarta, Indonesia

*E-mail: david.rumeser@pelindo.co.id

Abstract. Ports are a key element in maritime shipping as they play as vital hubs of international trade, comprising of 90% cargo being transported through the sea. Encouraging sustainability practices in maritime industry, however, has been an issue as ports innovate in various methods to promote sustainability, but are struggling in raising sustainability awareness of their leaders, employees, suppliers, customers, and stakeholders. As behavior change towards sustainability could not be achieved through formal education or training, an experiential learning method called serious or educational games are proposed in this study. This study firstly examines various sustainability practices from 4 (four) modern ports, namely the Hamburg Port Authority, Port of Amsterdam, PSA, and Pelindo. Furthermore, it proposes 3 (three) serious or educational games concepts that addresses critical sustainability issues, such as port infrastructure development, port function as a supply chain integrator, sustainable procurement, and terminal bottleneck that leads to energy inefficiency. These games are *VR Port Sustainability Game (VR PSG), Port Circular Integrator Game (PCIG), and Back-to-Basic Sustainable Terminal Design Game (b2b-STDG).*

Keywords. Sustainability, Environment, Climate Change, Circular Economy, Green Supply Chain, Energy Efficiency, Infrastructure Development, Port Management, Serious or Educational Games

1. Introduction

Maritime shipping is renowned for its economic and environmental efficiency. According to the United Nations Conference on Trade and Development (UNCTAD), global seaborne trade surpassed 10 billion tons by volume in 2015, marking a nearly 400% increase since 1970 (UNCTAD, 2016; ICS, 2024). This industry is responsible for transporting approximately 90% of global trade (Panayides & Song, 2012; Walker, et al., 2019). Ports, as vital hubs of international trade, play a crucial role in this system. They are essential for loading and unloading cargo and offer a range of value-added services, including storage, warehousing, packing, and organizing inland transportation (Hossain, Adams, & Walker, 2021). Although maritime transport is often

considered environmentally friendly in terms of emissions per kilometer, its significant role in global trade means that its environmental impact is still considerable (Lim, Pettit, Abouarghoub, & Beresford, 2019).
Research indicates that, despite substantial efforts by various stakeholders, numerous obstacles to pro-environmental behavior persist in the maritime industry, highlighting the need for further action to enhance maritime sustainability (Čulin, Bielić, & & Jakšić, 2019). Education plays a crucial role, as insufficient environmental knowledge is frequently cited as a barrier to environmental awareness and action. However, addressing knowledge gaps and increasing concern does not automatically lead to behavior change, as behavior is influenced by a range of interacting factors.

One of the change management or educational methods that can be utilized to develop awareness and behavior changes towards port sustainability is *serious games*. Serious games (SGs), also known as educational games, integrate the elements of a simulation designed for educational purposes with the engaging aspects of entertainment. These games aim to facilitate learning and behavior change (Connolly, Boyle, MacArthur, Hainey, & Boyle, 2012) while actively involving learners through their interactive features (Hendrix, Al-Sherbaz, & Victoria, 2016). SGs present a simplified and condensed version of real-world problems, allowing players to experiment with decisions and reflect on the results. Through multiple iterations or rounds, learners make choices, collaborate, and/or negotiate based on either set goals or personal interests (Mayer & De Jong, 2004).

This study aims to address a gap whereby ports innovate in various methods and systems to promote sustainability, but struggle in building sustainability awareness of their leaders, employees, suppliers, customers, and stakeholders. To achieve this aim, this study will review and analyze existing sustainability practices of 4 (four) modern ports, namely the Hamburg Port Authority, Port of Amsterdam, PSA, and Pelindo. On the basis of this review, the study will then recommend 3 (three) adjacent innovative serious games (SGs) concepts that can be utilized to raise sustainability awareness in ports, namely: *VR Port Sustainability Game (VR PSG), Port Circular Integrator Game (PCIG), and Back-to-Basic Sustainable Terminal Design Game (b2b-STDG).*

2. Recommendations of Innovative Serious Games Concepts

2.1 Virtual Reality Games for Sustainable Port Infrastructure Development

a. Existing Practices and Implementation

The Hamburg Port Authority (HPA) applies a 3-D Virtual Reality Port Model to simulate port infrastructure management (Zuesongdham, 2017). It is capable to model and visualize the entire port area and ship movements in real-time interaction.

The HPA has incorporated different digital building models (BIM models) into the VR port model to evaluate how they interact with the current infrastructure, as well as to conduct virtual construction inspections and design evaluations (Zuesongdham, 2017). The aim is to foster a shared understanding among planning participants and, importantly, non-expert stakeholders through 3D visualization, helping to identify planning issues early on.

b. Proposals for Innovative Serious Games: "VR Port Sustainability Game (VR PSG)"

In the long run, VR Port Model and Simulation has the potential to be improved and converted into a serious game that challenges its players to design and build port infrastructure that minimizes CO2 emission while also ensuring efficient operations. This study proposes a game called *VR Port Sustainability Game (VR PSG) whereby* players (i.e. port developers) decide on infrastructure design & materials and operational equipment allocation which will then affect CO2 emission and costs. Hence, raising awareness on how the choice of design, materials, and the use of energy can affect sustainability of ports.

Decision on materials has a significant effect on CO2 emission as shown in PSA's initiative to substitute ordinary cement with ground granulated blast furnace slag (GGBS) in the Tuas Project Phase 1. The initiative has reaped more than 53,000 tCO2e of carbon savings to-date and reduced construction carbon footprint by 72% (PSA, 2023).

Decision on equipment allocation, especially on the number of equipment and its energy source will certainly affect CO2 emission and costs. Port of Rotterdam, for instance, has a goal of which by 2030, most of container ships and passenger vessels (including cruise ships, ferries, and roll-on/roll-off ships) docked at the port will be connected to shore power. This will lead to a 2% decrease in CO2 emissions (equivalent to about 200,000 tons of CO2) from the port operations (Rotterdam, 2022).

2.2 Simulation Model for Sustainable Ports as a Part of Supply Chain

In the age of the global supply chain, ports have transformed from merely handling and storing cargo to becoming a crucial component of the worldwide supply network (Han, 2018). Hence, modern ports nowadays have envisioned themselves as an orchestrator of supply chain activities. For example, PSA International, which operates the world's largest transshipment hub, has the vision "to be a leading supply chain ecosystem orchestrator" (PSA, 2024). While, Pelindo, the 8th largest container terminal operator in the world, envision themselves to be "a world-class integrated maritime ecosystem leader" (Pelindo, 2024).

a. Existing Practices and Implementation
As a supply chain integrator, ports play a crucial part in "greening" the supply chain.
The Port of Amsterdam, for instance, aims to achieve complete circularity by 2050 and has introduced the "Upcycle Factory" platform to support the exchange of waste materials as secondary raw materials (Ardakanian, 2020).

This platform links diverse industries through a smart supply-demand matching algorithm, offering comprehensive services to both buyers and sellers. This initiative arises from the port's role as a hub for numerous industries that produce millions of tons of various waste streams annually.

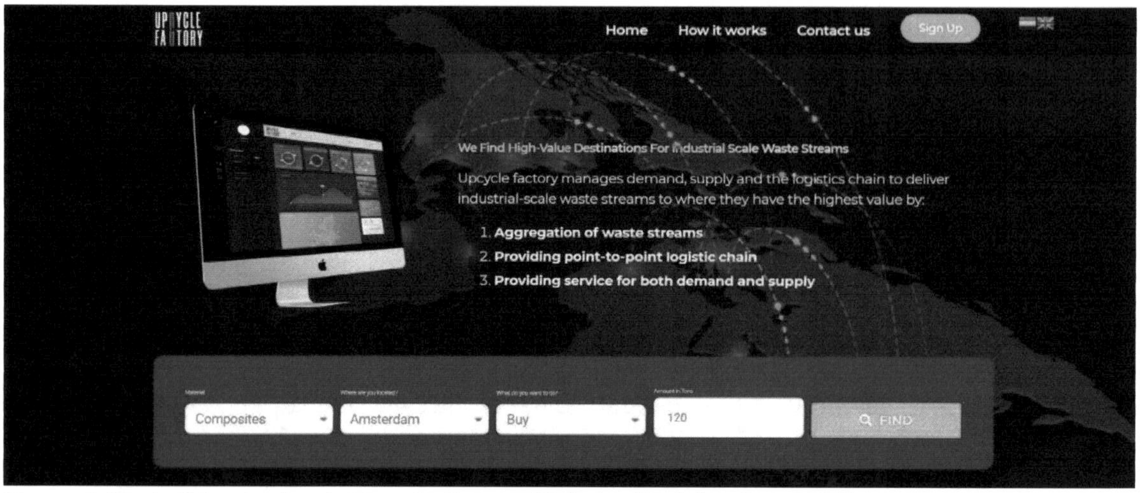

Figure 1. Port of Amsterdam's Upcycle Factory Website (Ardakanian, 2020)

Another example of how ports can affect their supply chain surroundings is shown in PSA's policy for sustainable procurement (PSA, 2023). The initiative ensures that PSA collaborates with suppliers who are aligned with PSA's goals of reducing carbon emissions and advancing ESG principles. PSA's suppliers are assessed based on the following criteria: energy efficiency, utilization of clean energy/fuels, reduced packaging, and utilization of recycled materials.

b. Proposal for Innovative Serious Games: "Port Circular Integrator Game (PCIG)"
Based on discussed existing practices (i.e. Port of Amsterdam's circular supply chain platform and PSA's sustainable procurement framework), this study proposes a serious games concept called the "Port Circular Integrator Game (PCIG)" whereby players are port top management and business executives who are tasked to make decisions on activities that are related to supply chain, business expansion, and procurement policy:

- *Suppliers selection* based on each supplier's profile, including pricing, product or service quality, as well as ESG compliance track record;
- *Business expansion by aggregating and matching the demand and supply of various industries* based on each industry's waste products and needs for secondary raw materials.

As shown in Figure 2 (i.e. the green lines), port managers can play a critical role in matching the needs of receiving/importing industries (e.g. Plastic Bricks Factory) with wastes as secondary raw materials from other industries who are the exporters hosted by their ports (e.g. Electronic Waste and Plastic Waste Producers). A serious games that simulate this interaction can help raise port managers' awareness and understanding of the "orchestrating" or "integrating" role of ports in circular economy.

Furthermore, decision makers should also consider each industry's ESG compliance track record as it will affect simulated carbon emission level which is one of the critical KPIs in the proposed game. This, in turn, will help raise awareness and improve understanding on the importance of sustainable procurement in ports.

IOP Conf. Series: Earth and Environmental Science **1461** (2025) 012019 doi:10.1088/1755-1315/1461/1/012019

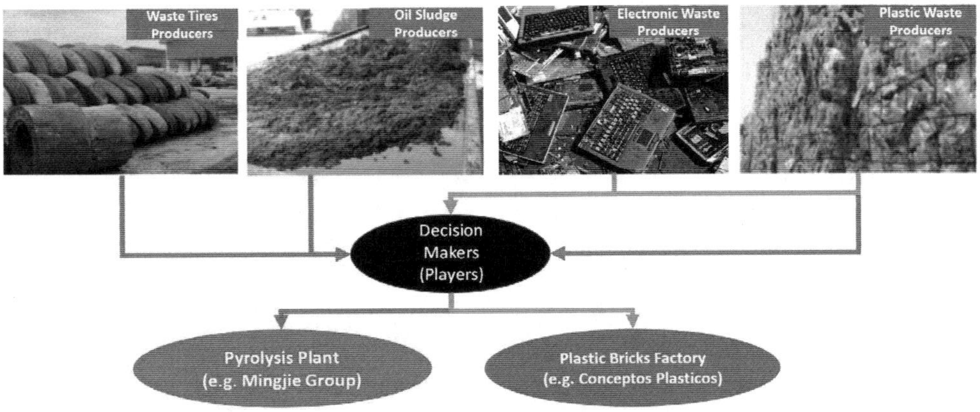

Figure 2: Illustration of Port Managers' Decision in Aggregating and Matching Supply and Demand of Waste Products as Secondary Raw Materials

2.3 Role-Play Simulation in Designing a Sustainable Container Terminal Operations

One of the approaches to improve sustainability in ports is by reducing energy usage which is a direct result of efficiency. This, however, cannot be achieved when there is a bottleneck between berth, yard, and gate operations as it limits container terminal capacity (van Battum, et al., 2023).

a. Existing Practices and Implementation

As a result of a merger of 4 largest port operator companies in Indonesia, PT Pelabuhan Indonesia (Persero) or Pelindo, the surviving entity, becomes the largest state-owned port operator in Indonesia and 8th largest container terminal operator in the world (Pelindo, 2024). The first priority after merger is to standardize port services across diverse terminals in Indonesia, and the biggest challenge is to enhance and standardize the skills of Pelindo's human resources which are varied significantly across different regions.

Amongst other skills, one of the basic skills which is an area of improvement is terminal design. Interestingly, the training methods applied by Pelindo, especially in the east region is through role-play simulation.

Figure 3: Terminal Design Role-Play Simulation in Sorong, East Indonesia

As depicted in Figure 3, teams are tasked to design container terminal by building a prototype that includes 3 aspects: berth, yard, and gate operations. The goal of this role is to improve the efficiency of the terminal by avoiding bottleneck, on the basis of the demand data provided by the training instructor.

b. Proposal for Innovative Serious Games: "Back-to-Basic Sustainable Terminal Design Game (b2b-STDG"

Sustainability aspect, however, is not discussed in the role-play. Consequently, this role-play can be further improved by calculating the indirect effect of bottleneck inefficiency when the terminal design is not balanced properly on CO2 emission. The role of training facilitator in this case is therefore critical to raise participants' awareness on how efficiency can lead to sustainability in ports.

4. Conclusion

This study seeks to fill a gap where ports tend to introduce various innovative methods and systems to advance sustainability but often overlook efforts to enhance sustainability awareness among their leaders, employees, suppliers, customers, and stakeholders. To address this, the study has examined and evaluated the current sustainability practices at 4 modern ports: the Hamburg Port Authority (HPA), Port of Amsterdam, PSA, and Pelindo.

The study then proposes related innovative serious game concepts:

- the VR Port Sustainability Game (VR PSG) that simulates sustainable port infrastructure design,
- the Port Circular Integrator Game (PCIG) that simulates circular supply chain and sustainable procurement, and
- the Back-to-Basics Sustainable Terminal Design Game (b2b-STDG) that simulates bottleneck inefficiency reduction in terminal design

as tools to improve sustainability awareness within ports.

References

Ardakanian, N. (2020). *IAPH World Port Sustainability Program*. Retrieved from https://sustainableworldports.org/project/port-of-amsterdam-upcycle-factory/

Connolly, T. M., Boyle, E. A., MacArthur, E., Hainey, T., & Boyle, J. M. (2012). A systematic literature review of empirical evidence on computer games and serious games. *Computers & Education*, 661-686.

Čulin, J., Bielić, T., & & Jakšić, K. (2019). Suggestions for improving the effectiveness of environmental education in the maritime sector. *Pomorstvo*, 232-237.

Han, C.-h. (2018). Assessing the impacts of port supply chain integration on port performance. *The Asian Journal of Shipping and Logistics*, 129-135.

Hendrix, M., Al-Sherbaz, A., & Victoria, B. (2016). Game based cyber security training: are serious games suitable for cyber security training? *International Journal of Serious Games*, 53-61.

Hossain, T., Adams, M., & Walker, T. R. (2021). Role of sustainability in global seaports. *Ocean & Coastal Management*.

ICS. (2024). Retrieved from Shipping and world trade: driving prosperity: https://www.ics-shipping.org/shipping-fact/shipping-and-world-trade-driving-prosperity/

Lim, S., Pettit, S., Abouarghoub, W., & Beresford, A. (2019). Port sustainability and performance: A systematic literature review. *Transportation Research Part D: Transport and Environment* , 47-64.

Mayer, I., & De Jong, M. (2004). Combining GDSS and gaming for decision support. *Group Decision and Negotiation*, 223-241.

Panayides, P., & Song, D. (2012). *Maritime Logistics : Contemporary Issues.* Bingley, U.K: Emerald Group Publishing Limited.

Pelindo. (2024). *Pelindo Company Website.* Retrieved from https://www.pelindo.co.id/

PSA. (2023). *PSA Sustainability Report 2023: Green Horizons - Going the Distance.* Singapore: PSA.

PSA. (2024). *PSA Company Website*. Retrieved from https://www.singaporepsa.com/about-us/our-vision-mission-values/

Rotterdam, P. o. (2022). *Port of Rotterdam*. Retrieved from The port of Rotterdam is working on shore power: https://www.portofrotterdam.com/en/news-and-press-releases/the-port-of-rotterdam-is-working-on-shore-power

UNCTAD. (2016). *Review of Maritime Transport 2016.* UNCTAD.

van Battum, C. H., Wiegmans, B., Atasoy, B., van Wingerden, E., de Waal, A., & Tavasszy, L. A. (2023). Performance improvements in container terminals through the bottleneck mitigation cycle. *Maritime Economics & Logistics*, 174-195.

Walker, T., Adebambo, O., Del Aguila Feijoo, M., Elhaimer, E., Hossain, T., & Edwards, S. (2019). *Environmental effects of marine transportation.* Academic Press.

Zuesongdham, P. (2017). *IAPH World Ports Sustainability Program*. Retrieved from Port of Hamburg – Virtual Reality for model-based port infrastructure management: https://sustainableworldports.org/project/port-of-hamburg-virtual-reality-for-model-based-port-infrastructure-management/

IOP Conf. Series: Earth and Environmental Science **1461** (2025) 012020 doi:10.1088/1755-1315/1461/1/012020

Analysis of the Maturity Level of Research and Innovation Management in Indonesian Port Companies Through Human Resources Expectations and Satisfaction Levels

Yudho Satrio Wirawan[1*], Febriandika Putra Anggia[1*]

[1] Department of Research and Innovation, PT Pelabuhan Indonesia (Persero), North Jakarta, Indonesia

*E-mail: yudho.satrio@pelindo.co.id, febriandika.anggia@pelindo.co.id

Abstract. Along with the rapid development of technology, innovation is the key to maintaining the competitiveness and operational efficiency of the company. This research aims to identify and evaluate the maturity level of the research and innovation management process at PT Pelabuhan Indonesia (Persero) or Pelindo. The survey was conducted involving 2,240 respondents covering six main pillars, namely strategy, culture, process, governance, expertise, and future foresight. The method used in this research involves the analysis of quantitative data generated from the survey. The results of the research show that the expertise pillar obtained the highest score, which is 82%, indicating that the ability and knowledge of employees are considered high enough to support research and innovation activities in the company. However, the governance and future foresight pillars still need to be improved, with each obtaining an average score of 78%. The findings highlight a gap between existing and ideal conditions, particularly in the governance and future foresight pillars, which requires targeted strategies to bridge these disparities. Although the average satisfaction with research and innovation management at Pelindo is 80%, a gap remains between current and ideal conditions. The findings highlight the need for Pelindo to develop future-focused innovation strategies, particularly improving governance and adopting a forward-looking approach. The discussion of this research provides insight for Pelindo management to design more effective strategies in improving research and innovation performance, as well as closing the gap that exists between current conditions and desired ideal conditions.

1. Introduction

1.1 Background

Innovation has an important role in encouraging long-term growth of economic activity in various industrial sectors. Innovation is a new product or process or any form of improvement of a previous product or process where the goal is to improve a certain state or situation in a company or organization [1] [2]. With the various advancements and widespread adoption of digital technologies today, such as IoT and AI, innovations both technological and non-technological have transformed the shape of products and services in most industries. This has

IOP Conf. Series: Earth and Environmental Science **1461** (2025) 012020 doi:10.1088/1755-1315/1461/1/012020

prompted a major change in the company [3]. The maritime and port sectors, as one of the industrial sectors, have also been affected by the development of innovation over the past decade [4]. For instance, Port of Busan in 2022 adopted cutting-edge innovations in automation, digital platforms, and AI-driven processes. These advancements were integrated not only into the port's operations but also in its organizational and business models, leading more efficient cargo handling, lower operational costs, and reduced environmental impact [5] . Moreover, initiatives such as the Marocco National Ports Agency's 2021 online Hackathon demonstrate the growing emphasis on digital innovation within the port sector. The port industry is starting to introduce many innovations that can help them in facing dynamic changes and more competitive competition [3].

1.2 Background of the Problem

PT Pelabuhan Indonesia (Persero) or Pelindo is a state-owned company that focuses on port services and is the largest port management company in Indonesia. Pelindo continues to encourage the development of innovation and improvement of corporate services, considering dynamic changes in various aspects in the port sector such as world trade flows, changes in economic trends, and changes in consumer demographic patterns [6]. Although various innovation initiations have been carried out by Pelindo, until now there has been no in-depth analysis of whether the implementation of the innovations carried out is in accordance with the framework used as a reference in carrying out innovation activities in the Company [7]. Therefore, it is necessary to conduct a reassessment to determine the level of maturity of the management process that is currently being carried out [8].

1.3 Research Objectives

This research aims to analyze the management of research and innovation at Pelindo, in order to find out the extent of the management process carried out and whether the implementation is in line with the framework that has been set. In addition, this research also focuses on assessing the maturity level of the process carried out by the Department of Research and Innovation as the main person in charge of innovation management at Pelindo. The results of this research are expected to assist Pelindo's Research and Innovation Manager in ensuring the sustainability of the implementation process, management, and suitability of the implementation of research and innovation with the established framework and corporate governance.

2. Research Description

2.1 Innovation in the Port Sector

Innovation in general is a product, process, or combination of the two that is new or has been improved and is different from the previous product or process and is made for potential users or used by a work unit or process unit [1]. Arduino specifically explains that innovation is a technological or organizational change (including culture and market as a separate subset) of a product, service, or product process that can reduce costs or improve the quality of the product/service to customers [9]. The OECD with its guidance in the Oslo Manual has made a classification related to the types of innovations developed into several types, as can be seen in Table 1.

Table 1. Types of innovations based on the Oslo Manual

Categories Innovation	Description
Product Innovation	New or improved products or services from previous company products and services.
Process Innovation	New or improved processes for one or more business functions performed by the company
Organizational Innovation	New or improved processes in terms of strategic management, corporate governance, human resources, and external stakeholder management from previous company processes
Marketing Innovation	New or improved processes in terms of marketing methods, pricing strategies, and sales and after-sales activities of the previous company

Although various innovations have been developed and researched in the port sector, completely technology-based innovations are rare [10]. This also applies to innovations related to management, organizational, and culture that do not use technological components in their development. Innovations with multi-disciplinary backgrounds are more commonly found in the port sector . In addition, changes to the market are common, with more innovations combining technology and management than technological innovations alone. This may be due to the large port industry, which makes the scope of influence of innovation wide, so that the change in the product market is not limited to a specific location and region [11]. However, a process, product, or innovation does not always depend on technology and its development as the main driver or to function. In fact, innovations that are not so dependent on technology are easier to achieve success in their innovation process [11]. Pelindo continues to strive to present innovations to improve services by developing in various lines, including the development of social media applications, correspondence (electronic office) as a standardization of scripts, centralization of IT Service Desk, and others [12].

2.2 Gap Analysis

Companies have objectives and goals, both short-term and long-term, that can be measured. The existing condition of the company can be assessed and compared with the ideal performance expected by the company in its objective [14]. The difference between the company's current achievement and the expected results will usually creates a gap between the two. Gap analysis is a method used to find out the reasons why the gap appears as well as efforts to develop steps to reduce or eliminate the gap [14]. Gap analysis can help companies improve their effectiveness in various business functions as well as assist managers in determining the steps that need to be taken to reduce the gap [15]. Based on, in conducting a gap analysis, the company initially needs to identify the main needs of the company's current state. Then, the company needs to establish the ideal circumstances expected by the company. Companies also need to pay attention to the gaps that arise and change and implement the company's plans to overcome these gaps [16].

There are different types of gaps, depending on what function is being analyzed. Explanations of the various types of gaps include:

- Expectation Gap: The difference between a company's performance expectations and the company's actual performance. Because the expectation gap refers to the difference in expectations for actual performance, the expectation gap includes various other types of gaps, namely the factual gap, the conformance gap, and the ideal gap.
    - Factual Gap: Involves a difference in the acceptance of facts related to the company's existing condition, both in terms of performance and behavior, between the company and its stakeholders.

IOP Conf. Series: Earth and Environmental Science **1461** (2025) 012020 doi:10.1088/1755-1315/1461/1/012020

  - Conformance Gap: The gap between what is actually happening in the company and what is expected according to the standards that have been set.
  - Ideal Gap: The difference between the actual condition of the Company and the ideal picture that the company expects.
- Legitimacy Gap: The difference between the general view of the behavior and actions of the company that should or is expected of the behavior and actions of the actual company.
- Performance Gap: The difference between a company's performance and the general view accepted as a company's performance, similar to the factual gap [18]

Gap analysis is used in this research to analyze the existing conditions of Pelindo's research and innovation management with the conditions expected by all workers in the company. Recent research on innovation maturity models, particularly those related to digital transformation in port management. For example, the Port of Rotterdam and Singapore Port Authority have implemented cutting-edge digital strategies, such as digital twin technology and blockchain for improving operational transparency, along with AI-driven predictive analytics and automated guided vehicles to boost efficiency [19]. These integrated approaches have greatly strengthened their competitive advantage. By comparing Pelindo's innovation framework and challenges with these global examples, it becomes clearer which areas need improvement, particularly in the adoption of digital technologies to enhance operational effectiveness.

2.3 Innovation Management and Innovation Measurement

Innovation management is necessary to ensure that the innovations developed are as planned. Innovation management is a procedure that allows companies to generate new ideas, practices, or products in a structured and systematic manner, positively impacting the company's performance. The success of implementing an innovation strategy can vary, so it is important for companies to measure and analyze innovation management and identify the improvements needed to achieve the desired performance[15]. Pelindo uses a special framework to manage innovation within its company, which can be seen in Figure 1.

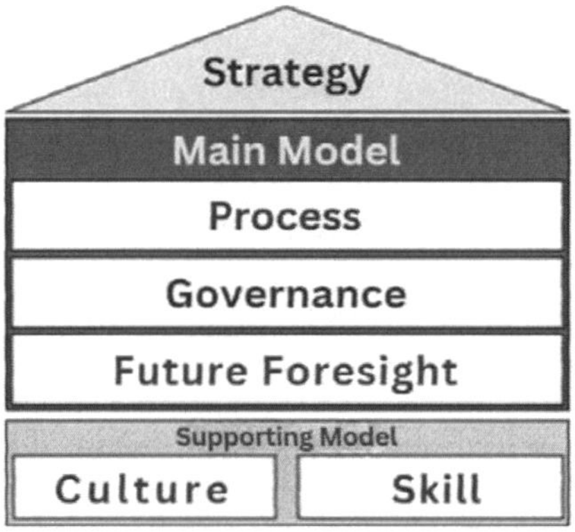

Figure 1. Pelindo's innovation framework

Pelindo has six dimensions based on the innovation framework, namely strategy, process, governance, foresight, culture, and expertise/talent, where these dimensions are used as a guide to regulate the company's overall innovation operation model. These dimensions are branched into 18 innovation sub-themes to provide a more focused view related to research and innovation. Dimensions and sub-themes can be seen in Table 2.

Table 2. Dimension pillars and sub-themes of innovation

Innovation Dimension	Innovation sub-theme
Strategy	Corporate strategy Impact measurement
Process	Innovation management Portfolio management Risk management Utilization of technology
Governance	Structure and network Roles and responsibilities Leadership involvement Performance management
Future Foresight	Digital and non-digital trends/forecasts
Culture	Internal collaboration and engagement Incentives and rewards External collaboration Value adoption
Talent	Innovation skills Recruitment management Learning and development

There are various methodologies that companies can use to measure and evaluate their innovation practices, some of which are the Oslo Manual which discusses standards for defining and measuring innovation [1], the Balanced Scorecard which although it seems unrelated to innovation, perspectives on internal processes and learning and growth perspectives. It can be used in analyzing existing innovations [4], as well as the Maturity Model used in assessing the maturity level of the company in managing its innovations. The researcher decided to use the innovation maturity model as a reference in assessing the maturity of the innovation process within Pelindo. This is because Pelindo has its own innovation maturity model that can be used to assess the company's maturity level.

2.4 Innovation Maturity Model in Measuring the Maturity of the Innovation Process

Various frameworks for assessing innovation management have been used in various industries, helping companies improve their innovations [13]. According to Acciaro [11], high innovation capability, namely the company's ability to continuously generate, develop, and implement new ideas, is important for the company's performance and competitive advantage. Maturity can be an indicator of a company's capabilities in the context of innovation [23]. The maturity model shows the extent to which a particular activity or process is managed and measured, which can improve performance. The innovation maturity model, a derivative of the capability maturity model, measures the level of innovation maturity within a company, indicating areas for improvement to achieve higher levels of maturity and competitive advantage [19]. In addition, the maturity model supports company benchmarking in comparing innovations and processes to ideal conditions, as well as providing guidance for improvement [24] Several innovation maturity models have been developed by entities such as Gartner Group [24], BERG Consulting [25], Blommerde [26], and Essmann [27]. A comparison of these models can be seen in Table 3.

Table 3. Identification of Innovation Maturity Model

Model Developer	Model Name	Model Characteristics
Gartner Group	Gartner Maturity Model for Innovation Management	The Gartner Maturity Model has 5 maturity levels and six dimensions of innovation management. Maturity level: level 1 Reactive, level 2 Active, level 3 Defined, level 4 Performing, level 5 Pervasive.6 dimensions of maturity model: Strategy and Intent, Processes and Practices, Culture and People, Organization and Infrastructure, Partnerships and Open Innovation, and Innovating How We Innovate.
Berg Consulting	The Innovation Maturity Model	The model consists of 5 maturity levels: Entry level innovation practices, Emerging innovation practices, Co-ordinated innovation practices, Innovation leadership, Industry innovation leadership.
Blommerde and Lynch, 2016	Service Innovation Capability Maturity Matrix	The model consists of 5 maturity levels and 4 dimensions of innovation management. Maturity level: Initial, Managed, Defined, Measured, Optimising. Dimensions: User involvement, Knowledge management, Strategizing, Networking.
Essmann, 2009	Innovation Capability Maturity Model	The model consists of 5 maturity levels: Ad-hoc innovation, Defined innovation, Supported innovation, Aligned innovation, Synergised innovation.

Based on the identification that has been carried out, the researcher will use the four maturity models as a reference in assessing the maturity level. Pelindo already has its innovation model in the process of managing research and innovation in their company. The innovation maturity model of Pelindo consists of five levels which can be seen in Table 4.

Table 4. Maturity level in Pelindo's Innovation Maturity Model

Maturity Level	Description	Innovation mindset
5 "Pioneering"	Innovation is the fundamental driver of the organization	Innovation Pioneer – "Innovation is a key driver of our organization"
4 "Leading"	Proven and continuously optimized innovation capabilities	Innovation Leader – "We are good innovators who know how to do it"
3 "Managing"	Innovation capabilities are being strategically developed across the organization and synchronized with organizational goals	Innovation Manager – "The whole organization is learning to innovate as a way to achieve our goals"
2 "Developing"	The organization accepts the need for innovation & capabilities and is being developed in a specific area	Innovation Developer – "Innovation is valuable and I see the areas where we"
1 "Resisting"	Ad-hoc innovation activities with limited support from the organization	Innovation Rejection – "Innovation is risky and we fear it could do more harm than good"

Pelindo has previously assessed the maturity of the research and innovation management process within its scope. The assessment was carried out by analyzing each pillar of the innovation framework dimension. The assessment was carried out to determine the level of maturity of research and innovation management at that time. Based on the assessment and gap between the current conditions and what Pelindo wants, which can be seen in figure 2, the maturity of the company's innovation can be categorized in the developing stage or level 2. Of course, the company wants to continue to improve the maturity of its process to be better than the previous assessment until it reaches the desired maturity with an average score of 4,72.

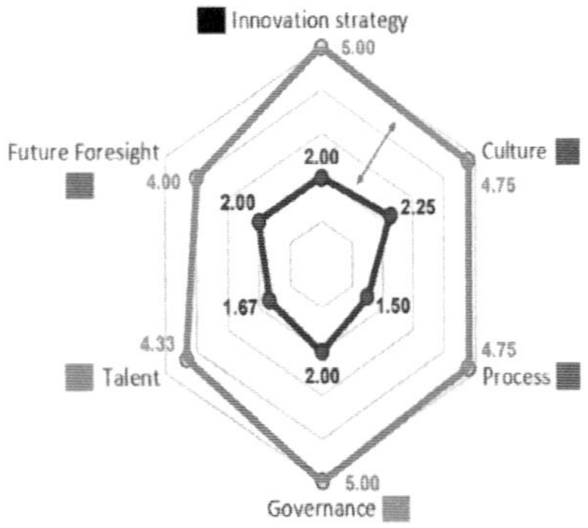

Figure 2. Overall score of the maturity of the company's innovation process

3. Methodology

The methodology for this research follows a structured approach, focusing on understanding and assessing the management of research and innovation at Pelindo. By employing a quantitative survey method, data was collected from approximately 12,000 Pelindo employees across Indonesia, resulting in 2,240 respondents. The survey questions underwent rigorous testing to ensure their validity and reliability, ensuring accurate measurement of innovation management dimensions and consistency in responses. The following steps outline the overall research process, from problem identification to the evaluation of maturity levels in research and innovation management, ensuring a comprehensive analysis of Pelindo's innovation practices:

- Identify the problem, The researcher identified problems related to the management of research and innovation at Pelindo to formulate problems to be solved.
- Literature Studies, looking for relevant references related to innovation management, gap analysis, innovation framework, and innovation maturity model as the basis for research.
- Survey Question Mapping, determining labels based on dimensions in Pelindo's innovation framework to facilitate visualization and assessment of maturity levels.
- Preparation of Data Collection Instruments, making survey questions and conducting interviews with related parties to collect information from workers within the Pelindo Group.
- Identify Existing and Ideal Conditions, analyze the existing conditions of research and innovation management and compare them with the expected ideal conditions.
- Data Processing and Maturity Level Measurement, processing survey data using Microsoft Excel and measuring the maturity level of research and innovation management based on existing models.
- Evaluation with the Innovation Maturity Model, analyzing the results of the survey according to the innovation maturity model to determine the current maturity level of research and innovation management at Pelindo.

These steps can be seen in figure 3:

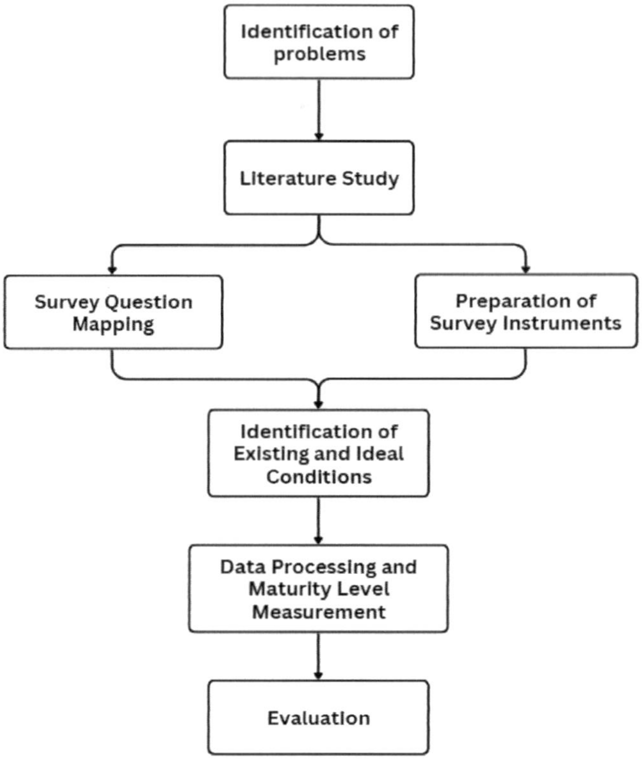

Figure 3. Research Methodology

The flowchart in figure 3 Research Methodology shows the systematic stages in this research, starting from problem identification to evaluation of survey results. Each stage is arranged with a clear goal to provide guidance in managing and improving the research and innovation process at Pelindo. The initial stage involves the identification of the problem and the research of the literature that forms the basis for building an initial understanding and formulation of the problem relevant to the research topic. Furthermore, the research involves the preparation of survey instruments, data processing, and evaluation using the innovation maturity model. Each of these stages is carried out in a structured manner to ensure that the data obtained is valid and can be interpreted properly. The final results of this stage provide recommendations that can be used by Pelindo to improve research and innovation strategies and management in the future.

4. Implementation

4.1 Design of Research Instruments

4.1.1 Designing Survey Questions

This stage involves the formulation of survey questions, as illustrated in Table 5, which will be disseminated to gather information regarding the existing conditions and the desired state of the research and innovation management process at Pelindo.

Table 5. Survey Question Design

Question	Estimated Dimension Pillar
The innovative products implemented at Pelindo come from ideas that are not only from internal parties but also from external parties	Strategy
Pelindo already has adequate technological resources to support research and innovation activities (seen from aspects such as IT hardware and software)	Process
Pelindo has evaluated research and innovation KPIs (as can be seen from the measurement of KPIs related to innovation)	Governance
Pelindo has often held trainings or workshops related to research and innovation to improve skills and knowledge for workers	Talent
Pelindo has provided a lot of support for workers to develop research and innovation in the workplace	Culture/Culture

4.1.2 Research Data

This research uses data obtained from the distribution of the survey. The survey was compiled using a google-form and distributed via email blast with the criteria that the research respondents were workers within the Pelindo Group. There are 2240 respondents (as of March 18, 2024) who can be used to process information related to research and innovation management. The following is the number of respondents and the origin of the respondent entity which can be seen in Table 6.

Table 6. Number and Origin of Respondent Entities

Corporate Entities	Count
Head Office (Holding)	174
Region 1 (including Branch or Sub Region 1)	118
Regional 2 (including Branch or Sub Regional 2)	199
Regional 3 (including Branch or Sub Regional 3)	48
Region 4 (including Branch or Sub Region 4)	123
SPJM (including the Company's Children/Grandchildren)	253
SPMT (including Company Children/Grandchildren)	182
SPSL (including the Company's Children/Grandchildren)	17
SPTP (including Company Children/Grandchildren)	857
Non-Cluster Subsidiaries	36
Other	233
Total	2240

5. Research Results

5.1 Identify Existing Conditions of the Research and Innovation Management Process

The identification of existing conditions is carried out to understand and analyze the current conditions of the research and innovation management process at Pelindo. The identification was carried out based on the results of a survey filled out by 2240 Pelindo Group workers. The following is an explanation of the quantitative analysis percentages of each pillar and subtheme it has.

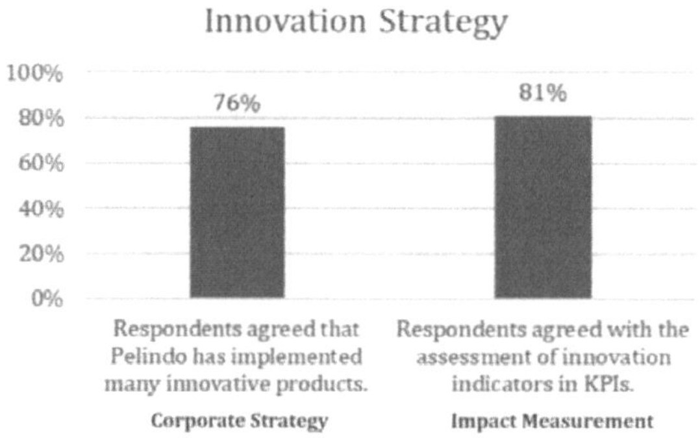

Figure 4. Innovation Strategy Survey Results

The survey results indicate that Pelindo has made significant strides in innovation, with 76% of respondents agreeing that the company has implemented numerous innovative products. This reflects a positive perception of Pelindo's corporate strategy in driving innovation. Additionally, 81% of respondents concurred that innovation is being effectively assessed through Key Performance Indicators (KPIs), highlighting the company's commitment to measuring the impact of its innovations and aligning them with organizational goals. This strong focus on innovation metrics suggests that Pelindo is actively monitoring its progress to ensure continuous improvement.

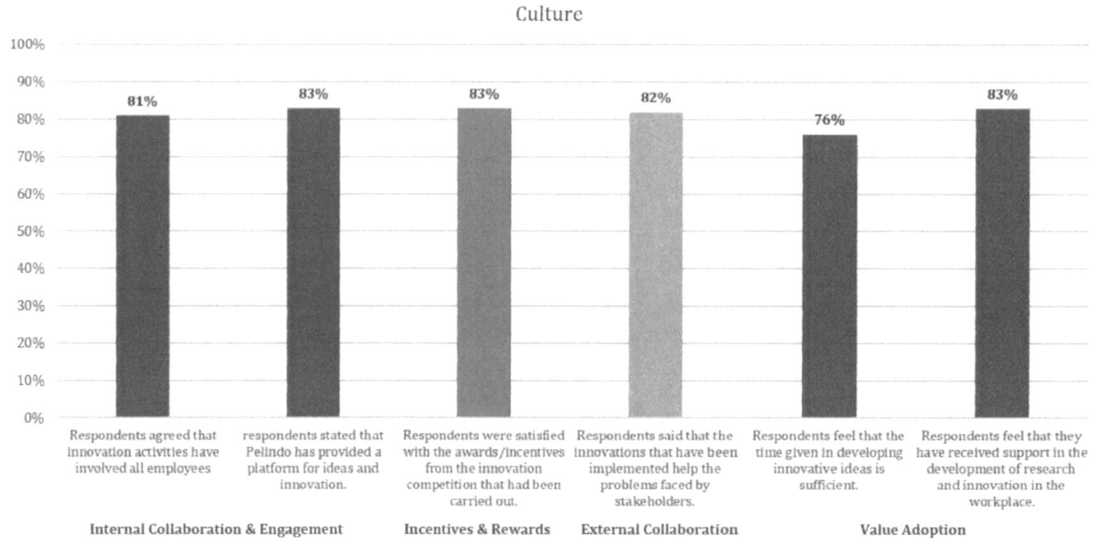

Figure 5. Culture Survey Results

The survey showed that Pelindo has built a strong innovation culture. 81% of respondents agreed that innovation activities involve all employees, and 83% confirmed that the company offers a platform to foster ideas and innovation. Regarding incentives, 83% were satisfied with the awards given for innovation competitions. External collaboration was also effective, with 82% of respondents acknowledging that the innovations implemented addressed stakeholder concerns. In addition, 76% believed that they were given enough time to develop innovative ideas, while 83% felt they received sufficient support for research and innovation in their workplace.

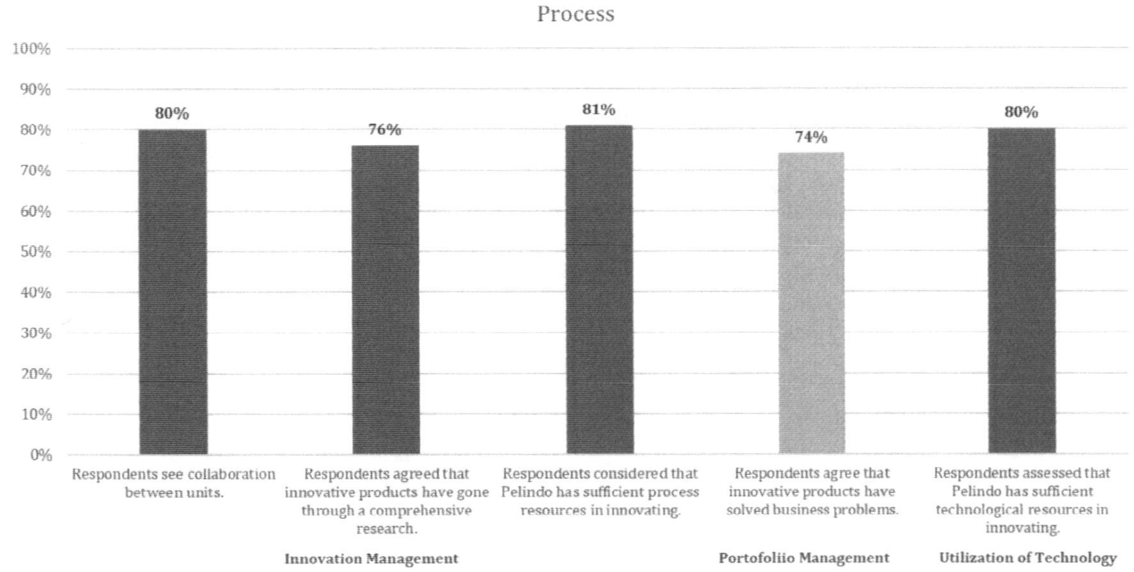

Figure 6. Process Survey Results

The survey results indicate that Pelindo's innovation process is well-structured. A notable 80% of respondents observe strong collaboration between units in managing innovation, while 76% agree that innovative products undergo thorough research. Furthermore, 81% believe that Pelindo has adequate resources to support the innovation process. In terms of portfolio management, 74% agree that the company's innovative products effectively address business challenges. Additionally, 80% of respondents feel that Pelindo utilizes sufficient technological resources to drive innovation, reflecting the company's commitment to leveraging technology in its operations.

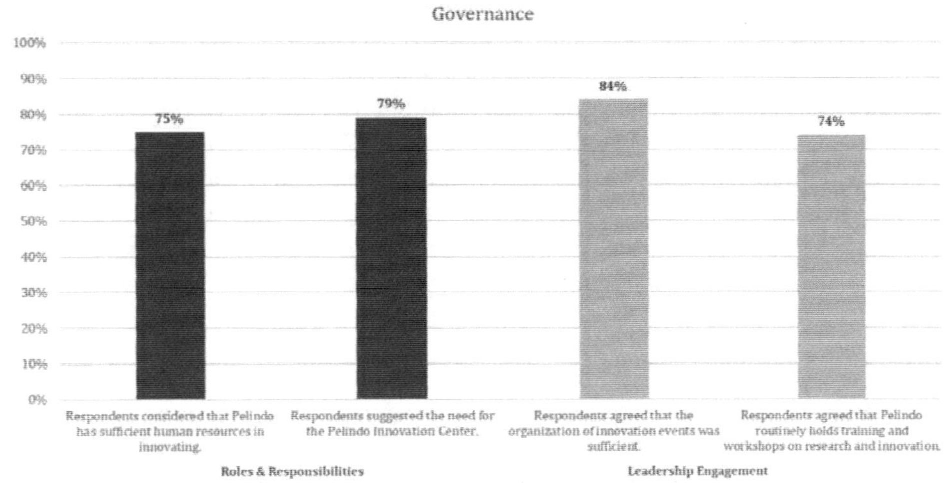

Figure 7. Governance Survey Results

The governance aspect of Pelindo's innovation framework demonstrates a generally positive response from participants. About 75% of respondents express satisfaction with the allocated budget for innovation initiatives, suggesting that financial resources are deemed adequate to support ongoing innovation efforts. Furthermore, 79% of respondents believe that external collaboration plays a crucial role in enriching the diversity and quality of innovative ideas, highlighting the importance of partnerships in Pelindo's innovation strategy. Additionally, an impressive 84% feel that Pelindo actively fosters a supportive culture for innovation, indicating a strong organizational commitment to encouraging creative thinking and initiative. Moreover, 74% agree that there is an acceptance of failure within the innovation process, which is essential for creating an environment conducive to experimentation and iterative learning. This collective feedback indicates that Pelindo is effectively cultivating a governance framework that not only supports innovation but also encourages growth through collaboration and a willingness to embrace challenges.

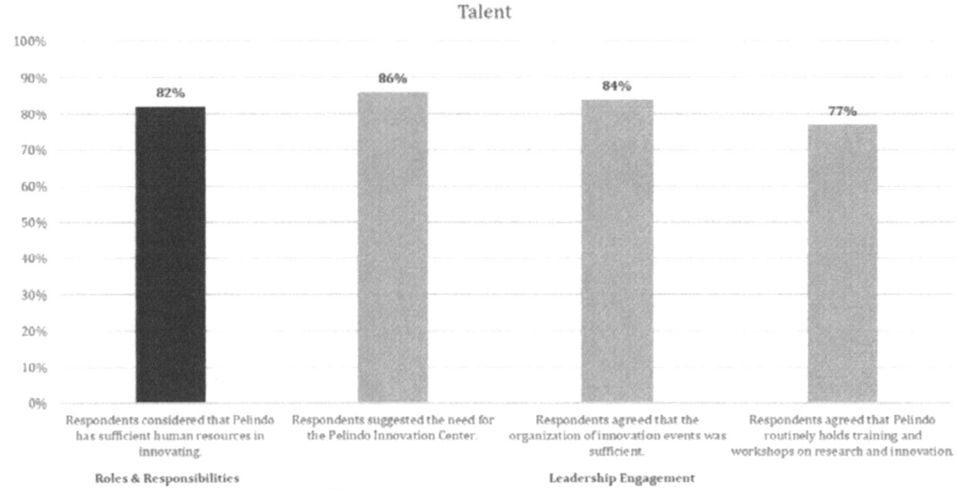

Figure 8. Talent Survey Results

The talent aspect of Pelindo's innovation framework reflects a strong belief in the organization's capabilities and commitment to nurturing its workforce. About 82% of respondents acknowledge that Pelindo possesses adequate human resources to drive innovation effectively. Additionally, there is a notable call for further development, with 86% of survey participants suggesting the establishment of a dedicated Pelindo Innovation Center, indicating a desire for a focused space to foster creative ideas. Furthermore, 84% agree that the organization of innovation events is sufficient, demonstrating that Pelindo is proactive in facilitating platforms for idea exchange and collaboration. However, 77% of respondents also noted that Pelindo regularly conducts training and workshops on research and innovation, underscoring the importance of continuous learning and skill development in enhancing the organization's innovative capabilities. Together, these insights suggest that Pelindo is on a positive trajectory in developing its talent pool to support innovation.

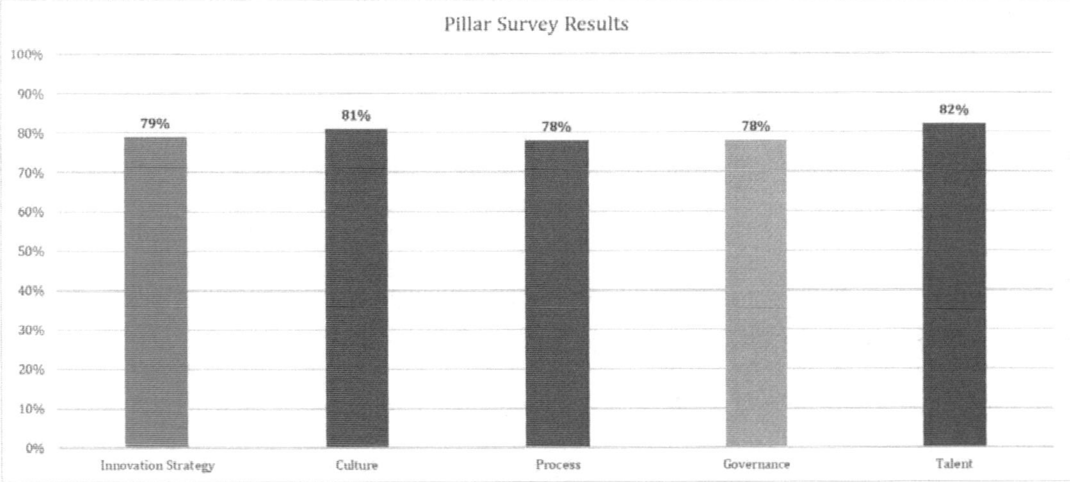

Figure 9. Pillar Survey Results

The survey covered five of the six pillars of innovation at Pelindo, yielding overall positive responses: innovation strategy received 79%, culture scored 81%, talent was rated at 82%, while both governance and process were the lowest at 78%. The relatively lower scores in governance and process may indicate potential areas for improvement. Specifically, while respondents expressed satisfaction with budget allocations and the encouragement of external collaboration, there might still be concerns regarding clarity in roles and responsibilities or the efficiency of internal processes. This could suggest a need for Pelindo to strengthen its governance structure and refine its innovation management processes to enhance overall effectiveness. By addressing these aspects, Pelindo can create a more robust environment that fosters creativity and innovation.

5.2 Identify the Ideal Expected Conditions

In the research that has previously been carried out by Pelindo, there are 18 results of subthemes or dimensions of the results of the innovation maturity assessment with the results of maturity at that time and also the desired maturity. The subtheme or dimension can be seen through figure 10 and the pillars that can be seen in figure 11 along with the color difference in the grouping of the maturity level status of the innovation as follows:

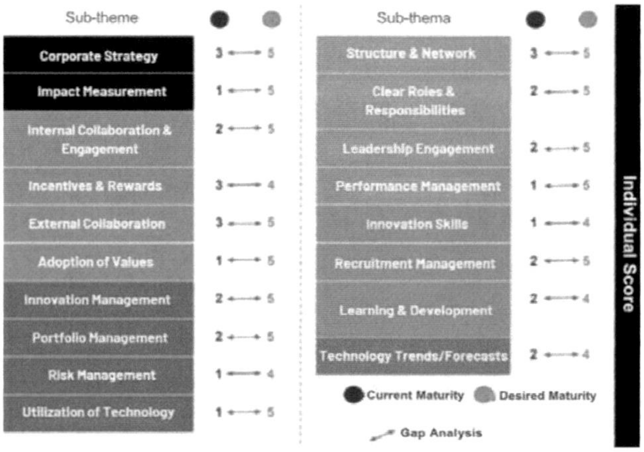

Figure 10. Summary Status of Maturity Level Innovation

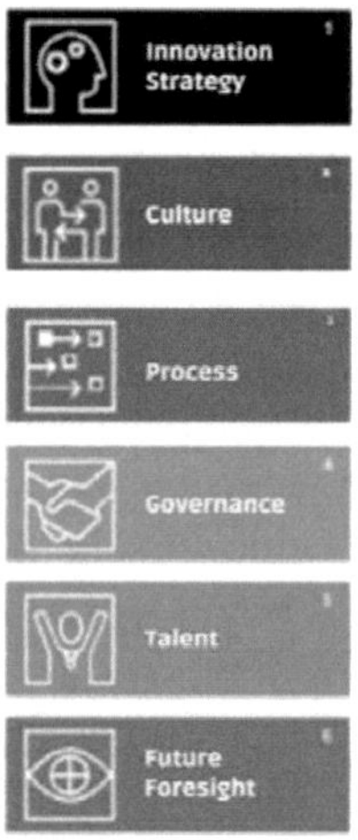

Figure 11. Pillar

Pelindo has conducted a research of the maturity level of research and innovation management as follows:

- In Strategy, the Company is in Level 3 and Level 1 in impact measurement. The results of the analysis show that Pelindo does not have a clear innovation strategy and has not aligned its innovation goals with the company's goals. The subtheme is included in the 'Innovation Strategy' pillar where the ideal condition is expected for the company to have a clear strategy and integrated with innovation, which supports long-term goals and can have a measurable impact on the company's performance. An effective impact measurement system allows for timely evaluation and adjustment, ensuring innovation contributes significantly to the achievement of the company's strategic objectives.
- Internal collaboration and engagement, currently at Level 2, incentives and rewards dimensions are currently at Level 3, external collaboration is at Level 1, and resources are currently at Level 2.
- The ideal condition expected in the 'Process' pillar is that an organization has a structured and efficient innovation management process, supporting the development and implementation of new ideas. An effective portfolio management system ensures investment in high-potential innovation projects and is in line with the company's

strategy. An efficient risk management framework is used to identify, assess, and manage risks related to innovation. The use of the latest technology is the key to supporting and accelerating the innovation process, as well as improving operational efficiency.

- The ideal conditions expected in the 'Governance' pillar include having a large number of innovation experts and dedicated resources across the unit to drive innovation strategies. The role of innovation must be defined, documented, and communicated to employees clearly, without overlapping duties and responsibilities. Leaders must be frequently engaged and provide full support to the entire organization to drive innovation. Additionally, innovation metrics should be adjusted at the individual level and used to determine performance management, ensuring that innovation becomes an integral part of performance evaluation.
- The ideal conditions expected in the 'Talent' pillar include that the company wants all staff to have good innovation capabilities with clarity of skills needed at each level, in accordance with the strategy and innovation needs. The necessary innovation skills must be clearly defined, documented, and included in the recruitment strategy. In addition, innovation learning and development programs must be embedded in the company's strategy and implemented through limited external procurement.
- In the 'Future Foresight' pillar, it can be seen that technology trends/forecasts are still at level 2 with ideal conditions expected to be at level 4. It is mentioned that the company is focusing on several digital and non-digital trends such as green port studies and the use of big data. However, a clear mechanism is still needed to keep up with the latest trends and inform innovation strategies. Pelindo also does not have adequate tools and programs for workers so that they can follow and apply these trends in the development of innovative products. Therefore, further actions to be able to enter the ideal condition expected at level 4 as "leading" require the creation of clear research mechanisms and programs, as well as the socialization of research results to employees to be able to monitor technological developments across industries.

5.3 Survey Question Mapping

The identification of existing conditions in the process of managing research and innovation at Pelindo is carried out based on the research instruments that have been prepared. Surveys are designed to collect in-depth data on various aspects of research and innovation management within a company. Identification of existing management conditions covers the entire innovation management process from the initial stage to implementation, including Innovation Strategy, Culture, Process, Governance, Talent/Expertise, and Future Foresight. The questions consist of 8 multiple-choice questions and 5 essay or opinion questions. The questions asked in the context of the analysis of the maturity level of Pelindo's research and innovation management process towards the measurement of expectations and satisfaction levels of its human resources have been grouped according to relevant subthemes to provide a comprehensive overview of various supporting aspects.

Regarding the justification for categorizing questions on pillars and subthemes with the aim of providing a clearer understanding of the aspects that support innovation activities in the company, can be seen in Table 7.

Table 7. Aspects of Innovation Activities in the Company

Pillar	No Subtheme	Justification
Strategy	1.1 Corporate Strategy	• Evaluate the implementation of innovative products as part of the company's strategy.
	1.2 Impact Measurement	• Assess the evaluation of research and innovation KPIs to measure the impact of innovation initiatives.
Culture	2.1 Collaboration & Internal Engagement	• This question evaluates the level of internal engagement and collaboration among workers in research and innovation programs. • Evaluate the availability of communication channels for workers to propose innovative ideas.
	2.2 Incentives & Rewards	• Evaluate the system of rewards and appreciation given to innovators and innovative ideas.
	2.3 External Collaboration	• Evaluate the extent to which Pelindo collaborates with external parties to identify opportunities and problems, which is an important aspect of the innovation culture.
	2.4 Value Adoption	• Assess management's support for workers' initiatives in developing research and innovation, which is part of the company's culture. Assess the allocation of time given to workers for the development of innovation, as part of the company's culture.
Process	3.1 Innovation Management	• Assess the existence of processes, frameworks, and procedures that support innovation management. • Evaluate the application of the principle of collaboration between work units in innovation management.
	3.2 Portfolio Management	• Evaluate the effectiveness of innovative products that have been implemented in answering business problems faced by Pelindo. Innovation portfolio management involves monitoring and assessing the success of various innovation projects, ensuring that each product developed and implemented is aligned with business needs and provides added value.
	3.3 Utilization of Technology	• Assessing the extent to which Pelindo has technology that supports research and innovation activities.
Governance	4.1 Structure & Network	• identify and evaluate key factors that affect Pelindo's ability to innovate. Companies can understand the important elements in internal and external structures and networks that need to be improved
	4.2 Clear Roles & Responsibilities	• Assess budget allocation for research and innovation, which is part of management responsibility. • Evaluate the extent to which Pelindo involves various stakeholders, both internal and external, in the innovation process
	4.3 Leadership Involvement	• Assess the extent to which the company is tolerant of failure in innovation development, reflecting leadership involvement in supporting innovation. • Assess the drive towards a culture of innovation within the company, which reflects leadership involvement.
Talent	5.1 Innovation Skills	• Assess the availability of human resources who have the necessary skills to support research and innovation. • This input is invaluable for developing training programs, workshops, and other initiatives that can strengthen

		workers' innovative capabilities, ensuring that they have the necessary skills to overcome challenges and capitalize on opportunities in an ever-evolving environment.
	5.2 Learning & Development	• This question assesses the extent to which Pelindo provides opportunities for workers to learn and develop through participation in events or events related to research and innovation. assess the need for physical facilities that support learning and innovation development. • Assess the frequency and quality of training and workshops held for research and innovation development
Future Foresight	6.1 Technology Trends/Forecasts	• Ask about innovative technologies that need to be implemented to overcome the company's business problems

5.4 Processing and Visualization of Survey Results

The report on the processing and visualization of survey results is of course taken based on questionnaire data that has been distributed and filled in by 2240 Pelindo Group workers who participate in the evaluation of research and innovation management and innovative product development at Pelindo with the stages of data processing in the form of diagrams as follows:

- Data Collection

  Data collection was carried out by making questionnaires/forms online to be distributed to respondents to various entities within the scope of Pelindo.
- Data Analysis

  The data obtained is categorized manually based on the mapping of innovation questions. Each of these categories includes several questions with different results. Then, the results obtained are converted into a percentage (%) to facilitate analysis and facilitate data visualization.
- Data Visualization

  Data visualization is a data representation process that is made to facilitate understanding or information obtained to help identify patterns, further variables, highlight anomalies and problems, and aid in decision-making. The spider diagram was chosen to visualize the data because it can identify strengths and weaknesses and analyze the balance of each variable can be seen in Figure 12. In this maturity level analysis that has been carried out, to visualize the data in the form of a spider diagram there are several stages:
  - Summing up the results of the values from the "Agree" and "Strongly Agree" options.
  - Classification by pillar and sub-theme.
  - Sum the data obtained on each sub-theme.
  - The amount of data they have is divided.
  - Average values were found.
  - The average value is entered into the spider diagram.

Figure 12. Employee satisfaction level in the management of research and innovation at Pelindo

- Interpretation of Results

  The results of the data analysis and visualization are then interpreted with the aim of providing a better understanding for research and innovation managers in strengthening the culture of innovation, development and management of innovation in the maritime sector globally which is highly competitive.

5.5 Analyze the Research and Innovation Management Process by Assesing the Level of Maturity
The survey results show that Pelindo has successfully implemented strategies and programs that support research and innovation, with an average satisfaction rate of 80% of employees with various aspects, including corporate strategy, collaboration culture, and governance. However, there is still room for improvement in innovation skill sets, workspace culture, and budget availability to encourage further innovation. While improvement-type innovation products are a priority with 47% of respondents, transformation towards modern technologies such as the Internet of Things (IoT), Cloud Data, Artificial Intelligence (AI), and Virtual Reality/Augmented Reality (VR/AR) is considered important to face the industrial revolution 4.0. An increased focus on continuous innovation and the development of future technological capabilities is a major concern for the company.

5.6 Evaluation of the Maturity Level of Existing Procss
Based on the analysis that has been carried out on the management of Pelindo's research and innovation, there are several gaps with the results of the previous assessment. The survey focuses more on managing current innovations and less on exploring the application of future foresight and questions that explore the future. The solution that can be done to find out the application of future foresight is to prepare a strategy to deal with future changes. In addition, Pelindo needs to implement smart manufacturing in the work environment, such as the use of IoT technology, Cloud Data, Artificial Intelligence, and Virtual Reality/Augmented Reality (VR/AR), to overcome the company's challenges. It is also important to improve three main aspects, namely Innovation Skill Set, Workspace Culture, and Budget Availability, to encourage sustainable innovation. This approach will

help ensure that the innovations implemented are relevant to the company's needs and contribute to future operational and strategic advancements.

6. Discussion

This research was conducted to identify and evaluate the maturity level of the research and innovation management process at Pelindo. In the era of globalization and digitalization, innovation is the key to maintaining the competitiveness and operational efficiency of companies. Therefore, it is important to understand the extent to which the company has managed research and innovation and how employees view the future of research and innovation at Pelindo.

The research for involved 2240 respondents and covered six main pillars, namely strategy, culture, process, governance, expertise, and future foresight. From the survey data obtained, the expertise pillar showed the highest score compared to other pillars by obtaining 82%. This indicates that the ability and knowledge of employees are considered high enough to support research and innovation activities in the company. However, the governance pillar that obtained an average score of 78% requires more attention. Employees feel that the budget allocated for research and innovation still needs to be increased and collaborate with external parties to gain more innovative insights. In addition, the future foresight pillar also needs more attention in long-term planning. Understanding these activities is important to prepare the right strategy, identify opportunities and challenges in the future, and encourage a future-oriented employee mindset.

The overall survey results showed an average satisfaction rate of 80% with various aspects of research and innovation management, indicating that the majority of employees were quite satisfied. However, there is still a gap between the existing conditions and the ideal conditions to be achieved, especially in terms of strengthening external cooperation and increasing resources for research and innovation.

This research provides important insights for Pelindo management to design more effective strategies in improving research and innovation performance in the future, as well as closing existing gaps. By comparing the results of this research with previous assessment studies, Pelindo can obtain perspectives related to areas that need to be improved, as well as use it as a foothold to optimize the research and innovation management process in the company.

7. Conclusion

Based on the results of the research that has been conducted, the maturity level of the research and innovation management process at Pelindo can be categorized as still at the "Developing" stage or level 2 on the company's Innovation Maturity Model scale. However, there has been significant progress in innovation management which indicates that Pelindo is undergoing a transition phase towards the "Managing" category in the Innovation Maturity Model. To move forward, Pelindo must prioritize the establishment of concrete key performance indicators (KPIs) to measure the impact and effectiveness of innovation over time. These KPIs will help track progress in innovation implementation and ensure sustained improvement. To achieve ideal conditions in the future, Pelindo's Research and Innovation Manager needs to strengthen several aspects that are currently not optimally mapped. One of the priorities is to improve the "future foresight" mindset by adopting a more future-oriented perspective to anticipate changes and opportunities to come. In addition, Pelindo needs to continue to facilitate the use of modern technologies such as IoT, Cloud Data, AI, and VR/AR in the innovation process to support digital transformation. From an industry-wide perspective, adopting these changes will not only support Pelindo's internal processes but also enhance its competitive edge in the broader market, enabling the company to better respond to industry trends and customer needs. With the implementation of these measures, Pelindo can build a stronger foundation in encouraging sustainable innovation, while increasing the company's competitiveness in the long term. Future research should focus on assessing the long-term impacts of these innovations, examining how they affect Pelindo's operational efficiency and market positioning as part of a larger industry trend towards digital transformation. This research provides insights that can be used as a guide in designing better research and innovation management strategies and processes in the future.

References

[1] OSLO/Eurostat, "Oslo Manual 2018: Guidelines for Collecting, Reporting and Using Data on Innovation, 4th Edition," *The Measurement of Scientific, Technological and Innovation Activities, OECD Publishing, Paris/Eurostat*, Oct. 2018, doi: 10.1787/9789264304604-en.

[2] T. Koukaki and A. Tei, "Innovation And Maritime Transport: A Systematic Review," *Case Stud Transp Policy*, vol. 8, no. 3, pp. 700–710, Sep. 2020, doi: 10.1016/j.cstp.2020.07.009.

[3] A. Bottalico, T. Vanelslander, and P. Verhoeven, "Innovation And Labor In The Port Industry: A Comparison Between Genoa And Antwerp," *Journal of Business Logistics*, vol. 43, no. 3, pp. 368–387, Sep. 2022, doi: 10.1111/jbl.12303.

[4] V. Carlan, C. Sys, T. Vanelslander, and A. Roumboutsos, "Digital Innovation In The Port Sector: Barriers And Facilitators," *Competition and Regulation in Network Industries*, vol. 18, no. 1–2, pp. 71–93, Jan. 2017, doi: 10.1177/1783591717734793.

[5] J. J. Lee, T. Y. Ham, and W. S. Jeong, "Economic Risk Assessment Of Ammonium Nitrate Explosions At The Busan Port By Determining The Building Damage Using A 3D Explosion Simulation," *Heliyon*, vol. 8, no. 12, Dec. 2022, doi: 10.1016/j.heliyon.2022.e12285.

[6] B. A. Setiono and S. Hidayat, "Effect of Organizational Transformation and Organizational Culture on Employee Performance of PT. Pelindo," *International Journal of Economics, Business and Management Research*, vol. 06, no. 10, pp. 255–265, 2022, doi: 10.51505/ijebmr.2022.61017.

[7] Kemenparekraf RI, "Grab Catat Lonjakan Pengiriman Barang dan GrabMart Selama Pandemi Corona," https://pedulicovid19.kemenparekraf.go.id/grab-catat-lonjakan-pengiriman-barang-dan-grabmart-selama-pandemi-corona/.

[8] H. S. Jaafar, M. L. A. Aziz, M. R. Ahmad, and N. Faisol, "Creating Innovation In Achieving Sustainability: Halal-Friendly Sustainable Port," *Sustainability (Switzerland)*, vol. 13, no. 23, Dec. 2021, doi: 10.3390/su132313339.

[9] R. Aronietis *et al.*, "How To Turn An Innovative Concept Into A Success? An Application To Seaport-Related Innovation," *Research in Transportation Economics*, vol. 42, no. 1, pp. 97–107, Jun. 2023, doi: 10.1016/j.retrec.2012.11.002.

[10] T. Vanelslander *et al.*, "A Serving Innovation Typology: Mapping Port-Related Innovations," *Transp Rev*, vol. 39, no. 5, pp. 611–629, Sep. 2019, doi: 10.1080/01441647.2019.1587794.

[11] A. Sahraoui, N. K. Tran, Y. Tliche, A. Kacem, and A. Taghipour, "Examining ICT Innovation for Sustainable Terminal Operations in Developing Countries: A Case Study of the Port of Radès in Tunisia," *Sustainability (Switzerland)*, vol. 15, no. 11, Jun. 2023, doi: 10.3390/su15119123.

[12] M. Acciaro *et al.*, "Are The Innovation Processes In Seaport Terminal Operations Successful?," *Maritime Policy and Management*, vol. 45, no. 6, pp. 787–802, Aug. 2018, doi: 10.1080/03088839.2018.1466062.

[13] PT Pelabuhan Indonesia (Persero), "Sustainability Report PT Pelabuhan Indonesia (Persero) 2022," Jakarta, 2022.

[14] M. L. A. M. Bogers, R. Garud, L. D. W. Thomas, P. Tuertscher, and Y. Yoo, "Digital innovation: transforming research and practice," *Innovation: Organization and Management*, vol. 24, no. 1, pp. 4–12, 2022, doi: 10.1080/14479338.2021.2005465.

[15] F. Driediger and V. Bhatiasevi, "Online grocery shopping in Thailand: Consumer acceptance and usage behavior," *Journal of Retailing and Consumer Services*, vol. 48, pp. 224–237, May 2019, doi: 10.1016/j.jretconser.2019.02.005.

[16] Yarty, H. Prabowo, Kurniawan, E. A. Kuncoro, R. B. Ikhsan, and J. Ohliati, "Consumer Acceptance in Grocery Shopping Mobile Applications," in *6th International Conference on Computing, Engineering, and Design, ICCED 2020*, Institute of Electrical and Electronics Engineers Inc., Oct. 2020. doi: 10.1109/ICCED51276.2020.9415854.

[17] G. Mortimer, S. Fazal e Hasan, L. Andrews, and J. Martin, "Online Grocery Shopping: The Impact Of Shopping Frequency On Perceived Risk," *International Review of Retail, Distribution and Consumer Research*, vol. 26, no. 2, pp. 202–223, Mar. 2022, doi: 10.1080/09593969.2015.1130737.

[18] C. Meyer, "Integration Of Baltic Small And Medium-Sized Ports In Regional Innovation Strategies On Smart Specialisation (RIS3)," *Journal of Open Innovation: Technology, Market, and Complexity*, vol. 7, no. 3, Sep. 2022, doi: 10.3390/joitmc7030184.

[19] R. Ridwan and M. A. L. Pambudi, "Singapore International Hub-Port And Its Effects On The Smooth Running Of The Sea Freight Transportation System In Indonesia," *Maritime Park Journal of Maritime Technology and Society*, pp. 111–120, Oct. 2023, doi: 10.62012/mp.v2i3.30980.

[20] F. D. Davis, "Perceived Usefulness, Perceived Ease of Use, and User Acceptance of Information Technology," *MIS Quarterly*, vol. 13, 1989.

[21] J. F. Hair, G. Tomas, M. Hult, C. M. Ringle, and M. Sarstedt, "Partial Least Squares Structural Equation Modeling (PLS-SEM) Using R: A workbook," 2021, doi: 10.1007/978-3-030.

[22] J. Henseler, C. M. Ringle, and R. R. Sinkovics, "The use of partial least squares path modeling in international marketing," *Advances in International Marketing*, vol. 20, pp. 277–319, 2009, doi: 10.1108/S1474-7979(2009)0000020014.

[23] Databooks, "Belanja di E-Commerce, Transaksi Paling Banyak Dilakukan Anak Muda," https://databoks.katadata.co.id/datapublish/2021/06/02/belanja-di-e-commerce-transaksi-paling-banyak-dilakukan-anak-muda.

[24] J. F. Hair, G. Tomas, M. Hult, C. M. Ringle, and M. Sarstedt, "A Primer on Partial Least Squares Structural Equation Modeling (PLS-SEM)," 2017. [Online]. Available: https://www.researchgate.net/publication/354331182

[25] R. S. Hamid and S. M. Anwar, "Structural Equation Modeling (SEM) Berbasis Varian: Konsep Dasar dan Aplikasi dengan Program SmartPLS 3.2.8 dalam Riset Bisnis," PT Inkubator Penulis Indonesia (Institut Penulis Indonesia).

[26] J. F. Hair, M. Sarstedt, L. Hopkins, and V. G. Kuppelwieser, "Partial Least Squares Structural Equation Modeling (PLS-SEM): An Emerging Tool In Business Research," 2018, *Emerald Group Publishing Ltd.* doi: 10.1108/EBR-10-2013-0128.

[27] A.-W. Chien, S. Kurnia, F. von Westarp, and von Westarp, "The Acceptance of Online Grocery Shopping," 2020. [Online]. Available: http://aisel.aisnet.org/bled2003/52

IOP Conf. Series: Earth and Environmental Science **1461** (2025) 012021 doi:10.1088/1755-1315/1461/1/012021

Sustainability in Port Development: Strategies for Environmental, Economic, and Social Resilience

Christino Boyke [1*], Siti Dwi Lazuardi[2], Hasan Iqbal Nur[3], Eka Wahyu Ardhi[4]

[1,2] Department of Marine Transportation Engineering, Faculty of Marine Technology, Institut Teknologi Sepuluh Nopember, Surabaya, Indonesia
[3] Department of Information System and Operation Management, University of Auckland, Auckland, New Zealand
[4] Department of Naval Architecture, Ocean And Marine Engineering, University of Strathclyde, Glasgow, UK

*E-mail: c .boyke@its.ac.id

Abstract. Ports are an essential infrastructure for global trade, with over 4000 seaports operating worldwide. As international trade continues to grow at a rapid pace, the development of new ports is increasing at a similar rate. During the construction and operating phase, ports are highly intersected with environmental effects, economic growth, and social impact on their surroundings. However, with the increasing pressure to mitigate their adverse climate and environmental impact, implementing sustainability frameworks offers a promising path forward. This paper studied the potential benefits of incorporating sustainability frameworks into port operation and development, focusing on environmental protection, economic sustainability, and social effects. The paper presented case studies showing the successful implementation of port sustainability frameworks, offering practical insights and best practices that can be directly applied by port authorities, legislators, and stakeholders engaged in port planning and operations.

1. Introduction

Ports are not just crucial; they are indispensable parts of the global commerce infrastructure. With over 4000 seaports operating globally, the construction of new terminals is expanding at a rapid pace to keep up with the growing volume of international trade. This global scale of port operations underscores the relevance of our research. During the construction and operating phase, ports are deeply intertwined with their surroundings' social impact, economic development, and environmental effects. The depletion of natural resources and the increasing complexity of ecological, financial, and development issues have brought about a pressing need to alter the existing policies on climate change mitigation. This need is urgent and cannot be overstated, as it directly impacts the future of port development. (Argyriou et al., 2022).

Simultaneously, technological advancements have led to significant changes in the shipping industry. These changes include implementing various information systems and technological modifications to ships, such as increasing their size to transport larger cargo loads. Additionally, there has been a trend towards specialization in specific markets, allowing for accommodating modern ships and new products (Hsu et al., 2023). Consequently, ports have evolved into intricate organizations with highly specialized operational frameworks. Therefore, they must continuously develop and secure sufficient capital to establish the necessary conditions for the provision of high-quality services and the adaptation to sustainability practices (Baccelli and Morino, 2020).

Many countries and regions actively encourage adopting environmentally friendly practices in the ports and shipping industry. This is done through various methods, including creating policies, researching and developing new technologies, and offering incentives. (Tong, 2022). However, establishing eco-friendly ports still faces obstacles to technological progress and infrastructure improvements. (Poulsen et al., 2018). Furthermore, converting the current port infrastructure to meet eco-friendly practices requires a considerable investment. (Becker et al., 2013). In addition, due to the different resource limitations faced by ports in developing countries, the implementation of green shipping shows differences in performance, with variations in environmental standards and requirements seen in various regions. (Lam and Notteboom, 2014). Hence, ports engaged in cross-border trade and international cooperation may need help with intricacies and difficulties related to compliance.

Prior research predominantly concentrated on analyzing environmental impact and sustainable port management, frequently neglecting thorough investigation of practical implementations, specifically regarding integrating cutting-edge technologies. Our paper examines approaches that improve sustainability and resilience in port operations and the opportunities and challenges facing sustainable port development today. In addition, we offer a thorough analysis of international best practices and derive practical recommendations for implementation and policy.

The paper is organized as follows: Section 2, titled "Literature Review," aims to critically evaluate relevant research, case studies, and best practices. It provides insights into the difficulties encountered by ports, the efficacy of various strategies, and ports' changing role in environmental preservation and community advancement. Section 3, "Sustainability Strategies in Port Development," expands on the literature review by explaining the strategies and actions taken to incorporate sustainability principles in port development. Section 4 focuses on case studies that examine successful examples of sustainability implementation in port development. These examples are presented, analyzed, and compared to the existing literature and goals outlined in the introduction. Section 4, "Analysis," compares the sustainability implementation strategies in the Port of Rotterdam and the Port of Shanghai. The "Conclusion" in Section 6 concisely summarizes the main findings, emphasizes their implications, and highlights their significant contribution to the progress of sustainable planning in ports.

2. Literature review

The evolution of port sustainability has made significant strides from the early days of minimal environmental consideration to today's comprehensive and strategic approaches. Initially, ports prioritized operational efficiency and economic growth with minimal regard for environmental consequences. In the 1980s and 1990s, environmental regulations in the United States, including the Clean Water Act and the Clean Air Act (U.S. Environmental Protection Agency, 1997), began encouraging ports to reduce their environmental impact and control pollution. The emergence of green port programs, designed to enhance energy efficiency, waste management, and emission reductions, directly resulted from the concept of sustainable development beginning to influence port policies in the 2000s. In the 2010s, sustainability initiatives were furthered by integrating green infrastructure, renewable energy, and smart technologies. Climate resilience, circular economy principles, sustainable finance, and decarbonization are becoming increasingly crucial to ports in the 2020s (Alamoush et al., 2021).Innovative solutions, regulatory evolution, collaborative efforts, and continued digitalization will likely influence port sustainability. These initiatives balance economic development, environmental stewardship, and social responsibility. Table 1 illustrates various initiatives in port sustainability that impact the shipping industry.

The International Maritime Organization (IMO) has implemented more stringent emission policies to decarbonize the maritime industry. (Ampah et al., 2021). The strategy, developed in April 2018, requires a 50% reduction in greenhouse gas emissions from shipping by 2050 compared to the levels observed in 2008 (Jimenez et al., 2022). In recent years, the MEPC meetings have documented the primary policies implemented to promote emissions reduction. MEPC's Ship Energy Efficiency Management Plan (SEEMP) has adhered to specific regulations for vessels of varying weight classes. MARPOL Annex VI's most recent amendment, set to take effect in 2022, includes modifications to two parameters: the Energy Efficiency Existing Ship Index (EEXI) and the Carbon Intensity Indicator (CII). By expanding the assessment scope and establishing new CII ratings, these revisions offer effective methodologies for reducing carbon intensity.

Table 1. Initiatives in Port Sustainability

Year	Organization	Description
1979	World Meteorological Organization (WMO)	During the year, an expert meeting examined the impact of atmospheric carbon dioxide on the ozone layer. The consensus was to establish a global plan of action to reduce atmospheric pollution. (World Meteorological Organization, 1979)
1988	United Nations Environmental Program (UNEP) and World Meteorological Organization (WMO)	The Intergovernmental Panel on Climate Change (IPCC) provides governments at all levels with the most current, reliable scientific information to assist them in developing climate policies.
1997	United Nations (U.N.)	The Kyoto Protocol was designed to restrict greenhouse gas emissions by a minimum of 5% below 1990 levels between 2008 and 2012. In 2023, 192 parties were involved. (Breidenich et al., 1998)
2005	The International Maritime Organization (IMO)	MARPOL Annex VI, adopted in 1997 and implemented in 2005, specifies limitations on primary air pollutants. (Van Roy et al., 2023)
2010	World Ports Climate Initiative (WPCI)	The WPCI recommended that ports enhance their GHG emissions inventories to reduce their contribution to climate change (Azarkamand et al., 2020).
2011	The International Maritime Organization (IMO)	IMO implemented the first mandatory GHG reduction regulations for an international industry by implementing binding energy efficiency measures (Joung et al., 2020).
2014	The World Association for Waterborne Transport Infrastructure (PIANC)	A guide was provided by the PIANC to port authorities to facilitate the introduction of the green port concept and to contribute to the transition from a reactive to a proactive approach.
2019	PIANC's Working Group 188	Working group 188 concentrated on the carbon footprint of activities associated with developing, maintaining, and operating port infrastructure and navigation channels, including managing dredged material (PIANC ENVIRONMENTAL COMMISSION, 2019).

Year	Organization	Description
2019	World Ports Climate Action Program (WPCAP)	Five working groups were established by the WPCAP: efficiency, policy, power to ship, fuels, and cargo handling equipment. Their initiatives encompass the development of models, pilots, research, and awareness.
2019	The Green Ship Technology Conference	The International Maritime Organization (IMO) has implemented a strategy to reduce greenhouse gas emissions from the shipping sector by 50% by 2050.
2020	PIANC Working Group 178	Working Group 178 presented a four-stage guideline framework to assist ports and waterway operators in effectively adapting to climate change (PIANC ENVIRONMENTAL COMMISSION, 2020).
2020	European Sea Ports Organization (ESPO)	The European Green Deal was the subject of a position paper presented by the ESPO. The paper outlined the goal of reducing CO2 emissions from ships at berth and in ports by 50% on average and across all shipping segments by 2030.

3. Sustainability strategies in port development

Sustainable port development is a critical endeavor that aims to balance economic growth, environmental preservation, and social well-being, thereby guaranteeing that ports remain viable and beneficial to all stakeholders in the long term. Figure 1 displays the port sustainability framework widely used in operation and management development.

Environmental stewardship is the foundation of this endeavor, and it entails a multifaceted strategy for minimizing the ecological impact of port operations. Emissions from cargo handling equipment, ships, and heavy vehicle traffic substantially contribute to port air pollution. To alleviate these effects, ports can implement strategies such as shore power (also referred to as cold ironing), which enables ships to connect to the local electricity grid while docking. This reduces the necessity for diesel engines to operate and significantly reduces greenhouse gas emissions. Furthermore, the implementation of cleaner fuels, including hydrogen and liquefied natural gas (LNG), in conjunction with the investment in electric and hybrid vehicles for port operations, can further enhance air quality and reduce emissions.

Another critical component of environmental stewardship in ports is waste management. Ports can reduce waste production, improve recycling efforts, and guarantee the proper disposal of hazardous materials by implementing circular economy principles. This not only mitigates the environmental impact but also enhances resource efficiency. Additionally, ports must establish stringent water quality management protocols to mitigate pollution caused by ballast water discharge, oil spills, and runoff. To safeguard marine ecosystems and preserve water quality, it is imperative to implement advanced technologies for oil spill response and promote ballast water treatment systems.

Preserving biodiversity is paramount, as port operations can disrupt local ecosystems and endanger marine life. Ports can mitigate these effects by regularly monitoring marine biodiversity, restoring degraded habitats, and establishing protected areas. These actions contribute to the overall health of the surrounding environment and reduce the ecological footprint of port operations.

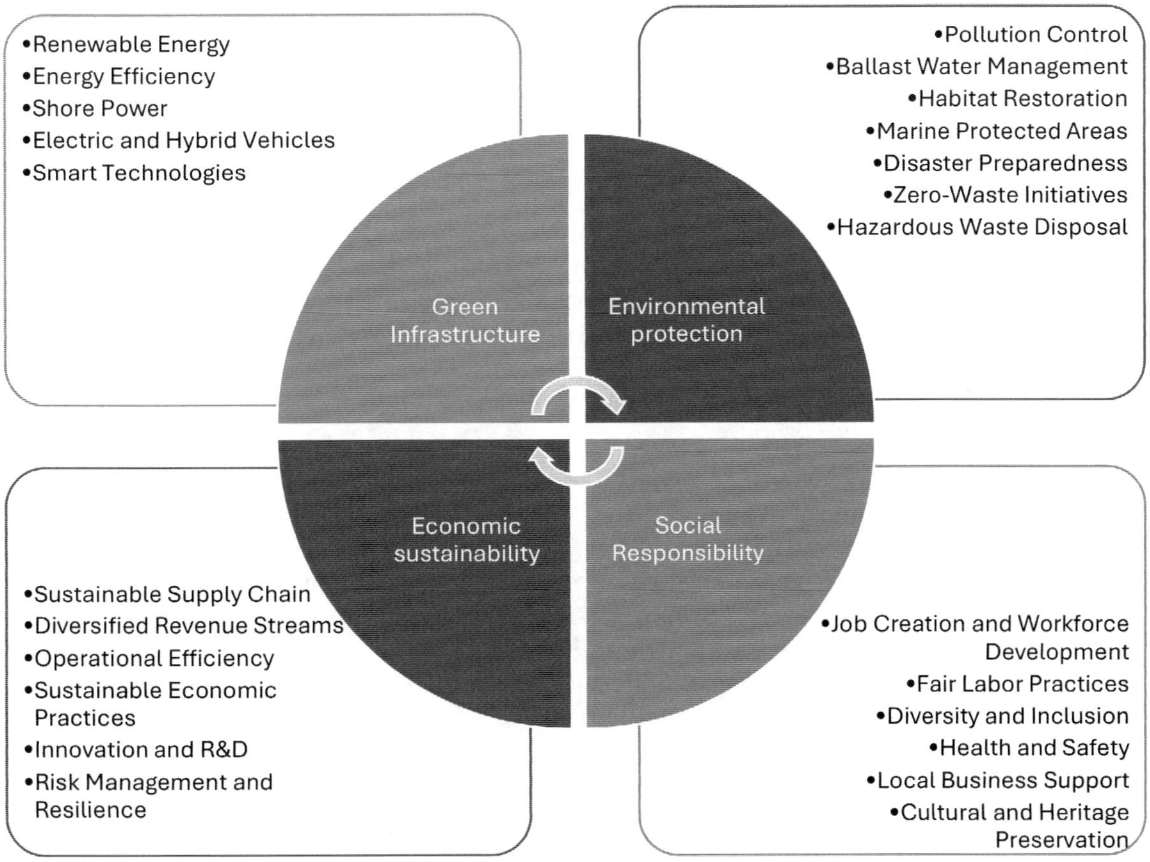

Figure 1. Port sustainability development framework

Economic sustainability is equally critical in developing ports and entails effectively utilizing energy, water, and land. Ports can enhance efficiency, reduce energy consumption, and lower costs by optimizing port infrastructure and operations. This can be accomplished by incorporating energy-efficient technologies, renewable energy sources, and automation and digitalization into port operations. Furthermore, ports must allocate resources to resilient infrastructure that can endure the effects of climate change, including the escalating frequency of storms, extreme weather events, and rising sea levels. With these obstacles, ports can guarantee long-term operational viability by constructing elevated platforms, flood defences, and flexible logistics systems.

Another essential element of economic sustainability is the establishment of sustainable supply chains. Ports are critical to global trade, and they can establish supply chains that prioritize low-carbon transport modes, reduce waste, and promote fair labour practices by collaborating with manufacturers, logistics providers, and shipping companies. This improves the port's sustainability and contributes to a more sustainable global economy.

Social responsibility is critical to sustainable port development, as ports frequently substantially impact nearby communities. Engaging with local communities to address noise, traffic, pollution, and land use concerns is imperative to establish trust and guarantee that port development is consistent with local values and needs. Community engagement programs can facilitate this process and help establish a positive relationship between the port and the community.

Additionally, sustainable port development should foster the development of the local workforce and generate employment opportunities. This entails providing training and educational programs that equip employees with the necessary skills to implement sustainable practices and new technologies in port operations. In doing so, ports can ensure that the workforce is prepared to meet the demands of modern, sustainable port operations while also contributing to the community's economic well-being.

Health and safety are also paramount in the context of social responsibility. Ports must implement strict safety protocols to protect workers and nearby communities from the risks associated with port operations, such as exposure to hazardous materials. Regular monitoring of health outcomes and proactive measures to reduce risks are essential components of a socially sustainable port.

Governance and policy frameworks are crucial in successfully implementing sustainability strategies in ports. Effective governance involves establishing clear sustainability goals, monitoring progress, and ensuring compliance with environmental regulations and standards. Ports must also engage in international collaboration, participating in initiatives like the World Ports Climate Action Program (WPCAP) and working with global organizations to standardize sustainable practices and share best practices and technologies.

Technological innovation is a critical factor in promoting sustainability in the development of ports. The transformation of ports into smart ports, which improve resource management, reduce emissions, and enhance operational efficiency, can be achieved by implementing digital technologies, including data analytics, the Internet of Things (IoT), artificial intelligence (AI), and automation. Energy storage solutions, renewable energy systems, and advanced waste treatment technologies are among the green technologies that can be invested in to mitigate the environmental impact of port operations further and contribute to a more sustainable future.

4. In-Depth Case Studies

4.1 Port of Rotterdam, Netherlands

Over 20% of the Netherlands' carbon dioxide (CO_2) emissions are attributed to activities involving fossil fuels in the Port of Rotterdam. The port's industrial and energy sector relies predominantly on fossil fuels. Hence, the port finds great significance in the Paris Agreement on climate change. The accord mandates promptly taking measures to decrease the release of greenhouse gases, including CO_2, significantly. (Port of Rotterdam, 2022).

The sustainability initiatives in Port of Rotterdam are displayed in Table 2. The Port of Rotterdam excels in renewable energy and sustainability by harnessing wind and solar power. Its offshore wind farms and solar panels efficiently produce environmentally friendly electricity. The heat network reduces dependence on fossil fuels by distributing residual industrial heat to nearby homes and businesses. The port employs cold ironing, which enables docked ships to replace diesel with shore power, thus reducing emissions. Furthermore, it promotes the development of blue hydrogen (converted through carbon capture) and green hydrogen (derived from renewable energy), facilitating the shift towards a low-carbon economy and reducing carbon emissions in shipping and heavy transportation sectors.

The Port of Rotterdam incorporates sustainability and environmental conservation as integral components through various significant initiatives. Effective flood risk management is paramount, achieved by implementing cutting-edge infrastructure such as storm surge barriers and elevated land to safeguard against escalating sea levels. To encourage environmentally friendly practices, the port provides incentives such as discounts for clean shipping and rewards vessels that use low-emission fuels or comply with environmental regulations. Furthermore, it promotes biodiversity by incorporating existing natural elements into the port, such as green areas and habitats, to sustain the local fauna. In addition, the port employs e-noses, a system of sensors that oversee air quality and identify hazardous emissions, thus supporting the reduction of environmental harm and enhancing public health.

The Port of Rotterdam prioritizes social responsibility in addition to its economic and environmental objectives. The port establishes a social agreement with local communities and stakeholders to guarantee inclusive development and employment opportunities. RDM Rotterdam (Research, Design, and Manufacturing) is an innovation hub encouraging collaboration among businesses, researchers, and students to advance sustainable technologies. The Port Welfare Committee is dedicated to the welfare of port workers, ensuring they have access to essential services and support. Furthermore, the Startbaan project is dedicated to assisting young people and unemployed individuals in obtaining skills and employment within the port, promoting social inclusion and economic growth.

Table 2. Port Sustainability Initiatives in the Port of Rotterdam

Categories	Initiative
Green Infrastructure	• Wind and solar power in the port. • Heat network • Cold Ironing • Solar Panel • Blue Hydrogen • Green Hydrogen
Environmental protection	• Flood risk management • Discount for clean shipping • Nature in the port • E-noses
Economic sustainability	• Waste to chemicals • Circular Economy • Efficiency in shipping
Social Responsibility	• Social agreement • RDM Rotterdam (Research, Design and Manufacturing) • Port welfare committee • The Startbaan project

(Port of Rotterdam, 2022).

4.2 Port of Shanghai, China

Table 3 categorizes the Port of Shanghai's initiatives into four key areas: Green Infrastructure, Environmental Protection, economic sustainability, and Social Responsibility. The port focuses on reducing its carbon footprint through shore power systems, enabling ships to use electricity instead of fuel while docked, and energy-efficient equipment like electric cranes. It also promotes sustainable construction with eco-friendly materials and seeks green building certifications for new infrastructure, contributing to lower environmental impact and operational efficiency.

The port has implemented emission control measures to safeguard the environment, promote low-sulfur fuels, and monitor air quality. It also prioritizes water quality protection through ballast water treatment and wastewater management. Additionally, the port supports waste management and biodiversity conservation projects, including restoring coastal ecosystems like mangroves to preserve local ecosystems.

The port drives economic growth by creating jobs and providing training programs for local workers. It also embraces digitalization and automation to improve efficiency, reduce costs, and support local supply chains. Investments in renewable energy and energy-efficient systems enhance the port's economic sustainability by reducing operational expenses.

Regarding social responsibility, the port prioritizes public health and safety through air and water quality monitoring. It engages local communities through public consultations and CSR initiatives like donations to schools and hospitals. The port also supports infrastructure development, improving transportation networks and public spaces, and ensures inclusive growth, offering opportunities for marginalized communities to benefit from port development.

Table 3. Port Sustainability Initiatives in the Port of Shanghai

Categories	Initiative
Green Infrastructure	• Shore Power Systems • Energy-Efficient Equipment • Sustainable Infrastructure Development • Green Building Certifications
Environmental protection	• Air Quality & Emission Control • Waste Management • Water Quality Protect • Biodiversity Conservation
Economic sustainability	• Job Creation & Skill Development • Digitalization & Automation • Supply Chain Development • Energy Cost Reduction
Social Responsibility	• Public Health & Safety Initiatives • Community Engagement & CSR • Inclusive Growth

5. Analysis

5.1 Success Factors

The Port of Rotterdam's success in sustainability and social initiatives is primarily attributed to the key factors in Table 4. Innovation is promoted through collaboration among businesses, government, and research institutions, while environmental leadership is fortified through investments in technologies such as renewable energy and hydrogen production. Sustainable practices are encouraged by incentives such as discounts for clean shipping, and a focus on social welfare supports the well-being of workers through initiatives like Startbaan. Finally, adaptive infrastructure for heat networks and flood risk management guarantees climate change resilience, reconciling sustainability with growth.

Table 4. Port Sustainability Initiatives Succes Factor in the Port of Rotterdam

Success Factor	Description
Collaboration and Partnerships	Cooperation between businesses, government, and research institutions fosters innovation (e.g., RDM Rotterdam).
Investment in Technologies	The port invests in hydrogen production, renewable energy, and air quality monitoring (e-noses) for environmental leadership.
Incentives for Sustainability	Programs like discounts for clean shipping encourage adopting sustainable practices in the shipping industry.
Commitment to Social Welfare	Projects such as Startbaan and the Port Welfare Committee support employment opportunities and worker well-being.
Adaptive Infrastructure	Flood risk management and heat networks ensure resilience against climate change and support long-term sustainability.

Table 5 outlines key success factors driving sustainability initiatives at the Port of Shanghai. Each success factor is accompanied by a description highlighting the specific actions or strategies contributing to the port's sustainability efforts. These factors cover a range of areas, from technology and energy use to regulatory compliance and workforce training. The goal is to reduce environmental impacts, improve operational efficiency, and comply with national and international sustainability standards.

This structured approach ensures that the Port of Shanghai remains competitive while minimizing its environmental footprint. By integrating these factors, the port has made significant strides in sustainable development, showcasing a comprehensive model for other ports globally.

Table 5. Port Sustainability Initiatives Succes Factor in the Port of Shanghai

Success Factor	Description
Green Infrastructure and Technology	Investment in shore power systems, electrified equipment, and energy-efficient vessels to reduce emissions.
Renewable Energy Integration	Use of solar and wind energy to power port operations, reducing reliance on fossil fuels.
Water and Air Pollution Control	Implement systems to control wastewater and reduce air emissions; comply with sulfur limits in marine fuel.
Automation and Digitalization	Automated container handling, intelligent logistics, and blockchain are used to improve efficiency and reduce environmental impact.
Waste Management and Recycling	Programs for waste reduction and recycling of shipping materials to minimize port waste.
Collaboration and Partnerships	Partnerships with shipping companies, government bodies, and international organizations for sustainable initiatives.
Regulatory Compliance and Incentives	Alignment with governmental policies and incentives to promote cleaner technologies and practices.
Sustainability Education and Workforce Training	Continuous training for employees on sustainable practices to foster a culture of environmental responsibility.

5.2 Barriers to Implementation

Several obstacles hinder the Port of Rotterdam's sustainability initiatives, such as the substantial initial expenses for renewable infrastructure and difficulties in expanding the use of emerging technologies like green hydrogen (Table 6). Intricate regulations impede the execution, while opposition from enterprises accustomed to conventional methods exacerbates the challenge. Rotterdam may face a temporary disadvantage due to global competition when disruptions in the supply chain can impede access to essential materials and technology. These facets complicate the port's transition to a more sustainable future.

Table 6. Port Sustainability Initiatives Implementation Barrier in the Port of Rotterdam

Barrier	Description
High Initial Costs	Significant upfront investments are needed for infrastructure like hydrogen production, renewable energy, and flood systems.
Technological Uncertainty	Challenges in scaling up innovative technologies (e.g., green hydrogen, cold ironing) and integrating them effectively.
Regulatory Complexities	Navigating diverse international and national regulations slows down implementation and creates planning uncertainty.
Stakeholder Resistance	Some businesses resist the shift to sustainable practices due to perceived economic risks and operational changes.
Global Competition	Competing ports may invest less heavily in sustainability, potentially placing Rotterdam at a short-term disadvantage.
Supply Chain Disruptions	Delays in obtaining materials and technology for projects like wind turbines and hydrogen facilities can impede progress.

Table 7 outlines the barriers to implementing sustainability initiatives at the Port of Shanghai, which may hinder progress toward environmental goals. These barriers range from financial challenges, such as the high costs of green technologies, to broader industry and regulatory issues, such as complex regulations and global competition. Overcoming these obstacles requires a concerted effort from the port authorities, governments, and industry stakeholders to make sustainability feasible, affordable, and widely accepted.

Table 7. Port Sustainability Initiatives Implementation Barrier in the Port of Shanghai

Barrier	Description
High Initial Costs	The upfront investment required for green technologies, renewable energy infrastructure, and automation can be substantial, which may deter immediate widespread adoption.
Technological Limitations	Some sustainable technologies still need to be developed or scalable for large-scale port operations, creating delays in full implementation.
Complex Regulatory Environment	Navigating varying local, national, and international regulations can complicate sustainability efforts, particularly when aligning with multiple regulatory bodies.
Stakeholder Resistance	There may be resistance from stakeholders such as shipping companies, suppliers, and employees who may be hesitant to adopt new sustainable practices due to perceived costs or operational changes.
Supply Chain Disruptions	Sustainability measures that alter existing operations may disrupt global supply chains, leading to logistical issues and increased costs.
Limited Expertise and Knowledge	A lack of technical knowledge or expertise within the workforce can slow the adoption of new technologies and practices, requiring extensive training and education.

Barrier	Description
Economic Pressures	Economic downturns or financial instability can force ports to prioritize cost-cutting over long-term sustainability investments, limiting resources for green initiatives.
Dependence on Fossil Fuels	Despite efforts to use renewable energy, the heavy reliance on fossil fuels for shipping and port machinery remains a significant barrier to achieving full sustainability.
Long Payback Periods	Many sustainability investments have long payback periods, which can discourage ports from committing to costly initiatives that may yield short-term financial returns.
Global Competition	Ports may fear that adopting expensive sustainable practices could make them less competitive globally if other ports continue to use cheaper, less sustainable methods.

5.3 Comparative analysis

a. Green Infrastructure Category

- Port of Rotterdam: This port focuses on large-scale renewable energy with offshore wind farms, solar panels, and an innovative heat network that redistributes residual industrial heat to nearby communities. It also promotes using blue and green hydrogen for clean energy, especially in the heavy transportation sector.
- Port of Shanghai: This port prioritizes shore power systems to reduce emissions by allowing docked ships to use electricity instead of diesel. It also emphasizes energy-efficient equipment (e.g., electric cranes) and sustainable construction using eco-friendly materials.

Comparison: Rotterdam has a more diverse green infrastructure, utilizing a mix of renewables (wind, solar) and hydrogen, while Shanghai focuses on energy-efficient technologies and shore power systems to reduce emissions.

b. Environmental Protection Category

- Port of Rotterdam: Implements comprehensive flood risk management through advanced infrastructure like storm surge barriers and elevated land. It also incentivizes clean shipping with discounts and integrates nature into the port to enhance biodiversity.
- Port of Shanghai: The port focuses on emission control through low-sulfur fuels and air quality monitoring. It also prioritizes water quality protection and waste management, including ballast water treatment and biodiversity conservation efforts, such as restoring coastal ecosystems.

Comparison: Rotterdam strongly emphasizes climate change resilience and biodiversity, whereas Shanghai's efforts focus more on air and water quality improvements and reducing port emissions.

c. Economic Sustainability Category

- Port of Rotterdam: This port supports a circular economy by turning waste into chemicals and improving shipping efficiency. It also invests in technologies that reduce operating costs and enhance economic efficiency.
- Port of Shanghai: This port embraces digitalization and automation to streamline operations, reduce costs, and boost efficiency in the supply chain. It also promotes job creation and skills development, offering training programs for the local workforce.

Comparison: While both ports prioritize economic sustainability, Rotterdam emphasizes circular economy initiatives and Rotterdam's energy investments, while Shanghai leverages digital tools and automation to improve efficiency and economic development.

d. Social Responsibility Category

- Port of Rotterdam: Engages in social agreements with local communities to promote inclusive development and employment. It also fosters collaboration between businesses,

IOP Conf. Series: Earth and Environmental Science **1461** (2025) 012021 doi:10.1088/1755-1315/1461/1/012021

researchers, and students through RDM Rotterdam. Initiatives like the Startbaan project provide skills and job opportunities, especially for youth and marginalized groups.
- Port of Shanghai: Focuses on public health and safety through air and water quality monitoring. It actively engages local communities in development discussions and supports Corporate Social Responsibility (CSR) initiatives, such as donations to schools and hospitals.

Comparison: Rotterdam has a more structured approach to social inclusion and innovation, involving public-private partnerships and workforce development. While emphasizing community engagement, Shanghai focuses more on public health and safety initiatives.

e. Barriers to Implementation Category
- Port of Rotterdam: This port faces challenges such as high initial costs for renewable infrastructure (hydrogen, flood management systems), technological uncertainties, and regulatory complexities. Potential resistance from stakeholders due to perceived risks and competition with other ports that may not prioritize sustainability also exists.
- Port of Shanghai: Faces similar challenges with high initial costs, technological limitations in scaling up sustainable infrastructure, and regulatory hurdles. Supply chain disruptions and economic pressures could also hinder investment in long-term green projects.

Comparison: Both ports struggle with high costs and regulatory challenges. Rotterdam's barriers focus more on integrating advanced technologies and global competition, while Shanghai's challenges include economic pressures and supply chain issues.

6. Conclusions and Recommendations
6.1 Conclusion

The transformation of major ports such as Rotterdam and Shanghai to sustainability emphasizes the critical role of ports in establishing a more environmentally friendly future for the global maritime industry. Ports are no longer merely centralized hubs for trade and logistics; they have evolved into critical contributors to economic efficiency, environmental protection, and social responsibility.

In general, the promotion of social welfare, the use of renewable energy, the adoption of green infrastructure, and the efficient management of waste are the primary factors that drive sustainability in port development. Renewable energy integration, automation, and shore power systems are among the initiatives assisting ports in reducing their carbon footprint, conserving resources, and enhancing overall operational efficiency. Simultaneously, ports acknowledge the significance of inclusive growth and community engagement, guaranteeing that local populations benefit from development while mitigating environmental impacts.

Nevertheless, the transition to sustainability is a difficult one. The most prevalent obstacles are the intricacies of navigating regulatory environments, technological limitations, and high initial costs. Furthermore, stakeholders may resist adopting novel, frequently expensive, sustainable practices. To guarantee that sustainability is economically viable and feasible over the long term, port authorities, governments, and private sectors must collaborate in overcoming these challenges.

In the final analysis, ports must strike a balance between environmental conservation, social responsibility, and economic growth. This comprehensive strategy is indispensable for the future of port development, as sustainability is not merely an objective; it is a prerequisite for preserving ecosystems, mitigating climate change, and preserving global supply chains. Ports can establish a more resilient and sustainable maritime industry by surmounting these obstacles and adopting innovation.

IOP Conf. Series: Earth and Environmental Science **1461** (2025) 012021 doi:10.1088/1755-1315/1461/1/012021

6.2 Recommendations

1. Increase investment in renewable energy and green technologies
   Ports should prioritize expanding renewable energy sources like solar, wind, and hydrogen to power their operations. By further developing shore power systems, ports can reduce greenhouse gas emissions from docked vessels. Government subsidies or incentives can help offset the initial high costs associated with these technologies.
2. Enhance collaboration and public-private partnerships
   Ports should foster collaboration with governments, research institutions, and private companies to drive innovation in sustainable technologies and infrastructure. Establishing public-private partnerships (PPPs) can facilitate the sharing of knowledge, risks, and financial resources, accelerating the adoption of sustainability initiatives.
3. Implement stronger incentives for sustainable practices
   Ports can introduce or expand programs that provide financial incentives for shipping companies and other stakeholders to adopt greener practices. Offering discounts for using low-emission fuels or adopting eco-friendly technologies can encourage broader compliance with sustainability goals.
4. Develop adaptive infrastructure to mitigate climate risks
   Ports should invest in climate-resilient infrastructure, such as flood barriers and elevated land areas, to address the increasing risks of rising sea levels and extreme weather. This ensures ports remain operational and secure while adapting to future environmental challenges.
5. Advance automation and digitalization for efficiency
   Ports should leverage automation, intelligent logistics, and digitalization (e.g., blockchain) to increase operational efficiency while reducing energy consumption and waste. Investing in intelligent port technologies can streamline processes, reduce bottlenecks, and minimize the environmental impact of port activities.
6. Promote stakeholder engagement and workforce development
   Sustainability initiatives should be accompanied by stakeholder education and engagement programs to reduce resistance and ensure all parties are aligned with sustainability goals. Investing in workforce training on green practices and technologies will build the necessary expertise for successful implementation.
7. Adopt circular economy and waste management solutions
   Ports should promote the circular economy by creating systems for recycling and reusing materials from shipping operations, such as packaging and fuel waste. Additionally, they can introduce waste-to-energy projects, turning waste into a resource that powers port operations and reduces environmental impact.
8. Align with international and local sustainability standards
   Ports must stay ahead of regulatory requirements by aligning with international and local sustainability standards. Compliance with regulations, such as those targeting emissions reductions and biodiversity protection, will ensure that ports remain competitive and avoid potential penalties or disruptions.
9. Long-term financial planning and cost management
   Ports should develop long-term strategies to manage the high initial costs of sustainability projects. Exploring financing mechanisms, such as green bonds or loans, can spread costs over time, making significant investments in renewable energy, infrastructure, and digitalization more feasible.
10. Monitor progress and adapt strategies
    Ports need to establish clear metrics and key performance indicators (KPIs) to monitor the success of their sustainability initiatives. Regular assessments will allow ports to adapt their strategies and ensure they meet environmental, social, and economic goals effectively.

References

Alamoush, A.S., Ballini, F., Ölçer, A.I., 2021. Revisiting port sustainability as a foundation for the implementation of the United Nations Sustainable Development Goals (UN SDGs). Journal of Shipping and Trade 6. https://doi.org/10.1186/s41072-021-00101-6

Ampah, J.D., Yusuf, A.A., Afrane, S., Jin, C., Liu, H., 2021. Reviewing two decades of cleaner alternative marine fuels: Towards IMO's decarbonization of the maritime transport sector. J Clean Prod 320, 128871. https://doi.org/10.1016/j.jclepro.2021.128871

Argyriou, I., Sifakis, N., Tsoutsos, T., 2022. Ranking measures to improve the sustainability of Mediterranean ports based on multicriteria decision analysis: a case study of Souda port, Chania, Crete. Environ Dev Sustain 24, 6449–6466. https://doi.org/10.1007/s10668-021-01711-7

Azarkamand, S., Wooldridge, C., Darbra, R.M., 2020. Review of Initiatives and Methodologies to Reduce CO2 Emissions and Climate Change Effects in Ports. Int J Environ Res Public Health 17, 3858. https://doi.org/10.3390/ijerph17113858

Baccelli, O., Morino, P., 2020. The role of port authorities in the promotion of logistics integration between ports and the railway system: The Italian experience. Research in Transportation Business & Management 35, 100451. https://doi.org/10.1016/j.rtbm.2020.100451

Becker, A.H., Acciaro, M., Asariotis, R., Cabrera, E., Cretegny, L., Crist, P., Esteban, M., Mather, A., Messner, S., Naruse, S., Ng, A.K.Y., Rahmstorf, S., Savonis, M., Song, D.-W., Stenek, V., Velegrakis, A.F., 2013. A note on climate change adaptation for seaports: a challenge for global ports, a challenge for global society. Clim Change 120, 683–695. https://doi.org/10.1007/s10584-013-0843-z

Breidenich, C., Magraw, D., Rowley, A., Rubin, J.W., 1998. The Kyoto Protocol to the United Nations Framework Convention on Climate Change. American Journal of International Law 92, 315–331. https://doi.org/10.2307/2998044

Hsu, C.-T., Chou, M.-T., Ding, J.-F., 2023. Key factors for the success of smart ports during the post-pandemic era. Ocean Coast Manag 233, 106455. https://doi.org/10.1016/j.ocecoaman.2022.106455

Jimenez, V.J., Kim, H., Munim, Z.H., 2022. A review of ship energy efficiency research and directions towards emission reduction in the maritime industry. J Clean Prod 366, 132888. https://doi.org/10.1016/j.jclepro.2022.132888

Joung, T.-H., Kang, S.-G., Lee, J.-K., Ahn, J., 2020. The IMO initial strategy for reducing Greenhouse Gas(GHG) emissions, and its follow-up actions towards 2050. Journal of International Maritime Safety, Environmental Affairs, and Shipping 4, 1–7. https://doi.org/10.1080/25725084.2019.1707938

Lam, J.S.L., Notteboom, T., 2014. The Greening of Ports: A Comparison of Port Management Tools Used by Leading Ports in Asia and Europe. Transp Rev 34, 169–189. https://doi.org/10.1080/01441647.2014.891162

PIANC ENVIRONMENTAL COMMISSION, 2020. PIANC REPORT N° 178 Climate Change Adaptation Planning for Ports and Inland Waterways. PIANC Secrétariat Général.

PIANC ENVIRONMENTAL COMMISSION, 2019. PIANC REPORT N° 188 Carbon Management for Port and Navigation Infrastructure. PIANC Secrétariat Général.

Port of Rotterdam, 2022. PERS application-December 2022.

Poulsen, R.T., Ponte, S., Sornn-Friese, H., 2018. Environmental upgrading in global value chains: The potential and limitations of ports in the greening of maritime transport. Geoforum 89, 83–95. https://doi.org/10.1016/j.geoforum.2018.01.011

Tong, H., 2022. The future development and restructuring of the international shipping industry: Conference report. Mar Policy 137, 104956. https://doi.org/10.1016/j.marpol.2022.104956

U.S. Environmental Protection Agency, 1997. The Benefits and Costs of the Clean Air Act, 1970 to 1990.

Van Roy, W., Scheldeman, K., Van Nieuwenhove, A., Merveille, J.-B., Schallier, R., Maes, F., 2023. Current progress in developing a MARPOL Annex VI enforcement strategy in the Bonn Agreement through remote measurements. Mar Policy 158, 105882. https://doi.org/10.1016/j.marpol.2023.105882

World Meteorological Organization, 1979. Proceedings of The World Climate Conference. Book On Demand Ltd.

Exploring Organizational Ambidexterity And Resilience In Integrated Port Network Development (Infrastructure & Sustainability Chapter)

Fajar Apriadi Darwita

Corporate Tranformation And Program Management Group, Pelindo Head Office
*E-mail: fajar.darwita@pelindo.co.id

ABSTRACT. This research explores the concepts of **Organizational Ambidexterity and Resilience** as key strategies in **the development of an Integrated Port Network**. In the context of the ever-evolving global supply chain, ports must simultaneously exploit existing capabilities and explore new opportunities to remain competitive and sustainable. As the market leader in Indonesia's port industry, **PT Pelabuhan Indonesia (Persero)** plays a significant role in advancing the sector, particularly in container services. O'Reilly & Tushman (2013) [9] describe Organizational Ambidexterity as an organization's ability to explore (seek new innovations) and exploit (maximize and improve efficiency of existing resources) simultaneously. This concept highlights that ambidextrous organizations can effectively operate in dynamic and uncertain environments by managing these often conflicting activities, which is crucial in addressing the operational complexities of modern ports. Similarly, Lam & Vis (2021) [6] define resilience as the ability of ports to adapt and recover from disruptions that could interrupt their operations. This research emphasizes the importance of strategic and well-implemented initiatives to enhance port resilience. By integrating these two concepts, the author recommends stakeholder collaboration, sustainable infrastructure development, process digitization and automation, enhanced intermodal connectivity, and a focus on sustainability and efficiency which is expected to be used as a reference and learning document for Pelindo in preparing the company's long-term plan and technical documents to the ministry of transportation in order to strengthen the national port strategy and the Ministry of State-Owned Enterprises as a corporate strengthening of the maritime and logistics sector.. **This research proposes a framework for implementing organizational ambidexterity and resilience through service standardization, synergy with shipping companies, and cooperation with industrial zones to develop a resilient Integrated Port Network**, which supports the company's long-term sustainable growth and enhances national competitiveness through improved national connectivity.

Keywords : Organizational Ambidexterity, Resilience, Sustainability, Sustainable Infrastructure development and Integrated Port Network

1. BACKGROUND OF THE RESEARCH

Port and logistics industry, as one of the critical components in the maritime sector, plays a vital role in the global economy, connecting producers and consumers worldwide. However, in recent years, this sector has faced significant challenges, particularly during the COVID-19 pandemic. In addition, as recovery processes have started, there remains ongoing economic uncertainty and geopolitical tensions, such as those in the Middle East and between Russia and Ukraine, which have impacted the global economy. Events like supply chain disruptions, port closures, and travel restrictions have resulted in shipment delays, increased costs, and heightened uncertainty. In this context, the development of an Integrated Port Network has become increasingly relevant to ensure smooth operations and sustainability within the maritime supply chain.

As a reference and example as a case study in this research, PT Pelindo plays a crucial role in Indonesia's economy as the largest state-owned port operator, managing multiple ports across the archipelago. One of Research Arief (2022) [1] explains approximately 90% of Indonesia's trade relies on maritime transport, Pelindo's efficiency directly influences national logistics costs and economic growth. The company's ability to adapt to disruptions is vital for maintaining trade flow and supporting local industries.

Within this context, the concepts of organizational ambidexterity and resilience become more crucial and relevant. Organizational Ambidexterity refers to an organization's ability to explore new innovations while continuing to exploit existing processes. This enables ports and logistics companies to quickly adapt to changing market conditions and demands. Meanwhile, resilience refers to the ability to recover from disruptions and operate effectively in unforeseen situations. Resilient organizations can manage risks and minimize the impact of disruptions on their operations.

1.1. Objective and Significance of the Research

This research objective to explore how organizational ambidexterity and resilience can accelerate the development of an Integrated Port Network. The focus of this research includes several key aspects:

Stakeholder Collaboration: Identifying how collaboration between ports, logistics providers, and government authorities can enhance operational efficiency and effectiveness.

Sustainable Infrastructure Development: Analyzing the importance of investing in environmentally friendly and efficient infrastructure to support the sustainability of port networks.

Process Digitization and Automation: Investigating the role of digital technology in increasing productivity and efficiency in port operations.

Intermodal Connectivity Enhancement: Examining the importance of integrating sea, land, and air transportation to create an efficient logistics network.

Focus on Sustainability and Efficiency: Assessing how the development of the Integrated Port Network can contribute to reducing emissions and more efficient use of resources.

By considering these factors, this research aims to provide a comprehensive conceptual understanding of how organizational ambidexterity and resilience can support the development of a more efficient and sustainable Integrated Port Network and offer strategic recommendations to stakeholders in the maritime industry..

2. LITERATURE REVIEW

2.1 Integrated Port Network

Presidential Regulation No. 18 of 2020 addresses the National Medium-Term Development Plan (RPJMN) for 2020-2024 [11], a strategic guide for Indonesia's national development. This regulation was drafted as part of the implementation of Law No. 25 of 2004 on the National Development Planning System. In this context, the RPJMN focuses on various sectors, including transportation, which is essential for enhancing national connectivity.

Integrated Port Network
(Main Port : Kuala Tanjung, Kijing, Tanjung Priok, Tanjung Perak, Makassar, Bitung Sorong)

Project Background	• Indonesia's Logistics Costs are the highest compared to a number of countries, including India, Malaysia, China, Thailand, and Vietnam • 28 major ports in the port master plan are not yet consolidated • 77% of shipping routes in Indonesia are dominated by port to port, while loop routes are only 23% • There is no standardization of the main ports (dock length, depth and stacking area), to be used as the main port of transhipment port.
Project Benefit	• Improve port performance with the standardization of the main port (maximum turn round time value of 24 hours) • Increase the efficiency of domestic shipping routes by forming loops regularly to 27% • As a supporting economic area integrated with the port
Implementation Time	2020-2025 (Five Years)

Image 2.1 Presidential Regulation Number: 18 of 2020 concerning RPJMN which contains the Integrated Port Network [11]

Presidential Regulation (Perpres) Number 18 of 2020 on the Medium-Term Development Plan (RPJMN) 2020-2024 [11] is a policy framework for Indonesia's national development that is valid for five years. Policy framework that describes the national strategic development plan that is valid for five years. One of the focuses of the RPJMN is the development of the transportation sector, including the port network.Key aspects of the IPN outlined in the RPJMN 2020-2024 include:

- National and International Connectivity: The development of a fully integrated network of major ports is intended to improve logistics efficiency, reduce costs, and facilitate more efficient distribution of goods.
- International Hub Ports: Strategic efforts to develop international hub ports will position Indonesia as a critical node in the global supply chain, increasing its competitiveness in the international market.
- Capacity Expansion: Increasing port capacity through infrastructure upgrades, modern equipment, and advanced technology is a priority. This includes the construction of new terminals and the modernization of shipping lanes.
- Sustainable Port Development: The RPJMN emphasizes the need for ports to adopt green technologies and sustainable practices to minimize environmental impact.
- Private Sector Collaboration: The government encourages private investment through partnerships to accelerate the development of port infrastructure and improve service quality.

The goal is to align the development of the Integrated Port Network with Indonesia's vision of becoming a Global Maritime Axis, which will enhance the nation's logistics and trade competitiveness.

2.2 Organization Amditexterity

There are various definitions of Organizational Ambidexterity that are referenced in this research including O'Reilly & Tushman (2013) [9] define as the ability of an organization to simultaneously explore (seek new innovations) and exploit (utilize and improve the efficiency of existing resources). O'Reilly & Tushman (2013) [9] emphasize that ambidextrous organizations can operate well in dynamic and uncertain environments by managing these two often conflicting activities.

Another definition is taken from Hill (2023) [3] which refers to the ability of an organization or organizational subsystem to perform two tasks with equal skill. The term "ambidexterity" is often metaphorically applied in the human body (i.e. having the skill of using both hands equally) to organizations. Although past research varies with respect to the duality applied to ambidexterity, researchers have now largely coalesced around organizational ambidexterity in reference to exploration and exploitation.

In general, organizational ambidexterity involves two main aspects:

- Exploration: The process of searching for and developing new innovations, including products, services, and processes.
- Exploitation: The use and optimization of existing resources to improve efficiency and effectiveness.
- These two elements are often in conflict with each other, so organizations must find the right balance to achieve long-term success.

Another researcher, Dolz (2019) [2] also stated that ambidextrous organizations can develop exploration and exploitation activities, where exploitation activities address the current survival of an institution. Meanwhile, exploration activities focus on future survival. These two activities are then known as ambidexterity (Pertusa-Ortega et al., 2020). Zhao el al., (2021) [16] conducted research between organizational ambidexterity and its effect on marketing performance which showed that organizational ambidexterity has a significant positive effect on marketing performance, highlighting the importance of the ability to adapt to market changes.

2.3. Resilience

The concept of resilience in the world of ports, as described by Lam & Vis (2021) [6] encompasses the ability of ports to adapt and recover from disruptions that may interfere with their operations. Port resilience from Shi, et all (2022) [13] can also be defined as the ability to face and overcome challenges that arise from various external factors, such as natural disasters, economic disruptions, or global health crises. Resilient ports are not only able to survive difficult situations but can also take advantage of opportunities that arise from these changes. Lam & Vis (2021) [6] also emphasized the importance of several things that ports should do to increase resilience. These practices include:

- Implementation of Sustainability Principles: Integrating environmentally friendly practices in port operations to mitigate climate change impacts and improve competitiveness.
- Sustainable Infrastructure Development: Investing in infrastructure designed to withstand natural disasters and climate change, so as to minimize damage and recovery time.
- Continuous Monitoring and Evaluation: Conducting regular evaluations of port performance and the effectiveness of strategies implemented to ensure that ports remain adaptive to changing conditions.

Resilience is becoming increasingly relevant and critical to the Integrated Port Network as ports need to be able to cope with challenges such as weather disruptions, accidents and demand surges. Turner (2013) [15] A resilient port can quickly restore operations and minimize the impact to the supply chains connected to on going process.

2.4. Relationship Between Ambidexterity, Resilience, and Pelindo's Role in Developing an Integrated Port Network

In this research, Author proposes to compile a research on the relationship between ambidexterity, resilience, and Pelindo's role as a port operator. Liu et al (2023) [7] found that resiliently-governed ports can guarantee higher port throughput while sustaining lower congestion when facing challenges from the global pandemic. Liu et al (2023) [8] built a port resilience index system based on the entropy weight method, integrating the port, the hinterland, and the local government as significant stakeholders. Their results suggest that resilient-governed ports are better able to maintain normal operations even in the face of unexpected global shocks like the pandemic.

Another research showed, Parola et al (2022) [10] examined how ambidextrous strategies enable ports to adapt to disruptions like COVID-19. They found that ambidextrous ports that pursue both incremental and radical innovations are more resilient. Investing in digital technologies while also optimizing existing processes allows ports to respond effectively to supply chain shocks.

Pelindo has faced various challenges, including the COVID-19 pandemic, which disrupted global supply chains and altered shipping patterns. Tukamuhabwa et al (2015) [14] The pandemic underscored the importance of resilience in port operations, prompting Pelindo to innovate its services and improve operational efficiencies. For example, Pelindo has implemented digital solutions to streamline cargo handling and enhance customer service, demonstrating its commitment to organizational ambidexterity by balancing exploration of new technologies with exploitation of existing processes.

The integration of Pelindo, which merged four state-owned port operators, aims to standardize operations across various port facilities. This standardization is crucial for enhancing operational efficiency and service quality, which are essential components of organizational ambidexterity. By adopting a unified operational framework, Pelindo can balance the exploration of innovative practices with the exploitation of existing capabilities, thereby improving overall performance

Pelindo plays a critical role in implementing the government's vision of making Indonesia a maritime hub. The development of an Integrated Port Network aims to enhance maritime connectivity across the archipelago, facilitating smoother logistics and transportation processes. This initiative aligns with global trends where ports are increasingly viewed as integral components of broader supply chain networks

According to the author perspective, Post Merger Pelindo creating opportunity become a more resilient entity within Indonesia's logistics sector. By consolidating resources and streamlining operations, Pelindo can better respond to market fluctuations and disruptions. The establishment of an Integrated Port Network (IPN) facilitates improved connectivity between ports and industrial zones, which is vital for maintaining resilience in supply chains. This integrated approach allows for more coordinated responses to logistical challenges, enhancing Pelindo's adaptability.

In this research, the author recommends the relationship between Organizational Ambidexterity, Resilience and Pelindo's Role in Developing Integrated Port Network that are relevant and very important including the following:

No	Exploration of Ambidexterity and Resilience	Scope Definition
1	Stakeholder Collaboration	Identifying how collaboration between ports, logistics providers, and government authorities can improve operational efficiency.
2	Sustainable Infrastructure Development	Analyzing the importance of investing in eco-friendly infrastructure to ensure sustainability.
3	Process Digitization and Automation	Investigating the role of digital technology in improving port operations.
4	Intermodal Connectivity Enhancement	Examining the integration of various transport modes to create an efficient logistics network.
5	Focus on Sustainability and Efficiency	Assessing how the development of the IPN contributes to resource efficiency and emission reductions.

Table 2.1 Relationship Between Ambidexterity, Resilience, and Pelindo's Role in Developing an Integrated Port Network

For a further analysis in this paper, the author recommends the detail recommendation about relationship between Organizational Ambidexterity, Resilience and Pelindo's Role in Developing Integrated Port Network especially in Inftrastucture and Sustainability Chapter that are relevant and very important including the following:

No	Infrastrucutre & Sustainabilty Chapter	Scope Definition And Implementation
1	Sustainable Infrastructure Development	Analyzing the importance of investing in eco-friendly infrastructure to ensure sustainability.
		Several Aspect for implementation : • Electrification of Equipment: Transitioning port machinery to electric power to reduce carbon emissions. This initiative aligns with Pelindo's commitment to environmental sustainability and helps mitigate climate change. • Integrated Waste Management: Implementing systems for managing ship waste effectively to minimize environmental impact. This includes recycling programs and proper disposal methods that adhere to environmental regulations. • Community Engagement: Implementation of Pelindo Program Of Social and Environmental Responsibility Such As Education Programs: Conducting workshops and outreach programs to educate local communities about sustainability practices and the importance of preserving marine ecosystems.
2	Focus on Sustainability and Efficiency	Assessing how the development of the IPN contributes to resource efficiency and emission reductions.
		Several Aspect for implementation : • Revitalization Projects: Ongoing efforts to modernize existing port infrastructure to accommodate larger vessels and improve logistics efficiency. This includes upgrading cargo handling systems and expanding storage capacities. • Sustainable Infrastructure Planning: Ensuring that new developments incorporate sustainable design principles, such as energy-efficient buildings and materials that minimize environmental impact. • Performance Metrics for Sustainability: Establishing clear metrics to evaluate the effectiveness of sustainability initiatives, ensuring accountability and transparency in reporting progress. • Feedback Mechanisms: Implementing systems for gathering feedback from stakeholders, including local communities and shipping partners, to continuously refine sustainability practices.

Table 2.2 Scope Definition and Implementation Of Sustainability And Infrastructure Between Organization Ambidexterity, Resilience, and Pelindo's Role in Developing an Integrated Port Network

2.5. Strategic Movemement Pelindo After Post Merger Integration

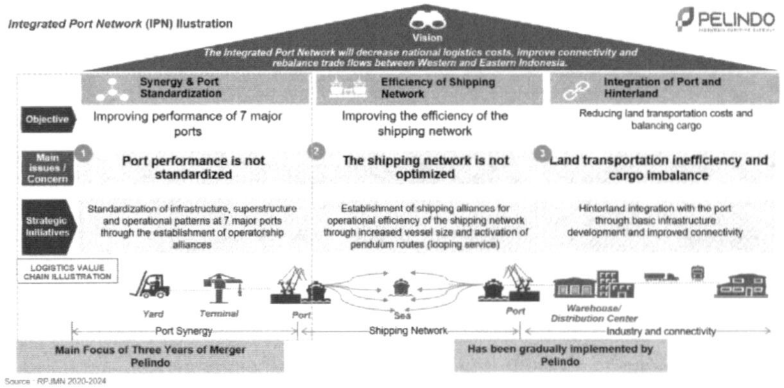

Image 2.2 Source : Post Merger Review & Evaluation Research and RPJMN 2020-2024 [5][12]

IOP Conf. Series: Earth and Environmental Science **1461** (2025) 012022 doi:10.1088/1755-1315/1461/1/012022

Pelindo's three-year transformation has focused on Phase 1 (One) of the Integrated Port Network, namely the standardization of key services, while the next year will focus on Phases 2 and 3, namely synergies with shipping and connectivity with ports and industrial areas : .

No	Category	Issue	Initiatives Taken	Next Step
1	Port Synergy and Standardization	Port performance has not been standardized	Standardization of infrastructure, superstructure and operational patterns at 7 major ports through the formation of operatorship alliances	Implementation at all other ports and complete the next stage of standardization, namely systemization and integration.
2	Shipping Network Efficiency	The shipping network has not been optimized	The formation of shipping alliances for operational efficiency of the shipping network through increasing the size of ships and activating pendulum roues (looping services)	Digital implementation synergies involving shipping and industrial estates
3	Port Integration with Hinterland	Inefficiency of land transportation and cargo imbalance	Hinterland integration with ports through basic infrastructure development and increased connectivity	Multiplication of port-based industrial estate development

Table 2.3 Implementation of Integrated Port Network and Follow-Up Action

3. Research Methodology

This research uses a qualitative approach with study case methodology and in depth interview. Interview was collected through several internal studies and internal exposure documents where questions and interview sessions of the port operations team consisting of Four Pelindo Subholding operating regions from Jakarta, Semarang, Surabaya and Jayapura where each team consists of 15-20 people from the operational team and operational support team through the interview method. In addition, researchers also analyzed company performance evaluations were conducted as well as document analysis related to port operations. This approach aims to gain an in-depth understanding of the dynamics of organizational ambidexterity and resilience in the Integrated Port Network Development.

The list of questions become key questions that become the source of the author's research related to the research methodology, especially for aspects of Organizational Ambidexterity and Post-Merger Pelindo Performance consisting of:

- Have there been improvements in service quality (e.g., BSH/BCH, TRT) following the Pelindo merger?
- Has productivity increased, and have costs been reduced?
- Can primary port customers introduce new services based on Pelindo's performance?
- Have unresolved operational issues persisted before and after the merger?
- How responsive are subholdings to customer requests and complaints?

In addition, for the aspects of organizational ambidexterity and resilience, there are several questions that form the basis of research recommended by the author as a basis for analysis and discussion, including:

- What are the stages of Sustainable Infrastructure Development that encourage the implementation of the Integrated Port Network?
- How is the impact of the development of Digitalization and Process Automation in realizing organizational ambidexterity?
- What are the efforts made by Pelindo in improving intermodal connectivity?
- How is Pelindo's focus on Sustainability and Efficiency?

The author recommends the Research Theory Model can be seen in the following model:

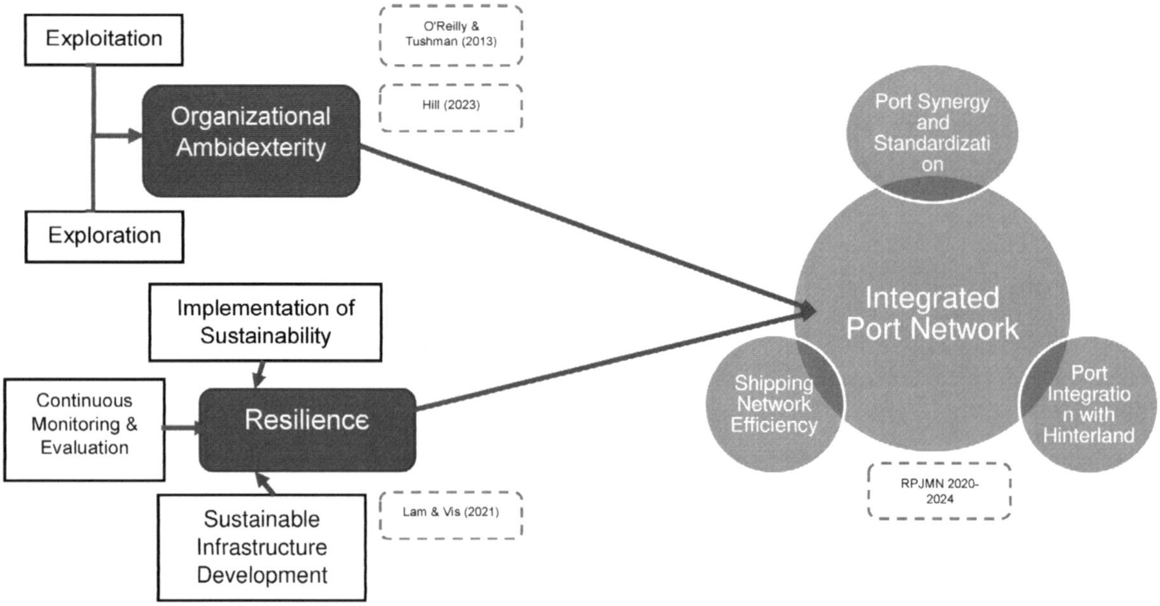

Image 3.1 Research Theory Model Recommendations

4. Discussion and Main Finding

4.1. Post Merger Pelindo Case Research

PT Pelabuhan Indonesia (Pelindo) Persero as a market leader in the port world, especially in container commodities, has a significant role in the development of the maritime world in Indonesia. Pelindo Post Merger received appreciation from customers for being able to maintain good operational performance without disruption and even improve the quality of operational performance in several areas that became the transformation pilot project which can be displayed with an operational performance matrix where in general there is an increase in operational performance productivity (BSH units) and a decrease in the waiting period of ships at the port (Vessel Port Stay units) with the following details :

Cabang	Productivity BSH (Box Ship Hour)*			Vessel Port Stay Hours*	
	Before	After		Before	After
		Average	Highest		Average
Belawan	20	38	61	55	32
Makassar	20	34	63	38	22
Ambon	12	26	35	37	24
Sorong	10	25	34	72	24
Nilam	25	35	58	21	18
Jayapura	18	36	52	36	17
Pantolan	22	25	35	16	15
Tarakan	12	15	21	31	27
Perawang	8	19	32	35	19

Tabel 4.1. Productivity BSH & Port Stay based on Optimal Performance (Internal Document Pelindo) [5]

According to the table and explanation above, it can be seen that in overall Pelindo's performance from the operational side is very positive and not interrupted by the merger and there are opportunities for new developments to strengthen Pelindo's performance which is more adaptive and solid through optimizing its resources regarding strengthening organizational ambidexterity and resilience which is the main theme of this paper. In response to this, the author provides some recommendations from the author regarding the exploration of organizational ambidexterity and resilience in the Integrated Port Network Development as follows:

4.1.1. Collaboration Between Stakeholders

In this aspect, the scope or focus that can be done is to identify how cooperation between ports, logistics providers, and government authorities can improve operational efficiency and effectiveness. Some initiatives that have been carried out by Pelindo include Intense Communication with Stakeholders and Customers regarding the Progress of the Pelindo Merger, Exploration related to Port Development through investment synergy and strategic partners and integrated Key Account Management. The things that are recommended for the writing to be the next stage according to the author include the implementation of regional expansion in the port business, especially container commodities, Government Support in the development of Port-based industrial areas, Determination of Competitive Schemes in the implementation of Cooperation with BUMN & private logistics players, Determination of Cooperation with industrial/logistics areas and Renegotiation of Cooperation Schemes for the Development of Port Operations Services owned by Ministries/Institutions.

4.1.2. Sustainable Infrastructure Development

In this aspect, the scope or focus that can be done is analyzing the importance of investment in environmentally friendly and efficient infrastructure to support the sustainability of the port network. Some initiatives that have been carried out by Pelindo include the development of Pelindo's existing business port infrastructure and equipment, gradual implementation of sustainable port operational infrastructure, implementation of steps related to the implementation of sustainability efforts in ports (green ports), rejuvenation and maintenance of port equipment. The recommended writing is the next stage according to the author, including the implementation of sustainable infrastructure planning in all port operational areas, multiplication of sustainability efforts in ports and collaboration on the implementation of sustainable infrastructure involving shipping and industrial areas.

4.1.3. Digitalization and Process Automation

In this aspect, the scope of Digitalization and Process Automation is optimizing the role of digital technology in increasing productivity and efficiency in port operations. Some initiatives that have been carried out by Pelindo include Standardization of operations at the port consisting of aspects of HR, business processes, technology and K3 as well as digitizing & systemizing port operations in the main Pelindo service areas. The things that are recommended in the writing as the next stage according to the author include Continuing Standardization and Systemization efforts throughout Pelindo's service areas by considering the market conditions served and the company's capabilities. Another thing that is recommended is to carry out digital implementation synergy involving shipping and industrial areas. This is crucial because to realize the Integrated Port Network, the role of the port is only one of the main domains outside the shipping and industrial area aspects. In addition, business processes that vary between industries can be a challenge in implementing the technology synergy that will be carried out

4.1.4. Improving Intermodal Connectivity

In this aspect, the scope of Improvement Intermodal Connectivity is explore the importance of integration between sea, land, and air transportation to create an efficient logistics network. Some initiatives that have been carried out by Pelindo include Cooperation with BUMN & private logistics players, Cooperation with industrial/logistics areas, Development of TUKS operational services with BUMN and the private sector, Development of Port operational services owned by Ministries/Institutions. The things that are recommended for writing as the next stage according to the author include port development through investment synergy and strategic partners, multiplication of port-based industrial area development [17]

4.1.5. Focus on Sustainability and Efficiency

In this aspect, the scope or focus that can be done is to assess how the development of the Integrated Port Network can contribute to reducing emissions and using more efficient resources. Some initiatives that have been carried out by Pelindo include Additional MEPS service offerings or additional service offerings from the core business for Pelindo's maritime services for Pelindo's key accounts, awareness stages or awareness of ESG implementation at Pelindo and sampling of monitoring the efficiency of Pelindo's main services. The things that are recommended for writing as the next stages according to the author include the implementation and commitment stages in implementing ESG at Pelindo, implementing regional expansion in the port business and comprehensive implementation of Pelindo's Main Service Efficiency Monitoring.

4.2. Key Findings of the Research

In addition, there are several author recommendations which are the main findings of the research in conducting the Exploration of Organizational Ambidexterity and Resilience in the Development of Integrated Port Networks, including:

4.2.1 Implementation of Pelindo Organizational Ambidexterity Post-Merger and Resilience

Some initial initiatives that could be "low hanging fruit" or early stage implementation programs in implementing organizational ambidexterity or balancing between exploitation and exploration at Pelindo include:

4.2.1.1 Synergy between Pelindo's Main Service Standardization and Service Commercialization

Since Three Years of the Pelindo Merger, a series of Pelindo's main service standardization programs have been carried out as written in the Roadmap of the Pelindo Head Office Operations Directorate (2023) [4] which consists of standardization, systemization, and integration in the majority of Pelindo's work areas and the main services provided.

The new author's recommendation from this stage is to synergize commercialization efforts and see it as a single stage of primary service standardization after the service standardization stage has been carried out. This is important to ensure operational performance that has been transformed by maintaining current customers and presenting new customers as a form of service commercialization.

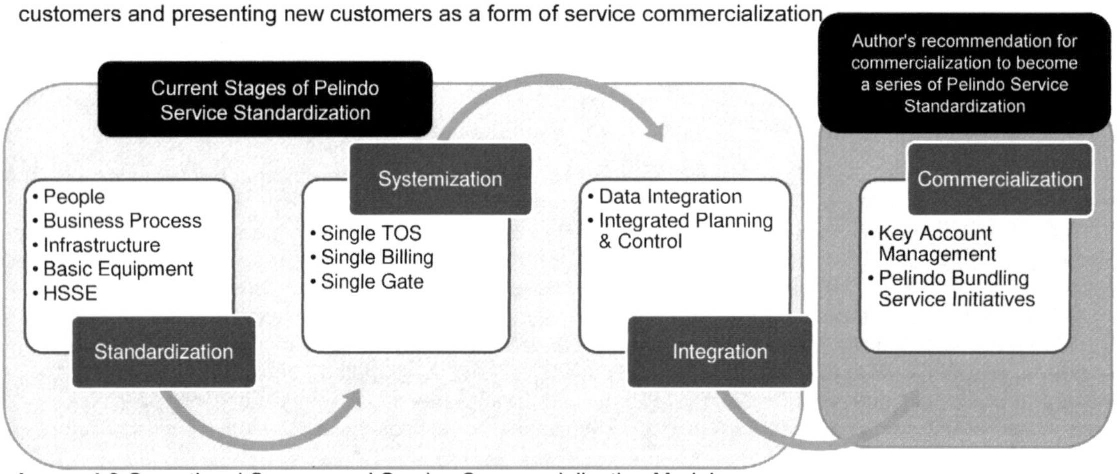

Image 4.2 Operational Synergy and Service Commercialization Model

4.2.1.2 Development of People Model (HR Profile Model) Enhancement Organizational Ambidexterity at Pelindo

The development of the People Model (HR Profile Model) which focuses on the implementation of organizational ambidexterity in Pelindo involves several strategic steps aimed at improving the capability and flexibility of human resources (HR). Here are some steps that must be taken.

I. Human Resources Competency Development

Pelindo has developed technical competency academies such as Operational and Commercial Academy and has been implemented in accordance with the placement of its employee. The next step is to conduct blended learning on the academy that has been carried out by combining operational and commercial modules to encourage Pelindo employee to apply ambidexterity even to the individual level and daily work.

II. Implementation of Organizational Culture

Pelindo continues to implement various initiatives that have been running previously that focus on supporting innovation and collaboration. This involves empowering employees to contribute to the innovation process and decision-making. Employees are encouraged to share ideas and participate in projects that can lead to process improvements that are operational and commercial in nature that drive the company's business expansion and synergy with shipping and industrial areas.

III. Selective and Flexible Organizational Structure and Competitive KPI Implementation

Related to the implementation of Ambidexterity Organization, there needs to be synergy and strengthening between operational and commercial functions so that a special team is needed as the front line or spearhead Pelindo implements a flexible organizational structure and competitive KPI Determination, especially those that produce profit centers in the form of Project Management Offices in the Operational and Commercial fields that allow the team to adapt quickly to market changes and customer needs. This includes the formation of cross-functional teams that can work together on innovative projects while remaining focused on operational efficiency.

With these steps, Pelindo seeks to create a People Model that not only supports HR development, but also increases the competitiveness of the organization, especially in terms of the company's business expansion aspect, which is currently facing increasingly sharp industrial challenges.

4.2.1.3 Joint Slot Operation Support between Shipping and Port Operator

The maritime industry, especially shipping, has begun implementing a joint slot scheme to serve the Eastern Indonesia Region and the outermost regions of Indonesia since 2014 (Bisnis.com) [20]. This scheme ensures that ship space is filled with goods that must be transported from and to these locations. In addition, the service is also part of encouraging the implementation of the sea toll idea.

The joint slot scheme is carried out in one to two and three companies in the arrangement of ship slots to be fully filled. For example, one ship is only filled up to 30% of the total ship capacity, while the rest is filled by other shipping companies and operated in one ship unit. Support from the port for the initiative of the world shipping joint slot is the implementation of blending services and cumulative rates for services managed by Pelindo.

4.2.1.4 Multiplication of Port-based Industrial Area Development (Port Centric Logistic)

The concept of Port-based Industrial Area Development, or known as Port Centric Logistic, refers to the integration between industrial areas and ports to improve logistics efficiency and connectivity. In Indonesia, the implementation of this concept is seen in several major projects aimed at driving regional and national economic growth.

There have been several success stories of Port-based Industrial Areas such as JIIPE in East Java which received an award as Best Industrial Special Economic Zone in July 2024 [18]. Which was obtained in a short time from the start of its operation since 2022.

As another example, the Integration of the Cibitung Cilincing Area in Tanjung Priok as the industrial backbone in the Jabodetabek area and currently the development of the Kuala Tanjung Industrial Area (North Sumatra), Batang (Central Java) and Kijing (West Kalimantan) as well as several other Industrial Areas that are being explored for implementation by Pelindo is being explored.

4.2.1.5 Government Support for Pain Points in the Indonesian Port and Maritime Logistics Industry.

The most important government support is optimizing existing ports and aligning marine logistics development strategies in accordance with the aspirations of port operators and strengthening the dominant industrial areas in the region.

Another thing to note is that there have been symptoms of Deindustrialization in Indonesia, which of course has an impact on the implementation of port-based industrial areas [19]. So that government efforts to increase industrialization start from downstreaming of natural resources and categorization of area-based industries.

From the Fiscal Support aspect, the Government is expected to provide various fiscal incentives to attract investment in the processing industry, such as tax holidays, tax allowances, exemptions from import duties on capital goods, and area-based incentives such as Special Economic Zones and Industrial Zones.

4.2.2 General Roadmap Implementation Organizational Ambidexterity Post-Merger and Resilience For Post Merger Pelindo

According to the analysis and key findings that have been presented previously, the authors provide recommendations General Roadmap For Pelindo Organizational Ambidexterity Post-Merger and Resilience

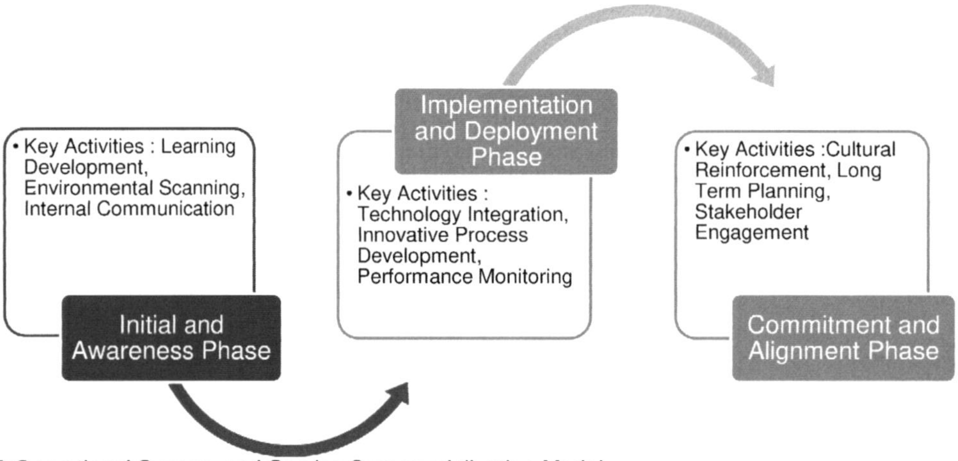

Image 4.3 Operational Synergy and Service Commercialization Model

which takes reference and guidance on the Pelindo Transformation Roadmap, Pelindo Human Capital Roadmap and Change Management implementation stages by Pelindo Internal Regulation where the

General Roadmap Organizational Ambidexterity Post-Merger and Resilience For Post Merger Pelindo consists of several stages including:

1) Initial and Awareness Phase

Objectives of the Initial and Awareness Phase are to ensure the establishment of cultivate awareness of ambidexterity within the organization and prepare employees for the forthcoming changes. There are key activities at this stage including :

a. Learning and Development Initiatives: Pelindo should implement training programs that elucidate the principles of organizational ambidexterity, emphasizing the balance between exploration (innovation) and exploitation (efficiency) and creating individual resilience.
b. Environmental Scanning: Conduct thorough analyses to identify external threats and opportunities within the maritime sector. This includes studying successful practices from Japanese ports, which have effectively integrated advanced technologies to enhance operational efficiency.
c. Internal Communication Campaigns: Launch campaigns to communicate the importance of innovation and adaptability, fostering a culture that embraces change.

2) Implementation and Deployment Phase

Objectives of the Implementation and Deployment Phase are aims to operationalize the concepts of organizational ambidexterity and resilience through practical applications across various units within Pelindo. There are key activities at this stage including :

a. Technology Integration : Implement systems such as the Terminal Operating System (TOS) Nusantara to streamline operations. This technology can facilitate real-time data sharing and improve cargo handling efficiency.
b. Innovative Process Development: Establish cross-functional teams tasked with developing innovative solutions tailored to local conditions, drawing inspiration from Japan's approach to continuous improvement (Kaizen).
c. Performance Monitoring: Develop key performance indicators (KPIs) to assess the effectiveness of ambidexterity initiatives, ensuring that both innovation and operational efficiency are measured.

3) Commitment and Alignment Phase

Objectives of the Commitment and Alignment Phase are focuses on solidifying commitment to organizational ambidexterity and resilience as a core organizational principle while ensuring alignment among all stakeholders.There are key activities at this stage including :

a. Cultural Reinforcement: Foster a culture that prioritizes innovation through recognition programs that reward creative contributions from employees.
b. Long-term Strategic Planning: Develop a strategic framework that encompasses both immediate operational goals and long-term growth objectives, including partnerships with international firms for knowledge transfer.
c. Stakeholder Engagement: Actively involve stakeholders in decision-making processes, ensuring their perspectives are integrated into strategic planning efforts.

4.2.3 Potential Long Term Impact Implementation Organizational Ambidexterity and Resilience For Post-Merger Pelindo

The roadmap for implementing organizational ambidexterity and resilience at Pelindo is structured to deliver both short-term outputs and long-term outcomes across three distinct phases. In the Initial and Awareness Phase, increased awareness sets the foundation for cultural shifts. The Implementation and Deployment Phase focuses on operational efficiencies that enhance competitiveness. Finally, the Commitment and Alignment Phase solidifies stakeholder engagement, fostering resilience.

Some points that the author suggests could be potential long term impact from the implementation of this roadmap include :

- Enhanced Organizational Readiness: Over time, Pelindo may cultivate a workforce that is more adaptable to change, equipped with skills necessary to navigate the complexities of Indonesia's maritime environment. This readiness is vital for maintaining operational resilience in a dynamic industry.
- Increased Competitiveness: By enhancing operational capabilities, Pelindo can improve its competitive position within the logistics sector, attracting more business from both domestic and international clients. This competitive edge is crucial for sustaining growth in a rapidly evolving market.
- Financial Performance Improvement: As operational efficiencies translate into cost savings and enhanced service offerings, Pelindo is likely to experience improved financial performance over time. Enhanced financial health will enable further investments in infrastructure and technology, reinforcing its market position.
- Organizational Resilience: With a robust framework for balancing exploration and exploitation, Pelindo will likely enhance its resilience against market fluctuations and disruptions. This resilience is particularly important for Indonesia as a maritime nation facing unique logistical challenges due to its geography.

By drawing lessons from global best practices while addressing Indonesia's unique maritime challenges, Pelindo can effectively navigate its post-merger landscape and emerge as a resilient leader in logistics. The anticipated impacts not only benefit Pelindo but also contribute positively to Indonesia's economic landscape by enhancing logistics efficiency and service quality for customers.

5. Conclusions and Implications for Next Research

5.1. Conclusion

This research explores the concept of organizational ambidexterity and resilience as key strategies in the Development of Integrated Port Network. Based on the case research and literature review conducted, there are several conclusions including the need for several strategic steps that have been explained previously, including:

- Collaboration Between Stakeholders
- Sustainable Infrastructure Development
- Digitalization and Process Automation
- Improving Intermodal Connectivity
- Focus on Sustainability and Efficiency

In addition, there are several "low hanging fruits" or the earliest implementation programs based on the author's proposals to encourage the development of the Integrated Port Network, including:

1. Implementation of Pelindo Organizational Ambidexterity Post-Merger
2. Formation of People Model (HR Profile Model) which focuses on implementing Organizational Ambidexterity at Pelindo
3. Joint Slot Operation Support between Shipping and Port Operator

4. Multiplication of Port-based Industrial Area Development (Port Centric Logistic)
5. Government Support for Pain Points in the Indonesian Port and Maritime Logistics Industry.

5.2. Implications and Further Research

Indonesia's mission to realize its ideals as the World Maritime Axis cannot but be carried out by multi-stakeholder and multi-entity collaboration which is formally summarized through Presidential Regulation (Perpres) Number 18 of 2020 concerning the National Medium-Term Development Plan (RPJMN) for 2020-2024 where one of the focuses of this RPJMN is the development of the transportation sector, including the Development of the Integrated Port Network. This research has explored Organizational Ambidexterity and Resilience in supporting these steps.

The next thing that needs to be explored is the Roadmap and Action Plan regarding port synergy between shipping and industrial areas that encourages a reduction in logistics costs in order to facilitate the flow of goods and passengers which is expected to increase national competitiveness.

In addition, in further research, there are several recommendations for researchers that can be the focus of next research including :

- Financial Viability of Organizational Ambidexterity and Resilience Initiatives at Pelindo which has focused on this study could analyze the cost-benefit implications of implementing ambidextrous strategies, including investments in technology and human resources. It could evaluate the potential returns on investment (ROI) from enhanced operational efficiencies and improved customer satisfaction.
- Comparative Analysis of Pelindo's Model with International Ports which has main objective doing study comparation Pelindo's post-merger strategies with those of successful international ports, such as those in Japan. It would explore how different operational models address similar challenges in logistics and customer service, highlighting best practices that Pelindo might adopt.
- Sustaining Competitive Advantage amid Technological Advancements which has main framework investigate how Pelindo can leverage emerging technologies (e.g., automation, digitalization) to maintain its market dominance. It would assess the impact of technological innovations on operational resilience and customer service.
- Govement Support Scheme Research : A detailed scheme of government support for social services to the community in general can be studied, both in terms of funding or service facilities. In addition, a broader analysis is needed to strengthen the "supply" aspect in the supply chain, namely extensive industrialization efforts according to the commodities of each region based on ports (Port Centric Logistic).

References

[1] Arief, O. H. R. (2022). Analisis Jumlah Armada Kapal Tunda Dengan Metode Simulasi Event Diskrit: Studi Kasus Pelabuhan Tanjung Priok. Fakultas Ekonomi dan Bisnis Universitas Indonesia.
[2] Dolz, C., Iborra, M., & Safón, V. (2019). "What explains the resilience of SMEs? Ambidexterity capability and strategic consistency." Journal of Business Research, 112, 1-10. https://doi.org/10.1016/j.jbusres.2019.01.026.
[3] Hill, S. A. (2023). Organizational Ambidexterity. In: Griffin, R. W. (Ed.), Oxford Bibliographies in Management. . New York, USA: Oxford University Press. doi: 10.1093/OBO/9780199846740-0214.
[4] Internal Document_Directorate of Operations Roadmap_ Terminal Services Group. Pelindo Head Office.
[5] Internal Document_Sharing Session Post Merger Pelindo. Corporate Transformation and Program Management Group. Pelindo Head Office.
[6] Lam, J.S.L., & Vis, I.F.A. (2021). "Resilient Port Operations: Strategies and Best Practices." Maritime Economics & Logistics, 23(2), 234-256.
[7] Liu, J., Gu, B., & Chen, J. (2023). Enablers for maritime supply chain resilience during pandemic: An integrated MCDM approach. Transportation Research Part E: Logistics and Transportation Review.
[8] Liu, J., Qi, Y., & Lyu, W. (2023). Port resilience in the post-COVID-19 era. Ocean Coast Manag. 2023 May 1; 238: 106565. Published online 2023 Mar 13.
[9] O'Reilly, C.A., & Tushman, M.L. (2013). "Organizational Ambidexterity: Past, Present, and Future." Academy of Management Perspectives, 27(4), 324-338.
[10] Parola, F., Pallis, A. A., & Rodrigue, J.-P. (2022). "Organizational ambidexterity in the post-COVID-19 era: A systematic literature review and future research agenda." Transportation Research Part E: Logistics and Transportation Review, 154,
[11] Presidential Regulation Number: 18 of 2020. RPJMN (National Medium-Term Development Plan).
[12] Post Merger Review & Evaluation. Internal Document. Pelindo Head Office. Corporate Transformation & Program Management Group.

[13] Shi, J., Chen, J., Xu, L., Di, Z., & Qu, Q. (2022). Improving the resilience of maritime supply chains: The integration of ports and inland transporters in duopoly markets. Maritime Policy & Management.
[14] Tukamuhabwa, B., Stevenson, M., & Busby, J. (2015). Supply chain resilience: definition, review and theoretical foundations for further research. International Journal of Production Research.
[15] Turner, N., & Kutsch, E. (2016). Understanding resilience in organizational operations-an ambidexterity perspective. Academy of Management Proceedings
[16] Zhao, Y., Bi, Y., & Han, Y. (2021). "The impact of organizational ambidexterity on marketing performance: Evidence from China." Journal of Business Research, 128, 358-367.
[17] https://www.antaranews.com/berita/3817239/kawasan-industri-dan-pelabuhan-terpadu-dinilai-tingkatkan-konektivitas
[18] https://jiipe.com/id/home/blogDetail/id/395#:~:text=JIIPE%20KEK%20Gresik%20Meraih%20Penghargaan,%ED%95%9C%EA%B5%AD%EC%96%B4
[19] https://apindo.or.id/index.php/media/pengusaha-sebut-gejala-deindustrialisasi-makin-nyata-beban-berat-pemerintahan-era-selanjutnya
[20] https://ekonomi.bisnis.com/read/20141120/98/274459/pelayaran-jalankan-skema-joint-slot

IOP Conf. Series: Earth and Environmental Science **1461** (2025) 012022 doi:10.1088/1755-1315/1461/1/012022

APPENDIX

No	Exploring Organizational Ambidexterity and Resilience	Scope implementation	Initiatives that have been undertaken	Next Steps
1	Collaboration Between Stakeholders	Identifying how collaboration between ports, logistics providers and government authorities can improve operational efficiency and effectiveness.	• Intense Communication with Stakeholders and Customers regarding Pelindo Merger Progress • Exploration related to Port Development through investment synergy and strategic partners • Integrated key account management	• Implementation of regional expansion in the port business for container commodities • Government support in developing port-based industrial areas • Establishing a competitive scheme in implementing cooperation with state-owned and private logistics players • Determination of cooperation with industrial/logistics areas • Renegotiation of Cooperation Scheme for Development of Port Operation Services Owned by Ministries/Institutions
2	Sustainable Infrastructure Development	Analyzing the importance of investing in environmentally friendly and efficient infrastructure to support the sustainability of the port network.	• Development of existing port infrastructure and equipment for Pelindo's business • Limited Implementation of Sustainable Infrastructure of port operations • Limited implementation related to the implementation of sustainability efforts at ports (green ports) • Rejuvenation and Maintenance of Port Equipment	• Implementation of Sustainable Infrastructure Planning in all port operational areas • Multiplication of sustainability efforts at the port • Collaboration on the implementation of sustainable infrastructure involving shipping and industrial areas
3	Digitalization and Process Automation	Examining the role of digital technology in improving productivity and efficiency in port operations	• Standardization of operations at the port • Digitalization & systemization of port operations	• Advanced Stage of Standardization and Digitalization of Port Operations • Synergy of digital implementation involving shipping and industrial areas
4	Improving Intermodal Connectivity	Examining the importance of integration between sea, land and air transportation to create an	• Cooperation with state-owned and private logistics players • Collaboration with industrial/logistics areas • Development of TUKS operational services with	• Multiplication of industrial area development based on ports

No	Exploring Organizational Ambidexterity and Resilience	Scope implementation	Initiatives that have been undertaken	Next Steps
		efficient logistics network.	state-owned and private companies • Development of Port Operation Services owned by Ministries/Institutions • Port development through investment synergy and strategic partners	
5	Focus on Sustainability and Efficiency	Assessing how the development of an Integrated Port Network can contribute to emission reductions and more efficient use of resources.	• Additional MEPS service offering for Pelindo key accounts • Stages of awareness regarding ESG implementation at Pelindo • Sampling of Pelindo's Main Service Efficiency Monitoring	• Implementation Stages and Commitment in the application of ESG at Pelindo • Implementation of regional expansion in the port business • Comprehensive implementation of Pelindo's Main Service Efficiency Monitoring

Enhancing Sustainability and Efficiency: Benefits and Challenges of Shore Power Connection Application in Indonesian Ports

Hasan Nur[1*], Achmad Mustakim[1], Christino Boyke[1] and Alwi Khaqiqi[2]
[1] Department of Marine Transportation Engineering, Sepuluh Nopember Institute of Technology, Surabaya, Indonesia
[2] Department of Shipbuilding Engineering, Shipbuilding Institute of Polytechnic Surabaya, Indonesia

*E-mail: hasaniqbaln@its.ac.id

Abstract. This paper investigates the benefits and challenges of implementing shore power connections (SPC) as part of green port initiatives in Indonesia. As highlighted in various studies, SPC offers a more sustainable solution for port operations by replacing a ship's marine fuel consumption with electricity, a more environmentally friendly alternative. Since 2016, at least 11 public ports in Indonesia have implemented SPC facilities. The benefits of SPC are substantial for shipping companies, port operators, and the government, as it reduces emissions and improves operational cost efficiency. The findings of this study show that SPC can lower sulfur dioxide (SO_2) and nitrogen oxide (NO_x) emissions by an average of 55% while reducing ship's operational costs by an average of 60% compared to using auxiliary engines while berthing. These results indicate that factors such as ship type, auxiliary engine specifications, and operating conditions significantly influence the outcomes. However, the current implementation of SPC still faces technical, operational, economic, and regulatory challenges. This paper concludes with recommendations for optimizing SPC applications in Indonesian ports to support the country's transition to greener port operations.

Keywords: infrastructure and sustainability, energy, shore power connection, port

1. Introduction

Ports are critical to international shipping, serving as essential nodes in transport networks. However, port activities also contribute to the global environment, mainly through emissions from ships during service. Ship emissions are produced not only during sailing due to the combustion of fossil-fuelled engines but also while ships are berthing at ports, requiring a continuous power supply typically generated by running auxiliary engines that emit exhaust gases. These emissions include harmful pollutants such as carbon dioxide (CO_2), nitrogen oxides (NO_x), sulphur oxides (SO_x), ammonia (NH_3), formaldehyde, particulate matter (PM), and particle number emissions (PN) [1]. In 2018, shipping activities contributed to 2.89% of global CO_2 emissions, approximately 15% of global NO_x emissions, and 13% of global SO_x emissions [2]. Without preventive measures, emissions from ships, especially in port areas, are likely to increase due to the growing demand for shipping, further escalating their environmental impact.

The potential environmental impact of port operations has drawn increasing attention from governments, industry, and academia, leading to the introduction of the green port concept to mitigate these adverse effects. The idea of a green port was officially proposed at the United Nations Climate Change Conference in 2009. This concept is characterized by a healthy ecological environment, reasonable utilization of resources, low energy consumption, and low pollution [3]. Port operations comprise activities in the water area and shore/terminal area. The water area operations encompass vessel operations, including ocean-going vessels (OGVs) and harbour crafts, while terminal activities comprise cargo/passenger handling equipment and heavy goods vehicle operations. Many studies have investigated the shares of emissions in port areas and showed that emissions of OGV operations are the main contributors to emissions in ports. For example, [4] highlighted that the emissions inventory of ocean-going vessels is above 60% of the total NO_x, PM, SO_2, and CO_2 in the Port of Oslo. Similar results yielded by [5][6] showed a significant contribution of emissions for OGVs in the Port of Los Angeles and Stockton Port, respectively. OGV operations in ports typically comprise anchoring, maneuvering, and berthing. The most emitted among these operations, according to [4] [7] [8], which comes from the auxiliary engines during berthing operations. Therefore, reducing emissions of ships at berth can decrease significant overall emissions in ports.

Shore power connection (SPC) has been proven to significantly reduce OGV's emissions at berth. This technology refers to utilizing electricity from the local grid to replace the power from ships' auxiliary engines during their berthing [9]. The first application of SPC was implemented at the Port of Gothenburg, Sweden, in 2000 to reduce emissions from ships while they are berthing [10]. Overall, this measure can reduce emissions of vessels by 48% - 70% of CO_2, 3% - 60% of SO_2, 40 – 60% of NOx, and 57 – 70% of black carbon in ports [11]. Many studies yielded various emissions reductions in specific ports studied worldwide, including the Port of Antwerp and Genoa, which reduce 50% of CO_2 and 97% of NOx emissions [12]. Furthermore, [8] estimated a 10% reduction of CO_2 emission for Ferry–Roro vessels and 5% for container ships in Port of Gothenburg. Also, several Asia ports have been investigated for emissions reduction, such as Kaohsiung port: NO_x (49.2%), SO_2 (63.2%), PM (39.4%), CO_2 by 57.2% and HC by 29.2% [13] and Shenzhen port: 20% of carbon emissions [14].

Apart from the benefit of emissions reduction, implementing SPC faces several challenges. Implementation of SPC turns into a "chicken and egg" dilemma, where ports are reluctant to build this facility unless enough vessels can connect, and ship owners are hesitant to retrofit their vessels without widespread SPC availability in ports visited [15] [16]. Investing in SPC requires high capital costs for ports and ships. High capital costs and high electricity prices, particularly in European countries, become high barriers for port operators to offer SPC at competitive rates. This is evidenced by the feasibility study conducted by [17] for the SPC investment at Southampton Port, UK, which concluded that it was not commercially feasible without public funding.

Global efforts such as the World Ports Climate Action Program (WPCAP), launched in 2018, with several ports collaborating to reduce greenhouse gas emissions through joint actions, including the accelerated development of SPC [18]. Furthermore, the EU established EU Fit 55, a set of binding targets to achieving net zero carbon by 2050, with an intermediate step committing to cutting emissions by at least 55% by 2030. Within these targets, a strategy for maritime transport under the FuelEU Maritime Regulation [19] will oblige container and passenger ships with over 5,000 gross tonnage (GT) at major EU ports to use SPC starting from 2030, and this will also apply to the rest of EU ports by 2035 [19]. A study by [20] demonstrates that ports, particularly in higher-income countries, have made significant strides in implementing environmental standards and shaping green port policies. They also highlight disparities between higher-income and lower-income countries. Ports in higher-income countries have been more effective in adopting enforcement-based approaches and influencing policy, while ports in lower-income countries face greater challenges in this regard.

As noted above, many studies have discussed the benefits and challenges of adopting SPC in higher-income countries' ports, including the USA, UK, and European countries. However, limited research has investigated the challenges and benefits of SPC in lower-income countries. Moreover, previous studies are solely concerned with reducing emission benefits, with a lack of discussion on additional benefits of SPC. To address this gap, this paper investigates SPC's benefits, challenges, and prospects and proposes recommendations for ports in lower-income countries, using Indonesia as a case study. A mixed-methods approach, combining qualitative and quantitative methods, was applied in this research. Our research contributes to the literature by understanding the adoption of SPC in the unique contexts of ports in Indonesia. This paper focuses on SO_2 and NO_x emissions due to the immediate impact on local air quality and public health in port areas.

This paper is structured as follows: Following the introduction, Section 2 describes the study's methodological approach, which combines both qualitative and quantitative methods. Section 3 provides an overview of the current state of SPC applications in Indonesian ports. Section 4 presents the results and discussion, focusing on SPC's benefits and challenges. Finally, Section 5 summarizes the main findings, discusses limitations, and offers recommendations for improving SPC applications.

2. Methodology

Mixed methods, combining qualitative and quantitative approaches, were applied in this research. The qualitative approach involved a site survey and focus group discussion (FGD). Furthermore, a quantitative approach was conducted to assess the potential benefits of SPC in reducing emissions and operational costs.

2.1 Survey and focus group discussion procedure

A site survey was conducted over seven days at three locations: Port of Tanjung Perak, Tanjung Emas, and Tanjung Intan, located in East and Central Java, Indonesia. These ports have been providing SPC services since 2018. The survey aimed to examine the actual application of SPC, including facilities and operations at different locations. Additionally, interviews were conducted with SPC operators and ship crews to explore the challenges associated with this service.

To gain further insight, a Focus Group Discussion (FGD) was held one week after the site survey, bringing together stakeholders, including representatives from shipping companies, port operators, SPC operators, and the government. A total of 18 participants attended the two-hour session. The FGD aimed to gather more detailed information from the perspective of management and policymakers, particularly regarding the benefits, challenges, and prospects of SPC implementation.

The FGD followed a two-stage procedure. In the first stage, presentations were given to engage the participants on key topics. For example, government representatives presented the fundamental regulations and policies related to gas emission control, and SPC operators discussed the current state of SPC, including facilities and operational aspects. The second stage involved a discussion among all participants, moderated to explore the potential benefits, challenges, and prospects of SPC. All invited participants, representing various Indonesian port stakeholders, were given the opportunity to share their experiences and suggestions, with a particular focus on input from shipping companies as the primary users of SPC.

2.2 Formulation of operational cost and emissions

2.2.1 Calculation of fuel consumption and costs
The cost of fuel oil consumption can be influenced by various variables, such as the size, shape, and condition of the hull, sailing conditions, speed, as well as the type and capacity of the main engine and auxiliary engines. [21]. The total fuel cost also depends on the prices of each type of fuel, such as Heavy Fuel Oil (HFO), Marine Diesel Oil (MDO), and Marine Fuel Oil (MFO). The fuel consumption for the vessels was estimated by using the following equation:

$$FC_t = (MCR\ (0.08 \times HP) \times SFOC \times t \times 10^{-6}) - (3 \times 1.05\ ton/hr) \quad (1)$$

Where: MCR (Maximum Continuous Rating) refers to the nominal power of the primary/auxiliary engine is calculated as 0.80 to 0.95 x *HP*, *SFOC* (Specific Fuel Oil Consumption) is calculated based on the condition of a new engine with a tolerance requirement of 180 g/HP/hour, *t* is the Operating time of the engine (hours), 10^{-6} is the Conversion from grams to tons, and 1.05 is the factor that accounts for fuel requirements due to leaks caused by engine aging. Equation 2 was used to convert the fuel consumption from tons to liter:

$$FC_l = FC_t \times 10^{3} \times \frac{1}{0.815} \quad (2)$$

Where 10^{-3} is the conversion from tons to kg, and 1/0.815 is the Weight/Specific Weight of HSD. Eventually, the fuel costs were estimated by the following equation:

$$FC = FC_l \times P \quad (3)$$

Where *FC* is the Fuel Cost (IDR), FC_l is the Fuel requirement (liters), and P is the fuel price (IDR/liters)

2.2.2 Calculation of cylinder engine oil consumption
The cylinder engine oil consumption for auxiliary engines on ships can be calculated by determining the lubrication needs for both the main and auxiliary engines and the cost of the lubricating oil itself. In this study, the cost of lubricating oil is sourced from the auxiliary engine.

Cylinder engine oil requirement :

$$FC = MCR\ (0{,}8 \times HP) \times 1{,}34 \times Number\ of\ AE\ Berthing \times Time/1000 \quad (4)$$

Where FC is the fuel cost (IDR), MCR (maximum continuous rating) is the nominal power of the main/auxiliary engine, which is calculated as 0.80 to 0.95 x *HP*, and *HP* is the electric power.

Cylinder engine oil cost :

$$FC = FC_c \times P \quad (5)$$

Where: FC is the fuel cost, FC_c is the cylinder engine oil requirement (liters), and *P* is the price of cylinder engine oil (IDR/liters).

2.2.3 Calculation of maintenance costs
Maintenance costs are used to determine the expenses required for operational activities. The following equation is an approach to estimating maintenance costs:

$$M_e = 0.03 \times \text{Investment} \quad (6)$$

Where: M_e is the maintenance cost, and investment is the investment cost.

2.2.4 Emission estimation
The methodology developed by the Puget Sound Maritime Air Forum, known as the OGV Emissions Estimating Methodology, is employed to estimate emissions produced by ships. This

methodology calculates emissions based on the activities performed by vessels in the area. Emissions are estimated using the following equations:

$$E_i = Energy \times E_F \times FCF \quad (7)$$
$$Energy = Load \times Activity \quad (8)$$

Where: E_i is the amount of emissions (by mode) (grams), Energy is the energy requirement (by mode), calculated as the Energy produced by the engine over a specified period (kWh), E_F is the emission factor (g/kWh), load is the maximum continuous rated engine, and activity is the activity level (hours).

2.2.5 Assumption

Calculations are also performed for the ship's auxiliary engine to calculate the total emissions produced. The emission factor for the auxiliary engine differs slightly from that of the propulsion engine. [22] To calculate the exhaust emissions produced by the auxiliary engine in the form of nitrogen oxides (NOx) and sulfur oxides (SOx), 9.8 g/KWh for NOx and 0.44 g/KWh for SOx were used. The assumptions used for fuel cost calculations include non-subsidized fuel at 13,700 IDR/liter [23], cylinder engine oil at 35,048 IDR/liter, specific fuel consumption at 252 ml/KWh, and a maximum continuous rating of 80%.

3. Current application

3.1 Regulatory aspect

Indonesia has ratified MARPOL ANNEX 6 on the Prevention of Air Pollution from Ships, which includes regulations limiting ship exhaust emissions of NOx and SOx. Additionally, Indonesia has committed to reducing greenhouse gas emissions by up to 29% by 2030, as outlined in its Enhance Nationally Determined Contribution across various sectors, including the port sector. This commitment is being pursued through initiatives that promote green port practices, including the use of SPC to control air pollution in and around port areas [24]. SPC is a promising approach for advancing the green port initiative, particularly in reducing ship emissions in port areas. The implementation of SPC in Indonesia is governed by several regulations, including Law No. 17 of 2008 on Shipping, Government Regulation No. 21 of 2010 on Maritime Environmental Protection, and Ministerial Regulation No. PM 29 of 2014 on Preventing Maritime Pollution. These regulations emphasize the importance of protecting the maritime environment by preventing pollution from ship operations, cargoes, and hazardous materials on ships. In response to these regulations, data from the Ministry of Transportation of Indonesia indicate that at least 11 ports and six terminals currently provide SPC facilities. This reflects the country's commitment to the Green Port program since its inception in 2016.

3.2 Technical aspect

SPC facilities are dispersed across several locations, including East and Central Java, Jakarta, North Sumatra, South Kalimantan, Bali, South Sulawesi, and West and East Nusa Tenggara, as shown in Figure 1. Based on the site survey, SPC facilities at Indonesian ports include high-voltage cables for electricity distribution from the local power grid to the dock, transformers to convert the electricity to voltages suitable for the ship's electricity to around 6-20 kV, and frequency to the appropriate frequency (50-60 Hz). Circuit breakers act as safety devices between the power source and the load side, automatically disconnecting in case of a disturbance. Inverters convert

IOP Conf. Series: Earth and Environmental Science **1461** (2025) 012023 doi:10.1088/1755-1315/1461/1/012023

direct current (DC) to alternating current (AC). Additionally, switchboard control panels manage electricity distribution, and housing protects transformers, inverters, circuit breakers, and switchboards from the weather. The frequency converter is placed on the dock inside a 20-foot container.

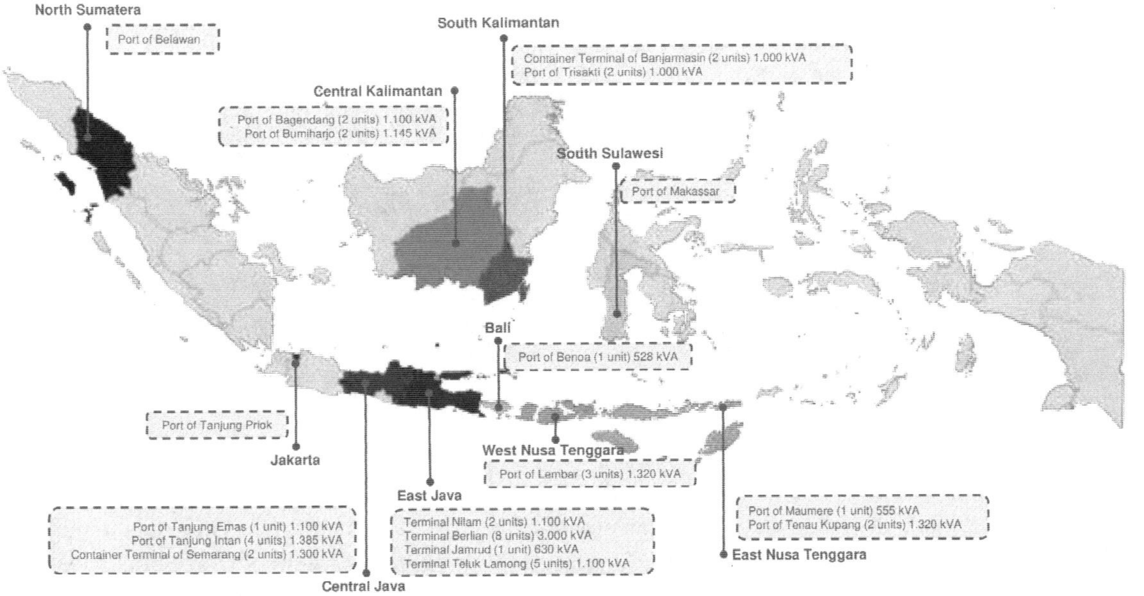

Figure 1. Location of shore power connection facilities across Indonesia

The electricity for SPC in the Indonesian port is mainly sourced from the national electricity power grid. The power capacity ranges between 525–3,000 kVA, with a frequency of 50/60 Hz and a voltage generally between 360-450 V. For instance, SPC services at Tanjung Perak Port have a power capacity of 630 kVA – 3,000 kVA with a voltage of 380 V and a frequency of 50 Hz. In contrast, at Tanjung Intan Port, the power capacity is 1,385 kVA with a voltage between 400 to 450 V and a frequency of 50 Hz or 60 Hz.

Figure **2** Illustrate the locations of SPC facilities at both the Port of Tanjung Perak and the Port of Tanjung Emas.

Port of Tanjung Intan

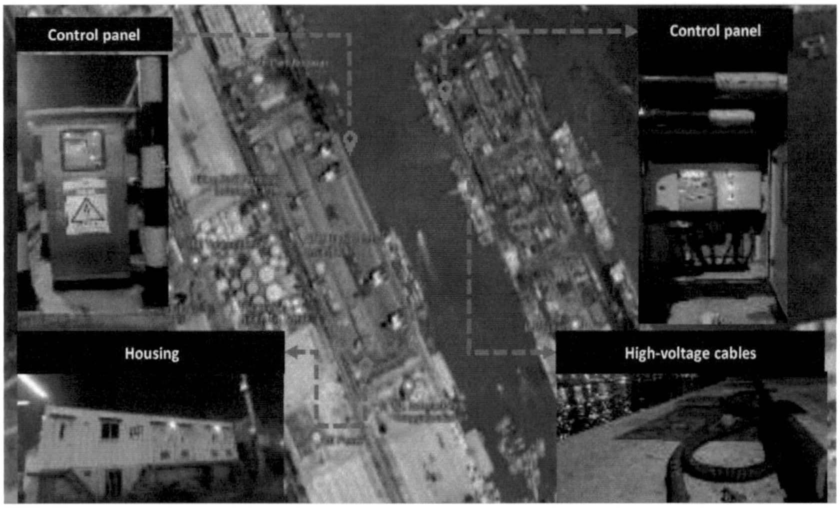

Port of Tanjung Perak

Figure 2. SPC facilities at the Port of Tanjung Perak and the Port of Tanjung Emas

Supporting equipment is also necessary on the ship's side. Most of this equipment is available on the vessel, as the SPC service concept at the port is similar to shipyards, where ships undergo periodic maintenance. Therefore, modifications to the ship's electrical system are only required if there is an issue with the system. Shipboard equipment includes a cable reel to connect the vessel to the dock's connectors while it is docking and a power grid to adjust the electrical voltage according to the ship's needs.

3.3 Operational aspect

Most public ports in Indonesia are operated by the country's state-owned port operator, with SPC services managed by their regional subsidiary companies. The standard operating procedure for SPC includes a ship compatibility survey and a trial connection. Typically, before a ship can receive SPC service, the SPC operator conducts a thorough survey to assess the condition of the ship's electrical system and determine the power load requirements while berthing. A trial connection is also performed. If the survey shows that the breaker does not trip and the system is deemed safe, the ship is approved to use the SPC service.

The use of SPC in Indonesian ports spans domestic and international vessels, including container ships, bulk carriers, tugboats, and pilot boats. However, container vessels dominate SPC usage. For instance, at Tanjung Perak Port, all SPC users are container vessels, with total electricity consumption reaching approximately 100,000 kWh in 2021. In contrast, at Tanjung Emas Port, the ships utilizing SPC include bulk carriers, tugboats, and pilot boats, with total electricity usage reaching 560,000 kWh. Similarly, at Tanjung Intan Port, bulk carriers, tugboats, and pilot boats are the primary users, consuming just over 250,000 kWh. The type of vessel plays a critical role in determining the overall electricity demand, as larger vessels like bulk carriers typically require more Energy than smaller vessels such as tugboats or pilot boats.

The tariff for SPC services ranges between 2,500 and 3,500 IDR/kWh, or approximately 0.15 USD/kWh. The price is determined based on the cost of electricity in each location, taxes, and specific cooperation agreements between the shipping company and the SPC operator. In some cases, SPC facilities are dedicated to certain shipping companies, with a minimum electricity consumption requirement per month or year. Additional costs include the connection and disconnection fees, which range from 500,000 to 3 million IDR per ship call, or approximately 30 to 200 USD per ship call. This is charged for physically connecting the ship to the SPC system and disconnecting it when it is ready to depart. The cost covers the work done by personnel and any necessary equipment.

4. Results and discussion

This section presents the findings from the site survey and FGD, which were conducted to gather insights from various stakeholders' perspectives. Additionally, it includes calculations and a comparative analysis of emissions estimates and operating costs to support the argument for the benefits of SPC. The discussion highlights the benefits of SPC and challenges related to technical, operational, economic, and regulatory aspects.

4.1 Benefits of SPC

4.1.1 Emission reduction

One of the primary benefits of implementing SPC is emission reduction, a benefit highlighted by many scholars, including [11], [12], and [13]. To contextualize this benefit within the application of SPC in Indonesian ports, we calculate the emission reductions based on the methodology described in the previous section. Our calculations include more than 500 vessels of various types and dimensions that have utilized SPC in three major ports: Tanjung Perak, Tanjung Emas, and Tanjung Intan. An example of ship specifications is provided in Table 1.

Table 1. Example of specifications for ships using shore connection services

	Ship-1 (Container vessel)	Ship-2 (Bulk carrier)	Ship-3 (Tugboat)
Main Dimension			
LOA (m)	120	96.5	36
DWT (Ton)	9,137	5,253	739
GT	6,285	2,99	386
Engine			
Main (Hp)	2,800	2,400	1,200
Auxiliary (HP)	448	163	108

The results indicate a significant reduction in emissions when using SPC, including a decrease of SO_2 by 15-89% and NO_x by 12-92%. These variations suggest that ship type, auxiliary engine specifications, and operating conditions are critical in determining emission reduction efficiency. A comparison of these results with previous studies is presented in Table 2, highlighting both the consistency and differences in emission reduction across various case studies.

Table 2. Comparison of emission reduction from SPC implementation

Author	Emission Reduction				Case study
	So_2	No_x	CO_2	PM	
T. Zis et al.	3% - 60%	40 – 60%	48% - 70%	-	-
Acciaro et al.	-	97%	50%	-	Port of Antwerp and Genoa
Chang & Wang	63.2%	49.2%	57.2%	39.4%	Port of Kaohsiung
This study	15-89%	12-92%	*Out of scope*	*Out of scope*	Port of Tanjung Perak, Tanjung Emas, and Tanjung Intan

4.1.2 Cost efficiency

Ship owners are primarily concerned with economic gains, so their decision to invest in an SPC facility and use SPC while potential cost savings influence berthing. To assess this, we calculated the cost of fuel consumption during berthing and compared it with the cost of using SPC. Ship owners can determine the most economical option by evaluating the costs with and without SPC. It is important to note that the cost of fuel consumed by auxiliary engines is affected by SPC use. Our results show that SPC is more economical than auxiliary engines, reducing costs by 50-65%. This outcome is supported by information from shipping companies that have been using SPC for several years, as shared during our FGD. The relatively inexpensive electricity significantly influences the cost efficiency in Indonesia. In contrast, countries in Europe and the US, which have higher electricity prices and greener grid sources, may experience a different level of cost-benefit. Thus, SPC offers notable advantages in countries where electricity costs are lower. Table 3 presents a cost comparison with and without SPC based on vessel type and characteristics, as shown in Table 3.

Table 3. Cost comparison with and without SPC

	Energy consumption (kWh per ship call)		Cost (Million IDR/Ship call)	
	Without SPC	With SPC	Without SPC	With SPC
Ship-1 (Container vessel)	3,258.0	291.0	13.98	1.20
Ship-2 (Bulk carrier)	2,411.0	129.0	10.34	0.78
Ship-3 (Tugboat)	1,821.0	95.0	7.46	0.46

4.2 Technical, operational, economic, and regulation challenges

Apart from the benefits of implementing SPC, several challenges remain. In the following section, we discuss the challenges shipping companies, port operators, and governments face. These challenges include technical and operational issues, such as the standardization of SPC infrastructure and the reluctance of ship crews. Economic challenges involve high initial investment and concerns about financial feasibility. Regulatory challenges focus on government participation and enforcement.

4.2.1 Technical and operational challenges

The lack of standardization in SPC systems presents significant technical and operational challenges, particularly concerning onboard facilities. To maximize the use of shore connections, it is crucial to standardize supporting equipment, such as main circuit breakers and other electrical installations. Alongside this, comprehensive crew training programs are needed to raise awareness of SPC's operational and environmental benefits, helping to overcome crew resistance through familiarity and certification incentives. Technical support will further facilitate the transition for existing ships, making the broader adoption of SPC systems more feasible. Additionally, conducting surveys and trial connections is essential to ensure the smooth operation of SPC services. The efficiency of ship service time at ports plays a vital role in the overall operational activities, especially for scheduled routes. SPC service times should be carefully considered, as the average connection and disconnection time is currently 90-120 minutes—longer than the average times at foreign ports, which are 82 minutes for container ships and 72 minutes for cruise ships [18]. Addressing these time inefficiencies is vital for improving SPC implementation in ports.

4.2.2 Economic challenges

One of the primary challenges in implementing Shore Power Connections (SPC) is the economic aspect, particularly concerning high investment costs and financial feasibility. [25] Highlight that, despite the ongoing development of SPC technology, economic concerns remain a significant barrier to its widespread adoption. Similarly, [26] Suggests that financial issues are the principal challenge to extensive SPC application. Establishing an SPC system requires significant investment from port operators and shipping companies. Ports must invest in substations with switch breakers, control panels, transformers, protection equipment, and communication systems to link ships to the shore. A frequency converter is often required to align the local electrical network's frequency with the vessel. According to information from the FGD session, the investment for SPC facilities in Indonesian ports ranges from 5 to 10 billion IDR, equivalent to 320,000 to 650,000 USD. Due to typically low utilization hours, port operators may face financial losses if usage falls below expectations. In some ports, the utilization rate is less than 25% of capacity, indicating low usage of these facilities. The uncertainty surrounding utilization rates further complicates investment decisions for port operators. Additionally, ongoing costs, including maintenance of SPC facilities, must be considered. For shipping companies, the investment includes additional electrical switchboards and cables to connect to the ship's main switchboard. The cost of SPC systems varies significantly depending on vessel type, size, and the need for an onboard transformer. This additional financial burden can be particularly challenging for smaller shipping companies.

4.2.3 Regulatory challenges

The government plays a significant role in developing regulations and enforcing green port initiatives, particularly in applying SPC. However, using SPC is optional in Indonesia due to several factors, including the availability of SPC facilities and financial concerns about the additional costs for shipping companies. Ship owners or shipping companies are primarily focused on the financial returns of their investments, while the government must balance economic and environmental considerations. The government must support port operators and shipping companies in promoting SPC implementation. In Indonesia, this role is typically carried out by the port authority. [27] highlight an example where government subsidies covered 30% of the SPC investment, ensuring maximum profits for all stakeholders, including shipping companies, port operators, government entities, and power grid companies. Financial support, such as subsidies, could incentivize shipping companies and port operators to invest in standardized SPC systems. As long as SPC use remains voluntary, ships will have the flexibility to continue relying on traditional fuels, which negatively impacts the utilization rates of SPC facilities and hinders the goals of green port initiatives. From a regulatory standpoint, it may be challenging for the government to mandate SPC usage, as it would require addressing the financial and practical concerns of various stakeholders.

5. Conclusion

Shore power connections (SPC) were introduced globally in 2000, and Indonesia has provided this facility since 2016. Currently, at least 11 ports in Indonesia offer SPC facilities, reflecting a solid commitment to the green port initiative and participation in the global emission reduction program. Implementing SPC is a promising strategy for mitigating pollutants and minimizing their environmental impact. This paper aims to investigate SPC's benefits, challenges, and prospects, providing recommendations for ports in lower-income countries through a case study of Indonesia. Our research contributes to the literature by examining the adoption of SPC in the unique context of Indonesian ports. However, this study may face limitations, including data availability, potential information bias due to limited stakeholder representation, and challenges in generalizing the results to other lower-income countries with varying port infrastructures, regulatory environments, and economic conditions.

The benefits of SPC extend to shipping companies, port operators, and the government, encompassing environmental impacts through reducing emissions and air pollution, economic impacts through cost savings, and enhanced reliability by reduced vessel downtime. Our findings indicate that the utilization of SPC, when compared to auxiliary engines, can reduce energy consumption (kWh) by an average of 60%, reduce cost by 50-65%, and decrease emissions of exhaust gases such as SO2 15-89% and NOx by12-92%. The efficiency is significantly influenced by the type of ship, size, and the auxiliary engine specifications. However, several challenges must be addressed: technical challenges related to the lack of standardization in SPC systems, which affects efforts to reduce connection and disconnection times; economic challenges involving high initial investment costs and financial feasibility; and regulatory challenges concerning compliance with policy development and enforcement.

To optimize the implementation of SPC in Indonesia, a coordinated approach involving the government, shipping companies, and port operators is essential. The government may develop a phased regulatory framework that mandates SPC use initially for larger ocean-going and international vessels, with plans to extend the requirement to domestic vessels based on size and type of vessel. Supportive policies, including subsidies and incentives, are crucial to mitigate high initial investment costs and encourage adoption. Standardization of SPC systems should be pursued to ensure compatibility and facilitate compliance. Monitoring and enforcement mechanisms must be established to ensure adherence to regulations and assess their effectiveness.

Shipping companies must consider investing in SPC technology on their vessel by conducting thorough cost-benefit analyses to understand financial advantages and operational savings. Early adoption is encouraged to capitalize on reduced operational costs and enhanced environmental performance. Comprehensive crew training programs should be implemented to familiarize crews with SPC operations and benefits, complemented by incentives for embracing the technology. Port operators must develop standardized SPC infrastructure to ensure compatibility with various vessels, investing in electrical installations and facilities that meet standardized specifications. Conducting surveys and connection trials will help ensure that SPC systems meet vessel needs, while service-level agreements should be established to guarantee reliable and consistent SPC service. Reducing connection and disconnection times will further enhance the efficiency of SPC services. Promoting SPC through awareness campaigns can increase utilization by highlighting its benefits, such as cost savings and environmental improvements. Additionally, implementing standard operating procedures (SOPs) for SPC operations, including safety protocols and emergency response training, will help manage risks and ensure smooth operations. This collaborative approach would help promote wider adoption of the SPC system to support the transition to greener port operations.

References

[1] P. T. Aakko-Saksa *et al.*, "Reduction in greenhouse gas and other emissions from ship engines: Current trends and future options," *Prog. Energy Combust. Sci.*, vol. 94, 2023, doi: 10.1016/j.pecs.2022.101055.

[2] IMO, "Fourth IMO GHG Study 2020 Full Report," *Int. Marit. Organ.*, vol. 6, no. 11, p. 524, 2021, [Online]. Available: https://wwwcdn.imo.org/localresources/en/OurWork/Environment/Documents/Fourt

h IMO GHG Study 2020 - Full report and annexes.pdf

[3] C. Wan, D. Zhang, X. Yan, and Z. Yang, "A novel model for the quantitative evaluation of green port development – A case study of major ports in China," *Transp. Res. Part D Transp. Environ.*, vol. 61, pp. 431–443, 2018, doi: 10.1016/j.trd.2017.06.021.

[4] S. López-Aparicio, D. Tønnesen, T. N. Thanh, and H. Neilson, "Shipping emissions in a Nordic port: Assessment of mitigation strategies," *Transp. Res. Part D Transp. Environ.*, vol. 53, no. 2017, pp. 205–216, 2017, doi: 10.1016/j.trd.2017.04.021.

[5] O. Cai, C. Peng, and X. Yu, "The development of port emissions inventory from decision-making perspective: Case of the port of Los Angeles," *IOP Conf. Ser. Earth Environ. Sci.*, vol. 675, no. 1, 2021, doi: 10.1088/1755-1315/675/1/012022.

[6] Starcrest, "Inventory of Air Emissions for Calender Year 2020 Port of Stockton California (Issue June)," 2023.

[7] R. A. Nguyen Khoi Tran, Jasmine Siu Lee Lam, Haiying Jia, "Emissions from container vessels in the port of Singapore," *Marit. Policy Manag.*, vol. 49(3), pp. 306–322, 2022.

[8] H. Winnes, L. Styhre, and E. Fridell, "Reducing GHG emissions from ships in port areas," *Res. Transp. Bus. Manag.*, vol. 17, pp. 73–82, 2015, doi: 10.1016/j.rtbm.2015.10.008.

[9] N. N. Abu Bakar, N. Bazmohammadi, J. C. Vasquez, and J. M. Guerrero, "Electrification of onshore power systems in maritime transportation towards decarbonization of ports: A review of the cold ironing technology," *Renew. Sustain. Energy Rev.*, vol. 178, no. May 2022, p. 113243, 2023, doi: 10.1016/j.rser.2023.113243.

[10] R. A. B. B. Review, "R-105-R-46-2-ABB-cable length on page 6 from substation to shore side transfomer," 2010.

[11] T. Zis, R. J. North, P. Angeloudis, W. Y. Ochieng, and M. G. H. Bell, "Evaluation of cold ironing and speed reduction policies to reduce ship emissions near and at ports," *Marit. Econ. Logist.*, vol. 16, no. 4, pp. 371–398, 2014, doi: 10.1057/mel.2014.6.

[12] M. Acciaro *et al.*, "Environmental sustainability in seaports: a framework for successful innovation," *Marit. Policy Manag.*, vol. 41, no. 5, pp. 480–500, 2014, doi: 10.1080/03088839.2014.932926.

[13] C. C. Chang and C. M. Wang, "Evaluating the effects of green port policy: Case study of Kaohsiung harbor in Taiwan," *Transp. Res. Part D Transp. Environ.*, vol. 17, no. 3, pp. 185–189, 2012, doi: 10.1016/j.trd.2011.11.006.

[14] L. Yang, Y. Cai, X. Zhong, Y. Shi, and Z. Zhang, "A carbon emission evaluation for an integrated logistics system-a case study of the port of shenzhen," *Sustain.*, vol. 9, no. 3, 2017, doi: 10.3390/su9030462.

[15] J. Yu, S. Voß, and G. Tang, "Strategy development for retrofitting ships for implementing shore side electricity," *Transp. Res. Part D Transp. Environ.*, vol. 74, no. August, pp. 201–213, 2019, doi: 10.1016/j.trd.2019.08.004.

[16] T. P. V. Zis, "Prospects of cold ironing as an emissions reduction option," *Transp. Res. Part A Policy Pract.*, vol. 119, no. November 2018, pp. 82–95, 2019, doi: 10.1016/j.tra.2018.11.003.

[17] S. Prosperity, F. Business, and C. Fbc, "FBC - Shore Power at Southampton Port," no. March, 2020.

[18] 2023 World Port Sustainability Program, "World Ports Climate Action Program," *2018*, 2018, [Online]. Available: https://sustainableworldports.org/wpcap/

[19] European Parliament, "Regulation (EU) 2023/1805 of the European Parliament and of the council of 13 September 2023 on the use of renewable and low-carbon fuels in maritime transport, and amending Directive 2009/16/EC (Text with EEA relevance)," *Off. J. Eur. Union*, vol. 66, no. July, pp. 48–100, 2023, doi: 10.2771/793264.22.9.2023.

[20] J. S. L. Lam and T. Notteboom, "The Greening of Ports: A Comparison of Port Management Tools Used by Leading Ports in Asia and Europe," *Transp. Rev.*, vol. 34, no. 2, pp. 169–189,

2014, doi: 10.1080/01441647.2014.891162.

[21] Saif Muizzadin Wadaullah, "Analisis Penerapan Layanan Shore Power Connection: Tinjauan Sektor Pelayaran Dan Pelabuhan," *Repos. its*, 2022, [Online]. Available: http://repository.its.ac.id/id/eprint/94429

[22] Starcrest Consulting Group, "Puget Sound Maritime Air Emissions Inventory," vol. 15, no. 2, pp. 1–23, 2018.

[23] Pertamina, "Daftar Harga Bahan Bakar Khusus / Non Subsidi TMT 10 Agustus 2024," 2024. https://www-pertamina-com.translate.goog/id/news-room/announcement/daftar-harga-bahan-bakar-khusus-non-subsidi-tmt-10-agustus-2024-zona-3?_x_tr_sl=id&_x_tr_tl=en&_x_tr_hl=en&_x_tr_pto=sc

[24] ENDC, "Enhance Nationally Determined Contribution Republic of Indonesia," *UNDP Clim. Promise*, pp. 1–17, 2022.

[25] J. Qi, S. Wang, and C. Peng, "Shore power management for maritime transportation: Status and perspectives," *Marit. Transp. Res.*, vol. 1, no. November, p. 100004, 2020, doi: 10.1016/j.martra.2020.100004.

[26] H. Li, "Analysis for Shore Power Economy in Preventing Air Pollution of Vessels are Docked at the Berth," *E3S Web Conf.*, vol. 118, pp. 0–4, 2019, doi: 10.1051/e3sconf/201911804020.

[27] X. H. Tianli Song, Yang Li, "Cost effective optimization analysis of shore to ship power system construction and operation," *IEEE Conf. Energy Internet Energy Syst. Integr.*, vol. EI2, pp. 1–6, 2017.

Comparison of Carbon Emission Estimation Results with Case Studies of Container Terminals

Muhammad Arif Budiyanto[1*], Eza Novia Putri[1] and Achmad Riadi[1]
[1] Naval Architecture and Marine Engineering, Universitas Indonesia, Depok – West Java, Indonesia. 16424
*E-mail: arif@eng.ui.ac.id
[1] ORCID ID 0000-0001-8118-7521

Abstract. Container loading and unloading activities are one of the lines that contribute the largest carbon emissions in the port area. Each port has begun to implement the Green Port concept as an effort to mitigate carbon emissions. This study aims to compare carbon emissions at container terminals in Indonesia, then analyze the calculation results to obtain a carbon emission reduction strategy. The calculation model in this study uses the bottom-up method which makes the fuel consumption value the result of calculations with a formula from the container movement theory. The case study uses container terminals spread across Indonesia. The results of this study indicate that the container terminal with the best container service efficiency based on CO2 emissions per TEU is case study 1 at 9.8 kg of CO2 emissions per TEU. The loading and unloading equipment that contributes the most CO2 emissions at each terminal is the Quay Crane. Terminal layout efficiency and Electrification of loading and unloading equipment are the best strategies based on the analysis of the results of this study.

Keywords : Carbon Emission, Container Port, Bottom-up model, electrification

1. Introduction

Sea transportation is still proven to be the main choice for industry players to distribute their goods, 90% of world trade is carried out by sea. At least more than 11 billion tons of commodities are distributed by ship each year (Andriessa, 2022). This is a challenge since the shipping industry also actively contributes to CO2 emissions which are harmful to the environment (Al Baroudi et al, 2021). Some of the negative impacts of high CO2 are increasing global temperatures, climate change, reducing the thickness of the ozone layer and acid rain. Around 940 million tons of CO2 are produced through shipping industry activities, equivalent to at least 2.5% of the world's total CO2 emissions (Innovation, 2021). The International Maritime Organization (IMO) seriously highlights the environmental problems caused by this exhaust gas. IMO is firmly committed to reducing carbon emissions with a target of reducing CO2 emissions by 40% by 2030 and 70% by 2050 compared to 2008. This commitment is realized by issuing MARPOL 73/78 Annex VI regulations which must be followed by all IMO members including Indonesia (IMO, 2023). This regulation contains regulations related to emissions of Nox, Sox, and substances that can damage the ozone layer as well as terms and conditions for the use of fuel used by ships and port

operations. In addition to regulations for the ships themselves, this regulation also targets port areas which are considered to also contribute to CO2 emissions.
According to the International Association of Ports and Harbors (Sahar Azarkamand, 2020), climate change that is occurring in this world is partly caused by increasing greenhouse gas (GHG) emissions due to port operational activities. The high CO2 emissions at the port come from a number of activities carried out at the Container Terminal (Shinoda, 2022). The container terminal itself is a place where the loading and unloading process of goods occurs, which is usually carried out with the help of loading and unloading equipment such as Quay Crane, Rubber Tyred Gantry, Reach Stacker, Side Loader, Truck and others (Shan, 2024).
The use of fuel in loading and unloading equipment is what affects exhaust emissions. In Indonesia itself, most container terminals still use diesel fuel to drive their loading and unloading equipment. Indonesia has begun targeting a reduction in CO2 emissions of up to 30% by 2030, this commitment is stated in the Nationally Determined Contribution (NDC) in the development sector both on land and at sea, one of the efforts to reduce it in the land sector is the start of the Green Port movement. Developed country have already implemented the Green Port program where loading and unloading equipment has switched from diesel to electric or hybrid (Du, 2019).
In 2020, Port Klang received an award from The APSN (APEC Port Services Network) Green Port Award System 2020 for its commitment and steps towards a sustainable and environmentally friendly port, this is the second award for Port Klang after receiving it in 2017 (Westports, 2021).
The purpose of this study is to compare carbon emissions at several container terminals in Indonesia, through comparative analysis of the calculation results, an analysis of the causal factors and conditions of each terminal will also be carried out so that an overview of the strategies that can be carried out to reduce CO2 emissions in the future is obtained.

2. Research Methods

This study uses CO2 emission calculations sourced from the Intergovernmental Panel on Climate Change (IPCC) Guideline 2006 and the Container Movement Theory. This theory uses the Bottom-Up method where the energy consumption value is positioned as the result of the calculation, not as an input value. The activity that is the input variable is the movement of move and ride containers. Ride containers are movements that use the help of trucks where in the calculation the distance variable is added which is calculated using the Manhattan-distance method. While for move containers, it is a movement that uses cranes, stacking cranes, RTGs or other tools. To use this theory, port operational data is needed such as Total Throughput of containers in one year, transportation modalities, Transhipment process and Terminal Layout (Budiyanto 2019, 2022). The theory of container movement is stated in equation 1.

$$W_x = \sum_{i=1}^{5} \sum_{J=1}^{1} ((V_{i,j} \times f_D) + (P_{i,j} \times f_E)) \quad (1)$$

Where Wx is total CO2 emissions produced by terminal X in kilograms, Vi,j is total consumption for one year from diesel device i to mode j in liters, fD is emission factor in kilograms of CO2/liters of diesel consumption, Pi,j is total consumption for one year from electrical device i to mode j in kWh and fE = emission factor in kilograms of CO2/kWh of electricity consumption.

The initial stage in this research is a literature study and discussion that continues to identify the problem, where this research raises the issue of CO2 emissions originating from the combustion of fuel for loading and unloading equipment at several container terminals in Indonesia. The results of this CO2 estimation will then be compared between container terminals in Indonesia. The calculation of CO2 emissions uses the Bottom-Up approach with the Container Movement Theory. An analysis is also carried out regarding the calculation results and CO2 emission reduction strategies. This research was conducted using 5 case studies in Indonesia with various container stacking capacities. The calculations in this study use the Bottom-Up approach based on operational activities at the terminal. The input values for the calculation are Throughput, container movement process, transportation modalities and terminal layout.

Loading and unloading equipment used in this study This data contains the type and number of loading and unloading equipment for each terminal and the fuel sources used are shown in Table 1.

Table 1. Container Handling Equipment of Case Study

	Case Study	Container Handling Equipment	Number	Energy Source
CT 1	Container Termina 1 Throughput 1.200.000 TEU's	Quay Crane	8	Electric
		Twin-lift Capable Crane	8	Electric
		E-RTG	24	Electric
		Prime Movers	44	Diesel
CT 2	Container Termina 2 Throughput 1.366.1962 TEU's	Quay Container Crane	12	Diesel
		Rubber Tyred Gantry	30	Diesel
		Reach Stacker	7	Diesel
		Sky Stacker	3	Diesel
		Forklift	29	Diesel
		Head Truck	157	Diesel
CT 3	Container Termina 1 Throughput 185.878 TEU's	Quay Container Crane	6	Diesel
		Rubber Tyred Gantry	18	Diesel
		Reach Stacker	5	Diesel
		Side Loader	1	Diesel
		Forklift	4	Diesel
		Terminal Tractor	28	Diesel
		Head Truck	4	Diesel
CT 4	Container Termina 1 Throughput 966.100 TEU's	Quay Container Crane	7	Diesel
		Rubber Tyred Gantry	25	Diesel
		Reach Stacker	2	Diesel
		Side Loader	1	Diesel
		Head Truck	48	Diesel
CT 5	Container Termina 1 Throughput 263.225 TEU's	Quay Container Crane	2	Diesel
		JIB Crane	2	Diesel
		Rail Mounted Gantry Crane	4	Diesel
		Reach Stacker	2	Diesel
		Forklift	4	Diesel
		Head Truck	11	Diesel

In the process of moving containers in the terminal area, it is divided into two parts, namely Stevedoring and Cargodoring. It can be seen in the picture that processes (1) and (2) are part of the unloading process and parts (3) and (4) are the loading process. In the unloading process, a Quay Crane is used to unload containers from ships to trucks, then the truck will take the containers to the stacking area while the loading process is the opposite of the unloading process. Energy consumption is not included in the input value but is the result of the calculation itself.

The results of the Energy Consumption calculation will be used as an input variable in the calculation of CO2 emissions. Loading and unloading equipment at container terminals has many varieties, so it is necessary to classify them based on their characteristics.

The calculation of the total CO2 emission estimate can use the IPCC Tier-1 default where the required data is the Total energy consumption for each type of equipment and fuel type and the default emission factor for each type of fuel. The following formula for total CO2 emissions is expressed in equation 2.

$$E_{CO2} = \sum_{i=1}^{n}((X_i \times f_{D,CO2}) + (Y_i \times f_{E,CO2})) \quad (2)$$

Where, E_{CO2} is the Total CO2 emissions produced by the terminal in kilograms, Xi is the Total energy consumption of equipment powered by diesel energy in liters, Yi is the Total energy consumption of equipment powered by electricity in kWh, $f_{D,CO2}$ is the CO2 emission factor from diesel energy sources (IPCC Guideline 2006 & Container Movement Theory, namely 2.67 kilogramsCO2/liter) $f_{E,CO2}$ is the CO2 emission factor from electricity sources (source: Ministry of Energy and Mineral Resources for Indonesian container terminals).

3. Result and Discussions

From the results of the total energy consumption of each container handling equipment, CO2 Emission calculations can be used by multiplying the total energy consumption by the emission factor for diesel and electricity to produce CO2 Emissions in Kilograms of CO2. The following are the results of the calculation of CO2 Emissions per type of equipment and total CO2 emissions at each terminal shown in Figure 1.

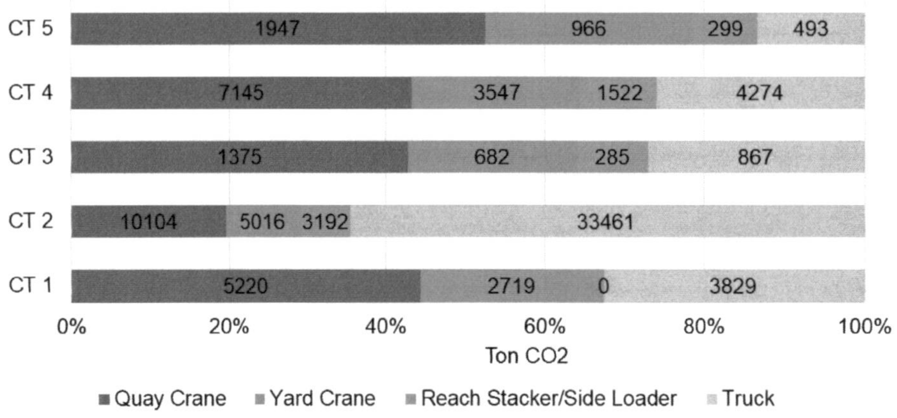

Figure 1. Carbon emissions of container handling equipment

IOP Conf. Series: Earth and Environmental Science **1461** (2025) 012024 doi:10.1088/1755-1315/1461/1/012024

The equipment that contributes the largest CO2 emissions in each terminal in Indonesia is the Quay Crane except in the CT2 case study. In the CT2 case study, it is the highest contributor of CO2 emissions, even the highest contributor of CO2 emissions of all equipment in all terminals, with a total emission of 33,461 tons of CO2. While the equipment with the next highest total CO2 emissions is the Quay Crane at 10,104 tons of CO2. From these results, it can be concluded in general that the equipment that contributes the highest CO2 emissions in both container terminals is the Quay Crane accompanied by the Rubber Tyred Gantry. The CT2 case study has a different case compared to other terminals where only this terminal has the result that the truck is the equipment that produces the highest CO2 emissions. This can be explained because the CT2 case study layout has a distance between the pier and the stacking yard that is still too far, coupled with the use of Truck fuel that still uses diesel.

After knowing the amount of CO2 emissions produced by each equipment, the next step is to obtain the total CO2 emissions produced by the terminal. The following is the total CO2 emissions for each container terminal in Indonesia which is shown in Figure 2. This result is comparable with the other study the CO2 emissions of container terminal is 11.58 kgCO2/TEU (Moya, 2019).

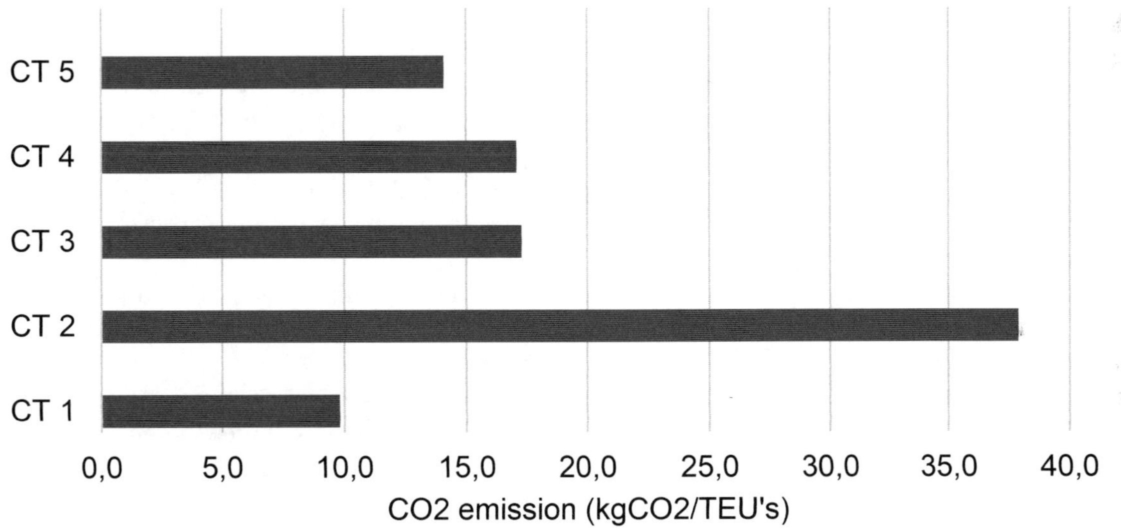

Figure 2. Total Carbon emissions of container terminal

Case study CT2 obtained a total CO2 emission at the container terminal of 51,773 tons of CO2, followed by Case study CT4 of 16,488 tons of CO2. Case study CT4 has a higher total CO2 emission compared to Case study CT1 even though the amount of Throughput is much lower, this is due to the use of different fuels, where Case study CT1 has used electric fuel for each Quay Crane and RTG operated in their terminals.

4. Conclusion

The results of this study indicate that the lowest carbon emissions are the CT2 case study of 9.8 kg of CO2 emissions per TEU. The loading and unloading equipment that contributes the most CO2 emissions at each terminal is the Quay Crane. Terminal layout efficiency and Electrification of loading and unloading equipment are the right strategies based on the analysis of the results of this study. Based on the results of the comparison of carbon emissions at each container terminal, the CT2 case study has a terminal with the highest total CO2 emissions per TEU, namely 37.9 kg of CO2 emissions per TEU. The higher the CO2 emissions per TEU, the greater the need for efficiency in handling each container. From the results of this comparison, it can provide a direction for future research, namely the use of electricity in loading and unloading equipment can be an alternative in efforts to reduce CO2 emissions caused by the combustion of fossil fuels in the equipment. Electrification of loading and unloading equipment has been shown to contribute to reducing CO2 emissions, with the use of electricity-based equipment being the most efficient terminal in this study.

Acknowledgements

The author would like to express his deepest appreciation and gratitude to the Directorate of Research and Development, Universitas Indonesia for the support and assistance provided in carrying out this research. Contract number: NKB-719/UN2.RST/HKP.05.00/2024.

References

Andriessa, R. (2022, April 18). Transportasi Maritim: The Backbone of International Trade Activities. Retrieved from CWTS: https://cwts.ugm.ac.id/2022/04/18/transportasi-maritim-tulang-punggung-dari-aktivitas-perdagangan-internasional/

Al Baroudi, H., Awoyomi, A., Patchigolla, K., Jonnalagadda, K., & Anthony, E. J. (2021). A review of large-scale CO2 shipping and marine emissions management for carbon capture, utilisation and storage. *Applied Energy*, *287*, 116510.

Budiyanto, M.A., Huzaifi, M.H., Sirait, S.J., 2019. Estimating CO2 Emissions in a Container Port based on Modality Movement in the Terminal Area. International Journal of Technology. Volume 10(8), pp. 1618-1625

Budiyanto, M. A., Habibie, M. R., & Shinoda, T. (2022). Estimation of CO2 emissions for ship activities at container port as an effort towards a green port index. Energy Reports, 8(Supplement 15), 229-236. https://doi.org/10.1016/j.egyr.2022.10.090

Du, K., Monios, J., & Wang, Y. (2019). Green port strategies in China. In *Green ports* (pp. 211-229). Elsevier.

IMO. (2023). 2023 IMO Strategy on Reduction of GHG Emissions From Ships. Retrieved from IMO: https://www.imo.org/en/OurWork/Environment/Pages/2023-IMO-Strategy-on-Reduction-of-GHG-Emissions-from-Ships.aspx

Innovation, U. R. (2021). Shipping Industry Reduces Carbon Emissions With Space Technology. UK: UKRI.

Martínez-Moya, J., Vazquez-Paja, B., & Maldonado, J. A. G. (2019). Energy efficiency and CO2 emissions of port container terminal equipment: Evidence from the Port of Valencia. *Energy Policy*, *131*, 312-319.

News, I. S. (2023, december 8). DP World targer 60% Renewable Power at Indian Terminals by 2026. Retrieved from India Shipping News : https://indiashippingnews.com/dp-world-targets-60-renewable-power-at-indian-terminals-by-2026/

Okşaş, O. (2023). Carbon emission strategies for container handling equipment using the activity-based method: A case study of Ambarlı container port in Turkiye. Marine Policy.

Pradana, A. S. (2023). Green port concept and its impact on port cost efficiency and green performance indicator in Tanjung Perak Port. International Conference On Sustainable Materials Science, Structures, And Manufacturing. AIP Conference Proceedings.

Prosertek. (2023). Which are the ports with the highest CO2 emissions. Amerika Serikat: Prosertek.

Shan, J., Zhang, M., Zhang, D., Zhou, X., Li, H., & Wang, B. (2024). Container Cranes. In *Handbook of Port Machinery* (pp. 451-678). Singapore: Springer Nature Singapore.

Sahar Azarkamand, C. W. (2020). Review of Initiatives and Methodologies to Reduce CO2 Emissions and Climate Change Effects in Ports. Int J Environ Res Public Health, 11.

Shinoda, T., Budiyanto, M. A., & Sugimura, Y. 2022. Analysis of energy conservation by roof shade installations in refrigerated container areas. *Journal of Cleaner Production, Volume(Issue)*, [Page numbers]. https://doi.org/10.1016/j.jclepro.2022.134402

Westports. (2021, Februari 10). Westports. Retrieved from Port Klang Awarded APSN Green Port 2020: https://www.westportsholdings.com/2021/02/10/port-klang-awarded-apsn-green-port-2020/

IOP Conf. Series: Earth and Environmental Science **1461** (2025) 012025 doi:10.1088/1755-1315/1461/1/012025

The Impact of Subsidized Maritime Transport Connectivity on Social Welfare in Eastern Indonesia: A System Dynamic Approach

Ilham Muliana Rahman[1*], Budisantoso Wirjodirdjo[2]

[1,2] Department of Industrial and Systems Engineering, Institut Teknologi Sepuluh Nopember, Surabaya, Indonesia

*E-mail: ilhammrt15@gmail.com

Abstract. The Indonesian government has implemented a maritime development strategy based on the "Transport Promotes the Trade" concept in the context of achieving Sustainable Development Goals, especially related to social welfare. This concept is expected to overcome welfare disparities between the Western and Eastern regions of Indonesia. Through the Subsidized Public Service Obligation (PSO) Maritime Transport program, which includes Pioneer Ships, Sea Tolls, and PSO Economic Passenger Ships, the government aims to improve inter-regional connectivity, open new trade routes, and enhance the welfare of communities in Eastern Indonesia. This research aims to evaluate the impact of subsidized maritime transport connectivity on social welfare in Eastern Indonesia. Using a system dynamics approach, this study analyses the changes in the connectivity index over times and its relationship with social welfare as measured by the Human Development Index (HDI). The findings of this research are expected to provide a deeper understanding of the interactions of welfare variables within the dynamic system, as well as the impact of changes in subsidized maritime transport connectivity on social welfare in Eastern Indonesia, increase in connectivity through the addition of ships results in an average increase of 0.009 in HDI.

1. Introduction

Development and growth in a region are generally intended to reduce welfare disparities between regions [1]. These disparities can arise due to variations and differences in resource potential and infrastructure facility [2]. This situation is evident in Indonesia, where almost all regions are scattered and surrounded by seas, leading to inevitable differences in culture, economy, and social welfare between regions [3].

Social welfare is a multidimensional concept that encompasses various aspects of welfare included in the Sustainable Development Goals in Indonesia. Social welfare is often measured using the Human Development Index (HDI) [4]. HDI also serves as an important indicator of success in efforts to improve quality of life. In Indonesia, significant disparities in HDI exist between regions, especially between Western and Eastern Indonesia [5]. The National Medium-Term Development Plan (RPJMN) 2020-2024 emphasizes the need to address regional disparities, aiming to equalize development across regions, enhance the competitive advantage of regional growth centres, and improve the quality and access to basic services [5].

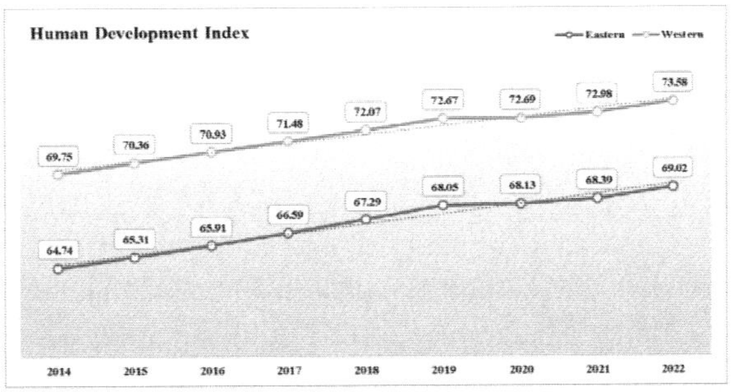

Figure 1. Human Development Index data for Western Indonesia and Eastern Indonesia [20]

Although the Human Development Index have improved overall, Eastern Indonesia still lags behind Western Indonesia across multiple aspects. Transportation connectivity significantly contributes to this gap. The geographical challenges in the region make transportation an essential element for development [2]. The economy is significantly boosted by transportation, with economic shifts further reinforcing the transportation sector, mainly maritime transport [1]. Maritime transport is the crucial role to enhance social welfare and efficiently transport of goods or passengers' mobility, the Indonesian government has introduced new policies, including the Sea Toll Program and subsidies for pioneer ships and Public Service Obligation (PSO) for economy class passengers. According to Presidential Regulation No. 70 of 2017, the Sea Toll Program aims to provide effective maritime connectivity with ships that sail regularly and on schedule from Western Indonesia to Eastern Indonesia [6]. The Sea Toll Program and the PSO are intended to reach and distribute logistics to remote, isolated, outermost, and border areas (3TP) and ensure the availability of goods and reduce price disparities to improve connectivity and social welfare [7].

Presidential Regulation Number 106 of 2015 about the Implementation of Public Service Obligations for Maritime Transport, it is also explained that the Government will finance the assignment to carry out Public Service Obligation for Maritime Transport through operational subsidies of Sea Toll Program, Pioneer Ships and PSO passenger economy ships, where pioneer ship is a feeder of Sea Toll and PSO Passengers Ships. This subsidy is given based on the gap between the revenue and operational cost. This subsidy will be included in the State Budget Plan (RAPBN) each year [8]. Subsidizing for Public Service Obligation Maritime Transport is crucial for social welfare in Indonesia, especially in Eastern Indonesia which is dominated by the 3TP region. According to [8], the role of maritime transportation as "Transport Promotes the Trade" is an option for maritime transportation in Indonesia which is fully supported by the Government. Maritime transportation become a role as a trigger for increasing connectivity and social welfare in region that changes over times due to feedback relationships.

The impact of transport infrastructure investment can be measured and evaluated based on societal welfare [9]. One effective method for such evaluation is System Dynamics Modelling (SDM), which is a simulation and modelling technique particularly suited for addressing complex, dynamic, and long-term management issues [10]. SDM captures the interplay of reinforcing and balancing feedback loops within a system, which can cause fluctuations and drive the system towards equilibrium or desired conditions. The approach relies on developing a dynamic model grounded in empirical evidence and rational hypotheses, allowing for a comprehensive understanding of the systemic impacts of transport connectivity [9].

This study aims to analyse the effect of subsidized maritime transport connectivity on social welfare in Eastern Indonesia, focusing on the Human Development Index (HDI). The Dynamic Simulation Method is used to model complex interactions between maritime transport system and social welfare. By using this approach, this study seeks to understand how the relationship and changes in maritime transport connectivity that supported by government subsidies can improve social welfare in Eastern Indonesia. By modelling various maritime transportation scenarios in term of connectivity and economics, this study offers a comprehensive understanding of the impact of subsidized policy. The simulation's findings are expected to provide better policy recommendations to achieve more balanced regional development and improve social welfare in Eastern Indonesia.

2. Literature Review

2.1 System Dynamics

Recent research has shown significant advancements in the application of System Dynamics Modelling (SDM) for evaluating complex and dynamic management issues, particularly in the context of transportation infrastructure. SDM is an effective tool for measuring the economic impact of transport infrastructure investments, as it allows for the analysis of feedback loops and interactions within a system to achieve desired outcomes [9]. SDM used to address challenges related to maritime transport policy and network capacity planning in a hub-and-spoke container shipping system, highlighting the model's capability to handle complex and non-linear planning processes [10]. Similarly, SDM can be used to evaluate alternative strategies for public transport in Surabaya, demonstrating its usefulness in assessing policy risks and supporting decision-making [11]. Additionally, SDM's intuitive nature makes it accessible and beneficial for understanding the structural dynamics of complex systems [12]. These studies illustrate the growing recognition of SDM as a versatile and robust tool in policy evaluation and strategic planning, providing valuable insights into the complexities of transport and urban management. Therefore, this study uses SDM to explore previously undiscussed relationships between PSO transport connectivity and social welfare using Anylogic Simulation.

2.2 Human Development Index

The concept of human development was initially developed as an alternative to measuring GDP per capita to assess poverty from various aspects that affect the quality of human life [13]. The HDI assessment was developed by United Nations Development Program (UNDP) which evaluates the achievement of human development in a country seen from three basic dimensions [5].

1. Long and Healthy Life dimension is measured using life expectancy at birth.
2. Knowledge dimension is measured by educational index through literacy index and mean length of schooling.
3. Decent Life dimension is measured by purchasing power parity through the average amount of expenditure per capita as an income approach that represents the achievement of development for a decent life.

Some indicators have been employed to assess both welfare and economic performance in previous studies, such as measuring welfare sub-criteria which described in the following approaches [14]:

1. Min Max Scale.
   The standardized value of an indicator with Min-Max transformation where the value of the value of X (indicator) and Z (Score Value [0,1]) changes every year.

$$Z = \frac{X - X_{Min}}{X_{Max} - X_{Min}}$$

2. Indicator Combination.
   The combination of several indicators is achieved by assigning weight to each Wellbeing Index indicator.

$$WBI = \sum_{i=1}^{n} W_i Z_i$$

Transportation facilities form a crucial part of human development, offering social and economic access to healthcare, education, economic resources and market [8]. This access will create opportunities for people in the 3TP area to develop regional resources, jobs and generating income. This process will increase purchasing power and standard of living.

2.3 Index Connectivity

Connectivity and mobility are indicator of the success of integrated transportation infrastructure development [15]. Mobility refers to physical movement consisting of the number of frequencies, speed and travelling distance for passengers and cargo. For that, the more and faster someone can travel to more destinations they want, the mobility of the area is considered good, so that increasing mobility will have an impact on increasing social welfare [16].

Routes of Sea Toll Program, Pioneer Ships and PSO Economic Passengers are liner shipping that connect each other port periodically. All nodes or ports in the networks represent direct or indirect connections. Based on UNCTAD (Nations Conference on Trade and Development) the Liner Shipping Connectivity Index (LSCI) aims to reflect the level of integration in the existing liner shipping network [14;15]. The parameters that can be used are as follows

1. Number of direct links
2. Transport Capacity, derivatives of the total cargo or passengers at the port.
3. Maximum ship capacity that calls at the port (GT or DWT).
4. Number of ship call or frequency of direct link.

LSCI serves as a measure of access to inter-region trade through its shipping network. The higher the index, the easier it is to access the transportation system for inter-region trade. A min-max calculation with determined weights according to this formula [14].

$$PCI_{mn} = \frac{\left(\frac{C_{mn} - C_{mn_{Min}}}{C_{mn_{Max}} - C_{mn_{Min}}}\right) \times B_m}{m}$$

Port Connectivity Index (PCI), Component of Connectivity (m), Number of Region (n), n - Component of Connectivity (C_{mn}), weight of component connectivity (B_m).

2.4 Overview PSO Maritime Transport

None of the studies on the impact of PSO Maritime Transportation Connectivity on social welfare have presented all maritime transportation program in Indonesia. So, this research will be conducted for three subsidized maritime transportation providers in the program.

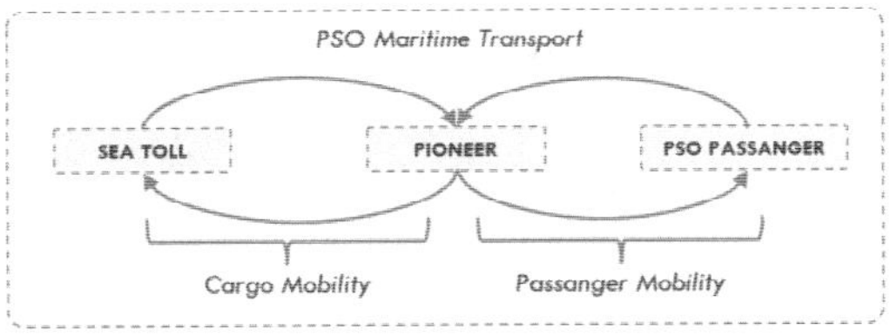

Figure 2. PSO Maritime Transport

The three maritime transportations of this program are integrated with each other and have the same goal. This program aims to improve the social welfare in the Eastern Indonesia. While the existing research only discusses it partially and has not considered the regional connectivity indicator based on maritime transportation infrastructure. This indicator is very important for assessing the movement of people and goods between regions and improving the social welfare in each region in Indonesia as archipelagic country.

2.4.1 Implementation of Pioneer Ships Program

Pioneer ships are vessels that serve isolated, remote, or underdeveloped regions within Indonesia, particularly those areas categorized as Lagging, Outermost, Remote, and Border (3TP) [18]. Based on the Pioneer Ships Program Decree from 2016 to 2022, there were 96 pioneer ship routes in 2016. This number increased to 113 routes in 2018 and 2019. In 2022, the number of pioneer ship routes had risen to 117 routes, the details as follows.

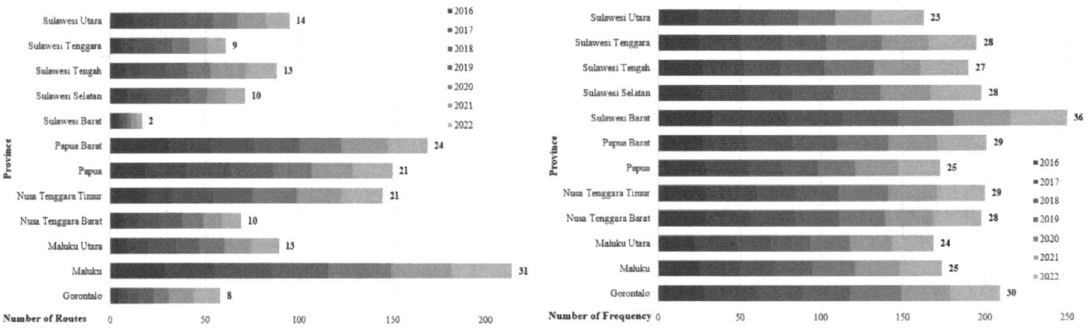

Figure 3. Number of routes and frequency Pioneer Ships in Eastern Indonesia [16].

Maluku has the highest number of routes, with a total of 32 routes in 2022, and the highest average number of routes, at 31 routes per year, compared to other provinces. The average frequency of each route for all provinces in Eastern Indonesia is 28 times per year.

2.4.2 Implementation of Sea Tolls Program

The Sea Toll is a government-initiated maritime logistics program launched in 2016 to enhance connectivity between Indonesia's main islands and its more remote regions. The program focuses on integrating remote areas (3TP) into the national economy, helping to drive down prices and facilitate trade by streamlining the distribution of goods [19]. The Sea Tolls transportation program has been implemented by the government since 2016. Figure 4 is the detail implementation of the Sea Toll transportation from 2016 to 2022 in Eastern Indonesia.

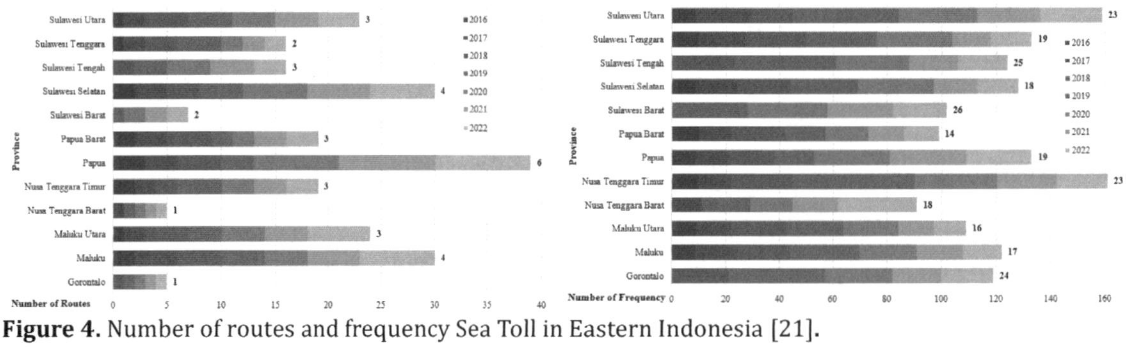

Figure 4. Number of routes and frequency Sea Toll in Eastern Indonesia [21].

IOP Conf. Series: Earth and Environmental Science **1461** (2025) 012025 doi:10.1088/1755-1315/1461/1/012025

Papua has the highest number of routes, with a total of 9 routes in 2022, and the highest average number of routes, at 6 routes per year, compared to other provinces. The average frequency of each route for all provinces in Eastern Indonesia is 20 times per year.

2.4.3 Implementation of PSO Economic Passenger Ships Program

PSO (Public Service Obligation) Passenger Ships are government-subsidized vessels operated by PT. PELNI that provide affordable passenger transport across Indonesia's archipelagic regions [18]. The implementation of the PSO Economic Passenger Ships program has had 26 routes from 2016 to 2022 based on PSO Economic Passenger Ships Program Decree. Figure 5 is the development of the PSO Economic Passenger Ship from 2016 to 2022 in Eastern Indonesia.

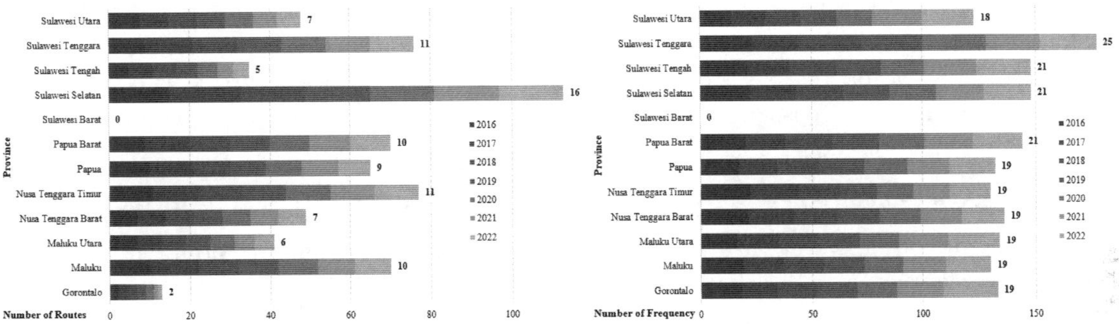

Figure 5. Number of routes and frequency PSO Economic Passenger in Eastern Indonesia [16].

South Sulawesi has the highest number of routes, with a total of 16 routes in 2022, and the highest average number of routes, at 16 routes per year, compared to other provinces. The average frequency of each route for all provinces in Eastern Indonesia is 25 times per year.

3. Data and Methodology

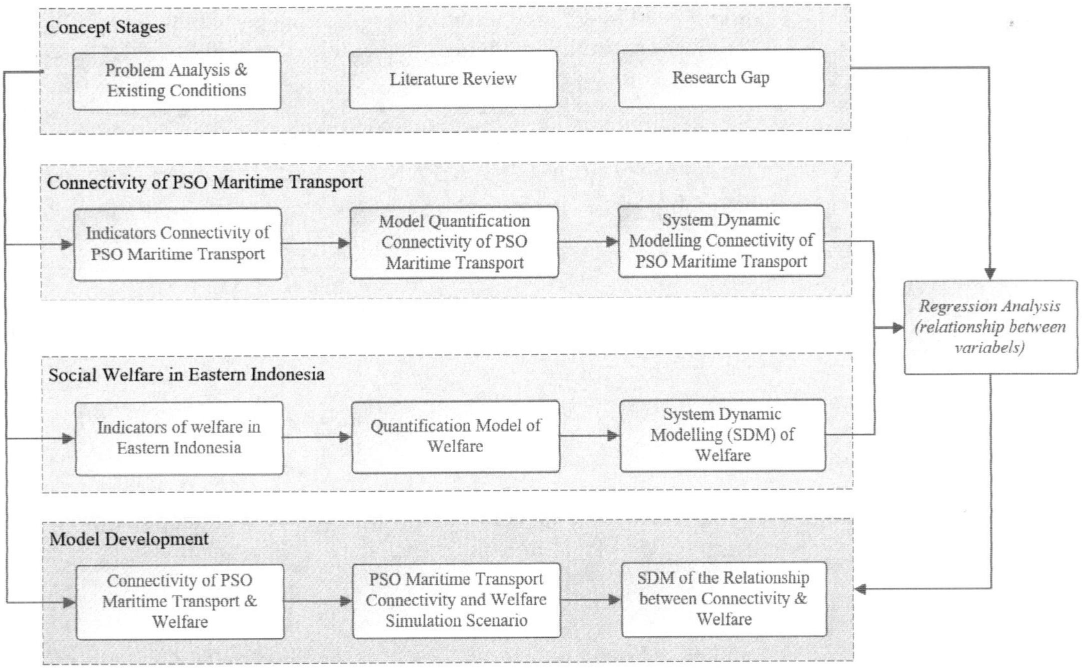

Figure 6. Research Stages

IOP Conf. Series: Earth and Environmental Science **1461** (2025) 012025 doi:10.1088/1755-1315/1461/1/012025

The approach used in this study is the integration of multiple linear regression analysis directly into system dynamic simulations. Considering the complexity of the relationships between regions in Indonesia, the use of multiple linear regression analysis allows for more detailed identification and quantification of factors affecting the Human Development Index (HDI). Integrating this analysis into system dynamic simulations enables more comprehensive understanding of how the connectivity of the Public Service Obligation (PSO) Maritime Transportation impacts the social welfare in Eastern Indonesia. The following are the stages of model development.

3.1 Indicators and Quantification Model

The connectivity od each region is measured based on 3 indicators, the number of direct links, the number of frequencies of direct link and the maximum ship size. After the connectivity value is obtained, a regression analysis is used to get an equation from relationship between connectivity and transport capacity (passengers and cargo). Transport capacity is depended on the number of ships, frequency and the number of segments in each region.

$$Y(Index\ Connectivity) = \beta_0 + \beta_1(Pax\ Transport\ Cap) + \beta_2(Cargo\ Transport\ Cap)$$

The indicators of social welfare are generally measured by the Human Development Index (HDI), which is categorized into three groups as follows.

Table 1. Indicator of Human Development Index.

Indicator	Description
Knowledge Dimension	Knowledge dimension is measured by educational index through literacy index and mean length of schooling
Health Dimension	Long and Healthy Life dimension is measured using life expectancy at birth.
Decent Life Dimension	Decent Life dimension is measured by purchasing power parity through the average amount of expenditure per capita as an income approach that represents the achievement of development for a decent life

To obtain the values of these indicators, a macro-level analysis is conducted using correlation and regression analysis based on socio-economic data from the Statistic Indonesia. The following data are used to assess the relationship between maritime transportation and social welfare.

Table 2. Relationship between Maritime Transportation and Social Welfare

ID	Dependent Variable	Independent Variable	
	Y	X1	X2
A	Cargo - BPS	GDRP	PSO Index Connectivity
B	Passenger – BPS	GDRP	PSO Index Connectivity
C	Price Consumer Index	Cargo – BPS	Salary
D	Poverty Population	HDI	Unemployment
E	Average Year of School	Per Capita Expenditure	Poverty Population
F	Life Expectancy Index	Per Capita Expenditure	Poverty Population
G	Literacy Index	Per Capita Expenditure	Poverty Population
H	Per Capita Expenditure	Price Consumer Index	Passenger - BPS
I	Salary	GDRP	HDI

Based on relationship maritime transportation and social welfare, a multiple linear regression model used to predict the value of a dependent variable based on several independent variables. the following equation can be derived and used as an input in the system dynamic simulation. $Y_i = \beta_0 + \beta_1 X_{1i} + \beta_2 X_{2i} + \cdots + \beta_k X_{ki} + \varepsilon_i, \quad i = 1,2,\ldots,n.$ where Y_i is the dependent

IOP Conf. Series: Earth and Environmental Science **1461** (2025) 012025 doi:10.1088/1755-1315/1461/1/012025

variable for observation i, representing the outcome that is to be predicted or explained. β_0 is the constant or intercept of the regression model, β_1, β_2 ... is the regression coefficients or slope that measure the impact of each independent.

3.2 Model Development

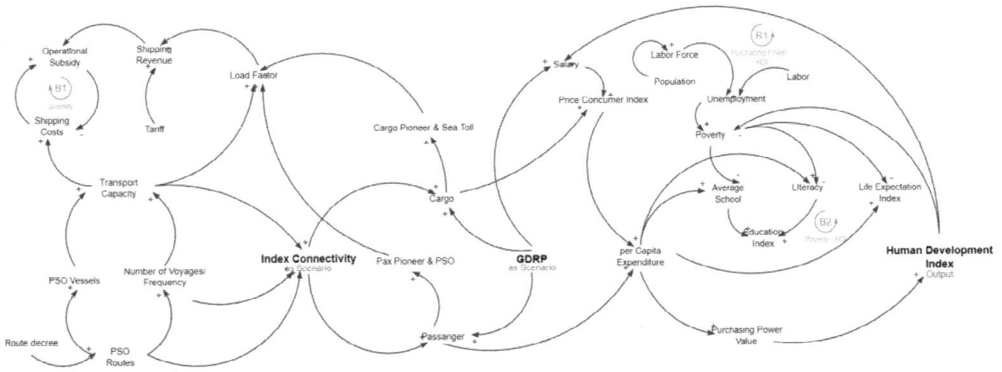

Figure 7. Causal Loop Diagram

Based on the previous indicators and model quantification, a causal loop diagram is constructed to illustrate the relationship between PSO Maritime Transportation connectivity and social welfare. The following is an explanation of some of the loops in the diagram.

Table 3. Description of Causal Loop Diagram

Loop		Order	Description
B1	Operational Subsidy	*Operational Subsidy – Shipping Costs*	Maritime transport subsidies are calculated based on the shipping costs; the lower the costs, the lower the subsidies
B2	Poverty - HDI	*Poverty – Education Index / Life Expectation Index – HDI - Poverty*	A reduction in poverty levels leads to an increase in the Education Index, Life Expectancy and improves the HDI and then decrease poverty.
R1	Purchasing Power - HDI	*HDI – Salary – PCI – per Capita Expenditure – Purchasing Power Value - HDI*	An increase in the HDI will also raise salary in each region, which will subsequently affect the Consumer Price Index and enhance the purchasing power value.

3.3 Verification and Validation

Verification and validation model is reviewing the developed model to determine if it accurately represents the actual data. Verification is conducted by checking for potential errors, including the consistency between variables and the cause-and-effect relationships. Validation is carried out by testing the model's structure, parameters, and behaviour to ensure they align with the actual data using RMSE and RMSPE.

3.4 Sensitivity Analysis

Sensitivity analysis is conducted on several variables that affect connectivity, including the number of routes or ships, and changes in regional GDP. These variables impact social welfare, which is represented by the Human Development Index.

4. Result and Analysis

4.1 Cost of PSO Maritime Transport

Based on the ship specifications, PM 55 of 2019 the Regulation on the Implementation of Pioneer Ships, PM 22 of 2018 the Regulation on the Implementation of Sea Toll Program and PM 10 of

2021 the Regulation on the Implementation of PSO Economic Passenger Ships, the shipping cost unit for each PSO Maritime Transportation is obtained as in the following table.

Table 4. Component Cost PSO Maritime Transport

Item	Unit	Pioneer	Sea Toll	PSO
Fixed Cost	Mil-IDR/Days	14.75	27.05	137.02
Bunker Cost	Mil-IDR/Nm	0.14	0.16	0.7
Port Charges	Mil-IDR/Call	1.51	2.27	6.45

The tariff of pioneer ship is represented in the equation $y = 1{,}365.1\, x^{0.6281}$ where, y is the tariff (IDR/Pax) and x is the distance in nautical mile (Nm). While the tariff of sea toll is represented in the equation $y = 4{,}835.2\, x + (3E + 06)$ where y is the tariff (IDR/TEU) and x is the distance (Nm) and the tariff for PSO Economic Passenger Ships is 600 *IDR/Pax.Nm.*

4.2 Index Connectivity

The calculation of the connectivity index for each region in each province uses a min-max scaling approach with the weights of each component being 60% for the number of direct links, 25% for the frequency of ships on each direct link, and 15% for the maximum capacity of ships (GT Max). The overall connectivity index values for each component in each region are as follows.

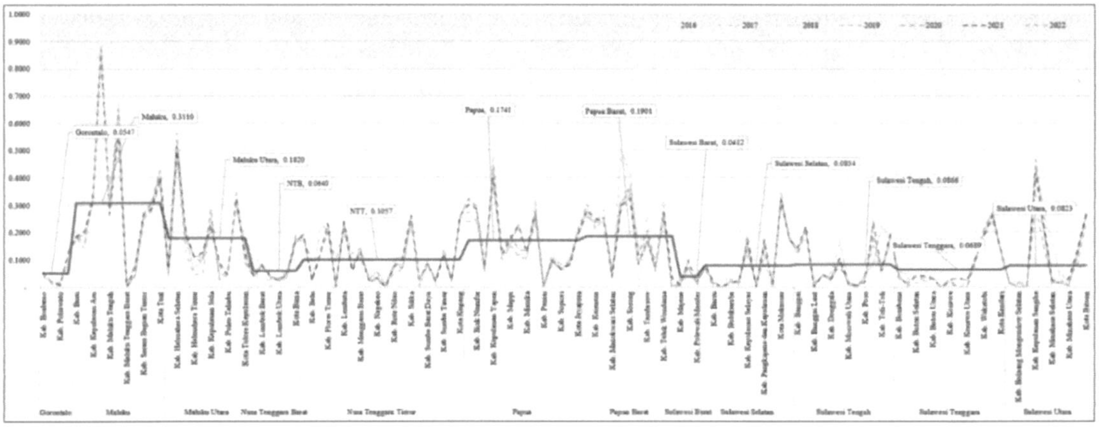

Figure 8. Index Connectivity in Each Region

Subsequently, a multiple linear regression model used to predict the value of a connectivity based on passenger and cargo transport capacity for each PSO Maritime Transportation. The results obtained are as follows.

Table 5. Regression Analysis of Index Connectivity and Transport Capacity

Component	GR	ML	MU	NB	NT	PA
Intercept (B_0)	3.1E-02	2.9E-01	9.1E-02	2.3E-02	1.0E-01	6.8E-02
Pax Transport Cap (B_1)	2.0E-08	2.7E-09	1.8E-08	5.7E-08	2.8E-10	2.8E-08
Cargo Transport Cap (B_2)	4.8E-08	1.4E-09	2.6E-08	6.4E-09	2.4E-09	3.4E-09
Multiple R	0.98	0.52	0.94	0.97	0.98	0.99
R *Square*	0.97	0.28	0.88	0.95	0.96	0.98
Component	**PB**	**SB**	**SS**	**SH**	**SG**	**SU**
Intercept (B_0)	3.9E-02	1.5E-02	3.1E-02	4.9E-02	5.8E-02	2.0E-03
Pax Transport Cap (B_1)	1.5E-08	1.1E-07	1.5E-08	1.9E-08	6.0E-09	7.1E-08
Cargo Transport Cap (B_2)	6.0E-08	2.7E-08	2.9E-08	6.1E-09	7.1E-10	1.5E-09
Multiple R	0.95	1.00	0.98	0.91	1.00	1.00
R *Square*	0.91	1.00	0.97	0.83	1.00	1.00

4.3 Relationship Between Connectivity and Welfare

Correlation and regression analysis are crucial steps in the system dynamic modelling and simulation process in this study. By understanding and utilizing the relationships between variables, we can build more effective models to predict system behaviour. The correlation value explains the extent to which the variability of one factor can be explained by its relationship with another factor. This value is calculated from 0 to 1 with the following criteria: Very high (0.7 – 1.0); High (0.5 - 0.7); Average (0.3 - 0.5); Weak (0.1 - 0.3). Based on relationship maritime transportation and social welfare, the results of the Regression Analysis and Analysis of Variance (ANOVA) to assess the relationships between variables for each province are as follows.

Table 6. Analysis of Variance

Province	Dependent	ID Var	B0	B1	B2	Mult R	F	Sig.
Gorontalo	Cargo - BPS	A	-1,525,537	0.05	46,323,538	0.77	2.89	0.17
	Passenger - BPS	B	33,670	0.00	103,897	0.96	12.29	0.08
	Price Consumer Index	C	85.647	0.000	0.000	0.99	41.25	0.11
	Poverty Population	D	48.065	-0.452	-0.420	1.00	2078.12	0.02
	Average Year of School	E	6	0.00	0	0.98	62.49	0.00
	Life Expetacy Index	F	74.274	0.000	-0.381	1.00	53521.70	0.00
	Literacy Index	G	96.994	0.000	-0.065	1.00	34245.49	0.00
	Per Capita Expenditure	H	5,276	0.00	0	0.99	80.67	0.00
	Salary	I	-20,417,560.796	-0.020	343,076.699	0.94	16.57	0.01
....	....	...	....	....	....	....	....	....

Based on the Table 6, the relationships between variables show a sufficiently strong level of correlation and significance. When the Sig-F ≤ α or the correlation is very high, the relationship is acceptable, and the values of the Intercept (β_0), Slope of Variable 1 (β_1), and Slope of Variable 2 (β_2) can be used as coefficients in the multiple linear equation that serves as input for dynamic system modelling.

4.4 Stock and Flow Diagram

Based on the Causal Loop Diagram, which considers the connectivity of Public Service Obligation (PSO) in maritime transportation with the social welfare model assessed by the Human Development Index, the stock and flow diagram in this model is divided into the following sub-models.

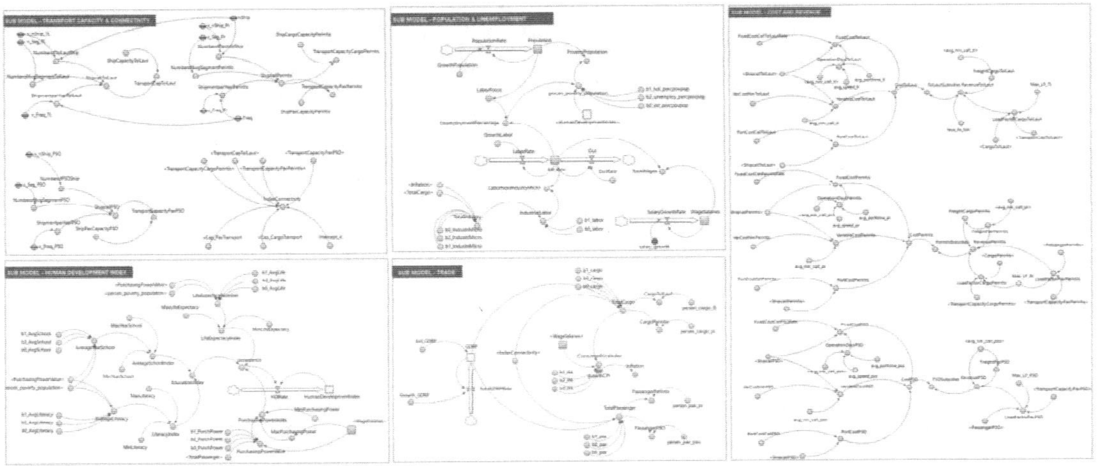

Figure 9. Stock and Flow Diagram

All regression equation components and input values are saved in a database and then used in the model with the help of table functions in AnyLogic. The table function in AnyLogic is used to define functions by providing a set of (argument, value) pairs, which are a set of base points on a plot. The following table function is used to support the modelling.

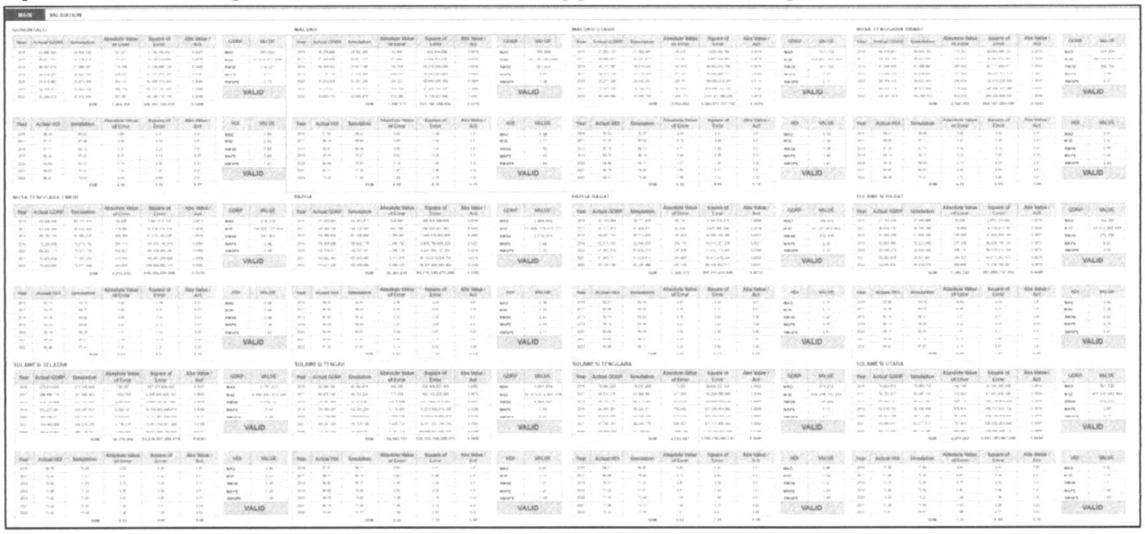

Figure 10. Verification and Validation

Verification and validation are essential processes in ensuring the accuracy and reliability of a simulation model. Verification involves checking whether the model is error-free and operates as intended by using tools "Build Problem" to identify any issues. Validation, on the other hand, involves confirming that the model's outputs match real-world data or behaviors. This is done by comparing the model's results to actual system performance through methods like algorithm and behavior testing by using MAPE (Mean Absolute Percentage Error) and RMSPE (Root Mean Square Percentage Error) to measure the accuracy of the model's predictions. If these metrics fall below a certain threshold (5%), the model is considered valid and acceptable for use

4.5 Sensitivity Analysis

Sensitivity analysis was conducted on the increase in the number of ships or routes in operation as parameters forming the connectivity index and the changes the economic conditions based on regional GDP that will impact social welfare, assessed by the Human Development Index. The following are the results of the sensitivity analysis of connectivity, subsidies and welfare.

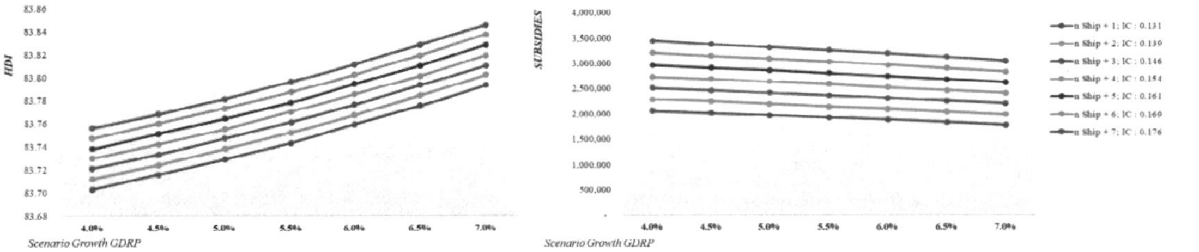

Figure 11. Sensitivity Analysis

Based on the sensitivity analysis results shown in the Figure 11, it can be concluded that adding one ship for each maritime transport results in an average increase in the connectivity index of 0.008 in Eastern Indonesia. This addition has an average impact on increasing the required subsidy value by 220,000 Mil-IDR, while each 0.05% increase in GDP growth has an average impact on reducing the subsidy by 55,000 Mil-IDR by the end of the simulation year. Regarding the Human Development Index, each increase in connectivity through the addition of ships results in an average increase of 0.009 in the Human Development Index. Additionally, each 0.05% increase in GDP growth results in an average increase of 0.015 in the Human Development Index by the end of the simulation year.

5. Conclusion

This study concludes that subsidized maritime transport connectivity positively impacts the social welfare in Eastern Indonesia. Sensitivity analysis shows that adding ships in each province and improving economic conditions lead to a higher Human Development Index (HDI) and reduced subsidy levels. These findings indicate that enhancing maritime connectivity reduces poverty, and mitigates regional disparities, aligning with the Sustainable Development Goals (SDGs) in Well-Being and Reduce Inequalities. This study also highlights the importance of a resilient sustainable maritime transport strategy in achieving long-term social welfare through causal loop diagram and sensitivity analysis in Eastern Indonesia.

References

[1] Tukan M, Achmadi T, Widjaja S. Seaport dimensional analysis towards economic growth in archipelagic regions. International Journal of Technology 2015;6:422–31. https://doi.org/10.14716/IJTECH.V6I3.243.
[2] Harto Widodo K, Kistia Purwadtya A, Soemardjito J. Developmnet of Sustainable Logsitics for Indonesian Remote and Rural Island Connectivity. Indonesian Journal of Geography 2021;53:118–25. https://doi.org/http://dx.doi.org/10.22146/ijg.57722.
[3] Rochwulaningsih Y, Sulistiyono ST, Masruroh NN, Maulany NN. Marine policy basis of Indonesia as a maritime state: The importance of integrated economy. Mar Policy 2019;108. https://doi.org/10.1016/j.marpol.2019.103602.
[4] Gunarto M, Wahasusmiah R, Indah Merina C. Determinants of Human Development Index in Indonesia. 2021.
[5] Dewi DM, Setyaningsih II, Romadhon A, Pusat B, Kabupaten S, Provinsi S, et al. Impact of Information and Communication Technology on The Welfare of Population in Eastern Indonesia. 2021.
[6] Badan Pusat Statistika. Konstruksi Dalam Angka Badan Pusat Statistika. 2022.
[7] Triantoro W. Comparative Cost Analysis of Domestic Container Shipping Network: A Case Study of Indonesian Sea-Toll Concept. Jurnal Penelitian Transportasi Laut 2020;22:33–46. https://doi.org/10.25104/transla.v22i1.1535.
[8] Kristini F, Bambang AN, Handoko W, Priadi AA. Efektivitas dan Efisiensi Program Tol Laut berbasis AHP (Studi Kasus: Pelabuhan Tahuna). Warta Penelitian Perhubungan 2019;31:75–82. https://doi.org/10.25104/warlit.v31i2.1269.
[9] Priadi AA, Laju IK, Nur HI, Achmadi T, Verdifauzi A. Evaluasi Penyelenggaraan Angkutan Perintis. Warta Penelitian Perhubungan 2021;33:77–86. https://doi.org/http://dx.doi.org/10.25104/warlit.v33i2.2004.
[10] Gunarto M, Wahasusmiah R, Indah Merina C. Determinants of Human Development Index in Indonesia. 2021.
[11] Chakrabartty SN. Wellbeing index avoiding scaling and weights. CULTURA EDUCACIÓN Y SOCIEDAD 2021;12:181–202. https://doi.org/10.17981/cultedusoc.12.2.2021.11.
[12] Kuswati, Herawati A. Konektivitas Transportasi Antarmoda di Kabupaten Tulungagung. 2017.
[13] Litman TA. Evaluating Accessibility for Transport Planning Evaluating Accessibility for Transportation Planning. 2023.
[14] Frazila RB, Zukhruf F. Measuring Connectivity for Domestic Maritime Transport Network. vol. 11. 2015.
[15] Bai X, Cheng L, Yang D, Cai O. Does the traffic volume of a port determine connectivity? Revisiting port connectivity measures with high-frequency satellite data. J Transp Geogr 2022;102. https://doi.org/10.1016/j.jtrangeo.2022.103385.
[16] Kementerian Perhubungan. Laporan Tahunan 2022 2022.

IOP Conf. Series: Earth and Environmental Science **1461** (2025) 012026 doi:10.1088/1755-1315/1461/1/012026

Optimizing Indonesian Pioneering Maritime Network through Port Clustering

Naning Yuliana[1*], Budisantoso Wirjodirjo[2] and Ahmad Rusdiansyah[3]

[1,2,3] Department of Industrial and Systems Engineering, Institut Teknologi Sepuluh Nopember, Surabaya 60111, Indonesia

*E-mail: naningyuliana13@gmail.com

Abstract. Indonesia's unique geography, characterized by scattered islands, presents challenges for maritime logistics, particularly in remote and border areas. To address these challenges, this study proposes a K-Medoids clustering model to optimize the pioneering maritime network, which is crucial for logistics distribution and population mobility in Eastern Indonesia. Unlike previous deterministic models, this approach groups ports based on their spatial characteristics, aiming to enhance network efficiency and resilience. By analyzing the coordinates of 512 ports, the model identified 10 optimal clusters, validated through gap statistics, silhouette analysis, and the Davies-Bouldin Index. The results indicate significant improvements in route efficiency, reduced voyage durations, and lower fuel consumption, leading to cost savings and environmental benefits. The optimized clusters could also enhance port infrastructure utilization and connectivity, supporting regional economic development. However, there is a trade-off with potential reductions in ship call frequency at individual ports. Overall, the K-Medoids model provides a robust framework for optimizing maritime logistics, balancing efficiency and service needs, while promoting regional growth.

1. Introduction

Indonesia, an archipelagic nation with over 17,000 islands, faces significant logistical challenges in its maritime transportation system. The dispersal of these islands necessitates a robust and efficient shipping network to ensure the continuous availability of goods and services, particularly in remote, outer, and border regions. To address these challenges, the Indonesian government provides subsidies and Public Service Obligations (PSO) to support logistics distribution and population mobility, especially in Eastern Indonesia. One such initiative is the pioneering maritime network, a regional shipping service aimed at improving connectivity. However, the current system struggles with inefficiencies due to overlapping shipping routes, long travel times, small cargo sizes, and high operational costs [1]. To enhance the efficiency and resilience of this network, a strategic optimization approach is required. This approach also supports the United Nations Sustainable Development Goals (SDGs) by promoting industry innovation and infrastructure (SDG 9), reducing inequalities (SDG 10), and fostering sustainable cities and communities (SDG 11).

Efforts to optimize maritime logistics in Indonesia have employed various methodologies, particularly focusing on clustering techniques to streamline operations. Previous studies [2] have utilized deterministic allocation models with R-tools to develop port clustering systems aimed at reducing overlaps in ship operational areas. Their findings indicated that while clustering can lead to significant improvements in operational efficiency, the choice of clustering method and the number of clusters formed critically influence the outcomes. Building upon these insights, this paper proposes an alternative clustering approach using the K-Medoids algorithm. Unlike traditional deterministic models, K-Medoids is well-suited for handling spatial characteristics and is robust against outliers, making it an ideal choice for the complex geographical distribution of Indonesian ports.

Further research has explored optimizing shipping networks through port clustering and network modelling. Clustering ports can potentially increase profits by reducing overlapping routes, though excessive clustering may decrease revenue [2]. Modelling studies have identified significant efficiency improvement potential, with up to 50% distance reduction for container transport [3]. The pioneering maritime network, crucial for connecting remote areas, has been found ineffective in the Maluku province, highlighting the need for improvement [4]. To address the optimal pattern of logistics movement network to support smooth distribution systems, clustering using Fuzzy C-means aims to group island clusters in adjacent locations [5], while using the minimum spanning tree genetic algorithm-based hybrid aims to get the optimal network pattern [6]. Clustering is also used in grouping passenger data and sea transportation vessels as a review material for the growth rate of sea transportation in each province in Indonesia [7]. Clustering analysis results in a 35% reduction in shipping costs [8].

These studies collectively highlight the potential benefits and challenges of optimizing shipping networks through advanced clustering and network modelling techniques. By implementing methods such as port clustering, Fuzzy C-means clustering, and genetic algorithm-based hybrid approaches, significant improvements in efficiency and cost reduction can be achieved. This research emphasizes the importance of strategic planning and optimization in Indonesian maritime transport system to enhance connectivity, reduce costs, and improve overall efficiency across the archipelago.

The clustering of ports can significantly reduce logistical inefficiencies by minimizing redundant routes and optimizing the allocation of resources. Traditional clustering techniques, while effective in certain contexts, often fall short when applied to highly irregular spatial distributions such as those found in Indonesia. The K-Medoids algorithm, by contrast, offers a more nuanced approach that can better accommodate the unique spatial dynamics of Indonesian ports.

This study specifically focuses on the remote, outer, and border regions in Indonesia, which are served by the pioneering maritime network. The analysis utilizes data from the year 2022 to ensure the most recent and relevant insights into the current logistical challenges and opportunities for optimization in these critical areas.

The research objectives include:

1. Developing a port clustering model: Utilizing the K-Medoids algorithm to create efficient and resilient port clusters based on spatial characteristics.
2. Evaluating clustering performance: Employing gap statistics, silhouette analysis, and the Davies-Bouldin Index to comprehensively validate the clustering outcomes.
3. Enhancing Network Efficiency: Formulating strategies to integrate the clustering model into the pioneering maritime network to minimize operational costs and maximize shipping service.

Through these objectives, the study aims to provide a strategic framework for optimizing the pioneering maritime network, thereby improving the overall efficiency and resilience of Indonesia's maritime logistics system. This optimization is expected to result in more cost-effective operations, reduced overlap in shipping routes, and improved service to remote and underserved regions.

2. Literature Review

2.1 The Role of Pioneering Shipping

The inter-island shipping network has been vital for trade within Indonesia, facilitating the movement of goods and passengers. However, it primarily focuses on profitable routes, leaving remote islands, despite their natural resources, without adequate transport. To address this, the Indonesian government mandates pioneering shipping routes, essential for regional connectivity but not commercially profitable. Both private and state-owned companies operate these routes, receiving subsidies to cover the gap between costs and government-set fares.

Pioneering shipping promotes economic development in isolated regions by enhancing market access and connectivity, integrating them into national and international trade networks, and improving access to goods and services. Currently, the system does not consider shipping distances and relies on local government recommendations. Sea Toll and Pelni ships transport goods from main ports to collector ports, while Pioneer ships carry goods to feeder ports, leading to inefficiencies and higher costs for Public Service Obligations (PSO).

Research has proposed an optimal port clustering model to eliminate overlap in ship operational areas. Optimizing port clusters is expected to reduce travel time, minimize costs, and maximize revenue [1]. However, the study found that more clusters increase operational costs and limit the number of ports served, reducing cargo capacity. This contradicts the goal of maximizing transportation service by reducing round-trip days and increasing frequency while minimizing costs. Additionally, the development of this clustering model has not yet incorporated the validation framework.

2.2 Clustering Analysis

Clustering analysis is an unsupervised learning method that groups data points based on their similarities [9][10], without prior labeling [11]. It assigns elements to groups to maximize similarity within each group [12]. Clustering algorithms include partitioning, hierarchical, density-based, grid-based, model-based [11], and fuzzy clustering, each with its own advantages and disadvantages [8]. Practical uses of cluster analysis include market research, pattern identification, data analysis, and image processing [13].

Spatial clustering abstracts the Earth's surface into two dimensions, incorporating location attributes explicitly or implicitly through partitioning. Partition algorithms divide data into subsets called partitions, each representing a cluster with specific characteristics: each cluster has at least one object, and objects belong to exactly one cluster. K-medoids is a partitioning algorithm that minimizes squared error by selecting data points as cluster centers (medoids) [14], rather than using the mean (as in k-means). It can handle outliers and work with any distance metric, though it is more complex and requires more iteration [15].

2.3 K-Medoids Clustering

The K-Medoids algorithm [16] clusters data points using representative objects called medoids, which are the most centrally located points within their clusters, minimizing average

IOP Conf. Series: Earth and Environmental Science **1461** (2025) 012026 doi:10.1088/1755-1315/1461/1/012026

dissimilarity. Unlike K-Means, K-Medoids is robust to noise and outliers because it uses medoids instead of means.

Advantages of K-Medoids include its robustness to noise and outliers, flexibility in using any similarity measure, and efficiency for small datasets. However, it does not scale well for large datasets, may yield different results with each run, is unsuitable for non-spherical clusters, and requires the number of clusters (k) to be specified in advance [17].

Algorithm steps [18]:

1. Initialization: Randomly select K data points as initial medoids.
2. Assignment: Assign each data point to the nearest medoid using distance metrics like Manhattan or Euclidean.
3. Update Medoids: Swap non-medoid points with current medoids if it reduces overall cost (sum of distances within clusters).
4. Iterative Refinement: Reassign data points to the nearest medoid, update medoids through beneficial swaps, repeat until no further improvement in cost or a stopping criterion is met.
5. Final Clustering: output the final set of medoids and corresponding cluster.

2.4 Transportation Cost

Transportation costs refer to the expenses that transport providers incur in delivering services, which include both fixed costs (such as infrastructure) and variable costs (such as operating expenses). The difference between these costs and the rates charged determines whether the service provider makes a profit or incurs a loss. Public transit rates are often set by political decisions, with subsidies partially covering the costs to maintain affordable mobility, even if this results in financial deficits. Several key factors influence transportation costs and rates, including [19,20]:

1. Geography: Distance and accessibility impact costs, with greater friction of distance leading to higher costs, varying by transport mode and route efficiency.
2. Type of product: Special handling, packaging, bulkiness, or perishability increase costs. Passenger transport also requires comfort and amenities.
3. Economies of scale: Larger transport quantities reduce unit costs.
4. Energy: Transport activities, particularly oil consumption, are significant energy consumers, making costs susceptible to energy price fluctuations.
5. Trade imbalance: Uneven outbound and return loads lead to empty movements and higher costs.
6. Infrastructure: Efficient and high-capacity transport modes and terminals lower costs, while poor infrastructure increases them.

3. Research Method

This research employs the K-Medoids algorithm for clustering and evaluates its performance using gap statistics, silhouette analysis, and the Davies-Bouldin Index.

3.1 Data Collection

Data for this study is sourced from the Ministry of Transportation's document as per KP.1028/DJPL/2021, comprising detailed port information and shipping routes within the pioneering maritime network.

3.2 Methodology

The clustering process follows these steps:

1. Data processing: Clean and prepare the dataset for analysis, removing any inconsistencies or missing values.
2. Initialization: Randomly select initial medoids from the dataset.
3. Clustering: Apply the K-Medoids algorithm to group ports based on their spatial characteristics.
4. Validation: Assess clustering performance using gap statistics, silhouette analysis, and the Davies-Bouldin Index.
5. Refinement: Iterate the process to optimize the clustering results, ensuring robustness and reliability.

3.3 The Gap Statistic

The gap statistic is a method for estimating the number of clusters in a dataset, applicable with any clustering algorithm like K-Medoids or hierarchical clustering. It compares the within-cluster dispersion to a null reference distribution to identify the optimal number of clusters where the observed dispersion is significantly lower than the expected dispersion [21].

Steps to compute the gap statistic:

1. Compute the observed data for k = 1, 2, ..., K clusters, and calculate the within-cluster dispersion measures W_k.
2. Generate B reference datasets using a uniform distribution over the observed data range or principal component-aligned box and calculate the within-dispersion measures $W^*{}_{kb}$ for each.
3. Compute the gap statistic:
$$Gap(k) = (1/B)\sum_b \log(W^*{}_{kb}) - \log\,(W_k)$$
4. Calculate the average log dispersion $\bar{l} = (1/B)\sum_b \log(W^*{}_{kb})$ and its standard deviation sd_k:
$$sd_k = [(1/B)\sum_b \{\log(W^*{}_{kb}) - \bar{l}\}^2]^{1/2}$$
Define $s_k = sd_k\sqrt{(1+1/B)}$.
5. Determine the optimal number of clusters:
$$\hat{k} = smallest\ k\ such\ that\ Gap(k) \geq Gap(k+1) - s_{k+1}$$

3.4 Silhouette Coefficient

The Silhouette coefficient is a measure used to evaluate the quality of clusters in a dataset, helping to estimate the optimal number of clusters. For a given observation i, let $a(i)$ represent the average distance to all other points within the same cluster (indicating cohesion), and $b(i)$ represent the average distance to points in the nearest neighbouring cluster (indicating separation). The silhouette statistic is then defined as [21]:

$$s(i) = \frac{b(i) - a(i)}{\max\{a(i), b(i)\}}$$

The value of $s(i)$ ranges from -1 to 1, where:

1. s(i) close to 1 indicates that the observation is well-clustered, with a large gap between $a(i)$ and $b(i)$.

2. s(i) around 0 suggests that the observation lies on the border of two clusters.
3. s(i) close to -1 indicates that the observation may have been assigned to the wrong cluster.

Table 1 provides the interpretation of the Silhouette Coefficient, with the largest index being 0.70 < s(i) ≤1.00, which indicates a strong clustering structure, and the lowest index being s(i) < 0.25, which suggests no discernible structure or poorly defined clusters.

Table 1. Interpretation of Silhouette Coefficient

Silhouette Coefficient	Interpretation
$0.70 < s(i) \leq 1.00$	Strong Structure
$0.50 < s(i) \leq 0.70$	Medium Structure
$0.25 < s(i) \leq 0.50$	Weak Structure
$s(i) < 0.25$	No Structure

3.5 Davies Bouldin Index

The Davies-Bouldin Index (DBI) evaluates the validity of clustering results by measuring both cluster compactness and separation. The index compares clustering algorithms to identify the best clustering scheme.

Criteria for clustering evaluation:

1. Compactness: Members of each cluster should be as close to each other as possible, commonly measured by variance.
2. Separation: Clusters should be widely separated.

Davies Bouldin Index is based on the similarity measure (R_{ij}), which depends on cluster dispersion s_i and cluster dissimilarity d_{ij}. The similarity measure must satisfy the following conditions [22]:

1. $R_{ij} \geq 0$
2. $R_{ij} = R_{ji}$
3. if $s_i = 0 \; and \; s_j = 0 \; then \; R_{ij} = 0$
4. if $s_j > s_k \; and \; d_{ij} = d_{ik} \; then \; R_{ij} = R_{ik}$
5. if $s_j > s_k \; and \; d_{ij} < d_{ik} \; then \; R_{ij} = R_{ik}$

Formula:

$$DB = \frac{1}{n_c}\sum_{i=1}^{n_c} R_i \, , where \; R_i = \max_{j=1..nc, i\neq j}(R_{ij}), i = 1 \dots n_c$$

The Davies-Bouldin Index balances cluster compactness and separation. A lower DBI value indicates better clustering, facilitating comparison across different clustering results.

4. Result and Discussion

The proposed K-Medoids clustering model demonstrates indicate significant improvements in the efficiency and resilience of the pioneering maritime network. Detailed results, including specific cluster formations and their performance metrics, are discussed below.

4.1 Clustering Analysis

The data utilized consists of the locations of ports for Pioneer shipping across Indonesia in 2022. A total of 512 ports were identified. The collected data includes port names, provinces, and geographic coordinates (latitude and longitude), which serve as attributes for the clustering analysis. Detailed information about these ports can be found in Table 2.

Table 2. Locations of Ports for Pioneer Shipping in 2022

No	Port	Province	Latitude	Longitude
1.	Adaut	Maluku	-8.12840241667	131.11053000000
2.	Aduwey	Papua Barat	-1.98617720746	129.90823155490
3.	Agats	Papua	-5.52359925000	138.13787900000
...	...	...	...	...

With the input consisting of unlabelled port coordinate data and the number of clusters set to be evaluated at 11, the gap statistic was employed for clustering using the K-Medoids algorithm implemented in MATLAB. The results of this clustering analysis are shown in Figure 1.

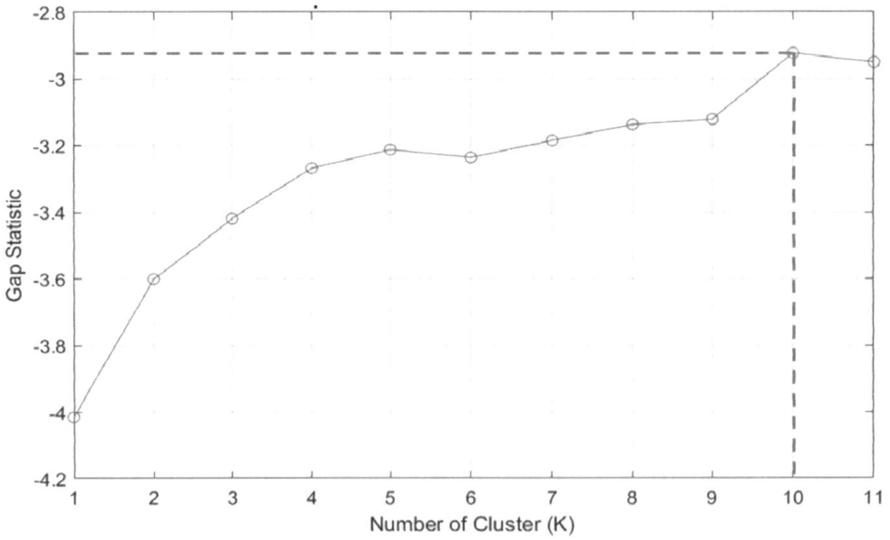

Figure 1. Gap Statistic

Based on the results of the gap statistic analysis, it was determined that 10 clusters yield the maximum gap with stable variability in the gap statistic. This indicates that 10 is the optimal number of clusters for the data. Additionally, it was observed that after reaching the maximum gap, if the gap statistic begins to decrease or the changes in gap values become minimal, it suggests that increasing the number of clusters further does not significantly enhance the clustering structure.

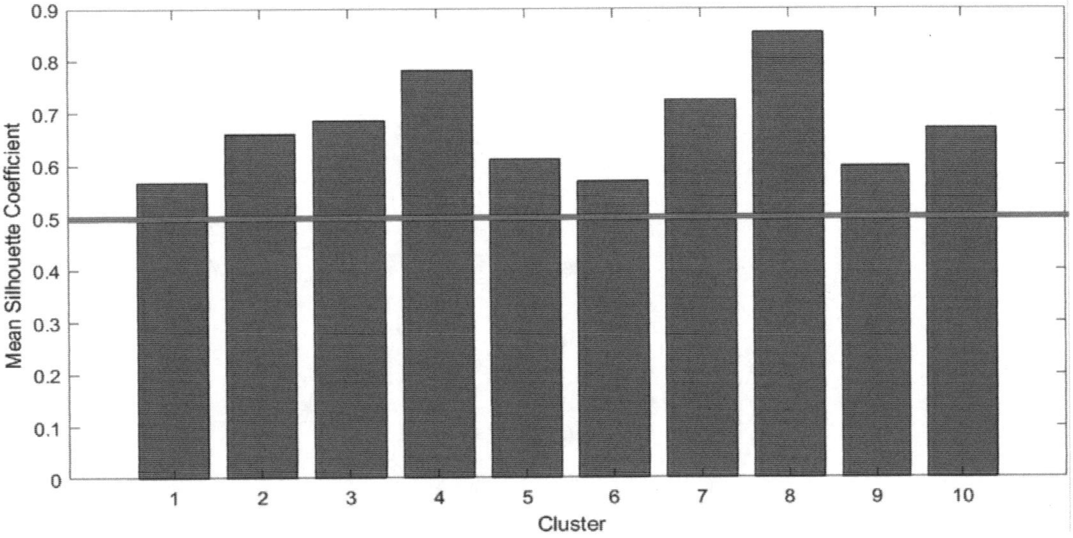

Figure 2. Mean Silhouette Coefficient of Each Cluster

To further assess the quality and coherence of the identified clusters, the silhouette coefficient is utilized. This metric evaluates how well each data point is assigned to its cluster relative to other clusters, offering a nuanced measure of clustering effectiveness. Applying the silhouette coefficient provides deeper insights into the internal consistency of the clusters determined by the gap statistic analysis. High silhouette coefficients within a cluster indicate that the cluster is cohesive and well-separated from other clusters, reflecting medium to strong structure. The results of this analysis are shown in Figure 2.

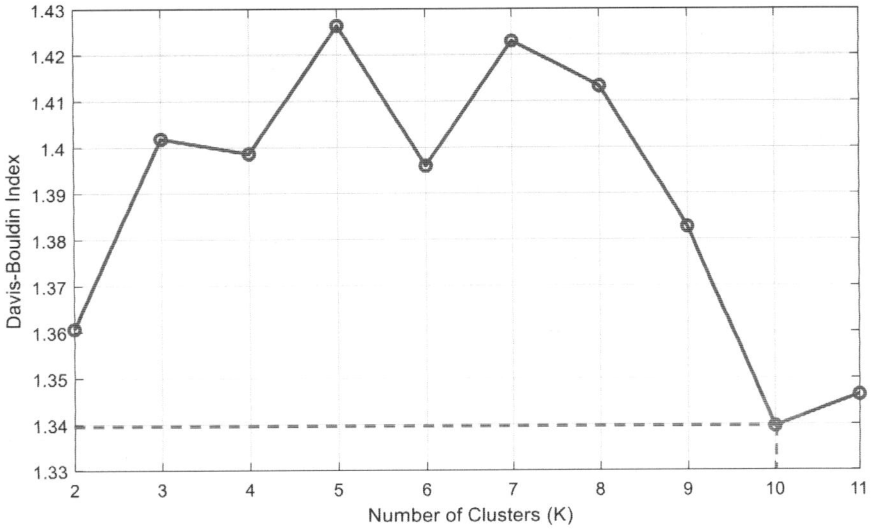

Figure 3. Davis Bouldin Index

The results, as indicated by the Davies-Bouldin Index (DBI) metric, affirm that the optimal number of clusters effectively captures the inherent structure of the dataset, ensuring both high intra-cluster homogeneity and inter-cluster heterogeneity. This enhances the reliability and validity of the clustering outcomes, contributing to a more nuanced understanding of the underlying data patterns. These findings are illustrated in Figure 3.

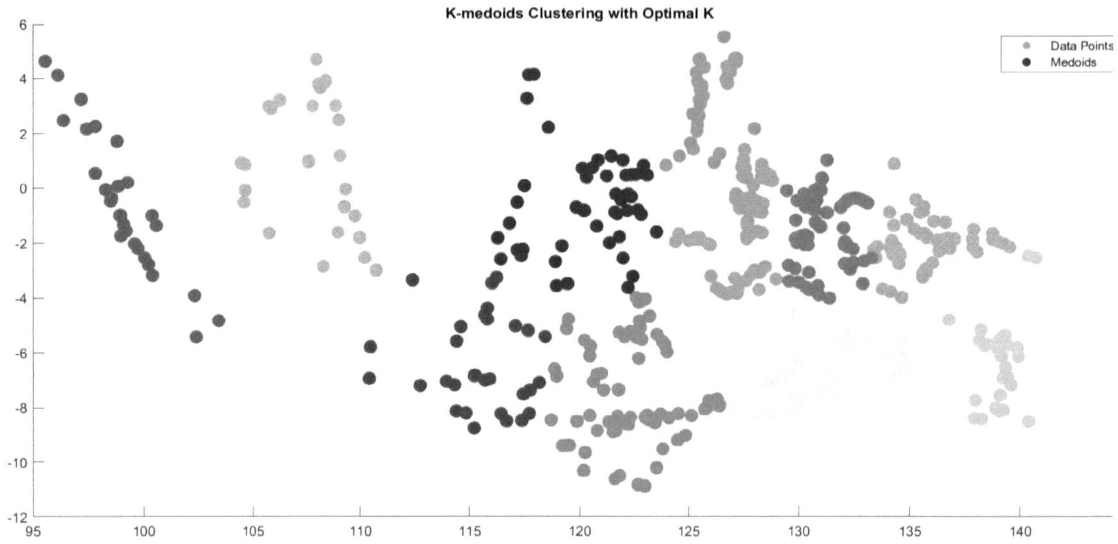

Figure 4. Clustering Result

Based on the results of the gap statistic, silhouette coefficient, and Davies-Bouldin Index, the clustering analysis identified 10 regions for the 512 ports, with the distribution shown in Figure 4. The cluster assignments for each port are detailed in Table 2.

Table 3. Cluster Assignment for Ports in 2022

No	Port	Cluster	Province	Latitude	Longitude
1.	Adaut	1	Maluku	-8.12840241667	131.11053000000
2.	Aduwey	2	Papua Barat	-1.98617720746	129.90823155490
3.	Agats	2	Papua	-5.52359925000	138.13787900000
...	...	...	...	...	...
509.	Yelu	2	Papua Barat	-2.05687122222	130.41409066667
510.	Yembekiri	8	Papua Barat	-2.36700799318	134.53112773338
511.	Yende	8	Papua Barat	-2.36704013889	134.53113844444
512.	Yenggarbun	8	Papua	-0.67195806318	135.61837349097

4.2 Implications for Maritime Logistic

The implementation of port clustering has significant and multifaceted implications for maritime logistics.

IOP Conf. Series: Earth and Environmental Science **1461** (2025) 012026 doi:10.1088/1755-1315/1461/1/012026

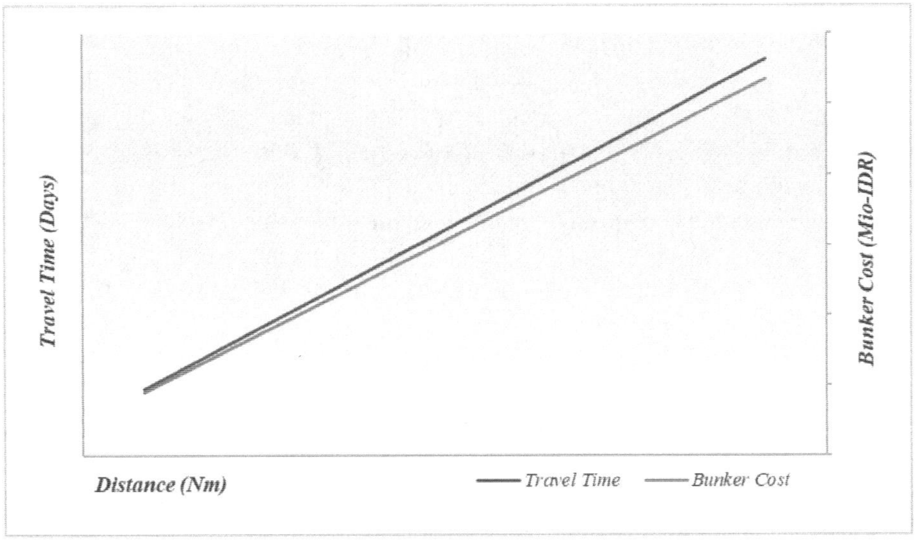

Figure 5. The Relationship Between Distance, Travel Time, and Bunker Cost

1. **Enhance Route Efficiency:**
   a. **Reduction in Route Overlaps**: Port clustering, along with the restriction of operational areas, allows for the assignment of vessels to specific clusters. This minimizes route overlaps and reduces redundancy, thereby maximizing fleet efficiency. Figure 5 illustrates the correlation between distance, travel time, and bunker costs.
   b. **Optimized Scheduling:** Clearly defined routes make scheduling more predictable and reliable, enabling ships to adhere to stricter schedules and improving overall service reliability.

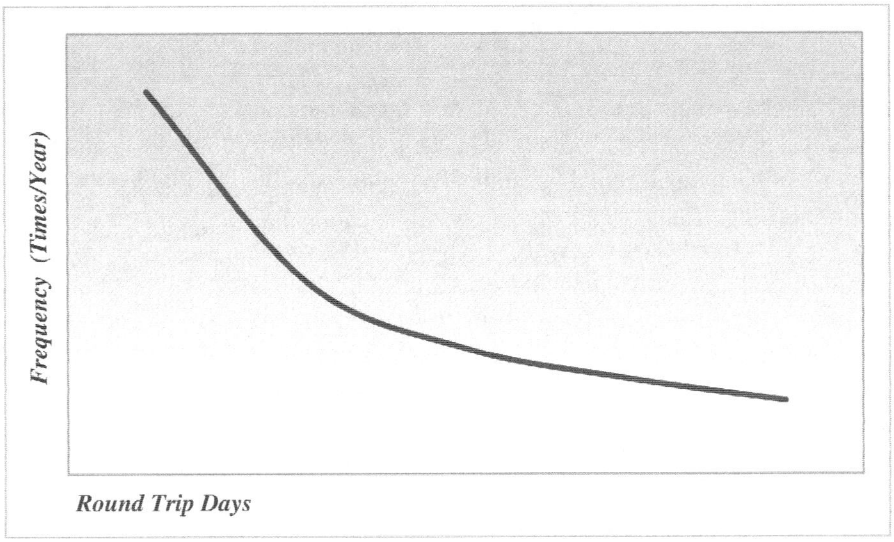

Figure 6. The Relationship Between Round Trip Days and Frequency

2. **Shortened Voyage Duration:**
   a. **Faster Turnaround:** Limiting vessels to smaller operational areas reduces the travel distance between ports, resulting in shorter voyage times. This allows for quicker turnarounds and more frequent trips within each cluster. Figure 6 shows the relationship between round trip days and frequency, demonstrating how reduced travel times lead to more frequent services.
   b. **Improved Service Frequency:** Shorter routes enable vessels to undertake more frequent trips, enhancing service frequency to each port within the cluster. This improvement leads to better connectivity and more regular service to remote and less accessible regions.

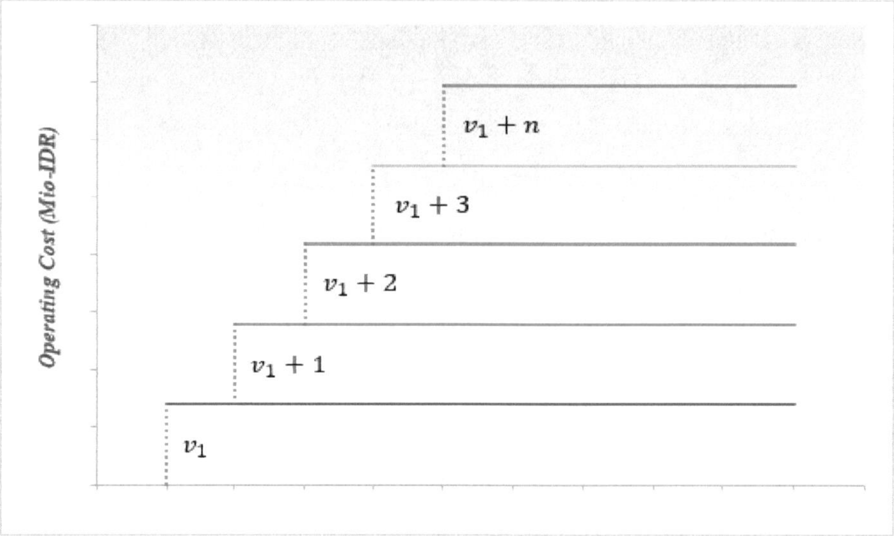

Figure 7. The Relationship Between Number of Ship and Operating Cost

3. **Cost Reduction:**
   a. **Lower Fuel Consumption**: By reducing the distances that vessels need to travel, fuel consumption is significantly decreased. This reduction not only lowers operational costs but also has environmental benefits by reducing the carbon footprint of maritime operations.
   b. **Efficiency in the Logistics System:** Overlapping routes can create inefficiencies, such as underutilized capacity and increased operational costs, when multiple ships call at the same ports. Port clustering and the definition of distinct service areas help to minimize these inefficiencies. Figure 7 illustrates the relationship between the number of ships and operating costs, showing how clustering can contribute to cost reduction.
4. **Enhance Port Infrastructure Utilization**
   Concentrating operations within clusters leads for better utilization of port infrastructure. Ports can invest in specialized facilities and services tailored to the specific needs of the vessels operating within their cluster, enhancing overall operational effectiveness.
5. **Regional Economic Development**
   a. **Boost to Local Economies:** Enhanced maritime connectivity and more frequent services can stimulate regional economic development by increasing trade, tourism, and investment in port regions.

b. **Support for Remote Areas:** Regular and reliable maritime services to remote and less developed areas support regional development goals, contributing to more balanced economic growth across different regions.

There is often a trade-off between transport capacity and the frequency of ship calls at various ports. This trade-off becomes particularly evident when considering overlapping routes and the implications of port clustering. While concentrating operations within port clusters can improve efficiency and reduce redundancy, it may also result in less frequent service for certain ports, particularly those not included in the clusters.

5. Conclusion

The K-Medoids clustering model significantly improves the efficiency and resilience of the maritime network by analyzing the coordinates of 512 ports across Indonesia. The model identified 10 optimal clusters, verified by the gap statistic, silhouette coefficient, and Davies-Bouldin Index. These clusters enhance route efficiency, reduce voyage durations, and lower fuel consumption, resulting in cost savings and environmental benefits.

The clustering also optimizes port infrastructure utilization and supports regional economic development by improving connectivity and service frequency, particularly in remote areas. However, the trade-off includes potential reductions in ship call frequency at individual ports, which could impact transport capacity and cargo handling. Overall, the K-Medoids clustering model offers a robust framework for optimizing maritime logistics, balancing efficiency with service needs, and promoting regional growth. Future research should explore dynamic cluster adjustments to adapt to changing conditions.

References

[1] Wahyono E, Tangkilisan YB, Marihandono D. Development of Inter-Island Shipping as a Bridge to Indonesia. Journal of Maritime Research 2016;13:29–38.

[2] Humang WP, Hadiwardoyo SP. Clustering on Freight Distribution System in Archipelagic Region with Deterministic Allocation Model. International Journal of Engineering Research and Technology 2019;12:2997–3005.

[3] Verhaeghe R, Halim RA, Tavasszy L. Optimizing the efficiency of the future maritime transport network of Indonesia. Freight Transport Modeling in Emerging Countries, Elsevier; 2021, p. 109–34. https://doi.org/10.1016/B978-0-12-821268-4.00006-X.

[4] Humang WP, Hadiwardoyo SP, Nahry. The Effectiveness of Pilot Marine Transport Services for Goods Distribution in Maluku Island Group, Indonesia. 2017 6th IEEE International Conference on Advanced Logistics and Transport (ICALT), IEEE; 2017, p. 165–9. https://doi.org/10.1109/ICAdLT.2017.8547034.

[5] Kismanti ST, Maidah AA. Penentuan Kelompok Jaringan Logistik pada Wilayah Kepulauan menggunakan Fuzzy C-Means. Borneo Engineering: Jurnal Teknik Sipil 2018;2:76–85.

[6] Tri Kismanti S, Ard Maidhah A, Amal Lama No J, Tarakan K. Penentuan Jaringan Logistik Menggunakan Fuzzy C-Means dan Minimum Spanning Tree Berbasis Genetic Algorithm. Jurnal INOVTEK POLBENG 2019;9:112–21.

[7] Surya Praja B, Daru Kusuma P, Setianingsih C. Penerapan Metode K-Means Clustering dalam Pengelompokan Data Penumpang dan Kapal Angkutan Laut di Indonesia Application of K-Means Clustering Method in Passenger and Ship Transport Data Grouping in Indonesia. e-Proceeding of Engineering, 2019, p. 1442–9.

[8] Atmojo AT, Siallagan M, Fanhas RS. Warehouse Location Optimization with Clustering Analysis to Minimize Shipping Costs in Indonesia's E-Commerce Case. International Journal of Current Science Research and Review 2023;6. https://doi.org/10.47191/ijcsrr/V6-i8-59.

[9] Fasulo D. An Analysis of Recent Work on Clustering Algorithms 1999.

[10] Patel D, Sarvakar K, Modi R. A Comparative Study of Clustering Data Mining: Techniques and Research Challenges 2014. https://doi.org/10.13140/2.1.1508.7042.

[11] Reddy BGO, Ussenaiah M. Literature Survey On Clustering Techniques. IOSR Journal of Computer Engineering (IOSRJCE) 2012;3:1–12.

[12] Milligan GW, Cooper MC. Methodology Review: Clustering Methods. Appl Psychol Meas 1987;11:329–54. https://doi.org/10.1177/014662168701100401.
[13] Velunachiyar S, Sivakumar K. Some Clustering Methods, Algorithms and their Applications. International Journal on Recent and Innovation Trends in Computing and Communication 2023;11:401–10. https://doi.org/10.17762/ijritcc.v11i6s.6946.
[14] Varghese BM, Poulose Jacob K, Jacob PK. Spatial Clustering Algorithms-An Overview. vol. 3. 2014.
[15] Omran MGH, Engelbrecht AP, Salman A. An overview of clustering methods. Intelligent Data Analysis 2007;11:583–605. https://doi.org/10.3233/ida-2007-11602.
[16] Kaufman L, Rousseeuw PJ. Finding Groups in Data. Wiley; 1990. https://doi.org/10.1002/9780470316801.
[17] Papakyriakou D, Barbounakis IS. Data Mining Methods: A Review. Int J Comput Appl 2022;183:5–19. https://doi.org/10.5120/ijca2022921884.
[18] Hossain A, Alam MA, Esa H, Islam KN, Assistant GR, Student G. Data Clustering: Prospects & Challenges 2023;11:2320–882.
[19] Hesse M, Rodrigue JP. The transport geography of logistics and freight distribution. J Transp Geogr 2004;12. https://doi.org/10.1016/j.jtrangeo.2003.12.004.
[20] Rodrigue J-P, Comtois C, Slack B. The Geography of Transport Systems. 3rd ed. New York: Routledge; 2013.
[21] Tibshirani R, Walther G, Hastie T. Estimating the Number of Clusters in a Data Set via the Gap Statistic. Source: Journal of the Royal Statistical Society Series B (Statistical Methodology) 2001;63:411–23.
[22] Patil SU, Nuli UA. A Review of Clustering and Clustering Quality Measurement. International Journal of Computer Engineering in Research Trends 2019;5. https://doi.org/10.22362/ijcert/2018/v5/i12/v5i1205.

IOP Conf. Series: Earth and Environmental Science **1461** (2025) 012027 doi:10.1088/1755-1315/1461/1/012027

Evaluating Transportation of Crude Palm Oil for GHG Emissions Reduction Using a Time/Cost-Distance Model: A Case Study of Special Terminal Policy for Crude Palm Oil Shipments from Central Kalimantan to East Java

Rengga Eka Putra Atmanegara[1*] and Noor Muhammad Ridha Fuadi[2]

[1] Department of Marine Transportation Engineering, Sepuluh Nopember Institute of Technology, Surabaya, Indonesia
[2] Research Center for Transportation Technology, , National Research and Innovation Agency, Jakarta, Indonesia

*E-mail: rengga.atmanegara@gmail.com
noor014@brin.go.id

Abstract. The palm oil industry is a significant contributor to Indonesia's economy, yet its environmental impact, including greenhouse gases (GHG) emissions from the transportation of Crude Palm Oil (CPO), poses significant challenges. This study employs the Time/Cost-Distance model to evaluate transportation scenarios used in the distribution of CPO from Central Kalimantan to East Java. By comparing transportation scenarios involving road and inland waterway networks, the study identifies the optimal configuration to minimize air pollution caused by GHG emissions while considering transportation time and cost. The results indicate that the "inland waterways network scenario," supported by the special terminal policy outlined in the Minister of Transportation Regulation Number 52 of 2022 on Special Terminals and Terminals for Private Use, provides a more environmentally friendly and cost-effective solution, despite covering a longer distance than the "road network scenario." The inland waterways network scenario results in total GHG emissions of 22.37 kg CO_2-eq/ton, a cost of $17.70/ton, and a travel time of 88.07 hours, compared to the road network scenario, which has total GHG emissions of 32.71 kg CO_2-eq/ton, a cost of $30.16/ton, and a travel time of 87.55 hours. One factor contributing to these differences is the presence of special terminals, which eliminate a leg of transportation in the inland waterways network scenario, thereby reducing both GHG emissions and transportation costs. This demonstrates that policy and law can enhance the sustainability of the palm oil industry and helping to address the issue of climate change through the reduction of GHG emissions.

1. Introduction

With a total of 46.22 million tons of palm oil produced (1) and 28.5 million tons of palm oil exported (2), Indonesia held the top position in terms of both the production and export of palm oil globally. Along with Malaysia, both countries accounted for nearly all of the world's palm oil production and export (3). The palm oil industry is important to Indonesia's Gross Domestic

IOP Conf. Series: Earth and Environmental Science **1461** (2025) 012027 doi:10.1088/1755-1315/1461/1/012027

Product (GDP), depending on the calculation method, palm oil contribution to Indonesia's GDP varies between 1.5% and 2.5% (4). Furthermore, the palm oil industry absorbed a huge workforce in Indonesia. The production of palm oil and its derivatives is estimated to impact around 50 million people in Indonesia as they depend directly or indirectly on it due to various multiplier effects (5).

In 2022, Central Kalimantan, a major palm oil-producing province was recorded to have 2,037,705 hectares of oil palm plantations, capable of producing 8,363,763 tons of CPO (1) and distributed a substantial portion of its CPO to various destinations, including East Java (6). The distribution process involving long-distance transportation from production areas to industrial centers or export ports contributes significantly to air pollution caused by greenhouse gases (GHG) emissions (7). These findings align with other studies indicating that the transportation sector is known to consume approximately 20% of the total final energy consumption nationally, with 97% of the sector relying on petroleum-based fuels (8). Considering this, efforts to reduce GHG emissions in the palm oil industry are essential, particularly in terms of its transportation sector.

The importance of reducing GHG emissions in the palm oil industry is also in line with the Sustainable Development Goals (SDGs) (9). Some SDGs relevant to this issue are:

- **SDG 13**: Climate Action: Taking urgent action to combat climate change and its impacts.
- **SDG 12**: Responsible Consumption and Production: Encouraging companies, including palm oil companies, to adopt sustainable business practices.
- **SDG 17**: Affordable and Clean Energy: Encouraging the use of cleaner and more efficient energy.

Efforts to reduce GHG emissions in the palm oil industry also play a role in helping to address the issue of climate change and enhancing the industry's sustainability. These efforts can be achieved through the implementation of policies and laws. This study discusses one such policy: the special terminal policy outlined in the Minister of Transportation Regulation Number 52 of 2022 on Special Terminals and Terminals for Private Use.

Based on Article 4 paragraph (a) of the Minister of Transportation Regulation Number 52 of 2022 concerning Special Terminals and Terminals for Private Use, it explains that Special Terminals or Terminals for Private Use can only be operated for ship traffic activities, passenger embarkation, and disembarkation, or loading and unloading of goods in the form of raw materials, production results, and production support equipment for own use (10). In the case of the palm oil industry, the development of special terminals for CPO processing will streamline the distribution process. This is particularly beneficial in parts of South and Central Kalimantan, where river transportation remains prevalent (11), offering multiple transportation options for CPO and providing greater flexibility and efficiency.

This study aims to compare GHG emissions from various modes of transportation used in the distribution of CPO from Central Kalimantan to East Java, while also considering transportation times and costs. Thus, efficient, and environmentally friendly transportation modes for long-distance CPO distribution in Indonesia can be recommended to reduce GHG emissions and support sustainable development goals.

IOP Conf. Series: Earth and Environmental Science **1461** (2025) 012027 doi:10.1088/1755-1315/1461/1/012027

2. Methodology

2.1. Time/Cost-Distance Model

To compare GHG emissions from various modes of transportation used in the distribution of CPO from Central Kalimantan to East Java, while also considering transportation times and costs, a transportation model will be developed. A transportation model aims to design a plan for transporting goods from sources to destinations. This model incorporates data on demand, budgets, and other relevant factors. The objective of the transportation model is to determine the quantity of goods to be delivered and the means of transportation to be used to achieve minimum costs. In other words, the transportation model seeks to establish an optimized plan for transporting goods from multiple sources to multiple destinations efficiently (12).

This paper will adopt the Time/Cost-Distance model as shown in Figure 1, which employs curve steepness to represent the cost changes of each transportation mode, with the slopes indicating the transport cost per distance and vertical surges displaying the cost steps of the multimodal transfer (13). The model provides intuitive and easily understandable graphical information on the route's cost-effectiveness and time efficiency by providing comparisons based on time, cost, and distance as demonstrated in previous research (14). This model and similar techniques have also been disseminated by the United Nations (15).

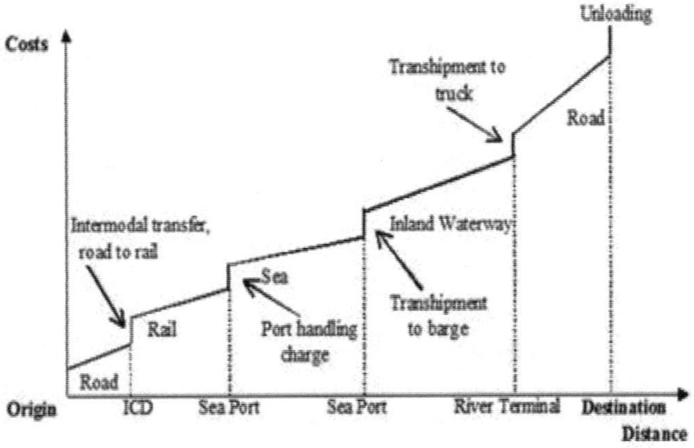

Figure 1. Time/Cost-Distance Model (13).

The Time/Cost-Distance model will be used as an analytical tool to evaluate transportation scenarios for Crude Palm Oil (CPO) distribution, focusing on time, costs, and environmental impact, to address the following research objectives:

1. How do different transportation modes compare in terms of environmental sustainability and economic efficiency for Crude Palm Oil (CPO) distribution in Indonesia?
2. How do special terminals influence the logistics chain, particularly in terms of transportation costs and GHG emissions?
3. Which transportation mode offers the most sustainable and cost-effective solution for long-distance CPO distribution in Indonesia, as evaluated using the Time/Cost-Distance model?

Hopefully, the findings from this analysis will provide valuable insights for policymakers and stakeholders in promoting sustainable and efficient transportation practices within the palm oil industry.

2.2. Identification of Transportation used to deliver CPO from Factory to Port

This research focuses on palm oil processing plants at PT. XYZ in Kotawaringin Timur Regency, Central Kalimantan. Alternative transportation modes proposed for CPO delivery include tank trucks, pipelines, railways, and barge tanker. However, the implementation of pipeline transportation is deemed unfeasible due to the significant distance between the processing plants and the port. Similarly, CPO transportation via railways is currently not viable because of inadequate infrastructure along the proposed route (12). Therefore, the feasible transportation alternatives are limited to tank trucks and barge tankers.

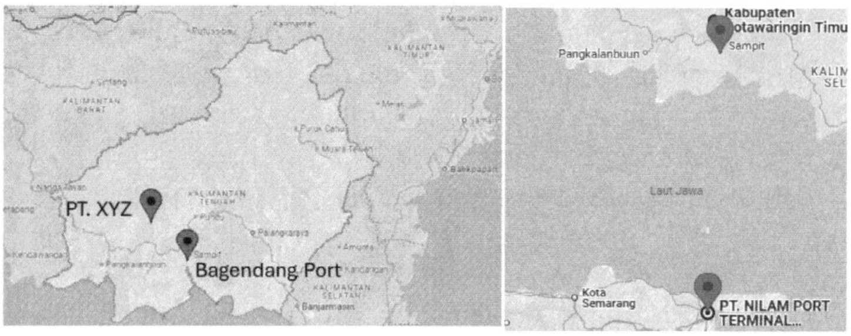

Figure 2. CPO Factory of PT. XYZ, Bagendang Port, and Nilam Terminal Location.

The selected plants were chosen as study objects due to their strategic locations, and because it has access to both road and inland waterways transportation. enabling a comparison between road and river transportation. Terminal Nilam at Tanjung Perak Port, Surabaya, was chosen as the final distribution destination. The selection of this location was based on the Road Network and Inland Waterways Network available in the location. So, as the starting point of distribution, this study focused on Bagendang Port. Meanwhile, Terminal Nilam at Tanjung Perak Surabaya was the final shipment destination. The selection of these two locations considered the specific characteristics as explained in previous research (6).

An important factor to consider in this research is that, for the purposes of this study, both scenarios are assumed to operate under conditions described in the previous study used as a basis. Consequently, this study does not introduce variations in CPO delivery using tank trucks or barge tankers. While this approach simplifies the analysis, it is not ideal, as palm oil transportation from plantations to factories via trucks can vary significantly due to factors such as changes in the distance traveled caused by non-specific transport instructions (16). Similarly, barge tanker operations could benefit from optimization efforts, such as reducing vessel speed to minimize emissions (17). However, such operational optimizations are not considered in this study, even though they have become a widely discussed theme due to the growing awareness of the shipping industry's impact on global climate (18) through increased greenhouse gas (GHG) emissions—a result of the rapid growth of global trade (19).

With the above assumptions, the distribution of CPO from the processing plant to the Hub Port involves three alternative transportation modes, each with distinct characteristics and associated transportation costs (20). Data related to these aspects will be sourced from previous research (7). which includes information on emissions, costs, and operational times for each mode involved in CPO distribution. This study will utilize the route along the Mentaya River, with its characteristics as shown in Table 1 (21), while also updating the data to reflect current conditions. Adjustments include setting the operational speed of barge tankers to 8–10 knots (22), and aligning operational costs from earlier research with current inflation rates. All the information is then summarized and presented in Table 2.

Table 1. River Information.

River Name	Length (Km)	Navigable (Km)	Depth (Meters)	Width (Meters)
Mentaya	400	270	6	350

Table 2. Transport Mode Information.

Transport Mode	Distances (km)	Average Speed (kph)	Max Capacity	Travel Cost ($/ton/km)	GHG Emissions (kg CO2-eq)
Light Tank Truck	88	52	7,1 T	0,125	13,811
Barge Tanker (River)	92	14,816	1265 Ton	0,017	4,814
Barge Tanker (Sea)	600	14,816	1265 Ton	0,017	15,980

This research will focus on two transportation alternatives: overland transportation using Light Tank Trucks to Bagendang Port, followed by transshipment via Barge Tanker, and direct Inland Waterways transportation using Barges Tanker to Tanjung Perak Port. The starting point of the multimodal transportation cost model in this study is Kotawaringin Timur Regency, Central Kalimantan Province. The CPO distribution modeling from the plant to the destination is divided into two models: via land to Bagendang Port and then by sea to Nilam Terminal (Tanjung Perak Port), and directly via river to Tanjung Perak Port as shown in Figure 3.

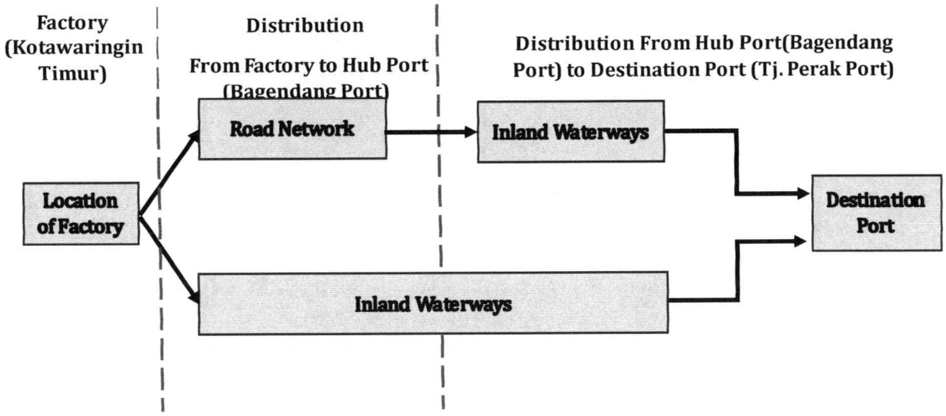

Figure 3. CPO Distribution Models.

3. Results and Discussions

3.1. CPO Transportation Simulation

Two transportation scenarios were projected for CPO transportation from the processing plant in Kotawaringin Timur to the Hub Port and subsequently to the destination port (North Nilam Terminal of Tanjung Perak Port). Time/Cost-Distance model was conducted for both the "road network scenario" and "inland waterways network scenario," involving land-sea and inland waterways-sea transportation. The multimodal simulations were divided into two scenarios: land-sea and river-sea, without considering possible variations in each scenario, and without considering weather factors on road and river conditions. Both simulations yielded calculations for transportation costs and GHG emissions for both road network scenario and inland waterway network scenario.

Table 3. Road Network Scenario Transportation Model of CPO Distribution from Central Borneo to East Java.

Leg	Distance (Km)	Cum. Network Distance (Km)	Cost ($/t)	Cum. Cost ($/t)	Operate Time (Hours)	Cum. Operate Time (Hours)	GHG Emissions (kg CO2-eq./ton CPO)	Cum. GHG Emissions (kg CO2-eq./ton CPO)
Loading on Factory	0,00	0,00	0,00	0,00	1,00	1,00	0,00	0,00
Kotawaringin Timur-Bagendang Port **(RN)**	88,00	88,00	10,99	10,99	1.69	2,69	13,81	13,81
Port Transshipment (Bagendang Port)	0,00	88,00	3.172	14,16	4,00	6,69	1,34	15,15
Bagendang-Surabaya **(RN)**	600,00	688,00	11.07	25,23	40,50	47,19	15,98	31,13
Seaport Transshipment (Nilam Terminal)	0,00	688,00	4,93	30,16	40,36	87,55	1,58	32,71
TOTAL		**688,00**		**30,16**		**87,55**		**32,71**

RN: Road Network

As shown in Table 3, in "road network scenario," for land transportation of 7,1 tons of CPO over 88 km using a single tank truck, the estimated travel time was approximately 1.69 hours. with a loading time of approximately 1 hour, the total transportation time was approximately 2,69 hours. The total loading and distribution cost was estimated at $10,99 per ton of CPO. The corresponding GHG emissions were calculated to be 13,81 kg CO_2-eq. per ton of CPO.

Upon arrival at Bagendang Port, a transshipment process was conducted at the inland port. Located approximately 60 km upstream of the Sampit River near Sampit City, the inland port incurred a transshipment cost of $3,17 per ton for transferring CPO from trucks to barges. The total loading and unloading time at the port, including waiting time, was 4 hours. Additionally, the total GHG emissions from this activity were 1.34 kg CO_2-eq. per ton of CPO (Domnik et al., 2019).

For the "inland waterways network scenario," involving the loading of CPO to the barge tanker at the plant and direct shipment to the North Nilam Terminal of Tanjung Perak Port, a 1,265-ton capacity oil barge towed by a tugboat was used. The barge traversed 92 km (49.68 nautical miles) of inland waterways from the processing plant to Bagendang Port, taking approximately 6.21 hours. Considering a loading time of 1 hour, the total transportation time was approximately 7.21 hours. The total loading and distribution cost was estimated at $1.7 per ton of CPO, with corresponding GHG emissions of 4,81 kg CO_2-eq. per ton of CPO.

At the destination port, a seaport transshipment process was carried out to transfer the CPO from the oil barge to storage tanks, with a cost of $4.93 per ton. The entire loading and unloading process, including waiting time, took 40.36 hours. This operation also produced GHG emissions of 1.58 kg CO_2-eq. per ton of CPO (Domnik et al., 2019). As a result, for the inland waterways network scenario, the total GHG emissions reached 22.37 kg CO_2-eq. per ton, with a total cost of $17.70 per ton and a total operation time of 88.07 hours.

Table 4. Inland Waterway Network Scenario Transportation Model of CPO Distribution from Central Borneo to East Java.

Leg	Distance (Km)	Cum. Network Distance (Km)	Cost ($/t)	Cum. Cost ($/t)	Operate Time (Hours)	Cum. Operate Time (Hours)	GHG Emissions (kg CO_2-eq./ton CPO)	Cum. GHG Emissions (kg CO_2-eq./ton CPO)
Loading on Factory	0,00	0,00	0,00	0,00	1,00	1,00	0,00	0,00
Kotawaringin Timur-Bagendang Port **(IWN)**	92,00	92,00	1,70	1,70	6.21	7,21	4,81	4,81
Bagendang-Surabaya **(IWN)**	600,00	692,00	11.07	12,76	40,50	47,71	15,98	20,79
Seaport Transshipment (Nilam Terminal)	0,00	692,00	4,93	17,70	40,36	88,07	1,58	22,37
TOTAL		**692,00**		**17,70**		**88,07**		**22,37**

IWN: Inland Waterways Network

3.2. Findings in the Study of CPO Shipment from Central Borneo to East Java in the Time/Cost-Distance Model

Based on the findings presented in Tables 3 and 4, the Time/Cost-Distance Model used for evaluating the transportation of Crude Palm Oil (CPO) from Kotawaringin Timur to the North Nilam Terminal at Tanjung Perak Port, Surabaya, has proven effective in modelling not only the time and cost associated with distance but also the GHG emissions relative to distance. This comprehensive model allows for a detailed analysis of how transportation time, cost, and environmental impact are interrelated. To further elaborate on these results, Tables 3 and 4 will be illustrated and discussed through graphical representations, as depicted in Figures 4 and 5. These figures will provide a visual summary of the data and offer deeper insights into the relationships and trends observed in the study.

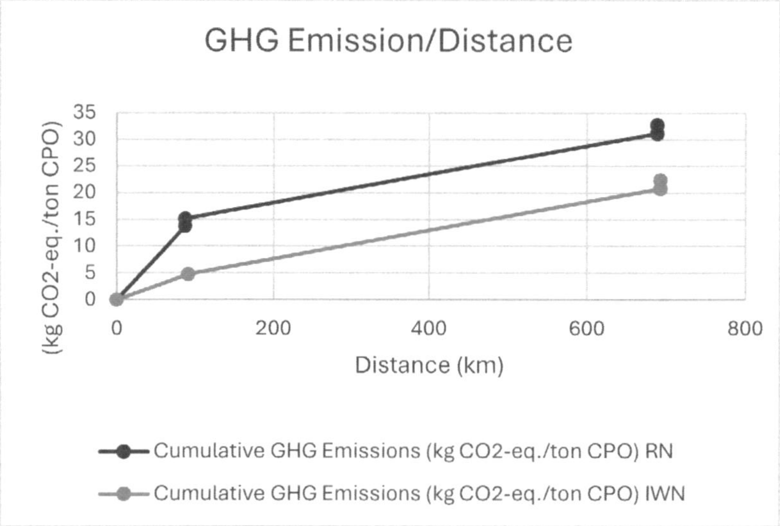

Figure 4. GHG Emissions/Distance Model of CPO Shipment from Central Kalimantan to East Java using Inland Waterways Network and Road Network.

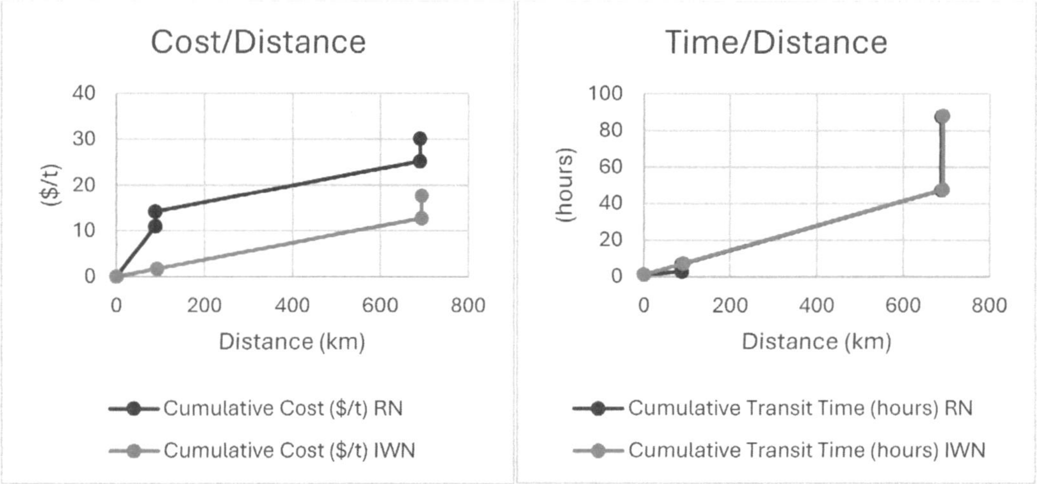

Figure 5. Cost/Distance Model (left) and Time/Distance Model (right) of CPO Shipment from Central Kalimantan to East Java using Inland Waterways Network and Road Network

Based on Figures 4 and 5, the model reveals that the choice of transportation mode plays a crucial role in determining both the environmental sustainability and economic efficiency of Crude Palm Oil (CPO) distribution. Inland waterways transport proves to be a more advantageous option due to its lower GHG emissions and typically lower transportation costs per kilometer, particularly over long distances.

A key benefit of the inland waterways transport scenario is the reduction in the logistics chain facilitated by specialized terminals. These terminals enable the elimination of one logistical leg, namely the transshipment operation connecting road network to inland port, which significantly reduces both transportation costs and GHG emissions. In contrast, the road network scenario does not offer this advantage, resulting in longer logistics chains and higher associated costs and GHG emissions.

The inland waterways network scenario's ability to cut out a segment of the logistics chain highlights its positive impact compared to the road network scenario. This streamlined approach not only enhances operational efficiency but also contributes to better environmental outcomes, demonstrating the substantial benefits of integrating special terminals into the transportation network.

Additionally, at the mode level, transportation modes operating on inland waterways demonstrate lower costs and GHG emissions compared to those operating on road networks. As shown in Table 5 below, this table highlights the comparison of costs and GHG emissions for each mode of transportation per ton and per kilometer transported. The inland waterway transport modes consistently show greater efficiency in terms of cost per ton and GHG emissions, making this option more environmentally friendly and more economically viable compared to road transport modes. This comparison reinforces the findings that river transport provides a more sustainable solution, especially for long-distance CPO distribution.

Table 5. Per Kilometer Calculation Based on Transportation Models of CPO Distribution from Central Borneo to East Java.

Leg	Cost ($/t) per Kilometer		Operational Time (Hours) per Kilometer		GHG Emissions (kg CO_2-eq./ton CPO) per Kilometer	
	RN	**IWN**	**RN**	**IWN**	**RN**	**IWN**
Kotawaringin Timur - Bagendang Port	0,125	0,018	0,019	0,067	0,157	0,052
Bagendang-Surabaya	0,018	0,018	0,067	0,067	0,027	0,027

RN: Road Network

IWN: Inland Waterways Network

4. Conclusion

In conclusion, this study highlights the significant impact that the choice of transportation mode has on the environmental sustainability and economic efficiency of crude palm oil (CPO) distribution. The inland waterways transport scenario emerges as the more favorable option, with total GHG emissions of 22.37 kg CO_2-eq./ton, a cost of $17.70 per ton, and a total operation time of 88.07 hours. In comparison, the road transport scenario results in higher GHG emissions of 32.71 kg CO_2-eq./ton, a cost of $30.16 per ton, and a total operation time of 87.55 hours. The use of specialized terminals in the inland waterways network further enhances its efficiency by eliminating a logistical leg, streamlining the supply chain, and reducing both costs and environmental impacts. In addition, this research is also demonstrating that the Time/Cost-Distance Model provides a comprehensive framework not only for analyzing time and cost but also for assessing GHG emissions, offering valuable insights for comparing transportation modes and scenarios. The findings emphasize the economic and environmental advantages of inland waterway transport, particularly for long-distance CPO distribution in Indonesia.

References

1. Directorate of Food Crops H and ECS. Indonesian Oil Palm Statistics 2022 [Internet]. Directorate of Food Crops H and ECS, editor. Vol. 16. Jakarta: BPS-Statistics Indonesia; 2023 [cited 2024 Nov 29]. Available from: https://web-api.bps.go.id/download.php?f=sNbflCCh1wHu7O0JscL6BGswRS9uTmZNVTY3ak9HV0I5eWVIL1ZjSzNtYXlwcHM0T280eVFzZ2RpUjAvUVRVbWN1MW5WMnFlYld4YlZ0RERuMkNCTzI0eE1wRnVPWDNEVTNSQ1YrbWNlS1hIcGRqY1hGczliTUlGQVloSEdHR1orbEwzTTZRYW9MT2NaVXBkb2pHNWhxSFErODlrY2Q3UnhZMUd0a2N1RkNWVWY2WlgwejB3UWhwL0dvZGJxbmd0czdSbUR0U2gyNnRCRzN2TnhsUlVJSmxMNTlZYU9tU3UxV3RLbkdNNWlFMVk4bGJUUExLbTFUU000aHVsKzVrWTdBUE1ackdnV3JkSG9Pcjc=&_gl=1*1jclj78*_ga*ODk1MjgwNTIxLjE3MzI4NjcwMzM.*_ga_XXTTVXWHDB*MTczMjg2NzAzMy4xLjAuMTczMjg2NzAzMy4wLjAuMA..

2. United States Department of Agriculture Foreign Agricultural Service. Oilseeds: World Markets and Trade [Internet]. 2023 Jan [cited 2024 Nov 29]. Available from: https://downloads.usda.library.cornell.edu/usda-esmis/files/tx31qh68h/k643cc49f/41688v39p/oilseeds.pdf

3. Siahaan M. Statista. 2024. Palm oil industry in Indonesia- statistics & facts. Available from: https://www.statista.com/topics/5921/palm-oil-industry-in-indonesia/#topicOverview

4. Indonesia-Investments. Indonesia-Investments. 2024. Palm Oil. Available from: https://www.indonesia-investments.com/business/commodities/palm-oil/item166

5. GAPKI. GAPKI. 2018. 50 Juta Penduduk Indonesia Sandarkan Hidupnya Pada Kelapa Sawit. Available from: https://gapki.id/news/2018/09/27/50-juta-penduduk-indonesia-sandarkan-hidupnya-pada-kelapa-sawit/#more-6025

6. Lailiah I, Hadi F, Wuryaningrum P. Analysis of Operation Patterns and Design of Crude Palm Oil (CPO) Transportation: A Case Study in Central Kalimantan – East Java [Internet]. [Surabaya]: Sepuluh Nopember Institute of Technology; 2018. Available from: repository.its.ac.id › 55802/1/04411440000048-Undergraduate

7. Dominik T, Kalber S, Leible L, Mahmudah N, Jahn C. Techno-economic assessment and GHG balance of the Indonesian palm oil supply path for europe-a case study of production and transport in Central Kalimantan. In: 27th European Biomass Conference and Exhibition. 2019. p. 27–30.

8. Purnomoasri RD, Handayani D. Analisis dan Mitigasi Emisi Gas Buang Akibat Transportasi (Studi Kasus Kabupaten Magetan). ENVIRO: Journal of Tropical Environmental Research. 2022 Mar 3;24(1):29.

9. Khairina E, Purnomo EP, Malawnai AD. Sustainable Development Goals: Kebijakan Berwawasan Lingkungan Guna Menjaga Ketahanan Lingkungan Di Kabupaten Bantul Daerah Istimewa Yogyakarta. Jurnal Ketahanan Nasional. 2020 Sep 18;26(2):155.

10. Kementerian Perhubungan Republik Indonesia. Peraturan Menteri Perhubungan Republik Indonesia Nomor PM 52 Tahun 2021 Tentang Terminal Khusus dan Terminal Untuk Kepentingan Sendiri. Jakarta; Jun 9, 2021.

11. Hardiani DP. Analisis Faktor-Faktor yang Mempengaruhi Pemilihan Angkutan Sungai di Banjarmasin. Jurnal Teknologi Berkelanjutan (Sustainable Technology Journal) Available on [Internet]. 2016 Apr 1 [cited 2024 Nov 29];5(1):35–41. Available from: http://jtb.ulm.ac.id/index.php/JTB/article/view/67

12. Atikah N, Sutopo W. Simulasi Model Dinamik Pengangkutan Crude Palm Oil (CPO) di PT. XYZ untuk Meminimalkan Biaya Transportasi Pengadaan Bahan. Vol. IX, J@TI Undip. 2014.

13. Beresford AKC, Dubey RC. Handbook on the Management and Operations of Dry Ports [Internet]. Geneva: UNCTAD; 1991. Available from: https://unctad.org/system/files/official-document/rdpldc7_en.pdf

14. Beresford A, Pettit S, Liu Y. Multimodal supply chains: iron ore from Australia to China. Potter A, editor. Supply Chain Management: An International Journal [Internet]. 2011;16(1):32–42. Available from: https://www.emerald.com/insight/content/doi/10.1108/13598541111103485/full/html

15. UN.ESCAP. Transit Transport Issues in Landlocked and Transit Developing Countries [Internet]. Landlocked Developing Countries Series. UN.ESCAP; 2003. Available from: https://hdl.handle.net/20.500.12870/2919

16. Tarigan U, Sidabutar RF, Tarigan UPP, Chen A. Analysis of Optimal Transport Route Determination of Oil Palm Fresh Fruit Bunches from Plantation to Processing Factory. J Phys Conf Ser [Internet]. 2018;1007(1):012028. Available from: https://dx.doi.org/10.1088/1742-6596/1007/1/012028

17. Yu H, Fang Z, Fu X, Liu J, Chen J. Literature review on emission control-based ship voyage optimization. Transp Res D Transp Environ [Internet]. 2021;93:102768. Available from: https://www.sciencedirect.com/science/article/pii/S1361920921000729

18. He Z, Lam JSL, Liang M. Impact of Disruption on Ship Emissions in Port: Case of Pandemic in Long Beach. Sustainability [Internet]. 2023;15(9). Available from: https://www.mdpi.com/2071-1050/15/9/7215

19. Balcombe P, Brierley J, Lewis C, Skatvedt L, Speirs J, Hawkes A, et al. How to decarbonise international shipping: Options for fuels, technologies and policies. Energy Convers

Manag [Internet]. 2019;182:72–88. Available from: https://www.sciencedirect.com/science/article/pii/S0196890418314250

20. Mahmudah N, Parikesit D, Malkhamah S, Priyanto S, Zuidgeest M. Modeling Freight Transportation for Crude Palm Oil (CPO) in Central Kalimantan. In: EACEF 2011 Designing and constructing in sustainability : proceedings of the 3rd international conference of European Asian Civil Engineering Forum [Internet]. Yogyakarta: Universitas ATMA; 2011 [cited 2024 Nov 29]. p. 49–54. Available from: https://www.researchgate.net/publication/274137924

21. BPS-Statistics of Kalimantan Tengah Province. Kalimantan Tengah Province in Figures 2023. BPS-Statistics of Kalimantan Tengah Province, editor. BPS-Statistics of Kalimantan Tengah Province; 2024.

22. Nugroho KR, Hasanudin H. Desain Kapal Multipurpose untuk Pelayaran Sungai Mentaya, Sampit, Kalimantan Tengah. Jurnal Teknik ITS [Internet]. 2020;9(1):G17–22. Available from: http://ejurnal.its.ac.id/index.php/teknik/article/view/51451

Weather-Based Ship Route Planning by Considering Added Resistance for Improving Fuel Efficiency

Yuda Apri Hermawan*, Lista Putri Adinda Rahmi, Raden Sjarief Widjaja, Totok Yulianto

Departement of Naval Architecture, Faculty of Marine Technology, Institut Teknologi Sepuluh Nopember

*E-mail: yuda.apri@its.ac.id

Abstract. Maritime authorities are advancing policies to reduce greenhouse gas emissions and environmental pollution by regulating fuels and lubricants. Enhancing ship operational efficiency is a key strategy for reducing fuel consumption and maintaining competitiveness in the maritime industry. This study addresses environmental concerns through route optimization based on fuel efficiency by incorporating weather conditions. The proposed model assesses ship resistance influenced by current, waves, and wind, integrating these factors to identify the most fuel-efficient routes. This study focuses on the implementation of weather routing in the eastern waters of Indonesia, with a case study involving ships navigating through adverse weather conditions. The research demonstrates that under bad weather, the potential increase in resistance can reach up to 18%, depending on the specific conditions. Additionally, the study proves that route optimization provides effective route recommendations based on the total resistance probability encountered. Modified Djikstra algorithm, reduced total resistance from 9.1 kN to 7.4 kN, reflecting a 10% improvement in fuel efficiency. By ranking routes based on minimized fuel consumption, the study provides a practical approach to achieving higher fuel efficiency and contributing to more sustainable maritime operations.

Keywords : Weather Routing, Fuel Efficiency, Ship Resistance, Climate Change, and Pollution

1. Introduction

International Maritime Organization (IMO), 90% of global trade is conducted via sea routes to support the export and import of goods, driving modernization efforts. This has led to an increase in the global fleet, with a reported 35% growth from 2015 to 2019 and an even larger increase of 51.5% compared to 2010 based on Equasis, 201. In response, the IMO and other maritime authorities are implementing fuel usage policies as a preventive measure to reduce greenhouse gas emissions and environmental pollution (1). Environmental pollution from the maritime industry is a critical issue (2), leading to the concept of "Green Shipping" or "Sustainable Shipping," which aims to use minimal energy for shipping goods as a contribution to improving current environmental conditions (3).

Regulatory developments are also being made to meet targets. In 2020, the IMO revised its emission reduction regulations, setting new targets to achieve a 20% reduction by 2030, a 70%

reduction by 2040, and net-zero emissions by 2050, all compared to 2008 emission levels. This replaced the previous target of a 70% reduction by 2050 and a 50% reduction in total annual emissions, first established in the 2018 GHG strategy based on IMO recapt in 2020. One solution continuously developed to address this issue is improving ship operational efficiency to reduce fuel oil consumption (4). Weather routing is a concept that involves planning a ship's route while considering weather conditions along the planned path (5). With a shorter distance, theoretically, fuel consumption and emissions are reduced. Weather-based route selection also helps predict accurate arrival times (6). By combining weather prediction data and optimization algorithms, weather routing opens opportunities for improved ship route management with a new technological perspective (1). The program's analysis includes several important calculations, such as ship resistance, hydrodynamic, and aerodynamic factors, considering speed and angle of sea currents and winds for accuracy (7).

This paper explores the use of weather routing to optimize ship fuel consumption on various shipping routes within Indonesia. The study aims to find the closest route that accounts for weather impacts on vessel resistance. A route with the lowest resistance will result in lower emissions, thereby contributing to environmental protection. The research selected a specific route as a case study due to its strategic importance and the challenges posed by varying weather conditions in the region. For the optimization process, the Djikstra algorithm was employed, known for its efficiency in pathfinding and route optimization.

2. Literature Review

Summarized from (8), multi-objective genetic algorithms are commonly used to solve ship routing optimization problems by optimizing routes and speed profiles while considering dynamic weather conditions. Although this technique is effective in generating Pareto-optimal solutions that help operators choose the best route, the computational complexity and the need for accurate data pose challenges. Three-dimensional (3D) dynamic programming is also used to optimize ship speed and heading angles, offering optimal solutions but with high algorithm complexity. Other weather-based optimization methods, such as shortest path algorithms and modified isochrone methods, are also applied to accommodate changing weather conditions, but they require significant computational resources and highly accurate weather data to provide optimal results. Fuel efficiency is a key factor in shipping to reduce costs and environmental impact. Research and practical experience indicate that fuel efficiency measures can offer varying levels of savings.

- Weather routing can reduce CO2 and black carbon emissions by 2% to 10%.
- Weather routing service providers often present more optimistic figures that may not always be achieved in practice.
- Alternative routes could reduce fuel consumption by up to 15% by avoiding high waves.

The variation in weather routing optimization results can range from 1% to 30% fuel and energy savings. To achieve optimal results, strategies such as employing advanced technologies, operational adjustments, continuous monitoring systems, and collaborating with reliable service providers are recommended (9).

IOP Conf. Series: Earth and Environmental Science **1461** (2025) 012028 doi:10.1088/1755-1315/1461/1/012028

3. Metodology

Weather routing optimization in shipping considers four main aspects: ship data (vessel parameters like speed and fuel consumption), weather data (current and forecasted conditions such as wind and waves), safety restrictions (navigational safety measures and exclusion zones), and navigational information (routes, ports, and aids). Combining these factors helps in planning efficient routes that improve fuel efficiency, reduce environmental impact, and ensure safety. To achieve an optimal route based on this optimization goals, several methods are considered in the execution of weather routing, as shown in **Figure 1.** below (9).

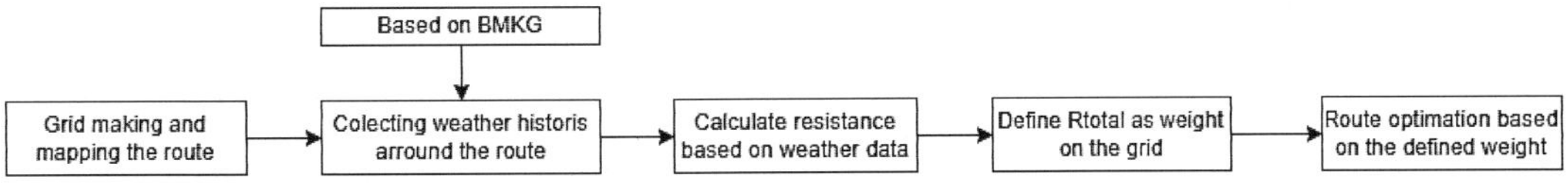

Figure 1. A schematic overview of the proposed approach

3.1 Weather Integration

The weather integration process in optimization involves collecting and preprocessing historical weather data, including significant wave height, wave mean period, primary swell, current direction, current speed, wind speed, and wind direction. This data is then mapped onto the geographical grid used in the optimization model, allowing the algorithm to dynamically adjust routes based on real-time or historical weather conditions. The goal is to find the most efficient and safe path, considering both the ship's capabilities and environmental factors, thereby enhancing route planning accuracy and reliability.

3.2 Fuel Consumption

Accurate weather forecasting can significantly reduce sailing risks and result in savings in terms of time and costs. The main reason is that bad weather can increase fuel consumption, especially on older and less efficient ships. Previous research has shown that Meteorology and Oceanography (METOC) systems can save up to 20% of fuel. Short-term forecasts are generally accurate for one to two weeks, while long-term predictions can cover periods of several years, with the impact of ocean current data often becoming apparent in the long term (10).

This formula is used to estimate the amount of fuel required for a ship to complete a voyage, taking into account the engine's power, fuel efficiency, distance to be traveled, and the ship's speed. It helps in planning and optimizing fuel usage, which is crucial for both cost efficiency and environmental impact in maritime operations. The key factor in calculating fuel consumption for routing in this program is the resistance encountered along the selected route. The following resistances are calculated in this program.

3.3 Resistance

$$R_{ts_{Calm,current}} = \frac{1}{2}\rho L d U_a^2 X'_{uu} + \frac{1}{2}\rho S U_a^2 C_w + \frac{1}{2}\rho S U_a^2 \Delta C_f \quad (1)$$

In his research on Weather Routing, (7) summarizes a concept for calculating resistance in calm waters, including the effects of currents. This equation is used due to its good accuracy in estimating resistance in still water conditions. The equation takes into account both the speed components and the angular direction of the variables involved in the calculation.

IOP Conf. Series: Earth and Environmental Science **1461** (2025) 012028 doi:10.1088/1755-1315/1461/1/012028

$$R_{wind} = \frac{1}{2}\rho_a A_T V_{wr}^2 C_{wx}(\chi'_{wind}) \tag{2}$$

Equation (2) is the calculation for wind resistance. The advantage of this method lies in its ability to provide accurate estimates of the impact of wind on a ship. By incorporating important factors such as air density, the area affected by the wind, and the square of the relative wind speed, this formula enables a detailed analysis of wind resistance (11).

$$\bar{R}_{aw} = 2\int_0^\infty C_{aw}(\omega_e)S_\zeta(\omega_e)d\omega_e \tag{3}$$

Equation (3) calculates the total value of added wave resistance experienced by a ship under irregular wave conditions based on regular condition (12). The integral in the equation sums the contributions to resistance from all wave frequencies present in the irregular wave spectrum, ranging from zero frequency to infinity (13). This provides a comprehensive view of how different waves affect the additional resistance encountered by the ship.

4. Study Case

4.1 Input Data

On March 14, 2024, there was a cancellation of the ferry sailing. This closure was due to a forecast from local Meteorology and Oceanography (METOC) systems from East Nusa Tenggara (NTB), that has been predicted the potential for high waves (2.50 to 4.00 meters) in areas such as the Southern Waters of Banten to Rote Island, the southern parts of the Bali-Alas-Lombok Strait, the western part of the Sumba Strait, and the southern part of the Sawu Sea. This research focuses on the ferry route across the Sawu Sea, which is well-known for its high waves, especially during the rainy season. The chosen destination for this study is the Waingapu Port (East Sumba) – Ende Port (Flores) route as shown in **Table 1** below, which crosses the Sawu Sea. This route is one of the key routes that connects and plays a vital role in the economic activities of East Nusa Tenggara. Given its moderate distance, this route is considered ideal for a study on route optimization to demonstrate the significant impact of high waves on recommended conventional route changes.

Table 1. Port's Coordinate

Port	Coordinate Reference System (CSR)	
	Latitude	Longitude
Start : Waingapu Port	-9.6662	120.2734
Finish : Ende Port	-8.8594	121.6356

[a] Coordinate Reference System (CSR) is a system used to define how the coordinates of a point on the earth are projected onto a map or digital model.

In addressing this case study, collecting historical data over a specific period is crucial. This historical data includes various information such as weather conditions, wave heights, wind speeds, and directional angles. This data will be the primary material used to analyze fuel consumption for ships encountering high waves. This information is necessary to evaluate how extreme sea and weather conditions affect fuel consumption. High waves and significant wind speeds can increase the resistance and operational load on a ship, which in turn affects fuel efficiency. This approach is also anticipated to provide positive and lasting effects, particularly concerning environmental issues (14).

Since the route used as the subject of the study is an economically significant route, a commercial vessel was chosen as the subject. The fuel consumption of the XYZ passenger ship is analyzed through calculations of the existing resistance. The ship has the following characteristics as shown in **Table 2.** below.

Table 2. Ship's Main Dimension

Main Dimension	Unit
LOA (Length of All)	55.5 m
B (Breadht)	12 m
H (Height)	3.2 m
T (Draft)	2.4 m

[a] Samples of the main dimensions that can be representative of the vessels on the route that selected for study

4.2 Route Optimization

In optimizing weather routing to minimize fuel consumption, constraints like Sea State are crucial for defining wave height and its impact on the ship's performance. Sea State provides a numerical measure of wave conditions, enabling the optimization algorithm to account for how different wave heights affect fuel efficiency and navigation. The selection of the appropriate algorithm is also vital, as it must effectively handle these constraints and data inputs.

4.2.1 Sea State

Table 3. Significant Wave Height by WMO

Sea state Code	Significant Wave Height (m)		Description
	Range	Mean	
0	0	0	Calm (glassy)
1	0.0-0.1	0.05	Calm (rippled)
2	0.1-0.5	0.3	Smooth (waveless)
3	0.5-1.25	0.875	Slight
4	1.25-2.5	1.875	Moderate
5	2.5-4.0	3.25	Rough
6	4.0-6.0	5	Very Rough
7	6.0-9.0	7.5	High
8	9.0-14.0	11.5	Very High
9	Over 14	Over 14	Phenomenal

[a] The sea state range is used as a parameter to define wave height

When a ship sails at sea, various external forces such as waves, currents, depth and area of the water, and wind impact its movement. These conditions are expressed through different levels of "Sea State," where higher Sea State levels indicate more severe wave conditions (15). Sea State conditions refer to standards established by the World Meteorological Organization (WMO). Sea

IOP Conf. Series: Earth and Environmental Science **1461** (2025) 012028 doi:10.1088/1755-1315/1461/1/012028

State is a numerical reference commonly used to measure and describe the severity of ocean conditions, it can be seen in **Table 3** above.

4.2.2 Djikstra Algorithm

In this study, the main focus is to determine the most fuel-efficient shipping route, aiming to minimize fuel consumption and operational costs. To achieve this goal, the study employs the Djikstra algorithm to optimize the route. This algorithm offers advantages in terms of processing speed and probability, making it suitable for complex and dynamic route calculations. The algorithm uses a queue to track path costs from the start node to neighboring nodes. It selects the node with the lowest path cost at each step, making it a "greedy algorithm." This greedy approach limits the number of nodes explored, giving Dijkstra an advantage in time complexity over the Bellman-Ford algorithm. While Dijkstra's algorithm cannot handle negative edge weights, this is not a significant issue for ship route optimization, as negative weights are not present in shipping costs. Therefore, Dijkstra's algorithm is more efficient and suitable for this study (16). The steps of the Dijkstra algorithm are as follows:

1. Set the distance d(node)d(node)d(node) to infinity for all nodes ∈N\in N∈N in graph GGG.
2. Initialize d(s)=0d(s) = 0d(s)=0 for the source node sss and add sss to the Queue with its distance value.
3. Extract the node uuu with the minimum distance from the Queue.
4. Perform edge relaxation at node uuu and add newly discovered nodes to the Queue.
5. Repeat steps 3 and 4 until the path to the goal node ttt is found.

Dijkstra's algorithm finds the shortest path in a graph by initializing all nodes with infinite distances, except the source node set to zero. It iteratively selects the node with the smallest distance, relaxes its edges, and updates distances for neighboring nodes. This systematic approach ensures efficient route optimization and is widely applicable for solving routing and optimization problems.

5. Result & Discussion

5.1 Mapping

Table 4. Mapping Syntax

Python syntax for mapping

```
# GeoDataFrame from latitude & longitude
    df = gpd.GeoDataFrame({'latitude': latitude,'longitude': longitude},
      geometry=gpd.points_from_xy(longitude, latitude))
# Plot grid in scale 1:10.000
    grid_size = 10_000  # 10 km grid
    grid_x = range(int(x_min), int(x_max) + grid_size, grid_size)
    grid_y = range(int(y_min), int(y_max) + grid_size, grid_size)
```

[a] The GeoDataFrame in this code is used to store location point data (longitude and latitude) in geospatial format. Then, the code forms a grid with a scale of 1:10.000, which can be used to carry out further analysis, such as mapping, grouping spatial data, or specific grid analysis of the data.

IOP Conf. Series: Earth and Environmental Science **1461** (2025) 012028 doi:10.1088/1755-1315/1461/1/012028

Mapping is a fundamental aspect of routing ships in relation to weather conditions. In practice, mapping ensures that geographic maps and weather data providers yield consistent interpretative results. In this study, a scale of 1:10,000 is used for route prediction. The choice of scale is crucial; a scale that is too coarse may result in lengthy computation times but high accuracy, while a scale that is too fine may expedite computations at the expense of precision. This scaling decision accounts for the integration with available weather data providers, with the assumption that each grid cell corresponds to a specific weather condition as predicted by Indonesian Meteorology and Oceanography (METOC) systems. Syntax that has been used for collecting weather data is shown in **Table 5** below.

Table 5. Syntax for collecting weather data

Python syntax for collecting weather data

```
data = {
        "model": "w3g_hires",
        "modelrun": "2024-08-29T00:00:00Z",
        "lon": -9.6662,
        "lat": 120.2734,
        "var": ["wspd", "wdir", "hs", "t01", "dir", "ptp00", "ptp01", "phs01", "phs00"]
```

[a] The Python syntax above is used to collect weather data, where the variables represent the desired data points: "wspd" refers to wind speed, "wdir" refers to wind direction, and other variables like "hs" represent significant wave height, "t01" is wave period, and "dir" is wave direction, among others.

5.2 Resistance Result

Shipping routes in the eastern part of Indonesia are known for their potential of high waves. The Indonesian Agency for Meteorology, Climatology, and Geophysics, the national weather forecasting agency, frequently issues warnings about this. The Sawu Sea, located between Timor, Sumba, and Flores islands in Indonesia, is known as a region prone to extreme weather, especially during the rainy season (October to March), when the southwest monsoon brings severe weather, including strong winds and high waves. The Sawu Sea is not exempt from the potential for high waves; however, during normal weather, several areas are actually quite safe for navigation, it is predicted to be around 0.7-1.5 meters.

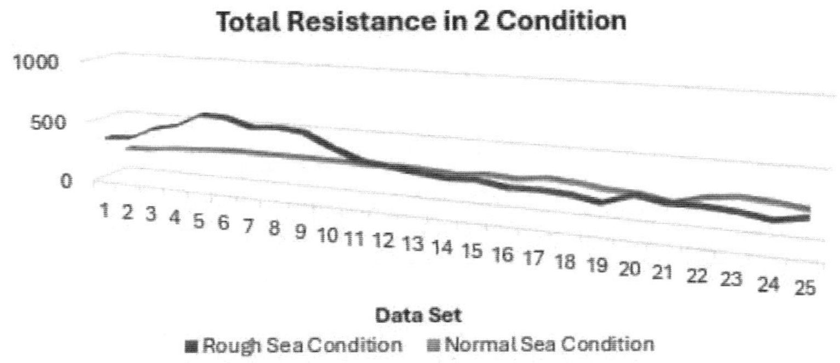

Figure 2. Total Resistance in Sea State

But apart from that, the case study raises the condition where the Sawu Sea is in a major storm condition (Rough Condition). From the weather data that has been obtained, several data represent a storm in the Sawu Sea. The graph in **Figure 2.** above clearly illustrates a significant difference in total resistance between stormy and normal sea conditions, with a notable discrepancy of 18%. The national weather forecasting agency consistently issues warnings during high wave conditions, as evidenced by the graph, where resistance due to adverse weather is markedly elevated. This presents serious safety concerns, as Indonesian vessels, particularly roll-on/roll-off (ro-ro) ferries, are not designed to withstand such severe wave conditions on a regular basis.

5.3 Optimization

Table 6. Optimation setting

Python syntax for the edges weight

```
for neighbor, dist in nearest_neighbors:
        Rtotal_i = G.nodes[str(row_i['Grid_ID'])]['Rtotal']
        Rtotal_j = G.nodes[neighbor]['Rtotal']
        weight = Rtotal_i + Rtotal_j
        G.add_edge(str(row_i['Grid_ID']), neighbor, weight=weight)
weight = Rtotal_i + Rtotal_j
next_node = min(unvisited_neighbors, key=lambda n: G[current_node][n]['weight'])
```

[a] The syntax calculates the edge weights between two nodes based on the sum of their Rtotal attributes and selects the next node with the lowest weight to build or update the graph.

In this research, a modified Dijkstra algorithm was implemented by incorporating the Rtotal value (Total Resistance) as an additional weight for each node, alongside the physical distance between nodes as shown in **Table 6**. The edge weight is calculated as the sum of the Rtotal values of the two connected nodes, prioritizing paths with lower Rtotal value. The algorithm selects the nearest neighbors based on a combination of distance and Rtotal, making it more efficient by considering factors beyond just distance when determining the optimal path. This adjustment to the traditional Dijkstra algorithm allows for a more nuanced approach to route optimization, particularly when additional constraints or costs are involved. It underscores the importance of integrating real-world factors into navigational strategies, such as environmental conditions or resource usage.

In conclusion, this study highlights the effectiveness of weather routing in optimizing ship navigation under adverse weather conditions. By analyzing the case of a vessel assumed to sail through bad weather, it was shown that navigating in such conditions poses significant challenges, particularly in terms of safety and increased resistance, requiring vessels to boost power to maintain speed. The findings reveal that route selection, informed by weather patterns, can offer valuable recommendations to minimize the resistance encountered. The visual results, where nodes are colored based on total resistance (Rtotal), clearly illustrate that darker nodes signify higher resistance.

The modified Dijkstra algorithm successfully provided route recommendations by connecting the nearest nodes with the least resistance, resulting in paths with the lowest total resistance. From a data set where the probability of encountering resistance was 9.1 kN, the optimized route reduced the total resistance to 7.4 kN, reflecting a 10% improvement in efficiency. This demonstrates the significant potential of weather-based route optimization in enhancing fuel efficiency and minimizing environmental impact. This improved route optimization is particularly relevant in scenarios where weather conditions or other factors contribute to resistance along the route. **Figure 3** show the visualization of the study's result.

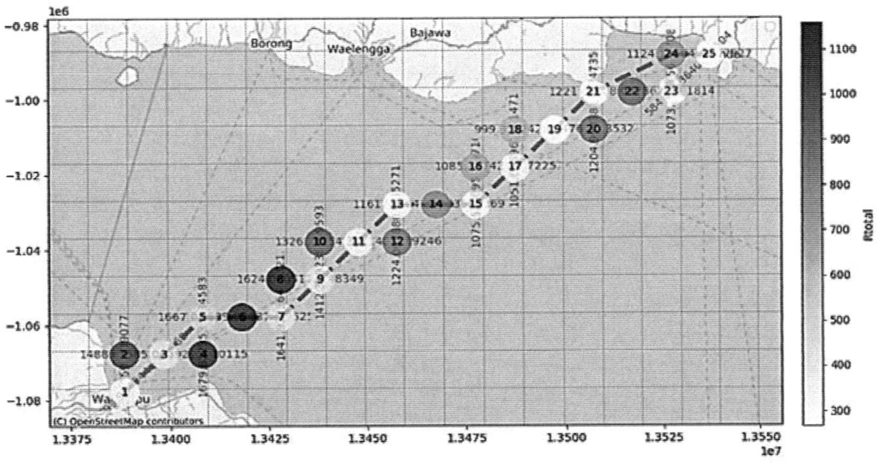

Figure 3. Optimation Result based on weather conditions

The route from Waingapu to Ende can be optimized using an algorithm that takes Sea State as a key parameter. The optimized route differs significantly from the conventional one, as it is not entirely straight. This is because the algorithm accounts for resistance, which is represented by a grid of the map. The shown nodes (labeled 1, 2, 3, untill 25) represent the center of gravity from the main grid, while the red numbers at the edges of the grid indicate the total resistance from one node to the next. The pattern formed represents the task that the optimization algorithm must complete. By choosing lower total resistance, the modified algorithm also translates directly to a reduction in the environmental impact of maritime operations. The resistance encountered on the route serves as a proxy for fuel consumption; therefore, a decrease in resistance equates to more efficient fuel usage. This underscores the substantial benefits of incorporating weather-based considerations into route planning, ultimately contributing to more sustainable and cost-effective navigation solutions.

6. Conclusion

In this study, weather routing was applied to Indonesian waters using the Dijkstra algorithm, with a case study focusing on routes experiencing adverse weather conditions. The selection of this route was intended to demonstrate the challenges of navigating during poor weather and to highlight the inefficiencies associated with such conditions. The findings revealed that resistance due to weather conditions could differ by up to 18%, significantly impacting operational aspects such as fuel consumption and safety.

The research confirmed that optimization through weather routing could provide route recommendations that minimize total resistance. The algorithm's adjustments led to a reduction in total resistance by 10% from the given probability, illustrating its effectiveness. This improvement underscores the potential of weather-based route optimization in enhancing operational efficiency, reducing fuel consumption, and improving safety. By integrating weather factors into route planning, the study highlights the benefits of optimizing navigation in challenging conditions, ultimately contributing to more cost-effective and safer maritime operations.

IOP Conf. Series: Earth and Environmental Science **1461** (2025) 012028 doi:10.1088/1755-1315/1461/1/012028

References

1. Perera LP, Soares CG. Weather routing and safe ship handling in the future of shipping. Ocean Engineering. 2017 Jan;130:684–95.
2. Acomi N, Acomi OC. Improving the Voyage Energy Efficiency by Using EEOI. Procedia Soc Behav Sci. 2014 Jul;138:531–6.
3. Tadros M, Ventura M, Soares CG. Review of current regulations, available technologies, and future trends in the green shipping industry. Ocean Engineering. 2023 Jul;280:114670.
4. Gkerekos C, Lazakis I. A novel, data-driven heuristic framework for vessel weather routing. Ocean Engineering. 2020 Feb;197:106887.
5. Simonsen MH, Larsson E, Mao W, Ringsberg JW. State-of-the-Art Within Ship Weather Routing. In: Volume 3: Structures, Safety and Reliability. American Society of Mechanical Engineers; 2015.
6. Grifoll M, Martorell L, Castells M, de Osés FXM. Ship weather routing using pathfinding algorithms: the case of Barcelona – Palma de Mallorca. Transportation Research Procedia. 2018;33:299–306.
7. Idomoto I. Study on Improving the Practicality of Optimal Route Selection System for Ships. [Kyushu]: Universitas Kyushu; 2014.
8. Pennino S, Gaglione S, Innac A, Piscopo V, Scamardella A. Development of a New Ship Adaptive Weather Routing Model Based on Seakeeping Analysis and Optimization. J Mar Sci Eng. 2020 Apr 10;8(4):270.
9. Walther L. Development of a Weather Routing System for Analysis and Optimization of Ship Voyages . 2021.
10. Zis TPV, Psaraftis HN, Ding L. Ship weather routing: A taxonomy and survey. Ocean Engineering. 2020 Oct;213:107697.
11. Fujiwara T, Ueno M, Futamura T. stimation of Wind Pressure Acting on a Ship's Hull. Journal of the Japan Society of Naval Architects and Ocean Engineers. 1998;(183).
12. Bhattacharya R. Dynamics of Marine Vehicles. 1978.
13. Liu S, Papanikolaou A. Improvement of the prediction of the added resistance in waves of ships with extreme main dimensional ratios through numerical experiments. Ocean Engineering. 2023 Apr;273:113963.
14. Zaccone R, Ottaviani E, Figari M, Altosole M. Ship voyage optimization for safe and energy-efficient navigation: A dynamic programming approach. Ocean Engineering. 2018 Apr;153:215–24.
15. Kusumah YW, Sulisetyono A, Putranto T. Analisis Pengaruh Gerakan Kapal Terhadap Added Wave Resistance Studi Kasus : Kapal Corvette. Jurnal Teknik ITS. 2018 Mar 29;7(1).
16. Shin YW, Abebe M, Noh Y, Lee S, Lee I, Kim D, et al. Near-Optimal Weather Routing by Using Improved A* Algorithm. Applied Sciences. 2020 Aug 30;10(17):6010.

IOP Conf. Series: Earth and Environmental Science **1461** (2025) 012029 doi:10.1088/1755-1315/1461/1/012029

Study of the Vehicle Loading System on the Ro-Ro Route Bajoe - Kolaka Ferry

Dahlia Dewi Apriani[1], Arnaldy Achmadita[2]

[1,2]Politeknik Pelayaran Barombong, Makassar, Indonesia

E-mail: [1]dahlia@poltekpelbarombong.ac.id, [1]dahliadewiapriani8@gmail.com, [2]achmadita.arnaldy@gmail.com

Abstract. Efforts to ensure safe ro-ro ferry shipping can be done by carrying out the correct loading process based on Ministry of Transportation Regulation Number: PM. 115 of 2016 concerning Procedures for Transporting Vehicles on Ships and PM. 30 of 2016 concerning Obligations for Fastening Vehicles on Ferry Transport Ships, that vehicles on board must be tied (lashed) and have a distance between the vehicle and/or the ship's walls. Vehicle tie-downs are intended to ensure that the vehicle does not move when the ship is sailing due to waves or strong winds so that the ship and its cargo can sail safely to their destination. The author examines the vehicle loading system on the ro-ro ferry Bajoe – Kolaka route. The Bajoe – Kolaka route was chosen as the research location because this crossing is one of the main commercial route connecting South Sulawesi with Southeast Sulawesi. This research uses a comparative descriptive method to describe the vehicle loading system on ro-ro ferries operating on the Bajoe - Kolaka route by comparing the existing conditions with loading procedures based on applicable regulations, and also examining the factors that influence the implementation of loading regulations . vehicle on board. From the results of the analysis it is known that the factors that influence the implementation of the vehicle loading system on ro-ro ferries based on established regulations are activity supervision, obstacles to implementing standard operating procedures, knowledge of ship crew and implementation of activity supervision . The better the supervision of activities, the avoidance of obstacles to the implementation of standard operating procedures the knowledge of ship crews improved, the more effective the implementation of the vehicle loading system will be.

1. Introduction

Efforts to realize the safety of ro-ro ferry shipping can be done by carrying out the correct loading process based on the Regulation of the Minister of Transportation Number: PM. 115 of 2016 concerning Procedures for Transporting Vehicles on Ships and PM. 30 of 2016 concerning Obligations for Vehicle Fastening on Ferry Transport Ships, that vehicles on board must be tied (lashing) and have a distance between vehicles and/or ship walls. Vehicle lashing is intended so that vehicles do not move when the ship is sailing due to waves or strong winds so that the ship and its cargo can sail safely to their destination. In addition to these national regulations, vehicle lashing on ships is also regulated in International regulations issued by the International Maritime Organization (IMO), including: Article 5 of the Safety of Life at Sea (SOLAS) Convention requires that vehicles transported on ships must be tied securely to prevent movement that could endanger the stability of the ship or the safety of passengers and the Code of Safe Practice for Cargo Stowage and Securing (CSS Code) which contains information on ensuring the safety of arranging and lashing cargo on ships, including vehicles. Sets standards for lashing equipment, securing procedures and checks during sea voyages, as well as IMO Resolution A.581(14) - Guidelines for Securing Arrangements for the Transport of Road Vehicles on Ro-Ro Ships which are specific guidelines for Ro-Ro ships that specifically deal with the lashing of land vehicles on this type of ship.

Several ship accidents that occurred due to the failure to have the vehicles lashed on board, namely the sinking of KM. Senopati Nusantara operated by the shipping company PT. Prisma Vista, this ro-ro ferry weighs 2,718 GT and has a passenger capacity of 1250 people. The initial suspicion of the sinking of this ship was due to bad weather and the release of vehicle lashings, especially large trucks, so that the vehicles were thrown over the maximum limit which caused the ship to capsize (Z Ariani, A Hendra, S Febriary, 2018), KMP. Saluang which capsized at the Sungai Ayak - Sungai Asam crossing pier, Sungai Ayak Village, Belitang Hilir District, Sekadau Regency On February 22, 2019, the capsizing of KMP. Bili which occurred on February 20, 2021 which caused all four-wheeled vehicles and 34 motorbikes to sink (Rizqi, Amalia, 2022), the most recent incident occurred in July 2023 in the waters of Ketapang, a truck loaded on the KMP ship. Dharma Ferry 1 capsized due to bad weather and high waves hitting the ship and in August 2023 a truck loaded with soybeans capsized while sailing from the Ketapang ferry port to the Gilimanuk ferry port using the KMP. Nusa Makmur, resulting in the crushing of 6 motorbikes because they were hit by the overturned truck due to not being lashed (detikbali, 2023).

In relation to the above, the author examines the vehicle loading system on the ro-ro ferry Bajoe – Kolaka route. The Bajoe - Kolaka route was chosen as the research location because this route is one of the main commercial crossings connecting South Sulawesi with Southeast Sulawesi. There are 9 (nine) ships operating on this route with operating hours starting from 10:00 to 21:00 WITA. This study uses a comparative descriptive method to describe the vehicle loading system on the ro-ro ferry operating on the Bajoe - Kolaka route by comparing the existing conditions with the loading procedures based on applicable regulations, and also examining the factors that influence the implementation of vehicle loading regulations on ships.

It is hoped that by knowing the existing conditions of loading vehicles on ships and the factors that influence the implementation of lashing vehicles on board, a priority government program can be taken to overcome ship accidents due to failure to tie up vehicles on board. It is also hoped that this research can add to the knowledge of the government, sailors, academics and other parties related to shipping safety, especially for ferry crossings using ro-ro ship in Indonesia.

2. Method

Based on the objectives to be achieved, the method used is a quantitative descriptive method with linear regression analysis, data processing is carried out using the SPSS application. This method was chosen to describe the existing conditions of the vehicle loading system on the ro-ro ferry on the Bajoe - Kolaka route and to analyze the dependent and independent variables to determine the factors that influence the implementation of vehicle loading regulations on the ro-ro ferry.

2.1 Data collection technique

The data used in this study are primary data and secondary data with the following data collection techniques:

2.1.1 Primary Data

Primary data is obtained using the following techniques:

a. Observation

Observation techniques are carried out by directly observing and recording the way the ship's crew loads vehicles, the equipment available on board and the way the ship's crew ties up vehicles (lashing) on board.

b. In-depth Interview(in-depth interview).

In-depth Interview Technique is conducted by using or preparing a questionnaire that has been prepared in advance to be answered by respondents. The questionnaire in this study is closed, meaning that respondents can only answer statements by choosing one of the available alternatives. The statements presented are qualitative, for data analysis purposes. The score is given 4,3,2,1 which is known as the Likert scale with score determination.

The scoring is determined by:

1) Answer Strongly Agree (SS) with a score of 4
2) Answer Agree (S) with a score of 3
3) Disagree (TS) answer with a score of 2
4) Strongly Disagree (STS) answer with a score of 1

2.1.2 Secondary Data

Secondary data is obtained by library study method. This data collection technique is to obtain theories and information that are closely related to the research material conducted by studying reference books, previous studies, regulations and other media.

2.1.3 Sampling Determination Techniques

The population in this study were ro-ro ferry ships operating on the Bajoe - Kolaka route, totaling 7 units and all crew members on duty on the ro-ro ferry ships whose positions consisted of Captain, First Mate, Second Mate, KKM, Second Engineer, Third Engineer, Helmsman, Oiler and Class. Because the number of ships operating was only 7 units, all of these ships became samples in this study.

The number of respondents to this study is as follows:

Table 1. Number of Respondents

No	Ferry Name	Number of Respondents
1	KMP. Masagena	17 people
2	KMP. Mishima	15 people
3	KMP. Earth City	16 people
4	KMP. Mandala Nusantara	16 people
5	KMP. Prime Nusantara	15 people
6.	KMP. Fais	16 people
7.	KMP. Muna City	19 people
Total		114 people

IOP Conf. Series: Earth and Environmental Science **1461** (2025) 012029 doi:10.1088/1755-1315/1461/1/012029

2.2 Analysis Design and Hypothesis Testing

Measurement of variables in this study using attitude measurement with the Likert type method that produces interval data. This Likert scale will measure and describe variables into variable indicators, then the indicators are used as benchmarks for compiling instrument items that can be statements or questions.

Table 2. Measurement of Variables

No	Variables	Variable Indicator	Total	Scale
1	Activity Monitoring (X1)	1. Ensure you get vehicle weight information 2. Ensure vehicles are weighed before loading 3. Ensure the vehicle is equipped with load information 4. Ensure that vehicles loaded onto the ship are in accordance 5. Ensure vehicles are positioned according to the loading plan 6. Ensure vehicles are secured in all conditions while on board. 7. Ensure that the vehicle is placed in the correct position	7	Intervals
2	Barriers to SOP Implementation (X2)	1. The occurrence of friction 2. Long and complicated organizational management or coordination flow 3. Organizational adherence to existing regulations 4. Increasing consumer needs 5. Bad weather or natural conditions 6. Lack of control 7. The assumption that SOPs make work more difficult or inflexible	7	Intervals
3	Crew Knowledge (X3)	1. Not yet understanding the purpose and benefits 2. Limited knowledge of existing regulations or systems 3. Inadequate capabilities and competencies	3	Intervals
4	Implementation of Vehicle Loading System (Y)	1. Technical calculation instructions 2. Loading limitations 3. Time limit for submitting information 4. Precise time instructions for loading 5. Possibility of oversized and inefficient loads 6. Proper placement of human resources	6	Intervals

Source: Processed data, 2023

3. Results and Discussion

The process of loading vehicles on board will affect the safety of the ship while sailing. Some studies that can provide an overview of the process of loading vehicles on board ferries in Indonesia are as follows:

IOP Conf. Series: Earth and Environmental Science **1461** (2025) 012029 doi:10.1088/1755-1315/1461/1/012029

Table 3. Research on Vehicle Loading on Ferries

No	Research Title	Research result
1	Evaluation of Vehicle Loading Procedures on Ferry Ships on the Bira-Pamatata Route, South Sulawesi Province (Febriansyah, et al., 2024, Multidisciplinary Journal of Science and Technology, Volume 02, No. 042024, pp.80-89, E-ISSN: 2988-1986)	a. The vehicle weighing equipment at the Bira Ferry Port is not working. b. Vehicle placement is not in accordance with the rules. c. Vehicles on board are not tied down
2	Vehicle Loading System on KMP. PORT LINK III for Sea Transportation Safety (C.I. Ilham, Katharina S, Yanti Desi, 2023, Inland Waterways Journal, Volume 5, Issue 1, ISSN. 2723-3642)	a. In the implementation of transportation services on board the vehicle is still found cargo handling is still not appropriate. b. The location of the vehicle should be longitudinal in the direction of the bow and stern but there are still vehicles that are located transversely. c. The condition of the loading space is suitable clean of oil and grease (grease).
3	Evaluation of the Implementation of Vehicle Binding on the Merak Bakauheni Crossing Ferry (FMD Widyarti, 2021, SDP Palembang Transportation Polytechnic)	a. The weighbridge at the Bakauheni Ferry Port is not operating b. Vehicle distance that does not comply with regulations. c. The non-compliance with Ministerial Regulation Number 30 of 2016 concerning the Obligation to Tie Down Vehicles on Ferry Transport Vessels, such as vehicles that are not lashed during the voyage.
4	Analysis of Vehicle Loading on the KMP. Kalibodri Ship on the Kendal – Kumai Crossing Route (MV. Fernanda, 2021, Poltektranssdp Repository)	a. There is no weighbridge at Kendal port yet b. Vehicle placement does not comply with applicable regulations. c. The number of lashing tools is less d. Not all vehicles on board are lashed.

Based on Table 3 above, it is known that there are still many Ro-Ro ferries that do not load vehicles on board according to the rules, namely:Regulation of the Minister of Transportation Number: PM. 115 of 2016 concerning Procedures for Transporting Vehicles on Ships and PM. 30 of 2016 concerning Obligations for Vehicle Binding on Ferry Transport Ships.

3.1 Vehicle Loading System on Ro-Ro Ferry Ships

The vehicle loading system on board the ship is as follows:

3.1.1 Vehicle weighing process

Bajoe ferry port has 1 (one) permanent scale unit which is used to weigh vehicles and their cargo before boarding the ship.

Based on the observation results, all goods vehicles to be loaded on the ship are weighed first. The weight of the vehicle and the type of goods carried are recorded by the weighing officer.

3.1.2 Vehicle Information System

Information on vehicle weight data and the type of cargo carried by vehicles that will enter the ship is informed by the weighing officer to the ship's land officer and then informed to the First Officer of the ship that will be loaded. The provision of this information aims to allow the First Officer to arrange the load according to the weight and type of load and to know the number of vehicles that can be transported based on the maximum weight of the load that can be loaded on the ship.

3.1.3 Loading of Vehicles on Board Ships

The arrangement and isolation of each vehicle entering the ship must comply with the regulations set out in the Regulation of the Minister of Transportation Number: PM. 115 of 2016 concerning Procedures for Transporting Vehicles on Ships and PM. 30 of 2016 concerning Obligations for Vehicle Binding on Ferry Transport Ships. The results of observations of 7 (seven) ships operating on the Bajoe - Kolaka route are as follows:

a. Availability of Lashing Gear

The types of vehicle lashing gear that can be used are rope automobile tiedown, sling tie with toothed lock (ratchet strap assembly), chain with reinforcement/tightener (chain with turnbuckle). 7 (seven) ships that were the objects of the study, namely KMP. Mishima, KMP. Masagena, KMP. Mandala Nusantara, KMP. Kota Muna, KMP. Faiz and KMP. Perdana Nusantara have sufficient lashing gear to lash vehicles on board.

b. Placement of Vehicles on Ships

Placement of vehicles on the ship as follows:

1) All vehicles are positioned lengthwise (longitudinally) in the direction of the bow or stern, in accordance with applicable regulations.
2) The location between vehicles and the ship's wall is given a distance for people's access. The distance between vehicles is at least 60 cm, between the front and back of the ship is at least 30 cm and the distance between the side of the vehicle and the ship's wall is at least 60 cm. Based on observations during loading at KMP. Muna City, there were several motorbikes placed next to the ship's wall and cars, thus blocking people's access. So not all ships meet the regulations for vehicle distance placement.

   Based on the results of observations, it is known that not all vehicle placements on ships are in accordance with established regulations.

c. Vehicle Lashing

Based on PM. 115 of 2016 concerning Procedures for Transporting Vehicles on Ships, vehicles that will be loaded on ships and have a minimum weight of 3.5 tons must be tied up both in good and bad sea and weather conditions. Based on the results of observations, it is known that lashing of vehicles with this weight was carried out on 7 (seven) ships that were samples in this study. The lashing was carried out by the ship's crew and supervised by port officers from the Land Transportation Management Center (BPTD) Class II of the South Sulawesi Region, Directorate General of Land Transportation.

3.2. Factors Influencing the Implementation of Vehicle Loading Systems on Ships

Factors affecting the implementation of vehicle loading system on board ship were analyzed using linear regression method and the calculation was done using Statistical Product and Service Solutions (SPSS) 25 for Windows. The variables in this study are factors affecting the implementation of vehicle loading regulation on board ro-ro ferry ships determined based on regulation and in-depth interview result with the ship crew. These factors are supervision of activities, obstacles to the implementation of regulation and knowledge of the ship crew regarding the implementation of lashing on board ship as independent variable, while the implementation of vehicle loading system as dependent variable.

The object of this study is 7 (seven) units of ro-ro ferry ships operating on the Bajoe - Kolaka route and the respondents in this study are the crew members who work on the ship. The characteristics of the respondents from the questionnaires that they have filled out and can be processed are as shown in Table 4.

Based on Table 4, it is known that the data amount to 114, the explanation of each data is as follows:

3.2.1. Activity Supervision has a minimum value of 20, a maximum value of 28 and an average value (mean) of 24.43 with a standard deviation of 2.300. The average value and standard deviation value of this activity supervision indicate that there is good data distribution because the average value is greater than the standard deviation X_1.

3.2.2. Inhibitors of SOP Implementation has a minimum value of 21, a maximum value of 28 and an average value (mean) of 24.47 with a standard deviation of 2.262. The average value and standard deviation value of the inhibitors of SOP implementation indicate that there is good data distribution because the average value is greater than the standard deviation X_2.

3.2.3. Crew Knowledge has a minimum value of 8, a maximum value of 12 and an average value (mean) of 10.07 with a standard deviation of 1.217. The average value and standard deviation value of crew knowledge indicate that there is good data distribution because the average value is greater than the standard deviation X_3.

Tabel 4. Respondent Characteristics

No	Information	Amount	Percentage (%)
1.	Gender		
	Male	114	100
	Female	0	0
	Total	114	100
2.	Level of Education		
	High School	44	38.6
	Diploma (D3)	57	50
	Bachelor (S1)	13	11.4
	Master (S2)	0	0
	Doctor (S3)	0	0
	Total	114	100
3	Position		
	Captain	7	6.1

	Deck Officer	16	14
	Merine Engineer	23	20.2
	Maritime Radio Operator	2	1.8
	Boss	4	3.5
	Helmsman	16	14
	Oiler	18	15.8
	Wipers	2	1.8
	Rating	20	17.5
	Ship's cook	6	5
	Total	144	100
4	Length of Work		
	1 - 5 Years	16	14
	6 - 10 years	29	25.4
	11 - 15 years	36	31.6
	16 - 20 years	15	13.2
	≥ 21 years	18	15.8
	Total	114	100

3.2.4. Implementation of Vehicle Loading System (Y) has a minimum value of 17, a maximum value of 24 and an average value (mean) of 19.65 with a standard deviation of 2.285. The average value and standard deviation value of the implementation of this vehicle loading system indicate that there is good data distribution because the average value is greater than the standard deviation.

3.3 Multiple Linear Regression Analysis Results

In this study, the hypothesis was tested using a multiple linear regression model to obtain a comprehensive picture of the influence of the variables of Activity Supervision, SOP Implementation Inhibitors and Crew Knowledge on the Implementation of the Vehicle Loading System which was carried out with the help of SPSS software version 25 for windows. The results of the multiple linear regression test can be seen in the following table.

Table 5. Multiple Linear Regression Results

Coefficients [a]

Model		Unstandardized Coefficients B	Standardized Coefficients Std. Error	Beta	t	Sig.
1	(Constant)	16.610	2.902		5.723	.000
	Total_X1	.214	.077	.216	2.799	.006
	Total_X2	.214	.078	.212	2.752	.007
	Total_X3	.138	.098	.127	2.049	.009

a. Dependent Variable : Total _Y

Based on the results of data management in Table 5 in the Standardized Coefficient column in the Beta section, the following multiple linear regression equation model was obtained :

$$Y = a + b1X1 + b2X2 + b3X3 + e$$
$$Y = 16.610 + 0.216X1 + 0.212X2 + 0.127X3 + e$$

The regression equation obtained can be explained as follows:

a. The constant value is 16.610, meaning that if the values of Activity Supervision, SOP Implementation Inhibitors and Crew Knowledge are assumed to be equal to 0, then the constant Vehicle Loading System Implementation will have a value of 16.610.
b. The regression coefficient value for the Activity Supervision variable (X1) is 0.216. This means that Activity Supervision has a positive effect on the Implementation of the Vehicle Loading System. This shows that the higher the Activity Supervision by 1 (one) unit, the higher the Vehicle Loading System Implementation value will be by 0.216 and vice versa.
c. The regression coefficient value for the variable Inhibitors of SOP Implementation (X2) is 0.212. This means that Inhibitors of SOP Implementation have a positive effect on the Implementation of Vehicle Loading Systems. This shows that the higher the Inhibitors of SOP Implementation by 1 (one) unit, the higher the Vehicle Loading System Implementation value will be by 0.212 and vice versa.
d. The regression coefficient value for the Crew Knowledge variable (X3) is 0.127. This means that Crew Knowledge has a positive effect on the Implementation of vehicle loading system. This shows that the higher the Crew Knowledge by 1 (one) unit, the vehicle loading system implementation value will increase by 0.127 and vice versa.

3.4 Risk Analysis

This risk analysis is carried out to determine the impacts that can occur due to not carrying out the process of loading vehicles on ships properly in accordance with the regulations in force in Indonesia, namely:Regulation of the Minister of Transportation Number: PM. 115 of 2016 concerning Procedures for Transporting Vehicles on Ships and PM. 30 of 2016 concerning Obligations for Vehicle Binding on Ferry Transport Ships.

This analysis was conducted based on the KNKT report published on social media, it was discovered that several accidents occurred due to the loading of vehicles on ships not being carried out properly, namely as follows:

a. KM. Musthika Kencana II In 2011 departed from Surabaya port to Tanjung Priok port, experienced a fire on the vehicle deck, but the crew could not extinguish the fire source with APAR because the distance between vehicles was narrow. This was caused by the arrangement of vehicle loading on the ship that was not in accordance with the rules, namelyThe location between vehicles and the ship's wall is given a distance for people's access. The distance between vehicles is at least 60 cm, between the front and back of the ship is at least 30 cm and the distance between the side of the vehicle and the ship's wall is at least 60 cm.
b. KMP. Munawar Ferry 2014 crossing Kayangan -Pototano, the vehicles on board shifted when the ship tilted because it was not lashed, thus making the stability of the ship poor and the ship.
c. KM. Bahari Indonesia. In 2020, there was a fire and it could not be extinguished because the vehicle was placed incorrectly, so the vehicle covered the hydrant or fire extinguisher and people's access road.
d. Dharma Rucitra III at Padangbai Port, Bali in 2020, experienced a tilt due to unattached vehicles on the deck shifting and piling up on the right side of the ship.

$$Y = a + b1X1 + b2X2 + b3X3 + b4X4\ e$$

4. Conclusion

The conclusion of this study is as follows:

a. The vehicle loading system on the ro-ro ferry on the Bajoe – Kolaka crossing is as follows:

No	Aspects according to Regulation	Implementation	
		In accordance	No
1	Vehicle weighing process	√	
2	Vehicle Information System	√	
3	Loading of Vehicles on Board Ships		
	Availability of Lashing Gear	√	
	Placement of Vehicles on Ships		X
	Vehicle Lashing	√	

b. Factors that influence the implementation of vehicle loading systems on ro-ro ferries based on established regulations are activity supervision, obstacles to the implementation of SOP, crew knowledge and implementation of pactivity supervision. The better the supervision of activities, the more obstacles to the implementation of SOPs can be avoided and the knowledge of the ship's crew can be improved, the more effective the implementation of the vehicle loading system will be.

5. Suggestion

Some things that are recommended so that the vehicle loading system on board the ship can be implemented properly, namely as follows:

a. Create a strict sanctions policy for ships that do not carry out lashing.
b. Improvement of training programs for ship crews.
c. Create an automatic vehicle lashing system.

Reference

[1] Ansori, M. 2020. Quantitative Research Methods 2nd Edition. Airlangga University Press
[2] C.I. Ilham, Katharina S, Yanti Desi, 2023. Vehicle Loading System on KMP. PORT LINK III for Sea Transportation Safety, Inland Waterways Journal, Volume 5, Issue 1, ISSN. 2723-3642.
[3] Febriansyah, et al., 2024. Evaluation of Vehicle Loading Procedures on Ferry Ships on the Bira-Pamatata Route, South Sulawesi Province, Multidisciplinary Journal of Science and Technology, Volume 02, No. 042024, pp.80-89, E-ISSN: 2988-1986.
[4] Fernandes, AAR 2018. Quantitative Research Methodology Systems Perspective: Revealing Novelty and Fulfilling Research Validity. Universitas Brawijaya Press.
[5] FMD Widyarti, 2021. Evaluation of the Implementation of Vehicle Binding on the Merak Bakauheni Crossing Ferry, SDP Palembang Transportation Polytechnic.
[6] Ghozali, I. 2013. Multivariate Analysis with IBM SPSS Program. Semarang: Faculty of Economics Publishing Agency, Diponegoro University.
[7] Insani, Katharina, Yanti, 2023, Vehicle Loading System on KMP. PORT LINK III for Sea Transportation Safety, Inland Water Ways Journal, Volume 5 Issue 1, April 2023, 1 – 6, ISSN 2723-3642.
[8] International Maritime Organization, 2014. International Safety Management (ISM) Code.
[9] Jaya, I. 2019. Application of Statistics for Educational Research. Jakarta: Prenada Media Group.
[10] Ministry of Transportation, 2016. Regulation of the Minister of Transportation Number: PM. 115 of 2061 concerningProcedures for Transporting Vehicles on Ships, 2016.
[11] Ministry of Transportation, 2016. Regulation of the Minister of Transportation Number: PM. 30 of 2016 concerningObligation to Tie Down Vehicles on Ferry Transport Vessels, 2016.

[12] MV. Fernanda, 2021. Analysis of Vehicle Loading on the KMP. Kalibodri Ship on the Kendal – Kumai Crossing Route, Poltektranssdp Repository.
[13] Priyatno, Duwi. 2017. Practical Guide to Data Processing Using SPSS. Yogyakarta: Andi Publisher.
[14] Rubianto., 2017. Ship Building, Ship Stability, Maritime Law, Ship's Aircraft. Publisher Maritim Djangkar.
[15] Rizqi, Amalia, 2022. Implementation of Vehicle Transportation Procedures on Ferry Ships on the Tebas Kuala – Parigi PIAI Crossing. Repository of SDP Palembang Transportation Polytechnic, 2022.
[16] Sarinah and Khairunas., 2015. Reducing Accident Rates on Ro-Ro Ships. Vol 1, No.3, May 2015: Journal of Transportation and Logistics Business Management.
[17] Subandrijo and Soegiyanto, 2016. Ship Stability for Nautical Study Program. Semarang Maritime.
[18] Sugiyono, 2018. Quantitative Research Methods. Bandung: Alfabeta.
[19] Z Ariani, A Hendra, S Febriary, 2018. Minimum Service Standards (SPM) and Lashing System on Ro-Ro Ships for Sea Crossing Transportation Safety, Gema Teknologi, Vol. 20, NO. 1 Period April 2018 – October 2018.
[20] https://www.detik.com/bali/berita/d-6819520/truk-terguling-di-kapal-gegara-gelombang-tinggi-di-selat-bali

IOP Conf. Series: Earth and Environmental Science **1461** (2025) 012030 doi:10.1088/1755-1315/1461/1/012030

BIOMETHANE-COMPRESSED NATURAL GAS (Bio-CNG) DISTRIBUTION MODEL IN SUPPORT OF THE LIQUEFIED PETROLEUM GAS (LPG) SUBSTITUTION PROGRAM

Dika Virginia D[1*], I G.N Sumanta Buana[1], Mohammad Kamil[1], Christino Boyke[1], Siti Dwi Lazuardi[1] and Pratiwi Wuryaningrum[1]

[1] Department of Maritime Transportation Engineering, Sepuluh Nopember Institute of Technology (ITS), Indonesia

[*]E-mail: dika.virginia@its.ac.id

Abstract. Indonesia's 2027 vision involves significantly reducing LPG imports to enhance energy independence. The aim is to promote alternative energy sources, such as Bio-CNG, and decrease reliance on imported fossil fuels. By developing and expanding domestic energy production, Indonesia intends to bolster its energy security and sustainability. Indonesia has been facing challenges to overcome the amount of energy imports, one of them being Liquified Petroleum Gas (LPG). Palm oil as one of the raw materials can be used to generate bioenergy in the future. On the other hand, it generates liquid waste, known as Palm Oil Mill Effluent (POME). Using POME to produce Biogas becomes the alternative and sustainable way to control the amount of POME as a waste of Palm Oil Production as well as providing economic satisfaction and benefits. The Ministry of Energy and Mineral Resources (ESDM) encourages the development of Biogas into Biomethane-Compressed Natural Gas (Bio CNG) on a commercial scale as a fuel to replace the LNG. In addition, The Bio CNG is purified from the Biogas that is generated from decomposing POME to replace the demand for LPG. The purpose of this research is to determine the demand value for Bio CNG, create a distribution model for transporting the Bio CNG from the Palm Oil Mill (PKS) to the unloading port as well as identify the transportation cost involved in distributing the Bio CNG. Several methods are used to obtain the purpose of the research such as Capacitated K-Medoids for clustering the area, TSP for determining alternative land transportation routes, VRP for determining modes and routes of the maritime transportation, Simulation, and Optimization for minimizing several storage capacities, and Sensitivity Analysis to access the impact of unit cost change over the component value. The amount of Bio CNG demand for achieving zero emissions in the palm oil industry is 2.18 million tons or 104,052 mmBtu, the amount of POME from the PKS to the Bio-CNG plant either directly or through a last-mile approach. After production, Bio-CNG is transported from the plant to the loading port via a direct route, and then from the loading port to the unloading port with a single CNG ship. The total cost and the unit cost are Rp455,303 million per year and Rp310,276 per mmBtu respectively.

IOP Conf. Series: Earth and Environmental Science **1461** (2025) 012030 doi:10.1088/1755-1315/1461/1/012030

1. Introduction

Indonesia is known worldwide as an agrarian country with a diverse range of natural resources, particularly in the agricultural sector. One of the well-known agricultural industries in Indonesia is the palm industry. According to the data from the Central Statistics Agency (BPS), Indonesia is the largest producer of palm oil (Crude Palm Oil/CPO) with a total plantation of 14 million hectares (1). The production process of palm oil has a positive impact on Indonesia's economy due to the high potential demand. However, this process generates liquid waste known as Palm Oil Mill Effluent (POME), which has environmental consequences. This waste, a byproduct of the fresh fruit bunches (TBS) used in the CPO production process, constitutes 70% of the total material. Currently, only a small portion of this waste, around 5%, is processed into Biogas that is utilized for power generation in palm oil mills (2). Biogas has various potential products, one of which is Biomethane Compressed Natural Gas (Bio-CNG) also often referred to as Compressed Biogas (CBG). Given the substantial potential and the need to reduce waste in the palm oil industry, accelerate the contribution of new renewable energy (EBT), and reduce the reliance on LPG import, which amounts to 6.34 Million Tones or about 75% of Indonesia's total LPG needs in 2021 (3). The Ministry of Energy and Mineral Resources (ESDM) is encouraging the development of biogas into Biomethane-Compressed Natural Gas (Bio-CNG) on a commercial basis. The advancement of Bio-CNG projects into new renewable energy sources offers a promising solution to decrease reliance on fossil fuels while being more environmentally friendly. Given the ample availability of raw materials, Bio-CNG holds long-term potential, particularly for fostering national economic growth and creating a positive ripple effect in local communities.

Currently, there is only one Bio-CNG plant in Indonesia, owned by PT. Dharma Satya Nusantara Tbk, located in Muara Wahau, East Kalimantan. In addition, PT. KIS Biofuels Indonesia plans to construct 25 other Bio-CNG plants in North Sumatra, which are expected to be completed in 2024The potential of POME is proposed for conversion into Bio-CNG due to its similar

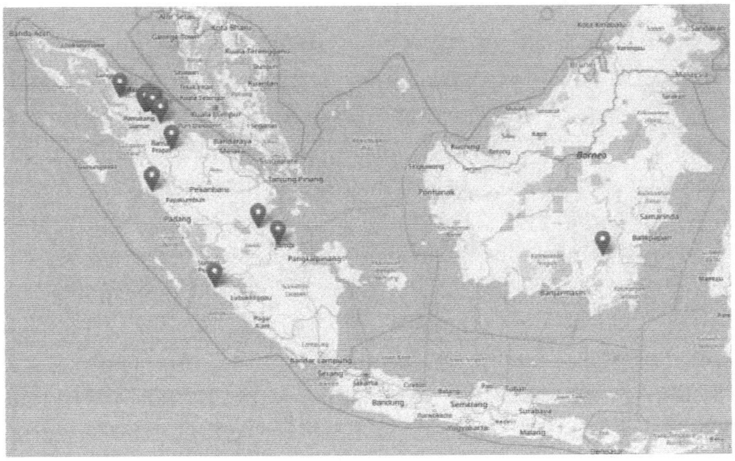

Figure 1. The distribution of Bio-CNG Plant across Indonesia.

characteristics to CNG (4). Bio-CNG has potential uses as a fuel, electricity generator, and heating source.

With the distribution of Palm Oil Mills (PKS) across Indonesia, there is a significant opportunity to utilize Bio-CNG on a larger scale, including for power generation, public vehicle

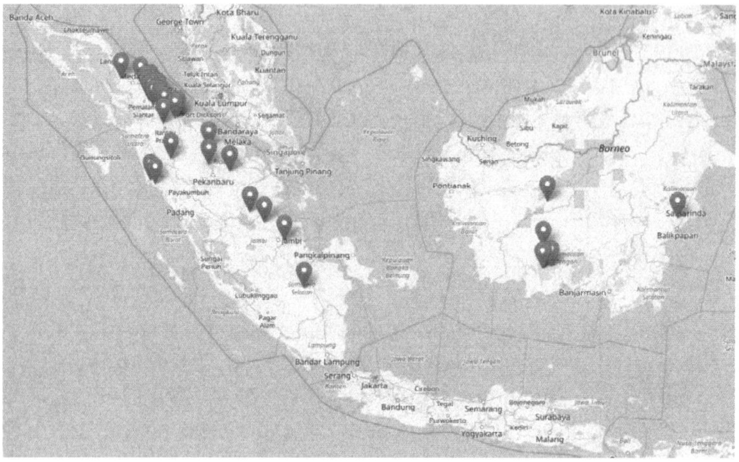

Figure 2. The distribution of Palm Oil Mills (PKS) across Indonesia.

fuel, and household needs. Therefore, this research aims to develop a transportation planning model to determine the most effective and efficient distribution patterns. To achieve this, the author employs K-Medoids for clustering, TSP for alternative land routes, VRP for route and maritime mode determination, and simulation for inventory management at each port.

2. Problem Definition

The optimal solution for transporting the Bio CNG through sea and land must be determined by the cluster of the Palm Oil Mills, Ships that comply for the shipping process, ship schedule, and also the inventory level at each port of destination while minimizing the total cost for the fleet, such as Capital Cost of the Fleet and Trucks, Operational Cost of the Fleets and Trucks, as well as the Capital and Operational cost of the Port Infrastructure.

According to the Operational, Two scenarios are provided, the first scenario is the trucks have a full commission day (365 days) and the second scenario is where the operations of the trucks follow the schedule of ship arrival at the loading port. Lastly, the cost analysis referred to in this stage involves calculating the expenses incurred from distributing Bio-CNG, including Capital and Operational Costs of the ships, trucks, and port infrastructure to determine the unit cost for each mmBtu, a unit energy used to measure the amount of energy contained in the Bio CNG.

3. Method

In this research paper, the author delves into the intricacies of mathematical expressions and its application to distribution problems. In this chapter, the main aims are to involve the formulation and explanation of mathematical expressions and the equations that capture the essential picture of the problem, starting with constructing, validating, and utilizing to advance our understanding of the problem.

3.1 Capacitated K-Medoids

The K-Medoids Algorithm is a partitional clustering technique similar to the K-Means, to partition the data set into N-clusters to minimize the distance between the data point or the demand points and their respective cluster or the supply points. Let N be the set of Palm Oil Mills, and M be the set of the Bio CNG plants available. We also have $i = \{1,2, \dots, N\}$, and $j = \{1,2, \dots, M\}$. We further let C_{ij} denotes the total cost of sending POME from Palm Oil Mills i to the Bio CNG plant j. Also, we introduced S_i and D_j as the amount of supply from Palm Oil Mills i and the demand at the Bio CNG plant j, respectively. The binary variable x_{ij} takes the value 1 if POME from the Palm Oil Mills $i \in N$ transported to the Bio CNG $j \in M$ using ship. While $CINF_j$ and Ct_{ij} denote the cost of building the infrastructure at plant j and the capital cost for trucks that ship the POME Plam Oil Mills $i \in N$ transport the POME to the Bio CNG $j \in M$, respectively. We also have several parameters regarding the data such as d_{ij} is given to explain the distance (Nm) between Palm Oil Mills i to Bio CNG plant j, nt_{ij} tells us about the number of trucks assigned from the Unloading port i to Bio CNG plant j, and fc_{ij} as the fuel cost spent from Palm Oil Mills i to Bio CNG plant j.

Using the notations, the model can be formulated as follows:

$$Minimize\ C_{ij} = \sum_{i=1}^{N}\sum_{j=1}^{M}((CINF_j \times x_{ij}) + (d_{ij} \times nt_{ij} \times fc_{ij} \times x_{ij}) + (Ct_{ij} \times nt_{ij} \times x_{ij}))$$

Subject to:

$$\sum_{i=1}^{K} S_{ij} \leq D_j, j = 1,2, \dots, M \qquad (4.1.1)$$

$$\sum_{j=1}^{M} x_{ij} = 1 \qquad (4.1.2)$$

$$x_{ij} \in \{0,1\} \qquad (4.1.3)$$

The objective function is to minimize the total cost which consists of the cost of building the infrastructure and capital cost of the truck required subject to the amount of supply for POME palm oil mills to cover the amount of Demand at the Bio CNG Plant.

3.2 Travelling Salesman Problem

This method is used to formulate the problem to obtain the number of trucks where multiple trucks are used to minimize the total distance. All the parameters are the same as written in Subchapter 4.1, here one additional parameter must be included in the TSP model, Q_k: The capacity of the truck k. Using the notations, the model can be formulated as follows:

$$Minimize \sum_{k=1}^{K}\sum_{i=1}^{U}\sum_{j=1}^{M}(d_{ijk} \times x_{ijk})$$

Subject to:

$$\sum_{k=1}^{K}\sum_{i=1}^{U} x_{ijk} = 1, \forall j \in M \qquad (4.2.1)$$

$$\sum_{j=1}^{M} x_{Ujk} = 1, \forall k \in K, \forall i \in U \quad (4.2.2)$$

$$\sum_{i=1}^{U} x_{iMk} = 1, \forall k \in K, \forall j \in M \quad (4.2.3)$$

$$\sum_{i=1}^{U} x_{ijk} - \sum_{j=1}^{M} x_{ijk} = 0, \forall i \in U, \forall k \in K \quad (4.2.4)$$

$$U_i - U_j + |U| + |M| \times x_{ijk} \leq |U| + |M| - 1, \forall i \in U, \forall j \in M, \forall k \in K \quad (4.2.5)$$

$$\sum_{i=1}^{U} S_j \times \left(\sum_{j=1}^{M} x_{ijk} \right) \leq Q_k, \forall k \in K \quad (4.2.6)$$

$$x_{ijk} \in \{0,1\}, \ \forall i \in U, \forall j \in M, \forall k \in K \quad (4.2.7)$$

This model is using the Nearest Neighbour to find the optimal solution for the distance of the trucks. We have the set of locations POME Plam Oil Mills $i \in N$ and the POME to the Bio CNG $j \in M$, respectively. Constraints 4.2.1 ensures that for each delivery point j in the set M, there is exactly one truck k assigned to handle it from a delivery point i in the set U. In other words, each delivery point must be served by one specific truck, ensuring full coverage without overlap. Besides, Constraints 4.2.2 ensures that each delivery i from the set of deliveries U is assigned to exactly one truck k. In other words, every delivery must be handled by one specific truck. Constraints 4.2.3 ensures that each truck k is responsible for one delivery point j. Each truck must serve a single delivery from the set of deliveries. Constraints 4.2.4 enforce flow conservation, meaning that the number of deliveries entering a point must equal the number of deliveries exiting it. This prevents deliveries from being unaccounted for. Then, Constraints 4.2.5 is designed to eliminate subtours. It imposes a penalty if a route connects deliveries i and j, it is ensuring that the route remains connected to the main path. The overall expression in 4.2.6 indicates that the total weighted assignments must not exceed a limit Q_k for each $k \in K$. This could represent resource limits, capacities, or budget constraints. The last constraint is the binary constraint, making this a combinatorial problem where each resource can either be assigned to a task of not.

3.3 Vehicle Routing Problem

This method is used to find the route and sea mode with the minimum cost (4). The compatibility considerations include port's depth, wave height, and the maximum economic distance of Bio-CNG. Let K, R,and P be the number of available fleets, routes and ports. In the set of fleets and routes, we have $k = 1,2, ... K, r = 1,2, ... R,$ and $p = 1,2, ... P.$ There are many parameters that matter in this problem:

- TC_{kr} : The total shipping cost of a ship k in the route r $(\frac{Rp}{year})$
- VC_{kr} : The total voyage cost of a ship k in the route r $(\frac{Rp}{year})$
- CHC_{kr} : Cargo handling cost of a ship k in the route r $(\frac{Rp}{year})$
- PC_{kr} : Port Charges of a ship k in the route r $(\frac{Rp}{year})$
- TCH_{kr} : Time charter hire of a ship k in the route r $(\frac{Rp}{year})$
- f_{kr} : The number of frequencies of a ship k in the route r (times)

- $cargo_{kr}$: Demand carried at the ship $i = k$ in the route r (mmBtu)
- D_r : The demand at Bio CNG r (mmBtu)
- nk_k : The number of ships assigned (Units)
- POL_p : Port of Loading p
- $Cport_p$: The amount of commodity loaded at port p (mmBtu)
- $Smax_p$: The maximum amount of Bio CNG at port p (mmBtu)
- price : Gas Proce per mmBtu (Rp/mmBtu)

Using the notations and parameters, the model can be formulated as follows:

$$Minimize \sum_{k=1}^{K}\sum_{r=1}^{R} f_{kr} \times TC_{kr}$$

where,

$$TC_{kr} = \left(\sum_{k=1}^{K}\sum_{r=1}^{R}(VC_{kr} \times f_{kr}) + (CHC_{kr} \times f_{kr}) + (PC_{kr} \times f_{kr}) + (TCH_{kr} \times nk_k)\right) + \left(\sum_{k=1}^{K}\sum_{r=1}^{R}(cargo_{kr} \times f_{kr}) \times D_r\right) \times price$$

Subject to:

$$\sum_{k=1}^{K}\sum_{r=1}^{R}(cargo_{kr} \geq D_r \tag{4.3.1}$$

$$\sum_{p=1}^{P} POL_p = 1 \tag{4.3.2}$$

$$\sum_{p=1}^{P} Cport_p = \sum_{p=1}^{P} Smax_p \tag{4.3.3}$$

The purpose of the objective function is to minimize the total cost across the fleets K and routes R. The first constraint ensures tha the total cargo assigned meets or exceeds the demand for each task in the route j. The second constraint specifies that exactly one point or origin (POL) must be selected from the available options. The last constraint ensures that the total capacity at ports matches the maximum available capacities, maintaining balance in resource distributions.

4. Analysis and Discussion

4.1 Design of the Bio-CNG Distribution Model

Bio-CNG does not yet have a distribution model because it is a very new commodity. Therefore, the author has designed a distribution model based on the flow of the Bio-CNG product itself and utilizes handling methods similar to CNG, as this gas has similar characteristics to Bio-CNG.

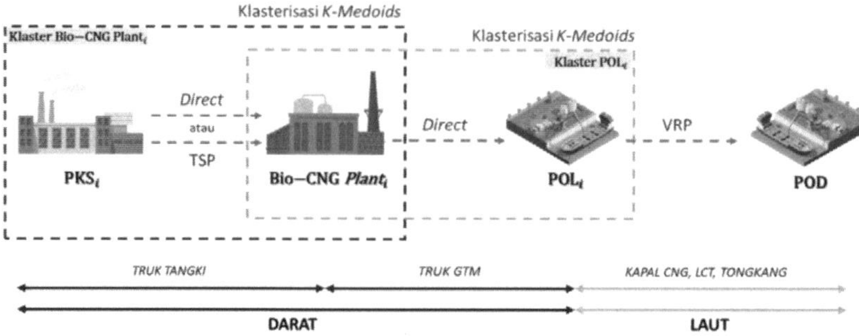

Figure 1. Author's Design: Distribution Model of Bio-CNG shipping from Palm Oil Mils to the Unloading Port

CNG and Bio-CNG share similar handling and storage methods, one of which involves using a Gas Transport Module (GTM). Through the GTM, Bio-CNG is compressed and then loaded into the module at an average pressure of 200-250 kg/m^2. After that, the GTM is transported using truck fleets or ships until it reaches its destination. For Bio-CNG to reach its destination, several stages must be followed, as shown in Figure 1.

Table 1. The Location of Bio-CNG Plants

Code	Bio CNG Plant	Province	Coordinate °*N*	Coordinate °*T*
PT A	PT BSP	Sumatera Utara	3,039	99,582
PT B	PT BSP Agrowiyana	Jambi	-1,077	103,11
PT C	PT BSP BPP	Sumatera Barat	0,22	99,47
PT D	PT BSP GLP	Sumatera Utara	2,58	99,78
PT E	PT BSP SNP	Jambi	-1,67	103,80
PT F	PT Mitra Puding Mas	Bengkulu	-3,13	101,60
PT G	PT Sawit Graha Manunggal	Kalimantan Tengah	-2,07	115,08
PT H	PT Tasik Raja	Sumatera Utara	1,66	100,16
PT I	PT United Kingdom Indonesia Plantation	Sumatera Utara	3,49	98,39
PT J	PTPN III Sei Mangkei	Sumatera Utara	3,13	99,34
PT K	PTPN III Sei Silau	Sumatera Utara	2,90	99,51
PT L	PTPN IV Bah Jambi	Sumatera Utara	2,98	99,22
PT M	PTPN IV Gunung Bayu	Sumatera Utara	2,99	99,27
PT N	PTPN IV Mayang	Sumatera Utara	3,02	99,33
PT O	PTPN IV Pasir Mandoge	Sumatera Utara	2,81	99,37

4.2 Clustering of Palm Oil Mills and Route Analysis from Palm Oil Mills to Bio-CNG Plant

By using the distribution of palm oil mills, clustering is conducted to group these mills with the nearest Bio-CNG plants (see Table 1), based on the maximum capacity of POME that can be accommodated at each plant.

IOP Conf. Series: Earth and Environmental Science **1461** (2025) 012030 doi:10.1088/1755-1315/1461/1/012030

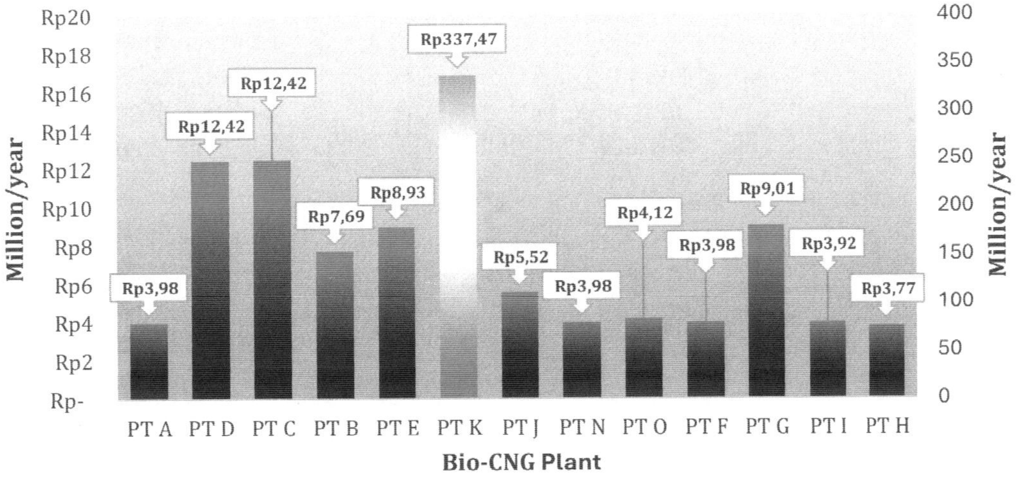

Figure 2. Total Shipping Cost from Palm Oil Mills to Bio CNG Plant (Rp = IDR)

As a result, From the clustering calculations, 2 (two) locations were not selected out of 15 potential sites for the construction of the Bio-CNG plant, namely PT Mitra Puding Mas and PT Tasik Jaya. The Bio-CNG cluster at PTPN IV Pasir Mandoge is supplied solely from its source Therefore, this cluster has no truck operational costs. On the other hand, PT Sawit Graha Manunggal has the highest truck operational costs, amounting to Rp337.47 million (see Figure 2).

The Bio-CNG cluster at PT Sawit Graha Manunggal has the longest route distance, measuring 2,049.2 km, requiring a minimum of 273 tanker trucks to serve this route (see Table 6). Consequently, this cluster incurs the highest costs at this stage, totaling Rp38.65 billion (see Figure 8). Based on Figure 9, there are 4 (four) clusters selected from the 6 (six) initial clusters. This is due to the lower costs incurred when using the last-mile scheme compared to the direct scheme. The selected clusters are PT BSP Agrowiyana, PT BSP SNP, PTPN III Sei Silau, and PT Sawit Graha Manunggal. Therefore, the total minimum cost from the palm oil mills to the Bio-CNG plant, including infrastructure costs, amounts to Rp110.3 billion.

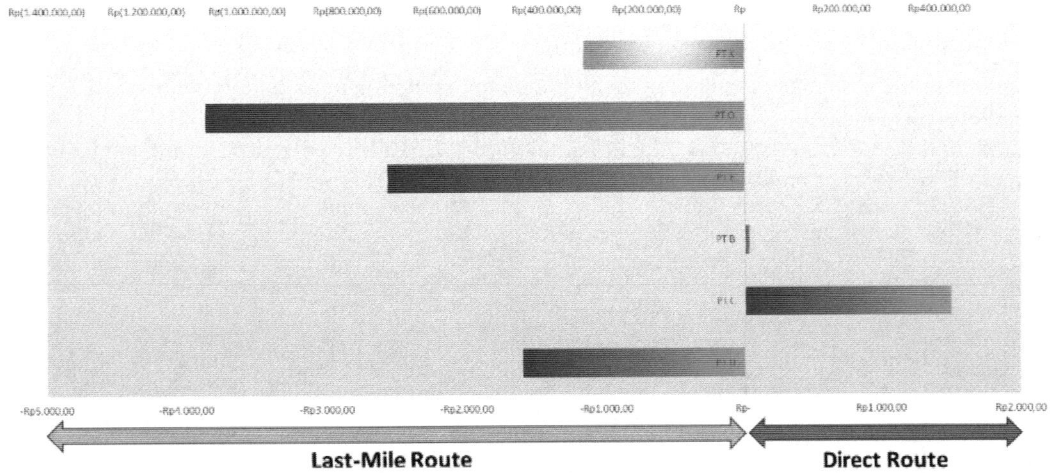

Figure 3. Cost Difference between the Direct Route Scheme and the Last-Mile Route Scheme (Rp = IDR)

Based on Figure 3, four clusters were selected from the six clusters mentioned. This is due to the lower costs incurred when using the last-mile scheme compared to the direct scheme. The selected clusters are PT BSP Agrowiyana, PT BSP SNP, PTPN III Sei Silau, and PT Sawit Graha Manunggal. Therefore, the total minimum cost from the PKS to the Bio-CNG plant, including infrastructure costs, amounts to IDR 110.3 billion.

4.3 Clustering of the Bio-CNG Plant

Similarly to the clustering of Palm Oil Mills, this clustering also employs the same method, namely k-medoids, using the selected Bio-CNG plant for construction based on the previous clustering phase. Belawan Port is the loading port that handles the largest Bio-CNG shipment, which is 879,711.3 mmBtu per year. The Bio-CNG supply is sourced from nine surrounding Bio-CNG plants (See Table 2).

Table 2. Group of Bio-CNG Plants

Port of Loading	Bio-CNG Plant	Total Bio-CNG Supply (mmBtu/yr)	Total Distance (km)
Belawan	PT A PT B PT J PT N PT O PT F PT G PT I PT H	879.711,30	3.368,9
Palembang	PT D PT E	362.641,30	1.488,4
Dumai	PT C	291.079,83	878,4
Semayang	PT K	399.532,41	680,3

4.4 The Route and Shipping Optimization

Optimization at this stage is carried out to find the combination of routes and maritime modes that results in the lowest costs. In selecting the route and mode, factors such as port depth, wave height, and the maximum economical distance for CNG cargo are considered. The selected vessel for this optimization is the CNG Jayanti Baruna. This ship has a length overall (LOA) of 110.92 meters, a beam of 17.6 meters, and a draft of 5 meters. Additionally, this vessel can transport 1,540 tons of Bio-CNG, equivalent to 73,370.59 mmBtu.

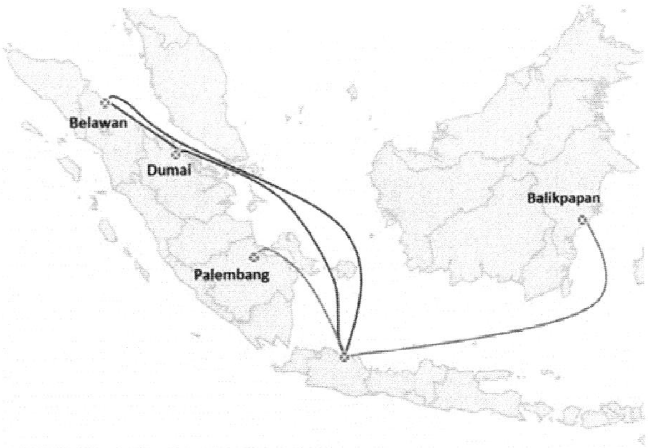

Figure 4. Selected Routes for CNG Vessel Each Year

In addition, three routes were chosen from the 64 possible route combinations: the Tanjung Priok-Belawan-Dumai-Tanjung Priok route with 11 frequencies per year, the Tanjung Priok-Palembang-Tanjung Priok route with 4 frequencies per year, and the Tanjung Priok-Semayang-Tanjung Priok route with 5 frequencies per year (See Figure 4).

4.5 Vessel Operational Analysis

The Operational Pattern Analysis aims to identify operational patterns for both the vessels and the trucks supplying the loading port. This phase includes two scenarios: Scenario 1, where the trucks operate year-round, and Scenario 2, where the trucks' operational days align with the arrival schedule of the vessels at the loading port. Although Scenario 2 has a significantly smaller storage capacity than Scenario 1, the number of trucks assigned is greater in Scenario 2. As a result, a cost analysis is needed to determine which scenario yields the lowest costs.

4.6 Cost Analysis

The total costs resulting from Scenario 1 and Scenario 2 are IDR 455 billion and IDR 494 billion, respectively. Therefore, the selected scenario in this scheduling simulation-optimization is Scenario 1. Although Scenario 1 incurs significant investment costs for storage capacity, the costs incurred are lower than the expenses related to additional trucks, their operational costs, and the holding costs at the Bio-CNG plant (see Figure 5).

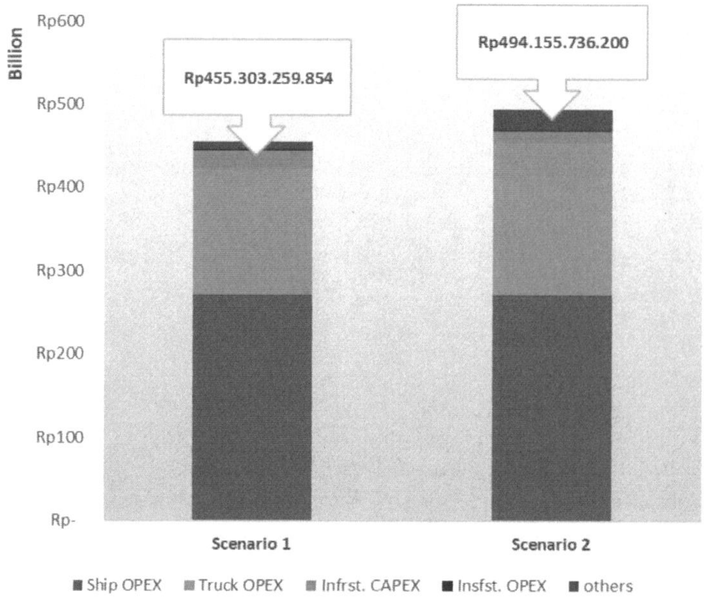

Figure 5. The Comparison of the Shipping and Trucking Total Cost (Rp = IDR)

The highest degree of slope in the cost components is attributed to ship opex and truck opex. Therefore, these cost components are sensitive to changes that can affect the unit cost. This indicates that if the values of ship opex or truck opex increase, the unit cost of Bio-CNG will be significantly higher compared to situations where the capex and opex of infrastructure rise (see Figure 6).

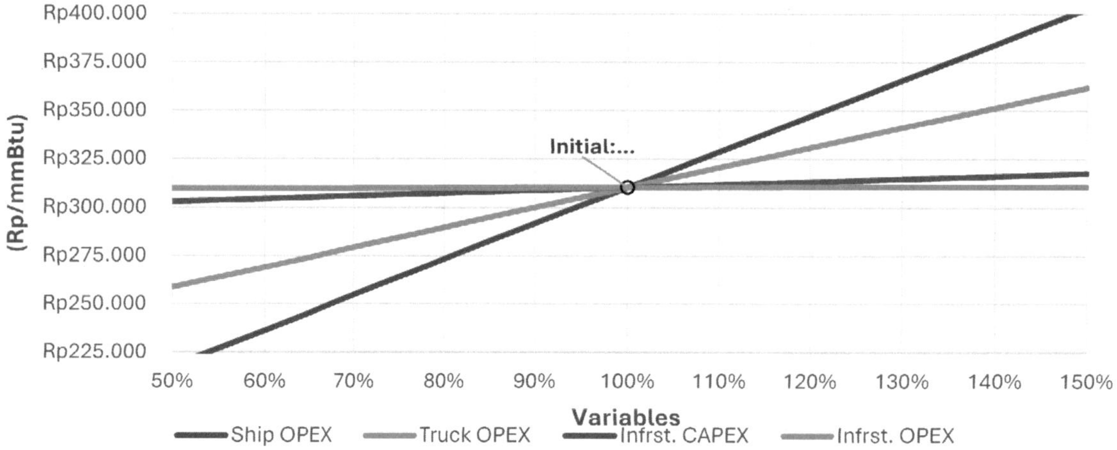

Figure 6. Sensitivity Analysis for the Component of the Costs (Rp = IDR)

The current unit cost of Bio-CNG, when compared to non-subsidized fuels available in the market, falls within the market price range. The highest price is held by CNG German at IDR 561,509/mmBtu, while the lowest is LPG UK at IDR 308,382/mmBtu. Additionally, compared to industrial LPG in the market, Bio-CNG is IDR 119,724 cheaper than 12 kg LPG and IDR 109,724 cheaper than 5 kg LPG (see Figure 7).

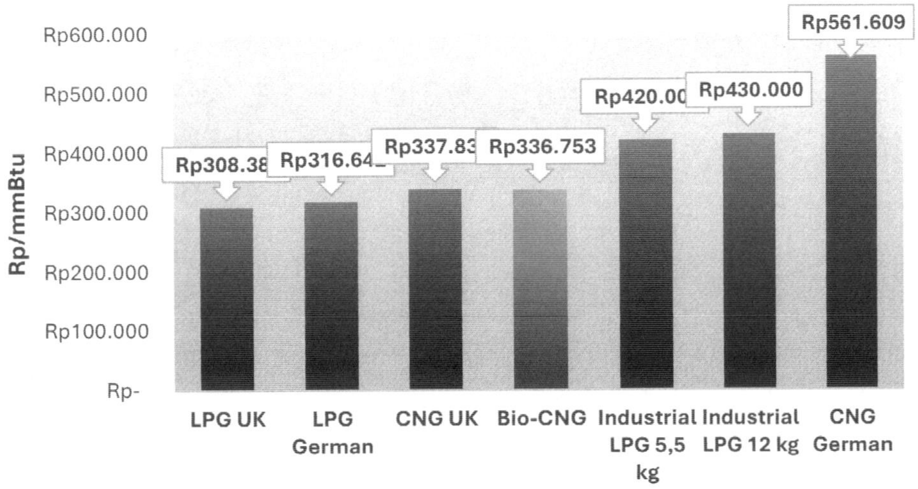

Figure 7. Comparison of Bio-CNG Unit Cost to Current Non-Subsidized Fuel Prices (Rp = IDR)

5. Conclusion

Based on the analysis and data processing conducted in this study, several conclusions can be drawn. First, the demand for Bio-CNG needed to achieve zero emissions in Indonesia's palm oil industry is estimated at 2.18 million tons, equivalent to 104.05 million mmBtu. Second, the distribution model developed includes the establishment of 13 Bio-CNG plants, with nine clusters utilizing a direct scheme and four clusters employing a last-mile scheme, all serviced by tank trucks. Additionally, four Points of Loading (POL) and selected clusters will be delivered using the direct scheme with dedicated GTM trucks throughout the year. The chosen vessel is one CNG ship, following the route from Tanjung Priok to Belawan (75%), then to Dumai (25%), with a total of 11 frequencies per year. Furthermore, there are four frequencies per year for the Tanjung Priok-Palembang-Tanjung Priok route and five frequencies per year for the Tanjung Priok-Semayang-Tanjung Priok route. Lastly, the minimum cost arising from this distribution model amounts to IDR 455 billion per year, or IDR 310,276 per mmBtu annually.

References

1. Luas Tanaman Perkebunan Menurut Provinsi - Tabel Statistik - Badan Pusat Statistik Indonesia [Internet]. [cited 2024 Sep 7]. Available from: https://www.bps.go.id/id/statistics-table/2/MTMxIzI=/luas-tanaman-perkebunan-menurut-provinsi.html (accessed Jul. 16, 2023).
2. Rahayu AS, Karsiwulan D, Yuwono H, Trisnawati I, Mulyasari S, Rahardjo S, et al. Buku Panduan Konversi POME Menjadi Biogas Pengembangan Proyek di Indonesia. Winrock Int [Internet]. 2015;100. Available from: https://www.winrock.org/wp-content/uploads/2016/05/CIRCLE-Handbook-INDO-compressed.pdf
3. Kementerian ESDM RI - Publikasi - Statistik [Internet]. [cited 2024 Sep 7]. Available from: https://www.esdm.go.id/id/publikasi/statistik
4. Ryckebosch E, Drouillon M, Vervaeren H. Techniques for transformation of biogas to biomethane. Biomass and Bioenergy. 2011 May 1;35(5):1633–45.
5. Li L, Ji B, Yu SS, Zhou S, Fang X. Branch-And-Price Algorithm for the Tramp Ship Routing and Scheduling Problem Considering Ship Speed and Payload. J Mar Sci Eng. 2022;10(12).
6. Nikfarjam A, Moosavi A, Rezaei N, Ebrahimnejad S. A green vehicle routing problem with time windows considering the heterogeneous fleet of vehicles: two metaheuristic algorithms. Eur J Ind Eng. 2019;13(4):507.

IOP Conf. Series: Earth and Environmental Science **1461** (2025) 012031 doi:10.1088/1755-1315/1461/1/012031

Alternative Safety Assurance for SPM Buoys Operating in Indonesian Waters by Implementing a New Verification Scheme

Ika Prasetyawan[1*]

[1] Global Offshore Technical Support Office, Lloyd's Register South Asia

*E-mail: Ika.Prasetyawan@lr.org

Abstract. Unlike FPSO where demands follow oil discovery, there are not many SPM buoy assets around being newly fabricated. SPM buoy is considered for offloading operation as an alternative where water depth is limited as compared to jetty terminal. Apart from nearshore operation, deep water SPM buoy has also been an option for offloading operation at location where the hydrocarbon fluid is transferred out either from PLEM or fixed offshore platforms.

Within Indonesian waters, SPM buoys are mandatory to be classed following new regulation PTK-015/SKKMA0000/2022/S9 issued by SKK MIGAS. This document provides both operational and maintenance guidance for special terminals[1]. However, some SPM buoys are in existence as non-classed assets (pre-PTK-15/2022) which would have been operating at reduced maintenance scope to lower operating cost. For this type of SPM buoys, an alternative method exists through verification scheme. A verification scheme introduces performance checking based on appropriate standards on safety-critical elements in terms of their availability, and survivability. Through this performance checking, SPM buoys operators (and most importantly the regulator) would have sufficient level of assurance for SPM buoys safe operation during the remaining of design service life.

The paper' objective is to illustrate the adoption of new verification scheme for SPM buoys operating within Indonesian waters, particularly applicable for assets built before 2022. Firstly, safety-critical elements are identified, then the performance standards are defined with agreed acceptance criteria based on international codes and standards. The safety-critical elements are then assigned with performance statement for basis of approval by the regulator, and where any deviation shall be deemed non-compliance for basis of rectification by the operator. The introduction of such verification scheme is an alternative to full classification scheme that would otherwise be challenging for existing SPM

[1] There are two definitions namely TERSUS (Fasilitas Terminal Khusus) and TUKS (Terminal Untuk Kepentingan Sendiri). Within this guidance document, the definition of "terminal" is associated with jetty terminal, whereas "floating terminal" could be subjectively interpreted as FPSO or SPM buoy.

buoys in operational for more than 10 years that could initially have been built without proper Class certificate.

1. Background

Early suggestion indicated that there are at least fourteen SPM offloading buoys operating in Indonesian waters by 2006 [1]. It is expected that the number of this type of assets might have increased in the last 15 years. Considering that Class certification is mandated by SKK MIGAS via PTK-015/SKKMA0000/2022/S9, many existing SPM buoys might be in operational without Classification. This could be the case for SPM buoys built prior to 2022 which are likely of having no Class certificate as they are mostly private offloading terminals. Furthermore, it seems to be common practices for terminal operators where abandoned SPM offloading infrastructures are re-purposed and / or interchanged with other spare SPM buoys, to facilitate offloading activities with shuttle tankers or new FPSO/FSO at reduced capacity. For these types of SPM buoys, the probability of mooring equipment failure is higher as they could be without active maintenance regime (as well as poor maintenance records); and where mooring system is regarded as barrier to hazards, the consequence of having fatal accident is also higher in comparison to SPM buoy with Classification scheme.

Although implementation of verification scheme is applicable to new built ranging from design, fabrication, installation, and operational stages; this paper discusses on options available for existing SPM buoys where acceptance into Classification scheme is challenging. There are three difficulty / challenging aspects to accept existing SPM buoys into class (Indonesia case), firstly is the material traceability which should come from approved manufacturer and be of acceptable offshore grade type material as required by IACS W22. Secondly, when the mooring chain is ageing with wear and corrosion, the determination of minimum breaking load as a function of chain diameter might not give accurate results unless confirming with either Finite Element Analysis or sample breaking test. Thirdly, experiment has shown that fatigue characteristic of ageing mooring chain may be influenced by wear and pitting corrosion where corrections are to be imposed on the S-N curve (research on this remains at early stage and correction factors are given for R-grade studless chain link only [2]).

In this paper, the existing asset would assume SPM buoy with minimum 10-year service life (with reasons that are explained in Section 3) during which period a substantial chain nominal diameter reduction would have occurred beyond API RP 2 I recommended value. The difficulty / challenging aspects elaborated above suggests that apart from quantitative analysis it is reasonable to argue that for this type of SPM buoy, the responsive actions and hazard mitigation plan is considered more urgent as assurance. It is also within the intention of this paper that the introduction of verification scheme within company level could at least contribute to the safety awareness for the operator.

In the sections to follow, introduction to verification implemented in UK HSE as regulator is provided, then some guides to develop performance standards for SPM buoys are given. The implementation of verification scheme specifically for existing SPM buoys in Indonesian waters is explained further with examples on SPM buoys performance standards and how to sign-off verification work instruction.

2. Introduction to Verification Scheme

2.1 Definition

Historically speaking, the verification scheme was meant to respond the loss of life and environmental damaged caused by major accidents such as Piper Alpha incident in North Sea in 1988 and Macondo blowout in the Gulf of Mexico in 2010. In these two incidents, the regulator (in this case UK HSE and EU respectively) impose the requirement to identify safety-critical elements (SCEs) of offshore assets to be verified as fit for purpose by independent verifier[2]. Within UK case example, following extensive public inquiry on Piper Alpha disaster, the verification scheme was introduced to replace long standing provisions of certification for fixed structures operating in UK waters.

Figure 1. Piper Alpha – before and after the disaster

Under certification scheme, certifying authority, which act as agent of regulator, enforces compliance of agreed international codes and standards for structural integrity of the installations; the certifying authority then issues certificate for fitness to the terminal operator (which is also referred to as the duty holder within context of a verification scheme) with certain validity (e.g. 5 years) subject to periodical inspection. On the other hand, verification scheme is required for the installations to have suitability in terms of appropriateness and effectiveness to manage hazards while maintaining their SCEs performing as intended. The parameters to measure suitability of such SCEs are through their specific performance of achieving goals identified during safety study. Such performance goals are to be verified by an independent competent person (ICP) based on his/her collective experience gathered from relevant cases. In principle, the generic steps involve in verification scheme can be elaborated as follows:

1. Identify major accident hazards from safety study;
2. Determine safety-critical elements (SCEs);
3. Develop performance standards;
4. Undertake verification activities and sign off verification report.

2.2 Contribution of Verification Scheme to Classification

2 There are geographical areas which implement different verification schemes, but only UK HSE version is taken as case example here. While the UK regulator perceive the safety case regulations also contain several additional regulations concerning hazard management, the implementation discussed in this paper is limited to a simple verification activity on SCEs applicable to a simple SPM buoy structure and mooring system which exclude complex processing plant and manning requirement.

IOP Conf. Series: Earth and Environmental Science **1461** (2025) 012031 doi:10.1088/1755-1315/1461/1/012031

The work in verification scheme provides favourable advantages for the duty holder where part of scope in the verification activities can be credited to class requirements on the basis that the identified SCEs can be demonstrated to be suitable in terms of maintaining structural integrity in addition to managing hazards. The verification activities by ICP in this case ensure that the verification scheme shall warrant, by largest part, compliance against class rules and regulations. Within Lloyd's Register the scope of design appraisal for SPM buoy shall cover buoy body structure and buoy mooring system including the use of class approved offshore grade materials (Rules and Regulations for Offshore Units), but to lesser parts of electrical and instrumentation, rigid piping, fire and safety, and flexible submarine hose (non-mandatory for class notation).

2.3 Performance Standards
Performance standards are criteria set to achieve suitability of SCEs in managing hazards. The suitability shall mean of having appropriateness in delivering its functions, and effectiveness when required. A generic approach to develop performance standards would be,

1. Adopt international codes / standards and / or IACS members rules and regulations;
2. Assess any hazards associated with SCEs;
3. Provide attributes to address element's functionality, availability, and survivability, based on following definitions:
    a. Functionality is the main duty that the safety-critical element, at a minimum level, is required to perform;
    b. Availability is the ability of safety-critical element to meet its intended function. This is often best described as a portion of the amount of time the element will be available to perform on demand;
    c. Survivability is the ability to remain function after a survival event such as cyclonic storm.

In point #2 above, the selection of SCEs by the duty holder must use as many available references as possible and consider experience in the past or previous projects to identify a major accident hazard (e.g. integrity failure could lead to riser rupture incident releasing hydrocarbon to the environment). When failure of an elements could cause such major accident hazard, or such element is required to mitigate or prevent against major accident hazard, then that element is considered safety critical (e.g. in SPM buoy case, the mooring system is safety-critical element).

Performance standards should be evaluated periodically to gather information for new learnings and to ensure they are still relevant as required by regulator[3]. Performance standards should also be reviewed when operating mode has changed (e.g. in SPM buoy case, this would be re-deployment to new location or whether the shuttle tanker capacity has changed). During service life, these performance standards may need to be amended with different criteria due to ageing installations which would have different probability to fail and consequences to accidents.

[3] The regulator for asset operating in Indonesian waters is SKK MIGAS.

3. Implementation of Verification Scheme on SPM Buoy

3.1 Preparing Performance Standards

As it appears (and technically so explained in section 2.3), performance standards are to be regularly evaluated when operating mode has changed, hence the implementation can be flexible. Especially for existing SPM buoys operating in Indonesian waters, which are initially not regulated under class, there is alternative acceptance criteria that can be imposed by the regulator to consider compliance via verification scheme on the basis that similar level of safety assurance can be achieved, at least during the remaining service life. It is valid to assume that existing SPM buoys are those where re-classification is challenging or less likely to be accepted especially for SPM buoys operating beyond 10-year service life. Such assumption is based on nominal chain diameter of 76mm ÷ 78mm normally found in most SPM buoys where within 10-year service life (with corrosion rate at 0.8mm/year at tropical waters[4]) the nominal diameter reduction would be around 10% (this is the limit value as per API RP 2 I). In addition to pitting corrosion that would be likely to occur within this period of service, a simple fatigue calculation could probably overestimate the fatigue life of the upper segment of mooring line since simple fatigue does not take effects of pitting corrosion into account. Within such situation, the compliance towards performance standards should be considered sufficient by the regulator and duty holder to ensure assurance against similar level of safety to Classification requirements, during service life. Furthermore, for this type of SPM buoy, the responsive actions and hazard mitigation plan is considered more urgent than assurance based on fatigue aspect for the reason explained above.

A route map for verification starts from scheme development which is then followed by the scheme execution itself. During the scheme development, references shall be made to relevant safety studies to appropriately select the worst scenario leading to potential major accidents[5]. Table 1 represents generic performance standards for SPM buoys where definitions of main designation functions are as follows,

Table 1. Generic Performance Standards of SPM Buoy Mooring System

Example - Performance Standards				<duty holder logo>
SCE Group/Asset: **Mooring System/SPM Buoy**				
SCE objective: **To prevent significant excursion that would lead to overstressing or rupture of riser**				
Performance Criteria	Technical Criteria	Evidence	Verification Task	Verification measure
The positional mooring system shall be capable of maintaining the SPM buoy at intended location.	The positional mooring system shall survive offloading operation under operating conditions, specific to the location	Metocean repot and positional mooring analysis report under site-specific environmental conditions Project specifications	To verify the offloading loads are within limitation of operating strength capacity	The mooring loads conform to have acceptable level `of safety factors for operating conditions The offloading loads conform to

[4] Based on Lloyd's Register survey experience in tropical water the corrosion rate of mooring chain is around 0.8mm – 1mm/year.

[5] Major accidents represent events caused by significant failures of the system and structural integrity that leads into loss of personnel, environmental damage, loss or major damages to asset, and disintegrated company reputation.

	The anchor line loads shall be within allowable limits The hawser load shall be within acceptable threshold load during operating conditions.	providing minimum key requirements for offloading operation		have acceptable limit to prevent overstressing of mooring line, hawser, riser and offloading hose.
The position of the shuttle tanker shall be continuously monitored to ensure integrity of anchor lines.	Monitoring and warning system shall be available to alert anchor line failure.	Operating guidelines to perform safe offloading operation.	To verify operating guidelines contains safety checklist for shuttle tanker berthing and responsive action plans are available in case integrity incidents.	Operating guidelines is in place and active to ensure safe procedures during offloading operation. Warning system conform at good working order.
Availability				
The mooring system shall be 100% available with minimum adverse effects of damaged line.	The positional mooring system shall remain survive under damage condition involving the loss of highest loaded mooring line (leading to highest excursion) and loss of second highest loaded mooring line (loading to highest line tension)	Metocean repot and positional mooring analysis report under damage condition (e.g. one line fails) In-service inspection plan (ISIP) document covering, periodical survey plan, damage assessment plan and post damage recovery plan to minimise lead time for procurement of spares	To verify the positional mooring system remain survive post incident/damage condition To verify that scope of in-water survey follows relevant codes or standards to maintain the unit availability within its design service life	The mooring system conform to remain survive post damage without leading to further line serial accident. The ISIP is active to maintain periodical survey in accordance with international codes and standards, as agreed by the flag The ISIP conforms to have limiting environment during damaged condition before recovery plan is initiated
The mooring system shall be 100% available with minimum adverse effect during the remaining service	The positional mooring system shall survive offloading operation under operating conditions,	Latest underwater report indicating physical condition of chain wear and corroded chain diameter	To verify minimum chain diameter is sufficient and offloading loads are within limitation of	The mooring loads conform to have acceptable level of safety factors for operating conditions, under

life under acceptable wear and corroded chain.	specific to the location with sufficient safety factor under acceptable wear and corroded chain.	measurement data. Risk assessment report incorporating assessment on pitting corrosion aspect with mitigating plan to be provided.	operating strength capacity under acceptable wear and corroded chain. To verify remaining fatigue life by taking pitting corrosion effect into consideration; in case this is not available mitigating plan to be verified and to be cross checked against duty holder safety management system.	acceptable wear and corroded chain. Mitigation plan is active to provide effective and efficient response in case integrity failure due to wear and corroded chain.
Survivability				
The positional mooring system shall survive its station keeping ability under extreme combination of site-specific wave, wind and current	The positional mooring system shall survive extreme conditions under 100-year return period (RP) wave and associated wind and current	Metocean repot and positional mooring analysis report under site-specific environmental conditions for 100-year RP environment Project specifications providing minimum key requirements for surviving 100-year RP environment	To verify the positional mooring system survival strength capacity under 100-year RP environment.	The mooring loads conform to have acceptable level of safety factors for survival conditions under 100-year RP environment.
The positional mooring system shall survive its station keeping ability under extreme combination of site-specific wave, wind and current, under acceptable wear and corroded chain.	The positional mooring system shall survive extreme conditions under 100-year return period (RP) wave and associated wind and current, under acceptable wear and corroded chain.	Latest underwater report indicating physical condition of chain wear and corroded chain diameter measurement data.	To verify the positional mooring system survival strength capacity under 100-year RP environment, under acceptable wear and corroded chain.	The mooring loads conform to have acceptable level of safety factors for survival conditions under 100-year RP environment, under acceptable wear and corroded chain.

Responsible technical authority name	:..................

3.2 Executing Verification Task

The ICP is considered as competent person and responsible to execute verification task. Such person could be nominated from a 3rd party assessor or even selected from discipline engineer of Classification bodies recognised by the regulator; this can be local Class (i.e. Biro Klasifikasi Indonesia or IACS members). It is arguably valid to consider that because the verification scheme is not directly related to issuance of Class Certificate (in this study the applicability is for SPM buoy pre-SKK MIGAS/PTK-15/2022), the provision of verification scheme should not necessarily warrant a dual-Class scheme as in the new-built or TOC cases under Indonesian's flag, in so far as the ICP is agreed as competent person working independently.

In this way, the verification scheme is likely to attract duty holders with flexible options to hire competent persons from either IACS or non-IACS member provided the competency of such persons is agreed by the regulator. The process of selecting competent persons to execute verification task is as simple as examining the curriculum vitae and checking involvement with past projects and be demonstrated as independent agent against conflict of interest.

The ICP shall work based on verification instruction and using performance standards as reference. The verification task should ensure the following key elements are addressed,

1. The appropriateness of performance standards for SCE, integrating key aspects of Classification rules and regulations and / or international codes and standards;
2. Provision to efficiently arrange way of communication between ICP and duty holder for review and recommendation, and for close out statement.

The performance standards as reference must be approved by the regulator and the ICP must work following a verification work instruction (VWI) document, as depicted in Table 2, acting as guidelines with step-by-step activities to ensure performance standards have been verified appropriately. Once the verification's steps have been completed, the ICP would need to sign-off the work instruction as part of the evidence document to ensure that the asset has been verified against performance standards. It should be noted that the VWI document would bear ICP's company logo although this is not Class Certificate but the signing-off would imply that all aspects of performance standards have been verified by independent person in the same manner as review scope by Classification (noting that all items in the performance standards need to be agreed by regulator and they are indeed extracted from requirements by Classification standards).

The difference with Classification scheme is on the freedom from duty holder to manage their own inspection regime and verification against non-offshore material. Although such verification scheme does not provide a system to bind the duty holder with responsibility to do routine maintenance of the asset, the performance standards dictate the requirement to have ISIP documentation where periodical survey plan must be active throughout the service life cycle. Because the performance standards are dynamic document, the regulator may impose mandatory to report periodical survey campaign annually or at time interval set-up in the ISIP document.

Table 2. Verification Work Instruction

Example – Verification Work Instruction	<duty holder logo> <ICP company logo>
SCE Group/Asset: **Mooring System/SPM Buoy**	
INSTRUCTIONS: The following activities shall be carried out to verify compliance of mooring system against performance standards: 1. Review the Basis of Design and confirms the document specifies and complies with appropriate codes and standards for the mooring system. 2. Review of the following aspects: a. SPM buoy mooring layout plan b. SPM buoy operation manual and in-service inspection plan 3. Review of material specifications (chains, connectors, tri-plates, anchors, stoppers) 4. Review on mooring analysis for the following aspects, a. Underwater report data to validate remaining MBL of otherwise wear and corroded mooring chain b. Environmental data (in case of redeployment this should be site specific) c. Design load report (for survival and offloading cases) for strength and fatigue, considering #a, for various critical load cases 5. Anchor holding capacity calculation (in case of redeployment this should be site specific soil condition)	
Completion / Signing-Off	
ICP Name :	
ICP Signature :	
Date :	

4. Conclusions

There are many SPM buoys in operational which may not be covered by Class rules. This would be typical SPM buoys in existence of pre-SKK MIGAS/PTK-15/2022. For this type of SPM buoys, implementation of Class rules is challenging due to the wear and corrosion state of the mooring chain in which determining MBL is not as accurate unless with sample breaking test and where the fatigue assessment must incorporate correction factors for wear and pitting corrosion. With these difficulty factors, an alternative and efficient means of providing safety assurance for SPM buoy is therefore required.

The introduction of new verification scheme can deliver strategic approach for SPM buoy terminal operator (the duty holder) providing alternative standard to have safety assurance for SPM buoys. In so far as Lloyd's Register experience involving verification-based projects, this is considered acceptable because the performance standards are basically derived from a combined rules and regulations from Classification, international codes, as well as incorporating some elements on risk assurance.

The VWI document is the deliverable from verification scheme signed-off by ICP which could be a competent person from respectable Classification bodies either IACS or non-IACS members consequently, this can be considered as legal document stating the level of safety assurance of assets (in this case SPM buoys). Such legal document can be regarded as valid reference for any legal business that may be conducted by the duty holder (e.g. obtaining marine insurance).

The regulator (SKK MIGAS) involvement in approving the performance standards would give added confidence that the scope of verification is sufficiently appropriate. The regulator, by examining performance standards against PTK-15/2022, would be able to communicate with the duty holder or ICP should there be a need to amend the content to include special attentions due to specific circumstances of the project/asset. This demonstrates the flexibility of performance standards document as flexible, appropriate, and efficient reference for safety assurance.

The implementation of verification scheme can be extended into other types of floating assets, as well as to incorporate fabrication and installation stages with the aim to encourage safety appreciation throughout the provision of operating floating offshore installation at fixed location.

Nomenclature

API RP	American Petroleum Institute Recommended Practices
FPSO	Floating Production, Storage, and Offloading
FSO	Floating Storage, and Offloading
HAZID	Hazard Identification
IACS	International Association of Classification Societies
ICP	Independent Competent Person
ISIP	In-service Inspection Plan
PLEM	Pipeline End Manifold
RP	Return Period
SCE	Safety Critical Element
SKK MIGAS	Satuan Kerja Khusus Kegiatan Minyak dan Gas Bumi
SPM	Single Point Mooring
TOC	Transfer of Class
UK HSE	United Kingdom Health and Safety Executive
VWI	Verification Work Instruction

References

[1] M. Irfan, D. R. Maulani and M. Tasrief, "Present Condition of Existing Single Point Mooring in Indonesia Waterways in View of BKI," *11th International Conference on Marine Technology,* 2018.

[2] P. M. V. L. V. S. Felipe A. Rezende, "A Practical Procedure for Fatigue Assessment of Mooring Line Chains With Pitting Corrosion," *ASME Journal of Offshore Mechanics and Arctic Engineering,* 2024.

[3] SKK MIGAS, Pelaksana Kegiatan Usaha Hulu Minyak dan Gas Bumi, Pedoman Tata Kerja, 2022.

[4] P. Campbell and H. Sewell, "Experience with Verification Schemes Under New UK Offshore Regulations," *Lloyd's Register Technical Association,* 1996.

[5] I. Prasetyawan, "Development of 3rd Party Verification and In-service Inspection Guidance Notes for SPM Offloading Buoy," Internal Lloyd's Register, 2021.

[6] American Petroleum Institute, Design and Analysis of Stationkeeping (API RP 2 SK), June 2015.

[7] American Petroleum Institute, In-service Inspection of Mooring (API RP 2I), June 2015.

[8] Noble Denton Europe Ltd, JIP FPS Mooring Integrity, 2005.

[9] E. Fontaine, A. Kilner, C. Carra, D. Washington, K. Ma, A. Phadke, D. Laskowski and G. Kusinski, "Industry Survey of Past Failures, Pre-emptive Replacements and Reported Degradations for Mooring Systems of Floating Production Units," *Offshore Technology Conference,* no. OTC-25273-MS, 2014.

[10] I. Prasetyawan, "Mooring System of Floating Offshore Units - Mooring Integrity (Internal Presentation)," Lloyd's Register, Kuala Lumpur, 2021.

Effect of triangle plate positioning on mooring line tension of single point mooring system

Murdjito[1], Madea Eka Silfiani[1], Ferdita Syalsabila[1*], Sujantoko[1], Santi Frestiqauli[1], Eko Budi Djatmiko[1]

[1] Department of Ocean Engineering, Institute of Technology Sepuluh Nopember, Surabaya, Indonesia

[*]E-mail: fsyalsabila@its.ac.id

Abstract. The marine transportation of crude oil from tankers to final terminals is facilitated using pipelines and oil terminals. The predominant form of the terminal is the Single Point Mooring system. The system is crucial as it anchors tankers and transfers crude oil from the ship to the system using floating hoses. Thus, periodic inspections are needed on the Single Point Mooring structure during its lifespan. In an inspection, there was a finding the Single Point Mooring 17,500 DWT vessel required rapid overhaul. Acquiring a substitute Single Point Mooring (SPM) system with 150,000 DWT in capacity is necessary to ensure the uninterrupted continuation of crude oil loading and unloading activities. However, a different mooring system was found between these two. Thus, a redesign of the mooring system is configured and analyzed in this paper. This paper investigates the strength of the mooring lines and the triangle plates. The different positions of triangle plates are also compared to give recommendations for the mooring system. The research findings indicate that the mooring system design satisfies the API RP 2SK criterion, both intact and damaged.

Keywords: *Damaged analysis, single point mooring, mooring line tension, triangle plate*

1. Introduction

Oil and gas are essential energy sources in transportation fuels. This energy source can be taken from oil and gas wells onshore or offshore. Oil and natural gas distribution from offshore can be done through transportation such as tankers or pipelines. Tankers are preferred because they can transport large quantities and are considered economical. Onshore or offshore terminals are the most used oil terminals. The most common type of offshore terminal is the Catenary Anchor Leg Mooring (CALM) system, a form of Single Point Mooring (SPM) [1]. This single-point molding (SPM) can anchor the tanker and channel petroleum from the tanker through a floating hose connected to the SPM, channelled ashore through an undersea pipeline. The SPM mooring system can rotate 360 degrees (weather vanning) so that in its operation, the tanker can rotate in the direction of the environmental load. Since introducing the CALM SPM system in 1958, it has operated 85% of the world's 700 oil terminals, with the most significant operating area in Southeast Asia, the Middle East, and West Africa [2].

Single Point Mooring (SPM) is used mainly in ship-shaped structures. This system allows the ship to move or rotate (weathervane) to minimize environmental loads, such as wind, currents, and waves that hit the ship's structure [3]. The catenary mooring line is moored on the seabed and connected to a floating substructure [4,5]. It has a slight flexural stiffness so that the tension in the mooring line holds the external load. External loads include the weight of the substructure, hydrodynamic tensile forces in the normal, tangential, and bi-normal directions, and additional inertial forces originating from the surrounding seawater [6].

The mooring line in the single-point mooring system (SPM) will receive a combination of tensile, bending, and fatigue [7]. Therefore, the dynamic modelling and analysis of the single-point mooring system is essential to ensure the safety and reliability of the mooring line [8]. In addition, due to the high cost of operating offshore facilities (including SPM) and their susceptible operation, reliability and risk assessment of their design and operation, particularly the operation of mooring lines, are complicated [9].

From the research that has been conducted regarding the reliability of the FSO mooring line, it is explained that the reliability of the mooring line is influenced by the pretension on the mooring line, MBL, and the environmental conditions where the FSO is located as well as the load from the FSO . Furthermore, the analysis of the reliability of the mooring system in SPM in extreme wave and current conditions has also been carried out by Eghbali et al. (2018). The study shows that increasing or decreasing the diameter of the mooring line will also affect the force that occurs on the mooring line. In addition, the behaviour of waves and currents shows that when the direction of the wave and current is closer to the direction of each mooring line, the force on the mooring line also increases [9].

Heraspatty (2016) also carried out an analysis of SPM. In another study, an analysis of the mooring system will also be carried out in SPM. The study concluded that the greater the tension that occurs on the mooring line, the smaller the reliability of the mooring line. In addition, it was concluded that there are three critical components in a mooring system: mooring line, triangle plate, and chain stopper [10].

Single Point Mooring (SPM) structure is inspected periodically during its operation. This inspection aims to check the condition of the Single Point Mooring (SPM) structure so that it can continue operating correctly. A study case is chosen for this research. At the time of the inspection, a SPM structure, SPM 17,500 DWT, was stated to be overhauled immediately. A replacement SPM is needed for the fuel oil loading-unloading operation at the site. The replacement SPM used to replace SPM 17,500 DWT during the docking process is Pertamina's SPM 150,000 DWT.

SPM 17,500 DWT during its operation was moored with six mooring lines with a diameter of 63.5 mm, while SPM 150,000 DWT was moored with four mooring lines with a diameter of 135 mm. Due to the difference in the number of mooring lines and the diameter of the mooring lines, it is necessary to modify the configuration from 6 to 4 mooring lines. The configuration modification combines the existing mooring line (63.5 mm) with a new one (135 mm) connected by a triangle plate. So, this research will discuss the modification design of the mooring system. The triangle plate will be varied to obtain an appropriate modification of the mooring system design. Local analysis is also carried out to determine the stress on the triangle plate locally.

2. Mooring System

2.1 Single Point Mooring

The mooring system has the function of limiting the movement of the ship due to wave loads so that it remains in position. In API RP 2SK (2005). In this research, an SPM mooring system is used. SPM consists of a buoy body and a mooring line component that connects the buoy to the seabed. This system allows the ship to move or rotate (weathervaning). It is usually used in ship-shaped structures. This system allows the ship to weathervaning to minimize environmental loads, such as wind, currents, and waves that hit the ship's structure. According to API RP 2SK (2005), there are many variations in Single Point Mooring (SPM), including Turret Mooring, a mooring system in which multiple mooring lines are attached to the turret, which includes bearings to allow the ship to rotate in the direction of the environmental load. The second one is CALM (Catenary Anchor Leg Mooring). It is a buoy equipped with several catenary chain legs moored to the seabed. Some mooring systems use synthetic rope hawsers and attach between the ship and the buoy. The last one is SALM (Single Anchor Leg Mooring). It consists of a structure permanently attached to the seabed utilizing piles. A bearing system allows a portion of it to rotate around a fixed geostatic part [3].

2.2 Triangle Plate

Triangle plate is a type of *connecting links*. According to API RP 2SK (2005), *connecting links* are used to connect the main components of a *mooring line*. *Connecting links* must be checked through testing in accordance with applicable standards [11]. The detail configuration of triangle plate is shown in Figure 1.

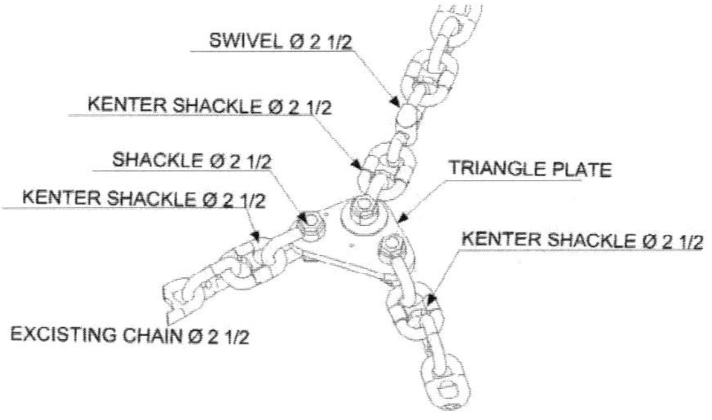

Figure 1. Triangle plate configuration

2.3 Mooring Line

In designing a mooring line so that the floating building can be in the position according to the purpose of its mooring, we must determine the appropriate length and pretension. The configuration of mooring line is shown in Figure 2. Equation 1 is a formula to determine the length of a mooring line [12]. l_{min} is minimum length of mooring rope (*chain line*), h is the vertical distance from *Fairlead* to *Seabed,* which is the summation of water depth and the fairleads height above waterline. W and T are the weight of mooring rope in water per unit length and maximum tension of mooring line (pre-tension), respectively.

$$l_{min} = h\left(2\frac{T_{max}}{wh} - 1\right)^{\frac{1}{2}} \quad (1)$$

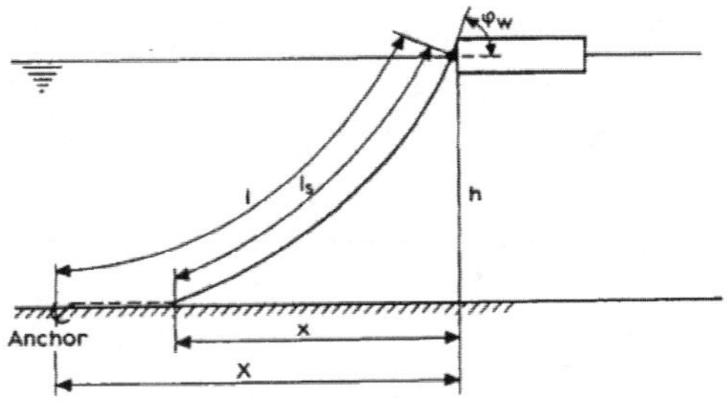

Figure 2. Mooring line length

The minimum distance of the mooring line is considered so that the heavy anchor chain does not experience too much stress so that the tension is distributed and still in the safe condition while operation [13]. The calculation of the minimum distance of the mooring line can be found by the Equation 2.

$$x = l - h\left(1 + 2\frac{a}{h}\right)^{0.5} + acosh^{-1}(1 + \frac{h}{a}) \quad (2)$$

The tension on the mooring rope must be in accordance with *the safety factor* criteria contained in the *rule. The safety factor* recommended by API RP 2SK (2005) for *intact condition* (ULS) is 1.67 while for *damage* condition (ALS) is 1.25. With the safety factor calculation is shown in Equation 3 [3].

$$Safety\ Factor = \frac{Minimum\ Breaking\ Load\ (MBL)}{Maximum\ Tension} \quad (3)$$

3. Response Amplitude Operator of Tanker

3.1 Tanker Model and Data

Based on the data of the 35,000 DWT tanker line plan, a hydrostatic analysis of the ship using to get RAO (Response Amplitude Operator), added mass, Damping forces and Wave Drift. The ship's hydrostatic analysis is used as an Input for mooring system analysis for the next Finite Element Method Analysis [14]. Analysis Tanker carried out under loading conditions full load and Ballast. Principal dimension Tanker used in this research is shown in Table 1.

Table 1. Principal dimension of Tanker

Parameters	Unit	Value	
		Full load	Ballast
Length Between Perpendicular (LPP)	m	175	175
Length Over All (LOA)	m	182.55	182.55
Breadth	m	27.34	27.34
Height	m	16.7	16.7
Draft	m	10.69	6.69
Displacement	MT	43,550.4	24,799.1

Response Amplitude Operator (RAO) is carried out under *full load* and *ballast* conditions with seven directions of wave loading, namely, $0°, 30°, 60°, 90°, 120°, 150°, 180°$ [15]. The RAO Movement Response is divided into two, namely, translation and rotation. The RAO movement response for translational movements (*surge, sway, heave*) is a comparison between the amplitude of the movement compared to the amplitude of the incident movement. Rotational RAO movement (*roll, pitch, yaw*) is a comparison between the amplitude of rotational movement and the slope of the wave or in degrees per meter [16,17]. The following is the RAO of the 35,000 DWT tanker at *full load* and *ballast conditions.*

3.2 RAO of Full Load Condition

Results of the analysis of the motion response of the RAO tanker 35,000 DWT condition full load can be shown in Figure 3. Surge RAO motion response on the condition full load occurs when the waves come from the direction of 0° at the frequency of 0.251 rad/s, which is 0.936 m/m. Meanwhile, the smallest surge motion occurs when the wave comes from the direction of 180° about 0.001 m/m. For the sway movement, the largest occurred in the direction of the wave 90° at a frequency of 0.251 rad/s, which is 0.951 m/m. Sway motion is getting smaller as the wave frequency increases. In the condition full load RAO Movement Heave has almost the same value at the lowest frequency in all directions of the coming wave which is about one. For the largest heave motion occurs from the 90° loading condition about 1.381 m/m at a frequency of 0.698 rad/s.

Rotational RAO motion responses (roll, pitch, yaw) are also presented in Figure 3. Roll motion with the highest degree occurs from the direction of 90°, at a frequency of 0.393 rad/s, which is 1.813 deg/m. The second motion is pitch which around 1.505 deg/m occurs from the direction 30° at a frequency of 0.524 rad/s. The last motion, yaw is the smallest motion than other rotational motions. Yaw movement, occurs around 0.3 deg/m at a frequency of 0.661 rad/s.

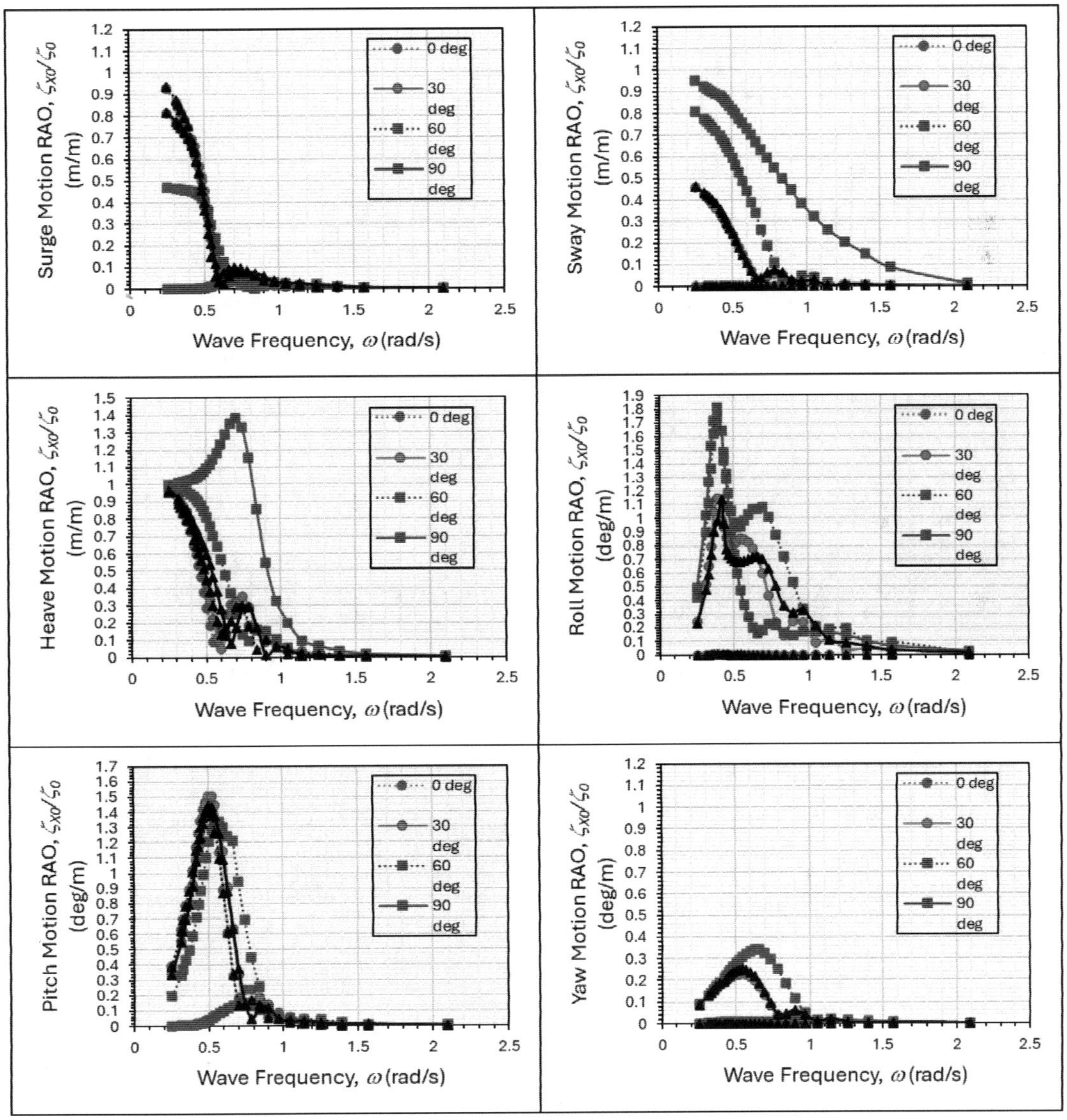

Figure 3. Response amplitude operator of 35.000 DWT tanker in full load condition

3.3 RAO of Ballast Condition

Figure 4 presents the RAO of tanker in ballast condition. Surge RAO motion response on the ballast condition occurs almost same with full load condition at the frequency of 0.251 rad/s, which is around 0.9 m/m. Furthermore, the sway and heave motions are also occur in the same condition as full load condition. However, the motions in ballast are slightly larger, which are 0.973 and 1.162 meter per meter for sway and heave, consecutively.

Rotational RAO motion responses (roll, pitch, yaw) are also presented in Figure 4. Similar with the translational motion, largest rotational motion responses are roll, pitch, and yaw. Roll motion response are five times higher which is 5.613 deg/m that occurs when the waves come from the direction of 90°, at a frequency of 0.785 rad/s. The second motion is pitch is smaller than full load condition which is 1.298 deg/m occurs from the direction 30° at a frequency of 0.546 rad/s. The last motion, yaw is the smallest motion than other rotational motions. Yaw movement, occurs around 0.4 deg/m at a frequency of 0.661 rad/s.

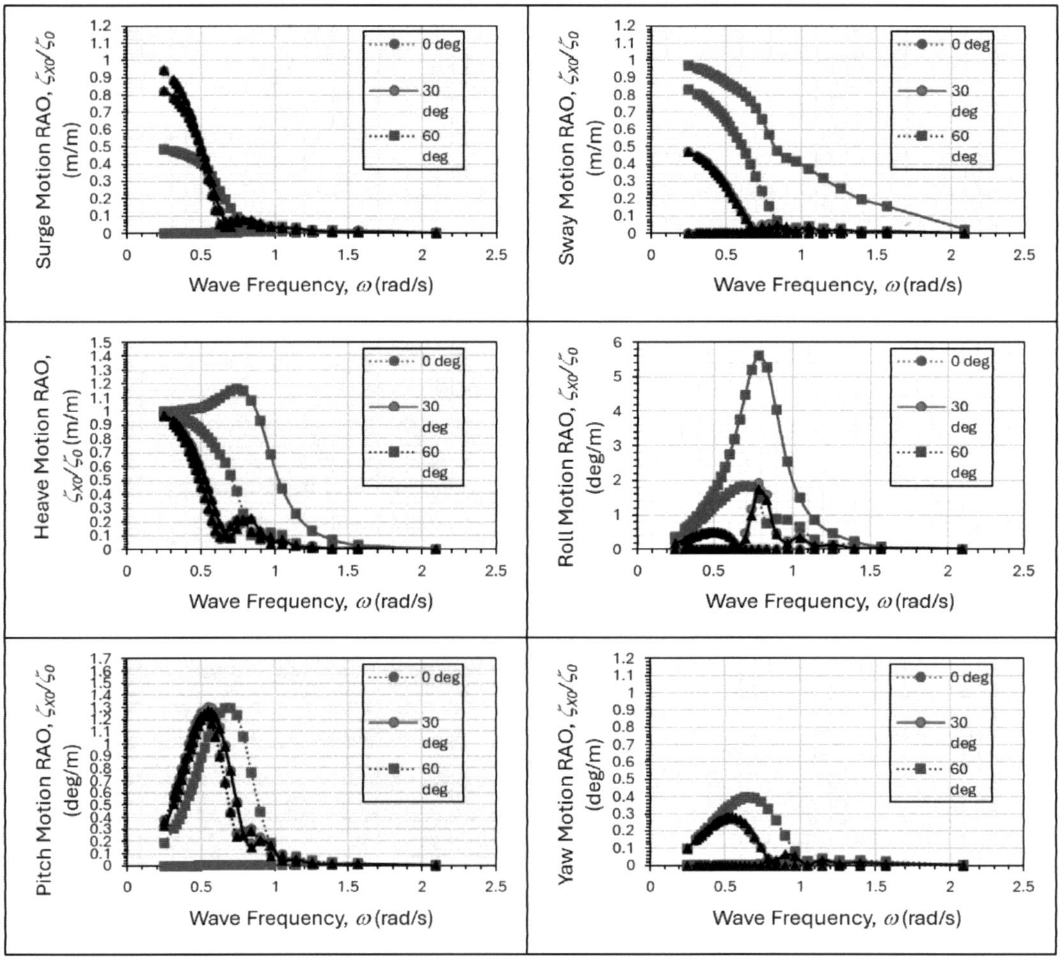

Figure 4. Response amplitude operator of 35.000 DWT tanker in ballast condition

IOP Conf. Series: Earth and Environmental Science **1461** (2025) 012032 doi:10.1088/1755-1315/1461/1/012032

4. Analysis of Mooring System

4.1 Data and Modelling

The SPM system is consisted of several parts, the structure of SPM itself, tanker, mooring line, hawser, and triangle plate. These components specifications are stated in Table 2 for SPM, Table 3 for mooring line, and Table 4 for Hawser. Mooring system analysis is conducted in finite element-based software. Mooring lines are analysed in two scenarios, intact and damaged condition. Intact condition is the condition which all the mooring lines are in their operational condition while damaged condition is analysed if a single line is snapped.

For the analysis of the condition of SPM (*Single Point Mooring*) is carried out with SPM conditions *Stand Alone* and operations. SPM Condition *Stand Alone* is when SPM is tethered with four *Mooring Line* without *Tanker* with a 100-year environmental charge or *Extreme*. Meanwhile, the operating condition at the time of SPM was tethered with four *Mooring Line* and there is *Tanker* who are tethered to SPM with environmental burden for 2 years or operating conditions. The environmental data is shown in Table 5. Figure 5 and Figure 6 presents the configuration of SPM model.

Table 2. SPM Specification

Parameter	Unit	Data
Outside Diameter	m	11
Inside Diameter	m	3.73
Height	m	4.8
Draft	m	2.62
Capacity	DWT	150.000
Mooring Chain	-	4 legs

Table 3. Mooring Line Specification

Description	Unit	New Mooring Line	Mooring Line Existing
Type	-	Chain, U2 Studlink	Chain, U2 Studlink
Mooring Chain	-	4 legs	6 legs
Chain diameter	mm	135	61.8
MBL	kN	8289.46	1848

Table 4. Mooring Hawser Specification

Mooring Hawser			Subsea Hose		
Parameter	Unit	Data	Parameter	Unit	Data
Type	-	Double Braided Grommet	Type	-	Double Carcass
Material	-	Rope/Nylon	Length	-	10.7
Size	Mm	96 dia	Nominal Bore	Inch	16
Length	m	54.86	MBR	m	1.6
			Weight in Air	Kg	1833

Table 5. Environmental data

Environment parameter	Return period	
	2 Year	100 Year
Significant wave height (m)	2.32	3.35
Peak period (s)	7.0	7.40
Maximum wave height (m)	4.0	6.0
Wind Speed (m/s)	18.01	29.84
Current speed at surface (m/s)	0.7	1.5
Current speed at near bottom (m/s)	0.6	1.2

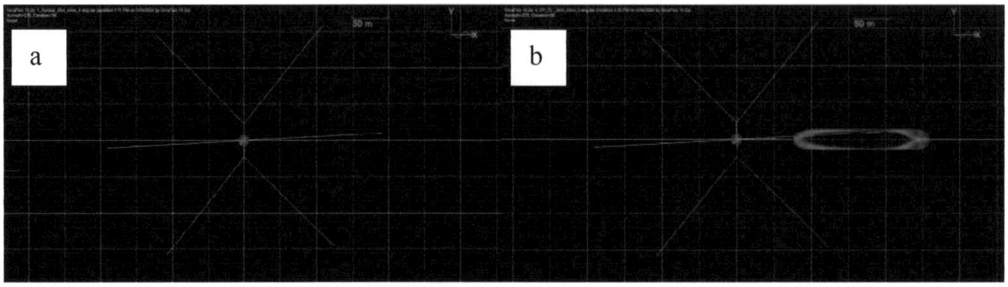

Figure 5. Configuration of stand-alone SPM model (a) and operational condition (b)

Analysis tension mooring line performed with a variety of positions triangle plate, namely at 26 m, 31 m, and 36 m from Chain Stopper. Mooring line that are connected to Triangle Plate be Mooring Line two and Mooring Line four. Triangle plate used to connect mooring line existing with Mooring Line new. Analysis of variations in triangle plate done to find out where the placement is triangle plate the most appropriate. On the analysis tension mooring line Environmental Burden is carried out from the direction of loading in line towards Mooring Line 1 (0°), between line (45°), and in line towards Mooring Line 4 (90°). The following Table 6 shows variations in the placement position triangle plate done for analysis Tension at Mooring Line. In the tension mooring line analysis, the safety factor criteria used are the criteria in API RP 2SK. In API RP 2SK it is explained that the minimum Safety Factor mooring line for intact conditions is 1.67 while for damaged conditions is 1.25. Thus, the maximum tension allowed based on API RP 2SK is the minimum breaking load (MBL) divided by the Safety Factor.

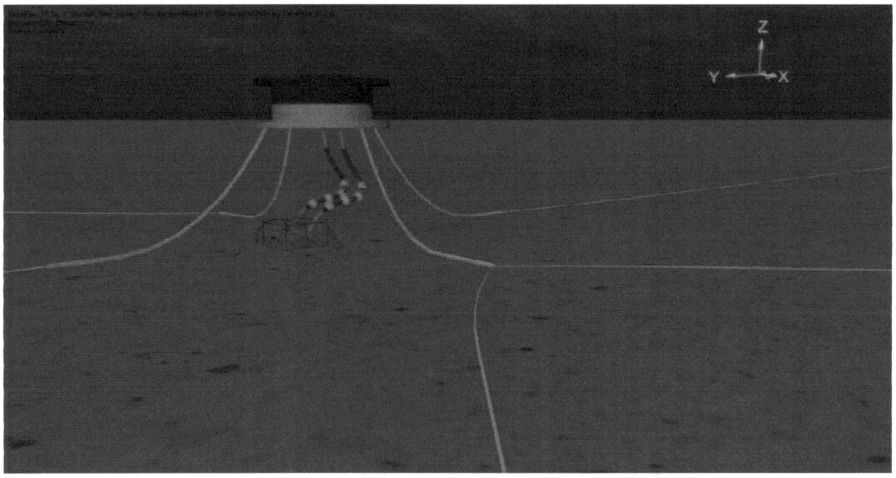

Figure 6. Model of SPM system in stand alone condition

4.2 Tension of mooring line in intact condition

There are nine analysis scenarios for each position variation *triangle plate*, so there are a total of twenty-seven analysis scenarios for all loading conditions *intact* on position variations *triangle plate*. Table 6 shows the loading scenario of each position variation *triangle plate.* Thus, the total variation in this research is 27.

Table 6. Loading condition of intact scenario

Load Case	Environmental Load	Heading (deg)	Vessel Draught
1	Extreme	0 (in-line mooring line 1)	No Vessel
2	Extreme	45 (between line)	No Vessel
3	Extreme	90 (in-line mooring line 4)	No Vessel
4	Operation	0 (in-line mooring line 1)	Full Load
5	Operation	45 (between line)	Full Load
6	Operation	90 (in-line mooring line 4)	Full Load
7	Operation	0 (in-line mooring line 1)	Ballast
8	Operation	45 (between line)	Ballast
9	Operation	90 (in-line mooring line 4)	Ballast

Under loading conditions Extreme SPM is tethered with four Mooring Line without being tethered to Tanker. The environmental loading condition used is environmental loading Extreme with 100 years of environmental data. The results of maximum tension are shown in Table 7. The influence of triangle positioning on mooring line is presented in Figure 7 and Figure 8.

Table 7. Maximum tension of each mooring line for extreme and intact condition

Case	Variations Position	Line1 [kN]	Line2(A) [kN]	Line3 [kN]	Line4(A) [kN]	Line2(B) [kN]	Line4(B) [kN]
1	26 m	283.77	90.76	90.51	85.21	41.21	37.64
2	26 m	203.38	97.56	100.73	170.68	50.94	126.09
3	26 m	110.67	120.46	106.14	269.48	85.67	236.29
10	31 m	275	134.97	90.43	121.9	85.4	71.82
11	31 m	211.84	106.14	91.08	243.11	58.09	203.88
12	31 m	109.49	119.81	101.81	332.13	72.9	306.09
19	36 m	261.24	285.53	97.33	295.18	238.81	250.36
20	36 m	207.24	364.13	89.95	385.18	324.88	347.14
21	36 m	108.28	372.15	101.93	462.53	328.26	435.65

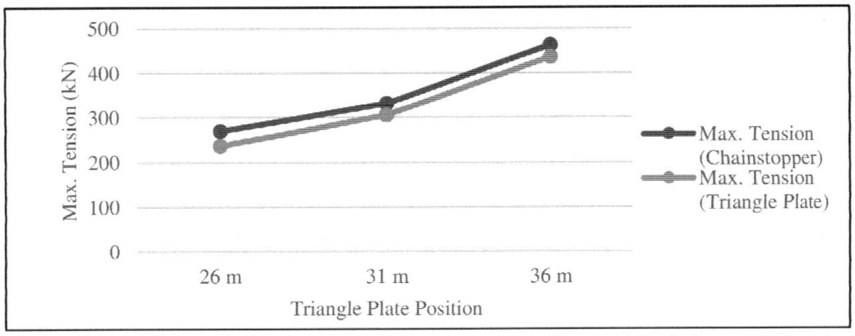

Figure 7. Effect of triangle position to maximum tension mooring line at chainstopper and triangle plate for extreme and intact condition

The condition of the SPM operation is tethered with four mooring lines moored with tankers. The environmental loading conditions used are the loading of the operating environment with 2-year environmental data. The tanker is modelled in two conditions, namely full load and ballast conditions.

Analysis for the effect of position variation triangle plate done on Mooring Line 4 with the direction of loading in line towards triangle plate (90°). Figure 8 shows the influence of variation triangle plate at Mooring Line 4. Meanwhile the detailed results shown in Table 8 and Table 9.

Table 8. Maximum tension of each mooring line for operational and full load condition

Case	Variations Position	Line1 [kN]	Line2(A) [kN]	Line3 [kN]	Line4(A) [kN]	Line2(B) [kN]	Line4(B) [kN]
4	26 m	224.24	80.01	111.02	80.87	41.21	37.44
5	26 m	184.71	93.40	83.37	155.49	51.65	113.30
6	26 m	107.64	81.46	100.38	236.01	50.59	199.65
13	31 m	212.15	248.46	62.41	237.75	205.23	192.95
14	31 m	167.29	218.47	87.52	308.61	176.38	266.04
15	31 m	93.69	195.91	76.47	336.88	153.41	298.82
22	36 m	209.81	273.88	127.35	263.28	231.31	216.78
23	36 m	176.76	249.90	73.66	350.13	208.23	311.40
24	36 m	100.38	222.59	92.40	405.20	180.66	371.06

Table 9. Maximum tension of each mooring line for operational and full load condition

Case	Variations Position	Line1 [kN]	Line2(A) [kN]	Line3 [kN]	Line4(A) [kN]	Line2(B) [kN]	Line4(B) [kN]
7	26 m	211.72	78.2	128.4	81.16	40.8	37.12
8	26 m	186.49	220.48	86.91	154.66	94.66	109.38
9	26 m	106.86	97.38	102.85	225.17	52.91	188.57
16	31 m	203.62	248.69	63.22	243.6	204.4	198.99
17	31 m	163.09	218.82	107.75	301.56	176.56	259.12
18	31 m	92.03	205.4	78.21	328.94	162.66	293.82
25	36 m	204.37	274.7	117.24	268.71	230.53	224.12
26	36 m	165.2	252.04	75.31	346.39	211.01	305.58
27	36 m	98.55	231.71	93.72	387.61	189.41	352.56

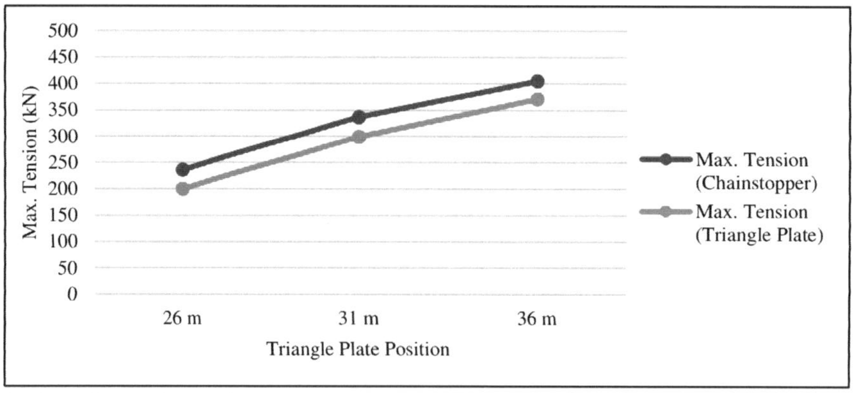

Figure 8. Effect of triangle position to maximum tension mooring line at chainstopper and triangle plate for operational and full load condition

4.3 Tension of mooring line in damaged condition

The effect of failure of one mooring line to the reliability of the damaged mooring system is significant [18]. Thus, the mooring line in damage condition is analysed using time domain. In this research, Mooring line number one is chosen to be damaged. There are six scenarios taken from the results Tension the largest of condition analysis intact on environmental loading Extreme and operation with a variety of positions triangle plate. Table 10 shows the loading scenario for the damaged.

Table 10. Loading condition of damaged scenario

Load Case	Environmental Load	Heading (deg)	Vessel Draught
3	Extreme	90 (in-line mooring line 4)	No Vessel
12	Extreme	90 (in-line mooring line 4)	No Vessel
21	Extreme	90 (in-line mooring line 4)	No Vessel
6	Operation	90 (in-line mooring line 4)	Full Load
15	Operation	90 (in-line mooring line 4)	Full Load
24	Operation	90 (in-line mooring line 4)	Full Load

In the condition damaged with environmental burden Extreme SPM is tethered with three Mooring Line without being tethered to Tanker. Environmental loading conditions Extreme with 100 years of environmental data. Table 11 shows the maximum tension of damaged scenario.

Table 11. Maximum tension of each mooring line for damaged scenario

Case	Variations Position	Line2(A) [kN]	Line3 [kN]	Line4(A) [kN]	Line2(B) [kN]	Line4(B) [kN]
3	26 m	452.22	92.21	286.02	345.4	258.26
12	31 m	181.28	93.82	460.43	122.98	445.36
21	36 m	381.31	93.2	617.32	331.62	588.17
6	26 m	160.75	585.3	603.01	91.22	582.82
15	31 m	299.83	549.43	613.67	263.2	591.57
24	36 m	387.78	568.36	724.25	353.67	698.15

From the Table 12, it shows that the placement of the triangle plate affects the magnitude of the tension that occurs on the mooring lines. From all the analysis of intact and damaged conditions, the largest tension mooring line was obtained at the triangle plate position of 36 m, then the triangle plate position was 31 m and the lowest tension mooring line was on the mooring line with the triangle plate position of 26 m. The tension on all mooring lines meets the criteria contained in API RP 2SK, namely the minimum safety factor for intact conditions of 1.67 and for damaged conditions of 1.25.

Table 12. Overall results of mooring line tension in intact and damaged condition

Case	Env. Load	Triangle Position	Line 4(A) [kN]	MBL [kN]	SF 4(A)	Status
Intact condition						
3	Extreme	26 m	374.6	8289.46	22.13	OK
12	Extreme	31 m	528.13	8289.46	15.7	OK
21	Extreme	36 m	643.36	8289.46	12.88	OK
6	Operating-Full Load	26 m	535.2	8289.46	15.49	OK
15	Operating-Full Load	31 m	588	8289.46	14.1	OK
24	Operating-Full Load	36 m	666.01	8289.46	12.45	OK
Damaged condition						
3	Extreme	26 m	286.02	8289.46	28.98	OK
12	Extreme	31 m	460.43	8289.46	18	OK
21	Extreme	36 m	617.32	8289.46	13.43	OK
6	Operating-Full Load	26 m	603.01	8289.46	13.75	OK
15	Operating-Full Load	31 m	613.67	8289.46	13.51	OK
24	Operating-Full Load	36 m	724.25	8289.46	11.45	OK

Conclusion

Analysis of mooring line system is conducted by varying the triangle plate position. In intact and damaged conditions and on all load cases, the smallest mooring line tension is at the modification of the triangle plate position of 26 m, then the triangle plate position is 31 m, and the largest tension mooring line occurs at the triangle plate position of 36 m from the chain stopper, which is 666.01 kN. Likewise with the damaged condition for all load cases, the largest tension mooring line also occurred at the position of the triangle plate 36 m from the chain stopper, which was 724.25 kN. All tensions in intact and damaged conditions also meet the criteria of API RP 2SK, for the minimum safety factor of intact condition is 1.67 and damaged condition is 1.25.

References

[1] Ju X, Victor C, Dong B, Meng X and Li J 2023 Numerical analysis of fishtailing motion , buoy kissing and pullback force in a catenary anchor leg mooring (CALM) moored tanker system *Ocean Eng.* **278** 114236 doi: 10.1016/j.oceaneng.2023.114236.

[2]Silfiani M E, Prastianto R W, Rosyid D M and Djatmiko E B 2023 Reliability Analysis on the Effect of Subsea Buoy to the Tension of Spread Mooring System **8** 802–9 doi: 10.12962/j25481479.v8i4.19588.

[3] American Petroleum Institute 2005 Design and Analysis of Stationkeeping Systems for Floating Structures *API RP 2SK* (Washington, D.C.: American Petroleum Institute)

[4] American Bureau of Shipping 2018 *Rules for Building and Classing Mobile Offshore Drilling Units* (Houston: American Bureau of Shipping)

[5] Karimirad M and Moan T 2011 Extreme dynamic structural response analysis of catenary moored spar wind turbine in harsh environmental conditions *J. Offshore Mech. Arct. Eng.* **133** 1–14 doi: 10.1115/1.4003393

[6] Vaz M A and Patel M H 2000 Three-dimensional behaviour of elastic marine cables in sheared currents **22** 45–53 doi: 10.1016/S0141-1187(99)00023-1

[7] Zhang Y, Zhang H, Xia R, Jiang S and Wang F 2024 Reliability analysis of carbon fi ber rod-reinforced umbilical cable under tension using an improved sampling method *Pet. Sci.* **21** 2769–78 doi: 10.1016/j.petsci.2024.02.016

[8] Jin Y, Xie K, Liu G, Peng Y and Wan B 2022 Nonlinear dynamics modeling and analysis of a marine buoy single-point mooring system *Ocean Eng.* **262** 112031 doi: 10.1016/j.oceaneng.2022.112031

[9] Eghbali B, Daghigh M, Daghigh Y and Azarsina F 2018 Mooring Line Reliability Analysis of Single Point Mooring (SPM) System under Extreme Wave and Current Conditions *Int. J. Marit. Technol.* **9** 41–9

[10] Heraspatty A 2016 *Analisa Operabilitas Dan Keandalan SPM 1134 CNOOC SES LTD. Terhadap Variasi Gelombang Seratus Tahunan* (Surabaya: Institut Teknologi Sepuluh Nopember)

[11] The Oil Companies International Marine Forum (OCIMF) 1997 *Mooring Equipment Guidelines 2nd Edition* (United Kingdom: Witherby Seamanship International)

[12] Faltinsen O M 1990 Sea Loads on Ships and Offshore Structures Cambridge

[13] Adaalah A H, Prastianto R W, Murdjito, Syalsabila F and Syarifudin M R 2022 Analysis of Mooring Line Tension in Floating Collar Net Cage *IOP Conf. Ser. Earth Environ. Sci.* **972** doi: 10.1088/1755-1315/972/1/012019

[14] Putra G R, Rosyid D M and Prastianto R W 2021 Mooring Lines Reliability Analysis Case Study : FSO Abherka *IPTEK J. Technol. Sci.* **32** doi: 10.12962/j20882033.v32i1.8628

[15] Chakrabarti S 2005 *Handbook of Offshore Engineering* (Illinois: Elsevier Ltd.)

[16] Bhattacharyya R 1978 *Dynamics of Marine Vehicles* (Wiley)

[17] Djatmiko E B 2012 *Perilaku dan Operabilitas Bangunan Laut Di Atas Gelombang Acak* (Surabaya: ITS Press)

[18] Hou H, Liu Y, Dong G and Xu T 2022 Reliability assessment of mooring system for fish cage considering one damaged mooring line *Ocean Eng.* **257** 111626 doi: 10.1016/j.oceaneng.2022.111626.

IOP Conf. Series: Earth and Environmental Science **1461** (2025) 012033 doi:10.1088/1755-1315/1461/1/012033

A Literature Review: Enhancing Maritime Risk Assessment through Advanced Fuzzy Approach

Diyah Purwitasari[1,2*], Ketut Buda Artana [2*], Dhimas Widhi Handani [2], and Fadilla Indrayuni Prastyasari[2]

[1] Department of Marine Engineering, Institut Teknologi Sepuluh Nopember, Surabaya, Indonesia
[2] Department of Sea Transportation, Politeknik Pelayaran Surabaya, Surabaya, Indonesia

*E-mail: purwitasari.diyah@gmail.com; kb.artana@gmail.com

Abstract. Maritime transportation played a vital role in global trade and economic growth but faced risks from weather, human error, and equipment failures. Traditional risk assessment methods often fell short in addressing the complexities and uncertainties of maritime operations, highlighting the need for more effective approaches. This review examined the application of Fuzzy Bayesian Networks (FBN) in maritime risk assessment, focusing on its integration with fuzzy logic and probabilistic tools to improve safety. Findings indicated that combining fuzzy frequency and consequence analysis provided a more flexible and accurate way to assess risks, helping to predict and prevent accidents by deepening insights into the likelihood and impact of risk events. This integrated model facilitated tailored risk mitigation strategies, promoting safer and more resilient maritime operations. As the industry expanded, incorporating advanced fuzzy methods became essential for enhancing safety and decision-making. The evolving potential of FBN, particularly with big data and machine learning, underscored its role in fostering efficient, sustainable maritime practices. Future research was encouraged to refine these models and apply them to new technologies in risk management, aligning with the Sustainable Development Goals to enhance resilience and sustainability in the maritime sector.

1. Introduction

Maritime transportation plays a crucial role in the global economy, accounting for over 80% of global trade, according to the UNCTAD Review of Maritime Transport (2023). This industry is especially significant in regions such as East Asia, where the shipping lanes contribute to 27.6% of the world's freight volume. However, with rapid growth comes a variety of challenges. The increase in maritime activities has led to more accidents, resulting in significant casualties, environmental damage, and economic loss [1]. In recent years, the maritime industry has faced heightened risks due to both traditional and non-traditional security threats. Issues like piracy and maritime terrorism are particularly pressing, and they compound existing problems such as natural disasters and operational errors [1]. Despite advancements in safety regulations, the maritime industry faces numerous risks, including adverse weather conditions, human error, equipment failures [2], and collisions, especially in congested or narrow waterways [3], and technological challenges, making the safety of maritime operations a priority [4]. Collisions at sea can result in substantial economic losses, environmental damage, and, in some cases, loss of life. The need for effective risk management and safety enhancement in maritime operations has prompted the development of new regulatory frameworks and risk assessment methodologies over the past few decades [5] [6] [7]. To mitigate these risks, international bodies such as the

International Maritime Organization (IMO) have introduced various conventions, such as the Safety of Life at Sea (SOLAS) and the Convention on the International Regulations for Preventing Collisions at Sea (COLREG), aimed at improving safety and navigation standards to reduce such risks, yet accidents like ship collisions continue to occur [8], especially in narrow waterways [3]. These incidents highlight the limitations of traditional risk management methods and underscore the need for more sophisticated tools [3] [9].

Risk assessment methods have evolved from traditional qualitative approaches to more sophisticated, data-driven models. Probabilistic Risk Assessment (PRA) has emerged as a valuable tool for identifying and mitigating risks in maritime operations. By reviewing existing literature, this study hoped to bridge gaps in understanding, highlight the shortcomings of traditional approaches, and emphasize the necessity for more advanced methods like Fuzzy Bayesian Networks (FBN). By exploring the application of FBN in risk analysis, the study sought to provide a comprehensive understanding of how fuzzy logic enhances risk predictions. Facing increasingly non-traditional security threats and operational challenges in maritime industries, a literature review provided insights into how FBN was applied in various high-risk environments, beyond its initial applications in chemical and process industries, and how these insights could be translated to maritime safetey. As in the bibliometric analysis conducted by Cao et al. (2023), that highlighted emerging trends in maritime risk research, revealing a growing interest in methodologies that combine fuzzy logic and probabilistic models with advanced technologies like machine learning [4], this article will perform bibliometric analysis and literature review using metadata from Scopus. Additionally, the review was driven by the need to reinforce the foundation for applying FBN in the author's dissertation, particularly in the Formal Safety Assessment (FSA) framework where FBN has yet to be utilized. By synthesizing insights from existing literature, this study aims to not only underscore the potential of FBN in bridging gaps in risk assessment but also laid the groundwork for advancing its application within FSA, aligning with the broader goals of improving maritime safety and operational resilience.

2. Method

Research on risk assessment had been ongoing for a long time. According to data from the National Transportation Safety Committee (KNKT) between January 2017 and August 2023, accidents still occurred, though less frequently. However, this data did not include ship accidents not investigated by the KNKT but still recorded by the harbourmaster's office or port. Therefore, further research on maritime transportation accident risks was still necessary to provide recommendations for reducing such accidents. This article utilized metadata from Scopus to conduct a bibliometric analysis, accompanied by a literature review that summarized various methods employed in risk research.

2.1 Bibliometrix Analysis

Based on 213 journals in Scopus with the keywords "Risk Assessment" and "Formal Safety Assessment," a trend topic was illustrated in Figure 1. Figure 1 shows a timeline of key terms related to risk assessment and safety research from 1994 to 2022. The size of the dots likely represents how frequently or prominently these terms appeared in research, while the horizontal lines show how long each topic has been relevant. In recent years, terms like risk controls, ship motion, health risks, and offshore oil well production have gained more attention, especially approaching 2022. These topics are now at the forefront of research, reflecting their importance in addressing current safety and risk management challenges. Likewise, Bayesian analysis, a vital tool in risk assessment, has become more significant, suggesting a shift toward data-driven, probabilistic approaches. Topics like transportation safety and maritime safety have been consistently researched over the past two decades, emphasizing their ongoing importance, particularly with the growth of global trade and maritime activities. Long-standing terms such as formal safety assessment, risk management, and safety assessments have also remained central to the field, with frequent mentions in research. While advanced approaches like Bayesian analysis and fuzzy logic are gaining traction, there seems to be a need for more research on areas like accident prevention, fishing vessel safety, and marine environment protection, which appear underrepresented or less developed in recent years.

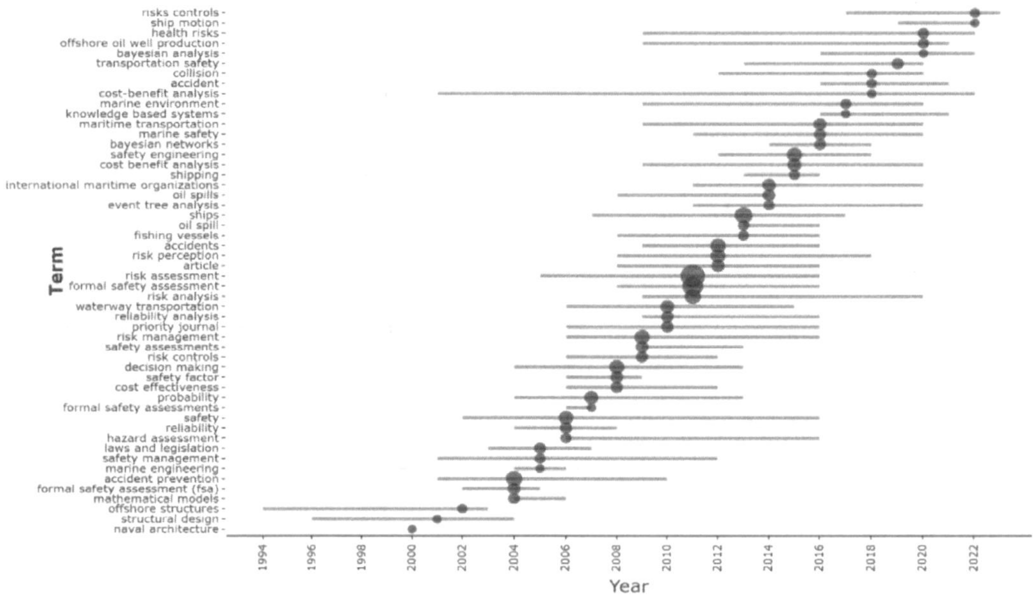

Figure 1. The trend of 213 Scopus journals on risk assessment and formal safety assessment from 1994 to 2022.

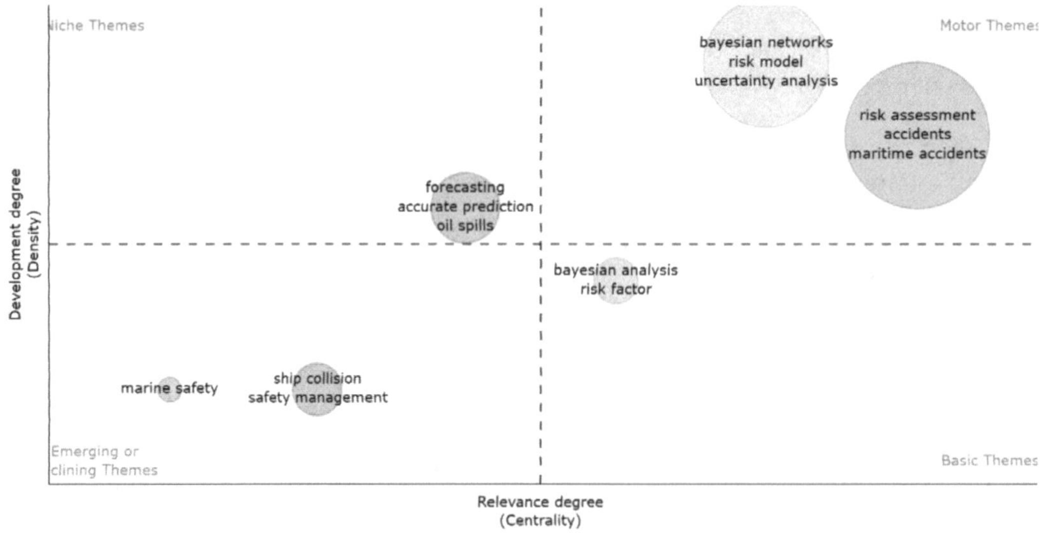

Figure 2. Topic Mapping with the Keywords "Risk Model" and "Maritime Accident" from 2013 to 2024.

Additionally, based on data from 21 journals in Scopus with the keywords "risk model" and "maritime accident" from 2013 to 2024, subtopics that need further development are found in the upper right quadrant due to their high density and relevance, including Bayesian networks, uncertainty analysis, risk models, risk assessment, and maritime accidents, as shown in Figure 2. In Figure 2, there are four quadrants: 1) The first quadrant, motor themes (Top-Right Quadrant), these themes are well-developed and also highly relevant. They act as drivers of the field and are often key research areas. Green and blue are areas driving research in maritime accidents, risk modelling, and safety networks. These are high-centrality, high-density themes, meaning they are well-established and play a pivotal role in the discourse of maritime safety and risk assessment. These should be the focus for stakeholders in the maritime industry because they are both relevant and well-established; 2) The second quadrant, basic themes (Bottom-Right Quadrant), these are central but underdeveloped themes. They are important in the field but may require further study or research to reach the level of motor themes. Orange and purple suggest areas like Bayesian analysis and risk factors that, while these topics are highly relevant but may still be emerging or lack the depth and advancement seen in the motor themes, indicating that topics like "Bayesian analysis" and "risk factors" are widely cited but not extensively researched (low density). Thus, the subtopics in that quadrant have potential for further development; 3) The third quadrant, niche themes (Top-Left Quadrant), these are well-developed but less relevant themes. The topics in it may be highly specialized or specific to certain subfields but do not connect broadly with other research areas. For instance, forecasting, accurate prediction, and oil spills, these topics are densely researched, but their broader relevance or applicability is lower, indicating they might be niche subjects within the maritime safety domain. Red, like oil spill prediction are more specialized but less central to the core research trends in this field; 4) The fourth quadrant, emerging or declining themes (Bottom-Left Quadrant), These are low-centrality and low-density themes, which means they are either in the early stages of development (emerging) or are becoming less important (declining). Brown and purple reflect areas like marine safety and ship collision safety management that could either be in their infancy or losing importance. These topics may benefit from revitalized interest or re-evaluation within the field.

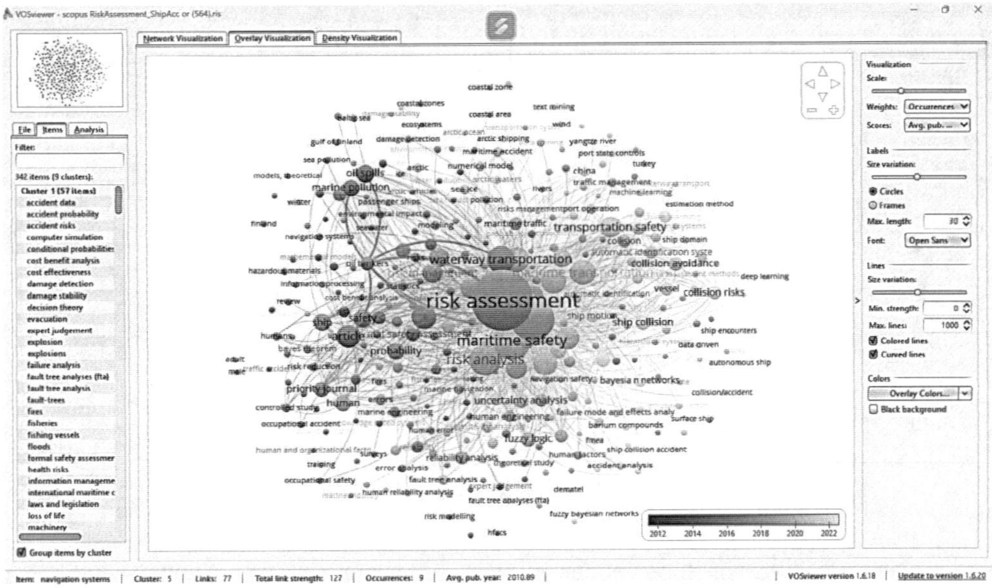

Figure 3. The overlay visualization bar display in VOSviewer for 9 clusters sourced from Scopus journal metadata with the keywords "risk assessment" and "ship accident".

Figure 3 display the overlay visualization bar in VOSviewer for 9 clusters sourced from Scopus journal metadata with the keywords "risk assessment" and "ship accident" from 1999 to 2023. The overlay visualization bar in figure 3, shows that risk assessment research still has a moderate impact, and there is a combination of methods used in risk analysis. The trend in risk assessment research shifts from green to yellow, connecting to items like Bayesian networks and fuzzy logic. Based on Figure 3, it is found that risk assessment research for ship accidents using FSA has been in use for a long time, but recent trends discuss risk assessment utilizing machine learning, AIS data, and the integration of multiple methods.
In risk assessment research, frequency or occurrence analysis and consequence analysis are essential components [18]. Various risk assessment methods employ different techniques for conducting frequency and consequence analyses.

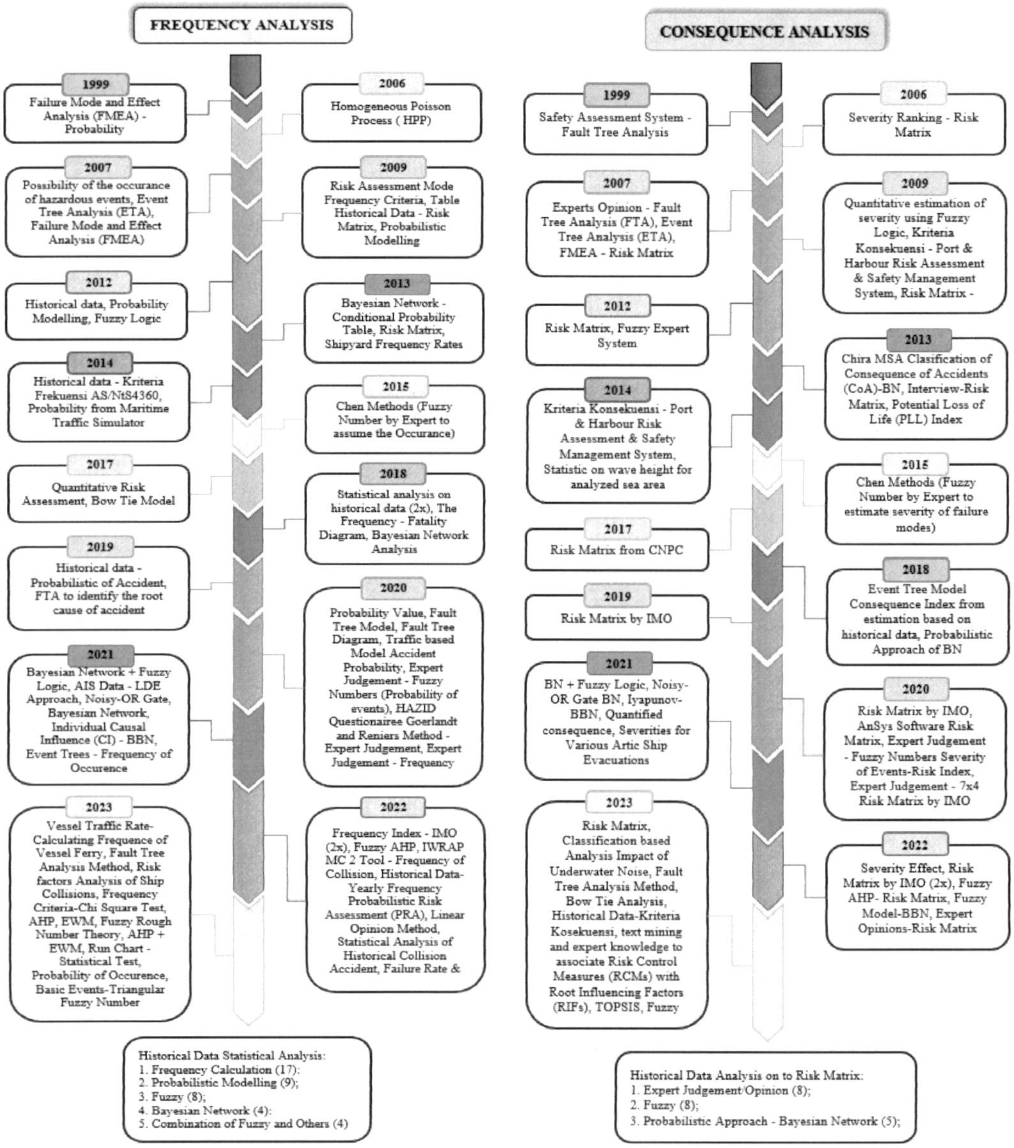

Figure 4. The milestones in risk assessment research from 1999 to 2023 related to frequency and consequence analysis.

Frequency analysis helps measure and visualize the importance and relationships between various terms in the field, providing insights into key topics, their interconnections, and how they have evolved over time, aiding researchers and practitioners in understanding the risk assessment research landscape. In Figure 4 (left side), a review of 50 journals on risk assessment reveals that recent research trends are moving towards the utilization of fuzzy methods combined with others, such as Fuzzy AHP, Fuzzy Bayesian Network, and Fuzzy Expert Systems. Fuzzy Bayesian Networks integrate Bayesian networks, which are probabilistic graphical models, with fuzzy logic to address uncertainty and incomplete data. Conducting frequency analysis using fuzzy Bayesian networks is a viable and effective approach for risk assessment, allowing for the integration of uncertain data and expert knowledge, making it a powerful tool for understanding and managing risks in complex systems [15].

On the other hand, consequence analysis in risk assessment is crucial for understanding and managing the potential impacts of risk events [16]. It enables organizations to prioritize risks, allocate resources effectively, and develop strong mitigation and response strategies. By thoroughly analysing consequences, organizations can enhance their resilience and preparedness for adverse events. Consistent with the findings from the review of risk assessment journals, Figure 4 (right side) shows that consequence analysis also trends towards the use of fuzzy methods, such as Fuzzy AHP, Fuzzy Bayesian Network, and Fuzzy Expert Systems.

2.2 Literature Review

The use of Fuzzy Bayesian Networks for frequency and consequence analysis in risk research is a growing trend that extends beyond the maritime sector, offshore risks to industrial applications, as demonstrated by the following studies.

2.2.1 Fuzzy Bayesian Network (FBN)

The Fuzzy Bayesian Network (FBN) is a probabilistic approach that merges Bayesian logic with fuzzy reasoning to model complex, uncertain maritime risk scenarios. Bayesian Networks (BN) are widely used to represent cause-and-effect relationships between risk factors, such as human error, environmental hazards, and equipment failures [3] [9]. In the context of a ship collision scenario, for instance, factors like miscommunication, poor visibility, and mechanical failures are interconnected. Bayesian Networks allow these variables to be mapped and analysed to predict the likelihood of accidents [9].

Liang and Liu (2024) demonstrated that FBN is particularly effective in maritime risk assessment due to its ability to model multiple risk nodes and their interactions, such as traffic congestion and adverse weather. By combining fuzzy logic with Bayesian reasoning, the model can process both quantitative data and qualitative expert opinions, allowing for more accurate risk predictions in uncertain environments [1]. Fuzzy logic is applied to account for uncertainties and ambiguities, which are common in maritime environments. In many situations, exact probabilities may be unknown, and expert opinions may be subjective. Fuzzy logic converts qualitative terms like "high risk" or "low risk" into numerical values, which can be incorporated into the Bayesian framework for analysis [3]. This approach allows for more flexibility in risk prediction, especially in dynamic environments like congested shipping lanes.

The methodology employed in Liang & Liu (2024) study, integrates the principles of Fuzzy Bayesian Networks (FBN) and a multilevel node relationship approach to assess maritime transportation risks. The FBN framework is a robust method for handling the uncertainty inherent in risk analysis. It combines fuzzy logic with Bayesian inference, enabling the inclusion of expert judgments and the handling of both crisp and fuzzy data [1]. The first step in the methodology involves the identification of key risk nodes, which represent various factors influencing maritime safety, such as ship traffic flow, extreme weather conditions, and terrorist threats. These nodes are discretized into several states (e.g., low, medium, high risk) using fuzzy logic theory, which allows the transformation of continuous and discrete data into probabilistic estimates [1]. This approach ensures that different types of data can be integrated within the Bayesian framework, overcoming the challenges of applying data-driven methods in cases where data may be incomplete or uncertain. Once the key risk nodes are identified and discretized, a Conditional Probability Table (CPT) is constructed to capture the relationships between these nodes. This process involves using rule-based methods to determine how certain factors (parent nodes) influence the likelihood of outcomes in the target nodes. The CPT enables the FBN model to aggregate inputs from various root nodes and calculate the overall probability of different risk levels [1]. In the final step, the FBN model performs a probabilistic inference to evaluate the transportation risk for different sea areas. The model aggregates the input data and calculates the likelihood of different risk outcomes, allowing for the identification of the most critical nodes that significantly impact maritime safety. This helps in determining which factors require targeted mitigation efforts [11].

In process industries, FBN is similarly applied to manage risks associated with hazardous materials and complex systems. For example, Yazdi and Kabir (2017) demonstrated that integrating fuzzy set theory with Bayesian networks improved the reliability of risk predictions for ethylene transportation, addressing the limitations of traditional fault tree analysis [10]. Research conducted by Luan (2023) and Sun (2023) integrated Fuzzy Bayesian Networks (FBN) in designing accident risk models in their studies on the risks of flammable liquid tankers (Luan, 2023) and maritime bauxite logistics (Sun, 2023). Luan and Sun (2023) believe that by using FBN, they can incorporate uncertainty and imprecision into the risk analysis process, making it more robust and accurate. By utilizing FBN, the studies successfully measured the level of risk and provided insights for risk mitigation strategies. This research demonstrates the potential for integrating FBN into ship accident risk models, enhancing their effectiveness in predicting and preventing maritime accidents [19] [20].

2.2.2 Fuzzy Reliability Analysis and Consequence Simulation

In contrast, the Fuzzy reliability analysis, Aydin et al. (2023) study, incorporates fuzzy reliability analysis with consequence simulation, a hybrid approach that not only estimates the probability of accidents but also simulates potential consequences [12]. This integration allows for a more detailed analysis of the impacts of accidents, considering factors such as ship type, crew performance, and environmental conditions. The fuzzy reliability method applies fuzzy sets to deal with uncertainties in reliability data, improving the prediction accuracy of system failures, while consequence simulation provides insight into the range of possible outcomes from a collision event. For example, a ship collision could result in anything from minor damage to a severe environmental disaster, depending on variables like ship speed and cargo [12].

Fuzzy reliability analysis is particularly effective in managing uncertainties within system failure data, providing a way to estimate failure probabilities even when historical data is incomplete. Consequence simulation then models the aftermath of accidents, offering a detailed picture of potential impacts on human life, property, and the environment [12]. This combination of predictive and consequence-based analysis allows for a more comprehensive risk assessment.

2.2.3 Bibliometric Analysis of Marine Accident Research

Cao et al. (2023) conducted a bibliometric analysis using tools like CiteSpace and VOSviewer to map the evolution of marine accident research over two decades. The study identified emerging trends, collaborative networks, and key areas of focus in maritime risk management. The analysis underscored the growing interest in advanced risk assessment methodologies, including FBN, and emphasized the increasing integration of big data and machine learning into accident prediction models. These emerging technologies are expected to further enhance maritime safety by improving the accuracy and speed of risk assessments [4]. This article conclusion is similar with the bibliometrix analysis done by the author.

3. Result and Discussion

3.1 Fuzzy Bayesian Network (FBN) Outcomes
The application of the FBN model in maritime risk assessment has produced significant results. For example, Aydin et al. (2021) demonstrated the effectiveness of the FBN model in predicting the likelihood of ship collisions. Key risk factors, including human error, communication breakdowns, and poor navigation, were identified and mapped using the FBN framework. The model predicted a 45% probability of collision under real-world conditions, outperforming traditional deterministic methods. By incorporating fuzzy logic, the FBN model effectively handled uncertainties, such as incomplete data or subjective expert judgments, allowing for more reliable risk predictions [3].

Similarly, Liang and Liu (2024) also applied the FBN framework to assess risks in congested maritime routes, finding that the model could effectively handle multiple variables, such as extreme weather, heavy traffic, and geopolitical instability. Their model identified critical risk nodes that require targeted interventions, such as enhanced navigation systems and improved communication protocols.

In process industries, Yazdi and Kabir (2017) showcased FBN's superiority in managing risks in ethylene transportation. The model allowed for a more accurate assessment of system unreliability by considering the dependencies among various risk factors, which traditional fault tree analysis methods failed to capture.

3.2 Fuzzy Reliability Analysis and Consequence Simulation Results
In a study combining fuzzy reliability analysis with consequence simulation, Aydin et al. (2023) evaluated both the likelihood of accidents and their potential impacts. The analysis revealed that communication failures and inadequate bridge team management were significant contributors to accidents. This dual analysis method allowed for a better understanding of both the likelihood and potential impact of accidents. The consequence simulation predicted that under certain conditions, the risk of collision could rise to 98%, depending on factors like ship type and environmental conditions. This method provided a deeper understanding of not only accident probabilities but also the full scope of potential consequences, from minor damages to severe environmental hazards [12].

3.3 Bibliometric Analysis Results
Cao et al. (2023) found that fuzzy logic-based models, including FBN, are increasingly recognized as vital tools in maritime risk assessment. Their bibliometric analysis highlighted human factors, autonomous ship operations, and environmental risks as emerging research hotspots, marking an important trend in the development of more accurate and reliable maritime risk assessments. The study also showed that international collaboration in developing these methodologies is increasing, particularly in the areas of big data and probabilistic risk assessment [4].

The integration of Fuzzy Bayesian Networks (FBN) into maritime risk assessment offers significant advantages over traditional methods. FBN models provide the flexibility to handle uncertain and incomplete data, allowing for more accurate predictions of accidents. Both the FBN and fuzzy reliability analysis models outperform deterministic approaches by incorporating expert judgment and accounting for dynamic risk factors in complex environments [10] [3] [12]. Liang and Liu's (2024) study further support the use of FBN models, especially in congested maritime routes where multiple risk factors interact. Their findings suggest that FBN can offer decision-makers a clearer understanding of where and how to focus safety interventions [1].

IOP Conf. Series: Earth and Environmental Science **1461** (2025) 012033 doi:10.1088/1755-1315/1461/1/012033

Similarly, Ren et al. (2024) found that FBN effectively managed offshore risks by incorporating both quantitative data and qualitative insights, demonstrating its adaptability across different maritime contexts [11].

Cao et al.'s (2023) bibliometric analysis confirms the growing importance of these probabilistic models in the maritime sector. As the industry advances with the introduction of autonomous ships and increased use of big data analytics, the ability to predict and manage risks more effectively will be crucial. The combination of fuzzy logic, Bayesian inference, and advanced data analytics provides a pathway toward safer and more efficient maritime operations. As technological advancements like machine learning continue to evolve, these methods will likely become even more integrated into maritime safety frameworks. Continued collaboration between researchers, regulatory agencies, and industry professionals will be essential to refining these tools and ensuring their widespread adoption [4].

This article provides a review of several studies that integrate Fuzzy logic with other methods to assist in risk analysis. The reviewed studies indicate that advanced fuzzy logic approaches are indeed highly effective and well-suited for use in maritime risk analysis. The literature review underscores the evolving landscape of maritime risk analysis and the increasing need for more sophisticated models capable of addressing uncertainty, ambiguity, and complexity inherent in maritime environments. Figure 1 illustrates the historical development and key milestones in maritime risk assessment. Traditional methods, such as Fault Tree Analysis (FTA) and Event Tree Analysis (ETA), are limited in their ability to handle the vagueness of real-world maritime risks. This gap has prompted the emergence of fuzzy-based models, which offer a more flexible approach to capturing and interpreting the inherent uncertainties in risk factors. Figure 2 highlights the various fuzzy logic methods that have been employed, such as Fuzzy Bayesian Networks (FBN), Fuzzy Failure Mode and Effect Analysis (FFMEA), and Fuzzy Cognitive Maps (FCM). These methods enhance traditional risk models by incorporating linguistic variables and subjective expert opinions, thus offering a more nuanced understanding of risks. Fuzzy systems' ability to process uncertain or incomplete data makes them particularly effective in maritime contexts where exact measurements are often impractical. Figure 3 provides a comparison between conventional risk assessment models and fuzzy-based approaches. The results from various studies indicate that fuzzy logic methodologies consistently outperform traditional models in terms of accuracy, reliability, and flexibility. Fuzzy models enable more precise predictions of accident likelihoods and provide deeper insights into potential risk mitigation strategies. The core methodology of the study (Figure 4) integrates frequency analysis and consequence analysis within a fuzzy logic framework. Frequency analysis estimates the likelihood of various risk events based on historical data, expert input, and probabilistic methods. When combined with fuzzy logic, the model effectively manages the uncertainties related to data gaps and subjective expert assessments, yielding more reliable likelihood estimates for rare but catastrophic events, such as oil spills or ship collisions. Consequence analysis, on the other hand, evaluates the potential impacts of these risk events on maritime operations, human safety, and the environment. The fuzzy logic approach enhances the traditional consequence analysis by incorporating imprecise input variables, which are typical in the maritime sector (e.g., the severity of environmental damage or the financial losses associated with a specific incident). By capturing these qualitative aspects, fuzzy models allow for a more holistic assessment of potential consequences.

4. Conclusion

By integrating frequency analysis with fuzzy logic, the estimation of risks and their consequences was significantly improved, particularly in the maritime sector. Frequency analysis utilized historical data and expert judgment to predict the likelihood of rare events such as oil spills or ship collisions, while fuzzy logic managed uncertainties in data and expert opinions, resulting in more reliable estimates. Similarly, in consequence analysis, fuzzy logic refined the evaluation of potential impacts on maritime operations, human safety, and the environment by addressing imprecise input variables, leading to a more comprehensive risk assessment.

The combination of fuzzy frequency and consequence analysis offered a highly adaptive and precise method for assessing maritime risks. This study demonstrated that this approach enhanced the ability to predict and prevent accidents by providing deeper insights into both the probability and impact of risk events. Furthermore, this integrated model facilitated the development of more effective risk mitigation strategies tailored to the unique challenges of the maritime industry.

This contribution was crucial not only for advancing maritime risk assessment but also for supporting the broader goals of sustainable development. By reducing operational risks, enhancing safety, and improving decision-making processes, fuzzy-based models helped safeguard marine ecosystems and promoted sustainable maritime practices. The use of Fuzzy Bayesian Networks (FBN) further strengthened this framework by integrating fuzzy logic with Bayesian inference, enabling the effective handling of uncertainties and outperforming traditional deterministic methods. As FBN models evolved with advancements in big data and machine learning, they played an increasingly vital role in ensuring the safety, efficiency, and sustainability of global maritime and process industries. Future research focused on refining these models to address emerging technological challenges, providing innovative solutions for risk management aligned with the principles of sustainability.

References

[1] J. &. L. J. Liang, "Maritime transportation risk assessment: A multilevel node relationship-based fuzzy bayesian network," *Ocean Engineering, 312, 119204,* 2024.

[2] L. M. X. a. C. L. Ma, "Risk evolution from causes to consequences of engine room fires on ships by mapping bow-tie into fuzzy Bayesian network," *Journal of Marine Engineering and Technology,* no. 10.1080/20464177.2024.2353955, 2024.

[3] M. A. E. T. O. &. A. O. Aydin, "Validation of risk analysis for ship collision using fuzzy Bayesian network approach.," *Ocean Engineering,* Vols. 231, 108973, no. https://doi.org/10.1016/j.oceaneng.2021.108973, 2021.

[4] Y. W. X. Y. Z. W. J. W. H. &. L. Z. Cao, "Research in marine accidents: A bibliometric analysis, systematic review and future directions," *Ocean Engineering,* Vols. 284, 115048, no. https://doi.org/10.1016/j.oceaneng.2023.115048, 2023.

[5] I. A. I. M. &. D. A. A. B. Baroroh, "Risk Analysis of Engine Room Module Installation with Integration of Bayesian Network and System Dynamics," *International Review of Mechanical Engineering,* Vols. 16, 6, pp. 299-308 10 p, 2022.

[6] W. e. a. Zhang, "A Fuzzy Bayesian Network-Based Method for Evaluating the Leakage Risk of STS LNG Bunkering," *ASCE-ASME Journal of Risk and Uncertainty in Engineering Systems, Part A: Civil Engineering,* vol. 10, no. 1, 2024.

[7] S. G. Y. Z. M. Z. &. J. W. Shanshan Fu., "Towards system-theoretic risk management for maritime transportation systems: A case study of the yangtze river estuary," *Ocean Engineering,* pp. 1 - 15., 2023.

[8] X. e. a. Meng, "Analyzing risk influencing factors of ship collision accidents: A data-driven Bayesian network model integrating physical knowledge," *Ocean and Coastal Management,* vol. 2561, no. 107311, 2024.

[9] I. D. A. K. B. K. H. P. E. &. N. M. F. Ratih, "A Bayesian Network for Classifying and Predicting Ship Collision," *Lecture Notes on Data Engineering and Communications Technologies. Springer Science and Business Media Deutschland GmbH,* vol. 191, pp. 403-414 12 p, 2024.

[10] M. &. K. S. Yazdi, "A fuzzy Bayesian network approach for risk analysis in process industries," *Process Safety and Environmental Protection,* vol. 111, no. https://doi.org/10.1016/j.psep.2017.08.015, pp. 507-519, 2017.

[11] J. J. I. W. J. X. D. L. &. Y. J. B. Ren, "An offshore risk analysis method using fuzzy Bayesian network," *Journal of Marine Engineering and Technology,* vol. 3(2), pp. 101-120, 2024.

[12] M. A. E. T. O. &. A. O. Aydin, "Integration of fuzzy reliability analysis and consequence simulation to conduct risk assessment," *Ocean Engineering,* Vols. 250, 111298, no. https://doi.org/10.1016/j.oceaneng.2023.111298, 2023.

[13] F. I. &. S. T. Prastyasari, "Study on Probabilistic Risk Assessment Model for Crossing Situation in Sunda Strait," *IOP Conference Series: Earth and Environmental Science,* Vols. 972, 1, 012043, 2022.

[14] S. Z. Z. a. L. G. Molin, "Uncertainty Analysis of the Estimated Risk in Formal Safety Assessment.," *Sustainability*, Vols. 10, 321, no. doi:10.3390/su1002032, pp. 1 - 16, 2018.

[15] N. A. S. F. A. &. P. D. A. Rahma, "Safety Risk Analysis Optimization Using Fuzzy: A Literature Review.," *Journal of Industrial Safety Engineering,* vol. 15(2), p. 152–160., 2023.

[16] L. D. Leoni, "Integration of fuzzy reliability analysis and consequence simulation to conduct risk assessment," *Journal of Loss Prevention in the Process Industries,* vol. 83, pp. 1 - 15, 2023.

[17] N. J. Van Eck, Manual for VOSviewer version 1.6.18, Universiteit Leiden, 2022.

[18] N. D. H. D. W. P. F. I. &. D. A. A. B. Wuryaningrum, "Frequency Analysis of Ship Collision and Its Impact on the Fulfillment of Supporting Facilities and Route Changes Due to Implementation of Sunda Strait TSS," *IOP Conference Series: Earth and Environmental Science.,* Vols. 557, 1, 012042, 2020.

[19] T. Z. Luan, "Dynamic risk analysis of flammable liquid road tanker based on fuzzy Bayesian network .," *Process Safety Progress,* vol. 42, pp. 737-751, 2023.

[20] J. W. Sun, "Risk Assessment of Bauxite Maritime Logistics Based on Improved FMECA and Fuzzy Bayesian Network .," *Journal of Marine Science and Engineering.,* 2023.

Analysis of Fuel Consumption on Ships during Auto Trajectory Operations Based on Fuzzy Logic

Aulia Siti Aisjah[1*], Syamsul Arifin[1], A.A. Masroeri[2], Nila Mauhibatin [1], Athalariq Yassirasyid Caesarani[1], Astried Ayuning Palupi[1], dan Muhammad Hagie Antasena[1]

[1] Department of Physics Engineering, Institut Teknologi Sepuluh Nopember, Surabaya 60111, INDONESIA
[2] Department of Marine Engineering, Institut Teknologi Sepuluh Nopember, Surabaya 60111, INDONESIA

*E-mail: auliasa@ep.its.ac.id,

Abstract. An auto trajectory ship motion control system has been developed on ships to control the ship's motion according to a predetermined trajectory. The appliance of a control system for maneuvering and moving ships is expected to respond quickly and be able to keep the direction and trajectory due to the sea environmental conditions. Changes in trajectory distance can be shorter or even longer than the recommended trajectory, which will affect fuel consumption. This paper studies and analyzes the effect of the ship's trajectory control system, the change in trajectory, and its relationship to the amount of fuel consumption throughout the voyage. The design of the control system uses the type 1 Mamdani fuzzy logic method. The research was carried out through a simulation with various tests, namely open loop and close loop, as well as close loop tests with an added wave disturbance factor. The tests were conducted with in accordance with IMO standards, and the test results were analyzed. The set point test of 10° and 20° in the rudder, resulted in an overshoot value of 0.7347° and 0.7355°. The simulation results of trajectory testing without disturbance and with disturbance have an error value that is still within the range of the permitted radius value or the circle of acceptance of the ship. The impact of the control system regarding to the fuel usage is its ability to reduce fuel consumption compared to when the control system is not applied.

Keywords: Auto Trajectory, Earth and Environmental Science, Fuel, Fuzzy Logic, Wave.

1. Introduction

Sea transportation in Indonesia is ranked first for the transportation of people and commodities around the islands of Indonesia. Sea transportation has been used as an option for trade actors for their economies and also an important component for the economic progress and efficiency of Indonesia's national logistics system. The island of Bali and Nusa Tenggara are an areas that is part of the Indonesian government's policy to develop the economy [1][2]. In supporting the process of optimization and effectiveness of the economic development, the flow of commodity distribution needs to be considered, hence transportation during the process is needed [3]. Efforts made by the government, especially the Ministry of Transportation, are to provide transportation for interisland mobilization and by opening new crossing routes using long-distance Ro-ro ferries [4]. This solution creates a positive impact in the sea transportation sector for interisland distribution that reduces traffic congestion on the island of Bali so that tourist comfort can also be maintained. Therefore, a new crossing route from Surabaya to Lombok was opened by the government, which is an alternative route for the distribution process to Nusa Tenggara.

Currently, the role of logistics is very important, especially in the economic sector, so it must be utilized to support the development of national logistics chain that can reach all regions of Indonesia, such as Bali and Nusa Tenggara [5]. One of the things to be considered on a ship's voyage is the maritime weather conditions [6]. Poor weather conditions can be an obstacle to a ship's operation, which can result in delays of ship's arrival and increasing the ship's operational costs [7].

High sea waves is one type of disturbance faced in the shipping process. Sea waves have an impact on changing the trajectory of the ship [8]. This change in trajectory can also affect the amount of fuel consumed during the shipping process. If the ship faces a high sea wave, it will look for a new trajectory and results in longer travel distance [9]. According to that statement, every ship company seeks to minimize the ship's fuel consumption through the selection of trajectories that have an impact on reducing fuel consumption while sailing [10].

This paper analyzes the effect of the auto trajectory control system on the length of the ship's trajectory, which is then associated with the fuel consumption of the ship. The analysis is carried out in two conditions, namely, without any disturbance occurring and with a disturbance occurring. The object of this research is the Dharma Rucitra VIII Ship. The auto trajectory system is obtained from the design of the control system on the rudder. The method of designing the auto trajectory system is using a fuzzy logic control system. The objective of the control system is for the ship to perform a track-keeping system from one point of departure to its arrival point. The research is conducted through a computer simulation using MATLAB R2024a software.

2. Methodology

2.1 Ship Dynamics Modeling and Actuators of Ship Steering System

Ship dynamics modeling in the form of mathematical equations in state space is derived based on the Davidson & Schiff approach [11]. The ship specification data are Length of ship: 137 meter, width (B): 25 meter, depth (T): 6.5 meter, displacement (Δ): 10793416.56 kilograms, height (H): 7.6 meter, mass of ship (M): 8549.527 tons, coefficient block (Cb): 0.473, rudder area (A_δ):16.318 m^2, radius of gyration (R): 30.825 meter, center of gravity (Xg): 4.04 meter, and ship service speed (U): 10.8 m/s . The dimensional specifications of the ship and the Davidson & Schiff model gives out the state space equation for sway and yaw motions, shown in Equation (1) below.

$$M\dot{v} + N(u_0)v = b\delta_R \tag{1}$$

With M is the inertia matrix, N is the damping matrix which is the sum of matrix D (damping) and matrix C (coriolis), $v = [v, r, \psi]^T$ is the state vector, b is the rudder force matrix and δ_R is the rudder angle.

$$\begin{bmatrix} \dot{v} \\ \dot{r} \\ \dot{\psi} \end{bmatrix} = \begin{bmatrix} -0{,}11839 & 4{,}13695 & 0 \\ -0{,}00034 & -0{,}02632 & 0 \\ 0 & 1 & 0 \end{bmatrix} \begin{bmatrix} v \\ r \\ \psi \end{bmatrix} + \begin{bmatrix} 0{,}15098 \\ -0{,}00068 \\ 0 \end{bmatrix} \delta_R \tag{2}$$

$$y = \begin{bmatrix} 0 & 0 & 1 \end{bmatrix} \begin{bmatrix} v \\ r \\ \psi \end{bmatrix} \tag{3}$$

The mathematical modelling of the actuator is a Van Amerongen-type steering system, with the following parameters: the rudder deflection range is between -35° to 35° and the opening rate variation is in the range of -7°/s to 7°/s.

2.2 Reference Trajectory Modelling and Disturbance Modelling

The reference trajectory shown in Figure 1 is a shipping trajectory from Tanjung Perak Port in Surabaya to Lembar Port in Lombok. The trajectory is divided into 16 positions/points. The initial point of departure became the reference point with UTM position of (0,0).

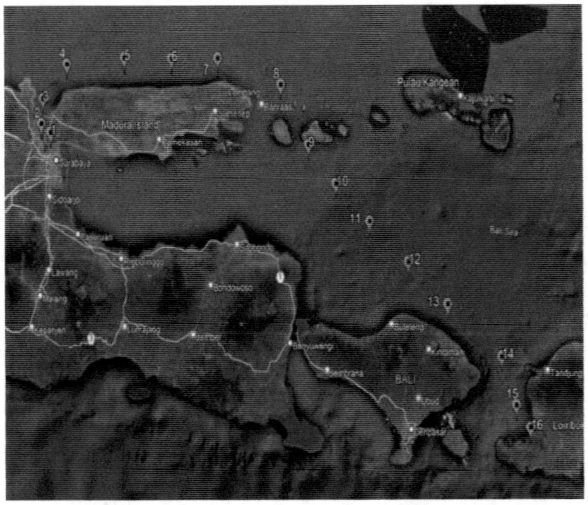

Point	X (m)	Y (m)
1	0	0
2	-7282,13	5219,33
3	-5908,13	19639,63
4	12999,17	39283,73
5	57539,21	42320,83
6	94099,77	41381,03
7	129028,27	41240,73
8	344012,37	25785,33
9	365133,17	-8607,97
10	386066,57	-31031,27
11	412167,77	-52594,67
12	441932,67	-75766,77
13	472319,27	-99971,57
14	513560,37	-130090,77
15	524975,77	-158069,27
16	535695,77	-170899,47

Figure 1. Reference trajectory

The trajectory of a ship along a voyage (Figure 1) is affected by the magnitude of wave disturbance in the sea of the trajectory. Disturbance factors are incorporated into ship dynamics modelling. Wave data was obtained from the Meteorology, Climatology and Geophysics Agency (BMKG). Modelling the motion of a ship above sea level by accommodating external disturbances is shown in Equation (4). The notations H and L in equation (4) are the values of ship motion variables at low frequency (L) and high frequency (H).

The coefficients of the state space equation shown in Equation (4) depend on the specifications of the ship and the hydrodynamic coefficients of the ship. By performing calculations according to the theory of Davison & Schiff [12], the equation of ship dynamics when sailing is shown in Equation (5).

$$\begin{bmatrix} \dot{x}_L \\ \dot{x}_H \end{bmatrix} = \begin{bmatrix} A_L & 0 \\ 0 & A_H \end{bmatrix} \begin{bmatrix} x_L \\ x_H \end{bmatrix} + \begin{bmatrix} B_L \\ 0 \end{bmatrix} \tau_L + \begin{bmatrix} 1 & 0 \\ 0 & 1 \end{bmatrix} \begin{bmatrix} w_L \\ w_H \end{bmatrix} \tag{4}$$

$$\begin{bmatrix} \dot{u}_L \\ \dot{v}_L \\ \dot{r}_L \\ \dot{x}_L \\ \dot{y}_L \\ \dot{\psi}_L \\ \dot{\xi}_x \\ \dot{\xi}_y \\ \dot{\xi}_\psi \\ \dot{x}_H \\ \dot{y}_H \\ \dot{\psi}_H \end{bmatrix} = \begin{bmatrix} 0 & 0 & 0 & 1 & 0 & 0 & 0 & 0 & 0 & 0 & 0 & 0 \\ 0 & 0 & 0 & 0 & 1 & 0 & 0 & 0 & 0 & 0 & 0 & 0 \\ 0 & 0 & 0 & 0 & 0 & 1 & 0 & 0 & 0 & 0 & 0 & 0 \\ 0 & 0 & 0 & -0{,}11836 & 4{,}14586 & 0 & 0 & 0 & 0 & 0 & 0 & 0 \\ 0 & 0 & 0 & -0{,}00034 & -0{,}02633 & 0 & 0 & 0 & 0 & 0 & 0 & 0 \\ 0 & 0 & 0 & 0 & 0 & 1 & 0 & 0 & 0 & 0 & 0 & 0 \\ 0 & 0 & 0 & 0 & 0 & 0 & 1 & 0 & 0 & 0 & 0 & 0 \\ 0 & 0 & 0 & 0 & 0 & 0 & 0 & 1 & 0 & 0 & 0 & 0 \\ 0 & 0 & 0 & 0 & 0 & 0 & 0 & 0 & 1 & 0 & 0 & 0 \\ 0 & 0 & 0 & 0 & 0 & 0 & -0{,}25044H^{-0{,}5} & 0 & 0 & -0{,}25044H^{-0{,}5} & 0 & 0 \\ 0 & 0 & 0 & 0 & 0 & 0 & 0 & -0{,}25044H^{-0{,}5} & 0 & 0 & -0{,}25044H^{-0{,}5} & 0 \\ 0 & 0 & 0 & 0 & 0 & 0 & 0 & 0 & -0{,}25044H^{-0{,}5} & 0 & 0 & -0{,}25044H^{-0{,}5} \end{bmatrix} \begin{bmatrix} u_L \\ v_L \\ r_L \\ x_L \\ y_L \\ \psi_L \\ \xi_x \\ \xi_y \\ \xi_\psi \\ x_H \\ y_H \\ \psi_H \end{bmatrix} + \begin{bmatrix} 0 \\ 0 \\ 0 \\ 0{,}15098 \\ -0{,}00068 \\ 0 \\ 0 \\ 0 \\ 0 \\ 0 \\ 0 \\ 0 \end{bmatrix} \delta_R + \begin{bmatrix} 0 \\ 0 \\ 0 \\ 0{,}15098 \\ -0{,}00068 \\ 0 \\ 0 \\ 0 \\ 0 \\ 0{,}79139H^{-0{,}5} \\ 0{,}79139H^{-0{,}5} \\ 0{,}79139H^{-0{,}5} \end{bmatrix} w \tag{5}$$

2.3 Auto Trajectory Control System Design

The design of the auto trajectory control system uses a fuzzy logic system, where this system consists of 3 units shown in Figure 2 below.

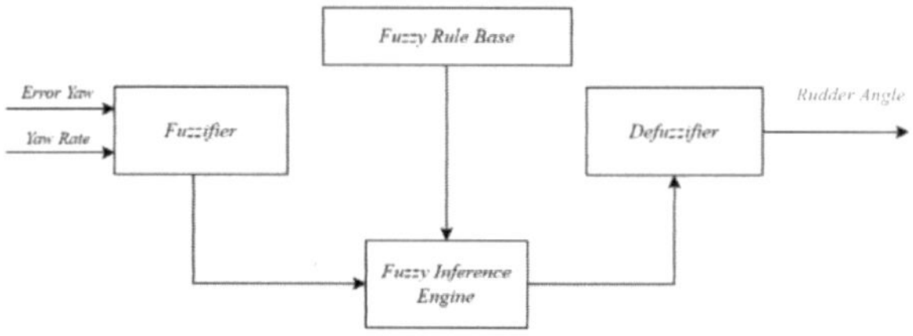

Figure 2. Fuzzy Logic System Block Diagram [13]

A fuzzy control system with 3 units, (i) fuzzifier, (ii) inference engine, and (iii) defuzzifier, which is a system that is able to process numerical (crisp) data into linguistics and by using rules in decision making, will produce a value that returns to its numerical value. The process is fuzzified, with the inputs being: yaw error and yaw rate, where both variables are divided into 7 (seven) membership functions. The membership functions used are seven fuzzy sets, including Negative Big (NB), Negative Medium (NM), Negative Small (NS), Zero (ZE), Positive Small (PS), Positive Medium (PM), and Positive Big (PB). The interval value for yaw error input is -35° to 35°, while the interval for yaw rate input is -7°/s to 7°/s. The decision making process selecting the right rule that has been planted in the inference engine unit. The rules embedded in the system are shown in Table 1 below. The fuzzification process for the yaw error and yaw rate variables is shown in Figure 3 (a) and (b), while the defuzzification process for the rudder angle variable is shown in Figure 3 (c).

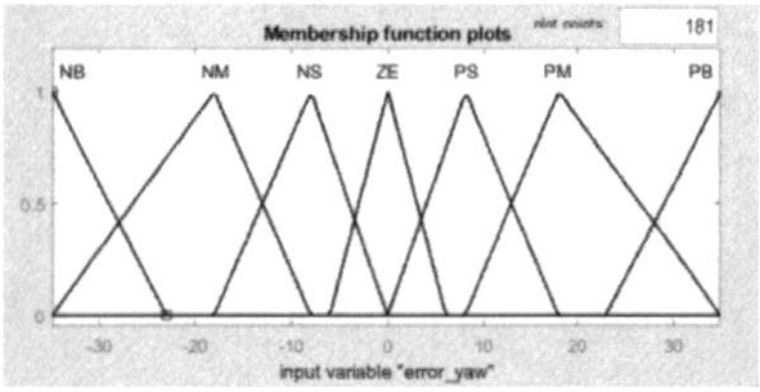

(a)

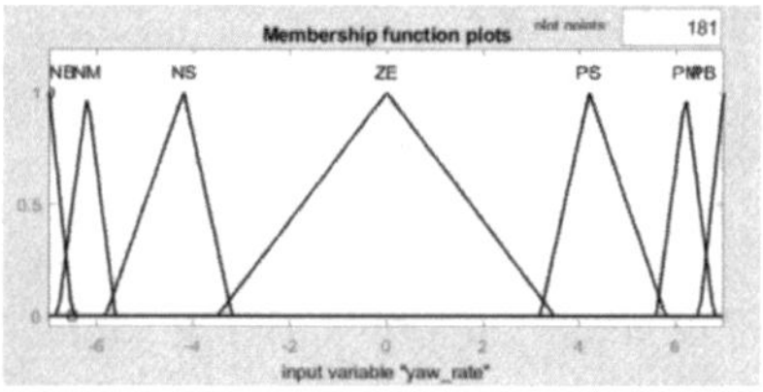

(b)

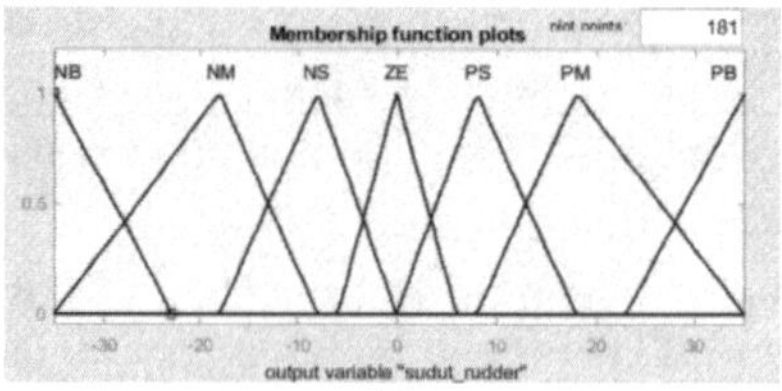

(c)

Figure 3. Membership Function Variables (a) Yaw Error, (b) Yaw Rate, and (c) Rudder Angle

The inference unit is a collection of rules used, and in this study 49 (forty-nine) were built. The rule design is carried out with various parameter variations in the fuzzy logic system, and one of these rules is shown in Table 1 below [14].

Table 1. Rule base on control system for rudder

r\e	NB	NM	NS	ZE	PS	PM	PB
NB	NB	NB	NB	NB	NM	NS	ZE
NM	NB	NB	NB	NM	NS	ZE	PS
NS	NB	NB	NM	NS	ZE	PS	PM
ZE	NB	NM	NS	ZE	PS	PM	PB

IOP Conf. Series: Earth and Environmental Science **1461** (2025) 012034 doi:10.1088/1755-1315/1461/1/012034

PS	NM	NS	ZE	PS	PM	PB	PB
PM	NS	ZE	PS	PM	PB	PB	PB
PB	ZE	PS	PM	PB	PB	PB	PB

2.4 Analysis of Fuel Consumption
The performance of the ship trajectory, when implemented by the autotrajectory system, will determine the length of the trajectory from the departure position to the destination. The length of the traverse resulting from the implementation of autotrajectory control will be the basis for determining the amount of fuel consumed. The results of the fuel consumption calculation are compared with the fuel consumption without the implementation of the autotrajectory control system. The vessel used is the Dharma Rucitra VIII, with the specifications shown in Table 2.

Table 2. Dharma Rucitra VIII Ship Engine Specification Data

Engine Type	Diesel
Engine Builder	MAN B&W
Model	8S50MC-C
RPM	127
Total *Power*	17185 kW
SFOC	117 g/kWh

The equation used to calculate the amount of fuel every hour is shown in equation (6) below [15].

$$Q = 0,85 \times 117 \times 17185 \times 10^{-3} = 1709\ kg/h \tag{6}$$

3. Result and Discussion

3.1 Analysis of Close Loop Test Results
In conducting close loop testing, it is carried out in accordance with the International Maritime Organization (IMO) standard using two set points, namely 10° and 20° [10].

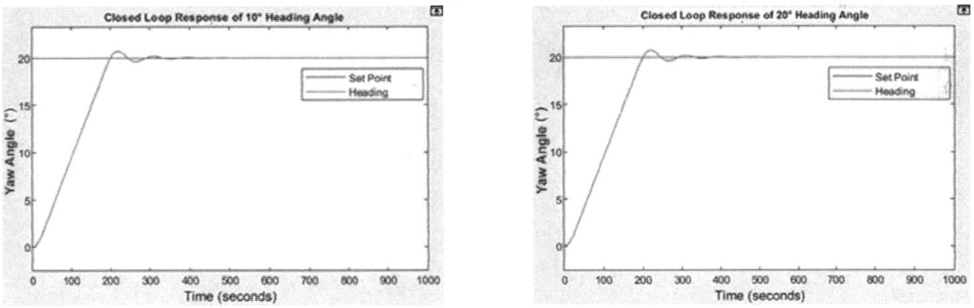

(a) (b)

Figure 4. (a). Close loop test response at 10° angle, (b). Close loop test response at 20° angle.

Figure 4 (a) shows the performance of the system when tested close loop with a set point of 10°, which shows an overshoot value of 0,7338°, a rise time of 107,338 seconds, and a settling time value of 452,735 seconds. The performance is in accordance with IMO criteria. The performance during the second test, when the rudder angle value is set to 20°, is shown in Figure 4 (b). The performance during the 20° set point test, is a rise time value of 201,649 seconds, a settling time of 546,369 seconds, and an overshoot of 0,7344°. Both performance test results showed compliance with IMO standards.

3.2 Analysis of trajectory test result without disturbance

The autotrajectory test for the track from Surabaya to Lembar, Lombok Island, was obtained by testing the ability of the control system to change the set point for the rudder, at each point of the track. The expected trajectory points correspond to the position of each point in Figure 1 above. The actual value of the ship's position, will provide a new calculation of the set point value of the rudder direction. And this will result in the position shown in Figure 5 below, with the zoom for position showing the difference between the actual trajectory and the expected trajectory.

Trajectory testing in the absence of interference showed errors between 9.9 - 23.27 meters (see zoom in Figure 5). This trajectory error value is still within the allowable error category [11].

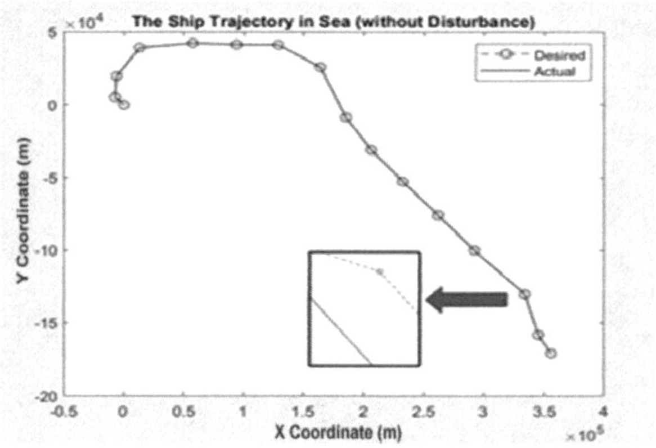

Figure 5. Trajectory test result without disturbance.

3.3 Analysis of trajectory test result with disturbance

The simulation when there is an external disturbance, namely a wave with a height of 1.5 metres, is shown in Figure 5.

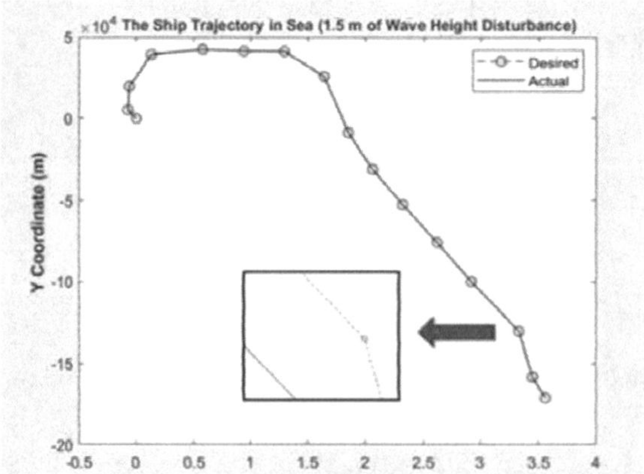

Figure 6. Trajectory test results with 1.5 meters of wave height disturbance.

Based on the results of the trajectory testing with a sea wave height disturbance of 1.5 meters of height shown in Figure 6, trajectory errors at each point are obtained. From the 16 points that are set as the desired waypoints, the average trajectory error is 26.09 meters, with the highest trajectory error value of 61.54 meters at point 14. The value of the trajectory error that is obtained still meets the circle of acceptance value, which is less than 0.5 of length of the ship.

IOP Conf. Series: Earth and Environmental Science **1461** (2025) 012034 doi:10.1088/1755-1315/1461/1/012034

3.4 Fuel Consumption Analysis

The simulation results of the control system without disturbance occurring and with a disturbance occurring able to show the trajectory traveled by the ship as well as the length of time of the voyage. Equation (6) is used to calculate the amount of fuel consumed by the ship during the voyage. For the case used in this study on the Dharma Rucitra VIII Ship, the analysis of the different fuel values used can be done, which are shown in Table 3 below.

Table 3. Fuel usage consumption on certain conditions.

Point	Fuel used in total without disturbance condition (ton)	Fuel used in total with 1.5 meter wave height disturbance condition (ton)
1	0,00	0,00
2	0,50	0,53
3	0,81	0,85
4	1,52	1,60
5	2,49	2,62
6	2,04	2,15
7	1,95	2,05
8	2,14	2,25
9	2,25	2,37
10	1,71	1,81
11	1,89	1,99
12	2,11	2,22
13	2,17	2,28
14	2,85	3,00
15	1,69	1,78
16	0,93	0,98
Total Fuel	27,06	28,48

Table 3 shows the difference in fuel consumption when the ship is implemented with the auto trajectory system. The value of fuel without disturbance condition in one voyage is 27.06 tons, while the 1.5 meters wave height disturbance condition requires more fuel, which is 28.38 tons, the difference between the two conditions is 0.52 tons.

4. Conclusion

- The fuzzy logic control system that has been designed can control the bow angle of the Dharma Rucitra VIII Ship in accordance with IMO standards and has an overshoot values of 0.7347° and 0.7355° for each test with a set points of 10° and 20° in the rudder angle. Those values are lower than the limit for the maximum value tolerance, which is 10° and 25°.
- The performance of the auto trajectory control system without disturbance has a trajectory error of 9.9 meters and can reduce fuel usage by 1.94 tons compared to without applying the auto trajectory control system.
- The performance of the auto trajectory control system with a 1.5 meter wave height disturbance has a trajectory error of 26.09 meters and can reduce fuel usage by 0.52 tons compared to without applying the trajectory control system.

References

[1] I. G. H. Saputra and H. A. Kurniawati, "Desain Kapal 3-in-1 Penumpang-Barang-Container Rute Surabaya – Lombok," *Jurnal Teknik ITS*, vol. 5, no. 2, pp. 92–97, 2016, doi: 10.12962/j23373539.v5i2.16878.

[2] Presiden Republik Indonesia, "Peraturan Presiden Republik Indonesia Nomor 32 Tahun 2011 Tentang Masterplan Percepatan dan Perluasan Pembangunan Ekonomi Indonesia 2011 - 2025," no. 1, pp. 1–9, 2011.

[3] A. Wirabrata and S. Aditua Fandhitya Silalahi, "HUBUNGAN INFRASTRUKTUR TRANSPORTASI DAN BIAYA LOGISTIK (The Linkage between Transportation and Logistic Cost)," *Jurnal Ekonomi & Kebijakan Publik*, vol. 3, no. NO. 1, pp. 79–90, 2012.

[4] G. D. Laksono, "ANALISA GETARAN RUANG KAMAR MESIN PADA KAPAL PENYEBRANGAN KETAPANG – GILIMANUK, KMP. DHARMA RUCITRA," *Revista CENIC. Ciencias Biológicas*, vol. 152, no. 3, p. 28, 2016.

[5] R. A. Kasengkang, S. Nangoy, and J. Sumarauw, "Analisis Logistik (Studi Kasus pada Pt. Remenia Satori Tepas-Kota Manado)," *Jurnal Berkala Ilmiah Efisiensi*, vol. 16, no. 01, pp. 750–759, 2016.

[6] L. Garcia-Alonso, T. G. Z. Moura, and D. Roibas, "The effect of weather conditions on port technical efficiency," *Mar Policy*, vol. 113, no. August 2019, p. 103816, 2020, doi: 10.1016/j.marpol.2020.103816.

[7] A. Hahn and D. Grunert, "ScienceDirect Planning in in ECDIS ECDIS to to Their Fully Applied for Fully Actuated Ships .," *IFAC-PapersOnLine*, vol. 49, no. 23, pp. 115–120, 2016, doi: 10.1016/j.ifacol.2016.10.330.

[8] A. Weintrit, "Marine Navigation and Safety of Sea Transportation," in *Marine Navigation and Safety of Sea Transportation*, Virginia: CRC Press, 2009, pp. 1–822.

[9] S. Y. Nurrani, *Akibat Rancangan Pengendali Trajectory Pada Kapal Container Dengan Lintasan Surabaya-*. 2017.

[10] A. Ansori, "Studi Pemanfaatan Internet of Things dan Data Mining untuk Pengawasan Bahan Bakar Minyak (Studi Kasus: Perusahaan Pelayaran Penumpang Nasional)," *Wave: Jurnal Ilmiah Teknologi Maritim*, vol. 12, no. 1, pp. 31–42, 2018, doi: 10.29122/jurnalwave.v12i1.2915.

[11] A. S. Aisjah, *Sistem kendali pada manuver kapal*. 2014.

[12] T. I. Fossen, *HANDBOOK OF MARINE CRAFT HYDRODYNAMICS*. Norway: John Wiley & Sons, 2011.

[13] D. Su and H. J. Wen, "APPLYING FUZZY LOGIC CONTROL TO PLEASURE CRAFTS".

[14] E. Omerdic, G. N. Roberts, and Z. Vukic, "A fuzzy track-keeping autopilot for ship steering," *J. Mar. Eng. Technol.*, vol. 2, no. 1, pp. 23–35, 2003, doi: 10.1080/20464177.2003.11020163.

[15] W. Du, Y. Li, G. Zhang, C. Wang, B. Zhu, and J. Qiao, "Energy saving method for ship weather routing optimization," *Ocean Eng.*, vol. 258, no. January, p. 111771, 2022, doi: 10.1016/j.oceaneng.2022.111771.

Observation of AIS signal coverage from Mount Sumbing compared to AIS data from VTS Semarang

Unggun Dahana[1] R O Saut Gurning[1] Titon Dutono[2]

[1] Department of Marine Engineering, Sepuluh Nopember Institute of Technology, Surabaya, Indonesia
[2] Department of Electrical Engineering, Surabaya State Electronic Polytechnic, Surabaya, Indonesia

* E-mail: unggun_dahono@yahoo.com, sautg@its.ac.id

Abstract. Automatic Identification System (AIS) is designed to operate on very high frequency (VHF) with the propagation characteristics limited to the principle of line of sight, where AIS receiver radio installed on coast radio station with a 100-metres-high antenna able to only monitor ships within 56 km. Considering such limitations, IMO has proposed a concept for positioning the AIS receiver antenna at the highest elevation to enhance the vessel surveillance range. To prove the concept, a survey of AIS reception has been conducted in Central Java in four different locations at different elevations, from 30 to 3,371 metres. The method employed was radio wave propagation prediction simulation at certain elevations, followed by AIS signal reception survey. The result indicated that from elevation of 3,371 metres, it could reach all coastal areas in Central Java to Surabaya Western Access Channel (SWAC). The implication of this study shows that it is potentially possible to construct support AIS stations to improve the performance of AIS coastal radio stations in Central Java in order to monitor vessels throughout this region with its 600 km of northern coastline and 300 km of southern coastline.

Keywords: AIS, Atmospheric Propagation, Elevation Effect, Mountain Resources

1. Introduction

Automatic Identification System (AIS) is a safety technology to avoid collides among vessels using the method of exchanging information of the ships' navigation autonomously, continuously and simultaneously based on the requirements of SOLAS and IMO since the end of 2024 [1]–[4]. Each vessel has a transmitter and receiver (transceivers) to send and receive navigation information from other ships at the same time. Since 2004, ships bigger than 300 GT and passenger ships are required to use AIS Class A with transmission power of 12.5 Watt, while fish ship and any other small boats are recommended to use AIS Class B with transmission power of 2.0 Watt voluntarily, since 2006.

Other than ship-to-ship transmission, AIS is also received by Vessel Traffic Service (VTS) in harbours or ship-to-shore and is used for ship traffic management [5]–[7]. AIS signal transmission is designed for short-range communication using VHF propagation wave range [8]–[10] and is not intended for any satellite communication [11]–[13].

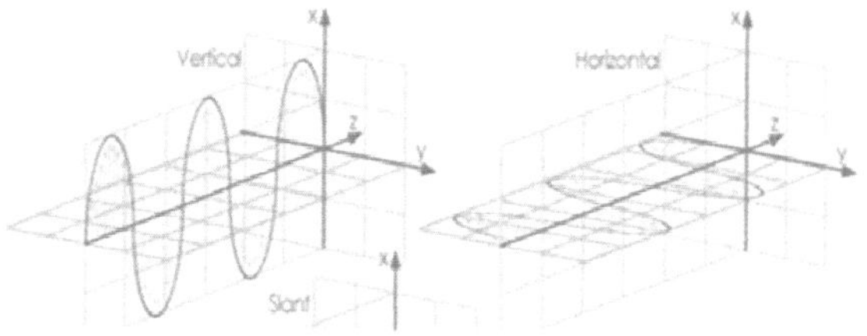

Figure 1. Vertical polarisation and horizontal polarisation with horizontal wave orientation. (https://jemengineering.com/blog-intro-to-antenna-polarization/)

This is due to the antenna installed are omnidirectional vertical polarisation type [14], that is positioned vertically using a vertical polarisation and transmits horizontally as in Figure 1, thus the signal received by Satellite AIS (S-AIS) that is located in perpendicular position to the ship is weak [15], as illustrated in Figure 2.

Figure 2. An illustration of AIS signal transmission strength with horizontal wave orientation received by terrestrial stations, plane, balloon, mountain, and satellite.

The maximum coverage of 56 km is a consequence of AIS station located on coastal areas with the antenna height of 100 metres [16]–[19]. In 2006, the concept of AIS Satellite (S-AIS) was proposed to fulfil the needs of longer surveillance range than the conventional coverage of Terrestrial AIS (T-AIS). In the publication of ITU-R M.2084, the coverage range up to 200 nm was required to anticipate the arrival of ships with dangerous goods, to improve safety, and to avoid offshore maritime illegal operation [12]. However, to receive AIS signals from satellite holds some fundamental challenges, as it unable to receive signals from thousands of ships at the same time as the signals will be overlapped [20]. In the perspective of satellite, some TDMA protocol clusters transmits at the same time, resulting in collision message [21]-[22], so that the AIS data collected are not complete. The satellite of LAPAN 2 orbiting in the altitude of 650 km only operates 11 hours per day and supports the module of AIS receiver ASR -100 Kongsberg Seatex that able to receive 104 AIS signal per minute only [23]. The collision message causes data corruption of up to 59.5%. [24]. Since satellites are always moving and earth rotates, to monitor the whole area of Indonesia simultaneously, it is required a constellation of eight units of satellites [25].

Based on the VHF propagation theory recommended by the IMO [16], it can be assumed that the AIS coverage range can be achieved at 200 nm if the location of the AIS receiving station is at an altitude of 6000 metres.

Table 1. Comparisons of coverage range between terrestrial station and elevated station.

Altitude	Coverage (km)	Power	Lifetime	Quality Latency	Cost
Satellite 700 km	5000	Battery	Years	Less Latency	Fantastic
Balloon 27 km	5000	Battery	Years	Less Delay	Very expensive
Airliner 10 km	800	Battery	Hours	Less Near real time	Expensive
Coastal 100 m	100	Fixed	All time	Normal Real time	Normal
Mountain 2 km	600	Fixed	All time	Normal Real time	Normal

Table 1 shows the comparisons among five various altitudes to AIS signal coverage, which is highly influenced by altitudes. From the cost perspective, mountain reaches more significant coverage at normal cost.

Besides, Indonesia as a country with a long coastline to 100,000 km is estimated to have 90s AIS Base Station (T-AIS) owned by the government. Compared to the U.S. that owns 20,000 km coastline only, it already has 200s T-AIS stations, as illustrated in Figure 3.

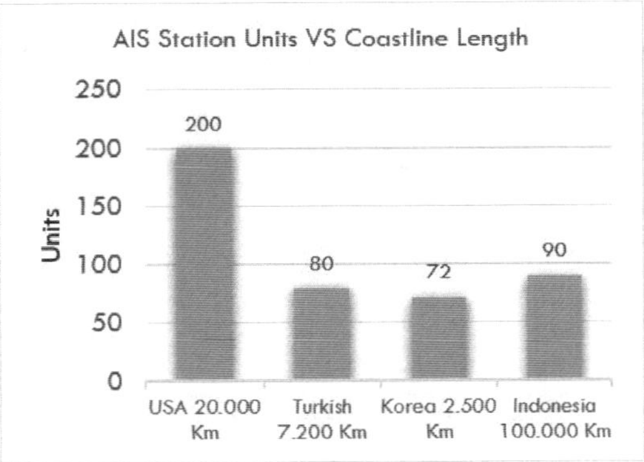

Figure 3. The numbers of AIS stations against coastline lengths.

AIS Surveillance Ratio (ASR) is the ratio of the number of AIS receivers against coastline length based on the characteristics of AIS signal propagation operated on VHF spectrum, as illustrated in Figure 4.

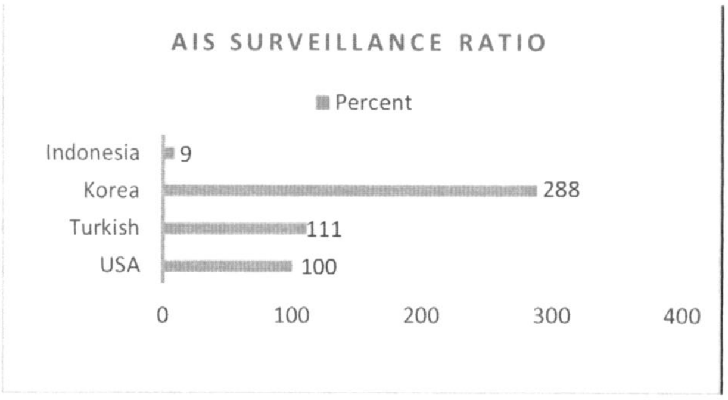

Figure 4. AIS Surveillance Ratio (ASR) of the number of AIS stations to coastlines.

Since the AIS coverage range is 56 km (radius) to the sea, one station is able to reach its left and right areas along the coast with the total of 112 km (diameter). The U.S. with its ASR ratio of 100% indicates that all of its coastal areas are covered by AIS receiver stations. Meanwhile, South Korea has a ratio of 288% indicating that all of its coastal areas are overlapped covered. Nevertheless, Indonesia's ratio is just about 9%, indicating that there are many coastal areas are not yet covered by AIS stations.

2. IMO Recommendations

The areas of ship monitoring are classified based on the characteristics of radio wave frequencies into four different areas, i.e., Sea Area A1, A2, A3, and A4. Sea Area A1 is categorised based on its VHF propagation wave up to the radius of 30 nm, while Sea Area A2 is categorised based on the characteristics of medium frequency (MF) propagation wave [3], [16] up to the radius of 100 nm.

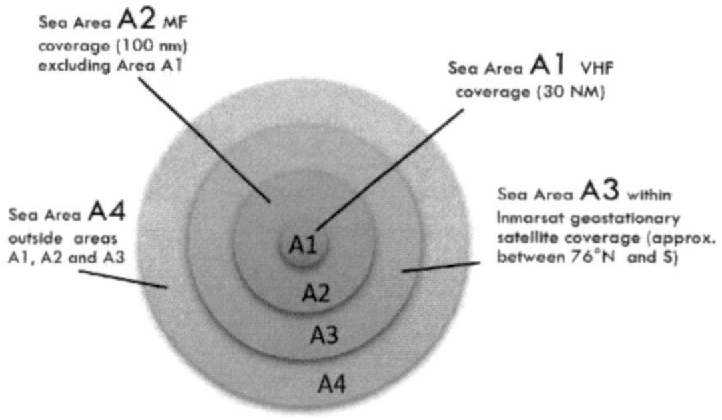

Figure 5. Sea Areas propagation coverage based on the IMO and GMDSS categories.

Figure 5 shows that Sea Area A1 is the circle where the radio station in its centre, with the maximum of VHF coverage as far as 30 nm. While the Sea Area 2 shows the MF frequency coverage as far as 100 nm. The AIS signal operates on the frequencies of 161.795 MHz and

162.025 Mhz, within the VHF spectrum in the propagation coverage of Sea Area A1. Sea Area 2 has the propagation characteristics of high frequency (HF) and medium frequency (MF), where the propagation able to cover up to thousand kilometres beyond the horizon due to the effect of mirror-like atmospheric reflection [26]. The AIS signal itself unable to be reflected by the atmosphere [27], but it transmits straight to the outer space. This phenomenon is in line with the Friis theory, which states that transmission power is negatively related to its frequency; the higher the frequency, the lower the transmission power.

IMO uses the formula (1) [11] to determine the effect of elevation towards AIS signal reception range. *A1* is a coverage range in nm. H (meter) is the height of receiver antenna in AIS receiver station in harbour; while h (meter) is the height of ship's antenna.

$$A1 = 2.5(\sqrt{H} + \sqrt{h}) \tag{2.1}$$

Based on the formula, the effect of elevation to AIS reception coverage range could be calculated as shown in the Table 2 below.

Table 2. IMO elevation effect to AIS signal coverage.

h/H	50 m	100 m
4 m	23 nm	30 nm

Source (IMO, 1995).

Table 2 explains that H (land station antenna) with elevation 50 m and h (ship antenna) 4 m, the maximum coverage range is 23 nm. If H 100 metres, the coverage range able to reach 30 nm. If the H is continued to 1000 metres, the range able to reach up to 84 nm, and when the H is at elevation 2000 metres, the range was able to reach 117 nm as shown in Table 3.

Table 3. IMO elevation effect extended.

Elevation (m)	1000	2000	3000	6000
Coverage (nm)	84	117	142	199

3.Theory of Radio Wave Propagation

The most fundamental in radio wave transmission formula is the Friis transmission formula [28], [29], also known as Friis free space formula, written as follows:

$$P_r(d) = P_t \frac{G_t G_r \lambda^2}{(4\pi d)^2} \tag{3.1}$$

Where:

P_r is a signal power received by receiver station (Watts)

P_t is a signal power transmitted by transmitter station (Watts)

G_r is a reveiver gain antenna

G_t is a transmit gain antenna

d is the distance between transmiter and receiver stations (metres)

λ is the operating wave length frequency (metres)

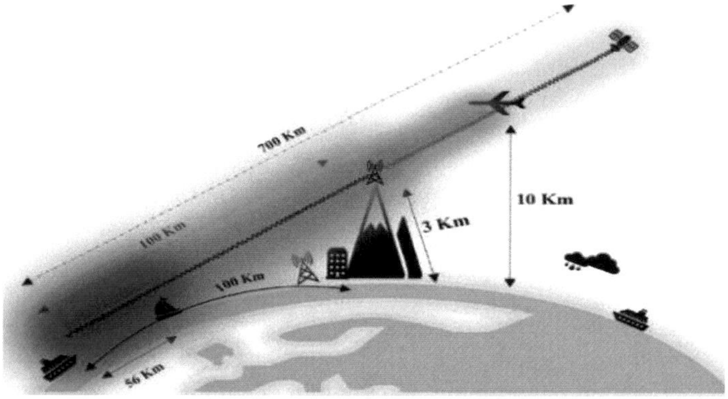

Figure 6. An illustration of line of sight of AIS signal.

Figure 6 shows an illustration of AIS signal transmission at the line of sight as the function of earth diameter and AIS receiver station at at a certain elevation. On a sea level, the distance of the AIS signal transmission from ships will be able to reach 56 km only, as it is affected by the curvature of the earth, but still detectable by AIS satellites at 700 km because it is not affected by the earth curvature. In addition, a mountain that is 100 km away is remained on the area line of sight, as its elevation reaches 3,000 metres. However, an AIS station located 100 km off-shore is *beyond line of sight* due to the earth curvature. Therefore, it should be assumed that the AIS signal strength received on a mountain 100 km away is greater than the signal strength received by AIS satellite station which is 700 km away.

4. Methods of Simulation and Survey

4.1 Simulation of VHF Radio Wave Propagation Prediction

The methodology employed in this study includes simulation and survey. To determine the locations of AIS signal receiver, it should be started by a simulation of propagation prediction on locations with elevations that are assumed to have strong propagation characteristics. This study conducted simulations on four different locations with various elevations. This simulation uses software that is operated on-line on the ve2db.com web site and is designed for amateur radio purposes. The simulations of four locations at different elevations are VTS Semarang, Gombel Hill, Mount Telomoyo and Mount Sumbing.

Table 4. Simulation results of propagation prediction using 6 dB antenna gain variable.

Location	Gain (dB)	Elevation (m)	Range (km)
VTS Semarang	6	30	60
Gombel	6	260	120
Telomoyo	6	1894	300
Sumbing	6	3371	320

The results of simulation show that from VTS Semarang, the area with strong signal propagation stretches along 60 km, which was shorter than Gombel Hill where the signal propagations showed strong characteristics within the radius of 120 km. Besides, the survey also found that areas with strong signal propagations in Mt. Telomoyo stretches along 300 km. Areas with weak signal propagations were carefully considered in this study as there might receive atmospheric disturbances and interference unwanted signals, thus creating noise which resulted in incomplete data content, such as ship name and GPS coordinate. The results of propagation prediction are summarized in Table 4 above.

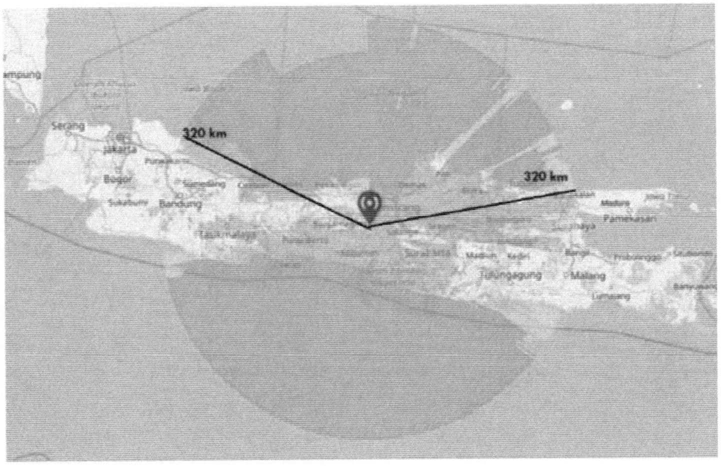

Figure 7. Result of propagation prediction simulation in Mt. Sumbing at 3371 m, shows 320 km (172 nm).

Figure 7 show the results of propagation prediction simulation which explained the area with the strongest signal propagation was on Mt. Sumbing at elevation 3371 metres where the area covered reached 320 km (172 nm).

The process for conducting survey and simulation is illustrated in the Figure 8 flow chart. It shows that the survey stage was influenced by several factors such as local weather, ocean weather, tidal conditions, topography. atmospheric propagation and vessels traffic density. Meanwhile, the surveyed parameters were also determined by adding some interventions such as signal strength, frequency, elevation, and maximum coverage.

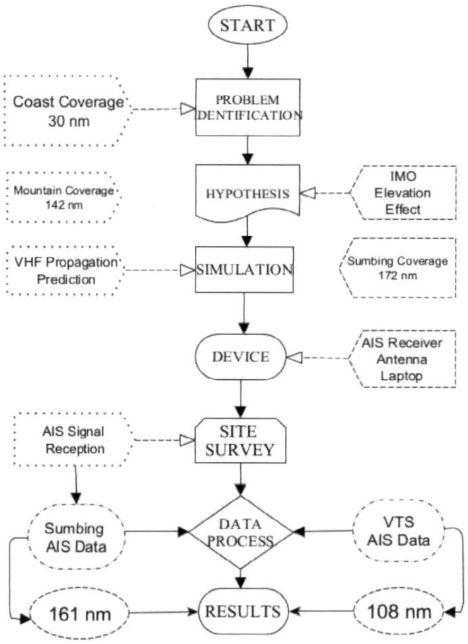

Figure 8. AIS signal observation flowchart .

Figure 8 shows the stage after AIS data was obtained from Mt. Sumbing with a range of 161 nm and then compared against AIS data from VTS Semarang with a range of 108 nm on the same date and time.

4.2 Survey of AIS Signal Reception

Following the results of propagation prediction simulations from the selected locations, the surveys of AIS signal reception were conducted on those places. The surveys were conducted in four different locations using longitudinal survey method, i.e. a survey conducted in different locations and times. In this study, the surveys were conducted at VTS Semarang lighthouse, Gombel Hill, Mt. Telomoyo, and Mt. Sumbing. The results showed that in VTS Semarang coverage range was 82.4 nm where it monitored 40 vessels. The broader area were covered in Gombel Hill and Mt. Telomoyo with coverage range, 121 nm and 171 nm respectively.

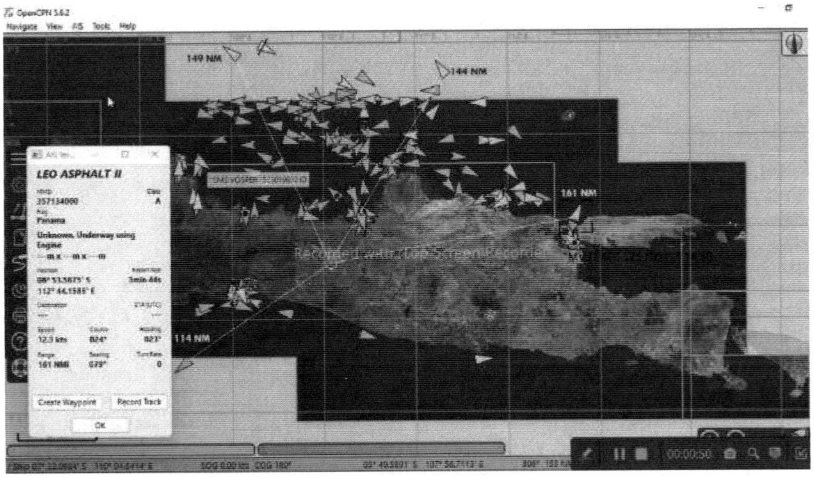

Figure 9. AIS survey result in Mt. Sumbing, coverage range 161 nm and 261 vessels density.

Furthermore, the survey results in Figure 9 shows that the coverage distance from Mt. Sumbing reached 161 nm and vessels traffic density reached 261 vessels. The results of all surveys are presented in Table 5 as follows.

Table 5. Survey results of AIS signal in four locations

Location	Gain (dB)	Elevation (m)	Range (nm)	Density (vessels)	Date	Time
VTS Semarang	6	30	82.44	40	March 18 2024	14:33:28
Gombel	6	260	121.50	65	March 17 2024	17:59:45
Telomoyo	6	1894	171.24	154	March 23 2024	11:44:37
Sumbing	3	3371	161.01	261	Dec 24 2023	07:25:39

Tabel 5 shows the effect of elevation on the AIS signal reception range, the higher the elevation, the longer the range of reception. It should be noted that the object of research is vessels movement, with distance and number of vessels changing each time. Thus, Mt. Telomoyo recorded a longer distance, even though it was lower than Mt. Sumbing, which received a larger number of vessels.

4.3 Simultaneous Survey for Data Comparison

A simultaneous survey is a survey conducted on two different location at the same time. In this study, the survey of AIS signal receiver was conducted on Mt. Sumbing at 3371 metres on December 24, 2023 at 07:25:39

The results were then compared to the AIS signal received by the VTS Semarang in December 24, 2023 at 07:05:36 using an antenna at 30 metres elevation. The furthest range of Mt. Sumbing was 161.01 nm with the density of 261 ships. In the meantime, VTS Semarang recorded its furthest range of 108.16 nm with the density of 84 vessels.

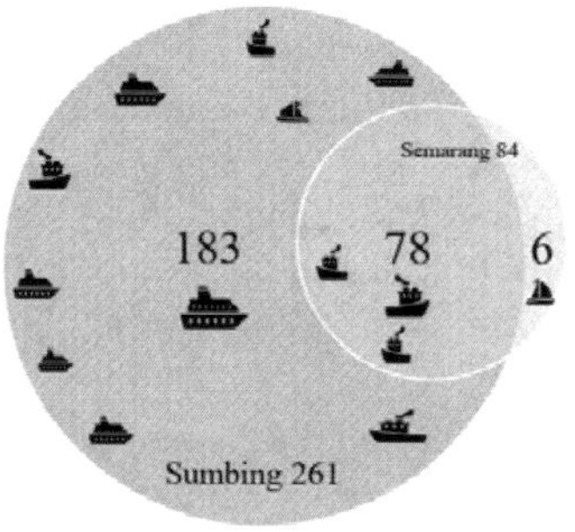

Figure 10. Venn diagram of the numbers of vessel received by Mt. Sumbing and VTS Semarang.

Figure 10 shows that there were 78 ships monitored by both AIS stations from Mt. Sumbing and VTS Semarang, while there were only 6 ships monitored by VTS Semarang but Mt. Sumbing, and 183 ships were monitored by Mt. Sumbing and were not by VTS Semarang. This means that 92.8 % of the ships monitored by VTS Semarang are monitored by Mt. Sumbing.

4.4 Variants Elevation Survey for Comparative Analysis

Some of the elevation variants that have been surveyed are Kamal Madura Port, Wonokitri, Penanjakan, Gantasan Ijen and Banyuwangi with varying results. However, these survey were not conducted simultaneously, so they cannot be compared directly on the same ship object at the detail level. Nevertheless, the results of this survey address the same objective, which is the density of vessel traffic in the waters of East Java and Bali.

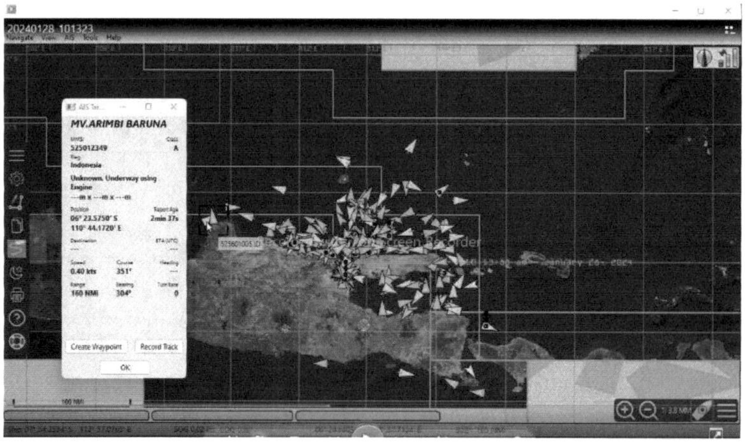

Figure 11 Survey results Penanjakan 2780 metress, 494 vessels, longest range 160 nm.

Figure 11 shows the results of the AIS signal reception survey at Penanjakan, which is one of a number of surveys with elevation variants targeting East Java waters. The results are then shown in the Figure 12 below, based on survey carried out at different elevations at different times.

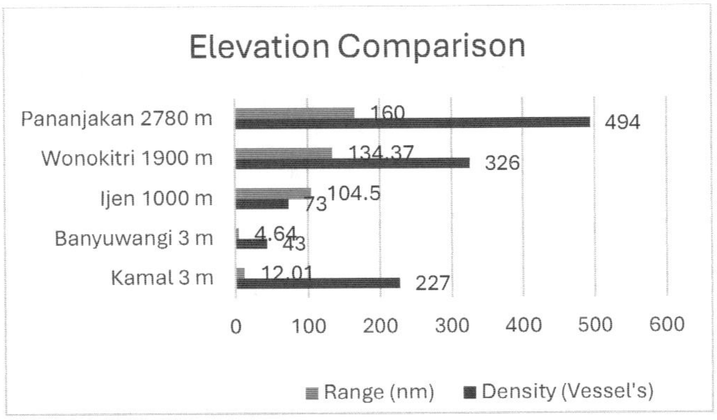

Figure 12. Comparison of elevation against range and traffic density.

Figure 12 shows the survey results from Kamal Port Madura with an antenna height of three metres and from Wonokitri at an elevation of 1900 metres is intended to observe SWAC Madura Straits waters. The survey from Banyuwangi coast with an antenna height of three metres and from Gantasan Ijen at an elevation of 1000 metres is intended to observe the Ketapang Gilimanuk crossing to the Lombok Strait.

From Figure 11 and Figure 12, it explained that from Penanjakan at an elevation of 2780 metres, the entire East Java waters can be covered, even from Jepara to the Lombok Strait, and the traffic density reaches 494 vessels in just 15 minutes of survey duration..

4.5 Propagation Challenges

In order to determine the difference in signal strength received between the coastal station and the mountain station, it is necessary to perform a simultaneous survey with a radio receiver that has a signal strength measurement feature. To generate detailed data on time and signal strength, a Software Define Radio device and a Raspberry data processor are required. The following data was obtained from a survey conducted simultaneously between Kamal and Bromo for one hour on November 30, 2023.

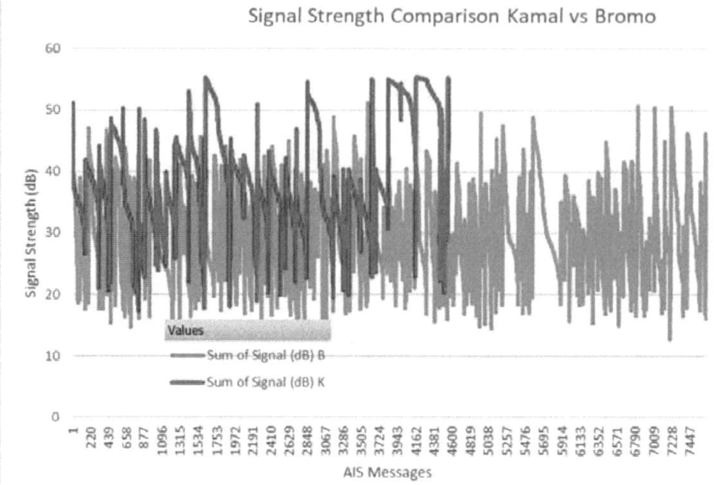

Figure 13. Signal Strength Comparison between Kamal and Bromo.

Figure 13 shows the accumulated received signal strength compared between Kamal and Bromo stations. The red line represents the accumulated received signal strength at Kamal, where 4550 AIS messages were received from 77 vessels within one hour.

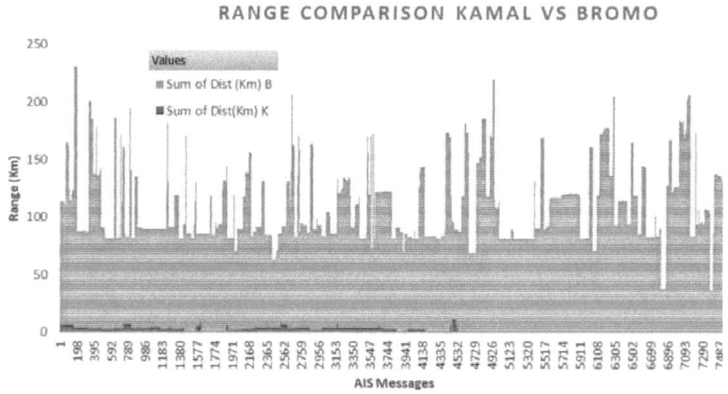

Figure 14. Range Comparison between Kamal and Bromo.

Figure 14 shows a comparison of the accumulated observed range between Kamal and Bromo stations. The blue line represents Bromo station which received 7469 AIS messages from 239 vessels within one hour.

4.6 TDMA Clusters Limitations
In order to manage the transmission turns, each AIS unit is equipped with the Time Division Multiple Access protocol, which uses the method of monitoring the frequency first before transmitting. Since the ability to monitor other vessels' transmissions is limited to line-of-sight, the TDMA protocol is limited to a 60km distance, called a TDMA cluster. In a TDMA cluster, each vessel transmits alternately and in harmony with the time slots allocated by the TDMA protocol. However, if it is outside the line of sight, TDMA will form other clusters, such as the Semarang, Cirebon and Cilacap clusters. These clusters are unable to hear each other, so the TDMA protocol will allocates the same time slot for transmission. This is not a problem for individual clusters, but it is a big problem if the AIS receiver location is able to listen to all clusters. This is the problem faced by the AIS satellite at an altitude of 700 km, which is capable of receiving hundreds of TDMA clusters, resulting in collision messages. Likewise, the AIS receiver station located at Mt. Sumbing at an altitude of 3371 metres also experienced collision messages originating from clusters in Central Java waters. However, due to the small number of TDMA clusters, the collision message did not have a significant impact.

4.7 Infrastructure Limitations and Weather Challenges
Unlike coastal radio stations, which are close to cities and easier to fulfil infrastructure needs, mountain AIS stations have fundamental limitations such as electricity, internet networks, road access, buildings, guards and maintenance. Power outages are a common occurrence, causing stations to become out of service. Bad weather with strong winds will cause the antenna to collapse, while horizontal lightning strikes will cause the vertical lightning rod, which is higher than the antenna, to fail. The characteristics of thin air at high elevations will cause a loss of dielectric capability, which will cause the switching power supply to short-circuit and burn out leading to complete failure of the AIS device.

4.8 Cost Comparison
According to Trong, a space-grade AIS receiver costs US$20,000 [31], but when equipped with auxiliary equipment such as antenna, UHF tranceiver, batteries, solar cells, power supplies, on-board computers and magnetorquers, the cost jumps to almost US$1,000,000. This cost does not include the cost of rocket launch, ground station infrastructure, operations and maintenance. Meanwhile, the cost of a mountain-based AIS station is estimated to be around US$2,000, using an AIS receiver costing US$1,000 for a single station.

5. Conclusion and Recommendation
This research has proved that the concept of IALA for placing antenna as high as possible to make it able to received AIS signals without any barriers. Whereas, the concept of IMO about the effect of elevation to AIS coverage range has also been proved through the survey of AIS signal coverage in nine different elevations from Central Java to East Java. Those two concepts was examined by the AIS survey in Mt Sumbing at the elevation of 3371 metres, resulting in coverage range to 161 nm, where the entire coastline from the north coast, south coast of Central Java to parts of West Java and East Java able to be monitored. While the AIS survey in Mt. Telomoyo at 1894 metres elevation resulted in coverage range of 171 nm, from Cirebon coastal areas to Surabaya Western Access Channel (SWAC). Then, AIS survey in Bromo at 2780 metres elevation was able to reach radius coverage to 160 nm.

Although AIS satellites have the ability to cover thousands of kilometres, but in terms of reception signal quality terrestrial AIS is superior, because it is closer to the source of the signal transmission and has fewer collision messages between TDMA clusters. Meanwhile, the cost of investing in single satellite AIS unit is enough to fund 500 AIS terrestrial stations.

The results of this research have the potential to contribute to maritime safety, protect the marine environment, prevent IUU fishing and improving maritime territorial surveillance. Indonesia, which is blessed with abundant mountainous resources, is a potential site for the establishment of an AIS stations network with the capability to cover the distances recommended by the ITU-R M.2084.

Acknowledgment
Researchers would like to express gratitude to Dian Nurdiana, S.T., M.T. as the Head of Second Class Navigation District of Semarang and all his staff, including seven Heads of Coastal Radio Station in Central Java, for their support in conducting this research about Automatic Identification System (AIS) propagation signal. Special thanks also to 2021 Marine Engineering student Rizki Azhari and the team for conduct the AIS expedition to Mount Sumbing.

References

[1] IMO, "Resolution A.917(22) GUIDELINES FOR THE ONBOARD OPERATIONAL USE OF SHIPBORNE AUTOMATIC IDENTIFICATION SYSTEMS (AIS), s.l.: INTERNATIONAL MARITIME ORGANIZATION," 2001.
[2] IALA, "IALA Guidelines on the Universal Automatic Identificaton System, s.l.: s.n," 2002.
[3] SOLAS, "International Convention for the Safety of at Sea 1974. s.l.:s.n," 2012.
[4] IALA-1082, "An Overview Of AIS Edition 2.0 | 1082, Saint German en Laye, France: International Association of Marine Aids to Navigation and Lighthouse Authorities," 2016.
[5] A.-L. S. Lapinski and A. W. Isenor, "Estimating Reception Coverage Characteristics of AIS," *J. Navig.*, vol. 64, pp. 609–623, 2011.
[6] E. Osekowska, "Design And Implementation Of Amaritime Traffic Modeling And Anomaly Detection Method, s.l.: Blekinge Institute of Technology," 2014.
[7] K. Wolsing, L. Roepert, J. Bauer, and K. Wehrle, "Anomaly Detection in Maritime AIS Tracks: A Review of Recent Approaches," *J. Mar. Sci. Eng.*, vol. 10, no. 112, 2022.
[8] S. Plass, R. Poehlmann, A. Dammann, and C. Gentner, "Investigations on AIS Signal Reception on Aircraft at Higher Altitudes, Oberpfaffenhofen, Wessling, Germany: IEEE," 2014.
[9] IALA, "An Overview Of AIS Edition 2.0 | 1082. 2.0 ed. Saint German en Laye, France: International Association of Marine Aids to Navigation and Lighthouse Authorities," 2016.
[10] Z. Vanche, A. Renaud, and A. Napoli, *Predicting AIS reception using tropospheric propagation forecast and machine learning*. United States., s.n: Denver, 2022.
[11] W. R. Cairns, "AIS and Long Range Identification & Tracking," *J. Navig.*, vol. 58, pp. 181–189, 2005.
[12] ITU, "Satellite detection of automatic identification system messages. s.l.:s.n," 2006.
[13] ITU, "Prediction procedure for the evaluation of interference between stations on the surface of the Earth at frequencies above about 0.1 GHz," in *s*, l.: International Telecommunications Union (ITU, 2021.
[14] IMO, "Guidelines For The Installation Of A Shipborne AIS." 2003.
[15] J. Carson-Jackson, "Satellite AIS -- Developing Technology or Existing Capability?," *J. Navig.*, vol. 65, pp. 303–321, 2012.
[16] IMO, "Provison of Radio Services for the Global Maritime Distress and Safety System (GMDSS). s.l.:s.n," 1995.
[17] P. Last, M. Hering-Bertram, and L. Linsen, "How automatic identification system (AIS) antenna setup affects AIS signal quality," *Ocean Eng.*, vol. 100, pp. 83–89, May 2015.
[18] X. Wang and S. Zhang, "Evaluation of Multipath Signal Loss for AIS Signals Transmitted on the Sea Surface," *Ocean Eng.*, vol. 146, pp. 9–20, 2017.
[19] H. Shuaiheng, Z. Shufang, W. Xiaoye, and Z. Jingbo, "A Novel Adaptive Noise Resistance Method Used for AIS Real-Time Signal Detection," *Chinese J. Electron.*, vol. 29, no. 2, Mar. 2020.
[20] A. Dembovskis, "AIS message extraction from overlapped AIS signals for SAT-AIS applications, s.l.: s.n," 2015.
[21] S. Mendes and others, "Satellite AIS - an End-to-End Simulation Approach," *Int. WS Simul. EGSE Facil. Sp. Program. SESP*, vol. 2010, pp. 28–30, Sep. 2010.
[22] D. S. Ilcev, "ANALYSES OF VHF-BAND MULTIFUNCTIONAL AIS ANTENNA ANTENNA," *IIOABJ*, vol. 8, no. 3, pp.

145–151, 2017.

[23] C. T. Judianto, "Spaceborne AIS: Teknologi Satelit LAPAN-A2/LAPAN ORARI Untuk Pengawasan Wilayah Maritim Indonesia," *Sos. Pus. Teknol. Satelit*, vol. 11, no. 1, Mar. 2016.

[24] A. Karim, R. Permala, M. Mukhayadi, and W. Hasbi, "(LAPAN-A2 AND LAPAN-A3 AUTOMATIC IDENTIFICATION SYSTEM (LAPAN-A2 AND LAPAN-A3 AUTOMATIC IDENTIFICATION SYSTEM (AIS) SATELIT DATA CORRECTION USING INTERPOLATION AND EXTRAPOLATION METHODE," *J. Teknol. Dirgant.*, vol. 7, pp. 159–167, Feb. 2019.

[25] M. Mukhayadi, A. Karim, W. Hasbi, and R. Permala, "Designing a constellation for AIS mission based on data acquisition of LAPANA2 and LAPAN-A3 satellites," *August*, vol. 17, no. 4, pp. 1774–1784, 2019.

[26] R. D. Straw and G. L. Hall, *Antenna Height and Communications Effectiveness*. Newington: The American Radio Relay League, Inc, 1999.

[27] K. F. Mathapo, "A software-defined radio implementation of maritime AIS, s.l.: s.n," 2007.

[28] H. T. Friis, "A Note on a Simple Transmission Formula," *s.l., s.n.*, pp. 254–256, 1946.

[29] O. Kurnaz and S. Helhel, "Near ground propagation model for pine tree forest environment," *AEU - Int. J. Electron. Commun.*, vol. 68, no. 10, pp. 944–950, 2014.

[30] S. Plass, R. Poehlmann, R. Hermenier, and A. Dammann, "Global Maritime Surveillance by Airliner-Based AIS Detection: Preliminary Analysis," *J. Navig.*, 2013.

[31] Trong, Thu Vu et al. "Constellation of small quick-launch and self-deorbiting nano-satellites with AIS receivers for global ship traffic monitoring." (2011).

IOP Conf. Series: Earth and Environmental Science **1461** (2025) 012036 doi:10.1088/1755-1315/1461/1/012036

The role of satellite image technology in maritime surveillance and national sovereignty using the Synthetic Aperture Radar (SAR) data approach

Rakha Asyrof[1*], Prasetyo Wibowo[2], Thai Duy Quy[3], Maulana Yafie Danendra[4], and Muhammad Riduwan[4]

[1] Faculty of Marine Science and Engineering, Hangtuah University, Indonesia
[2] Department of Informatics, Politeknik Elektronika Negeri Surabaya,Indonesia
[3] Department of Information Technology, Dalat University, Vietnam
[3] Department of Marine Transportation Engineering, Institut Teknologi Sepuluh Nopember, Indonesia

*E-mail: asyrofi@hangtuah.ac.id

Abstract. Synthetic Aperture Radar (SAR) technology has emerged as a crucial tool in maritime surveillance and national sovereignty. This study integrates deep learning techniques with SAR imaging, utilizing YOLOv8 to enhance ship detection accuracy. The proposed methodology involves a structured approach to object detection using SAR data. It begins with the acquisition of SAR images and their annotation, followed by pre-processing steps that include extracting relevant image information, converting annotations from COCO to YOLO format, and creating ground truth based on bounding box areas. The pre-processed data is then utilized for training a YOLOv8 model. After model training, the performance is evaluated using metrics such as precision, recall, F1-score, and mean Average Precision (mAP) to ensure the model's accuracy and reliability. This comprehensive framework aims to optimize object detection in SAR imagery by leveraging advanced techniques in data preparation and model evaluation.

1. Introduction

Synthetic Aperture Radar (SAR) technology has seen significant advancements in recent years, leading to enhanced capabilities in remote sensing applications. SAR is a crucial tool for various domains, including environmental monitoring, military reconnaissance, and disaster management. Recent research has focused on several key areas of SAR technology, reflecting its evolving nature and expanding applications [1]. One of the most significant breakthroughs in the field is the creation of specialized low-power SAR System designed to tackle specific monitoring tasks. Tasks, for example, the innovative work by Ustaili et al. in 2022 [2], who developed a low-power SAR system dedicated to continuous ship monitoring. This cutting-edge system utilizes a small satellite with reduced transmit power and high resolution, allowing it to detect illegal vessels with remarkable accuracy while skilfully managing azimuth ambiguities [2]. In the realm of maritime research, one of the most pressing challenges lies in detecting smalls ships. The task is complicated by the sidelobe effects and image blurring that often accompany these vessels [3]. However, Zhou et al [4]. In their 2023 study [4] have made significant strides by introducing a sidelobe-aware detection network. This innovative approach leverages dual-pooling for feature

IOP Conf. Series: Earth and Environmental Science **1461** (2025) 012036 doi:10.1088/1755-1315/1461/1/012036

extraction and a novel loss function based on Euclidean Distances, thereby enhancing the accuracy of small ship detection [4]. Radio Frequency Interference (RFI) poses a substantial challenge to SAR systems. Huang et al. (2022) introduced a semiparametric algorithm for efficient RFI suppression, which improves computational efficiency and performance in real-world applications [5]. Machine learning and artificial intelligence (AI) techniques are increasingly applied to SAR data processing. Zhou et al. (2021) [6] reviewed AI-based phase unwrapping methods in Interferometric SAR (InSAR) [7], highlighting both advancements and limitations in this area [6].

Similarly, Xu et al. (2022) [8] discussed the integration of compressed sensing and machine learning for sparse SAR imaging, focusing on improvements in super resolution and feature enhancement [8]. Change detection is a fundamental and crucial aspect of Synthetic Aperture Radar (SAR) imaging. This process involves analysing two or more images captured over the same geographical area at different times to identify changes [6]. Recent advancements in change detection technique have significantly improved the accuracy and efficiency of this process [4]. In 2022, Wang et al. [9] proposed a graph-based knowledge supplement network (GKSNet) that leverages contextual information from labelled datasets to enhance change detection performance. This approach utilizes the structural relationship within the data to improve the identification of changes, thereby increasing the overall accuracy of the process. Additionally, Qu et al. [10] introduced a dual-domain network (DDNet) in 2022. This network combines spatial and frequency domain features to enhance change detection accuracy and reduce noise. By integrating both spatial and frequency domain information, DDNet provides a more comprehensive analysis, leading to more reliable and precise change detection results [10]. In the realm of Synthetic Aperture Radar (SAR) imaging, several advancements have been crucial for achieving high-quality results. One of the most significant developments has been in motion compensation and autofocus techniques. These innovations are essential because they help mitigate the challenges posed by sensor movements, which can lead to deviations and aliasing errors in the imaging process. Schorlemer et al. (2021) [11] made a notable contribution by presenting methods for compensating sensor movements in Frequency Modulated Continuous Wave (FMCW) SAR Systems. Their techniques effectively correct these deviations, thereby reducing aliasing errors and enhancing the overall quality of the images obtained from these systems. Chen et al. (2022) [12] further expanded on this by providing a comprehensive review of motion compensation and autofocus techniques specifically for airborne SAR imaging. Their work highlighted the importance of these methods in ensuring that airborne SAR systems can produce accurate and reliable images, even under dynamic conditions. Target recognition and classification in Synthetic Aperture Radar (SAR) imagery are crucial for a variety of applications. For instance, researchers have explored the using of Inverse Synthetic Aperture Radar (ISAR) images for classifying automotive targets. A study by Pandey and Ram in 2022 [13] demonstrated high classification accuracy using ISAR images, highlighting the potential of this technology in identifying and categorizing targets in SAR imagery effectively. This advancement is significant as it enhances the ability to gather and analyse information from SAR data, which is particularly useful in environments where traditional optical images are less effective, such as at night or under heavy cloud cover.

Our contributions have significant practical implications for maritime operations, that state as below:

- **Low-power SAR Systems**: Developing efficient and compact SAR system for specialized monitoring tasks, such as ship detection.

- **Small Ship Detection**: Improving the accuracy of detecting small ships in SAR images through innovative technique like sidelobe-aware detection networks.
- **Target recognition and classification**: Improving the ability to identify and classify targets in SAR images, particularly using SAR techniques using YOLO model.

By addressing these critical challenges, our research provides a more resilient solution for maritime ship monitoring, ensuring the safety of personnel and the success of maritime operations.

2. Methodology

This methodology describes the application of deep learning techniques to Synthetic Aperture Radar (SAR) data for ship detection and position estimation, which is crucial for maritime surveillance and national sovereignty as shown in Figure 1.

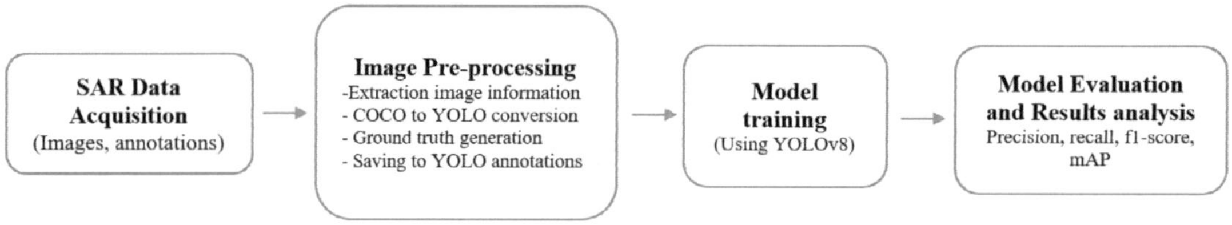

Figure 1. Methodology overview for SAR-based ship detection and position estimation.

2.1 SAR data acquisition

The primary goal of this data acquisition, there are how to gather relevant and high-quality satellite imagery that contains ships. This is crucial step because the accuracy of the model heavily depends on the quality and diversity of the training data. We utilize the SAR Ship Detection Dataset (SSDD), which is available for download from the official repository[1]. This dataset contains 1160 high-resolution SAR image, each with an average dimension of 500x500 pixels, and include 2358 ship instances. The SSDD is essential for maritime surveillance as it allows for continuous monitoring of shipping activities without the limitations imposed by weather or daylight conditions.

2.2 Image pre-processing

After data acquisition, the next step is to pre-process the SAR images and annotations to make them suitable for training the YOLOv8 model. There are many image augmentation methods as proposed in [14], our method includes several sub-tasks:

- **Extraction of Image Information**: Extract relevant details and metadata from the SAR images, which may involve filename, image id, bounding box coordinates, and area.
- **COCO to YOLO Conversion**: The annotations, initially in COCO format, are converted to the YOLO format. This step involves translating the bounding box coordinates and other annotation details into a format that YOLOv8 can interpret. The formula for converting from COCO to YOLO is given by:

[1] https://drive.google.com/file/d/1glNJUGotrbEyk43twwB9556AdngJsynZ/view

$$\text{YOLO bbox} = \left[\frac{x_{min}+\frac{width}{2}}{image\_width}, \frac{x_{min}+\frac{height}{2}}{image\_height}, \frac{width}{image\_width}, \frac{height}{image\_height}\right] \quad (1)$$

Where X_{min}, Y_{min}, *width*, and *height* are parameters from COCO format; *image_width*, *image_height* are the sizes of the original image.

- **Ground truth generation**: Ground truth data, based on the bounding box area, is created. This step is crucial for training the model as it defines the exact location of objects within each image.
- **Saving to YOLO annotations**: The pre-processed annotations are then saved in a format compatible with YOLOv8. This ensures that the model can correctly read and utilize the data during training.

2.3 Model training

Implementing a deep learning architecture capable of detecting ships in satellite imagery involves several key steps. Here's a detailed explanation of how to train a You Only Look Once (YOLO) model yolov8s.pt [15] for this task:

- **Implement YOLO Architecture**: We can use the YOLO architecture, which is a popular real-time object detection system. YOLO is known for its speed and accuracy in detecting objects within images. The architecture works by dividing the input image into a grid, where each cell in the grid predicts a certain number of bounding boxes and their associated class probabilities. This approach reframes object detection as a single regression problem, directly predicting bounding box coordinates and class probabilities from image pixels.
- **Using Keras**: To leverage the strengths of both frameworks, we can combine Keras with PyTorch. Keras provides a high-level interface for building neural networks, making it easier to implement complex models, PyTorch on the other hand, offers dynamic computation graphs and automatic differentiation, which are particularly useful for rapid prototyping and model optimization.
- **Optimize model parameters and layer configurations**: Optimizing the model's performance that involves adjusting hyperparameters such as the learning rate, batch size, and number of epochs. Additionally, we can fine-tune the layer configurations, including the number of convolutional layers and the number of filters. These adjustments help in optimizing the model's ability to detect ships accurately. For instance, in YOLOv4, the architecture uses a new CNN called Cross Stage Partial Network (CSPNet), which is a variant of the ResNet architecture designed specifically for object detection tasks. This architecture has a relatively shallow structure with only 54 convolutional layers but it cannot achieve- state-of-the-arts results on various object detection benchmarks.
- **Train the model**: After pre-processing the image dataset into the YOLO model, we need to train the model to learn identifying ships by predicting bounding boxes around them. The training process involves feeding the pre-processed images to the network and adjusting the model parameters to minimize the loss function. This process can be computationally intensive but it's essential for achieving high accuracy in ship detection.

2.4 Model evaluation and result analysis

To assess the performance of the trained model, we employ various metrics to gauge its accuracy and robustness. These metrics shows us how well the model performs in detecting ships, including true positives (correctly detected ships), false negatives (incorrectly detected non-ships as ships), and false negatives (missed actual ships). We utilize YOLO for advanced ship detection architecture. This approach ensures that the model is highly accurate in identifying ships by minizine the number of false positives. That dataset is created using SAR images annotated with ship locations. This dataset includes images with ships marked by bounding boxes, when helps the model learns to detect ships effectively. The model is trained to detect ships by minimizing a loss function that measures discrepancies between predicted and actual bounding boxes. Mean Average Precision at Intersection Over Union (IoU) thresholds ranging from 0.5 to 0.95 provides a comprehensive view of the model's performance. This metric indicates the model's robustness in detecting objects with varying levels of overlap, ensuring that it can accurately identify ships even when their bounding boxes have different levels of overlap with the actual ship locations.

3. Experiment results and discussion

3.1 Experiment Setup

This work presents the development of a classifier utilizing the YOLO architecture. The algorithm is implemented and coded using Keras with PyTorch. The algorithm is optimized by adjusting many variants of the number of layers and parameters. The Python platform is utilized for data manipulation and generating graphical representations. Additional modules utilized in these investigations include numpy, open cv, and pandas. The proposed methodology was evaluated and implemented on a personal computer equipped with an Intel(R) Core (TM) i7-8700 processor, NVIDIA GeForce RTX 3060 with 12 GB GDDR5X RAM, 16 GB of RAM, and running the Microsoft Windows 10 operating system.

3.2 Evaluation metric

The actual positive refers to the overall count of ships accurately identified by the algorithm based on satellite imagery. False positives refer to the instances where the algorithm incorrectly identifies non-ship objects as ships, whereas false negatives represent the cases where the program fails to detect ships. Precision, as defined by Equation (2), refers to the count of accurately identified objects in an image, excluding any mistakenly included objects. Recall is the proportion of accurate predictions to the number of objects in the image, as represented by Equation (3). For example, if a model accurately predicts 80 out of 100 genuine objects in an image, the recall rate is computed to be 80%. It is incorrect to assume that the model is accurate solely based on its high accuracy and recall. as the Equation (4), The *F1-score* is the mathematical average of precision and recall, calculated using the harmonic mean. Additionally, it represents the model's test correctness. The maximum attainable *F1-score* is 1, signifying flawless precision and recall, whereas the minimum attainable score is 0, indicating either no precision or recall. In image recognition, the mean Average accuracy (mAP) is the crucial parameter for evaluating the findings.

$$Precision = \frac{TP}{TP+FP} \tag{2}$$

$$Recall = \frac{TP}{TP+FN} \tag{3}$$

$$F1 - score = \frac{2*Precision*Recall}{Precision+Recall} \tag{4}$$

The Mean Average Precision (*mAP*) is a metric that calculates the average precision for each class and then takes the average of these values. It is represented by Equation (3), where q represents the number of queries and AP(q) represents the average precision for each query. After that, the

arithmetic progression (AP) average can be utilized to ascertain the mean average precision (mAP). *mAP50* and *mAP50-95* are assessment metrics commonly employed in object detection tasks. *mAP50* calculates the average precision by considering the intersection over the union (IoU) threshold 0.5.
On the other hand, *mAP50-95* calculates the average precision by considering IoU thresholds ranging from 0.5 to 0.95. These metrics are employed to evaluate the precision of object identification models by computing the precision and recall of the predicted bounding boxes to the ground truth. The *mAP50-95* metric offers a more extensive assessment by considering a broader spectrum of IoU thresholds, encompassing significant and minimal overlap between projected and actual bounding boxes. The metric selection to utilize is contingent upon the particular demands of the item identification task and the preferred equilibrium between precision and recall.

$$mAP = \frac{1}{Q}\sum_{q=1}^{Q} AP_q \quad (5)$$

3.3 Results

3.3.1 Ship detection performance

We employ YOLOv8 as a training optimizer, utilizing a learning rate 0.01 and a weight decay parameter of 0.0005. The training phase will consist of 50 epochs, each using a batch size 16. The image parameters for the original size are 800x800 pixels in height and breadth. Table 1 displays the outcomes of the YOLOv8 algorithm in identifying ship image collections. In summary, the precision is 75%, the recall is 76%, the F1 score is 75%, the mean average precision at 50% overlap (mAP50) is 83%, and the mean average precision from 50% to 95% overlap (mAP50-95) is 62%. These results demonstrate the model's strong detection capabilities, as evidenced by precision and recall scores exceeding 75%. The noteworthy aspect is the mAP50 outcome, indicating a significant value of 83%, signifying that many results exhibit an IoU surpassing 50. However, when contrasted with the mAP50-95, the results decline to 62%, implying that about 30% of the data exhibits an IoU over 95. This demonstrates the model's efficacy, as it exhibits nearly identical recognition of data findings between the training and testing data in the dataset.

Table 1. Ship detection performance metrics.

Class	Images	Precision	Recall	F1 Score	mAP50	mAP50-95
All	1962	75%	76%	75%	83%	62%
Small	1546	90%	70%	78%	87%	62%
Big	785	60%	82%	69%	78%	62%

According to image 2a, the f1 score curve demonstrates a 74% harmonization between large and small class labels, with a confidence level of 44%. Furthermore, the Precision-Recall curve for image 2b demonstrates an 83% value, while the Precision-confidence curve, derived from image 2c, stands at 94%. Figure 3a shows examples of detected ships in SAR images. The model effectively identifies ships even in challenging conditions, such as low contrast or cluttered backgrounds.

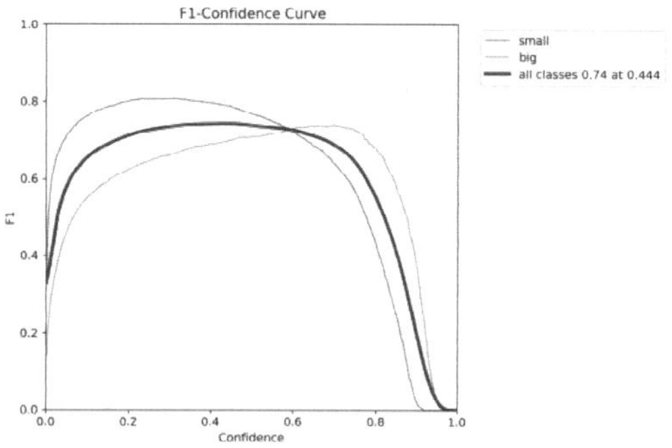

(a) F1-score curve

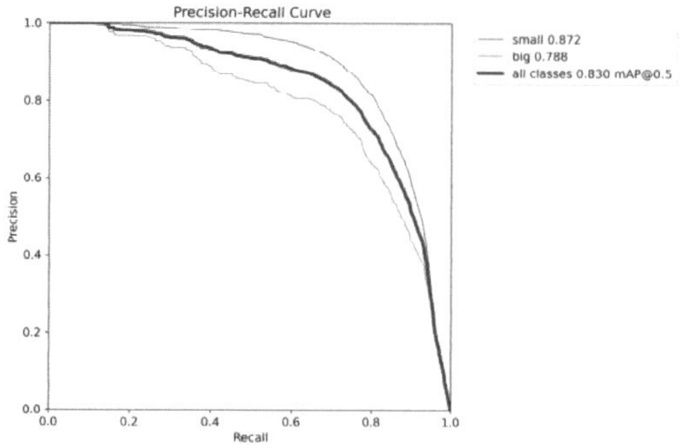

(b) Precision-Recall Curve

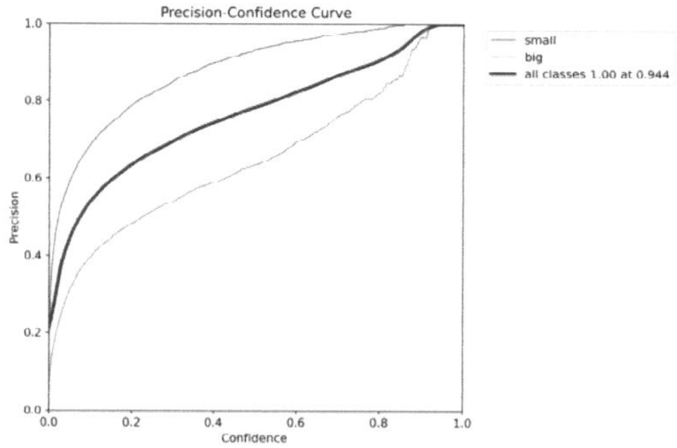

(c) Precision-Confidence Curve

Figure 2. Ship detection performance metrics (a) f1-Score curve, (b) Precision-Confidence curve, (c) Precision-Recall curve.

3.3.2 Position estimation accuracy

The accuracy of the position estimation was assessed by comparing the estimated ship positions with ground truth data from Automatic Identification Systems (AIS). The results are summarized in Table 2.

Table 2. Position estimation accuracy metrics. Mean Absolute Error (MAE) and Root Mean Square Error (RMSE) are calculated based on the distance between estimated and ground truth positions.

Metrics	Value
Mean Absolute Error (MAE)	1962
Root Mean Square Error (RMSE)	785

Figure 3b illustrates the estimated ship positions compared to the ground truth data. The estimation is generally accurate, with some discrepancies observed in high-density ship areas. The position estimation accuracy is satisfactory, with a mean absolute error of 1.2 km and a root mean square error of 1.8 km. These results show that the model can accurately estimate ship positions, which is crucial for maritime surveillance and national security. However, there are some areas where improvements can be made, particularly in high-density ship regions where overlapping or closely spaced ships may affect accuracy.

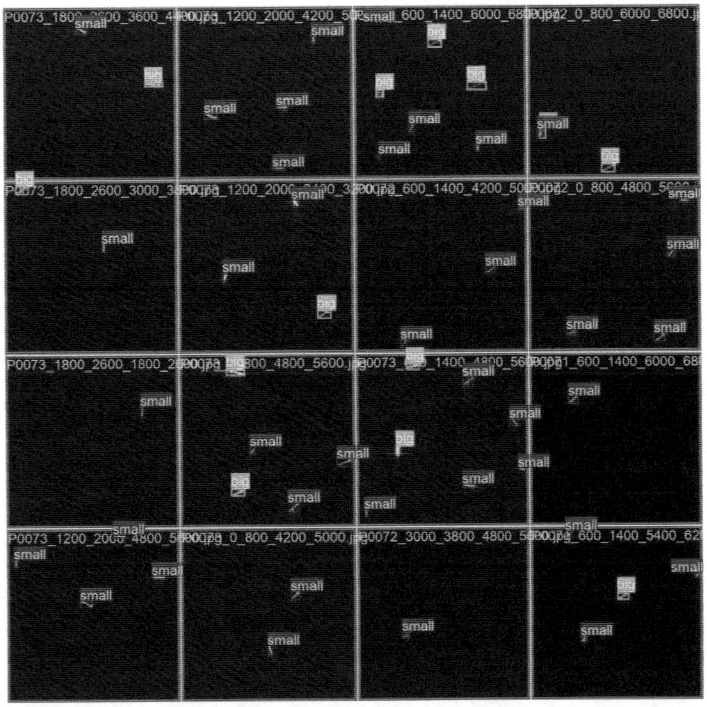

(a) Labels

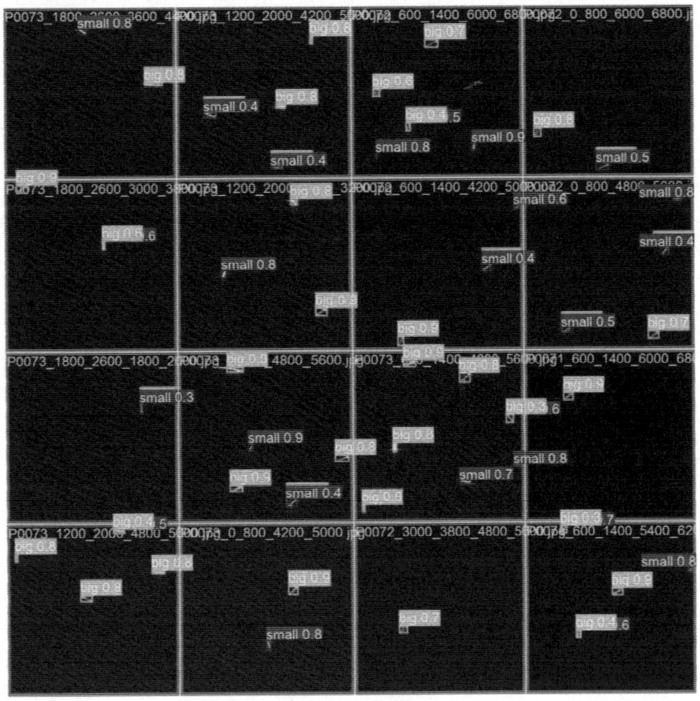

(b) Prediction

Figure 3. Comparison between ground truth label (a) and prediction result experiment (b).

Future work could focus on enhancing the model to handle such challenging scenarios better, possibly by incorporating more advanced techniques like multi-task learning or integrating additional data sources to improve overall performance.

4. Conclusion

In conclusion, this study has unequivocally shown the synergistic potential of integrating Synthetic Aperture Radar (SAR) data with deep learning techniques for enhancing maritime surveillance and national sovereignty. The application of YOLO model on annotated SAR images significantly improved ship detection and position estimation, validated through rigorous comparisons with ground truth data.

The high accuracy of these position estimates underscores the reliability and effectiveness of this approach for real-world applications. The implications are multifaceted and strategically significant, enhancing operational efficiency in maritime surveillance, protecting national sovereignty, and advancing remote sensing and artificial intelligence applications. Future research should focus on refining these models, exploring additional data sources, and adapting them to diverse scenarios, ultimately promising greater enhancements in maritime monitoring capabilities

Acknowledgments

We acknowledge the contributions of the various researchers whose work has been referenced in this manuscript. We are particularly grateful to Hangtuah University, Politeknik Elektronika Negeri Surabaya, Dalat University and Insitut Teknologi Sepuluh Nopember for our significant contributions, which have enriched our research project immensely.

References

[1] Kechagias-Stamatis O, Aouf N. Automatic Target Recognition on Synthetic Aperture Radar Imagery: A Survey. IEEE Aerospace and Electronic Systems Magazine 2021; 36:56–81. https://doi.org/10.1109/MAES.2021.3049857.

[2] Ustalli N, Krieger G, Villano M. A Low-Power, Ambiguous Synthetic Aperture Radar Concept for Continuous Ship Monitoring. IEEE J Sel Top Appl Earth Obs Remote Sens 2022; 15:1244–55. https://doi.org/10.1109/JSTARS.2021.3136935.

[3] Fracastoro G, Magli E, Poggi G, Scarpa G, Valsesia D, Verdoliva L. Deep Learning Methods for Synthetic Aperture Radar Image Despeckling: An Overview of Trends and Perspectives. IEEE Geosci Remote Sens Mag 2021;9:29–51. https://doi.org/10.1109/MGRS.2021.3070956.

[4] Zhou Y, Liu H, Ma F, Pan Z, Zhang F. A Sidelobe-Aware Small Ship Detection Network for Synthetic Aperture Radar Imagery. IEEE Transactions on Geoscience and Remote Sensing 2023;61. https://doi.org/10.1109/TGRS.2023.3264231.

[5] Huang Y, Chen Z, Wen C, Li J, Xia XG, Hong W. An Efficient Radio Frequency Interference Mitigation Algorithm in Real Synthetic Aperture Radar Data. IEEE Transactions on Geoscience and Remote Sensing 2022;60. https://doi.org/10.1109/TGRS.2022.3155068.

[6] Zhou L, Yu H, Lan Y, Xing M. Artificial Intelligence in Interferometric Synthetic Aperture Radar Phase Unwrapping: A Review. IEEE Geosci Remote Sens Mag 2021; 9:10–28. https://doi.org/10.1109/MGRS.2021.3065811.

[7] Wang C, Chang L, Wang XS, Zhang B, Stein A. Interferometric Synthetic Aperture Radar Statistical Inference in Deformation Measurement and Geophysical Inversion: A review. IEEE Geosci Remote Sens Mag 2024; 12:8–35. https://doi.org/10.1109/MGRS.2023.3344159.

[8] Xu G, Zhang B, Yu H, Chen J, Xing M, Hong W. Sparse Synthetic Aperture Radar Imaging from Compressed Sensing and Machine Learning: Theories, applications, and trends. IEEE Geosci Remote Sens Mag 2022; 10:32–69. https://doi.org/10.1109/MGRS.2022.3218801.

[9] Wang J, Gao F, Dong J, Zhang S, Du Q. Change Detection from Synthetic Aperture Radar Images via Graph-Based Knowledge Supplement Network. IEEE J Sel Top Appl Earth Obs Remote Sens 2022; 15:1823–36. https://doi.org/10.1109/JSTARS.2022.3146167.

[10] Qu X, Gao F, Dong J, Du Q, Li HC. Change Detection in Synthetic Aperture Radar Images Using a Dual-Domain Network. IEEE Geoscience and Remote Sensing Letters 2022;19. https://doi.org/10.1109/LGRS.2021.3073900.

[11] Schorlemer J, Schulz C, Pohl N, Rolfes I, Barowski J. Compensation of sensor movements in short-range fmcw synthetic aperture radar algorithms. IEEE Trans Microw Theory Tech 2021; 69:5145–59. https://doi.org/10.1109/TMTT.2021.3108399.

[12] Chen J, Xing M, Yu H, Liang B, Peng J, Sun GC. Motion Compensation/Autofocus in Airborne Synthetic Aperture Radar: A Review. IEEE Geosci Remote Sens Mag 2022; 10:185–206. https://doi.org/10.1109/MGRS.2021.3113982.

[13] Pandey N, Student Member G, Sundar Ram S, Member S. Classification of Automotive Targets Using Inverse Synthetic Aperture Radar Images. IEEE TRANSACTIONS ON INTELLIGENT VEHICLES 2022;7. https://doi.org/10.21227/hjw9-d428.

[14] Enkhbat A, Gochoo M, Shih TK, Cheewaprakobkit P, Aditya W, Quy TD, et al. Using Hybrid Models for Action Correction in Instrument Learning Based on AI. IEEE Access 2024. https://doi.org/10.1109/ACCESS.2024.3454170.

[15] Rakha Asyrofi, Yoni Azhar Winata. The Improvement Impact Performance of Face Detection Using YOLO Algorithm. Proceeding - 2019 International Conference of Artificial Intelligence and Information Technology, ICAIIT 2019, Institute of Electrical and Electronics Engineers Inc.; 2019, p. 6–9. https://doi.org/10.1109/ICAIIT.2019.8834526.

IOP Conf. Series: Earth and Environmental Science **1461** (2025) 012037 doi:10.1088/1755-1315/1461/1/012037

Determination of Hydrodynamic Coefficients for Rotary Derivatives with CFD Simulation

Yuda A. Hermawan[1*], Ketut Suastika[1], S. Al Farisy[1] and Rizky C. Ariesta[1]

[1] Department of Naval Architecture, ITS Surabaya, 60111 Surabaya, Indonesia

*E-mail: yuda.apri@its.ac.id

Abstract. Indonesia possesses rich natural resources, including extensive marine biodiversity and underwater infrastructure. To maximize its potential as a maritime nation, the development of underwater vehicles has emerged as a viable solution. In recent years, there has been significant growth in the development and use of these vehicles, particularly for exploring underwater environments. Hybrid Remotely Operated Vehicles (HROVs) combine the advantages of Autonomous Underwater Vehicles (AUVs) and Remotely Operated Vehicles (ROVs), offering versatile operational modes for industrial applications. UUV Segara Nauta (SN-01) is an HROV developed at the TKD Laboratory of FTK-ITS Surabaya. To ensure effective manoeuvering in various degrees of freedom, the motion abilities of HROVs are evaluated through manoeuver testing, which reveals the vehicle's motion characteristics and response to commands. These characteristics are represented by hydrodynamic coefficients derived from the forces and moments experienced by the vehicle, obtained through CFD numerical simulations. This study performs a virtual captive model test using the UUV Segara Nauta (SN-01) model to conduct rotating arm tests. The focus is on determining the linear hydrodynamic coefficients in the horizontal plane, specifically for sway and yaw motions. Test variations include different drift angles and radius values to assess the effects of lateral and angular velocities. The CFD simulation results, including force and moment values, are converted into non-dimensional values and analyzed using linear regression to obtain the hydrodynamic coefficients. These coefficients explain the model's behavior and underwater hydrodynamic performance, which is more responsive but tends to be unstable.

1. Introduction

The development of Unmanned Underwater Vehicles (UUVs) is crucial for supporting marine biodiversity and underwater infrastructure in Indonesia. To maximize its potential as a maritime nation, Indonesia must focus on developing and constructing its underwater vehicles, emphasizing technical aspects such as hydrodynamic performance, motion, and manoeuvrability. Unlike larger submarines, Unmanned Underwater Vehicles (UUVs) possess unique characteristics. There are various types of UUVs, including Autonomous Underwater Vehicles (AUVs) and Remotely Operated Vehicles (ROVs). However, recent advancements have led to the creation of a new category that combines the advantages of both, known as Hybrid Remotely Operated Vehicles (HROVs). These vehicles vary in size, and the environmental conditions of the water they operate significantly influence the forces and moments exerted on the vehicle's body during impact. These vehicles must maintain operational speeds without harming marine ecosystems and navigate varying depths effectively. Consequently, the vehicle's motion and manoeuvrability are largely determined by its body shape, making it essential to base the vehicle's design on a comprehensive analysis of its manoeuvring performance.

Manoeuvrability is a key factor in determining the limitations of an object's motion and control in water. For an underwater vehicle, manoeuvrability refers to its ability to adjust or maintain speed, depth, direction, and angle of motion, all of which are influenced by the vehicle's control mechanisms [1]. Manoeuvring analysis of underwater vehicles can be conducted using ship manoeuvring methods based on modular mathematical models. These models treat each hydrodynamic component involved in manoeuvring as a separate module, which is then integrated into a comprehensive numerical model for manoeuvring motion [2]. This modular approach allows for easier adjustments and modifications to individual parameters without affecting other components [3]. One such method, the Mathematical Model of Manoeuvring Group, has demonstrated high accuracy in predicting ship manoeuvring behaviour. Applying this modular mathematical model to underwater vehicles is particularly effective for predicting the manoeuvring motion of underwater tourist submarines. However, precise hydrodynamic coefficients are essential to calculate the manoeuvring performance of these vessels accurately.

Manoeuvrability can be evaluated through several methods, including empirical calculations, experimental testing, and numerical simulations. Among these methods, numerical simulation has become a primary approach for assessing the fundamental behaviour and manoeuvrability of a vehicle. Numerical simulations by utilizing computational fluid dynamics (CFD), offering cost-effective, accurate results into hydrodynamic coefficients and motion behaviours, are essential for evaluating UUV performance in six degrees of freedom. These simulations follow standard guidelines and procedures, such as those outlined in the ITTC captive model test procedure. Beyond addressing manoeuvring performance requirements through hydrodynamic coefficients, captive model tests also provide quicker results for assessing compliance with manoeuvring criteria set by IMO regulations. The ITTC procedure offers a detailed, step-by-step approach to simulating standard manoeuvring tests to obtain accurate hydrodynamic coefficients [4]. These coefficients, derived from manoeuvring equations, are crucial for describing and analysing spatial motion in six degrees of freedom (DOF) [5].

In previous studies, various approaches have been undertaken using captive model tests, particularly the rotating arm test. Toxopeus et al. [6] initiated a collaborative study using CFD to perform the calculations of the bare hull DARPA SUBOFF to investigate the capability of viscous-flow solvers for the prediction of the flow field around the hull. Either using static mesh or dynamic mesh, both methods give results that close to each other. Static mesh which non-inertial system is adopted to let the flow rotate around the stationary hull. Cao et al. [7] added source terms to those simulations in order to Coriolis and centrifugal forces will affect water particle to experience due to the rotation of the coordinate system. For different model shapes, Tang et al. [8] analysed the TUNA-SAND AUV, which has a complex block structure resembling a typical ROV, differing from streamlined shapes. They conducted both experimental and numerical simulations to validate their results, focusing on determining 16 viscosity-related damping coefficients and 12 potential flow-related inertia coefficients. Besides that, Hong et al. [9] assessed the hydrodynamic performance of the Underactuated Underwater Vehicle using PMM, oblique towing, and rotating arm tests. They employed dynamic and sliding mesh approaches, utilizing RANS and unsteady RANS equations, and validated the simulation results with experimental data, obtaining comprehensive coefficient values.

This study analyses the hydrodynamic performance of underwater vehicles by obtaining CFD-based rotational derivative hydrodynamic coefficients. The underwater vehicle used as the research object is a development Hybrid Remotely Operated Vehicle (HROV) model from the Digital Ship Technology Laboratory, Department of Naval Architecture, Institut Teknologi Sepuluh Nopember named UUV Segara Nauta (SN-01). The test was conducted virtually using CFD, applying a captive model test through a rotating arm test with radius variations and angular velocities. Determining the hydrodynamic coefficients of the underwater vehicle begins with solving the hydrodynamic forces and moments on the model in horizontal plane with three degrees of freedom (surge, yaw, and sway). The results are subsequently converted into nondimensional values, and hydrodynamic coefficients are determined using a linear regression approach.

2. Methodology

2.1 Coordinate system

The coordinate system adopted is the body-fixed coordinate system. In Figure 1, the body-fixed is located at the G-*xyz* coordinate right at the centre of gravity of the model while moving steady turn with respect to the O-*XYZ* coordinate as the global coordinate. The *x*, *y* and *z* axes define values for the forward direction, starboard direction, and downward direction, respectively. The body-fixed coordinates move in a steady turn rotation of radius R with angular velocity Ω. The use of a non-inertial reference frame introduces the effects of Coriolis and centrifugal forces in the momentum equation [10]. These forces are perpendicular to the velocity vector so they can act to change the direction of motion but not the kinetic energy of the flow.

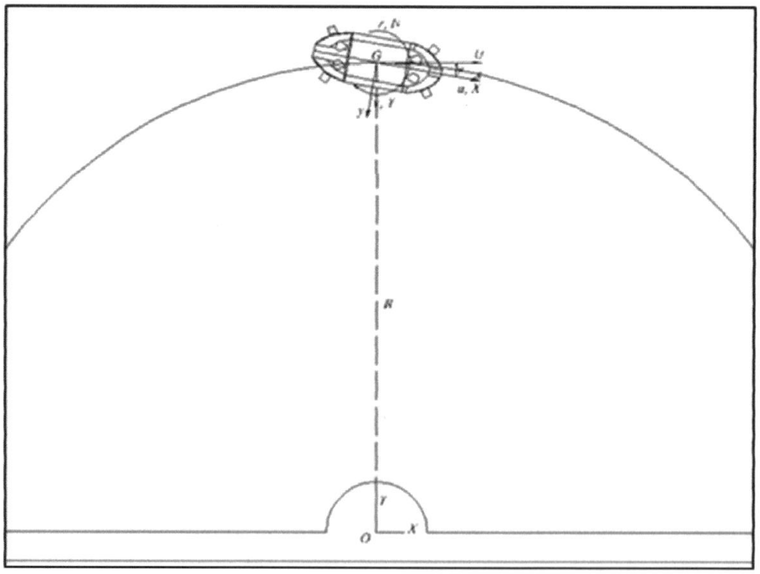

Figure 1. Coordinate system for rotational reference frame in the horizontal plane.

In a non-inertial frame of reference, it is essential to account for additional apparent forces, including the centrifugal and Coriolis forces. The centrifugal force acts radially outward from the axis of rotation, while the Coriolis force induces deflection in the path of moving objects relative to the rotating frame. These fictitious forces arise due to the acceleration of the reference frame and are crucial for accurate motion analysis. The forces will be incorporated as momentum source terms through the implementation of a User-Defined Function (UDF). In the horizontal plane, the source terms are specifically formulated to account for the effects of the rotating reference frame. The centrifugal force is proportional to the radial distance and the square of the angular velocity, while the Coriolis force is dependent on the velocity of the fluid and the angular velocity of the reference frame. By including these source terms, the simulation more accurately reflects the dynamics of the underwater vehicle in a rotating environment, enhancing the precision of the hydrodynamic performance analysis [7].

2.2 Parameter and equation motions

The resulting motion parameters include linear velocity U and angular velocity Ω. In body-fixed coordinates, the linear velocity U can be decomposed into three components: the longitudinal velocity u along the x-axis, the lateral velocity v along the y-axis, and the vertical velocity w along the z-axis. The same principle applies to the angular velocities Ω consisting of the roll angular velocity p about the x-axis, the pitch angular velocity q about the y-axis, and the yaw angular velocity r about the z-axis. For motion in the horizontal plane, it is adequate to consider the surge (longitudinal velocity u), sway (lateral velocity v), and yaw (angular velocity r).

The equations of motion of an underwater vehicle model for horizontal plane solved at specific degrees of freedom are described by the following notation [10, 11].

$$X = m[\dot{u} - vr + wq - x_G(q^2 + r^2) + y_G(pq - \dot{r}) + z_G(pr + \dot{q})] \quad (1)$$

$$Y = m[\dot{v} - wp + ur + x_G(qp + \dot{r}) - y_G(r^2 + p^2) + z_G(qr - \dot{p})] \quad (2)$$

$$N = I_z\dot{r} + (I_y - I_x)pq + m[x_G(\dot{v} - wp + ur) - y_G(\dot{u} - vr + wq)] \quad (3)$$

The model simulation measures the forces and moments acting on the vehicle body at the center of gravity (CG), which aligns with the centerline of the model. These forces include the longitudinal force X along the *x-axis* and the lateral force Y along the y-axis. The yaw moment considered is the yaw moment N around the *z-axis*. The hydrodynamic coefficients are derived from the total sum of hydrodynamic forces, external forces, and thrust forces. However, this study focuses solely on the hydrodynamic damping obtained from the simulation. Additionally, the model's motion is affected by the drift angle in the horizontal plane.

By conducting the rotating arm test, the model was manoeuvered to turn toward the starboard, tracing a circular path. This allowed for the determination of the hydrodynamic coefficients Y_r and N_r resulting from the rotational derivatives. The hydrodynamic coefficient X_r, influenced by rotational derivatives, are zero due to the vertical symmetry of the model. The linear velocity relation of the rotating model is defined by:

$$U = Rr \quad (4)$$

Where R is the radius of rotation and r is the angular velocity in rad/s. To determine the yaw rates r', which are nondimensional values representing angular velocity with respect to the z-axis, the following equation is provided.

$$r' = \frac{rL}{U} \quad (5)$$

The hydrodynamic coefficients are nondimensional quantities of the forces and moments obtained [12]. The process of nondimensionalizing these values is done as referring to (2).

$$X' = \frac{X}{\frac{1}{2}\rho L^2 U^2}; \quad Y' = \frac{Y}{\frac{1}{2}\rho L^2 U^2}; \quad N' = \frac{N}{\frac{1}{2}\rho L^3 U^2} \quad (6)$$

2.3 3D Model simplification

The model geometry is the Segara Nauta (SN-01) model, as shown in Figure 2. This model was developed at the Digital Ship Technology (TKD) Laboratory, Department of Naval Architecture, Institut Teknologi Sepuluh Nopember, featuring a hydrodynamic and modular fish line design [13]. The SN-01 has an overall length of 1.155 meters and can operate at depths of up to approximately 100 meters. Detailed specifications of the Segara Nauta SN-01 are provided in Table 1.

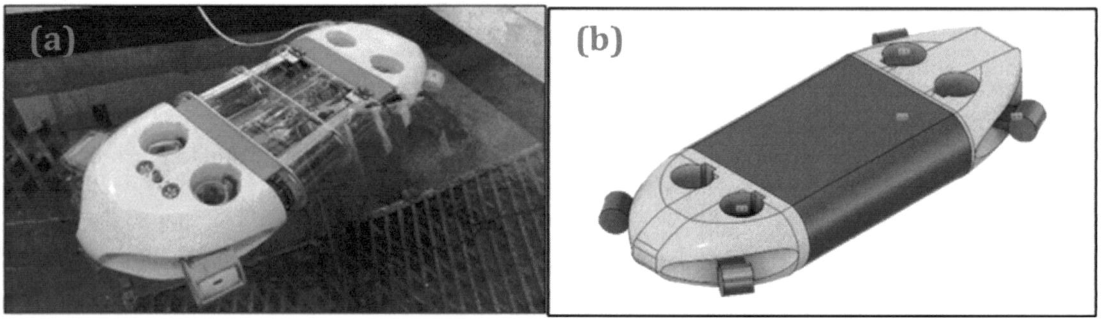

Figure 2. The test model used from the development of the TKD Laboratory FTK-ITS is named Segara Nauta (SN-01) (a); and its simplification model (b).

Table 1. Model specification of UUV SN-01.

Description	Symbol	Value	Unit
Length Overall	L_{OA}	1.155	m
Height	H	0.185	m
Breadth	B	0.450	m
Wetted Surface Area	S_{WA}	1.343	m^2
Center of Gravity	CG	0.582	m from fore

3. CFD Simulation

3.1 Simulation domain setup

The simulation domain is designed to represent the flow around the Segara Nauta (SN-01) model and is solved numerically. Proper domain selection is crucial to accurately depict how the model's shape affects the flow and to prevent boundary effects on the flow [14]. For the rotating arm test, the simulation domain is a half-cylinder, representing the shape of the rotating arm tank and experimental procedure that prevent the model from re-entering its generated flow. The half-cylindrical domain shape was selected to facilitate the clear definition of the flow's inlet and outlet, preventing any overlap that might occur in a full cylindrical domain. The simulation model is positioned at a specific radius (0, R) from the center of the simulation domain (0,0).

IOP Conf. Series: Earth and Environmental Science **1461** (2025) 012037 doi:10.1088/1755-1315/1461/1/012037

The simulation domain, created and discretized with a tetrahedral mesh, features a boundary layer value y^+ of 1 without a wall function. Tetrahedral mesh effectively captures the complex geometry of the model and the simulation domain; however, this comes at the cost of generating a significantly large number of elements. The simulation domain was designed in accordance with ITTC-2014 guidelines, which recommend positioning the boundaries at least one characteristic length L away from the geometry [15]. The boundaries in this simulation were defined by dividing the domain into six faces: the inlet, outlet, and walls [6]. The inlet, located in the positive *xz*-plane, is specified as inlet velocity with input angular velocity value of Ω. The outlet, in the negative *xz*-plane, uses a pressure outlet with default input to facilitate pressure-velocity coupling in RANS solving. The inner, outer, upper, and lower cylinder covers of the wall domain are specified as free-slip, while the wall model itself is specified as no-slip. The fore section of the model faces the positive *x*-axis, and the flow is generated in a circular pattern using a user-defined function (UDF).

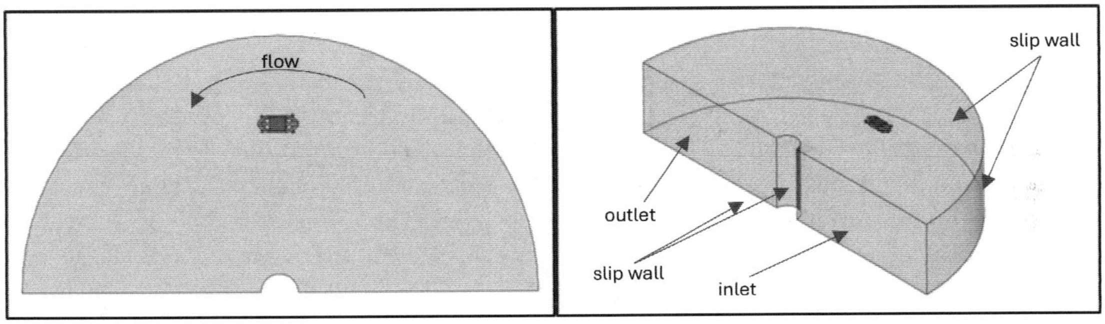

Figure 3. Schematics and simulation domain for the rotating arm test simulation.

3.2 Solver setup for rotating arm

In the steady-state simulation, the solver is configured as pressure-based and utilizes the $k-\omega$ SST turbulence model. This choice is based on the common practice of using a turbulence model to improve the accuracy of velocity profile analysis around the model without the need for wall functions. However, in some cases, the turbulence model alone may be sufficient for the simulation [6]. Within the cell zone conditions, a User-Defined Function (UDF) is employed to introduce source terms to the *x*-axis and *y*-axis momentum equations. The simulation employs the SIMPLE (Semi-Implicit Method for Pressure-Linked Equations) algorithm, which iteratively solves each variable to achieve convergence. Furthermore, the second-order upwind scheme is applied, offering enhanced result accuracy compared to the first-order upwind scheme. This approach ensures precise modeling of the fluid dynamics and turbulence characteristics within the simulation.

The test variations were conducted by altering the radius between the center of gravity (CG) of the model and the central point of the coordinate system, which is situated at the center of the simulation domain. The application of the captive model test adheres to the ITTC-2021 guidelines for rotating arm testing standards and parameters [4]. These variations, detailed in Table 2, aim to investigate the influence of different radial positions on the resulting forces and moments exerted on the model. The simulation results offer detailed data on the forces and moments acting on the model, particularly when it is subjected to rotational and lateral motion. This allows for a thorough analysis of the effects of radial displacement within the simulation domain.

Table 2. Model specification of UUV SN-01.

Linear velocity U (m/s)	Radius R (m)	Angular velocity Ω (rad/s)	Drift angle β
	20.000	0.12	
	15.000	0.16	
2.376	12.000	0.20	-12° s/d 12°, $\Delta\beta$ = 4°
	10.000	0.24	
	8.571	0.28	

3.3 Grid convergence index

To estimate the uncertainty of the simulation results, it is essential to perform a Grid Convergence Index (GCI) analysis. The GCI offers a systematic approach to quantify the error rate and assess the convergence of the simulation. This method involves evaluating three distinct mesh levels: Coarse, Medium, and Fine. Each mesh level is analyzed to determine the degree of numerical error and ensure that the solution is converging appropriately. The input data for this analysis are presented in Table 3, providing a basis for calculating the GCI and thereby validating the accuracy and reliability of the simulation results [16, 17].

Table 3. Detail inputs for Grid Convergence Index calculation.

Parameter	Fine (1)	Medium (2)	Coarse (3)
Surface mesh range (mm)	32 – 256	48 – 384	64 – 512
Total cell count	15,600,325	6,868,429	4,000,493
Dimensionless Y' ($\times 10^4$)	11.785	11.648	10.473

For the UUV SN-01 model, the Grid Convergence Index (GCI) procedure, as detailed in Table 4, was applied to assess the accuracy of the hydrodynamic coefficient calculations. The extrapolated value for the sway force coefficient (Y') was found to be 1.17% for the fine mesh quality level and 10.08% for the coarse mesh quality level. This indicates that the error estimation for the fine mesh is significantly lower, with an approximate relative error of 1.17% and an extrapolated relative error of 0.26%.

Additionally, incorporating a safety factor of 1.25 in the GCI calculation resulted in GCI values of 0.33% for the fine mesh level (GCI_{21}) and 1.04% for the coarse mesh level (GCI_{32}). These results demonstrate the reliability and precision of the simulation data [17]. The final estimated value of Y' is 11.816 with an uncertainty of 0.33%, underscoring the robustness of the fine mesh simulations and the effectiveness of the GCI methodology in ensuring the accuracy of the hydrodynamic coefficients for the UUV SN-01 model.

Table 4. Grid Convergence Index calculation.

Calculation	Symbol	Value
Refinement ratio	$r_i = r_{21} = r_{32}$	1.3
Difference of estimation	$\varepsilon_i = \varepsilon_{21} = \varepsilon_{32}$	$\varepsilon_{21} = -0.138$ $\varepsilon_{32} = -1.174$
Convergence ratio	R_i	0.117 (*Monotonic Convergence*)
Order of accuracy	p	9.4
Extrapolation	Y'^{i}_{ext}	$Y'^{21}_{ext} = 11.816$ $Y'^{32}_{ext} = 11.744$
Error estimation: Approximate relative error	e^{i}_{a}	$e^{21}_{a} = 1.17\%$ $e^{32}_{a} = 10.08\%$
Error estimation: Extrapolated relative error	e^{i}_{ext}	$e^{21}_{ext} = 0.26\%$ $e^{32}_{ext} = 0.82\%$
Factor of safety	F_s	1.25
Grid Convergence Index	GCI_i	$GCI_{21} = 0.33\%$ $GCI_{32} = 1.04\%$

4. Results and Discussion

4.1 Nondimensional force and moment

The results obtained for the UUV SN-01 model show that the sway force coefficient (Y') and yaw moment coefficient (N') increases as the yaw rates decreases, as shown in Figure 4. This aligns with Equation (5), where lower yaw rates suggest a decrease in the turning radius experienced by the model, while the model's characteristic length and linear velocity remain constant. The yaw rates (r') plays a significant role in the model's motion when assuming a circular turn. A high angular velocity at a low turning radius causes greater lateral forces due to centrifugal force, which acts as a reactive force to maintain the turning path. Consequently, during the design stage, it is crucial to consider the minimum turning radius while maintaining speed, as well as to ensure robust model design and controls to counteract the large drag forces experienced. Beside that, this trend indicates that larger drift angles result in increased lateral forces on the model due to a wider surface being exposed to the flow compared to the model's fore, which is designed to be more hydrodynamic.

Similarly, the yaw moment coefficient (N') exhibits a trend comparable to the sway force coefficient (Y'). The yaw moment coefficient (N'), related to the z-axis, impacts the rotational behavior of the model, where the moment experienced increases with higher drift angles. Notably, the slope of the N' curve is slightly steeper than that of the Y' curve, indicating a greater responsiveness of the yaw moment to changes in drift angle. This suggests that the yaw moment has a more significant influence on the model's drift angle, necessitating careful consideration in the design and control stages to ensure optimal manoeuvering performance, particularly in maintaining yaw motion.

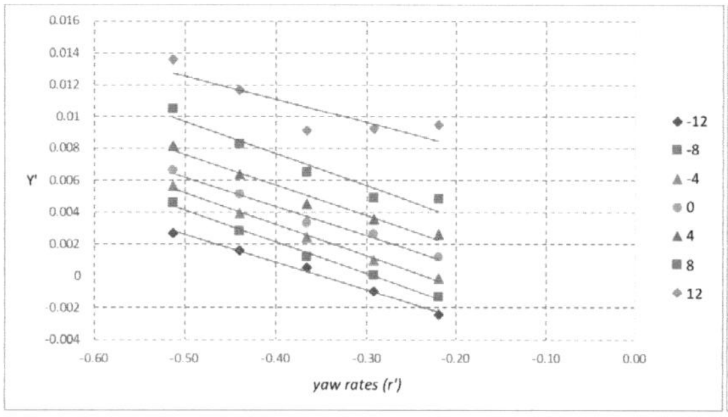

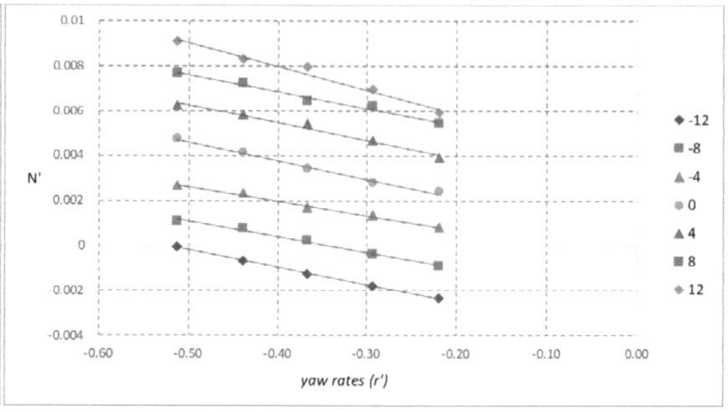

Figure 4. Force Coefficient Y' versus Yaw Rate Function r' (a) and Moment Coefficient N' versus Yaw Rate Function r' (b).

Figure 5 illustrates the pressure contours, revealing a higher-pressure gradient at the outer wall compared to the inner wall due to the centrifugal force in the rotational reference frame, consistent with the applied User-Defined Function (UDF). This pressure distribution highlights how the variation in drift angle significantly impacts flow separation, particularly at the aft and thruster sections of the model. As the drift angle increases, the separation point shifts, affecting the overall hydrodynamic performance. This analysis simulates a circular path with a constant radius, providing insights into the model's behaviour under these conditions. Additionally, variations in drift angle and radius were systematically examined to derive the corresponding force and moment values, enhancing the understanding of the model's dynamics. These findings are crucial for optimizing the design and control strategies of the UUV SN-01, ensuring robust performance in various operational scenarios.

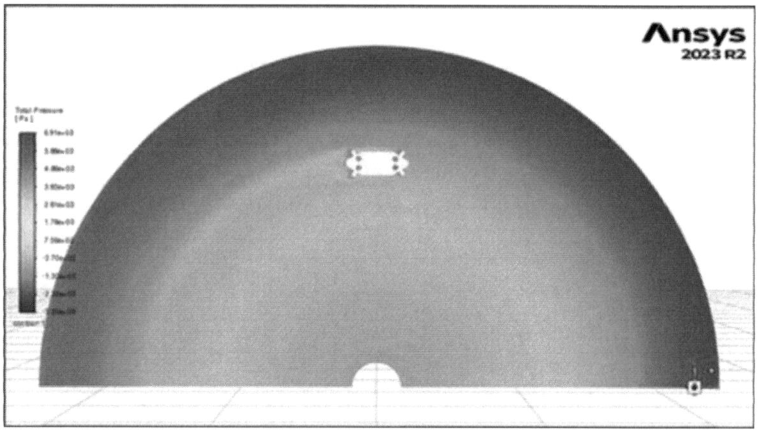

Figure 5. Pressure contour of UUV SN-01.

4.2 Hydrodynamic coefficients

The calculation of hydrodynamic coefficients involves using linear regression, where these coefficients are estimated from the force and moment values obtained through simulations. In this process, the known coefficient values are treated as response variables, while the nondimensional values of the motion components serve as regressor variables. By performing regression analysis on these variables, the relationship between the response and regressor variables is determined, with the slope of the regression model representing the hydrodynamic coefficient. To accurately determine the magnitude of these coefficients, it is essential to first convert the force and moment values into nondimensional forms, as outlined in equation (6). The calculation is shown in Figure 6.

Table 5. Hydrodynamic coefficients for rotary derivative.

Coefficient	Value $\times 10^3$
Y'_r	-14.441
N'_r	-6.841

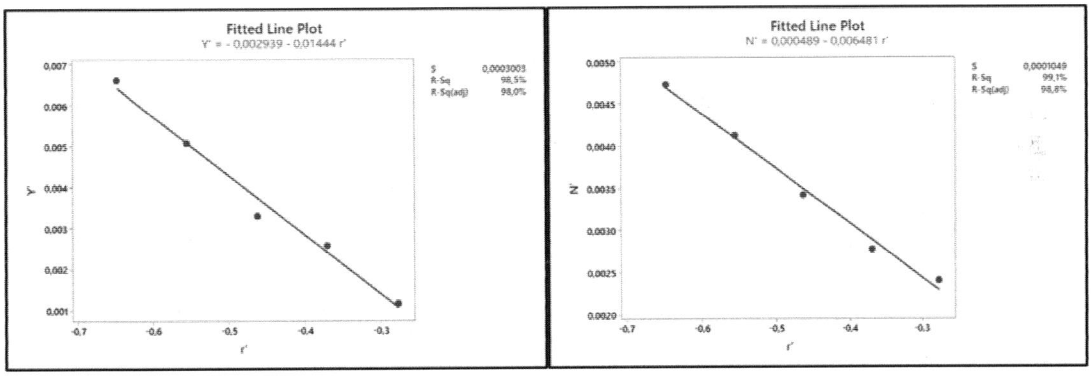

Figure 6. Linear regression function of yaw rates (r'): (a) Force Y' and (b) Moment N'.

5. Conclusion

The application of a virtual captive model test using CFD with rotating arm testing on the UUV SN-01 can be achieved by creating a simulation domain that mimics the actual test pool environment. In these simulations, variations in radius distance and drift angle were applied to the UUV SN-01 model. The resulting forces and moments from these simulations were then converted into nondimensional values for further analysis.

The hydrodynamic coefficients of the UUV, particularly those related to sway and yaw motion due to yaw rates (r') under steady conditions, can be derived from the CFD simulation results. This is accomplished through linear regression analysis, where the nondimensional values of forces and moments are regressed against the nondimensional values of the yaw rates (r') component. This approach allows for a detailed characterization of the UUV's hydrodynamic behaviour.

The CFD simulations revealed that the UUV SN-01 model exhibits a high sensitivity to changes in drift angle and yaw rate, leading to enhanced manoeuvrability. However, this increased sensitivity also results in greater instability in upper control, making it more challenging to maintain stable operation in varying underwater conditions. These findings underscore the need for careful consideration of control strategies to manage the vehicle's responsiveness while ensuring stability in the dynamic underwater environment.

Acknowledgments

This work was supported by the Institut Teknologi Sepuluh Nopember scientific research funding (No. 1209/PKS/ITS/2024).

References

[1] G. Dubbioso, R. Broglia, and S. Zaghi, "CFD analysis of turning abilities of a submarine model," *Ocean Eng.*, vol. 129, no. November 2016, pp. 459–479, 2017, doi: 10.1016/j.oceaneng.2016.10.046.

[2] S. Abdel-Latif, M. Abdel-Geliel, and E. Eldin Zakzouk, "Simulation of ship maneuvering behavior based on the modular mathematical model," *2013 21st Mediterr. Conf. Control Autom. MED 2013 - Conf. Proc.*, pp. 94–99, 2013, doi: 10.1109/MED.2013.6608704.

[3] Y. A. Hermawan and Y. Furukawa, "Coupled three-dimensional dynamics model of multi-component mooring line for motion analysis of floating offshore structure," *Ocean Eng.*, vol. 200, p. 106928, 2020, doi: 10.1016/j.oceaneng.2020.106928.

[4] ITTC, "ITTC - Recommended Procedures and Guidelines: Captive Model Test," *Int. Towing Tank Conf.*, pp. 1–20, 2021.

[5] P. Cardenas and E. A. De Barros, "Estimation of AUV Hydrodynamic Coefficients Using Analytical and System Identification Approaches," *IEEE J. Ocean. Eng.*, vol. 45, no. 4, pp. 1157–1176, 2020, doi: 10.1109/JOE.2019.2930421.

[6] S. Toxopeus *et al.*, "Collaborative CFD Exercise for a Submarine in a Steady Turn," in *Volume 5: Ocean Engineering; CFD and VIV*, Jul. 2012, pp. 761–772, doi: 10.1115/OMAE2012-83573.

[7] L. S. Cao, J. Zhu, and G. H. Zeng, "Numerical Simulation of Oblique Towing Tests and Rotating Arm Tests for a Submarine Model in Same Grid Topology," *Adv. Mater. Res.*, vol. 1083, no. November, pp. 190–196, 2015, doi: 10.4028/www.scientific.net/amr.1083.190.

[8] S. Tang, T. Ura, T. Nakatani, B. Thornton, and T. Jiang, "Estimation of the hydrodynamic coefficients of the complex-shaped autonomous underwater vehicle TUNA-SAND," *J. Mar. Sci. Technol.*, vol. 14, no. 3, pp. 373–386, Sep. 2009, doi: 10.1007/s00773-009-0055-4.

[9] L. Hong, X. Wang, D. Zhang, and H. Xu, "Numerical Study on Hydrodynamic Coefficient Estimation of an Underactuated Underwater Vehicle," *J. Mar. Sci. Eng.*, vol. 10, no. 8, p. 1049, Jul. 2022, doi: 10.3390/jmse10081049.

[10] O. M. Faltinsen, *Hydrodynamics of High-Speed Marine Vehicles*, vol. 9780521845. New York, USA: Cambridge University Press, 2006.

[11] T. I. Fossen, *Guidance and control of ocean vehicles*. Norway: John Wiley & Sons, 1994.

[12] J. Feldman, "DTNSRDC Revised Standarrd Submarine Equations of Motion," Bethesda, MD, 1979.

[13] K. R. Ramadhani, "Studi dan Rancang Bangun Unmanned Underwater Vehicle (UUV) Untuk Perairan Indonesia," Institut Teknologi Sepuluh Nopember, 2023.

[14] ITTC, "Practical Guidelines for Ship CFD Applications," *ITTC – Recomm. Proced. Guidel. ITTC*, pp. 1–8, 2011.

[15] ITTC, "ITTC – Recommended Procedures and Guidelines - Practical guidelines for ship CFD applications. 7.5-03-02-03 (Revision 01)," *ITTC – Recomm. Proced. Guidel.*, p. 19, 2014, [Online]. Available: https://ittc.info/media/4196/75-03-02-03.pdf.

[16] R. Akbar, I. K. Suastika, and I. K. A. P. Utama, "Numerical Study of Effects of Inhomogeneous Roughness on the Ship Friction Resistance," *BIO Web Conf.*, vol. 89, no. 3, pp. 17–27, 2024, doi: 10.1051/bioconf/20248913001.

[17] ITTC, "Uncertainty analysis in CFD Verification and validation methodology and procedures," *ITTC - Recomm. Proced. Guidel.*, pp. 1–13, 2021.

IOP Conf. Series: Earth and Environmental Science **1461** (2025) 012038 doi:10.1088/1755-1315/1461/1/012038

The Application of Artificial Neural Network - MISO (Multi Input Single Output) in Catamaran Ship Roll Motion Prediction

S Caramoy[1,*], W D Aryawan[1]

[1]Department of Naval Architecture, Institut Teknologi Sepuluh Nopember, Surabaya, Indonesia

*E-mail: 6018222003@student.its.ac.id and wasis@na.its.ac.id

Abstract. Artificial intelligence (AI) technology has become increasingly prevalent in various fields within the shipbuilding and marine industries. Typical applications of AI include production management, voyage analysis, and motion prediction. The main objective of this study is to develop a ship roll motion prediction tool for operational conditions that is fast, accurate, and computationally efficient. To achieve this, an AI algorithm based on Artificial Neural Networks (ANNs) with a Multiple Input Single Output (MISO) architecture was developed. The algorithm requires input parameters such as ship speed, wave height, wave period, and heading. The methodology for developing the predictive algorithm is presented, along with the ship motion data used for training and validating the ANN model. The ship motion data were generated using a frequency domain analysis based on the Boundary Element Method (BEM). The AI predictions were compared against 20% of the hydrodynamic analysis results to validate the model's performance. The results demonstrate that the developed AI tool effectively predicts the roll motion of a catamaran ship, capturing the non-linear behavior of ship dynamics under various operational conditions.

1. Introduction

The accurate prediction of ship motions, particularly roll motion, is of paramount importance in the marine and shipbuilding industries. Roll motion, a significant aspect of ship dynamics, affects vessel stability, operational efficiency, and the safety of onboard personnel and cargo [1]. This is especially crucial for catamaran ships, known for their unique hull configurations that can lead to pronounced rolling under certain sea conditions [2]. Traditional methods for predicting ship roll motion involve hydrodynamic simulations based on potential flow theory and frequency domain analysis, which require substantial computational resources and time. These methods, while accurate in ideal conditions, face challenges in providing real-time predictions for operational use.

Historically, ship motion prediction has relied on methods such as the Boundary Element Method (BEM) and Computational Fluid Dynamics (CFD) for understanding ship response in various sea states. BEM, as a frequency-domain approach, is widely adopted for its ability to analyze wave-structure interactions by solving boundary value problems under linear assumptions [3]. However, while effective for initial design assessments, these methods become computationally expensive when applied to more complex and non-linear conditions encountered in real-world operations. Moreover, they require simplifications that may limit their accuracy in capturing the full range of dynamic responses, particularly when dealing with non-linear behaviors in varied environmental conditions [4].

Recent advancements in artificial intelligence (AI) have shown promising potential to enhance predictive capabilities in the marine field. AI technologies, such as Artificial Neural Networks (ANNs), provide an alternative by leveraging data-driven approaches to predict complex phenomena [5], including ship motions, with less computational effort compared to traditional methods. As an example, Ao et al. [6] developed a Deep Learning algorithm capable to predict bare hull ship resistance. ANNs, with their ability to learn non-linear patterns from data, are increasingly being employed for various marine applications, such as voyage optimization, condition monitoring, and motion prediction. The adaptability of ANNs allows them to be trained on historical or simulated data to develop models that can predict real-time ship responses under diverse conditions, reducing the need for time-consuming hydrodynamic simulations [7].

While AI-based methods, particularly ANNs, have demonstrated potential in oceanography and marine applications [8], there remains a significant gap in their application for specific ship types, such as catamarans, under various operational conditions. The motivation for this study stems from the need to develop a fast, accurate, and computationally efficient predictive tool for catamaran roll motion that can operate in real-time environments. Unlike conventional hydrodynamic models that may struggle with computational demands and real-time applicability, an optimized ANN model can provide rapid predictions by learning from a variety of simulated ship motion data [9]. This study focuses on optimizing the ANN architecture, specifically using a Multiple Input Single Output (MISO) approach, to capture the complex, non-linear dynamics of catamaran roll motion across a range of speeds, wave heights, and wave frequencies.

2. Aim of this work

The primary objective of this study is to develop an ANN-based MISO model for predicting the roll motion of a catamaran ship under operational conditions. The proposed model aims to utilize ship speed, wave height, and wave frequency as input parameters to predict roll motion accurately. To achieve this, the study employs a frequency-domain method using BEM to generate a comprehensive set of ship motion data, which is then used to train the ANN model. The research also includes testing multiple ANN architectures to determine the optimum configuration that minimizes prediction errors, such as Mean Squared Error (MSE).

3. Methodology

3.1 Ship Response in Irregular Wave

The calculation of a ship's response in an irregular seaway is based on the linear superposition technique, which relies on two key assumptions. The first assumption is that the response of a vessel to any individual regular wave component is a linear function of that wave component's amplitude. This implies that the response would be the same as if the vessel were subjected to a continuous train of regular waves with the same amplitude and frequency over a long period. The response is assumed to be simple harmonic (i.e., sinusoidal) and to have a frequency equal to that of the encountered wave [3].

The second assumption underlying the linear superposition technique is that the response of a vessel to any given wave component is independent of its response to other wave components. This means that the total response of the ship in an irregular seaway is simply the sum of its responses to each individual wave component that makes up the seaway. In essence, the combined effect of the seaway on the ship can be obtained by summing up the effects of all the separate wave systems. The spectral density function of the ship's response $S_{\Phi}(\omega)$, determined by:

$$S_{\Phi}(\omega) = S_{\eta}|H(\omega)|^2 \qquad (1)$$

Where, $S_{\eta}(\omega)$ represents the spectral density function of the wave spectrum and $|H(\omega)|^2$ represents ship response amplitude operator.

The significance of the motion spectrum lies in its ability to fully describe the motion behavior of a ship in an irregular seaway. The motion spectrum provides valuable information about various aspects of the ship's movement, such as the average roll amplitude, the amplitude of one-third highest roll motions, and the highest possible roll amplitudes. For example, the average roll amplitude can be obtained from the area under the roll amplitude spectrum, denoted by m_0. The root mean square (RMS) value of the roll response is then given by:

$$(\Phi)_{avg} = 1.253\sqrt{m_0} \qquad (2)$$

3.2 ANN - MISO

A ANN-MISO is a specialized neural network architecture designed to process multiple input variables to generate a single output value. This network is structured with an input layer comprising multiple nodes that feed into a hidden layer, which then connects to a single output node [10]. The network's function can be expressed mathematically as:

$$y = f(\textstyle\sum_{i=1}^{n} w_i\, x_i + b) \qquad (3)$$

Where y represents the output, x_i are the input variables, w_i denote the corresponding weights, b is the bias term, and f is the activation function applied to the weighted sum of inputs. The principal advantage of the ANN-MISO lies in its capacity to capture and model intricate, non-linear relationships between multiple input variables and a singular output (Figure 1).

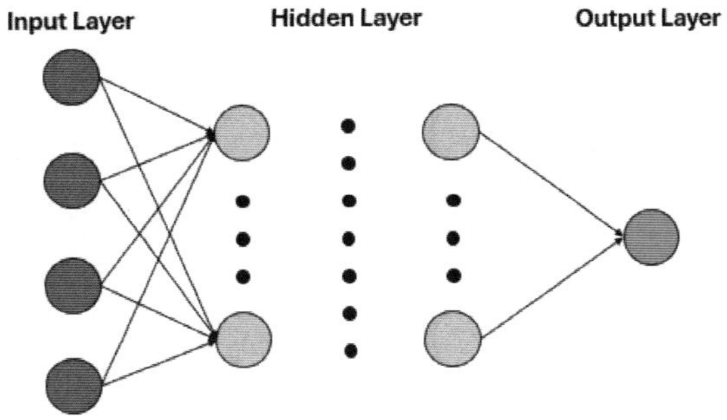

Figure 1. Example of ANN-MISO Architecture

To assess the performance of a MISO ANN, the MSE is commonly employed as an evaluation metric:

$$MSE = \frac{1}{N}\textstyle\sum_{j=1}^{N}\left(y_j - y^\wedge{}_j\right)^2 \qquad (4)$$

Where, N denotes the number of observations, y_j is the actual output value, and $y^\wedge{}_j$ is the predicted output value generated by the network. MSE quantifies the average squared deviation between the predicted and actual values, with lower MSE values indicating superior model [5].

IOP Conf. Series: Earth and Environmental Science **1461** (2025) 012038 doi:10.1088/1755-1315/1461/1/012038

4. Ship Roll Motion Simulation

4.1 Ship and Environmental Data

Accurate prediction of ship roll motion in varying sea conditions requires specific data related to the ship's dimensions, hydrostatic properties, and environmental conditions are provided as basis data to perform the ship motion simulation. The key parameters employed in this study include the ship's Length Between Perpendiculars (LBP), Breadth (B), Height (H), Draught (T), Displacement, Center of Gravity (CoG), and Radius of Gyration presented in Table 1. These parameters are fundamental in defining the ship's geometric configuration and mass distribution, which is crucial for simulating its hydrodynamic response and energy absorption during wave-induced motions.

The environmental conditions considered encompass variations in ship speed (2-8 knots), wave height (1-4 meters), and wave frequency (2-10 seconds, capturing a spectrum of sea states and the energy dynamics in operational scenarios the vessel may encounter.

Table 1. Catamaran Ship Particular and Environmental Data

Features	Value	Unit
LBP (Length Between Perpendicular)	45	meter
B (Breadth)	16	meter
H (Height)	7	meter
T (Draught)	3.7	meter
Displacement	990.3	ton
CoG		
x	21.23	meter
y	0	meter
z	3.3	meter
Radius of Gyration		
Kxx	6.4	meter
Kyy	11.25	meter
Kzz	11.25	meter
Ship Speed	2 - 8	knot
Wave Height	1 - 4	meter
Wave Period	2 - 10	second

This data is used to perform simulations of the ship's roll motion using the BEM. The BEM allows for a detailed hydrodynamic analysis of the ship's response to waves and their energy, considering different combinations of ship speed, wave height, and wave frequency. These simulations provide the roll motion and energy interaction data needed for training and validating the ANN-based predictive model. This approach enables the evaluation of the vessel's dynamic energy response under diverse conditions.

4.2 Ship Simulation Using BEM

In this study, the prediction of ship roll motion is conducted using a frequency domain method, which requires the calculation of the Response Amplitude Operator (RAO). The RAO is a critical parameter that represents the ship's response to wave excitation in the frequency domain, enabling the assessment of roll motion under various sea states and operational conditions.

To obtain the RAO data necessary for the frequency domain analysis, a ship simulation is performed using the BEM. The BEM is a numerical technique widely used for solving problems related to wave-structure interaction. It provides a robust framework for calculating the hydrodynamic forces and moments acting on the vessel, which are essential for determining the RAO.

The inputs for the BEM simulation are derived from the ship particulars and environmental data outlined in Subsection 4.1, including the ship's geometric and hydrostatic properties (e.g., length, breadth, draught, and displacement) and environmental parameters (e.g., ship speed, wave height, and wave frequency). By incorporating these parameters, the BEM simulation accounts for the various factors influencing the ship's roll motion.

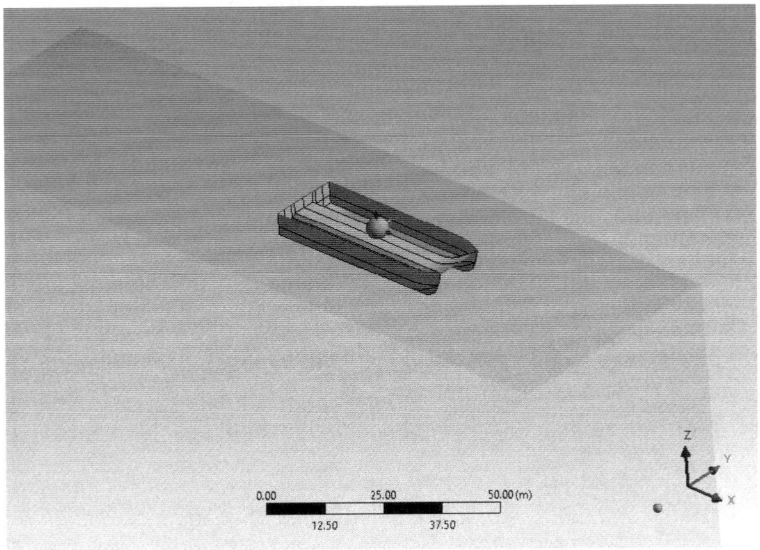

Figure 2. Ship Motion Simulation Using BEM

Figure 2 illustrates simulation setup of a catamaran ship using the BEM. The model showcases the computational domain in which the ship's hydrodynamic performance is analyzed under various sea conditions. The simulation environment includes the catamaran's hull geometry and surrounding fluid domain, where wave interactions and resulting ship motions are computed.

The output of the BEM simulation is a set of RAO data corresponding to different ship speeds and wave frequencies (Figure 3). This variation in RAO data across different speeds provides insight into how the ship's roll motion changes under different operating and environmental conditions. The RAO data serves as the foundation for roll motion calculation (1) in a range of the wave height in irregular wave sea conditions.

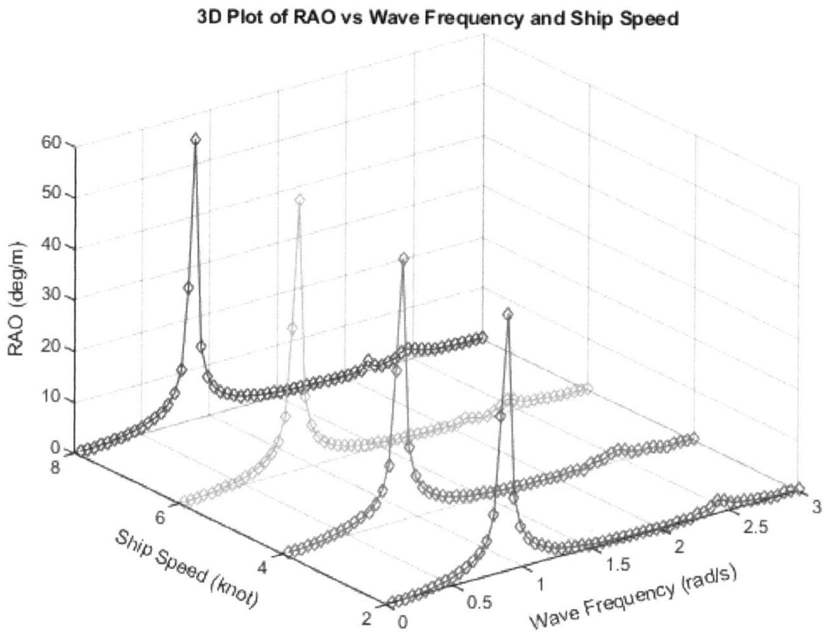

Figure 3. Ship Roll RAO Result in Different Ship Speed

5. ANN Architecture

In this study, an ANN-MISO architecture is employed to predict the roll motion of a catamaran ship based on varying sea conditions and operational parameters. The ANN model is designed to process three input parameters, representing key environmental and operational conditions, and generate a single output, which corresponds to the predicted roll motion.

Table 2. ANN Model Hyperparameter

Parameter	Value
Input Layer	3
Hidden Layer	1-2-3
Output Layer	1
Neuron	5-10-15-20
Data Ratio	
Training Data	80%
Test Data	20%
Optimizer	Adam
Activation Function	ReLu

Table 2 present the assumption of the hyperparameter used in this study. The ANN model consists of an input layer with 3 neurons, representing three input parameters which is ship speed, wave height, and wave period. The network has a single output layer with 1 neuron to predict the roll motion.

To determine the optimal ANN MISO model for predicting roll motion, multiple variations of the ANN architecture were trained and evaluated [11]. These variations were based on different configurations of hidden layers (ranging from 1 to 3) and the number of neurons in each hidden layer (5, 10, 15, and 20). The performance of each configuration was assessed to identify the best combination that provides accurate predictions with minimal computational overhead.

The training process involved adjusting the weights and biases of the network using the Adam optimizer to minimize the error between predicted and actual roll motion data [12]. The variation in the number of hidden layers and neurons allowed for an exploration of the trade-off between model complexity and performance, ultimately aiming to achieve a balance between predictive accuracy and computational efficiency.

6. Result and discussion

6.1 Ship Roll Motion Data

In this study, the ship roll motion data is obtained by calculating the ship response spectrum, as described in Subsection 3.1. The ship response spectrum is derived by multiplying the JONSWAP wave spectrum [13], with the square of the RAO (1) (2), which was generated using the BEM simulations outlined in Subsection 4.2.

Table 3. Ship Roll Motion Data

Speed (kn.)	Hs (m)	Tp (s)	Roll (deg.)
2	2	2	0.325
2	2	4	0.758
2	2	6	14.132
2	2	8	11.569
2	2	10	2.485
2	3	2	0.487
2	3	4	1.137
2	3	6	21.197
.	.	.	.
.	.	.	.
.	.	.	.
8	4	6	30.015
8	4	8	22.172
8	4	10	4.364

The resulting ship roll motion data comprises several sets of inputs and corresponding outputs. As presented in Table 3, the input parameters include ship speed, wave height, and wave period, which reflect different operational and environmental conditions. The output is the predicted roll angle of the ship under these conditions. This roll motion data is used for developing the ANN model, serving as both training and testing datasets.

6.2 ANN – MISO Model Development

The development of the ANN-MISO model in this study involves training 12 different variations of ANN architectures to identify the most accurate configuration for predicting ship roll motion. Each ANN model variation was constructed with different combinations of hidden layers and neuron counts, as outlined in Section 5. The performance of these models was evaluated based on the MSE, a common metric used to measure the average squared difference between the predicted and actual values (4).

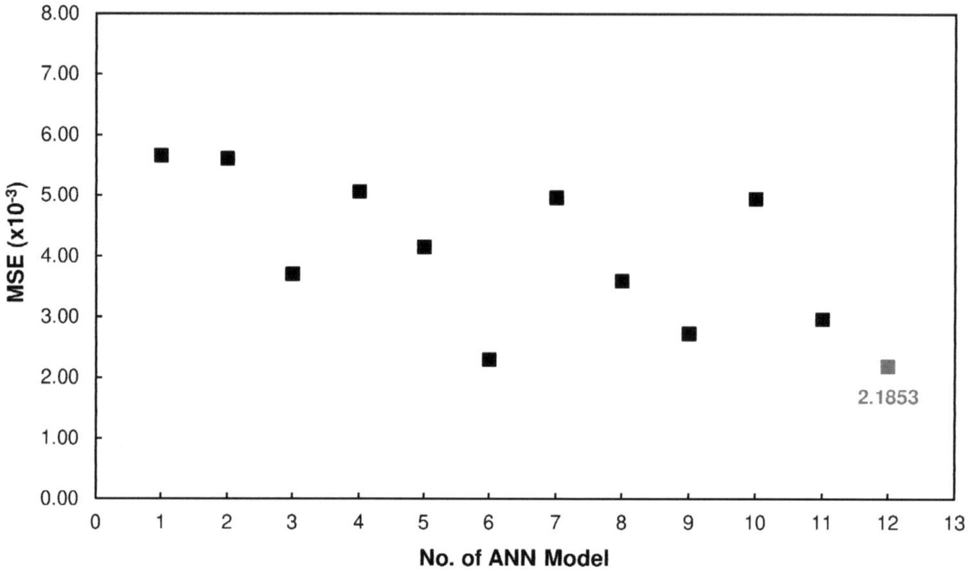

Figure 4. MSE Result of ANN Model Variation

Figure 4 illustrates the MSE values obtained for each of the 12 ANN models. It can be observed that the MSE values vary across the different model architectures, indicating differences in their prediction accuracies. The goal is to select the ANN model with the lowest MSE, as it represents the most accurate prediction capability [14]. Among all the models, Model No. 12, marked in red, achieved the smallest MSE value of 2.1853×10^{-3}. This makes Model No. 12 the most optimal choice for predicting ship roll motion in this study, as it demonstrates the highest level of prediction accuracy with the lowest error rate.

Model No. 12 was selected as the optimal model, further testing and validation were performed to evaluate its performance in capturing the non-linearities present in ship roll motion data. The results, shown in Figure 5, illustrate the comparison between the actual and predicted roll motion data. The model effectively captures the non-linear behavior of the ship's roll motion, indicating that it is capable of accurately modeling the complex dynamics involved. This result suggests that the developed ANN-MISO model works well in predicting ship roll motion under varying operational and environmental conditions.

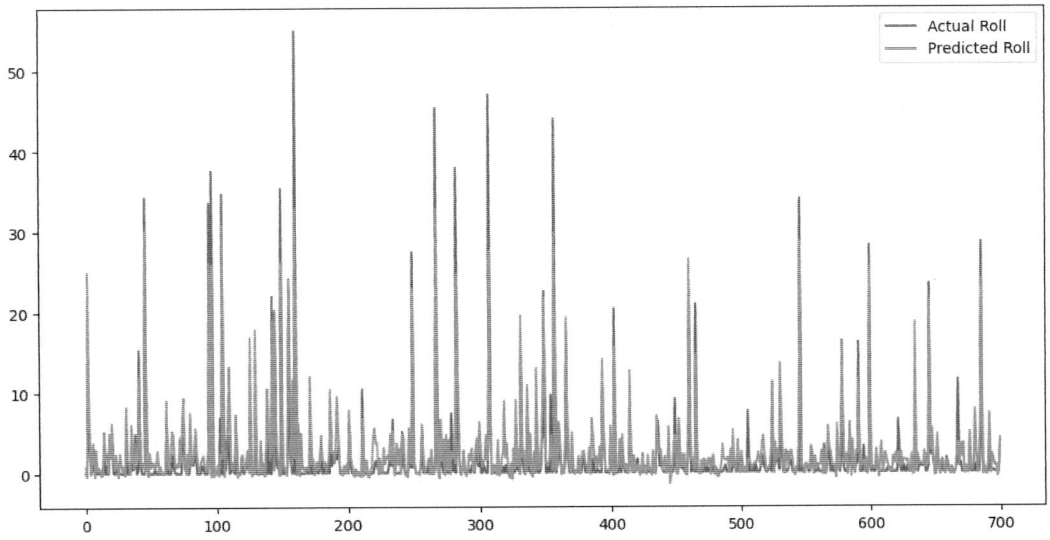

Figure 5. Validation of ANN Model No.12

6.3 Catamaran Ship Roll Motion Prediction

In this subsection, the selected ANN-MISO model applied to predict the roll motion of a catamaran ship under operational conditions. The model was tested using a set of input assumptions to simulate real-world scenarios. The input features for the model include varying ship speeds ranging from 0.5 to 8 knots, a constant wave heading of 90 degrees, a wave height of 1 meter, and a wave period of 6 seconds as presented in Table 4.

Table 4. Ship Operational Data

Features	Value	Unit
Ship Speed	0.5-8	knot
Wave Heading	90	deg
Wave Height	1	meter
Wave Period	6	second

The prediction results, as shown in Figure 6, indicate that the roll motion of the catamaran ship decreases as the ship speed increases. This trend is consistent with the general understanding that higher ship speeds can lead to reduced roll amplitudes due to increased stability and less time for wave-induced forces to act on the vessel [15]. Notably, there is a small spike observed at a speed of 5.5 knots, which suggests that the model can capture non-linearities present in the ship roll motion data. This spike could be attributed to resonance effects or specific wave-ship interactions at this speed.

Overall, the prediction results demonstrate that the developed ANN-MISO model effectively captures the non-linear behavior of catamaran ship roll motion, providing valuable insights into ship stability under varying operational conditions.

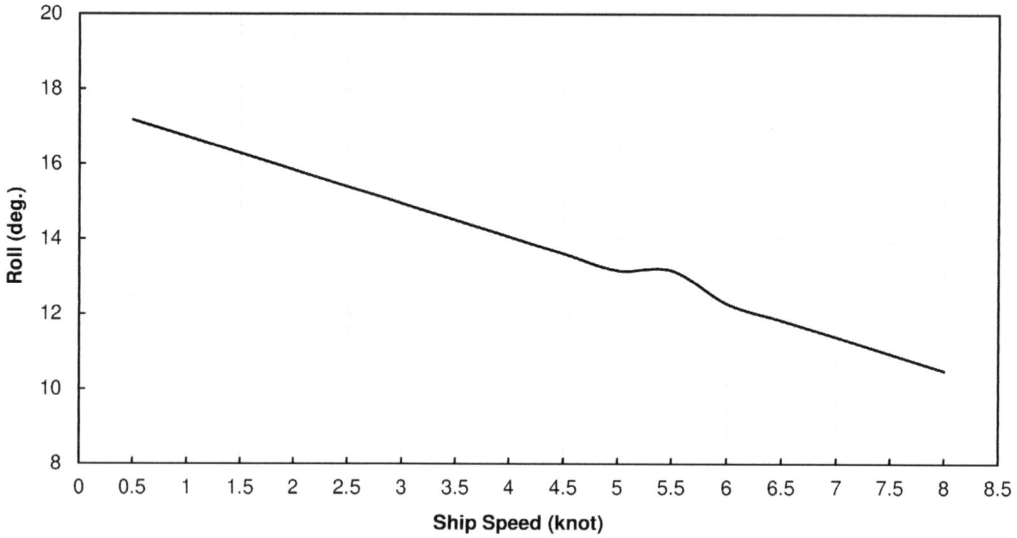

Figure 6. Catamaran Ship Roll Motion Prediction Using ANN-MISO

7. Conclusion

This study successfully developed an ANN model using a MISO architecture for predicting the roll motion of a catamaran ship under varying operational conditions. The model trained using ship motion data derived from hydrodynamic simulations conducted in the frequency domain using BEM. The ANN model demonstrated strong predictive capabilities, capturing non-linear characteristics of ship roll motion with minimal MSE. The results indicate that the ANN model can effectively simulate real-world scenarios, identifying critical points such as resonance effects. This capability is essential for enhancing operational decision-making and improving the safety of marine operations.

References

[1] Sharma, A. K., Sharma, A., & Singh, S. (2022). Artificial neural networks in maritime applications: A review. Journal of Marine Engineering and Technology, 11(1), 1-17.

[2] Ogilvie, T. F. (1976). The influence of hull form on the roll motion of a catamaran. *Journal of Ship Research*, 20(3), 121-132.

[3] Bhattacharyya, R. (1978). *Dynamics of marine vehicles*. Wiley.

[4] Park, G.I., Choi, K.,Y., Lee, J.H., & Kim, M.S. (2004). Introduction of Optimum Navigation Route Assessment System Based on Weather Forecasting and Seakeeping Prediction. Journal of Korean Navigation and Port Research, 28(10), 833-841. https://doi.org/10.5394/KINPR.2004.28.10.833

[5] Russell, S., & Norvig, P. (2021). *Artificial intelligence: A modern approach* (4th ed.). Pearson Education.

[6] Ao, Y., Li, Y., Gong, J., & Li, S. (2021). An artificial intelligence-aided design (AIAD) of ship hull structures. *Journal of Ocean Engineering Science*, doi: 10.1016/j.joes.2021.11.003

[7] Zhou, T., Hu, Q., Hu, Z., & Zhen, R. (2022). An adaptive hyper parameter tuning model for ship fuel consumption prediction under complex maritime environments. *Journal of Ocean Engineering Science*, 7, 255–263. doi:10.1016/j.joes.2021.08.007

[8] Sayli, A., Dursun, A., & Ganiler, O. (2010). Nonlinear meta-models for conceptual sea-keeping design of fishing vessels. *Ocean Engineering*, 37(7-8), 730–741. doi: 10.1016/j.oceaneng.2010.02.005

[9] Kim, Y.R., Park, J.-B., & Moon, S.-B. (2018). Prediction of Ship Roll Motion Using Machine Learning-based Surrogate Model. Journal of Korean Navigation and Port Research 42(6), 395-405. https://doi.org/10.5394/KINPR.2018.42.6.395

[10] Goodfellow, I., Bengio, Y., & Courville, A. (2016). Deep learning. Cambridge, MA: MIT Press.

[11] Sapp, C. E. (2017). Preparing and Architecting for Machine Learning. *Gartner Technical Professional Advice, January*, 1–38.

[12] Yan, X., Wang, L., & Zhang, X. (2012). Artificial neural networks: A survey of architectures, learning algorithms, and applications. Neural Networks, 37, 58-78.

[13] Hasselmann, K., Barnett, T., Bouws, E., Carlson, H., Cartwright, D., Enke, K., Walden,H. (1973). Measurements of Wind-Wave Growth and Swell Decay during the Joint North Sea Wave Project (JONSWAP).

[14] Li, Y., Wang, Y., Zhang, Z., & Wang, J. (2022). Seakeeping performance prediction of ship in service using artificial intelligence. Ocean Engineering, 244, 109996.

[15] Lloyd, A. (1989). *SEAKEEPING: Ship Behaviour in Rough Weather.* Chichester: Ellis Horwood Limited.

Improvement of Remote Control Monitoring System on Yanmar TF-65 Engine Using PLC Method Based on Digital Twin

Dwi Priyanta[1], M Badrus Zaman[1], Ari Yuda Dwitama Hamdono[1], Nurhadi Siswantoro[1], Trika Pitana[1], and Hari Prastowo[1]

[1] Department of Marine Engineering, Faculty of Marine Technology, Sepuluh Nopember Institute of Technology, Surabaya 60111, Indonesia

E-mail: priyanta@its.ac.id

Abstract. Technological advancements not only affect physical and virtual objects in the industrial sector. Technology also acts as a solution to improve safety in industrial environments, with the aim of reducing the risk of accidents for human workers. The engine is a vital component in every vehicle, including ships. To monitor engine performance and ensure the safety of passengers, goods, and crew, a reliable monitoring system is required. Digital Twin enables real-time monitoring and control of engine parameters, such as RPM and engine throttle control. The technology provides easy-to-understand digital visualization and enables quick identification of potential problems. In this case, the Progammable Logic Controller (PLC) and Digital Twin become a system that can be used to monitor and control the engine in real-time. In this research, a proximity sensor equipped with infrared light and a stepper motor that acts as an engine throttle controller are used to monitor RPM. The experimental results show that the RPM value measured through the proximity sensor compared to the use of a tachometer has a low error value of 4.31%. And the utilization of stepper motor in controlling the engine throttle is also effective. These two components are connected to the PLC for the process of receiving data which then the value of the results is transmitted to the server to be displayed through mobile apps. This research contributes to the application of Digital Twin technology in the maritime industry by increasing safety and efficiency in ship propulsion systems.

Keywords. Digital Twin, Monitoring, Remote Controller, Sensor, Stepper Motor

1. Introduction

Industry 4.0 technology has gained significant traction across various sectors, including data science, information technology, and manufacturing. This new era is underpinned by essential infrastructure, notably the Internet of Things (IoT). In this context, IoT acts as a vital planning and control system, enabling organizations to leverage data for informed decision-making. Additionally, the Cyber-Physical Production System (CPS) represents an integrated framework that facilitates both control and monitoring, with the capability to make autonomous decisions based on the data it processes. Thus, Industry 4.0 embodies a multidisciplinary revolution that amalgamates concepts, methodologies, and technologies aimed at advancing business practices. However, to fully realize its potential, there remains a pressing need for more comprehensive research across diverse sectors and scales [1].

Today's technological advancements extend beyond mere physical and virtual objects within the industrial sector; they play a crucial role in enhancing safety in industrial environments, ultimately reducing the risk of accidents for human workers. The application of Industry 4.0 safety technologies has proven notably successful, particularly in fostering collaboration between humans and robots. In this regard, these technologies merge virtual models with real-time data obtained from physical objects, utilizing sensors applied to robots, machines, and human operators [2] [3]. The digital twin system exemplifies this integration, as it incorporates data from the physical environment into a virtual model, thereby increasing the realism of implementations. Through digital twins, safety training can be simulated in a virtual environment that mirrors actual conditions, allowing for the adaptation of information regarding machine functions, robotic actions, and human behavior. Consequently, the digital twin serves as an effective tool for conducting safety training, performing risk analyses, and planning safety measures [2].

Initially, digital twin technology found its primary applications in military and aerospace sectors. However, research on this technology has surged significantly over the years. This increasing scholarly interest has led Gartner to recognize the digital twin as one of the ten strategic technology trends for four consecutive years, from 2017 to 2020. Looking ahead, the relevance of digital twin technology is anticipated to continue escalating over the next 5 to 10 years. The digital twin comprises five key dimensions: physical entity, virtual model, connection, data, and service [4]. This methodology proves invaluable for integrating real-time modeling, simulation, machine learning, prototyping, optimization, and big data management [5].

These technologies facilitate a seamless connection between the physical and digital worlds through CPS, leveraging IoT sensors [6]. A prime example of such technology is the remote-control system for electronic devices, which typically utilizes push buttons as control inputs. In a remote-control setup, two main components are essential: the local controller and the remote controller. The operator engages with the local controller, which provides access to control mechanisms, while the remote controller is directly linked to the equipment being managed. In this research, a control system can be designed using a Programmable Logic Controller (PLC) as the central control unit, along with a means to transmit data to the Human-Machine Interface (HMI), serving as the user interface.

Table 1. Yanmar TF-65 Diesel Engine Digital Twin Development Research Roadmap in Digital Marine Operation and Maintenance Laboratory, Marine Engineering Department - ITS

Digital Twin for Yanmar TF-65 Diesel Engine Development Road Map		
1st Generation	2nd Generation	3rd Generation
▪ Design of Website for Engine Performance and Fuel Consumption Monitoring ▪ RPM and Exhaust Gas Temperature Monitoring ▪ Database Design ▪ Cooling Water Flow and Fuel Consumption Monitoring ▪ Engine Throttle Remote Controller Design ▪ Telemetry System Design	Engine Start Remote Controller Design and Improvement for The Database	▪ Mobile Application Front End Design Based on Diesel Engine Performance for Digital Twin Development ▪ Backend Database Design on Digital Twin Mobile Application ▪ Analysis of Fuel Consumption in Diesel Engines Using the Development Method of Monitoring and Controlling via Digital PLC Platform on Digital Twin Technology ▪ Design of Temperature Monitoring System of Diesel Engine Yanmar TF-65 Based on Programmable Logic Controller (PLC) for Digital Twin Development ▪ Improvement of Remote Control Monitoring System on Yanmar TF-65 Engine Using PLC Method Based on Digital Twin

The key point of this research is to develop the existing digital twin of Yanmar TF-45 engine to be improved in terms of sensor readings, data migration from the engine to the database so that it can be presented in the HMI as a user interface. Table 1 shows the Yanmar TF-65 Diesel Engine Digital Twin

IOP Conf. Series: Earth and Environmental Science **1461** (2025) 012039 doi:10.1088/1755-1315/1461/1/012039

Development Research Roadmap in Digital Marine Operation and Maintenance Laboratory, Marine Engineering Department – ITS. A different sensor optimization and controlling system will be designed so as to realize a better digital twin concept on existing assets. In addition, because this technology provides real-time monitoring and controlling capabilities of asset operational conditions, it is expected to help prevent the risk of engine failure and improve safety on ships.

2. Basic Theory

Applying digital models to the function of remotely controlling and monitoring industrial systems is part of the current digital transition of industry. Industry 4.0, which represents the latest technology, features innovative control strategies, including data driven PID control created to modify stability control.

2.1. Digital Twin Concept

The Digital Twin (DT) dates back to 2003, when Grieves introduced the concept for the first time in his course on "product lifecycle management". Although the concept was not specific enough at the time, the initial form of DT was proposed to include three parts: the physical product, the virtual product, and the connection between them. DT-enabling technologies have experienced exponential growth since then. In 2012, the concept of DT was updated by the National Aeronautics and Space Administration (NASA), which defined DT as a multiphysics, multiscale, probabilistic, and ultrafidelity simulation that reflects, in a timely manner, the state of the corresponding twin based on historical data, real-time sensor data, and physical models. DT is becoming a popular research topic. According to Gabor et al., DTs are specialized simulations, built on expert knowledge and real data collected from existing systems, to produce more accurate simulations in different time and space scales [7]. According to Yildiz et al., DT is a digital representation that can describe the production process and product performance. The meaning of DT has become more concrete since then, leading to several specialized concepts such as airframe digital twin (ADT) and experimental digital twin (EDT) [8].

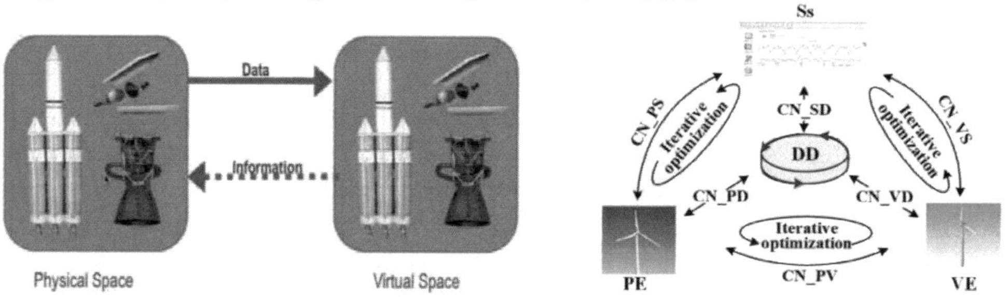

Figure 1. Three dimension DT and five dimension DT models [4]

There are different understandings of the Digital Twin. Some researchers believe that DT should be focused on simulation development. However, there are also those who argue that DT consists of three elements, namely, physical, virtual, and connection parts [9]. Figure 1 shows the basic framework where virtual space is mapped to physical space through connection parts that exchange data and information. Based on the three-dimension model by Tao, Fei et al., propose that a complete DT should include five dimensions, namely, physical, virtual, connection, data and service parts [4]. The basic framework in Figure 2 shows that with the data model included in the DT dimension section, it expands the definition of DT to be an integration of several model elements from the previous dimensions.

2.2. Digital Twin Framework

The Digital Twin used to monitor and control the Yanmar TF-65 engine consists of 3 aspects, namely data analysis, data acquisition and data processing [10] [11]. This model involves five dimensions including physical assets, digital models, connections, data and services provided. In addition, supporting the connection dimension at Digital Twin currently uses Haiwell Happy and PLC to connect data from sensors to the database and app platform. One of the objectives of writing this report is to improve the system that has been built by creating a more optimal server with better sensor support so that the data migration process can run according to the directions given. The designed digital twin framework can be seen in the Figure 2.

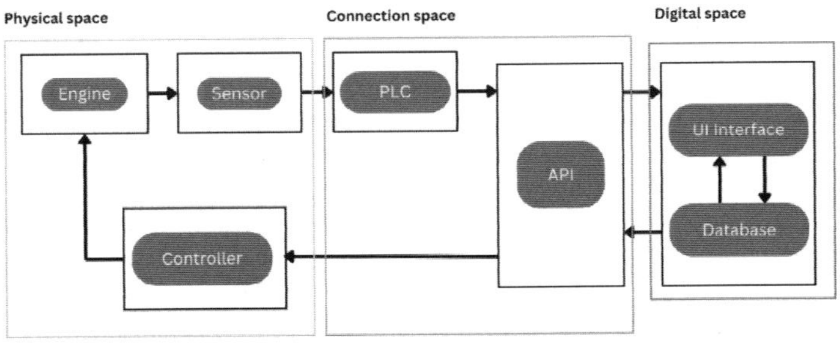

Figure 2. Digital Twin framework design

2.3. Sensor

Sensors are equipment that functions to detect emitted signals that come from changes in energy, such as physical energy, chemical energy, electrical energy and so on. In its own classification, sensors are generally divided into 3 types, namely Mechanical Sensors, Optical Sensors, and Thermal Sensors. Mechanical sensors are sensors that detect and convert mechanical motion such as position displacement, straight motion, circular motion and so on. Examples of mechanical sensors are in strain gauge, proximity, load cell, etc. While optical sensors are sensors that detect changes in light from a direct light source or through refraction of light in a room or object. Examples of optical sensors are photo cell, photo multiplier, photo voltaic. The thermal sensor has a function to detect changes in temperature in certain dimensions of space such as thermocouples, thermistors, infrared pyrometers, etc. Sensor networks provide a challenging scope of programming and computing, because the devices tend to be small and prone to crashes and in its operation also the device does not play a role to overcome in case of failure. So that failure is overcome by repairing other devices. Due to these limitations, it is necessary to have a data collection system in the sensor network to support the performance of each other's devices [12] [13].

2.4. Data Migration

Data migration is the process of moving historical data or existing data archives to new storage, new systems, or new file formats. Although it sounds simple, it actually involves many steps that must be seriously considered. Some of the things that need to be considered when doing data migration is planning. Before data migration is carried out, of course, planning is carried out first, by determining the objectives, data sources, and migration methods that will be used. Next is the creation of data backups where in the pre-migration process all existing data should be backed up (backup data) to anticipate loss of information and data in the event of a migration failure. After the migration process is complete, verify that the data has been successfully moved and the new system can function properly.

2.5. Programmable Logic Controller (PLC)

A Programmable Logic Controller (PLC) is a programmable electronic device that is widely used to control industrial processes. It mainly consists of a processing unit and I/O modules to acquire and interact with sensors and actuators. Although PLC architecture and programming are defined in the IEC 61131 standard, there are minor differences in implementation from various manufacturers [14]. In this research, focus on Haiwell PLCs, because to unify the sensors and actuators in other studies so that they can be well connected to the same PLC. So, the data migration process can run as instructed.

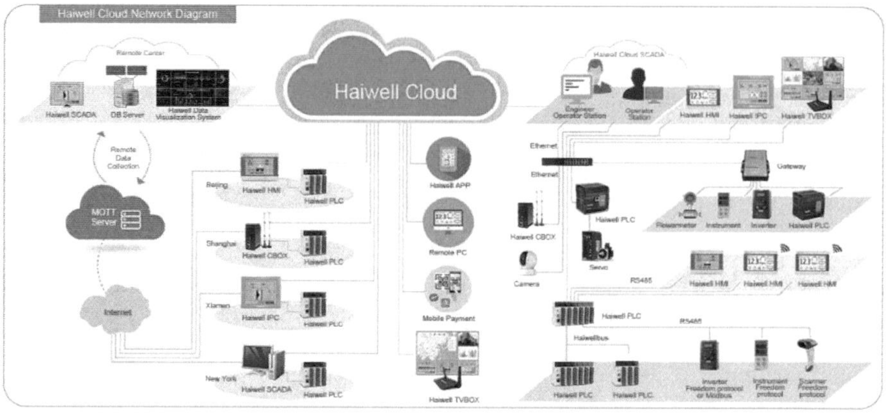

Figure 3. Haiwell Cloud Network Diagram (*Haiwell Manual Book*)

PLCs utilize programmable memory to store instructions and implement logic functions [15]. These functions include combinational, sequential, timing, enumeration, and arithmetic logic to control industrial machines and processes. PLCs are designed to be operable by engineers with limited knowledge of computers and programming languages. The PLC interface is designed in such a way that not only computer programming experts can create or change its programs. Therefore, PLC designers have provided an initial program in the device (called a pre-program) that allows the control program to be entered using a simple and easy to understand programming language. In this research, Haiwell PLC is used because it has the advantages of performance flexibility, affordable price, and easy to make changes and tracking in case of problems / failures. In addition, all instructions in it are integrated, making it easier to learn.

2.6. Database

The main purpose of a database system is to allow users to compose an abstraction view of data. Thus, the interaction between the user and the system becomes simpler. Databases can also present different views to users, programmers, and administrators. Since not all database users have the same training and users are divided into different levels, the complexity of the database is hidden from users through multiple levels of data abstraction. The purpose of data organization, among others, is to provide data storage to be used in the present and in the future. It is also to control data every cycle so that it can always be up to date and can reflect specific changes that occur in each system. Securing data against possible additions, modifications, theft or other disturbances can play a role in the data organization process.

3. Methodology

Following the background of the rapid development of industry 4.0 and digital twin research, supports the making of this research. The digital twin itself requires data from physical assets, where the asset used for this research is the Yanmar TF-65 engine. Sensors and controllers for this engine starter already exist for now, but their performance or role is still not optimal and very limited. Therefore, it is necessary to improve this part. The sensor will be placed on the engine and by using PLC to help output the signal from the program calculation to the actuator (stepper) - the controlling signal. This improvement factor is carried out to ensure that the command given to the machine can be more precise and it is hoped that the output given is appropriate.

3.1. Problem Identification

The digital twin itself requires data from physical assets, where the asset used for this research is the Yanmar TF-65 engine. Sensors and controllers for this engine starter already exist for now, but their performance or role is still not optimal and very limited. Therefore, it is necessary to improve this part. The sensor will be placed on the engine and by using PLC to help output the signal from the program calculation to the actuator (stepper) - the controlling signal. This improvement factor is carried out to ensure that the command given to the machine can be more precise and it is hoped that the output given is appropriate.

3.2. Sensor, Stepper and PLC Design
This research used a Programmable Logic Controller (PLC). PLC is an electronic device that plays a role in assisting the monitoring and controlling process. In this research to support the monitoring and controlling process there are 3 sensors and one stepper motor. These sensors will later read important parameters on the Yanmar TF-65 diesel engine. Stepper motor is one type of motor that is widely used in various industrial applications. With a construction that only has coils on the stator while the rotor is a permanent magnet. In this research used the Haiwell PLC model AT12M0T (see Figure 4).

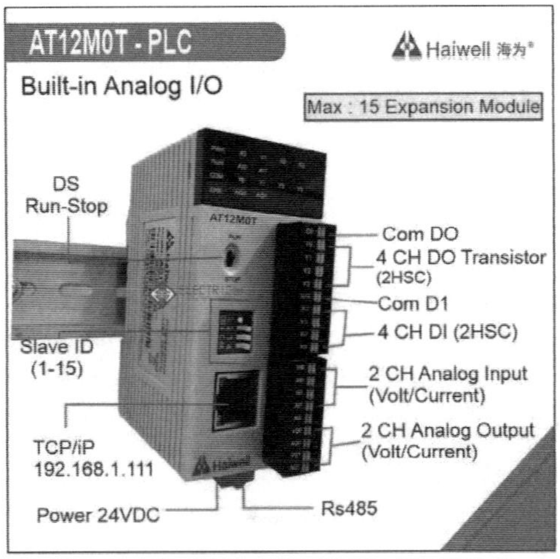

Figure 4. Haiwell AT 12M0T PLC

A sensor is a device intended to detect the presence of an event or change in value around the equipment. The result of the change or detection provides a response in the form of an output. The output given by the sensor in this research is the change in rotation value (RPM) on the Yanmar TF-65 engine. The E18-D80NK sensor used in this research is known as an infrared obstacle detection sensor, which is a sensor that uses infrared as an indicator of detecting objects in front of it (see Figure 5). Objects that can be detected are between 3 cm and 80 cm away.

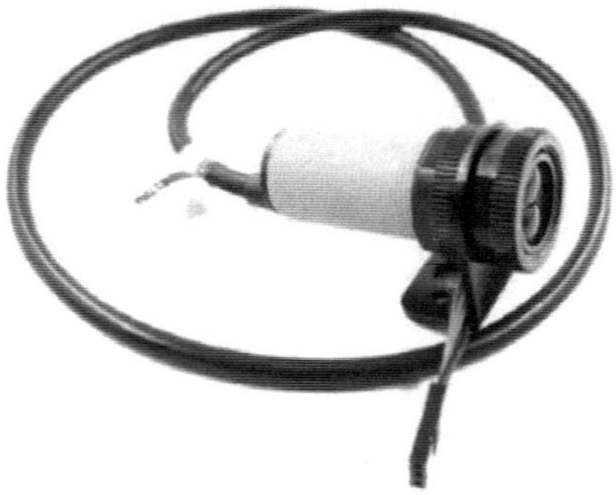

Figure 5. E18-D80NK infrared sensor

Stepper motors have the characteristic of moving based on the sequence of pulses given to the motor. The pulse sequence is related to the direction of rotation of the motor so that the rotational speed of the motor is continuous with the frequency given. In this stepper motor application is supported by using a stepper motor driver. Microstep driver DM542 is the driver used in this research. Microstep driver

DM542 is an easy-to-use stepper motor driver, where this driver has a DC 24-60 (V) power input. Stepper drivers function to support speed and direction control.

Figure 6. Stepper motor nema 23 & Driver DM542

3.3. Programming

After assembling the sensor, motor, and PLC, then create a program to operate the motor. The programming language uses ladder logic on the Haiwell Happy platform as a serial monitor. If there are no errors in programming, the program can be uploaded, and the sensor and motor can work according to the command. Because the purpose of this research is to be able to remote controlling through mobile apps, it is necessary to ensure that the transfer of data from sensors and motors to the serial monitor is successful so that it can reach the server.

In reading the parameters on the RPM sensor, programming needs to be done first. The programming used in this research is using the Haiwell Happy platform. The programming language used is ladder logic. In the ladder arrangement there are several components to form one program. As in the ladder for the RPM sensor using instructions on High-speed control, namely SPD (Speed Detection). In this RPM monitoring, the TnP>0 rule is used because to get the engine RPM value based on detection time not detection pulse. Where for input the TnP value is 1000 and the sensor input to the PLC is entered into slot X0 (external input) (see Figure 7-9).

The stepper motor control process also requires programming first. Haiwell Happy is the platform used to do programming in this research. As with programming on the proximity sensor, the stepper motor itself also has a ladder program that functions to set the motor so that it can rotate clockwise or counterclockwise according to the operator's wishes.

Figure 7. RPM infrared proximity sensor program ladder (*Haiwell Happy*)

IOP Conf. Series: Earth and Environmental Science **1461** (2025) 012039
doi:10.1088/1755-1315/1461/1/012039

Figure 8. Ladder and stepper motor simulation process (*clockwise*)

Figure 9. Ladder and stepper motor simulation process (*counter-clockwise*)

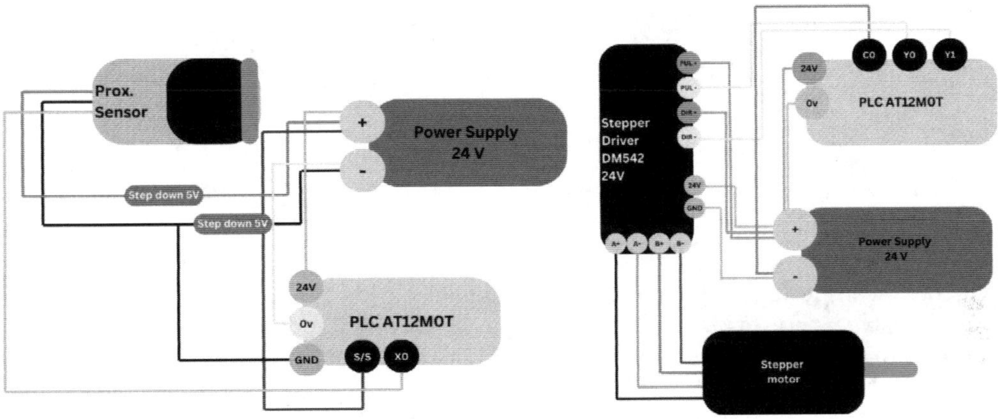

Figure 10. Wiring diagram of proximity sensor infrared RPM & Wiring diagram of stepper motor (*author*)

3.4. Data Migration and Validation

This stage aims to identify the form of data migration and select the software to be implemented for this project. The information obtained from the previous steps was used to determine the scope of the project, create an initial migration plan, record initial data issues, and break down the project into smaller, manageable tasks. If the data is not successfully read or sent to the server, it is possible that the input data is invalid, so it is necessary to reassemble the sensors, steppers and PLCs and retest until the data is successfully transmitted.

4. Results and Discussion

4.1. Results of RPM Sensor Readings and Throttle Controlling

The results of the sensor readings are as follows which can be seen in the Table 2-4. Data is taken every 2 minutes in each experiment using a lamp load equipped with a power meter sensor.

Table 2. 1st RPM Data Collection Experiment

Taken On	RPM (Digital)	Tachometer
26/06/2024 13:10	1620	1651
26/06/2024 13:10	1620	1667
26/06/2024 13:11	1620	1668
26/06/2024 13:11	1620	1665
26/06/2024 13:11	1620	1665

Table 3. 2nd RPM Data Collection Experiment

Taken On	RPM (Digital)	Tachometer
26/06/2024 13:14	1560	1625
26/06/2024 13:15	1560	1624
26/06/2024 13:15	1560	1622
26/06/2024 13:15	1560	1626
26/06/2024 13:15	1560	1625

Table 4. 3*rd* RPM Data Collection Experiment

Taken At	RPM (Digital)	Tachometer
26/06/2024 13:17	1500	1603
26/06/2024 13:18	1500	1603
26/06/2024 13:18	1500	1602
26/06/2024 13:18	1500	1602
26/06/2024 13:18	1500	1602

The experiment shows that the presence of a lamp load affects the performance of the engine RPM speed. As in each experiment, the RPM data sequentially gets lower as the load increases. This is in accordance with the power formula theory, namely:

$$Power\ (P) = Torque\ x\ RPM \tag{1}$$

The equation shows that the power value is directly proportional to the torque. When the torque is at its maximum, the output power from the engine to the load will be greater. However, this is inversely proportional to the RPM value. Where when the torque value is getting bigger, the RPM value will be lower. This is common in diesel engines due to their design characteristics, namely small space, optimal air induction system and the use of high heat diesel fuel. So that it causes the RPM value at the maximum torque condition to be low. It can also be seen in the experimental results above, the higher the load given, the RPM value will decrease.

To be able to control stepper motors through mobile apps, first the ladder program data that has been created in Haiwell Happy must be sent to Haiwell Scada. From Haiwell Scada, data can be sent to enter the HMI (tablet) and monitoring through mobile apps. The controlling process can be done by starting the machine first. After the engine starts, in the mobile apps there is a controlling section. There is a throttle button (+) marked in green and (-) marked in dark red and both functions to increase and decrease the angle of rotation of the stepper motor. The following is the engine throttle calibration value through a stepper motor in various angular conditions, such as:

a. Stepper motor calibration on engine throttle in 0° angle condition shows an RPM value of 0 RPM.

b. Calibration of the stepper motor on the engine throttle in the 40 ° angle condition shows an RPM value of ± 780 RPM.
c. Stepper motor calibration on engine throttle at 80° angle condition shows an RPM value of ± 1200 RPM.
d. Stepper motor calibration on engine throttle in 130° angle condition shows an RPM value of ± 1860 RPM.
e. Calibration of the stepper motor on the engine throttle in the 180° angle condition shows an RPM value of ± 2040 RPM.

After the proximity sensor and stepper motor data are obtained through Haiwell Happy, the next step is to transmit the data to Haiwell Scada. The data can be displayed on the local platform, namely the HMI (Human Machine Interface) in the form of a tablet and can be a component to be sent to the server later so that it can be accessed through the application (see Figure 11).

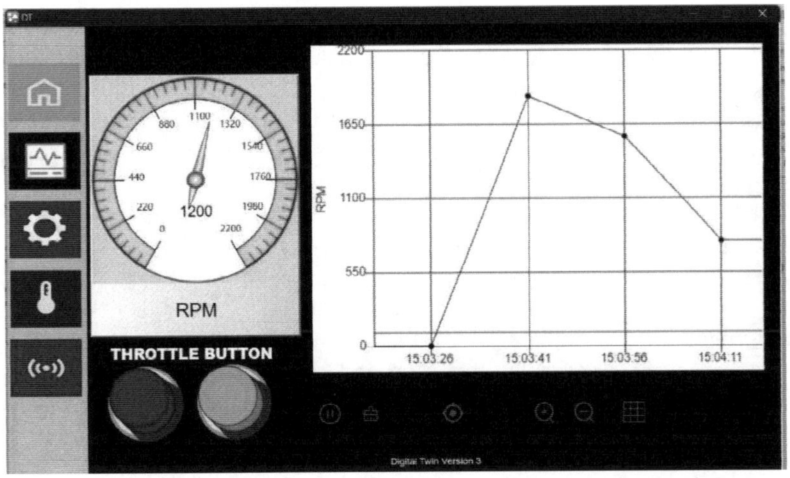

Figure 11. Design view of the RPM section on the HMI that presents the RPM meter, *throttle button* and comparison graph between RPM value and operating time (Haiwell Scada).

In accordance with the picture above, when the engine automatically operates the RPM meter will work and present the RPM value in real-time. In addition, there is a graph that shows the running RPM value and is compared to the time every 15 seconds.

4.2. Validation of Sensor Reading Results
After comparing the RPM value between the proximity sensor and the tachometer, there is a difference in results that is not too far away. Therefore, to find out how many percent of the difference in value it is necessary to calculate the error value. The results of the error calculation in each experiment are presented in the tables below as follows:

Table 5. Average Error of the Whole Trial

Trial Number	Average Error (%)
1	2.60
2	3.96
3	6.39
Overall Average	4.31

After recapitulating the average error above, the overall average value is 4.31%. The error value shows that proximity sensors that use infrared sensors have varying values with varying error values. From the above factors, it causes difficulties in knowing whether the engine is experiencing too high or too low rotation. Rotation on an engine that is too high can cause an increase in temperature and low rotation can occur due to a load on the engine that is too large. Therefore, the error value must be corrected to know the condition of the engine directly, moreover the goal in this Digital Twin research is to be able to monitor directly and accurately (real time).

5. Conclusion

According to experiment, the reading of RPM parameters can be done by placing the proximity sensor under the flywheel area of the Yanmar TF-65 engine. The findings indicate 4.31% average error in RPM readings, it is crucial to consider the broader implications of this improvement for engine safety and efficiency. A reduction in error directly enhances the accuracy of engine monitoring, which is vital for ensuring optimal performance. More precise RPM readings can lead to better throttle control, resulting in improved fuel efficiency and reduced emissions. Additionally, enhanced accuracy contributes to safer engine operation by minimizing the risk of mechanical failures or accidents caused by incorrect data interpretation. In the process of sending data, both monitoring and controlling from the PLC for stepper motors and proximity sensors are connected through the Herza server so that the data can be accessed through the application. A NEMA 23 stepper motor with a torque of 3.6 Nm can be used to control the Yanmar TF-65 engine throttle. This stepper motor is connected to the DM542 motor driver circuit and PLC so that the motor can be controlled through the HMI tablet or application. To ensure the performance of the stepper motor is in good condition, engine throttle calibration is carried out through the stepper motor in various angular conditions including angle 0°, 40°, 80°, 130°, 180°.

6. Acknowledgments

The authors thank the Digital Marine Operation and Maintenance Laboratory, Department of Marine Engineering, and Institut Teknologi Sepuluh Nopember for supporting facilities and funding this research.

7. References

[1] Lemstra, M. A. M. S., & de Mesquita, M. A. (2023). Industry 4.0: A tertiary literature review. Technological Forecasting and Social Change, 186, 122204.

[2] Pitkäaho, T., Kaarlela, T., Pieskä, S., & Sarlin, S. (2021). Indoor positioning, artificial intelligence and digital twins for enhanced robotics safety. IFAC-PapersOnLine, 54(1), 540-545.

[3] Söderäng, E., Hautala, S., Mikulski, M., Storm, X., & Niemi, S. (2022). Development of a digital twin for real-time simulation of a combustion engine-based power plant with battery storage and grid coupling. Energy Conversion and Management, 266, 115793.

[4] Tao, F., Zhang, M., & Nee, A. Y. C. (2019). Digital twin driven smart manufacturing. Academic press.

[5] Sharma, A., Kosasih, E., Zhang, J., Brintrup, A., & Calinescu, A. (2022). Digital twins: State of the art theory and practice, challenges, and open research questions. Journal of Industrial Information Integration, 30, 100383.

[6] Škraba, A., Stanovov, V., & Semenkin, E. (2020). Development of control systems kit for study of PID controller in the framework of cyber-physical systems. In IOP Conference Series: Materials Science and Engineering (Vol. 734, No. 1, p. 012105). IOP Publishing.

[7] Gabor, T., Belzner, L., Kiermeier, M., Beck, M. T., & Neitz, A. (2016). A simulation-based architecture for smart cyber-physical systems. In 2016 IEEE international conference on autonomic computing (ICAC) (pp. 374-379). IEEE.

[8] Yildiz, E., Møller, C., & Bilberg, A. (2020). Virtual factory: digital twin based integrated factory simulations. Procedia CIRP, 93, 216-221.

[9] Nezhad, M. M., Neshat, M., Sylaios, G., & Garcia, D. A. (2024). Marine energy digitalization digital twin's approaches. Renewable and Sustainable Energy Reviews, 191, 114065.

IOP Conf. Series: Earth and Environmental Science **1461** (2025) 012039 doi:10.1088/1755-1315/1461/1/012039

[10] Priyanta, D., Prastowo, H., Nurrahman, R., Siswantoro, N., Pitana, T., Zaman, M. B., ... & Parluhutan, Y. H. (2024). Digital Twin Monitoring System for Diesel Engine Based on Arduino, Case Study: RPM and Exhaust Gas Temperature. In Marine Technology (pp. 75-82). Cham: Springer Nature Switzerland.
[11] Billey, A., & Wuest, T. (2023). Energy digital Twins in smart manufacturing Systems: A literature review. Manufacturing Letters, 35, 1318-1325.
[12] Vetelino, J., & Reghu, A. (2017). Introduction to sensors. CRC press.
[13] Santos, J. L. (2021). Optical sensors for industry 4.0. IEEE Journal of Selected Topics in Quantum Electronics, 27(6), 1-11.
[14] Adiego, B. F., Darvas, D., Viñuela, E. B., Tournier, J. C., Bliudze, S., Blech, J. O., & Suárez, V. M. G. (2015). Applying model checking to industrial-sized PLC programs. IEEE Transactions on Industrial Informatics, 11(6), 1400-1410.
[15] Alphonsus, E. R., & Abdullah, M. O. (2016). A review on the applications of programmable logic controllers (PLCs). Renewable and Sustainable Energy Reviews, 60, 1185-1205.

IOP Conf. Series: Earth and Environmental Science **1461** (2025) 012040 doi:10.1088/1755-1315/1461/1/012040

Design Configuration of Horizontal Motion Control System for Unmanned Underwater Vehicle

Yuda Apri Hermawan[1*], Muhammad Ikhsan Suryadarma[1], and Totok Yulianto[1]

[1] Department of Naval Architecture, Institut Teknologi Sepuluh Nopember, Surabaya, Indonesia

*E-mail: yuda.apri@its.ac.id

Abstract. Unmanned underwater vehicles (UUVs) represent a significant advancement in underwater technology, offering a promising alternative for conducting underwater operations. One of the most prominent concepts is the hybrid UUV which provides many operational advantages. Nevertheless, this concept necessitates the implementation of an interplanetary energy system that is constrained in its capacity. Consequently, the propulsion system must exhibit optimal efficiency while simultaneously ensuring the requisite levels of manoeuvrability and controllability. To meet these requirements, one way that can be done is to use an adjustable thruster angle configuration. However, the change in configuration affects the dynamics model of the UUV, which in turn affects the control system. To overcome this problem, an electronic module will be designed to be able to accommodate thruster changes for all motions in the horizontal plane with on-the-fly parameter changesSTM32F4111 will be used as the computing centre to control 4 motors in the horizontal plane. To be able to make parameter changes, the Robot Operating System (ROS) is used as communication between the control unit and the microcontroller. The control system will be analysed by observing the yaw angle output from the GY-25 sensor during forward motion. The parameter values for each angle will be obtained through tuning, resulting in a maximum decrease in the error value of 3° for the 15° angle, 2° for the 30° angle, and 1° for the 45° angle. The discrepancy in the manoeuvrability capabilities of the UUV for each angle configuration in forward movement is 15°, with a 23% difference, in backward movement it is 15°, with a 24% difference, in right movement it is 45°, with a 59% difference, and in rotation movement it is 45°, with a 33% difference. These discrepancies suggest that the 45° configuration is the optimal choice when considering all movement types.

Keywords : Unmanned underwater Vehicle, Configuration, Thruster, and Horizontal

1. Introduction

The current state of underwater robot technology it is not only developed for the needs of the petroleum industry but can also be used for several other purposes. This development introduces a new concept, the Hybrid Underwater Robotic Vehicle (HURV), which combines the advantages of the Remotely Operated Vehicle (ROV) and the Autonomous Underwater Vehicle (AUV) in a single system that can perform both exploration and intervention tasks simultaneously(1). However, HRUVs also present a combination of the requirements and limitations of both types of UUVs. As a vehicle with high utility requirements, a remotely operated vehicle (ROV) must be capable of precise control and free movement. To achieve this, the propulsion system on ROVs is typically arranged in different directions (vector layout) and comprises more than one thruster (2). In contrast, AUVs are designed to achieve high endurance, which is achieved at the expense of manoeuvrability due to the use of a single propulsion system. Even though UUVs have been designed with a low-drag hull form and efficient propulsion in mind, the travelling distance of a UUV is largely contingent upon the number of internal energy sources that it incorporates, which can significantly impact the overall volume (3). Consequently, the HRUV must possess the capacity to move in a multitude of degrees of freedom (DOF) while simultaneously maintaining the lowest possible power consumption.

The adjustable configuration affords the UUV a high degree of flexibility, thereby enabling it to customise the mission route. However, modifying the thruster angle configuration also influences the overall dynamic model of the UUV, which in turn affects the control system that has been established. Various types of control systems have been tried to be applied to maintain the direction of the unmanned underwater vehicle to match the desired input. A method with simple logic through a joystick has been applied, but it has poor performance.(4). Control systems such as PID have been widely used by underwater vehicles and have good performance results with low computational value(5–8). Not only that, PID control systems have also been found to survive disturbances from the environment well(9). However, with the change in thruster configuration that also changes the dynamics of the vehicle, the linear PID control system will be difficult to achieve steady state error with previously tuned parameters. This can be overcome by the application of nonlinear control systems or adaptive control systems. With the addition of an adaptation system to the control system, changes in dynamics and external disturbances can be predicted to change the parameters of the control system. Adaptive systems with fuzzy methods can be applied as control systems (10)or applied with other control systems such as PID(11). In addition, the adaptive control system with backstepping has also been proven to control the position of the underwater vehicle despite the uncertainty of the vehicle dynamics and external environmental disturbances(12–15). However, the use of complex control systems can increase computational costs, making it difficult to apply to compact underwater vehicles.

Therefore, it is essential to adjust the control system and to understand its effects, as these are important factors in UUVs with different configurations. This paper aims to propose a design of control system module that can run the control system in the horizontal plane on the UUV used and has a control unit that can change the control system on the fly based on the configuration used. The effect of the configuration on the control system must also be analysed to maximise the performance of the control system used.

2. Control System Design

2.1 Electronic Component

In the development of an unmanned underwater vehicle (UUV) with three degrees of freedom (DoF) in its movement capabilities, the selection of appropriate electronic components is important. The factors considered and reasons for component selection are related to the

IOP Conf. Series: Earth and Environmental Science **1461** (2025) 012040 doi:10.1088/1755-1315/1461/1/012040

reliability, compatibility, capability and suitability of the components for UUV operation. As the main processing tool, the STM 32F411 microcontroller is used, which has high processing capabilities with speeds up to 100 MHz, efficient power, and sufficient RAM and flash memory. Meanwhile, the GY-25 MPU6050 sensor is used to obtain heading measurements by offering a 6 DoF inertial sensor consisting of a 3-axis accelerometer and a 3-axis gyroscope at an affordable

Table 1. Electronic components used for control systems

No.	Component	Component Specification
1	Microcontroller	STM32F411
2	Inertial Meassurement Unit Sensor	GY-25 MPU6050
3	DC Converter	Mini 560 Buck Converter
4	Electronic Speed Controller	ESC Littlebee 30A
5	Battery	Li-Po 4S 4500mAh
6	Controller	Logitech F310
7	Thruster	APISQUEEN U1
8	Cable Theter	Zimmlink Ethernet CAT 5e Cable

cost. The results of the component selection can be seen in **Table 1**.

2.2 Electronic System Design

This electronic system is designed to control the movement of the robot using 8 ESCs connected to 8 Brushless Direct Current (BLDC) motors. Each BLDC motor is responsible for driving one axis or actuator on the robot, thus enabling the robot to perform movements in 6 DOF which include translation and rotation. The combination of the 8 ESCs and BLDC motors, along with the MPU6050 sensor, allows the robot to perform movements in 6 DOF. The robot can perform translations in three axes (x, y, z) as well as rotations in three axes (roll, pitch, yaw). Precise control of each BLDC motor through the ESC and the use of MPU6050 sensor data allows the robot to perform precise, goal-directed movements. overall diagram can be seen in **Figure 1**.

The designed circuit is also equipped with several additional features such as the presence of diodes to avoid reverse current, besides that there are also several LED indicators that can facilitate problem solving. Printed Circuit Boards are also designed to enable more compact and tidy circuits, especially for electronic devices with limited space such as UUVs.

2.3 Underwater Vehicle Dynamic

The thruster modelling is based on Kadiyam (2021) (16). Based on this configuration we can define $F_f^i = \begin{bmatrix} F_f^A & F_f^B & F_f^C & F_f^D \end{bmatrix}^T$, the vector represents the force of each thruster in the horizontal plane. The general moment and force matrices in vector form $\tau_f^i = \begin{bmatrix} F_f^X & F_f^Y & M_f^Z \end{bmatrix}^T$ are expressed in the {V} coordinate system. With the following notation, the force vector F_f^i of the six thrusters is mapped to the vector τ_f^i. With this relationship, the following equation is formed.

$$\tau_f^i = BF_f^i$$

$$B = \begin{bmatrix} -\cos\theta & -\cos\theta & \cos\theta & \cos\theta \\ -\sin\theta & \sin\theta & \sin\theta & -\sin\theta \\ -(l_H + a_H)\sin\theta & (l_H + a_H)\sin\theta & -(l_H + a_H)\sin\theta & -(l_H + a_H)\sin\theta \end{bmatrix} \quad (1)$$

The selection of the deviation angle is based on numerical simulations conducted before (2)showing that the angle 15° has the largest total force on the longitudinal axis while 45° is the smallest and the angle 30° has performance between them. The thruster configuration of the UUV can be seen in **Figure 2**.

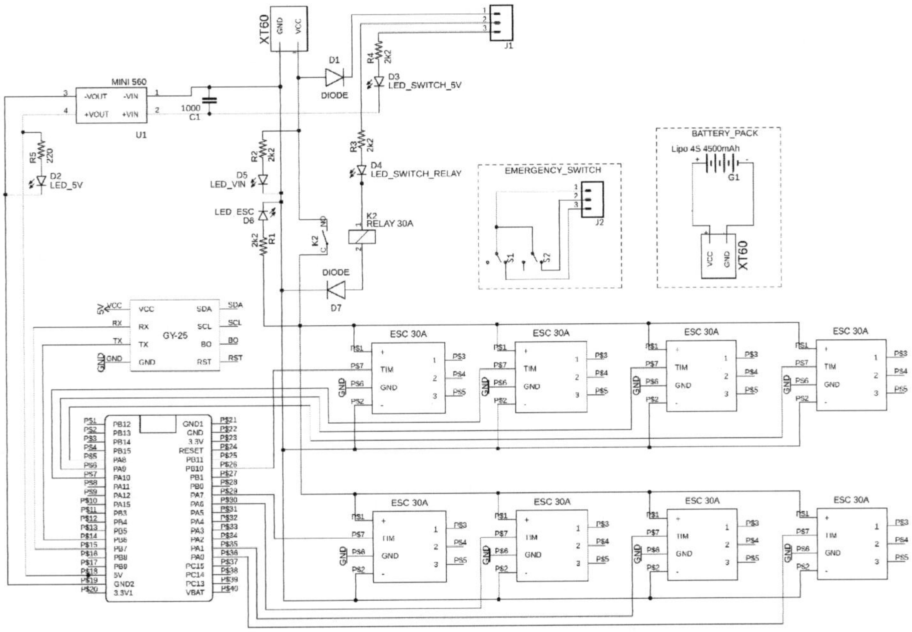

Figure 1. Overall Electronic Schematic of the control system with 8 ESCs and STM32F411 microcontroller.

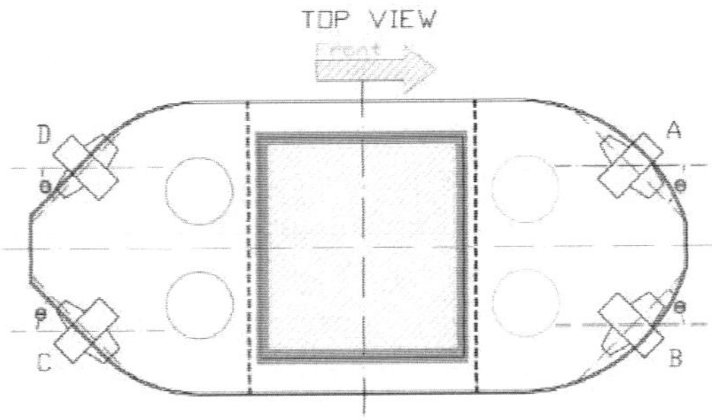

Figure 2. Overall Electronic Schematic of the control system with 8 ESCs and STM32F411 microcontroller.

IOP Conf. Series: Earth and Environmental Science **1461** (2025) 012040 doi:10.1088/1755-1315/1461/1/012040

2.4 Control System Algorithm

The embedded motion control system within the microcontroller is responsible for translating the input from the controller into pulse-width modulation (PWM) signals for each motor. The PID control method is employed for the purpose of regulating the movement of the UUV along each respective axis in horizontal plane. The structure of the control system diagram can be seen in **Figure 3**.

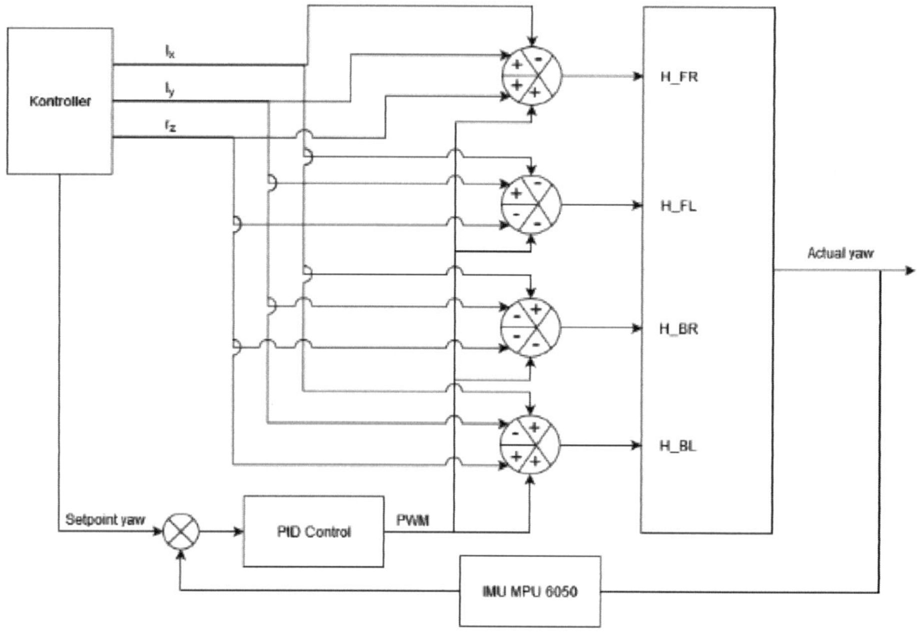

Figure 3. Overall diagram of the UUV PID control system in the horizontal plane

The ROV manufacturing programme, which employed the STM32CubeIDE tool, was executed in accordance with the following sequence of steps: initial configuration, setting of timers and PWM units, utilisation of the MPU6050 IMU sensor, construction of a PID control algorithm, implementation of the main loop of the programme, compilation, upload to microcontroller, validation test, and reference to the official STM32F411 and MPU6050 documentation [8].

The ROS (Robot Operating System) framework serves as a conduit between programs in the control unit. The ROS framework has been designed to develop large-scale robotic systems with high flexibility and includes a collection of tools, libraries, and conventions that can simplify the creation of complex robotic systems(17). Furthermore, ROS offers a structured communication framework. This is done to facilitate the development of the system on the control unit, allowing the addition of programs in a modular fashion while maintaining communication between them. Furthermore, this system enables the programme to run more efficiently in parallel. The ROS framework that has been created can be seen in **Figure 4**, there are several programs that communicate with each other.

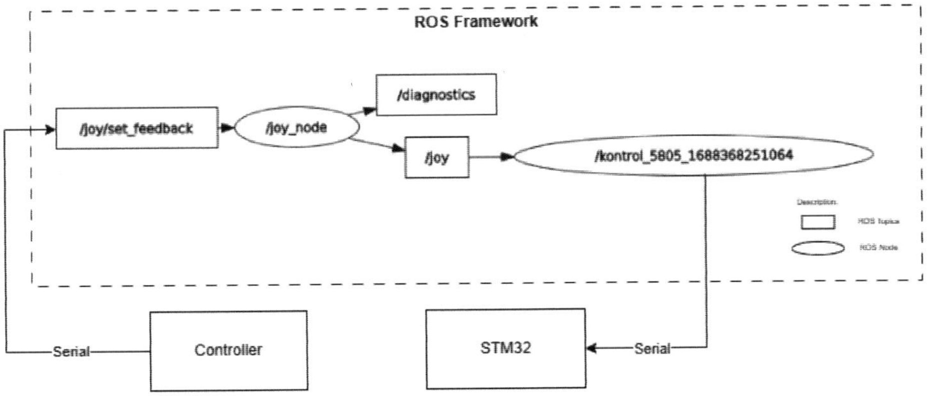

Figure 4. Overall ROS Framework of the control system to connect the control unit with the microcontroller

3. Result and Discussion

3.1 UUV Motion Testing for System Identification

The data retrieval method is done by sending an input signal in the form of a throttle value from the controller with a range of -500 to 500. The output value taken is the yaw heading angle data reading from the GY-25 sensor based on the input value given. Testing is repeated until good data is obtained. In identifying the system, additional data is needed to validate the data model that has been taken. This additional data is in the form of the same input and output values by varying the input values given.

Figure 5 shows the results of taking additional movement data to validate the data model that will be performed during system identification at the angle configurations 15^0, 30^0, and 45^0. The time interval between input values is different when compared to the previously captured movement data. The resulting maximum deviation angle values also have differences due to different input intervals. The system will later identify using the data by simulating the input given with the predicted data model whether it remains appropriate if given different inputs.

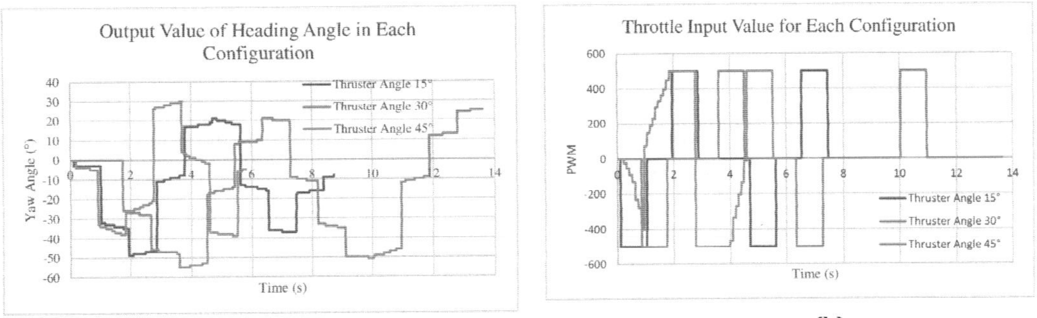

(a) (b)

Figure 5. Graph of movement test results for system identification; (a) output value, (b) input value.

3.2 PID Tuning With PID Tuner

In effective system control, proper tuning of PD (Proportional-Derivative) parameters is essential. One approach is to use the PID Tuner tool in MATLAB, which provides an intuitive interface and useful tools for PID tuning. This tool will simulate the ship's movement using the ship's dynamic model that has been obtained previously through the system identification method. The MATLAB automated tuning algorithm selects PID gains with the objective of achieving an optimal balance between performance (response time, bandwidth) and robustness (stability margins). The algorithm is preset to design for a 60-degree phase margin. The default value is used to see how the thruster angle differs with respect to the controller gain value. The tuning result of the controller gain value can be seen in **Table 2**.

Table 2. Electronic components used for control systems

Configuration	Kp	Kd	Rise Time	Settling Time
15°	10,8	8,75	0,48	2,12
30°	9,8	8,6	0,58	2,35
45°	13	2,6	0,75	3,4

3.3 Movement Testing With and Without Control System

This test is carried out to see the performance of the PD control system in each configuration and angle to see the performance of the manoeuvre produced at each angle configuration. Data collection begins with forward motion without PD and then data is taken with PD to analyse the difference. The performance testing method of the PD control system was conducted in a swimming pool with 5 m of calm water for forward movement by recording the IMU sensor angle reading to measure the direction angle of the UUV.

It can be observed in **Figure 6**, that the configurations exhibit disparate movement patterns in the absence of the PD control system. The thruster angle configuration of 15° tends to move towards a positive angle, whereas the thruster angle configuration of 30° has a tendency to move towards a negative angle. Meanwhile, the thruster angle configuration of 45° has the greatest heading angle deviation at a positive value, making it the most significant deviation value observed in any thruster angle configuration. The discrepancy in error values observed across the various configurations can be attributed to the thrust force exerted by the thrusters on the transverse axis, which is greater in the UUV.

The application of PD that illustrate in **Figure 7** at the angle configuration of 45° yields more pronounced outcomes than in the configuration of 30°. In this configuration, the error value can be reduced to -1° from a previous maximum of -2°. This reduction in maximum error provides an improvement in stabilisation performance, with a difference of 1 degree compared to the configuration of 30.This is attributable to the enhanced yaw manoeuvrability of the 45° configuration, which enables the system to rectify the angle with greater ease in yaw conditions.

The results of the three tests indicate that the PD control system exhibits a differential in its error correction capability, with the greatest capability observed in the configuration with the greatest angle.

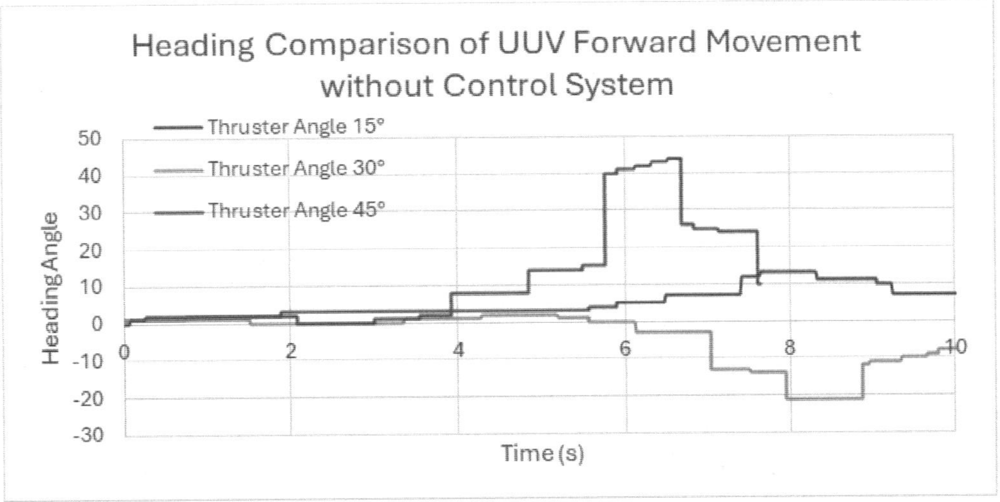

Figure 6. Overall diagram of the UUV PID control system in the horizontal plane

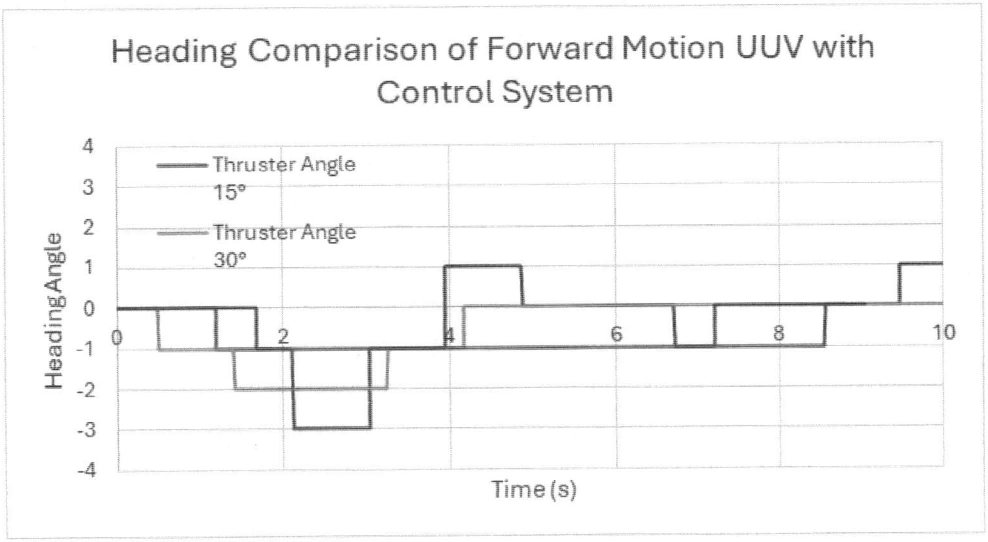

Figure 7. Overall diagram of the UUV PID control system in the horizontal plane

3.4 Movement Performance Testing

Tests were conducted to measure the speed of the UUV with different angular configurations as well as to see the difference in speed when using and without using a PD control system. The test method was to give a full signal to move the UUV up to a distance of 5 metres for the longitudinal axis and 1 metre for the transverse distance and 180° for the rotational movement.

IOP Conf. Series: Earth and Environmental Science **1461** (2025) 012040 doi:10.1088/1755-1315/1461/1/012040

The results shown in **Table 3** show the characteristics of each thruster angle configuration on the UUV motion performance when operating. The 15° angle gives the best surge performance at the expense of sway which, due to the angle, cannot overcome the drag. While the 30° angle gives a slightly slower surge motion, the UUV can still perform other maneuvers so it would be suitable for the type of remote operations that still require detailed monitoring. If the operation to be carried out really requires great maneuverability, a 45° angle is the best choice considering the capability of all movements. Although it has the lowest surge performance, the sway and yaw movements are the best by a considerable margin.

Table 3. Electronic components used for control systems

No.	Movement	Configuration	Time (s)	Difference	Percentage
1		15°	4,7	0,0	0%
2	Forward 5 m	30°	5,3	-0,6	-13%
3		45°	5,8	-1,1	-23%
4		15°	5,5	0,0	0%
5	Backward 5 m	30°	6,2	-0,7	-13%
6		45°	6,8	-1,3	-24%
7		15°	-	-	-
8	Right 1 m	30°	5,4	-2,0	-59%
9		45°	3,4	0,0	0%
10		15°	2,8	-0,7	-33%
11	Rotation 180°	30°	2,3	-0,2	-10%
12		45°	2,1	0,0	0%

4 Conclusion

The control system designed has a design of control system module that can run the control system in the horizontal plane while changing the configuration has been done. The control system unit uses the ROS framework that can input the desired control system parameters er and PD control system is used as a control to reduce errors based on a specified setpoint, namely the yaw angle value. The results of the analysis demonstrate a notable alteration in the capacity of the PD control system, in accordance with the modification in thruster angle configuration. This is substantiated by the maximum error of 3° for thruster angle 15°, 2° for thruster angle 30°, and 1° for thruster angle 45°. The difference in UUV manoeuvrability for each thtuster angle configuration in forward motion is greatest with the optimal configuration at 15°, with a difference of 23%. In backward motion, the optimal configuration is 15°, with a difference of 24%. In right motion, the optimal configuration is 45°, with a difference of 59%. In rotation motion, the optimal configuration is 45°, with a difference of 33%. This discrepancy suggests that the 45° configuration is the optimal choice, taking into account all motion types.

IOP Conf. Series: Earth and Environmental Science **1461** (2025) 012040 doi:10.1088/1755-1315/1461/1/012040

References

[1] Yu C, Xiang X, Maurelli F, Zhang Q, Zhao R, Xu G. Onboard system of hybrid underwater robotic vehicles: Integrated software architecture and control algorithm. Ocean Engineering. 2019 Sep 1;187.

[2] Zhou B, Zhao M. Numerical simulation of thruster-thruster interaction for ROV with vector layout propulsion system. Ocean Engineering. 2020 Aug 15;210.

[3] Fernandes PG, Stevenson P, Brierley AS, Fernandes PG, Stevenson P, Brierley AS. AUVs as research vessels: the pros and cons. In: ICES ASC 2002 Theme Session J: Use of marine research vessels in ICES . Copenhagen; 2002. p. 1–11.

[4] Dukan F, Sørensen AJ. Joystick in Closed-loop Control of ROVs with Experimental Results. IFAC Proceeding. 2012;45(5).

[5] Allotta B, Pagliai M, Pugi L. Design and Modelling of Innovative Propulsion Layouts with Pivoted Thrusters For Underwater Vehicles. In: IFAC-PapersOnLine. Elsevier B.V.; 2016. p. 454–9.

[6] Soylu S, Proctor AA, Podhorodeski RP, Bradley C, Buckham BJ. Precise trajectory control for an inspection class ROV. Ocean Engineering. 2016 Jan 1;111:508–23.

[7] Budak S, Tekin MH, Durdu A, Sungur C. Kalman Filter and PID Application on Underwater Vehicles. Turkish Journal of Forecasting. 2022 Aug 31;06(1):27–33.

[8] Bayusari I, Mario Alfarino A, Hikmarika H, Husin Z, Dwijayanti S, Yudho Suprapto B. Position Control System of Autonomous Underwater Vehicle using PID Controller. Semarang; 2021 Oct.

[9] Walker KL, Stokes AA, Kiprakis A, Giorgio-Serchi F. Impact of Thruster Dynamics on the Feasibility of ROV Station Keeping in Waves. In: 2020 Global Oceans 2020: Singapore - US Gulf Coast. Institute of Electrical and Electronics Engineers Inc.; 2020.

[10] Li M, Yu C, Zhang X, Liu C, Lian L. Fuzzy adaptive trajectory tracking control of work-class ROVs considering thruster dynamics. Ocean Engineering. 2023 Jan 1;267.

[11] Kong F, Guo Y, Lyu W. Dynamics Modeling and Motion Control of an New Unmanned Underwater Vehicle. IEEE Access. 2020;8(8):30119–26.

[12] Ohrem SJ, Amundsen HB, Caharija W, Holden C. Robust adaptive backstepping DP control of ROVs. Control Eng Pract. 2022 Oct 1;127.

[13] Zhang X, Jiang K. Backstepping-based adaptive control of underactuated AUV subject to unknown dynamics and zero tracking errors. Ocean Engineering. 2024 Jun 15;302.

[14] Hosseinnajad A, Mohajer N, Nahavandi S. Novel barrier Lyapunov function-based backstepping fault tolerant control system for an ROV with thruster constraints. Ocean Engineering. 2023 Oct 1;285.

[15] Gong H, Er MJ, Liu Y, Ma C. Three-dimensional optimal trajectory tracking control of underactuated AUVs with uncertain dynamics and input saturation. Ocean Engineering. 2024 Apr 15;298.

[16] Kadiyam J, Mohan S, Deshmukh D, Seo T. Simulation-based semi-empirical comparative study of fixed and vectored thruster configurations for an underwater vehicle. Ocean Engineering. 2021 Aug 15;234.

[17] Serrano D. Introduction to ROS-Robot Operating System [Internet]. 2015 [cited 2024 Oct 8]. Available from: https://www.sto.nato.int/publications/STO%20Educational%20Notes/STO-EN-SCI-271/EN-SCI-271-06.pdf?__cf_chl_tk=5l76CD6c3bsj8ATBURXi84VUK7NaqmO0FlhoQgMTF4w-1728378834-0.0.1.1-5503

IOP Conf. Series: Earth and Environmental Science **1461** (2025) 012041 doi:10.1088/1755-1315/1461/1/012041

Thermal Performance of Immersion Cooling On Li-Ion Battery with Paraffin Wax-based Phase Change Material (PCM) under Ship Rolling and Pitching Angles

Yogyandini Mega Kirana[1], Sutopo Purwono Fitri[1*] and Makoto Shibahara[2]

[1] Department of Marine Engineering, Institut Teknologi Sepuluh Nopember, Surabaya, Indonesia
[2] Graduate School of Maritime Sciences, Kobe University, Kobe, Japan

*E-mail: sutopopf@its.ac.id

Abstract. Battery Thermal Management System (BTMS) onboard electric ships are crucial for optimising battery performance and longevity. By regulating the temperature, these systems ensure that batteries operate within safe limits, thus reducing the risk of thermal runaway. The present work aimed to determine the thermal performance of immersion cooling BTMS using paraffin wax-based PCM as a coolant on a LiFePo4 Li-Ion battery pack at discharging rates of 1C, 2C, and 3C. The immersion coolant rates were set to 30%, 50%, and 90%, and the inclination angles of the simulated ship's pitching and rolling conditions were 15° and 35°, respectively. It was found that direct immersion cooling by paraffin PCM significantly affects the lowering of the battery pack's discharging temperature. Discharging the battery without a coolant resulted in a battery temperature of 45.38 °C, above the safe limit temperature of 40°C for LiFePo4. Meanwhile, discharging the battery pack immersed in paraffin PCM reduced the temperature by approximately 6 -7 °C at 30% immersion, 8 -10° C at 50% immersion, and 11-12° C at 90% immersion. The effect of the 35° inclination angle of the battery pack showed a slightly higher temperature distribution than the 15° one at the low immersed battery's wall. However, the pitching and rolling conditions gave the battery pack a uniform temperature distribution.

1. Introduction

Ships heavily rely on fossil fuels, which contributes to pollution. To create a cleaner shipping environment, the International Maritime Organization (IMO) aims to reduce a minimum of 50% of greenhouse gas (GHG) by 2050 compared to the level in 2008 by adopting the resolution of MEPC.304(72). One of the alternative methods of ship propulsion is the fully electric drive, where the battery is the foremost essential component for power sources [1]. The types of batteries commonly used are lithium-ion (Li-ion) batteries, which are a type of rechargeable battery whose working principle is based on the movement of lithium ions from negative (anode) to positive (cathode) to charge and discharge the battery [2][3]. This type of battery is usually manufactured in the form of small packages of batteries, which are then grouped into a battery pack. The two major types of lithium-ion batteries are Nickel-Manganese-Cobalt (NMC) and Lithium Iron

Phosphate/LiFePo4 (LFP). Lithium-ion batteries are often chosen due to their relatively high specific energy density, long lifecycle, good lifespan, and reasonable cost [4].

When batteries are charged and discharged, thermal energy may be generated by exothermic chemical reactions and the Joule effect, which cause heat generation that, if not removed properly, will increase the operating temperature, causing performance degradation and eventually resulting in thermal runaway [5]. The higher the discharge rate, the higher the speed of the battery capacity degradation will also be [6]. Zhang et al. (2023) reported experimental and computational simulation of battery discharging up to 5C on a single prismatic LFP battery [7]. He confirmed that the battery discharging without any cooling could reach 61.3° C, beyond the safe working temperature range of Li-ion batteries and cause a significant risk of thermal runaway.

To ensure that the battery operates in optimal condition, Battery Thermal Management Strategies (BTMS) is required [8][9][10]. An effective BTMS should have the capability of ensuring that Lithium-ion batteries perform within acceptable and appropriate temperature ranges. Due to the limitations of air cooling, liquid cooling is an alternative for large-scale BTMS [11]. Compared to air cooling, liquid cooling has a higher specific heat capacity and better thermal conductivity [12]. Immersion cooling has gained attention for its efficient thermal management in electronic devices and electric vehicles.. In this immersion cooling, the entire battery cell is immersed and will be in direct contact with a non-conductive dielectric liquid, thus creating homogeneous conditions and high thermal energy transfer capacity for the heat dissipation process and reducing thermal contact resistance [13]. Kim and Pesaran (2007) performed a single-cell simulation study comparing three thermal management systems: air cooling, indirect water/glycol jacket cooling, and direct mineral oil cooling.. Direct cooling exhibited the lowest maximum cell-surface temperature difference and the best heat transfer coefficient among the three cooling methods, especially with smaller cooling channel diameters and higher flow rates [14]. Choi et al. (2024) experimentally and numerically observed the temperature performance of a polymer intercell heat exchanger with paraffin-based PCM and demonstrated the best cooling performance under the 3C charging condition [15]. Paraffin-based PCM became one of the good alternatives for direct battery cooling because it is inexpensive, safe, reliable, and non-irritating [16].

This study developed a direct immersion cooling method using paraffin wax-based (C22-C45) PCM coolant for LFP battery packs at 30%, 50%, and 90% immersion rates. The battery pack's thermal performance was experimentally measured and investigated based on rapid 1C-3C discharge rates and the variations of 15° and 35° of the ship's pitching and rolling inclination angles.

2. Experimental Apparatus Setup

The schematic diagram and configuration of the experimental setup for this research are shown in Figures 1 and 2, respectively. The experimental apparatus consisted of several equipments, including an 8 LiFePo4 pouch cell battery (3.2V 20Ah) in series with a 5 mm gap space in between the batteries, Battery Management System (BMS) (JK-BD6A17S8P, JK BMS), battery meter/analyser (PZEM-015, Peacefair), battery charger (C600, GuangZhou Electrony Corporation), load bank consisting of 50-100 W 1-4Ω braking resistors, inclinometer, switch, data logger (T7-Pro, LabJack), and computer laptop. According to the immersion rate parameter based on the percentage of the immersed height of the battery cell, the coolant of paraffin wax PCM was directly poured at the above melting point condition into the acrylic chamber of the battery pack, as shown in Figure 3. The steel platform is adjusted according to the inclination angle parameters, as seen in Figure 4. The experimental parameters are listed in Table 1.

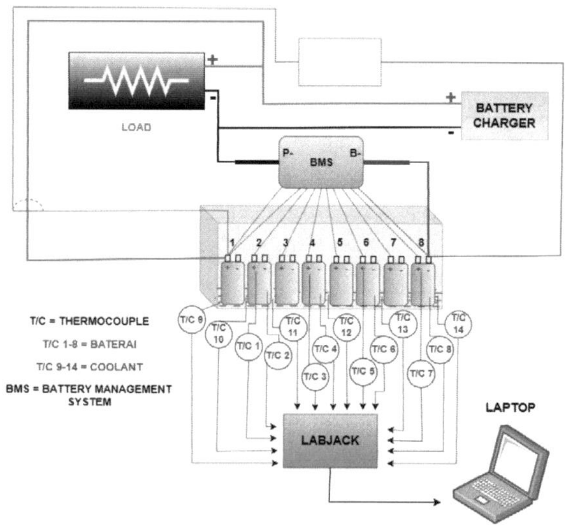

Figure 1. Schematic diagram of experimental apparatus

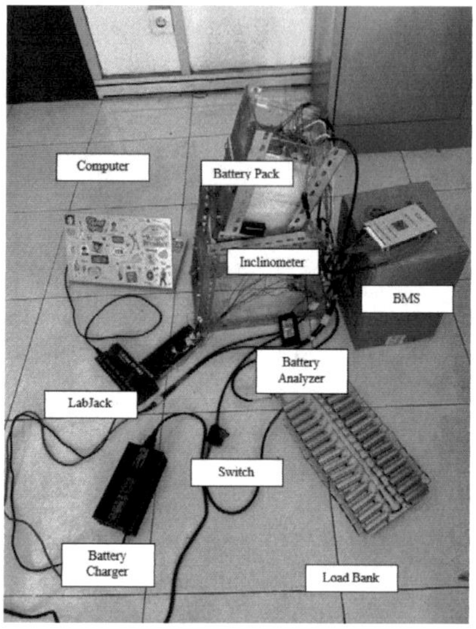

Figure 2. Configuration of the experimental apparatus

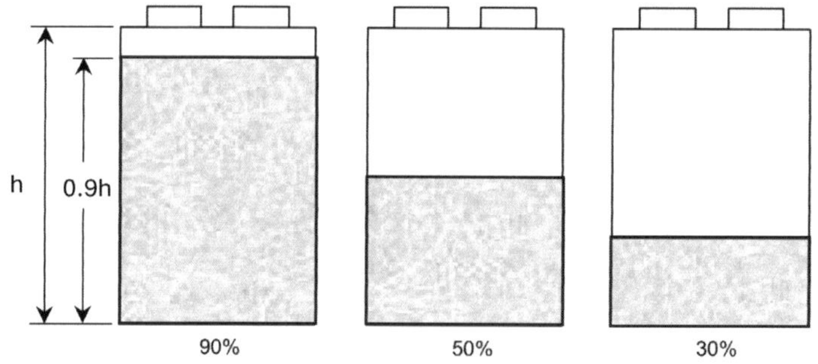

Figure 3. Schematic of immersion rate

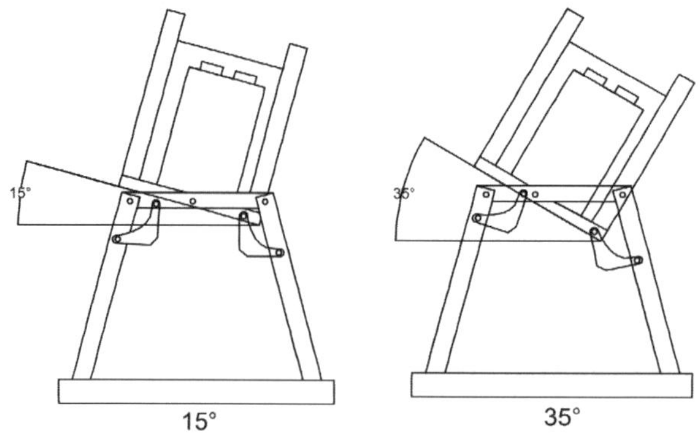

Figure 4. Schematic of inclination angle

Table 1. Experimental parameters

Variable	Description
Battery type	LFP Li-Ion 3.2v 20Ah
Battery cell configuration	8S
Discharge rate	1C, 2C, and 3C
Type of PCM coolant	Paraffin C22-C45
Melting point	58 – 60 °C
Heat capacity	2500 kJ/kg °C
Density	920 kg/m^3
Rolling/pitching inclination angles	15° and 35°
Coolant immersion rates	30%, 50%, and 90%

3. Results and Discussion

Figures 5(a) and 5(b) show the LFP battery average temperature distribution without PCM coolant at discharge rates 1C, 2C, and 3C for 60', 30', and 20', respectively. In the duration of 20 minutes, the highest temperature rise was achieved by 3C discharge rate with 17.53 °C temperature increase, followed by 2C discharge rate with 11.8 °C temperature increase, and by 5.8 °C temperature increase at 1C discharge rate. The battery pack operated unsafely by performing an over 40 °C safe limit temperature for the LFP battery type. As shown in Fig. 5 (b), it is also observed that the battery's wall of the 4th cell has the highest temperature at almost average of 45 °C.

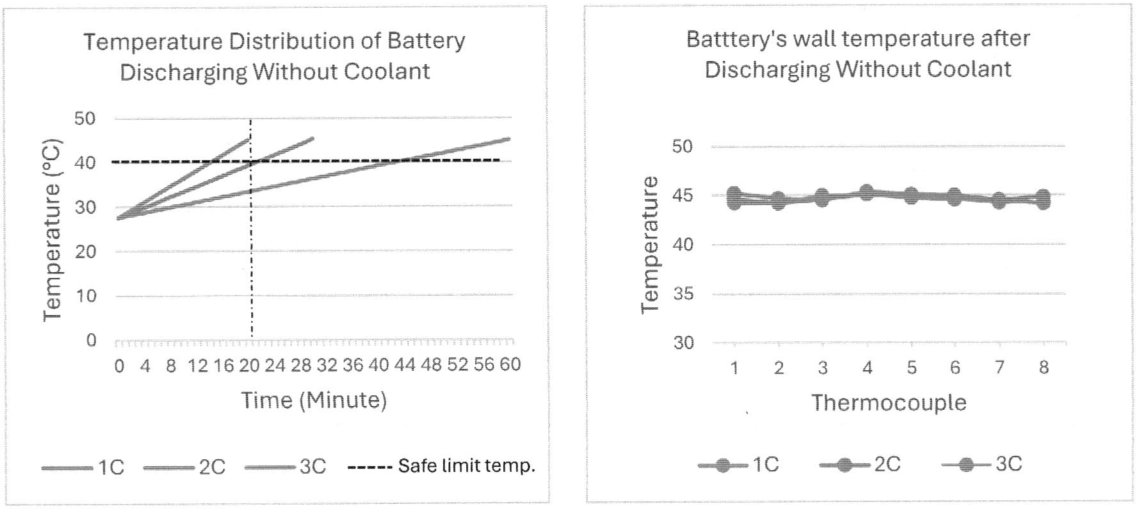

(a) (b)

Figure 5. (a) T-history of Temperature Increase of Battery Discharging Without Coolant, (b) Battery's Wall Temperature Distribution After Discharging

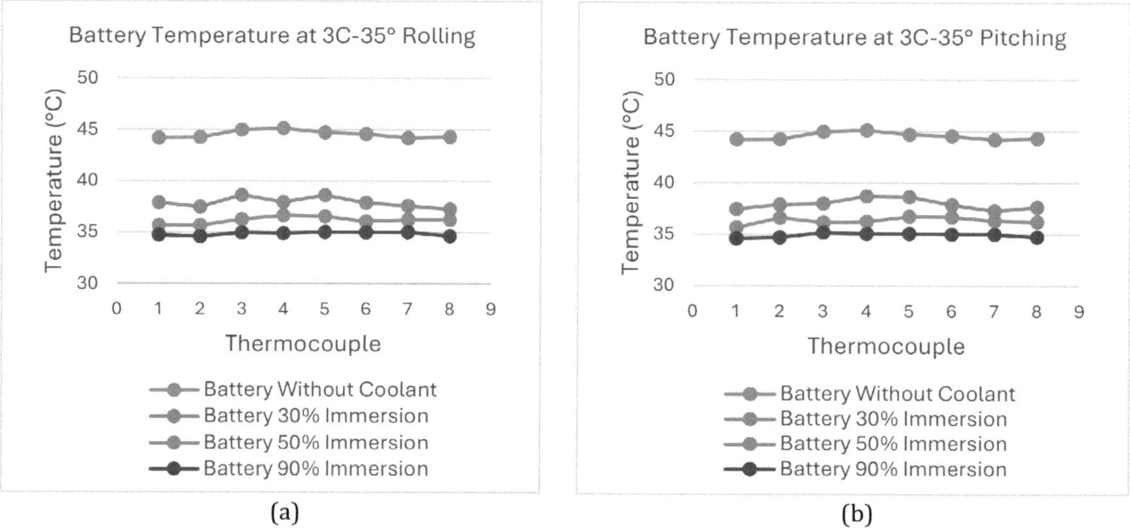

Figure 6. Temperature Distribution of Battery Pack at (a) 3C Discharge Rate & 35° Rolling Angle, (b) 3C Discharge Rate & 35° Pitching Angle

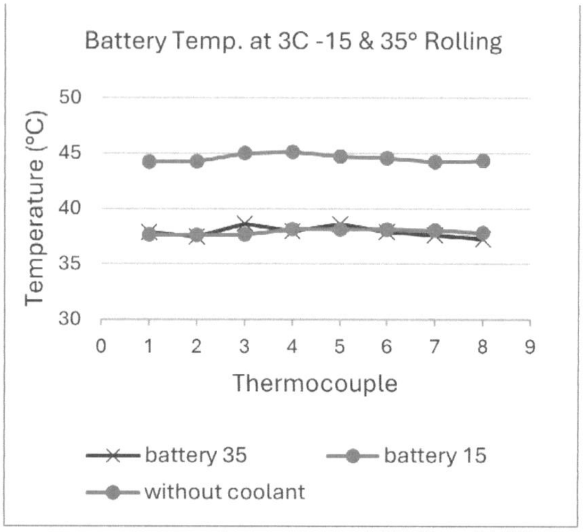

Figure 7. Comparison of Temperature Distribution of Battery at 3C Discharge Rate and 30% Immersion Rate for 15° and 35° Rolling Angles

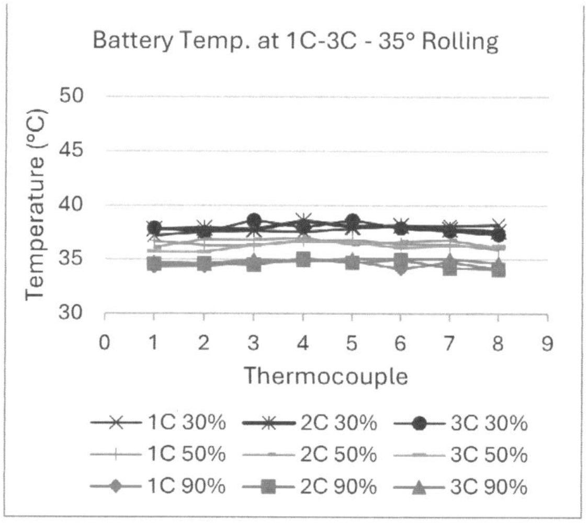

Figure 8. Comparison of Temperature Distribution of Battery at all Discharge and Immersion Rates for 35° Rolling Angle

As shown in Fig. 6(a) and 6(b), it is found that at both 35° of rolling and pitching inclination angle, the 30% immersion rate showed the highest average of the battery's wall temperature up to 38.7° C. The 50% immersion rate performed temperature distribution up to 36.7° C, and the 90% immersion rate performed the lowest temperature distribution up to 35.1° C. It was also confirmed that all the immersion rate cooling using Paraffin PCM showed a battery-safe operation temperature below 40° C.

Fig. 7 compares the battery's temperature distribution at 15° and 35° rolling angles at a 3C discharge rate and 30% PCM immersion. The average temperature distribution of both angles has no significant differences. The 15° angle has an average battery wall temperature of 37.9° C, while the average coolant temperature is 36.9° C. The 35° angle has an average battery wall temperature of 37.9° C, while the average coolant temperature is also 36.9° C. Figure 8 shows the temperature distribution of the battery on a 35° rolling angle at all range discharge rates of 1C, 2C, and 3C; and all range of immersion rates of 30%, 50%, and 90%. The immersion rate of 30% for all discharge rates showed higher values of the battery's temperature than at the other 50% and 90% ones. However, all immersion rates of cooling by paraffin PCM showed a battery-safe operation temperature below 40° C.

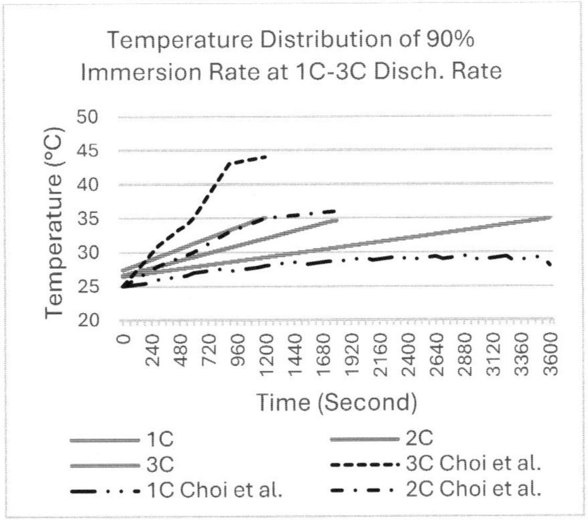

Figure 9. Comparison of T-history-based Temperature Distribution of Battery at All Discharge Rate and 90% Immersion Rate with the Results from Choi et al. (2024)[15]

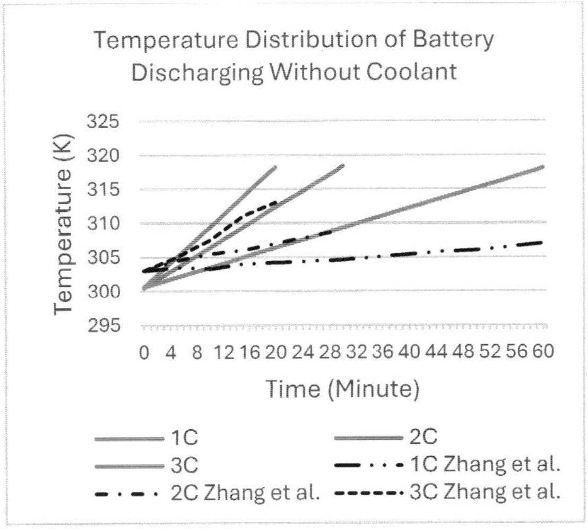

Figure 10. Comparison of T-history-based Temperature Distribution of Battery Discharging at All Discharge Rates without Coolant with the Results from Zhang et al. (2023)[7]

Figure 9 compares the temperature distribution of the battery based on the T-history method for 90% immersion rate cooling at all discharge rates with the results from [15]. In the present work, the immersion cooling of the battery using paraffin wax PCM shows that the highest temperature of around 35° C was achieved at all discharge rates, while Choi et al. reported the performance of 8-series Lithium nickel manganese cobalt (LiNiMnCo) type of Li-Ion battery cooling using Polymer Intercell Heat Exchanger with paraffin-based PCM ($C_{18}H_{38}$, n-octadecane and $C_{20}H_{42}$, n-eicosane) that reached different temperatures at each charge rate, i.e., 1C at around 28° C, 2C at around 34° C, and 3C at around 40° C [15]. The results of PCM cooling of Li-Ion battery were almost agreed upon at specific charge/discharge rate, despite the different types of paraffin PCM and Li-Ion battery types.

In Fig. 10, it can also be seen that the results of temperature distribution of battery discharging without coolant from 1C to 3C rates are compared with the results from [7]. The present work showed a 45° C of the highest temperature at all discharge rates, which almost agreed with the 41° C of the highest temperature at 3C discharge rate from Zhang et al.'s result. The authors performed experimental and computational evaluation on a single prismatic 3.2 V, 50 Ah LFP battery [7].

4. Conclusions

This study experimentally investigated a direct immersion cooling method using paraffin-based (C22-C45) PCM on a Li-Ion LFP battery to overcome the potential thermal runaway during rapid discharging and dynamic motion effect of the ship's rolling and pitching. The major conclusions were summarised as follows:

1. The direct immersion cooling using paraffin wax-based PCM coolant successfully demonstrated a significant lowering of the maximum temperature of the battery below 40° C of safe limit operation temperature of LFP battery at all 1C – 3C immersion rates, while the discharging without coolant could reach 45° C of potential risk of thermal runaway;
2. The ship's rolling and pitching inclination angles at 15° and 35° had no significant effect on battery temperature distribution at all discharge and immersion rates. It was evident that the immersion cooling by paraffin-based PCM has the capability for the high-duty and dynamic motion application of Li-Ion battery onboard ships;
3. The paraffin-based PCM battery immersion cooling could provide better cooling or thermal performance under high immersion rates and rapid discharging and showed promising potential BTMS technique for significant improvements in battery performance and safety in the application onboard electric ships.

Acknowledgements

It is gratefully acknowledged for the financial support received from ITS Research Fund 2024 in the scheme of Penelitian Keilmuan (Contract No.: 1183/PKS/ITS/2024).

References

[1] Jeong, B., Jang, H., Lee, W., Park, C., Ha, S., Kim, D. K., & Cho, N.-K. (2022). Is electric battery propulsion for ships truly the lifecycle energy solution for marine environmental protection as a whole? Journal of Cleaner Production, 355, 131756. https://doi.org/10.1016/j.jclepro.2022.131756

[2] Haizhou, Z. (2017). Modeling of Lithium-ion Battery for Charging/Discharging Characteristics Based on Circuit Model. International Journal of Online Engineering (iJOE), 13(06), 86. https://doi.org/10.3991/ijoe.v13i06.6799

[3] Olabi, A. G., Abbas, Q., Shinde, P. A., & Abdelkareem, M. A. (2023). Rechargeable batteries: Technological advancement, challenges, current and emerging applications. Energy, 266, 126408. https://doi.org/10.1016/j.energy.2022.126408

[4] Zhao, G., Wang, X., Negnevitsky, M., & Li, C. (2023). An up-to-date review on the design improvement and optimization of the liquid-cooling battery thermal management system for electric vehicles. Applied Thermal Engineering, 219, 119626. https://doi.org/10.1016/j.applthermaleng.2022.119626

[5] Al-Zareer, M., Dincer, I., & Rosen, M. A. (2019). A novel approach for performance improvement of liquid to vapor based battery cooling systems. Energy Conversion and Management, 187, 191–204. https://doi.org/10.1016/j.enconman.2019.02.063

[6] Zhu, J., Su, P., Dewi Darma, M. S., Hua, W., Mereacre, L., Liu-Théato, X., Heere, M., Sørensen, D. R., Dai, H., Wei, X., Knapp, M., & Ehrenberg, H. (2022). Multiscale investigation of discharge rate dependence of capacity fade for lithium-ion battery. Journal of Power Sources, 536, 231516. https://doi.org/10.1016/j.jpowsour.2022.231516

[7] Zhang, Y., Liu, H., Liu, S., Pan, S., Tian, C., & Hu, J. (2023). Prediction model of thermal behavior of lithium battery module under high charge-discharge rate. Journal of Energy Storage, 74. https://doi.org/10.1016/j.est.2023.109366

[8] Roe, C., Feng, X., White, G., Li, R., Wang, H., Rui, X., Li, C., Zhang, F., Null, V., Parkes, M., Patel, Y., Wang, Y., Wang, H., Ouyang, M., Offer, G., & Wu, B. (2022). Immersion cooling for lithium-ion batteries – A review. Journal of Power Sources, 525, 231094. https://doi.org/10.1016/j.jpowsour.2022.231094

[9] Chen, D., Jiang, J., Kim, G.H., Yang, C., Pesaran, A. (2016) Comparison of Different Cooling Meth-ods for Lithium Ion Battery Cells. Applied Thermal Engineering, 94, 846-854. https://doi.org/10.1016/j.applthermaleng.2015.10.015
[10] Zhao, C., Cao, W., Dong, T., Dong, T., Jiang, F. (2018) Thermal Behavior Study of Discharg-ing/Charging Cylindrical Lithium-Ion Battery Module Cooled by Channeled Liquid Flow. International Journal of Heat and Mass Transfer, 120, 751-762. https://doi.org/10.1016/ j.ijheatmasstransfer.2017.12.083
[11] Shang, Z., Qi, H., Liu, X., Ouyang, C., Wang, Y. (2019). Structural optimization of lithium-ion battery for improving thermal performance based on a liquid cooling system, Int. J. Heat Mass Tran. 130. 33–41. https://doi.org/10.1016/j.ijheatmasstransfer.2018.10.074
[12] Karthik, C. A., Kalita, P., Cui, X., Peng, X. (2020). Thermal management for prevention of failures of lithium-ion battery packs in electric vehicles: a review and critical future aspects, Energy Storage 2, https://doi.org/10.1002/est2.137
[13] Wu, W., Wang, S., Wu, W., Chen, K., Hong, S., Lai, Y. (2019). A critical review of battery thermal performance and liquid based battery thermal management, Energy Convers. Manag. 182. 262–281, https://doi.org/10.1016/j.enconman.2018.12.051
[14] Kim, G.-H and Pesaran, A. (2007). Battery thermal management design modeling, World Electr. Veh. J. 1. 126–133, https://doi.org/10.3390/wevj1010126
[15] Choi, H., Hong, J., Lee, S., Kim, D., & Lee, H. (2024). A novel battery thermal management system for cooling/preheating utilizing a polymer intercell heat exchanger with phase change material. Applied Thermal Engineering, 238, 122248. https://doi.org/10.1016/j.applthermaleng.2023.122248
[16] Vakhshouri, A. R. (2020). Paraffin as Phase Change Material. In Paraffin: An Overview (p. 65). BoD - Books on Demand.

IOP Conf. Series: Earth and Environmental Science **1461** (2025) 012042 doi:10.1088/1755-1315/1461/1/012042

Maintenance Improvement on Feed Water System of Steam Power Plant: A Lesson Learned of the Application of Reliability-Centered Maintenance 3

Dwi Priyanta[1*], Ruhul Sunan Jazirah[1], Hari Prastowo[1], Nurhadi Siswantoro[1], Muhammad Badrus Zaman[1], and Trika Pitana[1]

[1] Department of Marine Engineering, Faculty of Marine Technology, Institut Teknologi Sepuluh Nopember, Surabaya, Indonesia

*E-mail: priyanta@its.ac.id

Abstract. The feed water system is one of the most important components of steam power plants because it defines the system's performance. Quick turnaround time is an important factor in operations that should be achieved to avoid causing interruptions. Maintenance techniques should, therefore, be effectively used to prevent the failure modes that lead to system failure. Specifically, this research aims to implement the Reliability-Centered Maintenance (RCM) 3 on the feed water system to improve its maintenance. The study assesses the important failure modes from failure history and develops new maintenance tasks to address these issues. The following have been integrated upon the application of RCM-3: 15 activities are planned by condition, eight are restored, and two are discarded. These improvements minimize the probability of system failures and, at the same time, improve the reliability factor. This research offers useful knowledge and efficient ways to enhance the maintenance approaches in steam power plants, thus ensuring more reliable operations.

1. Introduction

The feed water system, a pivotal component of a steam power plant, directly influences its overall performance. Its main function is to supply water to the boiler, which is heated to produce steam that powers the turbines. Any malfunctions in this system can significantly impact the plant's performance, potentially leading to reduced power generation and serious safety risks.

Reducing downtime is one of the main issues with feed water system maintenance. It is crucial to complete repairs swiftly to prevent power production interruptions. Significant operational delays might result from inadequate maintenance, which can be expensive. The necessity for targeted maintenance techniques, known as Reliability-Centered Maintenance (RCM), that minimize operating disruptions while preserving system reliability was first recognized by Nowlan and Heap (1978)[1].

RCM has experienced significant growth since its introduction. The RCM methodology has been updated to RCM2 (1997) by Moubray with the help of Nowlan for non-aviation commercial industry applications [2]. The American Bureau of Shipping (ABS) has released RCM guidance notes for marine applications[3]. The guidelines have been developed following the RCM method devised by Nowlan and Heap. The RCM2 approach has been further developed and upgraded to RCM3 (2018), also known as Risk-Based Reliability-Centered Maintenance by Basson. This update incorporates risk analysis into the original RCM2 method[4].

The major objective of this study is to utilize Reliability-Centered Maintenance (RCM) 3 to enhance the maintenance of the feed water system in steam power plants. Reliability Centered Maintenance (RCM) is a systematic methodology that identifies potential failure modes, assesses their consequences, and recommends proactive maintenance actions to improve system reliability. The RCM technique helps businesses prioritize actions that prevent major failures by enabling the creation of a maintenance plan based on actual operating data. RCM 3, a more sophisticated version, enhances these concepts by integrating predictive and condition-based maintenance.

2. Literature Review

2.1 Overview of maintenance strategies in power plant

Maintenance in steam power plants is essential to ensure continuous operation, reduce downtime, and extend the lifespan of critical components like the feed water system. Traditional maintenance strategies, such as preventive and corrective maintenance, have been widely employed. Preventive maintenance involves scheduled inspections and repairs to avoid unexpected failures, while corrective maintenance deals with repairs after a failure occurs.

Proactive maintenance strategies enhance efficiency and save costs in electric power plants. However, poor execution can lead to increased downtime, higher costs, safety risks, and regulatory issues, highlighting the importance of proper maintenance[5].

Power generation systems must be maintained effectively to satisfy daily electricity requirements. Numerous strategies, including reliability-centered maintenance, business-centered maintenance, and total productive maintenance, have been implemented in response to the growing complexity of apparatus. A study revealed that nearly all power stations utilize the SAP™ information system, and they employ various maintenance strategies, including run-to-failure, time-based, and condition-based maintenance.[6].

A recent study explores effective maintenance tactics in South African power stations, focusing on preventive, predictive, and corrective strategies. It also explores human factors and computerized systems to enhance maintenance practices. The research aims to support scholars and practitioners in South Africa and developing nations, ensuring reliability and safety in power plants[7].

2.2 RCM in power plant

The application of RCM in power plant industries was promoted by the Electric Power Research Institute (EPRI) [8]. Reliant Energy investigated from July 1998 to August 2000. They implemented a cost-effective maintenance program by conducting a multi-plant analysis with Streamlined Reliability Centered Maintenance (RCM). This study examined a total of 223 systems from 11 power facilities. Reliability analysis and maintenance history reviews are combined in the Streamlined RCM method, which is based on classical RCM, to generate maintenance plans. The initiative's feasibility was demonstrated through a pilot project at the W.A. Parish plant, leading to more extensive implementation. The process transformed Reliant Energy's maintenance approach from reactive to proactive, with a greater emphasis on predictive maintenance technologies. The analysis pinpointed equipment that was underperforming and proposed new tasks to resolve these concerns, thereby improving the performance of the facility and the allocation of resources.

Since the initial initiation of the application of RCM in power-generating industries, many studies have been conducted regarding the implementation of RCM. A case study was conducted

IOP Conf. Series: Earth and Environmental Science **1461** (2025) 012042 doi:10.1088/1755-1315/1461/1/012042

to introduce a novel methodology that integrates Reliability-Centered Maintenance (RCM), Risk-Based Maintenance (RBM), and Condition-Based Maintenance (CBM), aiming to improve plant availability in thermal power plants. Maintenance programs and component reliability significantly impact the availability of thermal power plants. Consequently, the Maintenance Priority Index (MPI) is employed to prioritize components, while a Failure Defense Task (FDT) is used for monitoring and assessment [9].

Reference [10] provided another study of RCM implementation in a power plant. Despite utilizing short- and long-term maintenance, the power plant facility was not yet operating optimally. It was recommended that RCM II be implemented to enhance maintenance and efficiency. Many incompatibilities in the existing maintenance strategies were discovered during the assessment. The final analysis indicated that certain enhancements to maintenance strategies were suggested. It was also recommended that the study be broadened to incorporate RCM2 completely.

A guide for implementing, revising, or modifying a maintenance strategy for industrial equipment, such as a power transformer, based on RCM may refer to [11]. The strategy is designed to plan for catastrophic failures, ensuring immediate availability and reliability in an industrial system.

The other case study demonstrates the implementation of RCM in thermal power plants to improve maintenance schedules and reduce long downtimes, which may refer to [12]. The seven-step RCM process involves defining, analyzing, and acting, with Pareto analysis prioritizing critical components. The study also demonstrates RCM's enhanced reliability benefits by developing an ANN (artificial neural network) model. The results show a significant improvement in predicted reliability values, highlighting the importance of RCM in various industrial sectors.

Another benefit of implementing a maintenance RCM strategy, based on FMECA analysis, has been shown to reduce maintenance costs significantly, demonstrating the feasibility of this approach in optimizing equipment maintenance management [13].

RCM is essential for maintaining the availability of critical systems and guaranteeing their reliability. RCM enhances equipment availability and reduces maintenance orders by optimizing preventive maintenance strategies. Case studies were examined in Reference[14] to illustrate the economic advantages of RCM, with a particular emphasis on its ability to optimize costs and meet annual production objectives. Organizations could guarantee the availability and dependability of their equipment through the implementation of RCM, which would subsequently stimulate productivity and enhance efficiency.

2.3 RCM3 in power plant

Since its introduction in 2019, few studies have been conducted on implementing RCM3, particularly in the power generation sector. Initial research contributing to RCM3 may consult reference[15]. This study introduces a new method, Reliability and Risk Centered Maintenance (RRCM), which combines Risk-Based Maintenance (RBM) and Reliability-Centered Maintenance (RCM). RRCM assists organizations in developing maintenance strategies that prioritize risk and reliability. A case study in a hydroelectric power plant illustrates how well RRCM works in developing maintenance schedules that prioritize cost, risk, and dependability.

3. Methodology

The first step in applying the RCM3 methodology to the feed water system of a steam power plant is gathering comprehensive historical data on system failures. Historical failure records serve as a crucial foundation for identifying failure modes and assessing the reliability of various components within the system. Other data relevant to the RCM3 analysis are also collected.

The RCM3 methodology is founded on eight critical questions that guide the analysis process[4]. These questions ensure that all potential failure modes are thoroughly evaluated, and appropriate maintenance strategies are developed. The eight questions are:

1. What are the operating conditions (how is the equipment or system being used)?
2. What are the functions and associated performance standards of the asset in its present operating context?
3. In what ways does it fail to fulfill its functions (failed states)?
4. What causes each failed state (failure modes)?
5. What happens when each failure occurs (failure effects and consequence severity)?
6. What are the risks associated with each failure (inherent risk quantified)?
7. What must be done to reduce intolerable risks to a tolerable level (using proactive risk management strategies)?
8. What can be done to reduce or manage tolerable risks in a cost-effective way?

The first question related to the operating context of the equipment. The operating context is the basis for developing the risk management strategies and maintenance program. RCM3 Information Worksheet is designed to document the answers to questions number 2 to number 6. The technically feasible and worth-doing maintenance plan in RCM3 is determined by answering questions 7 or 8. The answers to those questions are documented in the RCM3 Decision Worksheet. RCM3 Decision Diagram, which represents the framework of the decision process, shall be used together with RCM3 Decision Worksheet.

4. Results

4.1 Feed water system

The primary function of the feed water system in a steam power plant is to deliver high-pressure feed water to the boiler, which is crucial for the boiler's efficient and safe operation and the entire power generation process. Table 1 displays the list of key equipment for the analyzed feed water system.

Table 1. Key equipment of feed water system

Equipment	No. of unit	Brand and Type
HP BFP	4	Sulzer – Type GSG 100-290/9
LP BFP	4	Ensival – Type CN 80-32
HP BCP	2	Ensival Type PRNR 200/40A
LP BCP	2	Ensival Type CN 100-20
Deaerator	1	Mitsubishi
Economizer	1	CMI Energy
Evaporator	1	CMI Energy
CEP	2	Torishima P67740
Condensate Preheater	1	CMI Energy

4.2 Operating context

Developing the operating context is a prerequisite for conducting a proper RCM analysis. It answers the first eight questions of the RCM3 foundation. It is also the initial step in RCM3 and is often the process's most overlooked aspect. Operating context can be defined as the circumstances in which a physical asset or system is expected to operate. Table 2 shows the sample of the operating context of one of the key equipment listed in Table 1.

Table 2. Operating context of the high-pressure boiler feed pump

Key equipment	Operating context
HP BFP	The high-pressure (HP) steam generation system, positioned after the high-pressure economizer, utilizes gas turbine exhaust to produce steam with specific pressure and temperature qualities. This steam is supplied to the high-pressure steam drum and the main HP steam system, maintaining temperature limits for the steam turbine during operation. It also provides high-pressure spray water for the attemperator and discharges/temperate HP steam to the cold reheat piping system. The intermediate pressure (IP) steam generation system, located between the IP economizer and intermediate superheater steam system, generates intermediate pressure steam and supplies feedwater to the IP steam drum. Superheated IP steam is fed into the intermediate section of the steam turbine via the reheat system in this specific feed water system. It is required to have 4 High-Pressure BFPs, of which three are operational and one is on standby. Three HP BFPs must also operate 24 hours a day and 7 days a week to produce sufficient electrical energy. HP BFP must transfer water from the deaerator into the HP drum at not less than 578 tonnes/hour. The failure of this equipment would result in reduced electrical energy production caused by the lack of feed water entering the drum, resulting in major cost loss. The shutdown of this equipment would result in an electricity blackout.

4.3 Operating context

Developing the operating context is a prerequisite for conducting a proper RCM analysis. It answers the first eight questions of the RCM3 foundation. It is also the initial step in RCM3 and is often the process's most overlooked aspect. Operating context can be defined as the circumstances in which a physical asset or system is expected to operate. Table 2 shows the sample of the operating context of one of the key equipment listed in Table 1.

4.4 Information worksheet

The Information Worksheet is designed to document the analyzed item's functions and associated failed states, failure modes, failure effects and consequence severity, and inherent risk. The worksheet documents the answers to questions number 2 to number 6. The information worksheet may also be called a failure mode and effect criticality analysis (FMECA).

4.4.1 Function and failed state

One of the primary objectives of the RCM3 analysis is to preserve the critical equipment's function. From this function identification, the RCM3 analysis grew to determine the maintenance tasks that are technically feasible and worth doing.

Every identified function may experience functional failure or a failed state. A failed state occurs when the item is unable to fulfill a function to a standard of performance that is acceptable to the user. Table 3 shows the documentation of the functions and their associated failed state of the key equipment of the feed water system.

4.4.2 Failure mode

RCM3 defines failure mode as any event that causes a failed state and the mechanisms leading up to the failed state. Table 4 displays the findings of failure mode in the feed water system.

Table 3. Function identification of key equipment

Key equipment	Function	Functional failure
HP Boiler Feed Pump	To transfer water from the deaerator into the HP drum at not less than 578 tonne/hours	See notes A, B, and C.
LP Boiler Feed Pump	To transfer water from the deaerator into the LP drum at not less than 200 tonne/hours	See notes A, B, and C.
HP Boiler Circulating Pump	To Circulate water inside the HP drum at not less than 578 tonne/hours	See notes A, B, and C
LP Boiler Circulating Pump	To Circulate water inside the LP drum at not less than 200 tonne/hours	See notes A, B, and C
Condensate Extraction Pump	To transfer water from the deaerator into the HP drum at not less than 227 tonne/hours	See notes A, B, and C.
Deaerator	To remove air and other dissolved gasses from feed water and to remove oxygen up to 7ppb	Unable to evaporate demanded 174 tonne/hours
Economizer	To heat feed water at not less than 238 Celsius before transferring it into the evaporator	Unable to remove air and other dissolved gasses from feed water and to remove oxygen up to 7ppb
Evaporator	To evaporate the feed water at not less than 174 tonne/hour before evaporating it into the drum	Unable to heat feed water at no less than 238 Celsius before transferring it into the evaporator
Condensate Preheater	To heat condensate water from condensate extraction pumps at a temperature of not less than 104 Celsius.	Unable to heat feed water at not less than 104 Celsius before transferring it into the deaerator

A. the pump is unable to supply the demanded water of at least X Tonne/Hour and Feed the Water System in the absence of a standby redundant system (total failure)
B. the pump is unable to supply the demanded water of at least X Tonne/Hour and Feed Water System in the presence of standby redundant (partial failure)
C. the pump is unable to supply the demanded water of at least X Tonne/Hour and Feed Water System in the absence of standby redundant (hidden failure)

4.4.3 Inherent risk

RCM3 uses a 5x5 risk matrix to assess the risk of a string of analyses, from function identification to risk identification. Risk rating is evaluated based on a score derived from the probability of failure and the consequence of failure of the item in question. A risk score between 1 and 5 is considered low risk, between 6 and 12 is medium risk, between 13 and 20 is significant risk, and between 21 and 25 is high risk. Both significant and high risks are deemed intolerable risks. Table 5 summarizes the inherent risk determination of failure modes listed in Table 4.

4.5 Task selection

The Decision Worksheet is a crucial tool in selecting maintenance tasks in RCM3. It guides the selection process by answering questions about the asset's functions and potential failure modes. These questions help quantify the risks associated with each failure mode and select between proactive tasks like preventive and predictive maintenance and default actions like reactive maintenance or "run to failure" modes. This systematic approach prioritizes tasks based on risk and cost projections, optimizing resource allocation and enhancing asset reliability and safety. The Task Manager displays this selection logic. Table 6 summarizes maintenance tasks identified for the feed water system.

Table 4. Failure modes found in feed water system

Failure Mode Number	Failure Mode	Failure Mechanism	Found in
1	Normal wear and tear on impeller and casing and other rotary part	current standby changeover strategy leads to both pumps wearing out in a close time frame	Pumps
2	abnormal acceleration aging process on sealing apparatus	Improper Installation of sealing system parts	Pumps
3	Motor Winding Temperature overheat	Improper cooling system cleaning and varnishing insulation	Pumps
4	Mech Seal Leakage	Misalignment, Abrasion, Improper Installation	Pumps
5	Bearing Seize	Insufficient Lubrication and Wearout	Pumps
6	Impeller Drift	Abrasion	Pumps
7	Flexible Coupling Breakage	Wearout, Misalignment, Hydraulic Pressure Drop	Pumps
8	Tube Leakage	Corrosion, Aging, Welding Cracks	Evaporator
9	Unstable Drum Level	Cracked joint welding, abnormal steam and water level, a leak at the indicator level	Evaporator, Economizer, condensate Preheater
10	Level Control Valve Abnormal	Aging	Deaerator, Condensate, Preheater
11	Recirculation Valve Abnormal	Aging	Deaerator, Economizer, Condensate Preheater

Table 5. Summary of the inherent risk and residual risk of failure modes found in the feed water system

Failure mode number	Key equipment group	Failed state*	Inherent risk score	Inherent risk rating	Residual risk score	Residual risk rating
1	Pump	A	10	medium risk	8	medium risk
		B	12	medium risk	6	medium risk
		C	8	medium risk	4	low risk
2	Pump	A	10	medium risk	8	medium risk
		B	12	medium risk	6	medium risk
		C	8	medium risk	4	low risk
3	Pump	A	10	medium risk	8	medium risk
		B	12	medium risk	6	medium risk
		C	8	medium risk	4	low risk
4	Pump	A	10	medium risk	8	medium risk
		B	12	medium risk	6	medium risk
		C	8	medium risk	4	low risk
5	Pump	A	10	medium risk	8	medium risk
		B	12	medium risk	6	medium risk
		C	8	medium risk	4	low risk
6	Pump	A	10	medium risk	8	medium risk
		B	12	medium risk	6	medium risk
		C	8	medium risk	4	low risk
7	Pump	A	10	medium risk	8	medium risk
		B	12	medium risk	6	medium risk
		C	8	medium risk	4	low risk

* Refer to Table 3

Table 5. Summary of the inherent risk and residual risk of failure modes found in the feed water system (Continued)

Failure mode number	Key equipment group	Failed state*	Inherent risk score	Inherent risk rating	Residual risk score	Residual risk rating
7	Pump	A	10	medium risk	8	medium risk
		B	12	medium risk	6	medium risk
		C	8	medium risk	4	low risk
8	HRSG Equipment	as in Table 3	12	medium risk	4	low risk
9	HRSG Equipment	as in Table 3	12	medium risk	4	low risk
10	HRSG Equipment	as in Table 3	12	medium risk	4	low risk
11	HRSG Equipment	as in Table 3	12	medium risk	4	low risk

* Refer to Table 3

Table 6. Maintenance task identified for the feed water system

Maintenance task type	Number of tasks
Scheduled on condition	15
Scheduled restoration	8
Scheduled discard	1

4.6 Revised risk rating

When risk management strategies or recommended maintenance tasks are implemented, a revised risk rating is an expected or residual risk rating.

5. Conclusion

The RCM3 method has been used to analyze the feed water system of a steam power plant. The following are the conclusions that can be drawn from the study's results.

Pumps in feed water systems are categorized as protected function-type assets and are equipped with a protective device, namely a standby redundant pump. The currently implemented pump switching strategy carries a risk of failure within a relatively short timeframe; therefore, a new strategy for switching pumps is recommended.

Three technically feasible and worth doing-tasks can be performed to mitigate failure modes: On-Condition Tasks, Scheduled Restoration, and Scheduled Discard Tasks. The number of maintenance tasks in each category is 15, 8, and 1, respectively.

To address the hidden failure modes in pumps, such as those in standby pumps, it is essential to implement measures that make these failure modes evident. One effective strategy is the installation of a Switching System Readiness Check Circuit on the asset.

Equipment such as evaporators, deaerators, economists, and condensate preheaters were found to have minimal issues, with only two failure modes being detected for each category. Conversely, seven failure modes were observed in pump equipment, including HP BFP, LP BFP, HP BCP, LP BCP, and CEP, in three distinct states. Due to the presence of redundant failsafe pumps, managing these pumps is more intricate, necessitating the implementation of complicated handling procedures.

Acknowledgments

The authors thank the Department of Marine Engineering, Institut Teknologi Sepuluh Nopember (ITS), Surabaya, Indonesia, for supporting this research.

References

[1] Nowlan F S and Heap H F 1978 *Reliability-Centered Maintenance* (Washington: U.S. Department of Defense)
[2] Moubray J 1997 *Reliability-Centered Maintenance II* (Industrial Press)
[3] American Bureau of Shipping 2018 *RELIABILITY-CENTERED MAINTENANCE* (Houston: American Bureau of Shipping)
[4] Basson M 2019 *RCM3: Risk-based Reliability Centered Maintenance* (Industrual Press, Inc.)
[5] Al Ameeri T A, Ab Rahman M N and Muhamad N 2023 Analysing Effective and Ineffective Impacts of Maintenance Strategies on Electric Power Plants: A Comprehensive Approach *Energies (Basel)* **16**
[6] Ndjenja L and Visser J K 2015 DEVELOPMENT OF A MAINTENANCE STRATEGY FOR POWER GENERATION PLANTS *SAIEE Africa Research Journal* **106** 132–40
[7] Nkosi M 2023 Assessing the Maintenance Strategies and Techniques Used in South African Power Plants: A Review *Proceedings of the 6th European Conference on Industrial Engineering and Operations Management* (Lisbon: Industrial Engineering and Operations Management (IEOM) Society) pp 550–60
[8] Toomey G, Brunette T and DeCoster M 2002 *Streamlined Reliability Centered Maintenance at Reliant Energy* (Walnut Creek,California)
[9] Pariaman H, Garniwa I, Surjandari I and Sugiarto B 2015 Availability Improvement Methodology in Thermal Power Plant *Science and Engineering* **2** 43–52
[10] Singgih M L, Prasetyawan Y, Sutikno, Hartanto D, Kurniawan F R and Wicaksana W T 2019 Maintenance management improvement based on reliability centered maintenance II in energy generating industries *IOP Conference Series: Materials Science and Engineering* vol 528 (Institute of Physics Publishing)
[11] Martinez-Monseco F J 2020 An approach to a practical optimization of reliability centered maintenance. Case study: power transformer in hydro power plant *Journal of Applied Research in Technology & Engineering* **1** 37
[12] Bhangu N S, Pahuja G L and Singh R 2017 Enhancing reliability of thermal power plant by implementing RCM policy and developing reliability prediction model: a case study *International Journal of System Assurance Engineering and Management* **8** 1923–36
[13] Yssaad B, Khiat M and Chaker A 2014 Reliability centered maintenance optimization for power distribution systems *International Journal of Electrical Power and Energy Systems* **55** 108–15
[14] A.A. Bubshait and Basurrah A 2021 Reliability Centered Maintenance: Case Studies *Transactions on Engineering Technologies* ed Sio-Iong Ao, Len Gelman and Haeng Kon Kim (Singapore: Springer) pp 203–16
[15] da Silva R F, Melani A H de A, Michalski M A de C and de Souza G F M 2023 Reliability and Risk Centered Maintenance: A Novel Method for Supporting Maintenance Management *Applied Sciences (Switzerland)* **13**

IOP Conf. Series: Earth and Environmental Science **1461** (2025) 012043 doi:10.1088/1755-1315/1461/1/012043

Calculation for Energy Potential of Circular Sector Pendulum for Wave Energy Generator

IS Arief *[1], AZM Fathalah[1], B Cahyono[1], GG Pratama[1], A. Baidowi[1], Amiadji[1]

[1] Department of Marine Engineering, Sepuluh Nopember Institute of Technology, Surabaya, Indonesia
*E-mail: irfansya@its.ac.id

Abstract. Indonesia, as the most archipelagic country in the world, offers an abundant ocean energy resource, one of which is wave energy, especially on the shore. One concern regarding wave energy converters is the value of the torque that can be produced by relatively small devices that can provide sufficient energy for a small community on an island. This paper provides the study results of a wave energy converter that uses experiments and analytics to utilise the circular sector pendulum. This research examines the potential torque and power generated by a pendulum in a sector-type Wave Energy Converter (WEC) system with frame tilt angle and pendulum mass variations. The results show that the maximum exploitable torque is 1235.64 Nm at a frame tilt angle of 25°, a mass of 409.8 kg, and a deflection angle of 175°. At different frame tilt angles, the generated torque varies, with maximum torques of 999.99 Nm (20°), 756.73 Nm (15°), 507.71 Nm (10°), and 254.82 Nm (5°). The highest potential power generated was recorded at a frame tilt angle of 25°, a mass of 409.8 kg, and a deflection angle of 175°, amounting to 12.82 kW. This potential power decreases with the reduction of the frame tilt angle, respectively to 10.37 kW (20°), 7.8 kW (15°), 5.26 kW (10°), and 2.6 kW (5°), with an increase in the potential power of 1.2% to 2% for each increase in frame tilt angle.

1. Introduction

In the vast ocean, waves constantly move, carrying enormous amounts of energy. Ocean wave power generation technology with pendulum systems harnesses this natural motion to generate electricity[1]. The system consists of a pontoon that floats on the ocean's surface and is equipped with a pendulum that moves with the motion of the waves[2]. When ocean waves hit the pontoon, the pontoon moves up and down and sways. This movement is then passed on to the pendulum mounted inside the pontoon[3]. The pendulum swings with the movement of the pontoon, and this swing is converted into mechanical energy[4]. This mechanical energy is then converted into electrical energy through a generator connected to the pendulum.

Several researchers and industries have developed ocean wave energy with pendulum systems, such as SEAREV [5] and Penguin Wave Energy [6]. As technology advances and our understanding of the marine environment deepens, the potential of wave energy through pendulum systems for renewable energy solutions in the future is promising[7]. Pendulum systems installed on pontoons as the main driver of electricity production from ocean waves will provide significant benefits in terms of cost-effectiveness and scalability.

Energy demand in Indonesia is increasing along with the growing population, which has reached approximately 274.9 million people. Electricity consumption in July 2020 reached 138.6 TWh and is expected to rise to 468 TWh by 2040. Per capita electricity consumption has steadily increased from 1991 to 2022, with a 4% increase in 2022 compared to 2021. The Ministry of Energy and Mineral Resources (ESDM) plans to boost electricity consumption in 2023 by ensuring a 24-hour power supply nationwide, targeting electricity consumption of 1336 kWh per capita).

The rising energy demand leads to an increase in greenhouse effects due to dependence on fossil energy. The use of fossil energy not only raises greenhouse gas emissions but also accelerates climate change, depletes energy supplies, and causes geopolitical conflicts over oil reserves. Dependence on fossil energy for electricity generation is the primary cause of climate change. The Meteorology, Climatology, and Geophysics Agency (BMKG) released data on average air temperature anomalies in Indonesia, showing a normal air temperature of 26.8°C for 1991-2020, while in 2022, it was 27.0°C. Annual air temperature anomalies are almost all positive, with a maximum in Sentani-Jayapura of 0.8°C and a minimum in Karel Sadsuitubun-Southeast Maluku of -0.7°C.

Approximately 76.94% of Indonesia's territory consists of the sea, which gives it great potential to utilise new renewable energy (EBT) from the ocean. This potential energy includes tidal power, wave energy, and ocean thermal energy. This energy can be generated from the movement of seawater due to wind forces that create ocean waves. Wind forms due to pressure differences in two locations caused by temperature differences due to air heating by the sun. Therefore, wave energy is categorised as renewable energy. Besides wave energy, Indonesia also has tidal energy potential. Coastal areas in Indonesia have an average tidal current speed of 1.5 meters per second, with some areas reaching current speeds of 2.5 to 3.4 meters per second, such as the straits in Bali, Lombok, and East Nusa Tenggara. The strait between Taliabu Island and Mangole Island in the Sula Islands, North Maluku Province, has the strongest tidal currents, reaching speeds of 5 meters per second. Based on this explanation, it can be concluded that the development of Wave Energy Converters (WEC) has great potential in Indonesia. Many studies have been conducted discussing WEC, and one exciting system to apply is the Pendulum Wave Energy Converter. This pendulum system uses a horizontal pendulum (vertical axis) that moves according to the direction of the ocean waves. This wave energy technology can address the electricity energy issues in Indonesia as an archipelago and maritime country. Considering the population distribution in Indonesia, which consists of many islands, the government needs to actively research and develop ocean energy, especially wave energy. This is because WEC technology has the potential to be a solution to the electricity energy problems in an archipelagic country like Indonesia.

It is very important to understand the torque that can be harnessed from the pendulum and the potential power produced from the pendulum system; this paper studies the potential energy of the circular sector pendulum of a wave energy system using a calculation approach.

2. Method

2.1 Theoretical Approach

Data processing involves performing several calculations such as torque and potential power based on calculations, creating torque graphs for each variation, and creating potential power graphs. This research uses basic equations for torque and Newton's equations. This is because torque and force have a strong relationship in rotating objects. The equations used are as follows, as shown in Figure 1: the free-body diagram of the circular pendulum and the forces acting on its system can be modelled to obtain the wave energy. The equation used in this research is:

$$\tau = Ft \times Rcg$$
$$Ft = Fb(1-\cos\alpha)$$
$$Fr = Fb\cos\alpha$$
$$Fb = mg\sin\beta \quad (1)$$

The equations for power, torque, and angular velocity are in the calculations. This is because the pendulum shaft is a rotating object. Torque is a measure of the force that causes an object to rotate. Angular velocity, on the other hand, is the rate at which the angle changes over time and measures how quickly an object is rotating. This can be expressed with the following formulas:

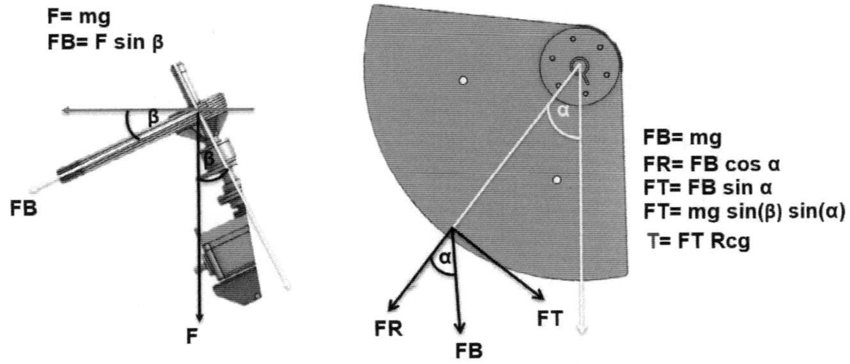

Figure 1. The free-Body diagram of the circular pendulum

$$P = \tau\omega \quad (2)$$

Angular velocity (ω) is the ratio of speed to distance. In this study, the distance used is from the center of the pendulum's axis to the center of gravity of the pendulum (r). Thus, the equation is obtained as follows:

$$\omega = v_{\mathrm{maks}} / rcg \quad (3)$$

The maximum velocity referred to here is the distance traveled by the pendulum swing per unit of time. This velocity is obtained based on the principle of conservation of energy. With the same mass, the velocity equation is given as follows:

$$v_{\mathrm{maks}} = \sqrt{2gh} \quad (4)$$

The distance (h) referred to is the product of the distance from the pendulum axis to the centre of gravity of the pendulum and the cosine of the pendulum's displacement angle:

$$h = r_{cg}\,(1 - \cos\alpha) \tag{5}$$

Table 1. Parameter and variable for to know the product of energy by a circular sector of pendulum.

Parameter	Amount of	Unit
Mass of Pendulum (m)	68.3 (1 disk) 136.6 (2 disk) 204.9 (3 disk) 273.2 (4 disk) 341.5 (5 disk) 409.8 (6 disk)	kg
Frame tilt angle	5 10 15 20 25	Degree
Initial deviation (Amplitude)	5; 10; 15 20; 30; 45 60; 75; 90 105; 120; 135 150; 165; 175	Degree

2.2 Data and Scrnario

The analytical approach to calculating the energy generated by a sector pendulum depends on several parameters and variables in the system. Several parameters affect the amount of power generated, such as the pendulum's weight, the angle of inclination of the frame, and the initial deviation that occurs.

3. Result and Discuss

The results obtained from the calculations in this study are the amount of torque and power produced based on the difference in pendulum weight, frame tilt and initial angle (deviation) of the pendulum.

Based on the calculations performed, a comparison graph between torque and deflection angle has been created. Variations in frame tilt angles lead to increased data variations. In this case, the frame tilt angle variation graph is divided into two categories for more accessible analysis: one with the same mass across different frame tilt angles and the other with the same frame tilt angle but varying masses, as shown in Figure 2.

IOP Conf. Series: Earth and Environmental Science **1461** (2025) 012043 doi:10.1088/1755-1315/1461/1/012043

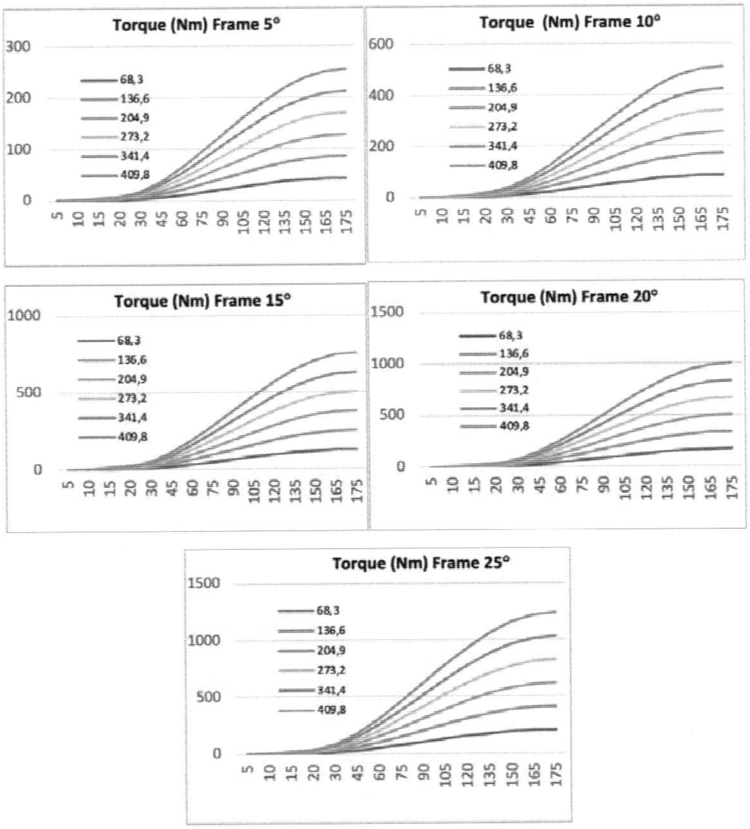

Figure 2. Torque every different mass pendulum, angle tilt frame and initial angle (deviation)

The maximum torque that can be utilised from the pendulum is 1235.64 Nm, which occurs at a frame tilt angle variation of 25° with a mass of 409.8 kg and a deflection angle of 175°. The maximum torque appears at the most extensive mass and deflection angle at each frame tilt angle. Each variation of mass and maximum deflection angle at a frame tilt angle of 20° produces a torque of 999.99 Nm; at a frame tilt angle of 15° produces 756.73 Nm; at a frame tilt angle of 10° produces 507.71 Nm; and at a frame tilt angle of 5° produces 254.82 Nm. This torque will subsequently be input for other Wave Energy Conversion System components. Maximum torque will be the reference for calculating the transmission system.

The pendulum velocity value is needed to obtain the power value with various initial deviation angles. The calculation shows (Figure 3) that the angular velocity has the same value relative to the centre of the pendulum rotation angle; this is shown in equation (4), which is only influenced by the height of the pendulum at the time of the initial deviation angle.

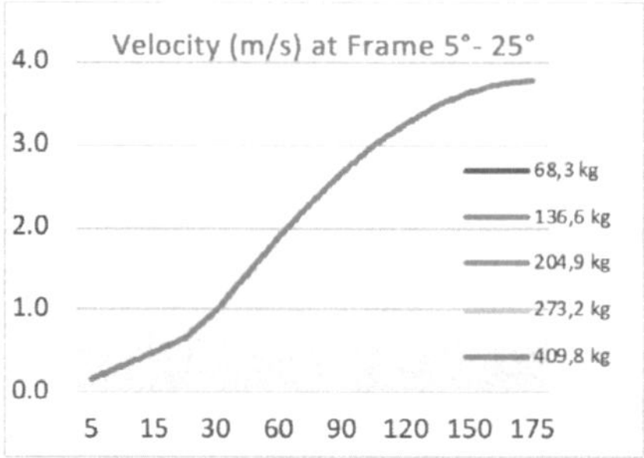

Figure 3. The value of velocity in a circular sector pendulum depends of the height of the radius Center of Gravity Pendulum

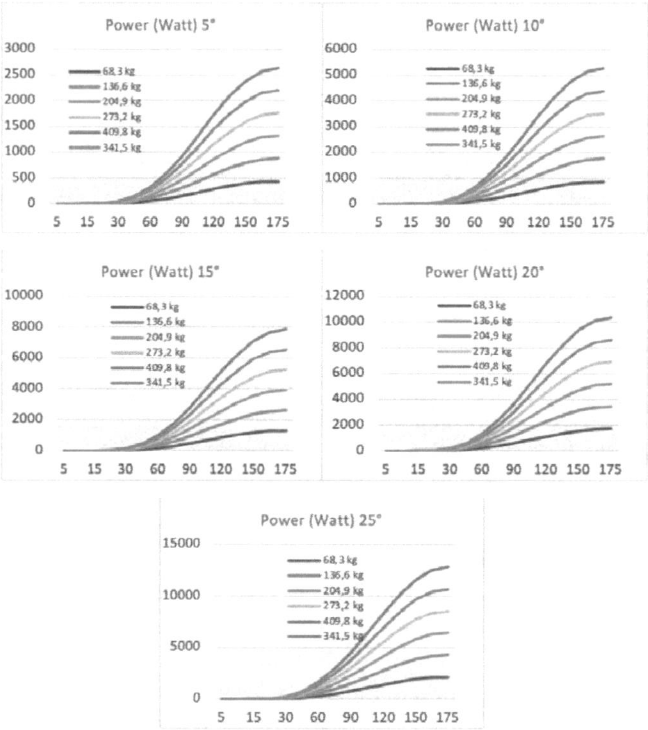

Figure 4. Torque every different mass pendulum, angle tilt frame and initial angle (deviation)

The two graphs above show the maximum speed and power potential at a frame tilt angle of 25° for the Wave Energy Conversion System used. Both graphs exhibit a trend where the values tend to increase with larger deflection angles. In the maximum speed graph, variations in mass do not affect the results, so there is no variation in the graph for different pendulum masses. The

IOP Conf. Series: Earth and Environmental Science **1461** (2025) 012043 doi:10.1088/1755-1315/1461/1/012043

highest point is at a deflection angle of 175° with a value of 3.78 m/s. In contrast, the power potential graph at a frame tilt angle of 25° shows that the highest power is achieved at a mass variation of 409.8 kg and a deflection angle of 175°, with a power output of 12817.17 Watts or 12.82 kW.

4. Conclusion

The conclusion of the above calculations provides a picture as below:

Based on the graph, it can be concluded that for a mass of 409.8 kg, the highest torque occurs at a deflection angle of 175° with a frame tilt angle of 25°, resulting in 1235.64 Nm. This is because torque is closely related to the frame tilt angle. The resulting torque increases as the frame tilt angle deviates further from 0°. Therefore, a tilt angle of 25° produces a higher torque than other frame tilt angles.

The maximum power potential generated in this research occurs at a frame tilt angle of 25° with a mass variation of 409.8 kg and a deflection angle of 175°, producing 12817.17 Watts or equivalent to 12.82 kW. With the same mass and deflection angle variation, a frame tilt angle of 20° produces the maximum power potential of 10.37 kW, a frame tilt angle of 15° produces 7.8 kW, a frame tilt angle of 10° produces 5.26 kW and a frame tilt angle of 5° produces 2.6 kW. With each increase in frame tilt angle, there is an increase in power potential of 1.2% to 2% from the previous power potential.

References

[1] I. S. Arief *et al.*, "Response to Pontoon and Pendulum Motion at Wave Energy Converter Based on Pendulum System," in *E3S Web of Conferences*, 2018. doi: 10.1051/e3sconf/20184301022.

[2] I. S. Arief, I. K. A. P. Utama, R. Hantoro, J. Prananda, T. R. Arvisa, and R. F. Kusuma, "Mooring Experimental Study of Motion Response for Pendulum Wave Energy Converters," in *IOP Conference Series: Materials Science and Engineering*, 2019. doi: 10.1088/1757-899X/462/1/012010.

[3] I. S. Arief, T. R. Arvisa, and A. Baidowi, "Uji Eksperimen Respon Gerakn Ponton Trimaran Untuk Pembangkit Listrik Tenaga Gelombang," Surabaya, 2018.

[4] A. F. de O. Falcão, "Wave energy utilization: A review of the technologies," *Renewable and Sustainable Energy Reviews*, vol. 14, no. 3, pp. 899–918, Apr. 2010, doi: 10.1016/J.RSER.2009.11.003.

[5] J. Cordonnier, F. Gorintin, A. De Cagny, A. H. Clément, and A. Babarit, "SEAREV: Case study of the development of a wave energy converter," *Renew Energy*, vol. 80, 2015, doi: 10.1016/j.renene.2015.01.061.

[6] Wello, "The Penguin Wave Energy Converter." Accessed: Dec. 19, 2022. [Online]. Available: https://wello.eu.

[7] X. Jiang, H. Shi, F. Cao, Z. Zhao, M. Li, and Z. Chen, "System analysis and experimental investigation of a pendulum-based wave energy converter," *Ocean Engineering*, vol. 277, p. 114300, Jun. 2023, doi: 10.1016/j.oceaneng.2023.114300.

IOP Conf. Series: Earth and Environmental Science **1461** (2025) 012044 doi:10.1088/1755-1315/1461/1/012044

Exploring the Maritime Passenger Transport in Algeria: Opportunities and Challenges for Tourism Growth

Mouloud KHELF[1*]

[1] Laboratory of Transportation Engineering and Environment (LITE), Department of Transport Engineering, University of Constantine 1, Campus Zarzara, Constantine 25000, Algeria.

*E-mail: mouloud.khelf@umc.edu.dz

Abstract. The Algerian government plans substantial investments in several transportation infrastructures to reduce pollution and establish a sustainable transportation system. Among these initiatives is the maritime passenger transport, aimed at enhancing the tourism sector. This paper aims to analyze the status of maritime passenger transport in Algeria, evaluate its performance, and address future challenges, particularly in terms of tourism. The methodology used in this work is based, firstly, on an analysis of the passenger flow recorded between 1998 and 2020 at the Algerian ports. Then, secondly, it is based on the evaluation of the Algerian maritime passenger transport company between 2011 and 2020. These permit us to analyze the exact situation and to know the several constraints that influence the performance of this transportation system in Algeria. The results show that the highest flow of passengers in the Algerian ports was between 2004 and 2007. In addition, the number of passengers transported by the Algerian company decreased by an average of 74.61% between 2011 and 2014 to an average of 50.88% between 2017 and 2019. This paper concludes with some suggestions to improve the performance of maritime passenger transport in Algeria.

1. Introduction

Tourism is as important as industry, playing a pivotal role in driving the development of a country's economy [1]. Nowadays, the availability of diverse transportation options, particularly sustainable modes aimed at reducing pollution, has enabled tourists to explore destinations across the globe [2]. Maritime transport is an essential mode for the economic development of countries, thanks to its large transport capacity, its prestige and importance in the tourism sector [3, 4]. Algeria is among the countries that have given a lot of value to this sector, especially in recent years [5]. This is through the investment, renovation and construction of infrastructures, such as the construction of the port of El HAMDANIA, Cherchell, Tipaza. This new infrastructure will cost from 5 to 6 billion dollars (USD), and it will make it possible to process 25.7 million tons of goods per year [6].

In addition, the Algerian maritime passenger transport company has strengthened its fleet through the purchase of a new ship named BADJI MOKHTAR III, which has a capacity of 1800 passengers and 600 vehicles. This new ship entered service on November 1, 2021. This new acquisition is the result of negotiations between the Algerian company for Maritime Passengers'

IOP Conf. Series: Earth and Environmental Science **1461** (2025) 012044 doi:10.1088/1755-1315/1461/1/012044

Transport and a Chinese company specializing in shipbuilding called Guangzhou Shipyard International Company; it cost 175 million dollars [7]. The length of the Algerian maritime facade extends over 1622 km [8, 9]. Algeria is a country that has a large facade on the Mediterranean area, which allows an important commercial exchange with the partners of this zone, mainly France, Spain, and Italy [10]. Furthermore, its scenic Mediterranean coastline has garnered attention from international tourists, including those from the United States, who are drawn to explore Algeria via cruise expeditions [11].

Otherwise, the Algerian government has launched the objective of opening maritime passenger transport lines to several destinations in Africa and Asia in the future [12]. This decision will elevate the number of tourists in the next years. To achieve this objective, the Algerian government has decided to allow the private company to invest in the passengers' maritime transport and to speed up this procedure in order to create competition with other maritime companies of the other countries in the Mediterranean zone and to improve the service quality of this transportation system [13, 14].

The following section presents the Algerian passengers' transport company, fleet, and ports, which ensure the exchange of travelers between Algeria and the other countries in the Mediterranean Sea.

2. Case study of the Algerian maritime transport company

The Algerian Company for Maritime Passengers' Transport was created by decree (N°87-155) on July 14, 1987. On April 7, 1990, the company obtained the status of an economic public company under the aegis of Holding Services. The activities of the company are mainly the maritime transport of passengers and vehicles, freight transport, and rolling freight. As well as activities ancillary to maritime transport, in particular the consignment of passenger ships and the sale of tickets. The company fleet consists of four ships [15]:

Table 1. The fleet characteristics of the maritime transport company in Algeria.

Fleet	Construction date	Construction place	Capacity (Passengers)	Capacity (Vehicles)
Tariq Ibn Ziyad	1995	Spain	1276	500
Tassili II	2004	Spain	1320	300
El Djazair II	2005	Spain	1320	300
Badji Mokhtar III	2021	China	1800	600

Previously, the Algerian Ferries Company had 5 other ships named El-Djazair, Tassili, Tipasa, Zeralda, and Hoggar; the first 4 were sold between 2000 and 2005, and the last one is currently out of service. The four ships in service regularly provide voyages from the Algerian ports of Algiers, Oran, Annaba, Bejaia, Skikda, and sometimes Mostaganem to the destinations of Marseille (France), Barcelona and Alicante (Spain), and Genoa (Italy). As well, there are other destinations from time to time added to the program for the cities of Naples (Italy), Toulon (France), and Valencia (Spain) (Figure 1) [16-17].

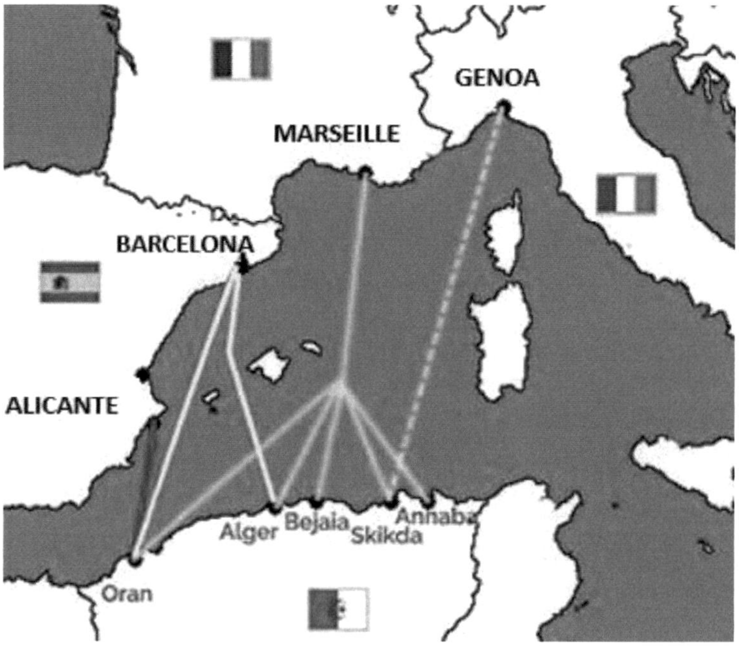

Figure 1. The Origin destination of ships between Algerian and European ports [15].

3. Method

The methodology followed in this paper to assess the maritime passengers' transport in Algeria is divided into 2 sections.

The first section focuses on the assessment of the passengers' departing and arriving at the Algerian ports to know the passengers' traffic frequency per year from 2000 to 2020.

The second section of the results presents an assessment of the Algerian maritime passenger transport company to recognize the performance of this important company and to know in the conclusion the causes of its performance.

The data used in this work were recorded mainly from the annual reports provided by the National Office of Statistics (ONS, Algeria), from 2000 to 2020 [17].

4. Results

4.1 Assessment of the of passengers arriving and departing at the Algerian ports

Figures 2, 3, and 4 present the number of passengers' frequency at the Algerian ports for 20 years. The number of passengers increased from 2000 to 2007, and then obviously decreased from 2008 until 2015. In the last 4 years, it has increased again, but the COVID-19 pandemic has affected this augmentation at the end of 2019 and into 2020.

IOP Conf. Series: Earth and Environmental Science **1461** (2025) 012044
doi:10.1088/1755-1315/1461/1/012044

The highest passenger number departing and arriving at the Algerian ports is recorded between 2004 and 2007, reaching 46,726 passengers in 2006. The most significant fluctuation in passenger traffic (both arrivals and departures) at Algerian infrastructures was observed in 2016, exceeding 20%.

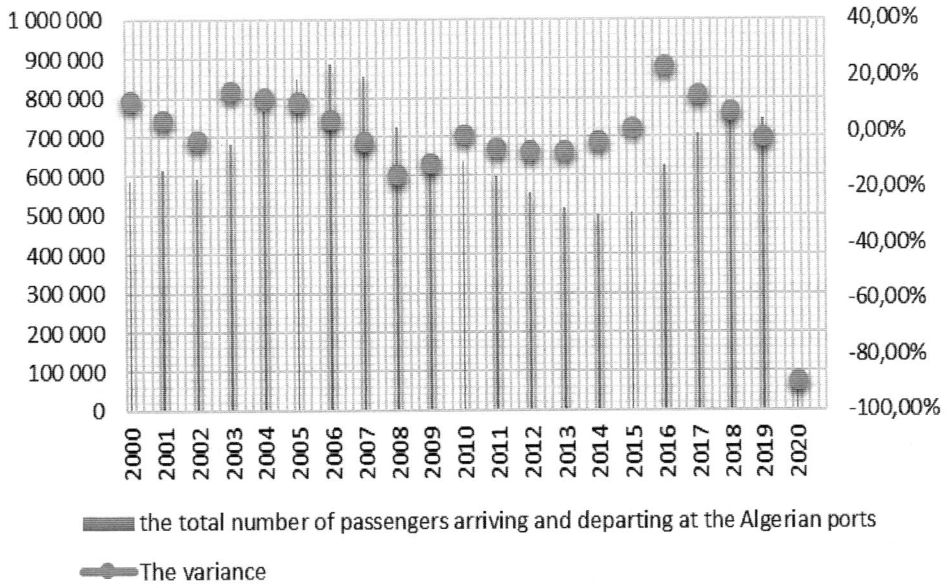

Figure 2. The total number of passengers' arriving and departing at the Algerian ports per year.

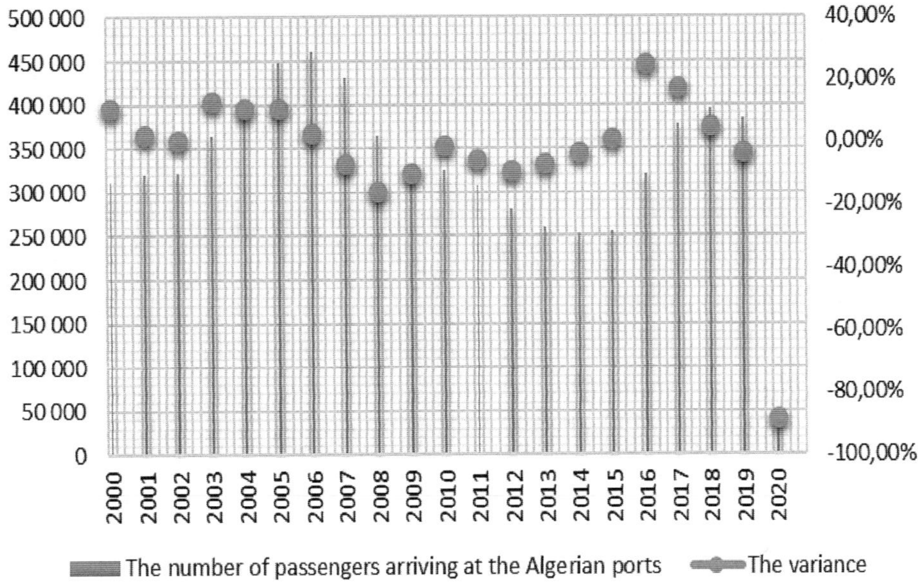

Figure 3. The number of passengers arriving at the Algerian ports per year.

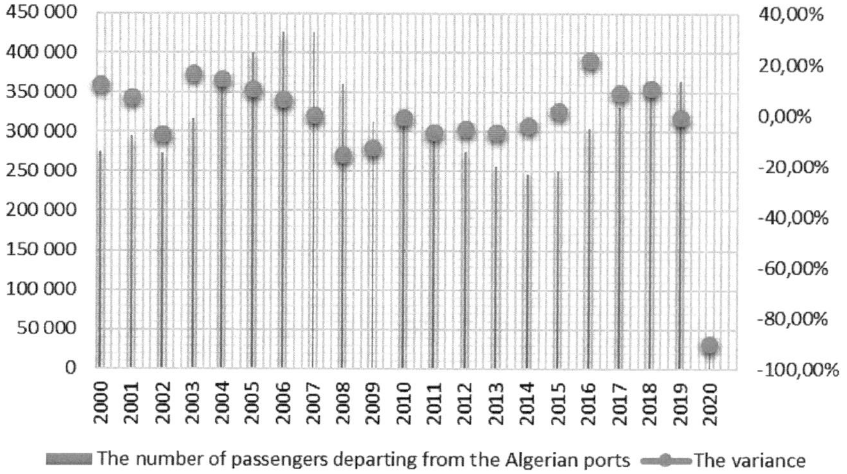

Figure 4. The number of passengers departing from the Algerian ports per year.

4.2 Assessment of the of passengers traveled with the Algerian company

Figure 5 shows the number of passengers who traveled with the Algerian company of maritime passenger transport per year. The highest number of passengers is recorded in 2012, with 447 323 passengers. The number of passengers decreased from 2013 to 2015 and then increased until the COVID-19 pandemic. Figure 6 represents the number of passengers' cruises made per year. The highest number of cruises made is in 2018 and 2019, with 641 cruises. The number of cruise deceased between 2014 and 2015 because the passengers' demand is reduced in this period. The most notable fluctuation in passenger numbers traveling with the Algerian company occurred during the eight-year period from 2012 to 2020, peaking in 2016. It is worth mentioning that the highest count of cruises was recorded in 2017.

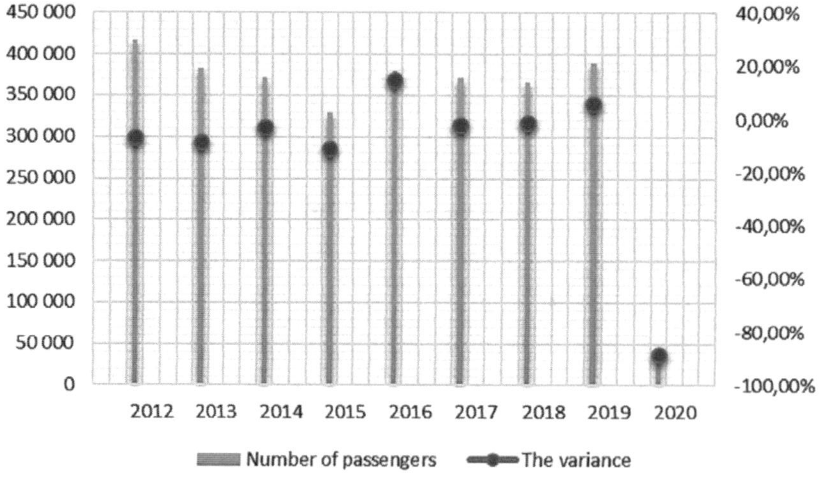

Figure 5. The number of passengers who traveled with the Algerian company per year.

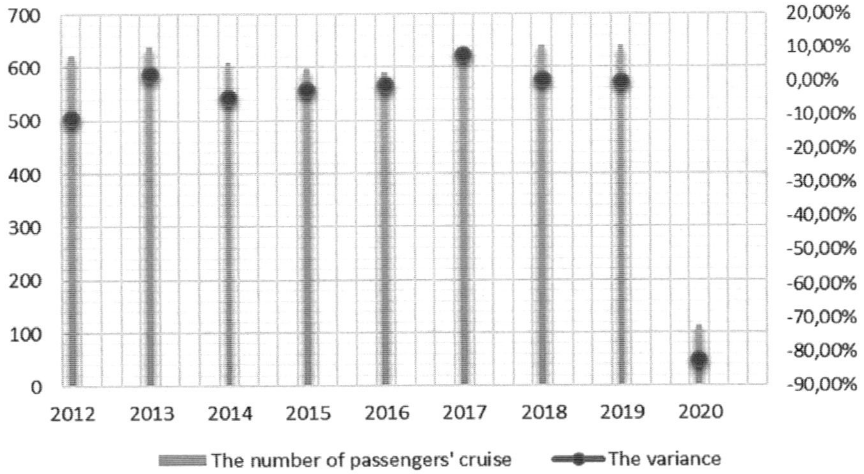

Figure 6. The number of passengers' cruise per year.

Figure 7 illustrates the number of passengers recorded with the number of cruises made per year. In 2012, the number of passengers transported by the company is 416,972 using 622 cruises, but in 2018 and 2019, the company transported 366,747 and 388,820 passengers, respectively, with 641 cruises per year. It is obvious that the number of passengers is less in these two years compared to 2012, and the number of cruises in this year is higher than in 2018 and 2019. This can prove the bad management of the company in that period.

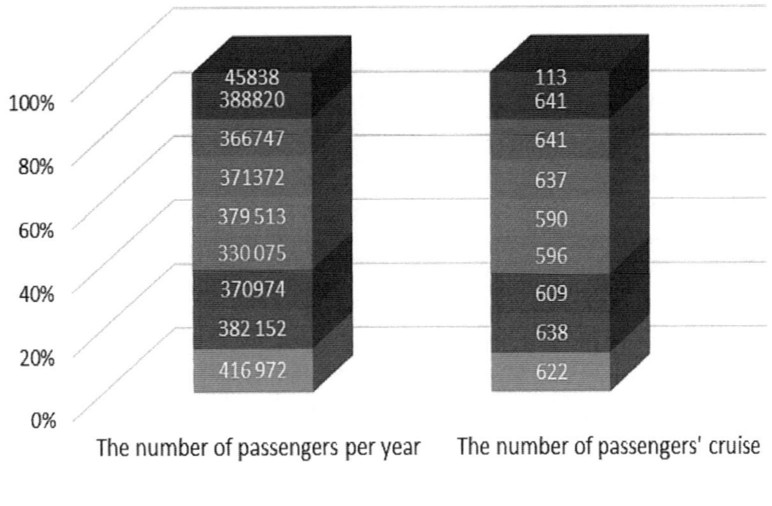

Figure 7. The number of passengers recorded with the number of cruises made per year.

Figure 8 indicates the rate of passengers transported by the Algerian Company compared to the total passengers' frequency at the Algerian ports.

From 2011 to 2014, the rate of passengers transported by the company was almost 75% of the total passengers recorded at the different Algerian infrastructures. After this period, the rate of passengers transported by the company decreased for all years, reaching 48.14% in 2018. The other international companies have benefited from the percentage lost by the Algerian company in these last few years. This is due to several problems in the Maritime Algerian Company, but between 2019 and 2020 the number of passengers transported by the company increased even if in under the Covid 19.

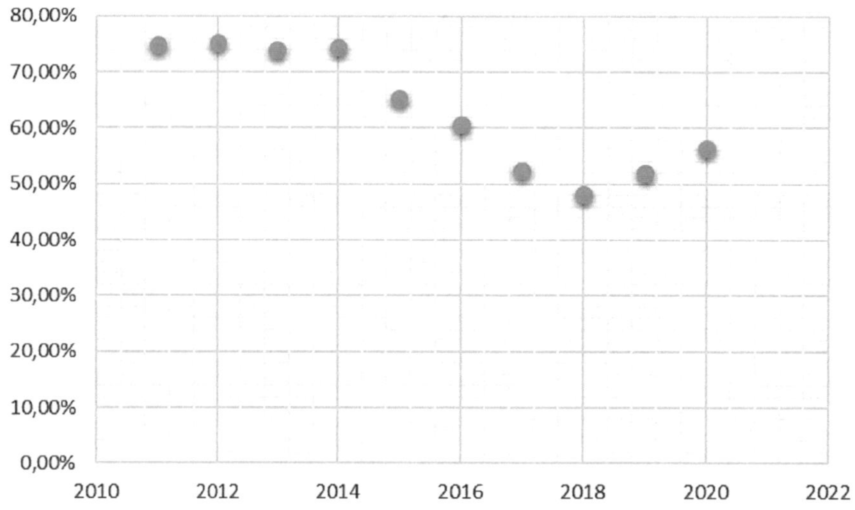

Figure 8. The rate of passengers transported by the Algerian Company compared to the total

Overall, these findings underscore the mismanagement within the company. This was particularly evident when the Algerian ship Badji Mokhtar arrived in Algiers from Marseille, France, carrying only 72 passengers and 25 vehicles. Despite the ship's capacity to accommodate 1800 passengers and 600 vehicles, many passengers were stranded in Marseille after being informed that the ship was fully booked [18]. Consequently, the implementation of a new ticketing strategy is important to prevent the recurrence of such issues.

5. Conclusion

The objective of this paper was to show the importance of transportation to the tourism, this by analyzing the situation of maritime passenger transport in Algeria. To realize it, an analysis of passenger traffic data at Algerian ports and the performance of the Algerian maritime passenger transport company has been done.

The paper reveals that it exists an important passengers' traffic flow at the Algerian ports and tourists are interested by visiting Algeria by cruise. The results of this study prove that the passengers' maritime transport is not well managed in Algeria. The Algerian company lost between 15% and 25% of passengers to the other companies of the Mediterranean Sea from 2015 to 2020, compared to 2014. This bad management can discourage tourists to visit the country. The government is investing in this important transportation mode to contribute efficiently to the economic development of Algeria, but this investment and effort should be made with efficient management of the Algerian Maritime company, because Algeria has a strategic position in the Mediterranean Sea and its economy has grown, especially in the last two years.

The main recommendation to improve the performance of the Algerian company and to encourage tourists to visit Algeria by maritime transportation is to use an electronic system for selling tickets online. The capacity of transport offered by the Algerian company should be in equilibrium with supply; this could not be realized without a statistical study of the previous years in several periods and especially in summer. In addition, the company should offer good service quality to the tourists to attract them and make good profits. Finally, the company should offer an interesting cruise for tourists along the Algerian coast, because this will improve its profits.

References

[1] Chebli, A. & Ben Said, F., 2020. The impact of covid-19 on tourist consumption behaviour : A perspective article. *Journal of Tourism Management Research, 7 (2)*, pp. 196-207. DOI: 10.18488/journal.31.2020.72.196.207.

[2] Gay J.C., 2006. Transport et mise en tourisme du Monde. *In: Collection EDYTEM. Cahiers de géographie, Transport et tourisme. 4.* pp. 11-22. https://doi.org/10.3406/edyte.2006.960.

[3] Beriatos, E. & Papageorgiou, M., 2011. Maritime and coastal spatial planning: the case of Greece and the Mediterranean. *Sustainable Development and Planning V, WIT Transactions on Ecology and the Environment, 150*, pp. 3–17. 10.2495/SDP110011.

[4] Fratila (Adam), A., Gavril (Moldovan), I.A., Nita, S.C., and Hrebenciuc, A., 2021. The Importance of Maritime Transport for Economic Growth in the European Union: A Panel Data Analysis. *Sustainability, 13(14)*, p.7961. https://doi.org/10.3390/su13147961.

[5] La redaction., 2020. Algérie : un plan de développement pour le transport maritime. Le Journal de la Marine Machande. https://www.journalmarinemarchande.eu/filinfo/algerie-un-plan-de-developpement-pour-le-transport-maritime. Accessed on: 29 March 2023

[6] Anyango, M., 2020. ALGERIA: 5 to 6 billion USD for the port of El Hamdania. ze-africanews. https://ze-africanews.com/en/algeria-5-to-6-billion-usd-for-the-port-of-el-hamdania/. Accessed on: 29 March 2023

[7] Oumakhlouf, Y., 2020. ENTMV : Acquisition d'un bateau neuf pour le transport des voyageurs. DZ Entreprise, La Portail de l'Entreprise Algérienne. https://www.dzentreprise.net/entmv-bateau-neuf-transport-voyageurs. Accessed on: 29 March 2023

[8] Kacemi, M., 2011. Protection and Development of the Coastal Areas in Algeria: Legislation and Instruments. The Case of Oran's Coastal Areas. *OpenEdition Journals : Etudes Caribéenne.* https://doi.org/10.4000/etudescaribeennes.5959

[9] Aroua, M., Boulahdid, M., Radakovitch, O., Benhalima, M., Guendouzi, Y., Fowler, S.W., and Ambrosi, J-P., 2023. Sediment pollution: An assessment of anthropogenic and geogenic trace element contributions along the central Algerian coast, *Marine Pollution Bulletin, 186*, p.114398.

[10] MOHAMED-CHERIF, F.Z., 2012. Shipping Linkages between Algeria and the Euro-Mediterranean Area: Realities and Perspectives. *OpenEdition Journals : Territory in movement Journal of geography and planning,* pp.86-99. https://doi.org/10.4000/tem.1907

[11] Salima, A., 2022. Algérie : après Oran, les touristes américains explorent Alger. Dzairdaily. https://www.dzairdaily.com/algerie-apres-oran-touristes-americains-explorent-alger/ . Accessed on: 13 May 2023.

[12] Aps., 2022. Accélération des discussions pour l'ouverture de la ligne maritime commerciale entre l'Algérie et le Qatar. Algérie Press Service. https://www.aps.dz/economie/136644-acceleration-des-discussions-pour-l-ouverture-de-la-ligne-maritime-commerciale-entre-l-algerie-et-le-qatar. Accessed on: 29 March 2023

[13] Derkouche, K., 2022. Bientôt une compagnie maritime privée en Algérie. Tourism&voyages. https://tourismetvoyages.dz/bientot-une-compagnie-maritime-privee-en-algerie/. Accessed on: 29 March 2023

[14] Aris, A., Transport maritime en Algérie : feu vert pour les compagnies privées. algerie-expat. https://www.algerie-expat.com/tourisme-et-voyages/transport-maritime-en-algerie-feu-vert-pour-les-compagnies-privees/15642/. Accessed on: 29 March 2023.

[15] Algerie Ferries., https://www.algerieferries.com/compagnie/presentation. Accessed on: 09 April 2023.

[16] MEZIANE, Y., 2017. Essai d'analyse de l'état des lieux et perspective du transport maritime urbain des voyageurs en Algérie assuré par l'ENTMV. *Master's Thesis : UNIVERSITE MOULOUD MAMMERI DE TIZI-OUZOU.* pp. 1-62.

[17] Office National des Statistiques (ONS)., Bulletins trimestriels des statistiques (2000-2020). https://www.ons.dz/spip.php?rubrique267. Accessed on: 29 March 2023.

[18] Bounabi, S., 2022. Algérie Ferries : Mauvaise gestion et soute à scandales. *REPORTERS, Quotidien National d'Information.* https://www.reporters.dz/algerie-ferries-mauvaise-gestion-et-soute-a-scandales/. Accessed on: 29 March 2023.

Fuzzy Delphi Method (FDM) Analysis On The Significance Of Collaboration Elements In Oil Spill Preparedness And Response

Ismila Che Ishak[1*], Aminuddin Md Arof[2], Md Redzuan Zoolfakar[3], Mohd Fairoz Rozali[4], Ahmad Shahrul Nizam Isha[5], Hayatul Safrah Salleh[6], Muhammad Kasffi Ramli[7]

[1,2,7] Maritime Management Section, Universiti Kuala Lumpur Malaysian Institute of Marine Engineering Technology, 32200 Lumut, Perak, Malaysia
[3] Centre For Industrial Linkages, Universiti Kuala Lumpur 1016 Jalan Sultan Ismail, 50250 Kuala Lumpur, Malaysia
[4]International Maritime Organization, Atase Maritim, High Commission of Malaysia, 45-46 Belgrave Square, London
[5]Management and Humanities Department, Universiti Teknologi Petronas, 32610 Seri Iskandar, Perak, Malaysia
[6]Faculty of Business, Economics, and Social Development, Universiti Malaysia Terengganu, 21030 Kuala Terengganu, Malaysia

*E-mail: ismila@unikl.edu.my

Abstract. Once the oil spills have occurred it has introduced acute effects on the environment, marine life, human life, society, communities, economics, aquatic ecosystems, and shoreline societies. This research analyses the ranking of the significant elements of collaboration that are essential in oil spill preparation and responses. The oil spill statistical incidents reported by the Department of Environment (DOE) from 2014 to 2023 have been analysed. Malaysia has encountered 140 oil spill cases. This research has utilized the questionnaire survey from the Modified Delphi Survey Method which involved the purposive expert respondents from 11 related agencies in Malaysia. The agencies are the Department of Environment (DOE), Marine Department, MMEA, Fishery Department, Meteorology Department, Royal Malaysian Navy, Air Operation Force, PIMMAG, PETRONAS, Fire and Rescue Department, and Marine Park. The data has been examined by Statistical Package for the Social Science (SPSS) version 26.0 analysis tools such as Frequency, Mean Analysis, Standard Deviation, and Cronbach's Alpha gathered from 59 expert respondents. The ranking of the significant elements has been examined utilizing the Fuzzy Delphi Method (FDM). The results show the top five significant elements are: i) Coordinated decision-making at 0.914, ii) Adequate communication among agencies at 0.913, iii) Good response time at 0.912, iv) Good leadership link at 0.908, and v) Large teamwork cooperation at 0.907. Nevertheless, as recommended by the experts the other significant element is efficient budget allocation and suitable equipment which are also essentials to be anticipated.

Keywords: pollution, environmental, fuzzy delphi method, oil spill, collaboration

1. Introduction

The oil spills are generated through several sources such as incidental leakages from vessels and offshore oil platforms and the oil spills regularly findings in excessive economic costs and destructive maritime environmental degradation as mentioned by *Shi, Wang, Luo & Zhang (2019)[17].* The oil spill beginning tankers or other ships, which comprise the carriage of liquid or bulk such as crude oil, fuel oil, or heating oil a substantial source of hydrocarbon in going into the seas, lakes, and rivers as stated by *Doshi, Repo, Heiskanen, Sirviö & Sillanpää (2017) [5].* Once the oil spill incident occurs, it will harm the marine ecosystem and the sea population as agreed by *Ishak, Ishak, Ali & Isha (2020) [9].* Thus, this research goal is to analyse the important ranking of substantial elements for collaboration in oil spill preparation and response in Malaysia. Collaboration is repeatedly assumed to be one approach to effectively sharing insufficient resources while progressing cooperation by establishing inter-organization as suggested by *Thomson, Perry & Miller (2009) [18].* As for the response to oil spill preparation and response activity, collaboration is anticipated the utmost dependent in oil spill preparation and response. The expert respondents of this research comprise eleven related organizations such as the Department of Environment (DOE), Marine Department, MMEA, Fishery Department, Meteorology Department, Royal Malaysian Navy, Air Operation Force, PIMMAG, PETRONAS, Fire and Rescue Department and Marine Park. The result of this research is to be conveyed among the response agencies as a reference and suggestion.

2. Methodology

FDM indicates to the Fuzzy Delphi Method that is labeled as a multi-criteria decision-making method for complex and unstructured problems as proposed by *Saaty Thomas (2008) [12].* This research associations the qualitative methods through the preceding literature review process and the quantitative method involves closed-ended and open-ended questions in the initial Delphi Survey by *Sekaran & Bougie (2016) [13] and Sekaran & Bougie (2013) [14].* The Delphi is identified as a numerous-criteria decision-managerial instrument which has been depleted in all concentrations linked to decision making and it has been broadly selected by students and practitioners by *Vaidya & Kumar (2006) [20].*

2.1 Reliability and Validity Test of The Instruments

This research utilizes reliability and validity tests to test the reliability of the measurement within research, whilst validity refers to the generalizability of the findings by *Hasson & Keeney (2011) [7] and Hasson, Keeney & McKenna (2000) [8].* The content validity has been conducted by selected certified experts and pilot testing. The criterion-related validity has been applied by using a measurement that could distinguish between individual judgments. Both Delphi and Fuzzy must count on the feedback acquired from the experts. Thus, to devise priority scales, the targeted expert respondent has been established among those who have detailed knowledge and a minimum of 5 years of working experience linked to oil spill preparedness and response by *Arof & Nair (2017) [1].*

IOP Conf. Series: Earth and Environmental Science **1461** (2025) 012045 doi:10.1088/1755-1315/1461/1/012045

2.2 Pilot Test

This research involved a pilot test among little research to test research procedures, data gathering methods, sample enrolment policies, and other practices in provision for research analyses [6]. The pilot evaluation was piloted among five expert samples of the related organizations and authorization experts which were from the Marine Department, Department of Environment, PIMMAG, Maritime Academy, and Academician University.

2.3 Delphi Questionnaire Survey

The Delphi questionnaire survey comprises Section A on basic information about respondents or experts and their organization exposure. Section B involves a survey question for 14 elements covering situational analysis, communication, decision-making, leadership, teamwork collaboration, response time, stockpile management, knowledge, expertise, and technology sharing, media management, standard SOPs, regular exercise, government support, suitable equipment used, and weather and meteorological conditions. The targeted experts are invited to convey their opinions on each of the questions.

2.4 Theoretical Framework

The anticipated theoretical framework in this research encompasses independent and dependent variables which have been adapted and edited as shown in Figure 1 by *Crichton, Lauche & Flin (2005). [3].*

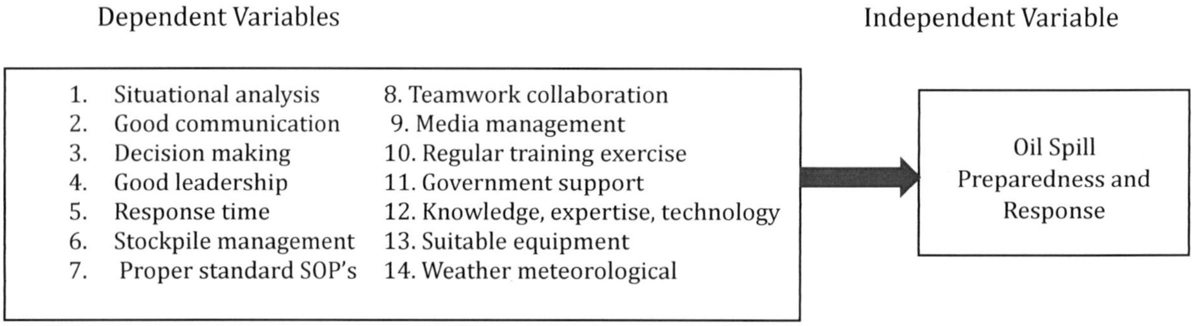

Figure 1. Proposed Summary of Conceptual Theoretical Framework

2.5 Likert Scale

A Likert Scale is utilized as an essential device in attitude, social surveys, and the collection of attitudinal data. The Likert Scales force respondents to select among the 7-point level of agreement for that argument established on their opinion as exhibited in Table 1. To evaluate the respondent's intensity of conformity regarding the statement prepared in the questionnaires utilizing an application of the Likert Scale [4].

Table 1. Likert Scale Score of 7 points [9]

1	2	3	4	5	6	7
Least Important	Less Important	Slightly Important	Moderate Important	Important	Very Important	Most Important

3. Results

3.1 Respondents Details

The questionnaire was conducted online to reach the respondents, followed up by email, phone calls, and WhatsApp messages. The respondent's details show that generally of the respondents were male with 45 respondents or 76.3%. This is due to the biggest gender tasks in the marine arena being dictated by a male as this field demands male staff. The age of 31-40 years old is high among the respondents or 32 respondents or 54.2%. This age is the greatest of the respondents who have finished their studies by receiving bachelor credentials from numerous qualifications to facilitate them to perform in the maritime industry. It indicates the greatest of the respondents have an education at the bachelor level 34 respondents or 57.6%. In addition, the current position shows that most of the respondents were in middle management with 25 respondents or 42.4%. Thus, obtaining a rank in the central organization encourages the staff to have a good quality of working attitude. In addition, the experience of the oil spill in 5 years of 38 respondents, or 64.4% influenced the greatest responses. This is for some of the expert respondents who have more than 5 years of working experience and are exposed to oil spill incidents directly. A length of more than 5 years to 15 working experiences is sufficient to evaluate the expertise of their effort. It is also from working experience that the workers have expanded a lot of experience in operational situations and intersect the benchmark of work corresponding with the salary gained by *Wiesenfeld, Brockner & Trope (2017) [22]*. Lastly, the familiarity of the respondents with the oil spill cases reveals an adequate quantity of 33 respondents or 55.9%. Consequently, it is predicted that this respondent presents honesty and trustworthiness to the data accumulated during the questionnaire distribution.

3.2 Mean Score Interpretation

The average mean score from 14 items is 6.34 and, it is revealed as Very High by an interpretation from *Moidunny (2009) [10]*. The Cronbach's coefficient of alpha internal consistency reliability is applied to measure the consistency of questionnaires as presented in Table 3 by *Ursachi, Horodnic & Zait (2015) [19]; Sekaran & Bougie (2016) [13] and Vijayalaxmi, Anuradha, Sekaran, Meqdad & Kadry 2020) [21]*. Table 3 displays a degree of internal reliability which is valuable for recognizing the degree to which the rankings from a group of judges held simultaneously to evaluate an ordinary attribute by *Osborne & Dillon (2008) [23]* and by *Sekaran & Bougie (2016)* [13]. Hence, Table 5 shows a coefficient alpha of 0.937 from a maximum of 1 simply indicates the scale scores attained from the expert respondents are realistically reliable by *Sekaran & Bougie (2016)* [13]; *Sekaran & Bougie (2013) [14], and Green & Salkind (2007) [24]*. It displays that Cronbach's alpha reliability statistic of the significant elements for the success of the coordination efforts for the oil spill response is at 0.937 from the 14 elements. Table 6 shows the mean, median, and standard deviation (SD) results for 14 significant elements. The average Mean result ranges from 5.80 to 6.54. Meanwhile, the average Median is 6.34 ranging from 6 to 7, and the SD ranges from 0.626 to 1.047. All 59 items are valid and sufficient for data analysis. The SD for the judgment on element factor 9 of the promotion of oil spills through media is at 1.047 which is high and has exceeded 1.0 points on a 7-point Likert scale.

Table 2. Mean Score Interpretation [10]

Mean Score	Interpretation
1.00-1.80	Very Low
1.81-2.60	Low
2.61-3.20	Medium
3.21-4.20	High
4.21-5.00	Very High

Table 3. Cronbach Alpha Interpretation [10]

Cronbach Alpha	Internal Consistency
$\alpha \geq 0.9$	Excellent
$0.9 > \alpha \geq 0.8$	Good
$0.8 > \alpha \geq 0.7$	Acceptable
$0.7 > \alpha \geq 0.6$	Questionable
$0.6 > \alpha \geq 0.5$	Poor
$0.5 > \alpha$	Unacceptable

Table 4. Case Processing Summary (Source: SPSS 26.0)

		N	%
Cases	Valid	59	100.0
	Excluded	0	0.0
	Total	59	100.0

a. Listwise deletion based on all variables in the procedure.

Table 5. Reliability Statistics (Source: SPSS 26.0)

Cronbach's Alpha	N of Items
.937	14

Table 6. Mean, Median, and Standard Deviation (Source: SPSS 26.0)

N	Valid	Element 1	Element 2	Element 3	Element 4	Element 5	Element 6	Element 7
Mean		6.34	6.51	6.54	6.51	6.51	6.29	6.37
Median		6	7	7	7	7	6	6
Std Deviation		0.734	0.626	0.678	0.751	0.653	0.696	0.717
N	**Valid**	**Element 8**	**Element 9**	**Element 10**	**Element 11**	**Element 12**	**Element 13**	**Element 14**
Mean		6.49	5.80	6.22	6.41	6.31	6.44	6.10
Median		7	6	6	7	7	7	6
Std Deviation		0.774	**1.047**	0.744	0.698	0.836	.702	.977

3.3 Fuzzy Delphi Method (FDM) Analysis

There are nine steps concerned in applying FDM by *Ridhuan &Nurul Rabiah (2021) [11].*

a) Step 1: Determine the Research Title and Research Construct.

b) Step 2: Determine the Experts' Respondent: There are several suggestions for the sample size of the respondents involved in Fuzzy. The amount of experts in the Delphi method is 10–15, if there is high uniformity among experts by *Adler & Ziglio (1996) [25].* Meanwhile, the assortment of experts is 10-50 by *Jones & Twiss (1978) [26].* This research utilized respondents 53 who were taken from 11 different organizations.

c) Step 3: Determination of the Fuzzy Likert Scale: This research has applied the Likert scale of 7 points for Fuzzy decision-making (FDM) to produce concise and correct fuzzy score values. The anticipated highest level of agreement such as Likert scale point no 7 is most important to the lowest level of agreement as Likert scale no 1 respectively.

IOP Conf. Series: Earth and Environmental Science **1461** (2025) 012045 doi:10.1088/1755-1315/1461/1/012045

d) Step 4: The Questionnaire Distribution to the Experts and Process of Entering the Likert Scale Value into Item Space for Each Expert: This step involves the converting process from the Likert scale to the fuzzy scale.
e) Step 5: Process of Determining the Value Of 'D' (Threshold Value): These values show the consensus among the experts. The table also shows whether to accept all three items or not. This is according to the law of the Fuzzy Delphi Method (FDM) for Threshold value, (d) is equal to 0.2 or less.
f) Step 6: The Process of Determining the Number of Experts in the Red Box Below According to The Number Of Experts In The Research.
g) Step 7: The Proses of Determining the Percentage of Consensus for Each Element and Overall Elements: This is the second condition involved in the FDMv20 analysis. The required consensus percentage that is to be accepted is 75.00% or more by *Chu & Hwang (2008) [2]. The* other suggested consensus among the experts must be the same or more than 75.00% by *Murry &Hammons (1995) [27].*
h) Step 8: Fuzzy Evaluation Process - The Process of Determining Score (Item Position / Priority).
i) Step 9: Results of Analysis of Research Data Findings: Those experts' consensus of 75% and above are accepted, and those below 75% are rejected by *Chu & Hwang (2008)* [2]; by *Murry &Hammons (1995)* [27], and by *Subhashini, Li, Zhang & Atukorale (2022) [16].*

Table 7. The Expert Consensus Percentage

Link Of System	
Percentage	Results
0%	Reject
75%	Accept

4. Conclusion

Table 8 shows the rank of significant elements of oil spill preparation and response. On top of the top five ranking of the significant elements that need to be highlighted, the experts also suggested significant elements of efficient budget allocation and equipment. This could assist the related agencies in establishing what significant elements need to be prioritized. The result shows that the top five ranking needs in significant collaboration during oil spill preparedness and response are: i) Coordinated decision-making by 0.914, ii) Good response time by 0.912, iii) Good leadership links by 0.908, iv) Adequate communication among agencies by 0.913, and v) Large teamwork cooperation by 0.907. It is believed that a consolidative coordination range thru a serious core on consistent multi-stakeholder incident organization cooperation is substantial as agreed by *Sabela-Rikhotso, Van Niekerk & Nemakonde, (2022) [28].* In addition, a good formation among local municipalities has mainstreamed the oil spill response collaboration including sufficient budgets for inter-organizational planning and preparedness as stated by *Sabela-Rikhotso, Van Niekerk & Nemakonde, (2022) [28].*

Table 8. The Summary of Expert's Consensus for 14 Significant Elements

No	Significant Element (SE)	Condition of Triangular Fuzzy Numbers		Condition of Fuzzy Evaluation Process				Experts consensus	Accepted Element	Ranking
		Threshold value, d	Experts Consensus Percentage d ≤ 0.2r, %	m1	m2	m3	Fuzzy score (A)			
1	Adequate situational analysis	0.110	98.3%	0.768	0.920	0.985	0.891	0.9	Agree	9
2	**Adequate communication among agencies**	0.128	93.2%	0.802	0.944	0.993	**0.913**	0.9	Agree	**2**
3	**Coordinated decision making**	0.132	93.2%	0.808	0.944	0.990	**0.914**	0.9	Agree	**1**
4	**Good leadership links**	0.094	93.22%	0.802	0.937	0.985	**0.908**	0.9	Agree	**4**
5	**Good response time**	0.126	89.83%	0.802	0.942	0.992	**0.912**	0.9	Agree	**3**
6	Stockpile management support at the primary period	0.139	89.83%	0.758	0.915	0.986	0.886	0.9	Agree	10
7	Harmonization of managerial procedures among agencies	0.102	89.83%	0.775	0.925	0.986	0.895	0.9	Agree	7
8	**Large teamwork cooperation**	0.137	88.14%	0.798	0.937	0.985	**0.907**	0.9	Agree	**5**
9	Promotion of oil spill preparation and response	0.152	88.14%	0.659	0.832	0.937	0.810	0.9	Agree	14
10	Oil spill preparation and response efficiency	0.114	88.14%	0.744	0.903	0.981	0.876	0.9	Agree	12
11	Government agreements to support oil spill issue	0.133	89.83%	0.781	0.929	0.988	0.899	0.9	Agree	8
12	Oil spill preparedness and response service quality	0.138	91.53%	0.761	0.908	0.976	0.882	0.9	Agree	11
13	Suitable equipment in responses	0.098	91.53%	0.788	0.932	0.988	0.903	0.9	Agree	6
14	Weather & meteorological conditions	0.131	93.22%	0.720	0.878	0.959	0.853	0.9	Agree	13

Acknowledgment

This research is conducted for University Kuala Lumpur Malaysia Malaysian Institute of Marine Engineering Technology with support from the UNIKL academic staff, and other academic staff from outside establishments collected with additional associated respondents from numerous organizations. With no reliable commitment and assistance from the groups, it is challenging for the research to be completed on schedule. This research monetarily attained a grant from the

Ministry of Higher Education Malaysia through the Fundamental Research Grant Scheme (FRGS) with grant number FRGS/1/2020/SSI03/ UNIKL/02/1.

References

[1] Arof A M. and Nair R. (2017). The identification of key success factors for interstate Ro-Ro short sea shipping in Brunei-Indonesia-Malaysia-Philippines: a Delphi approach. International Journal of Shipping and Transport Logistics. no. 9 (3): 261-279. https://doi.org/10.1504/IJSTL.2017.083470

[2] Chu H C. and Hwang G J. (2008). A Delphi-based approach to developing expert systems with the cooperation of multiple experts. Expert systems with applications, no. 34(4): 2826-2840. *https://doi.org/10.1016/j.eswa.2007.05.034*

[3] Crichton M T., Lauche K. and Flin R. (2005). Incident Command Skills in the Management of an Oil Industry Drilling Incident: a Case Research, no. 13(3):116–128. https://doi.org/10.1111/j.1468-5973.2005.00466.x

[4] Dittrich R., Francis B., Hatzinger R. and Katzenbeisser W. (2016). A paired comparison approach for the analysis of sets of Likert-scale responses. Statistical Modelling. https://doi.org/10.1177/1471082X0600700102

[5] Doshi B., Repo E., Heiskanen J P., Sirviö J A. and Sillanpää M. (2017). Effectiveness of N, O-carboxymethyl chitosan on destabilization of Marine Diesel, Diesel, and Marine-2T oil for oil spill treatment. Carbohydrate polymers, no. 167: 326-336. https://doi.org/10.1016/j.carbpol.2017.03.064

[6] Hassan Z A., Schattner P. and Mazza D. (2006). Doing pilot research: why is it essential? Malaysian family physician: the official journal of the Academy of Family Physicians of Malaysia, no.1(2-3): 70.

[7] Hasson F. and Keeney S. (2011). Enhancing rigor in Delphi technique research. Technological Forecasting and Social Change, no. 78(9): 1695-1704. https://doi.org/10.1016/j.techfore.2011.04.005

[8] Hasson F., Keeney S. and McKenna H. (2000). Research guidelines for the Delphi survey technique. Journal of Advanced Nursing, no. 32(4): 1008-1015. https://doi.org/10.1046/j.1365-2648.2000.t01-1-01567.x

[9] Ishak I C., Ishak N A. L., Ali N M. and Isha A S N. (2020). Research on preparedness and response to oil spill. In Journal of Physics: Conference Series. vol.1529, no. 3: 032088. IOP Publishing. https://doi.org/10.47750/pnr.2022.13.S10.621

[10] Moidunny K. (2009). The effectiveness of the National Professional Qualifications For Educational Leader (NPQEL) (Unpublished doctoral dissertation), Bangi: The National University of Malaysia. https://doi.org/10.1016/j.techfore.2011.04.005

[11] Ridhuan M J, Nurul Rabiah M N. (2021). Kepelbagaian Methodologi Dalam Penyelidikan Reka Bentuk dan Pembangunan. Cetakan Ketiga. Qaisar Prestige Resources. ISBN 978-967-18106.

[12] Saaty Thomas L. (2008). How to make and justify a decision: The Analytic Hierarchy Process, System Research, and Information Technologies, Issue 1, 95-108.29.

[13] Sekaran U. and Bougie R. (2016). Research methods for business: A skill-building approach. John Wiley & Sons.

[14] Sekaran U. and Bougie. R. (2013). Research Methods for Business: A Skill-Building Approach.

[15] Shi X., Wang Y., Luo M., and Zhang C. (2019). Assessing the feasibility of marine oil spill contingency plans from an information perspective. Safety Science, no.112: 38-47. https://doi.org/10.1016/j.ssci.2018.09.014

[16] Subhashini L D., Li Y, Zhang J. and Atukorale A S. (2022). Integration of fuzzy logic and a convolutional neural network in three-way decision-making. Expert Systems with Applications, no.202:117103. https://doi.org/10.1016/j.eswa.2022.117103
[17] Shi X, Wang Y, Luo M. and Zhang C. (2019). Assessing the feasibility of marine oil spill contingency plans from an information perspective. Safety Science, 112 (December 2016), 38–47. https://doi.org/10.1016/j.ssci.2018.09.014
[18] Thomson A M, Perry J L. and Miller T K. (2009). Conceptualizing and measuring collaboration. Journal of Public Administration Research and Theory, no. 19(1): 23-56. https://doi.org/10.1093/jopart/mum036
[19] Ursachi G, Horodnic I A. and Zait A. (2015). How reliable are measurement scales? External factors with indirect influence on reliability estimators. Procedia Economics and Finance, no. 20: 679-686. https://doi.org/10.1016/S2212-5671(15)00123-9
[20] Vaidya O S. and Kumar S. (2006). Analytic hierarchy process: An overview of applications. European Journal of Operational Research, no.169(1): 1-29. https://doi.org/10.1016/j.ejor.2004.04.028
[21] Vijayalaxmi B, Anuradha C, Sekaran K, Meqdad M N. and Kadry S. (2020). Image processing-based eye detection methods a theoretical review. Bulletin of Electrical Engineering and Informatics, no. 9(3):1189-1197. https://doi.org/10.11591/eei.v9i3.1783
[22] Wiesenfeld B M, Reyt J N, Brockner J. and Trope Y. (2017). Construal level theory in organizational research. Annual Review of Organizational Psychology and Organizational Behaviour, no.4: 367-400. https://doi.org/10.1146/annurev-orgpsych-032516-113115
[23] Osborne J. and Dillon J. (2008). Science education in Europe: Critical reflections. Vol.13. London: The Nuffield Foundation. https://efepereth.wdfiles.com/local--files/science-education/Sci_Ed_in_Europe_Report_Final.pdf
[24] Green B S. and Salkind J N. (2007). Using SPSS for Windows and Macintosh (5th ed.). New Jersey, Prentice Hall. https://dergipark.org.tr/en/pub/ekuad/issue/74727/1227281
[25] Adler M. and Ziglio E. (1996). Gazing into the oracle: The Delphi method and its application to social policy and public health. Jessica Kingsley Publishers.
[26] Jones H. and Twiss B C. (1978). Forecasting technology for planning decisions. 658.4 J6. https://cir.nii.ac.jp/crid/1130282270039005440
[27] Murry Jr J. W. and Hammons J. O. (1995). Delphi: A versatile methodology for conducting qualitative research. The review of higher education, no. 18(4): 423-436. https://doi.org/10.1353/rhe.1995.0008
[28] Sabela-Rikhotso, P. T. Z., van Niekerk, D., & Nemakonde, L. D. (2022). A conceptual model for marine oil spill management in South Africa. *Disaster Prevention and Management: An International Journal, 31*(4), 457-474. https://doi.org/10.1108/DPM-08-2021-0241

Implementing Buffer Zones Around Nusa Penida Island in the Lombok Strait to Enhance Maritime Safety & Environmental Protection

Joe Ronald Kurniawan Bokau[1*], Park Young-Soo[2] and Kim Dae-Won[2]

[1, 2] Department of Navigation Sciences, National Korea Maritime and Ocean University, Busan, South Korea

*E-mail: joeronald.kb@g.kmou.ac.kr

Abstract. Lombok Strait connects Indian Ocean with Bali Sea and Makassar Strait as an adjoining of archipelagic sea lanes and becoming one of the chokepoints in the maritime supply chain. As one of the most important alternative routes for seaborne trades, the trade volume by weight and by value in the Lombok and Makassar Strait will increase if the flow through the Malacca Strait or Sunda Strait in closure. Lombok Strait also one of the major routes in Asia-Pacific region beside Alas Strait, connecting Australia to Asia. As an example, 82% of iron ore exported to China came from Australia and most of the port located in the western part of Australia where Port Hedland as the major port for exporting natural resources such as Iron Ore to the world, exported 474 million iron ore in 2,716 ships to China in 2023 alone. Nusa Penida Island is one of the areas within Coral Triangle in the Lombok Strait area that is expected to be designated as Particularly Sensitive Sea Areas (PSSA) by Indonesian Government. Through this study, we examine the relation of the marine traffic in the vicinity since the area is prone to adverse impacts emerging from an increase volume of shipping as of 2023, there were 15,099 passing ships and 21,547 crossing ships in the Lombok Strait area through the Traffic Separation Scheme (TSS). This study used AIS data and spatial approach according to PIANC, CESMA and IMO Ship Manoeuvring Standard mimicking turning circle of the model ship according to actual manoeuvring table obtained to determine "buffer zone" that consist of safety zone, reservation area, safety margin and manoeuvring margin. In terms of safe distances and vulnerable area related to nearest shore was observed 2.3km with the closest distance to the to the core zone was 1,2km. This study makes a valuable contribution to the subject of maritime traffic management by providing a realistic method to enhancing safety and environmental protection in accord with IMO Resolution A.982(24). The findings offer useful perspectives for maritime authorities and policymakers in developing and executing safe TSS, guaranteeing the most effective management of high-traffic maritime routes.

Keywords: Infrastructure and sustainability, environmental risk assessment, maritime safety, marine traffic, particularly sensitive sea areas (PSSA), Lombok strait, environmental protection

1. Introduction

Lombok Strait is a busy waterway with high shipping traffic density, ranking as the third highest in Indonesia [1]. The implementation of the Traffic Separation Scheme (TSS) in the Lombok Strait has been pivotal in enhancing maritime safety and managing traffic in the region [2]. However, the Malacca Strait, another critical waterway in the region, faces challenges related to navigation safety and collision risks due to high-density traffic [3]. Similarly, the Istanbul Strait, known for its narrowness and heavy international shipping usage, poses environmental challenges due to maritime traffic [4]. Recent accident in Suez Canal and threat of security around Hormuz made this chokepoint become so important in case similar thing happen in Malacca Strait or Sunda Strait where most of the traffic will mostly opted Lombok Strait [5]. This raise of concern and possibilities needs to be addressed by carrying out comprehensive assessment in the area related to increasing traffic and the impact to the regional environment.

Impact on the regional marine environment extends to marine radioecology, with studies focusing on the distribution of radionuclides in the Lombok Strait where it served as an outflow and crucial pathway for water masses from the Pacific Ocean to the Indian Ocean through the ITF (Indonesian Through Flow) and plays a vital role in the broader oceanic system [6]. This strait is also essential for interocean water exchange in the Indonesian Seas [7]. The ITF significantly impacts the global climate system by influencing heat and freshwater transport between the oceans [8]. Hence due to its nature, Lombok Strait has become one of the highest difficulties in ship handling that ship master should concentrate on, particularly when leaving and arriving at the port in Padang Bai area [9]. Moreover, Lombok Strait is one of the Indonesia Archipelagic Sea Lane (IASL) passageways, further highlighting its importance in maritime transportation routes [10]. With 15,136 passing ships and 21,603 crossing ships in the area alone on 2023, the danger of navigation to the surrounding environment become profoundly imminent.

Therefore, Government of Indonesia through Ministry of Transport and Directorate General of Sea Transportation as the recognized administration by International Maritime Organization (IMO) proposes Nusa Penida Island and Gili Island as Particularly Sensitive Sea Area (PSSA). Figure 1 is the location of the Nusa Penida island and its current Marine Protected Area and also its location in the proximity of the Lombok Strait where the marine traffic is heavily concentrated along the Traffic Separation Schemes (TSS).

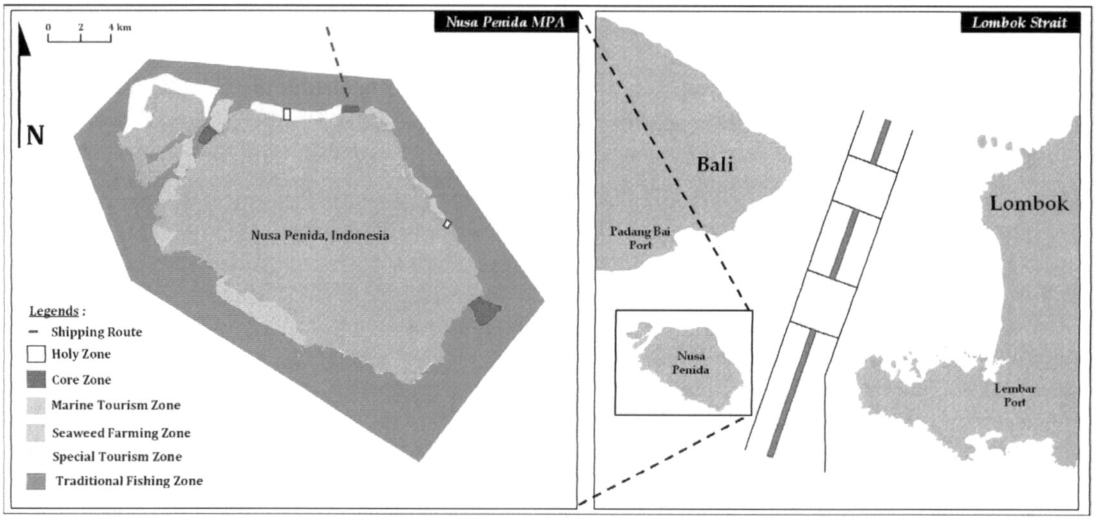

Figure 1. The location of Nusa Penida Island in the proximity of Lombok Strait with Traffic Separation Scheme (TSS) with the Marine Protected Area (MPA) zones [11].

The increase of traffic in the vicinity will creates threat in the future, therefore in this study we will examine the latest traffic and searching for any "rogue ship" where it violates MPA. The new routing measures according to the finding also discussed with the creation of buffer zone in the proximity of the island for the future protection of sensitive area.

2. Literature Review

The complexity of marine traffic, especially in busy waterways, increases the risk of accidents, collisions, and oil spills, further threatening marine environments [12, 13]. This busy waterway causes significant environmental impacts, including air and water pollution, disturbances in marine ecosystems, and air quality issues in port areas and coastal regions, posing adverse effects on human health and the environment [14]. The discharge of pollutants from marine traffic, such as oil spills, ballast water discharges, and antifouling paints, can also lead to marine pollution and ecosystem degradation [15, 16]. These pollutants can harm marine organisms, disrupt food chains, and degrade water quality, impacting biodiversity and ecosystem health. Furthermore, the risk of ship strikes on marine mammals, such as whales, poses a significant threat to these species' survival [17]. Efforts to mitigate the environmental impacts of marine traffic include the development of emission reduction technologies, implementation of traffic separation schemes, and monitoring of shipping activities were undertaken [18,19]. Strategies to reduce air and water pollution from ships, enhance navigation safety, and protect marine ecosystems are essential for promoting sustainable maritime practices. Additionally, the use of remote sensing technologies for maritime surveillance can aid in monitoring and managing marine pollution, ensuring the safety and sustainability of marine environments [20].

Particularly Sensitive Sea Areas (PSSAs) are crucial in marine governance to protect ecosystems and resources from international shipping activities. Designating these regions allows for specific protective measures and regulations, with the International Maritime Organization (IMO) emphasizing resilience-based governance principles in managing these sensitive sea areas [21]. UNCLOS Article 192 reads: "States have the obligation to protect and preserve the marine environment", therefore Indonesia have clear foundation to implement such preventative measure. The sensitivity of marine ecosystems to environmental changes, such as sea ice variations, underscores the significance of core zones within Particularly Sensitive Sea Areas [22]. Understanding core zones within PSSAs is crucial for assessing climate change and human activities' impact on marine environments as seen in the result of research in similar area show high sensitivity to climate variations, requiring conservation and management strategies while underwater noise pollution also threatens marine species' communication, navigation and feeding behaviours [23,24]. Another study highlighted the construction and operation of offshore wind farms can increase ship traffic, potentially affecting the vocalizations and habitats of marine mammals such as the Taiwanese humpback dolphin [25]. To address these concerns, it is vital to establish effective buffer zones that consider the needs of protected species and minimize disturbances. Some researcher did propose buffer zones, which are responsible for maintaining and protecting core areas within national park ecosystems in the land and concluded that this buffer zone play a critical role in preserving biodiversity and ecological balance [26]. A so-called "transition zones" have been proposed as a management approach to reduce the environmental impact of global shipping traffic on marine ecosystems [27,28].

The Confederation of European Shipmasters' Associations (CESMA) collaborates with stakeholders, policymakers, and researchers to optimize spatial planning for offshore wind farms and also promotes best practices in impact assessment, mitigation, and compensation to

IOP Conf. Series: Earth and Environmental Science **1461** (2025) 012046 doi:10.1088/1755-1315/1461/1/012046

minimize environmental impacts and ensure the long-term sustainability of offshore wind farm operations aligning with the objectives of the Maritime Spatial Planning Directive 2014/89/EU (MSPD) at the EU level [29,30,31]. These best practices include spatial planning whenever country decide to develop Offshore Wind Farm (OWF) with adding zone called "buffer zone", that consist of safety zone, reservation area and safety margin. By creating this safe boundary zones around maritime routes, the spread of shipping-related impacts can be controlled, thereby decreasing risks to marine species and ecosystems. These zones act as a protective barrier, minimizing the direct effects of shipping activities on vulnerable marine habitats and species, aligning with conservation and sustainability goals. By doing similar approach in the research area, there will be a safety guard to the policy-maker regarding the PSSA proposal.

3. Methodology

Since 2004, the International Maritime Organization (IMO) has mandated the use of the Automatic Identification System (AIS) for maritime safety. This study uses AIS data from Benoa Vessel Traffic Services (VTS) for high-detailed ship movement recording, with an average time of 6 seconds for moving ships compared to 10 seconds for 0-14 knots ships [32]. The research flow in Figure 2.

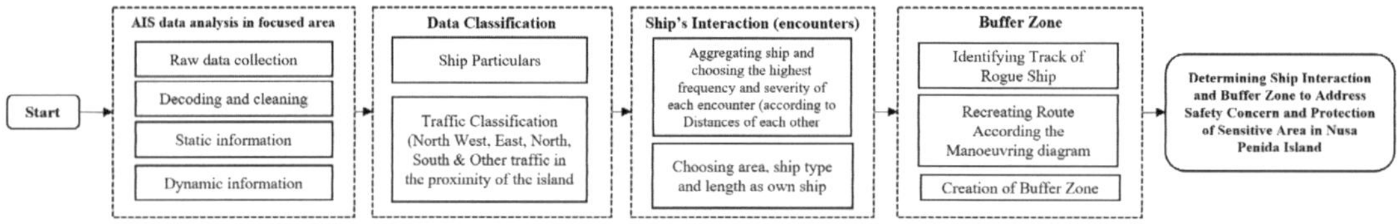

Figure 2. Research Flow.

AIS data enabled to model ship density, track vessel/ship movements, and simulate marine traffic scenarios with a high degree of accuracy [33,34]. Using 7 days of AIS data [35,36] retrieved from the Benoa VTS, the vessel track movement (dynamic data) in Figure 3 shows the example of raw AIS data visualized using Geographic Information System (GIS) and the vessel movement in

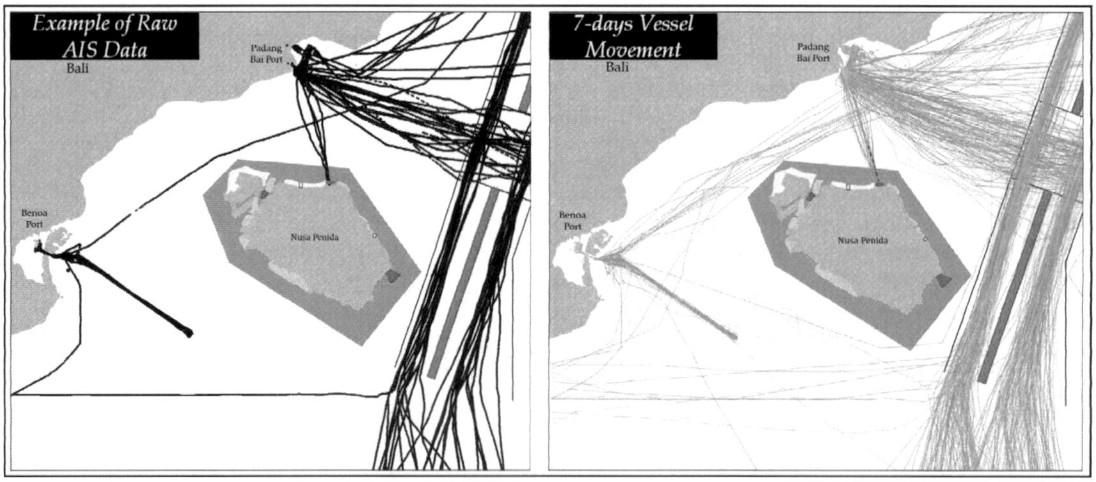

Figure 3. The visualization of AIS Data in GIS of the research area.

the proximity of the Nusa Penida Island. This AIS data were further analyzed to obtain the detail of the traffic such as ship name and Maritime Mobile Service Identity (MMSI) and retrieve each of their particulars (static data). Buffer zone was determined according to CESMA in OWF recommendation and the bulk carrier was chosen as model ship of the rogue ship[37]. This rogue ship was based on assumption that violators (rogue) constitute about 10% of the traffic [38].

4. Result & Discussion

The result were systematically arranged according to the Annex of IMO Resolution A.982(24)-5- where it stated that one of the consideration factors prior to designation of PSSA is 'vulnerability to impacts from international shipping' with vessel traffic characteristics that consisting of:

4.1 Operational factors

Maritime activities in the proximity of the Nusa Penida island according to the AIS data derived from the analysis were mostly consisting of passing by traffic. We divided based on the region with the visualization depicted in the Figure 4 below:

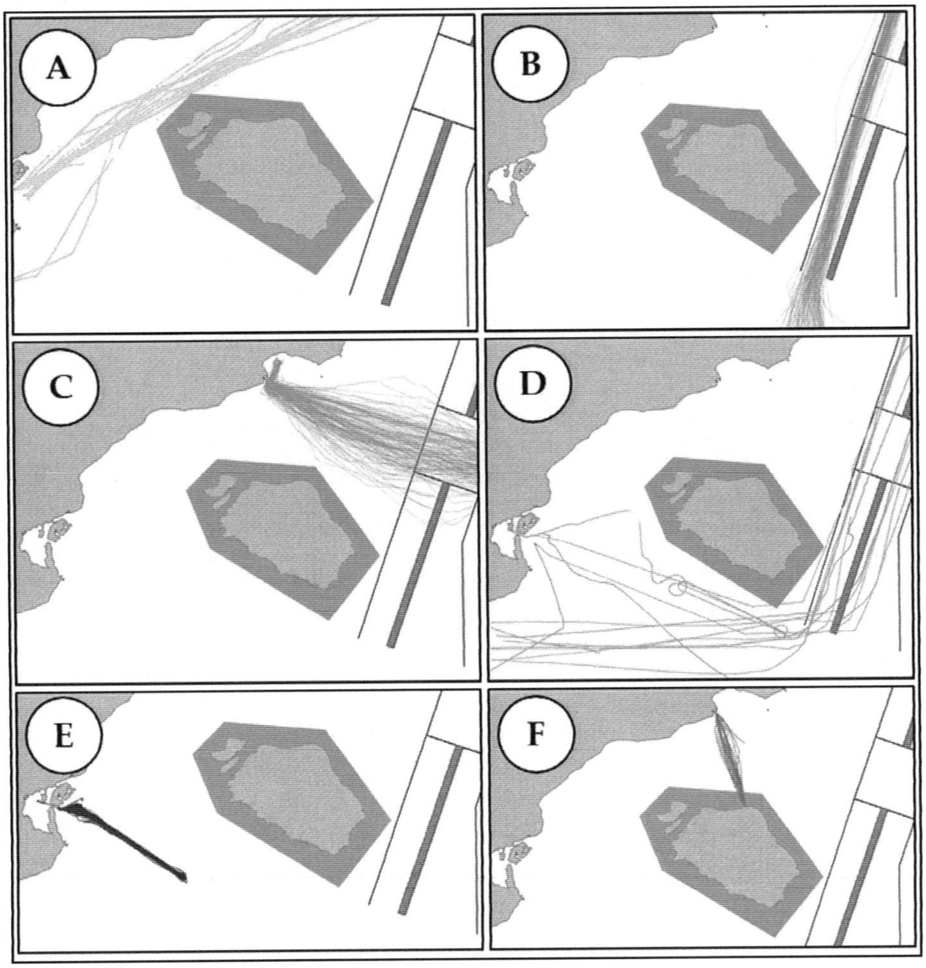

Figure 4. North west inshore traffic (A), east traffic inside TSS which is southbound traffic (B), north traffic which is crossing TSS traffic (C), south traffic which is end of the TSS direct to or from Benoa or southern Bali (D), offshore Benoa Port traffic (E), and Nusa Penida ferries (F).

The marine traffic visualized above only consists of ships that equipped with AIS. No AIS data retrieved from small fishing boats, pleasure craft along the coastal area from Benoa to Sanur in Bali Island and fast boat serving between Sanur and Nusa Penida. A Floating Storage Regasification Unit (FSRU) docked in Benoa Port serves ship-to-ship (STS) transfer of LNG to another ship (SS. Triputra). Two ships (motor hopper) frequently discharge sediments from dredging works in Benoa Port (Figure 3E).

4.2 Ship types

The types of ship in the passing through or adjacent to the Nusa Penida Island according to 7-days AIS data analysis given in the Table 1 as follow.

Table 1. Ship types.

Description	Total traffic (7-day) [a]	Total ships [b]	Ship type [c]
North West Traffic	17	12	9 Oil Tanker, 3 Passenger Ship
East Traffic (TSS southbound)	108	108	88 Bulk Carrier, 3 General Cargo, 2 Offshore Support Vessel, 7 LNG Tanker, 1 LPG Tanker, 7 Oil Tanker
East Traffic (TSS northbound)	132	128	111 Bulk Carrier, 4 General Cargo, 1 Dredger, 5 LNG Tanker, 2 LPG Tanker, 5 Oil Tanker
North Traffic (crossing TSS)	104	18	17 Ro-ro Ferries, 1 LPG Tanker
South Traffic	12	12	5 Oil Tanker, 1 LNG Tanker, 2 Bulk Carrier, 1 General Cargo, 3 Fishing Vessel,
Offshore Benoa Port	12	2	2 Motor Hopper / Cargo
Nusa Penida Ferries	7	1	1 Ro-ro Ferries

[a] Traffic means number of ships observed during 24 hours (1-day) for 7-days

[b] Ships means number of each particular ship in the area (for example 1 ship only serving Nusa Penida Ferries route in 7 days)

[c] Ship type classified based on AIS data & ship database

Overall, there were 281 ships that recorded in the AIS data in the surrounding waters of the Nusa Penida Island, 72% is bulk carrier, 9% is oil tanker, 7% is ro-ro/passenger, 6% is LNG/LPG tanker, 3% is general cargo ships, 2% is other ships (dredger, OSV, motor hopper) and 1% is Fishing Vessel. Among the 240 ships on the east traffic (TSS southbund and northbound), 199 of them are Bulk Carrier with average length of 285 meters and 89,000 GT, therefore the closest is Cape Size Bulk Carrier which will be the base of our model. The author also had years of experience navigating similar ship in the past and the manoeuvrings diagram was taken from the particular ship.

4.3 Traffic characteristic

The volume and concentration of the marine traffic relatively adequate since the width of both TSS lane equals to 5.40nm – separation zone (0.50nm) / 2 = 2.45nm (4,537m) per lane. This value is still within the IMO recommendation (routes should be designated based on the number of vessels passing annually through the target route considering the maximum length of the vessels that pass through the route). The current or forthcoming largest vessel, which is currently 400 m long against the current width of the lane and number of ships that using the lane annually still in adequate numbers and with the current status able to accommodating up to 18,000 ships annually. Table 2 is the data derived from the Benoa VTS to grasp a better understanding of the Lombok Strait marine traffic.

Table 2. Annual traffic in the Lombok Strait.

Classification	Description	2021	2022	2023
Traffic	Passing	14,965	15,092	15,136
	Crossing	14,649	18,302	21,603
	Total	29,614	33,394	36,739
	Annual Increase %age	-	12.7%	10.01%
Type	Cargo Ship	11,411 (38.53%)	11,228 (33.62)	11,469 (31.22%)
	Passenger/Ferry	13,649 (46.09%)	17,255 (51.67%)	20,652 (56.21%)
	Tankers	3,427 (11.57%)	3,542 (10.61%)	3,497 (9.52%)
	Others	1,127 (3.81%)	1,369 (4.10%)	1,121 (3.05%)
Flag	Indonesian	17,510 (59.20%)	21,385 (63.64%)	24,313 (66.18%)
	Foreign	12,070 (40.80%)	12,213 (36.35%)	12,426 (33.82%)

Sources: Annual Traffic according to Benoa VTS (https://app.disnavbenoa.id/vtsdashboard)

From the Table 2 data, indicates that there has been a noticeable increase in maritime traffic over three years (2021 to 2023). Total marine traffic rose with 12.7% increase in 2022 and an 10.01% increase in 2023. Passenger/Ferry ships saw the most significant growth, increasing from 46.09% of total traffic in 2021 to 56.21% in 2023 or 6,913 traffic alone, this also indicates crossing traffic become denser over the year. Cargo ships remained relatively stable, accounting for approximately 31-38% of traffic, while tankers and other ships saw slight decreases in their proportion of the total. Additionally, the proportion of ships flying the Indonesian flag increased significantly, from 59.2% in 2021 to 66.17% in 2023. The share of foreign-flagged ships decreased from 40.8% to 33.82% over the same period and this shift indicates a growing dominance of Indonesian-flagged vessels in the region.

IMO A 24/Res.982 only stated the vulnerability to impacts from "international shipping", therefore, we will focus on the international shipping traffic located nearest to the core zone which is the southbound TSS traffic depicted in Figure 3B. In the observed ship type data (Table 1), the majority of the vessels were Bulk Carriers, making up 88 of the totals, which represents a significant 81.48% of the fleet. This was followed by Oil Tankers and LNG Tankers, each with 7 vessels, accounting for 6.48% of the total fleet respectively. General Cargo ships were less common, with only 3 vessels, representing 2.77% of the fleet. Offshore Support Vessels made up an even smaller proportion, with just 2 vessels or 1.85%, and LPG Tankers were the least represented, with only 1 vessel, making up 0.92% of the fleet. This distribution highlights the dominance of Bulk Carriers in the fleet composition in the area, with other vessel types comprising a much smaller share.

4.4 Ship interaction – vulnerable area

The ship interaction within the TSS is primarily overtaking, with one ship overtaking another. However, as ships approach the precautionary area, the situation changes to crossing. The number of crossing ships is expected to increase due to an increase in passenger/ferry traffic. Table 2 and Table 1 show similar numbers of crossing ships and ferries in the northern part of Nusa Penida island (southbound TSS traffic and crossing traffic in northern part of the island) with the visualization in the Figure 5.

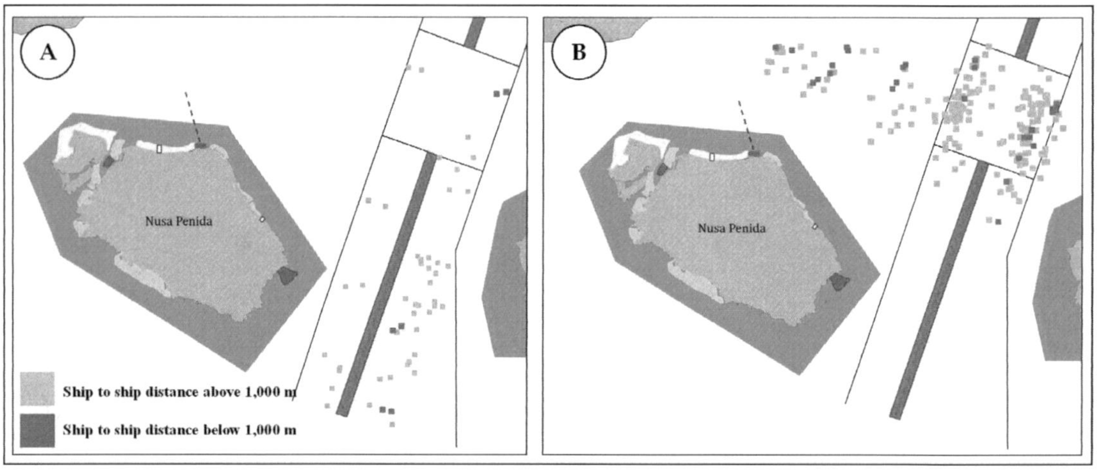

Figure 5. Ship interaction in the TSS related to International Shipping inside TSS (A); Ship interaction related to the crossing ships involving ferries (B).

From the AIS data analysis revealed that there were 213 encounter situations between ships (distances of 0m-3,500m), with 70 overtaking and 143 crossing. The total traffic was 344 (Table 1 - northbound TSS, southbound TSS, and crossing), indicating a 61% chance of encounters during 7-day observation. The densest area for encounters was in the precautionary area, with the nearest distance involving ships observed in crossing situation between ferry and another ferry at 280m while there was a Bulk Carrier also in crossing with ferries with 570m distances. The higher number of ships encounters also observed in the northbound ships.

4.5 Distance offshore or other dangers to navigation

The distance of the traffic from the nearest shore in the area ranging from 2,300 to 12,000 meters. The nearest distances were the north west traffic but since the core zone in a very close proximity to the southbound TSS, we will further examine towards the creation of 'buffer zone' according to

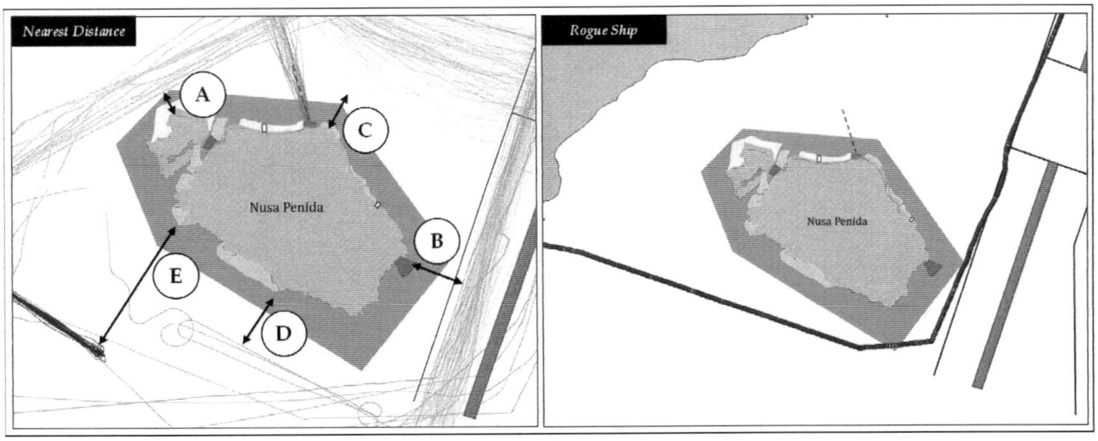

Figure 6. Left figure - The nearest ship from North west inshore traffic is 2,3km (A), east traffic inside TSS which is southbound traffic is 4.5km (B), north traffic which is crossing TSS traffic is 3.5km (C), south traffic which is end of the TSS direct to or from Benoa or southern Bali is 4.5km (D), offshore Benoa Port traffic is 12km (E). Right figure – track of rogue ship observed went outside TSS and in a close proximity to the core zone (1,2km).

the southbound traffic and the rogue ship. There is one rogue ship observed being in close proximity to the core zone with only 1.2km distances as seen in the Figure 6.

4.6 Buffer Zone in the Core Zone

As stated in the IMO Resolution A.982(24), associated protective measures and how it directly contributes to the adequate protection of the core area will be delivered in this section. The best practices derived from CESMA in spatial planning of the OWF is taken. According to CESMA, buffer zone derived from PIANC (World Association for Waterborne Transport Infrastructure) where they define this zone as total distance from the edge of ship fairway (or in this case the outer line of the TSS) to an offshore installation implemented to ensure navigational safety and it may vary along the length of a shipping route. It is fixed based on the width of safety margin, reservation area and safety zone. In this research we consider the outermost island in Nusa Penida as the limit of the buffer zone to the TSS boundary limit.

Safety zone in the offshore installation translated to 500m radius around the offshore installations as mandated by UNCLOS. Reservation area is an area reserved for the future use by shipping or future installation. Safety margin means an area reserved solely for the ship manoeuvre particulary in the case of emergencies to ensure the safety of navigation and it may be determined qualitatively (perception of stakeholders/expert) or quantitatively (based on ship manoeuvring characteristic and static and dynamic of the vessel). Some other define this safety margin according to the IMO ship manoeuvering standard (MSC.137(76)) and ship size (6*L for port & starboard turns, where L is 98.5% of length over all (LOA) of the largest ship operating in the area and can also be designated as a no-go zone for ships except for emergencies. This safety margin also consist of 0.3 NM (Nautical Mile) of manoeuvre margin that translates for starboard turns, the ship require additional distance to manoeuvre. The result of the spatial visualization as seen in the Figure 7. The steps of visualizing are :

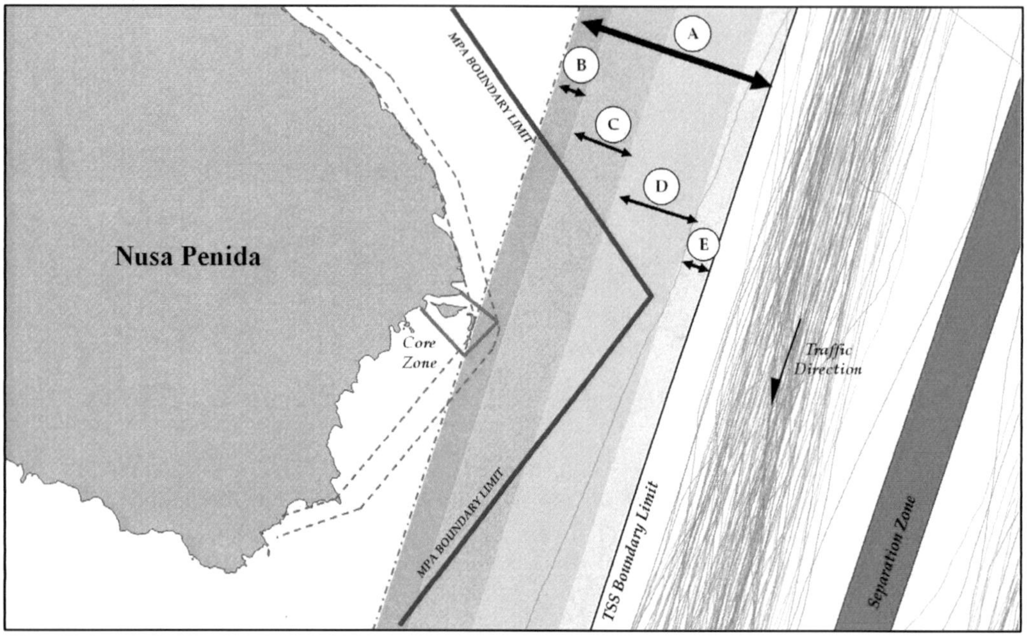

Figure 7. The visualization of the "buffer zone" (A), safety zone (B), reservation area (C), safety margin (D) and manoeuvre margin (E).

a) Using measurement tool in GIS we determine the distance of 500m from the outermost island perpendicular to the TSS line as the limit in determining the the safety zone ('B' in figure 3).
b) Then from the TSS limit we measure 0.3nm of line perpendicular with the TSS line towards the island. This 0.3 NM translates for starboard turns where a ship required as additional distance to manouvre, hence the name is manoeuvre margin ('E' in figure 3).
c) The safety margin as the reserved area solely for the ship manoeuvre taken into account the largest ship in the area was a Very Large Ore Carrier (VLOC) with Length Over All (LOA) is 360 meter. L = 98.5% of 360 is 354.6m. So 6*L = 2,127m is the safety margin ('D' in figure 3).
d) Now remain 'C' which is the reservation area. Since this reserved area were intended for future shipping, we can use IMO recommendation where if the intended traffic <4,400 vessel/year : two ship side-to-side where the L is the LOA of the largest ship intended to use the reservation area or the inshore traffic zone in the future, in this case we take example of L=225m panamax bulk carrier, therefore the formula is 2 ship side-to-side × 1 (one-way route) × L = 2 x 1 x 225 = 450m width of the reservation area. This scenario seems to be appropriate in open waters where there are no obstruction or marine protected area. In this Nusa Penida PSSA, the reservation area will be determine by calculating first the distance from the TSS outer line with the shore line which in this case is 3,900m.
e) Buffer zone = B + C + D + E

3,900 = 500m + ? + 2,127m + 555m (0.3nm)

= 3,900 – 3,182

= 718m for the reservation area

From the steps above we are able to determine the buffer zone of 3,900m that include of 718m of reservation area which more than enough compare with the IMO recommendation for the future shipping. We then recreate the rogue ship manoeuvring characteristic taken from one of the ship type with the highest frequencies traveling using Lombok Strait which is Cape Size Bulk Carrier (LOA=292m) and visualize it with the result in the Figure 8. The starboard turn of the model ship according to the manoeuvring characteristic surpass the manoeuvre margin, safety margin and reaching the reservation area with approximately 800m to the core zone. The model ship which is the rogue ship manoeuvre is based on ballast condition since every southbound bulk carrier mostly in ballast condition when sailing to the loading port in Australia.

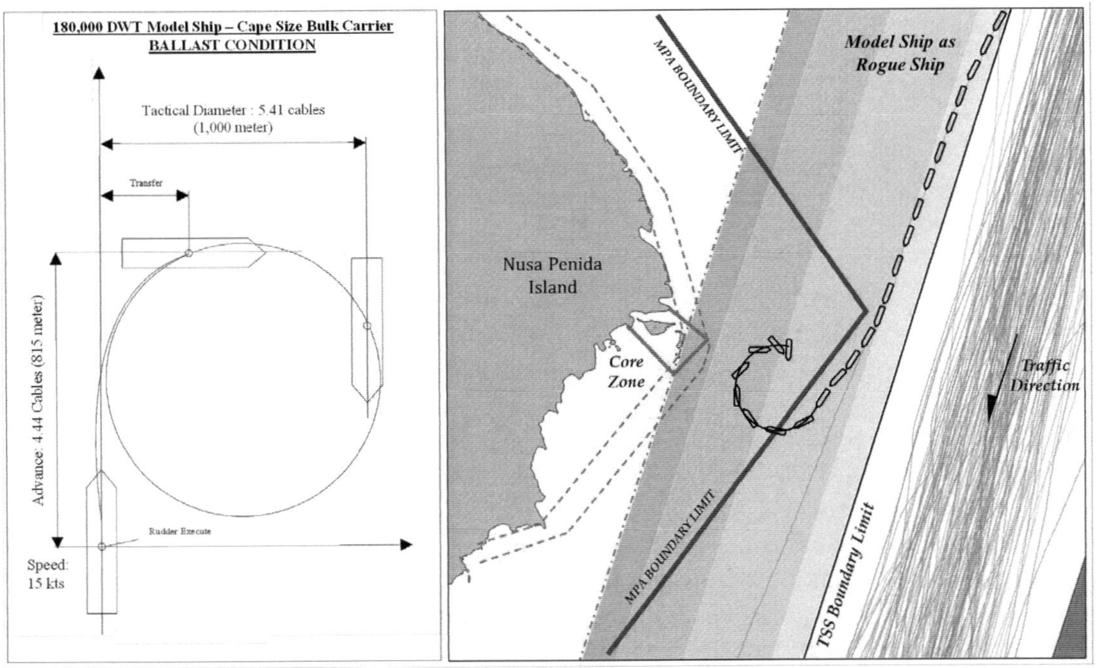

Figure 8. Manoeuvring diagram of the model ship and visualization of the model ship manoeuvre inside the buffer zone.

As seen in the figure 8 that shows the example of rogue ship deviate from the route and assuming taking critical emergency manoeuvering that can lead to potential close proximity to the core zone inside MPA. By implementing steps in figure 7, the buffer zone proposal will further increased the protected area by 300 meters related to the TSS lane.

4.7 PSSA area & TSS modification

Using buffer zone as a parameter in Figure 7, we propose PSSA area surrounding the Nusa Penida island as seen in the Figure 9 with the modification of southern part of TSS. The proposed PSSA area is larger than existing MPA by applying buffer zone in whole Nusa Penida Island and due to the wider area, the southern part of TSS needs to modify. This modification did not change the overal width of the channel but rather following the northbound channel entrance as seen in Figure 3A. By modify the TSS, the distance of the traffic and the core zone will be much farther (6.5km compared with the existing 4.5km).

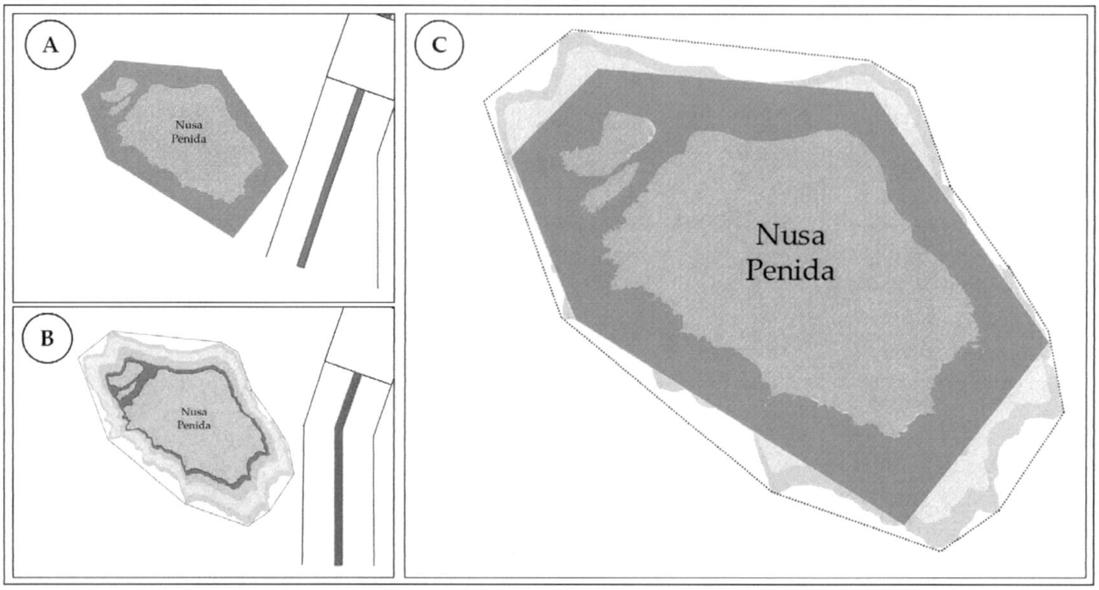

Figure 9. Existing MPA and TSS (A); Buffer zone being implemented with the modification of TSS (B); Comparison of existing MPA and proposed PSSA area based on buffer zone (C).

5. Conclusion

This study highlights the need for careful management and protection of this sensitive area especially when establishing it as a Particularly Sensitive Sea Area surrounded by dense traffic with consideration as follow:

a. The marine traffic around Nusa Penida Island is predominantly composed of passing ships and crossing ships as per AIS data analysis. The traffic consists largely of vessels equipped with AIS, excluding small fishing boats, pleasure crafts, and fast boats, which become the limitation of this study. Significant maritime activities include ship-to-ship LNG transfers and sediment discharge from dredging operations also observed;

b. AIS data over a 7-day period identified 281 ships in the waters surrounding Nusa Penida. The majority (72%) were bulk carriers, followed by oil tankers (9%), ro-ro/passenger ships (7%), LNG/LPG tankers (6%), general cargo ships (3%), other ships (2%), and fishing vessels (1%). This highlights the dominance of bulk carriers in the area, with other vessel types comprising a smaller portion of the traffic;

c. The traffic volume and lane width in the Traffic Separation Scheme (TSS) are adequate according to IMO standards. The largest vessels in records, measuring 400m, are currently accommodated within the existing TSS, with capacity for up to 18,000 ships per lane annually;

d. Ship interactions within the TSS primarily involve overtaking, with crossing becoming more prevalent near precautionary areas. A significant number of crossing situations are expected to increase with the rise in passenger/ferry traffic. AIS data revealed 213 encounters, predominantly in the precautionary area, where close-quarters situations were frequently observed, indicating a 61% chance of encounters during 7-day observation, this indicates that this area is vulnerable with regards to marine traffic;

e. The offshore distances of maritime traffic range from 2,300 to 12,000 meters, with the closest proximity being on the northwest side of Nusa Penida. The close proximity of the core zone to the southbound TSS raises concerns, especially since one example of 'rogue ship' has been observed, necessitating the consideration of a buffer zone;

f. A buffer zone around Nusa Penida Island was proposed based on best practices and guidelines from CESMA and PIANC, ensuring safe navigation around the island. The buffer zone was calculated to be 3,900 meters with the existing TSS, incorporating a safety margin, manoeuvre margin, and reservation area to safeguard against potential navigational hazards.
g. The PSSA area and the TSS were modified based on the established buffer zone, ensuring enhanced protection of the core area around Nusa Penida Island with the distance of the traffic and the core zone become farther (6.5km compared with the existing 4.5km). This modification is crucial for the designation of Nusa Penida as a PSSA, balancing the needs of international shipping with the protection of the island's sensitive marine environment and safeguarding the local fishing community;

References

[1] A. Octavian, T. Trismadi & P. Lestari 2020 The importance of establishing particularly sensitive sea areas in Lombok Strait: Maritime security perspective, *IOP Conf. Ser.: Earth Environ Sci.* **557**, no. 1, pp. 1–12, https://doi.org/10.1088/1755-1315/557/1/012013

[2] Suryadi, A. 2024 The effect of the implementation of Traffic Separation Scheme (TSS) in maintaining maritime environment safety of Lombok Strait. *Dinamika Bahari*, **5**(1), 50-56. https://doi.org/10.46484/db.v5i1.521

[3] Zaman, M. 2019 Navigation safety for marine traffic in the Malacca Strait using AIS data. *Asian J of Appl. Sci.* **7**(4). https://doi.org/10.24203/ajas.v7i4.5827

[4] Birpınar, M., Talu, G., & Gönençgil, B. 2008 Environmental effects of maritime traffic on the istanbul strait. *Env. Mon. & Ass.* **152**(1-4), 13-23. https://doi.org/10.1007/s10661-008-0292-8

[5] Wang, S., Jia, H., Lu, J., & Yang, D. 2023 Crude oil transportation route choices: A connectivity reliability-based approach. *Rel. Eng. & Sys. Saf.* **235**, 109254. https://doi.org/10.1016/j.ress.2023.109254

[6] Putra, D. and Suseno, H. 2017 Present status of Marine Radioecology in Bali and Lombok waters: Distribution study of 137cs and natural radionuclides in Lombok Strait. *Omni-Akuatika*, **13**(2). https://doi.org/10.20884/1.oa.2017.13.2.251

[7] Chonnaniyah, N., Karang, I. W. G. A., & Osawa, T. 2021 Internal solitary waves propagation speed estimation in the northern-part of Lombok Strait observed by Sentinel-1 SAR and Himawari-8 images. *IOP Conf. Ser.: Earth Environ Sci.* **944** (1), 012042. https://doi.org/10.1088/1755-1315/944/1/012042

[8] Sprintall, J. and Révelard, A. 2014 The Indonesian throughflow response to Indo-Pacific climate variability. *J. of Geophy. Res. Oceans*, **119**(2), 1161-1175. https://doi.org/10.1002/2013jc009533

[9] Priadi, A. A. 2012 Assessing Safety of Ferry Routes by Ship Handling Model through AHP and Fuzzy Approach. *Int. Info. Manag.* **02**(25), 277–283. https://doi.org/10.4236/iim.2012.225039

[10] Prastyasari, F. I., Dinariyana, A. a. B., Nugroho, T. F., & Sukmana, A. S. 2023. Formal Safety Assessment for determining sensitive area due to traffic of tourism vessels at Lombok Strait. *IOP Conf. Ser.: Earth Environ Sci.* **1166**(1), 012052. https://doi.org/10.1088/1755-1315/1166/1/012052

[11] Pemkab, K., & KNP, P. 2011 Zonasi Kawasan Konservasi Perairan (KKP) Nusa Penida, Kabupaten Klungkung, Bali. Nusa Penida: Pemkab Klungkung.

[12] Lee, J. and Cho, I. 2022 Extracting the maritime traffic route in Korea based on probabilistic approach using automatic identification system big data. *App. Sci.* **12**(2), 635. https://doi.org/10.3390/app12020635

[13] Daranda, A. and Dzemyda, G. 2020 Navigation decision support: discover of vessel traffic anomaly according to the historic marine data. *Int. J. of Comp. Comm. & Cont.*, **15**(3). https://doi.org/10.15837/ijccc.2020.3.3864

[14] Langella, G., Iodice, P., Amoresano, A., & Senatore, A. (2016). Ship engines and air pollutants: emission during fuel change-over and dispersion over coastal areas. *Int. J. of En. & Env. Eng.* **7**(3), 307-320. https://doi.org/10.1007/s40095-016-0211-7

[15] Eichler, P. (2024). Evaluation of seasonal anoxia in florianópolis north bay through environmental indicators. *J. of Aquaculture & Mar. Bio.*, **13**(1), 31-35. https://doi.org/10.15406/jamb.2024.13.00393

[16] Baghdad, K., Bendraoua, A., & Hasnaoui, A. (2015). Assessment of marine pollution at bay of oran, algeria. *J. of Wat. Res. &Prot.*, **07**(13), 1010-1016. https://doi.org/10.4236/jwarp.2015.713082

[17] Redfern, J., McKenna, M., Moore, T., Calambokidis, J., DeAngelis, M., Becker, E. & Chivers, S. 2013 Assessing the risk of ships striking large whales in marine spatial planning. *Cons. Bio.*, **27**(2), 292-302. https://doi.org/10.1111/cobi.12029

[18] Jalkanen, J., Brink, A., Kalli, J., Pettersson, H., Kukkonen, J., & Stipa, T. 2009 A modelling system for the exhaust emissions of marine traffic and its application in the Baltic sea area. https://doi.org/10.5194/acpd-9-15339-2009

[19] Pitana, T., Kobayashi, E., & Wakabayashi, N. 2010 Estimation of exhaust emissions of marine traffic using Automatic Identification System data (Case study: Madura Strait area, Indonesia). https://doi.org/10.1109/oceanssyd.2010.5603866

[20] Melillos, G., Themistocleous, K., Danezis, C., Michaelides, S., Hadjimitsis, D., Jacobsen, S. & Tings, B. 2020 The use of remote sensing for maritime surveillance for security and safety in Cyprus. https://doi.org/10.1117/12.2567102

[21] Hillmer-Pegram, K. and Robards, M. 2015 Relevance of a particularly sensitive sea area to the Bering Strait region: A policy analysis using resilience-based governance principles. *Eco. and Soc.*, **20**(1). https://doi.org/10.5751/es-07081-200126

[22] Chadwick, M., Crosta, X., Esper, O., Thöle, L., & Kohfeld, K. 2022 Compilation of Southern Ocean sea-ice records covering the last glacial-interglacial cycle (12–130 ka). https://doi.org/10.5194/cp-2022-15

[23] Larayedh, R. 2024 Numerical investigation of shipping noise in the red sea. *Sci. Rep.*, **14**(1). https://doi.org/10.1038/s41598-024-56523-2

[24] Luttenberger, L., Slišković, M., Ančić, I., & Boljat, H. 2022 Environmental impact of underwater noise. *J. of Mar. & Trans. Sci.* **4**(4), 45-54. https://doi.org/10.18048/2022.04.02

[25] Hu, W., Siddagangaiah, S., Chen, C., & Pieretti, N. 2022 Impact of vessel transit on vocalizations of the taiwanese humpback dolphin. *Diversity*, **14**(6), 426. https://doi.org/10.3390/d14060426

[26] Jabrayilov, E. 2021 Exploring environmental connection: ecological network opportunities in the shahdagh national park, azerbaijan. Rudn *J. of Eco. & Life Saf.* **29**(1), 92-102. https://doi.org/10.22363/2313-2310-2021-29-1-92-102

[27] Pirotta, V., Grech, A., Jonsen, I., Laurance, W., & Harcourt, R. 2018 Consequences of global shipping traffic for marine giants. *Frontiers in Eco. & the Env.*, **17**(1), 39-47. https://doi.org/10.1002/fee.1987

[28] Sèbe, M. and Gourguet, S. 2021 Opinion: to save whales, look to the sky. *Proc. of the Nat. Acad. of Sci.*, **119**(1). https://doi.org/10.1073/pnas.2121360119

[29] Bergström, L., Kautsky, L., Malm, T., Rosenberg, R., Wahlberg, M., Capetillo, N., Wilhelmsson, D. 2014 Effects of offshore wind farms on marine wildlife—a generalized impact assessment. *Env. Res. Let.*, **9**(3), 034012. https://doi.org/10.1088/1748-9326/9/3/034012

[30] Lüdeke, J. 2017 Offshore wind energy: good practice in impact assessment, mitigation and compensation. *J. of Env. Assess. Pol. and Man.*, **19**(01), 1750005. https://doi.org/10.1142/s1464333217500053

[31] deCastro, M., Costoya, X., Salvador, S., Carvalho, D., Gómez-Gesteira, M., Larruga, F., ... & Gimeno, L. (2018). An overview of offshore wind energy resources in Europe under present and future climate. *Annals of the New York Acad. of Sci.* **1436**(1), 70-97. https://doi.org/10.1111/nyas.13924

[32] IEC61993-2. 2001 "Part 2: Class A shipborne equipment of the universal automatic identification system (AIS) Operational and performance requirements, methods of test and required test results," *Int. Elec. Committee*

[33] Liu, Y., Zou, L., Zou, Z., & Guo, H. 2018 Predictions of ship maneuverability based on virtual captive model tests. *Eng. App of Comp. Fluid Mec.*, **12**(1), 334-353. https://doi.org/10.1080/19942060.2018.1439773

[34] Miyake, R., Fukuto, J., & Hasegawa, K. 2015 Procedure for Marine Traffic Simulation with AIS Data. TransNav the International *J. on Mar. Nav. & Saf. of Sea Trans.*, **9**(1), 59–66. https://doi.org/10.12716/1001.09.01.07

[35] Yoo, S., Park, S., Park, Y., & Jeong, J. 2015 A Study on the Observation Days of Maritime Traffic Investigation. *J. of the Korean Soc. of Mar. Env. & Saf.*, **21**(4), 397–402. https://doi.org/10.7837/kosomes.2015.21.4.397

[36] Kundakci, B. and Nas, S. 2018 Mapping marine traffic density by using AIS data: an application in the Northern Aegean Sea. *Polish Maritime Research*, **25**(4), 49-58. https://doi.org/10.2478/pomr-2018-0131

[37] Ma, X., Fan, S., Blanco-Davis, E., Shi, G., & Yang, Z. 2024 Bulk carrier accident severity analysis in Australian waters using a data-driven Bayesian network. *Oc. Eng.*, **310**, 118605. https://doi.org/10.1016/j.oceaneng.2024.118605

[38] Maritime Transportation: Safety Management and Risk Analysis. (n.d., p. 168). Routledge & CRC Press.

IOP Conf. Series: Earth and Environmental Science **1461** (2025) 012047 doi:10.1088/1755-1315/1461/1/012047

Stakeholder dynamics in Whale Shark Conservation: A social-economic analysis of marine tourism in Botubarani Beach, Gorontalo, Indonesia

Tatak Setiadi[1*], Jauhar Wahyuni[2], Iin Susilawati Lantu[3]

[1,2] Department of Communication Science, State University of Surabaya, Surabaya, Indonesia
[3] Department of Fish Processing Technology, Gorontalo State University, Gorontalo, Indonesia

*E-mail: tataksetiadi@unesa.ac.id

Abstract. Indonesia's marine biodiversity is highlighted by the presence of whale sharks (*Rhincodon typus*), the largest fish species averaging 9.7 meters in length and 9 tons in weight, which appear in only 4-5 locations nationwide. Botubarani Beach in Gorontalo, Indonesia has recently become one such site, hosting around six whale sharks since 2018. Since 2013, the existence of whale sharks is protected under the Decree of the Minister of Marine Affairs and Fisheries of the Republic of Indonesia. Their appearances in Botubarani Beach creates a new model of marine tourism, which includes activities like diving, snorkelling, and surfing that are managed by various stakeholders. As human activity around the whale shark increases, the interests of the various stakeholders in managing Botubarani Beach as both a tourist destination and a whale shark conservation area have the potential to conflicting with each other. This research is conducted by field observation and in-depth interview with several key stakeholders in whale shark's conservation area. Qualitative data are analysed using stakeholder analysis by mapping each level of interest and power of influence. And then, by participation analysis resulting there are social dynamics driving the growing of whale shark tourism. The findings reveal several contradictory viewpoints among various stakeholders to sustain the conservation area with or without the appearance of the whale sharks themselves.

1. Introduction

The whale shark, the largest fish species in the ocean, appeared and interacted with human firstly known in Ningaloo Marine Park on Western Australia in 1989 [5]. And then, as of 2014, whale shark tourism can be found at numerous places worldwide including the eastern Gulf of Mexico, Honduras, Belize, the Philippines, Mozambique, Seychelles, and the Maldives [6]. As filter feeders [8], they help regulate nutrient cycles and contribute to the overall balance of the ecosystems they inhabit, making them a keystone species and an important indicator of environmental health. Indonesia, with its rich marine biodiversity and extensive archipelagic habitats, is home to these gentle giants [1,17,25]. The presence of the whale shark has been confirmed in various locations across Indonesia, including Probolinggo, Derawan, Talisayan, Botubarani, Saleh Bay, Cenderawasih Bay, and Kaimana [9]. However, whale sharks face numerous threats, including habitat degradation and climate change, shipping activity, and overcrowding tourist which have contributed to a decline in their populations [4,12,18]. Listed as endangered on the IUCN Red List [15], their protection, especially in areas like Gorontalo, is essential for local conservation efforts.

At the same time, the global rise of marine tourism, particularly whale shark tourism, has emerged as a significant economic activity, including in Indonesia [3]. While this offers substantial benefits to local communities, such as economic growth and job creation [9,21], it also presents significant challenges for conservation efforts. The diverse range of tourism activities such as diving, snorkelling, fishing, boating, and marine wildlife watching cater to a broad spectrum of tourists, from adventurers to eco-tourists. However, these activities can lead to environmental challenges such as coral reef degradation [27], pollution, and disturbances to wildlife, emphasizing the urgent need for sustainable tourism practices [22]. Balancing the economic potential of whale shark tourism with the need to protect this endangered species is a complex issue [28], requiring careful priorities consideration of both conservation and tourism goals [23].

This balance is particularly crucial in Indonesia, where the cultural and economic significance of whale sharks is closely tied to their ecological importance [2,13,16,21]. Botubarani Beach in Gorontalo, Indonesia, has recently gained recognition as a whale shark tourism destination. The seasonal appearance of whale sharks in Gorontalo's coastal waters makes it an ideal location for eco-tourism and geo-tourism [1]. However, understanding the dynamics among the various stakeholders involved in whale shark conservation and tourism at Botubarani Beach is vital to ensuring sustainable tourism practices. In the context of whale shark conservation and marine tourism, stakeholders include all individuals, groups, and organizations with a vested interest in the management, protection, and use of marine resources. These stakeholders typically consist of local communities, government agencies, conservation organizations, tourists, and private sector actors such as tour operators [10,24].

The dynamics between key stakeholders at Botubarani Beach are crucial for creating social-economic value [17] and implementing effective whale shark conservation strategies. The coexistence of various stakeholders at the beach presents opportunities for collaboration, as well as potential conflicts. In Botubarani Village, there are various types of participation, including decision-making, implementation, benefits, and evaluation, as proposed by Cohen & Uphoff [7]. However, it's important to address potential conflicts that may arise between local communities due to these types of participation. For instance, conflicts might occur between those who depend on fishing for their livelihood and others who are involved in conservation efforts or government bodies regulating fishing activities to protect whale sharks [11]. The economic dependency of the local community on traditional fishing can create tension when conservation measures restrict access to fishing grounds, affecting their income and food security directly. Additionally, unequal distribution of tourism revenue [14], where larger external operators may benefit more than local residents, can exacerbate feelings of inequality and lead to social tensions [19]. Therefore, this study aims to analyze the interests and power dynamics of different stakeholders, identify potential conflicts and synergies, and explore opportunities for collaborative conservation.

2. Materials and Methods

2.1. Data collection

The study was conducted from July 2024 to September 2024 in Botubarani Beach, Gorontalo, Indonesia. Data were collected qualitatively by field research and in-depth interview towards several stakeholders in Botubarani Beach, specifically of those related to whale shark tourism activities. Key informants during the research are including local residents, local Diving Management, Village Owned Enterprise (BUMDesa), Village Authorities, Coastal and Marine Resources Management Department (BPSPL) of Gorontalo Area, and the Department of Tourism Gorontalo. All informants are the stakeholders who play their significant roles in the development of the Botubarani beach, Boturabrani village and the marine conservation in which whale sharks live in Gorontalo.

Table 1. Tourism Expenses in Whaleshark Tourism.

Type of Service	Price (USD)	Capacity / Description
Boat Rental	$6.67	Capacity for 3 people to explore the Hiu Paus waters.
Glass Bottom Boat Rental	$16.67	Provides the experience of viewing Hiu Paus up close.
Snorkeling	$2.33	Additional fee for snorkeling activities.
Snorkeling Equipment Rental	$3.33	Rental fee per set for snorkeling equipment.
Swimsuit Rental	$3.33	Rental fee per person for swimsuits.
Whale Shark Feed	$1.33	Cost to purchase shrimp as whale shark feed.
Underwater Camera Rental	$16.67	For taking photos underwater.
Glass Bottom Boat with Drone	$16.67	Rental of a glass bottom boat equipped with a drone.
Homestay Rental	$16.67	Includes breakfast; can accommodate 8 guests per month.
Keychain Souvenir	$1.67	Souvenir in the form of keychains.
Souvenir Shirt	$10.00	Souvenir in the form of shirts.
Souvenir Bag	$3.33	Souvenir in the form of bags.

Source: Research Data Collection (2024)

2.2. Data analysis

The collected data was transcribed and analyzed using the DPSIR (Driving forces, Pressures, State, Impact, and Response) framework, which is commonly utilized to integrate applied science and human use management of marine environments [20]. The framework breaks down the analysis into driving forces, which are fundamental social processes shaping human activities impacting the environment, pressures resulting from these forces, the current state of the environment, the impacts on human well-being, and responses from institutions to address environmental changes. Additionally, data analysis involved descriptive assessment based on Cohen & Uphoff's participation types [7], focusing on decision-making, implementation, benefits, and evaluation. The results will be thoroughly discussed in light of current literature and accompanied by graphics and supporting visual material.

Table 2. Data Related to DPSIR Analysis

DPSIR	Approach	Botubarani Beach	Whale Shark	Stakeholders
Driving forces	Fundamental social processes shaping human activities impacting the environment	Botubarani has become one of the latest tourist destinations in Gorontalo	Visitors are curious about the presence of Whale Sharks	• Community • Investors
Pressures	Resulting from these forces	Botubarani Beach has developed into an increasingly popular tourist destination, attracting visitors from outside the region	The sea conditions are becoming increasingly crowded due to the growing number of visitors	• Ministry of Tourism and Creative Economy at Gorontalo Province. • Ministry of Maritime and Fisheries Affairs (KKP)
State	The current state of the environment	The beach area is supported by the development of tourist facilities	There are no strict regulations regarding visitor restrictions for Whale Sharks	• Investors • Ministry of Maritime and Fisheries Affairs (KKP)
Impact	Impacts on human well-being	The people of Botubarani Village earn a significant income through the independently managed tourism sector	The people of Gorontalo and visitors from outside Gorontalo have a basic understanding of Whale Sharks	• Community • Tourists
Response	Responses from institutions to address environmental changes	Evaluations of the development of the tourism sector are conducted regularly	Efforts are being made to ensure the comfort of Whale Sharks and to establish a conservation area around them	• Ministry of Tourism and Creative Economy at Gorontalo Province • Ministry of Maritime and Fisheries Affairs (KKP)

Source: Research Data Collection (2024)

IOP Conf. Series: Earth and Environmental Science **1461** (2025) 012047 doi:10.1088/1755-1315/1461/1/012047

3. Results and Discussion

3.1. Type of Whaleshark in Botubarani Beach, Gorontalo

The whale shark, known as Rhincodon typus, is found in tropical and subtropical oceans around the world and is the largest living fish in the world today. Whale sharks are known to feed on a variety of planktonic organisms. Whale sharks prefer plankton and able to travel long distances in search of enough food to support their large size. The maximum size of a whale shark is estimated to be 20 metres [10]. A good environment with optimal foraging conditions will tend to attract whale sharks to congregate.

Whale sharks have been sighted in several areas of Indonesia, including Probolinggo, Gorontalo, Teluk Saleh, Kaimana and Teluk Cendrawasih [10]. However, at Botubarani in Tomini Bay, whale sharks' surface to search for larval aggregations of seasonal migratory fish known locally as nike [10]. Nike are also known as the larvae of a species of goby (thought to be an Awaous species) that spend most of their lives in freshwater but have larval stages in estuarine or coastal seawater [10]. This natural situation means that whale sharks are often easily spotted at Botubarani Beach, even in close proximity to tourist boats, and this unique whale shark behaviour has led to the rapid establishment of the tourism sector at Botubarani Beach.

From 2016 to 2024, a total of 60 Rhincodon typus whale sharks were tagged and all of the tag showed that whale sharks are all males [29]. Visual identification results conducted by the Makassar Coastal and Marine Resources Management Centre, Ministry of Marine Affairs and Fisheries (KKP) showed that the sex of the whale shark in Botubarani Gorontalo was male. The data was compared with data from whale sharks stranded in Java, Bali and Nusa Tenggara and all showed the same data. Based on the results of direct observation research conducted by Whale Shark Indonesia (WSID) together with BPSPL Makassar, WWF-Indonesia and IPB in 2016, male whale sharks actively forage in coastal waters and move far away, in contrast to female whale sharks. Satellite-tagged whale sharks in Botubarani move towards Bitung, Lembeh Island, Pagimana, Togian, Tojo Una Una, Marisa, Boalemo and Botubarani, all areas within Gorontalo province, towards North Sulawesi. In May 2024, two orca whales (killer whales) were spotted in the waters off Botubarani. Orca is a whale that is considered as a top predator, hunting fish down to large blue whales. That's why orcas are called killer whales. After the appearance of the orca whales in the Botubarani beach area, three whale sharks disappeared [30].

3.2. Driving forces of whale shark presence and the emergence of tourism in Botubarani

Based on the seasonal calendar of whale shark occurrence throughout 2024, whale sharks appear in Botubarani waters almost every day, in numbers ranging from one to four per day. The appearance of whale sharks in Botubarani waters can be attributed to the geographic and environmental conditions of the region. Botubarani, located within the geographic confines of Tomini Bay, represents a critical migratory corridor for a multitude of marine species. This includes not only the whale shark, but also dolphins, orcas, and other fauna. In addition to typical migratory patterns, the availability of food sources in the area is believed to influence the presence of whale sharks [10]. The region's waters provide an abundance of plankton and small fish, which constitute a significant part of the whale shark's diet. Furthermore, the establishment of a shrimp processing factory within the area has contributed to this phenomenon, as the factory's waste products have provided a supplemental food source for the whale sharks. The termination of shrimp production has led local residents to procure shrimp waste from external sources to continue attracting the whale sharks to the area.

Figure 1. Record of whale shark appearances in 2024.

The coastal village of Botubarani was historically dependent on fishing as a primary source of livelihood. The unexpected appearance of whale sharks in the vicinity of the village initially prompted local interest. Over time, the presence of the whale sharks gained widespread attention, which led the local community to recognize the potential for developing a tourism-focused initiative centered around these distinctive marine creatures. In order to capitalise on this opportunity, the community has adapted to this new economic reality by leveraging the whale shark presence to develop sustainable tourism initiatives, thereby supporting the local economy.

3.3. Pressures on the Environment

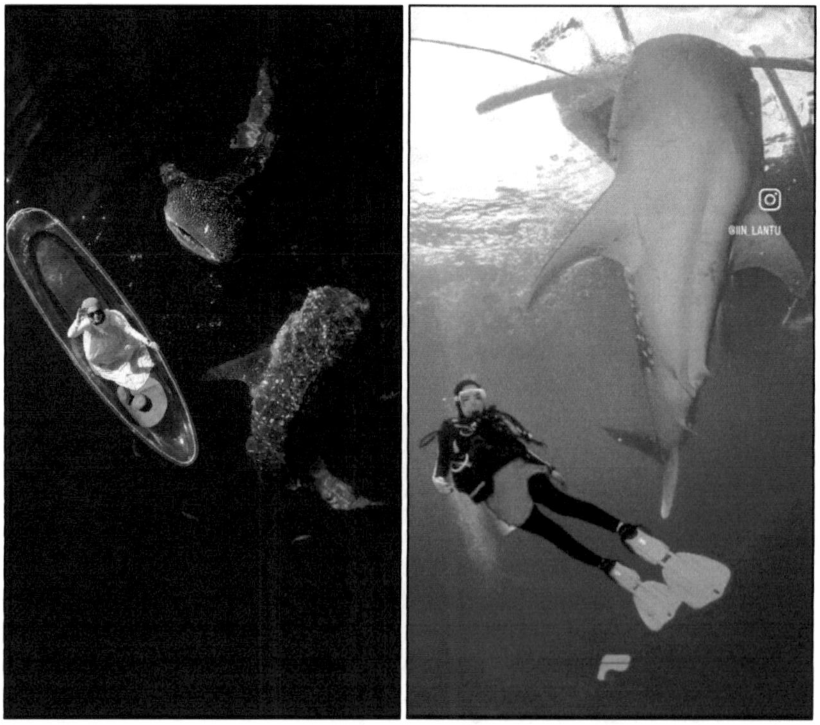

Figure 2. Tourist activities and interactions with whale shark.

IOP Conf. Series: Earth and Environmental Science **1461** (2025) 012047 doi:10.1088/1755-1315/1461/1/012047

Pressure to Increase Visitor Numbers
The appearance of whale sharks has become a significant draw for tourists eager to witness these natural phenomena in person. One of the benefits of visiting Gorontalo for whale shark tourism is the relatively low cost compared to other destinations. While the majority of visitors originate from outside the region, the affordable price point offers tourists an unforgettable experience of observing whale sharks in their natural habitat. The affordability of Gorontalo has contributed to its popularity as a destination for nature enthusiasts and whale shark admirers [19], resulting in an increased pressure on the site to accommodate the rising number of visitors. The challenge is to achieve a balance between the economic benefits to be derived from tourism and the protection of the whale sharks and their habitat.

Impact of High Visitor Traffic on Marine Habitat
As whale shark tourism grows in popularity, the influx of visitors to the coastal area, coupled with the practice of underwater activities, has the potential to disrupt marine life. In order to mitigate this, a number regulations have been implemented by the central government, in particularly by the Ministry of Marine Affairs and Fisheries (KKP). These include a prohibition on direct interaction with the whale sharks and a requirement for tourists to maintain a safe distance from the animals. Furthermore, the local government in Gorontalo has introduced additional regulations specific to the whale shark tourism site in Botubarani Village. The objective of these regulations is to minimise the impact of human activity on the natural environment [17], thereby ensuring that the presence of visitors does not have an adverse effect on the delicate marine ecosystem.

Human Activity and Its Impact on Whale Shark Behavior
The interaction of tourists with whale sharks, whether through the utilisation of boat tours or swimming in their proximate vicinity, must be managed with the intention of avoiding any disturbance to the animals. It is of the utmost importance to ensure the maintenance of an appropriate distance, as well as to refrain from any contact with the whale sharks. It is of the utmost importance to respect the natural behavior of whale sharks in order to preserve their ecosystem. Tourists must refrain from engaging in any actions that could cause excessive distress or disruption to the animals, as this could have an adverse effect on their overall well-being and behaviour. Adherence to these guidelines facilitates the responsible practice of whale shark tourism, thereby ensuring the long-term sustainability of the species and contributing to the positive management of the marine environment.

3.4. State of the Environment

Table 3. Whale Shark Sighting Seasonal Calendar 2024

Month	Average Numbers of Sighting	Not Sighting
January	1 to 4 Whale Sharks per day	
February	1 to 3 Whale Sharks per day	
March	1 to 3 Whale Sharks per day	During March 25th – 27th No whale shark sightings for 3 days straight

IOP Conf. Series: Earth and Environmental Science **1461** (2025) 012047 doi:10.1088/1755-1315/1461/1/012047

Month	Average Numbers of Sighting	Not Sighting
April	1 to 5 Whale Sharks per day	
May	1 to 4 Whale Sharks per day	During May 25th – 31st No whale shark sightings for 7 days straight
June	1 to 3 Whale Sharks per day	During June 1st – 9th No whale shark sightings for 9 days straight
July	1 to 3 Whale Sharks per day	
August	1 to 4 Whale Sharks per day	During August 20th – 22nd No data recorded
September	1 to 4 Whale Sharks per day	
October	1 to 2 Whale Sharks per day	During October 22nd – 24th No whale shark sightings for 3 days straight
November	-	-
December	-	-

Source: Research Data Collection (2024)

Biodiversity Conditions
The waters of Botubarani Village are particularly conducive to the presence of whale sharks. The presence of natural food sources, such as shrimp waste and plankton, in addition to the coastal geography, render the area highly attractive to the species. The beach exhibits a distinctive feature in the form of a deep-sea trench situated in close proximity to the shoreline. Research findings that this trench represents a particularly favourable habitat for whale sharks. The deep waters provide a secure environment for the sharks, particularly during the night when they are likely to remain in the trench area.

Impact of Increased Visitors on the Beach Environment
Despite an increase in visitor numbers, the condition of the beach remains well maintained. This is a consequence of the collaborative efforts of local management, the government, the community, and related agencies, all of which are dedicated to the preservation of the sustainability of the tourism area. The commitment to environmental stewardship by these stakeholders has ensured the preservation of the natural environment, thereby enabling both the coexistence of tourism and marine life in harmony.

Tourism Facilities and Environmental Preservation
The existing tourism facilities in Botubarani have had a beneficial impact on environmental conservation efforts. The implementation of government programmes designed to maintain the cleanliness of the whale shark tourism area has proved to be an effective strategy for the protection of the surrounding ecosystem. Furthermore, local small and medium-sized enterprises (SMEs) have engaged in recycling activities, transforming waste materials from the area into souvenirs. Furthermore, tourists are encouraged to contribute to environmental sustainability by

adhering to government initiatives that promote the preservation of the whale shark ecosystem. This collective endeavour ensures that the tourism industry is able to facilitate both economic growth and environmental conservation.

Potential Migration or Habitat Shift of Whale Sharks
It is documented that whale sharks have been known to temporarily leave the Botubarani waters for approximately two weeks, likely due to the presence of orcas in the area. The temporary relocation of the whale sharks was managed by local stakeholders, who devised strategies to encourage their return. One method involved hanging sacks of shrimp in the water to attract the whale sharks back to the area, as they are drawn to the scent of shrimp. This adaptive response by the local community demonstrates a proactive approach to maintaining the presence of whale sharks, ensuring the continuation of tourism activities. The potential for whale shark migration, which has occurred in the past, is in accordance with the DPSIR framework. This is despite the fact that the migration only occurred over a few weeks and had a significant impact on the local community at Botubarani beach. In previous research, the DPSIR framework has been employed as a tool for analysing issues pertaining to the marine environment. The most crucial aspect is the management of assets that are inherently risky. When such assets have an adverse impact on human, economic and safety assets, they become a risk for the community [20].

3.5. Impacts: The Influence of Environmental Changes on Human Well-being
The Physical Location for Human Activities and Habitat
The coastal area of Botubarani, which is known for its frequent appearances of whale sharks, has become a site of significant human activity and tourism. However, the definitive boundaries of the area where whale sharks are commonly sighted remain uncertain, given the inherent unpredictability of their movements. The frequent appearance of these animals in designated tourist spots has transformed the area into a popular destination for visitors. However, this phenomenon also presents challenges in managing and regulating their habitat to ensure the sustainability of the whale shark population.

Human Resources, Natural Resources, Infrastructure, and Its Impact on Community Welfare
Despite the expansion of whale shark tourism, Botubarani Village is facing challenges related to the availability of resources, both human and physical aspects [18]. The village has initiated several community programmes with the objective of developing whale shark tourism. However, these programmes have not been fully utilized due to a lack of skilled and trained human resources. Furthermore, the village's tourism facilities are limited, with notable gaps in essential equipment such as a monitoring tower, life jackets, and other mandatory safety items for visitors. The lack of suitable infrastructure and the absence of essential facilities, such as accommodation and restaurants, also represent significant obstacles to the development of tourism in the village. In order to enhance the overall visitor experience and stimulate tourism, it is essential to reinforce the capabilities of local human resources and invest in infrastructure upgrades.

Social and Economic Impact: Local Prosperity and the Effects of Whale Shark Tourism
The emergence of whale shark tourism has resulted in a notable transformation in the economic activities and livelihoods of the local community. Many former fishermen have switched occupations to become whale shark tour operators, thereby highlighting the economic advantages that tourism has brought to the area. This shift has resulted in the creation of additional income

opportunities for the community, as local residents have capitalized on the tourism boom by selling a range of products and services, including food, beverages, souvenirs, clothing, and transportation services [24]. The economic prosperity generated by whale shark tourism has led to an improvement in the overall well-being of the community. However, the absence of whale sharks for extended periods, such as during migration, represents a threat to this newfound prosperity. The challenge for stakeholders is to develop strategies that will mitigate the negative economic impact of the whale sharks' temporary absence, thereby ensuring the long-term sustainability of the local economy.

Economic Issue: Coordination between Whale Shark Experts and Tourism Managers
Effective coordination between whale shark experts and tourism managers is essential to ensure the optimal utilisation of the economic benefits of whale shark tourism in Botubarani. This collaboration ensures that the tourism activities are in alignment with the ecological needs of the whale sharks, while also providing income-generating opportunities for the local community. The tourism sector has become a significant source of income for the residents, thereby facilitating improvements in their livelihoods. Community participation in the tourism industry, for instance through the sale of goods and services in the vicinity of the whale shark site, serves to enhance the local economy while also facilitating local economic growth.

Figure 3. Tourism operator activities.

3.6. Responses to Prioritize and Address Environmental Changes

Decision Making
In response to the growth of whale shark tourism in Botubarani, decision-making processes have increasingly focused on achieving a balance between economic gains and the conservation of the species and its habitat. Local authorities, in collaboration with relevant stakeholders, including marine conservation organisations and tourism operators, are responsible for making decisions that address both the immediate needs of the community and the long-term sustainability of whale shark populations. This requires a considered approach to the assessment of environmental impacts and the formulation of tourism management strategies that are aligned with the stated of conservation goals.

Money-Oriented or Whale Shark Sustainability?

Although the economic benefits of whale shark tourism are evident [21], particularly in the form of financial gains for local residents, the community is also actively engaged in the preservation of the whale sharks' well-being. While financial profit is a significant motivating factor, it is not the sole focus of the initiative. The local community is aware of the importance of maintaining the whale sharks' presence in the area in order to ensure the continued availability of tourism opportunities through conservation initiative [1]. It is imperative to strike a balance between economic advancement and ecological sustainability [13], as the long-term viability of tourism is contingent upon the continued wellbeing and safety of the whale sharks. The community's dedication to the conservation of the species serves to illustrate the necessity of pursuing both economic development and environmental protection in parallel.

Policy

In order to guarantee the security and longevity of whale shark tourism, regulations have been put in place to control the interactions between tourists and the whale sharks [12]. The local government in Botubarani has established collaborative arrangements with law enforcement agencies, including the local police (Polsek), military (Babinsa), and the Ministry of Marine Affairs and Fisheries (KKP). The objective of these partnerships is to ensure the safety of both visitors and whale sharks, while enforcing guidelines designed to protect the animals. The policies include restrictions on direct human-whale shark interactions, the maintenance of appropriate distances, and the prevention of actions that could harm the animals or disrupt their natural behaviors. These measures guarantee that whale shark tourism will continue to be a sustainable and environmentally conscious practice in the village.

3.7. Characterizing the Participation of Stakeholders

Table 4. Characteristic of Stakeholders' Participation.

Decision-Making	Implementation	Benefits	Evaluation
Local Communities (Those who have transitioned from fishing to tourism-related activities, provide input into tourism development strategies based on their experience and local knowledge)	**Local Residents** (Actively contribute by becoming tour guides, selling local products, and participating in waste recycling programs to support environmental preservation)	**Local Community** (Benefits economically from the influx of tourists, with opportunities for increased income from providing tourism-related services and products)	**Local Community** (Provides feedback on how tourism affects their livelihood and the environment, highlighting the importance of preserving whale sharks for future sustainability)
Local Government (Sets policies and regulations governing tourism activities, balancing economic growth with the conservation of whale sharks)	**Local Small and Medium Enterprises (SMEs)** (Engage in recycling activities, turning waste into souvenirs, and providing services that contribute to tourism)	**Tourism Operators - Investors** (Profit from managing whale shark tours and related activities)	**Marine Conservation Organizations - KKP** (Assess the environmental impact of tourism and propose adjustments to tourism activities to minimize harm to whale sharks)
Marine Conservation Organizations - KKP (Provide expertise and recommendations on how to preserve whale sharks and their habitat, influencing local policies)	**Local Government** (Implement policies to regulate tourist behavior and protect the whale sharks)	**Local Government** (Gains economic and social benefits by enhancing the village's reputation as a tourist destination and improving local infrastructure)	**Local Government** (Evaluates the success of policies and initiatives related to whale shark tourism and adjusts them to ensure long-term sustainability)
Tourism Operators - Investors (Work with the community and authorities to ensure that tourism activities align with conservation efforts while maximizing economic benefits)	**Marine Conservation Organizations - KKP** (Implement policies to regulate tourist behavior and protect the whale sharks)	N/A	N/A

*Data analyzed by participation stages frameworks of Cohen dan Uphoff (1980).

In the context of whale shark tourism at Botubarani Beach, Gorontalo, Indonesia, two primary stakeholders are most likely to compete for resources and power: tourism operators and marine conservation organisations. Both parties have a significant interest in the utilisation of resources related to whale shark tourism. From the perspective of the tourism operators, the resources at

IOP Conf. Series: Earth and Environmental Science **1461** (2025) 012047 doi:10.1088/1755-1315/1461/1/012047

issue include access to the marine environment, as well as the capacity to attract visitors. The ability to do so directly influences the revenues they are able to generate and the number of local employment opportunities they are able to create. Conversely, marine conservation organisations are focused on the preservation and protection of the marine environment. This involves the securing of funding for research, conservation efforts and the gaining of public support for their initiatives. The investigation of stakeholder dynamics in Goorontalo regarding the management of the presence of whale sharks and the expansion of tourism potential in the region can be conducted through the utilisation of the DPSIR approach, particularly in relation to marine ecosystems. The application of the DPSIR approach has been demonstrated to a range of stakeholders, including researchers, academic institutions, policy and decision-makers at the national and local levels, environmental non-governmental organisations (NGOs), and the general public [20].

There is considerable competition between these stakeholders in several key areas. Firstly, there is the allocation of resources, including funding, research grants and government support. It is often the case that tourism operators possess greater economic power than other stakeholders due to the revenue-generating activities they undertake and their influence on local economies. In contrast, conservation organisations, while potentially having less financial power, possess considerable moral authority and public support, particularly when advocating for environmental protection. Secondly, both stakeholders seek to influence policy in a manner that aligns with their respective objectives. Tourism operators may seek to influence policy in a way that facilitates the expansion of their activities and the relaxation of regulatory constraints. In contrast, conservation organisations may advocate for the implementation of more stringent regulatory measures with the objective of safeguarding marine ecosystems. Such competition for influence over policy can give rise to tensions and conflicts. A further area of potential conflict is the perception of the public. Both stakeholders are engaged in efforts to secure favourable media coverage and public support, which can impact policy decisions and resource allocation. Tourism operators may emphasize the economic benefits and local job creation, whereas conservation organisations may prioritize the importance of environmental preservation and the long-term sustainability of whale shark tourism.

4. Conclusion

The analysis of whale shark tourism in Botubarani Beach, Gorontalo, highlights the complex interplay between environmental and socio-economic factors that influence both the presence of whale sharks and the growth of tourism. The region's favourable habitat for whale sharks, along with the presence of a shrimp processing factory, has contributed to the attraction of these marine giants. The expansion of the tourism industry, driven by the attraction of whale sharks, has created both opportunities and challenges. The increased visitor numbers have exerted pressure on the local marine environment. Local stakeholders have embraced whale shark tourism as a means of economic growth, transitioning from traditional fishing to tourism-related activities. Nevertheless, challenges remain in the form of limited resources, infrastructure deficiencies and competition between tourism operators and marine conservation organisations. To guarantee the long-term sustainability of whale shark tourism, it is vital to foster collaboration between stakeholders, establish transparent regulations, promote shared objectives and facilitate the effective communication through the open dialogue.

Acknowledgments

The authors would like to thank the Directorate of Research and Community Service Universitas Negeri Surabaya for supporting the research. And the authors would also like to thank all the community and stakeholders in Gorontalo who involved in this research.

References

[1] Adisubroto U and Pardede T S 2021 Whale Shark in The Development of Geopark and Sustainable Tourism in Gorontalo, Indonesia *Jurnal Perencanaan Pembangunan: The Indonesian Journal of Development Planning* **5** 363-371.

[2] Anna Z and Saputra D S 2017 Economic valuation of whale shark tourism in Cenderawasih Bay National Park, Papua, Indonesia *Biodiversitas* **18** 1026-1034.

[3] Barry C, Legaspi C, Clarke T M, Araujo G, Bradshaw C J A, Gleiss A C, Meyer L, and Huveneers C 2023 Estimating the energetic cost of whale shark tourism *Biological Conservation* **284** 1-10.

[4] Cárdenas-Palono N, Mimila-Herreraa E, Trujillo-Córdovaa J A, Abunaderb I V, and Herrera-Silveirab J 2024 Whale shark abundance, spatial distribution and effective tourism areas in the northern Mexican Caribbean *Journal of Ecotourism* 1-17.

[5] Catlin J and Jones R 2010 Whale shark tourism at Ningaloo Marine Park: A longitudinal study of wildlife tourism *Tourism Management* **31** 386-394.

[6] Cauga E F, Collins N, Hancock J, and Rees R 2014 Whale shark economics: a valuation of wildlife tourism in South Ari Atoll, Maldives *PeerJ* 1-17.

[7] Cohen J M and Uphoff N T 1980 Participation's Place in Rural Development: Seeking Clarity through Specificity *World Development* **8** 213-235.

[8] D'Antonio B, Barry C, and Beck A 2024 Whale shark (Rhincodon typus) observed gulping on the seafloor at Ningaloo reef aggregation site *Marine and Freshwater Behaviour and Physiology* 1-5.

[9] Djunaidi A, Jompa J, Nadiarti N, Bahar A, and Tilahunga S D 2021 Benefit sharing from whale shark tourism in Botubarani *IOP Conf. Series: Earth and Environmental Science* **763** 012063.

[10] Djunaidi A, Jompa J, Nadiarti N, Bahar A, Tilahunga S D J, Lilienfield D, and Hani M S 2020 Analysis of two whale shark watching destinations in Indonesia: status and ecotourism potential *Biodiversitas* **21** 4911-4923.

[11] Djunaidi A, Jompa J, Nadiarti, Bahar A, Sianipar A, Hasan A W, Alaydrus I S, and Erdmann M 2019 Potential tourism development for whale shark (Rhincodon typus) watching in eastern Indonesia *IOP Conf. Series: Earth and Environmental Science* **253** 012043.

[12] Freya C Womersley, Christoph A Rohner, Kátya Abrantes, Pedro Afonso, Shin Arunrugstichai, Steffen S Bach, Shir Bar, Adi Barash, Peter Barnes, Adam Barnett, Ginevra Boldrocchi, Noemie Buffat, Tom Canon, Clara Canovas Perez, Metavee Chuangcharoendee, Jesse E M Cochran, Rafael de la Parra, Stella Diamant, William Driggers, Christine L Dudgeon, Mark V Erdmann, Richard Fitzpatrick, Anna Flam, Jorge Fontes, Gemma Francis, Beatriz Eugenia Galvan, Rachel T Graham, Sofia M Green, Jonathan R Green, Ya'ara Grosmark, Hector M Guzman, Royale S Hardenstine, Maria Harvey, Jessica Harvey-Carroll, Abdi Wunanto Hasan, Alex R Hearn, Jill M Hendon, Mochamad Iqbal Herwata Putra, Mahardika Rizqi Himawan, Eric Hoffmayer, Jason Holmberg, Hua Hsun Hsu, Mohammed Y Jaidah, Ashlee Jansen, Christy Judd, Baraka Kuguru, Emily Lester, Bruno C L Macena, Kirsty Magson, Rossana Maguiño, Mabel Manjaji-Matsumoto, Stacia D Marcoux, Travis Marcoux, Jennifer McKinney, Mark Meekan, Alejandra Mendoza, Muhammad Moazzam, Emily Monacella, Brad Norman, Cameron Perry, Simon Pierce, Clare Prebble, Dení Ramírez Macías, Holly Raudino, Samantha Reynolds, David Robinson, David Rowat, Mudjekeewis D Santos, Jennifer Schmidt, Chad Scott, Sian Tian See, Abraham Sianipar, Conrad W Speed, Ismail Syakurachman, Julian A Tyne, Kelly Waples, Chloe Winn, Ranny R Yuneni, Irthisham Zareer, Gonzalo Araujo 2024 Identifying priority sites for whale shark ship collision management globally *Science of The Total Environment* **934** 172776.

[13] Habibi P, Azizurrohman M, and Novita D 2021 Whale Shark Tourism and Well-Being: A Case Study of Labuan Jambu *Jurnal Ilmiah Hospitality* **10** 1-8.

[14] Indradi M T K and Ristiawan R R 2019 Partisipasi Masyarakat dalam Pengelolaan Atraksi Wisata Hiu Paus di Pantai Botubarani, Gorontalo *Unpublished Thesis of UGM* https://etd.repository.ugm.ac.id/penelitian/detail/183943 Accessed on 5 September 2024.

[15] IUCN 2024 The IUCN Red List of Threatened Species Version 2024-1 https://www.iucnredlist.org Accessed on 5 September 2024.

IOP Conf. Series: Earth and Environmental Science **1461** (2025) 012047 doi:10.1088/1755-1315/1461/1/012047

[16] Meyers M M, Francis M P, Erdmann M, Constantine R, and Sianipar A 2020 Movement patterns of whale sharks in Cenderawasih Bay, Indonesia, revealed through long-term satellite tagging *Pacific Conservation Biology* **26** 353-364.
[17] Monoarfa S F, Yulianda F, Taryono, and Fahrudin A 2020 Economic Value of Whale Shark Tourism in Botubarani Village, Bone Bolango Regency, Gorontalo Province *J. Ilmu dan Teknologi Kelautan Tropis* **12** 779-790.
[18] Nijman V 2023 Illegal Trade in Protected Sharks: The Case of Artisanal Whale Shark Meat Fisheries in Java, Indonesia *Animals* **13** 1-15.
[19] Pakaya R, Handayani O W K, KS Soegiyanto, and Sulaiman S 2023 Daya Tarik Snorkeling terhadap Ekowisata Whale Shark Botubarani sebagai Ikon Sport Tourism Kabupaten Bone Bolango *Prosiding Seminar Nasional Pascasarjana UNNES* 692-696.
[20] Patricio J, Elliott M, Mazik K, Papadopoulou K-N, and Smith C J 2016 DPSIR - Two Decades of Trying to Develop a Unifying Framework for Marine Environmental Management? *Frontiers in Marine Science* **3** 1-14.
[21] Rahayu S, Nurhidayati S, Wijayanti N, Ayu I W, and Edrial 2023 Analisis Dampak Ekonomi terhadap Kesejahteraan Masyarakat Wilayah Pesisir Wisata Hiu Paus Desa Labuhan Jambu Kabupaten Sumbawa *Entrepreneur: Jurnal Bisnis Manajemen dan Kewirausahaan* **4** 48-58.
[22] Sino A R, Kasim F, and Hamzah S N 2016 Evaluasi Ekowisata Hiu Paus di Desa Botubarani *Nikè: Jurnal Ilmiah Perikanan dan Kelautan* **4** 132-139.
[23] Sholeh M and Farid A 2021 Policy analysis of marine tourism of Gili Iyang Island, Madura, Indonesia *IOP Conf. Series: Earth and Environmental Science* **674** 012018.
[24] Stacey N E, Karam J, Meekan M G, Pickering S, and Ninef J 2012 Prospects for Whale Shark Conservation in Eastern Indonesia Through Bajo Traditional Ecological Knowledge and Community-based Monitoring *Conservation and Society* **10** 63-75.
[25] White W and Cavanagh R D 2007 Whale shark landings in Indonesian artisanal shark and ray fisheries *Fisheries Research* **84** 128-131.
[26] Wijayanti N, Nurhidayati S, Rahayu S, Ayu I W, and Edrial 2024 The Direct Effect of Whale Shark Ecotourism on The Social Economy of The Community in Sumbawa Indonesia *Journal of Global Sustainable Agriculture* **4** 160-166.
[27] Wong C W, Conti-Jerpe I, Raymundo L J, Dingle C, Araujo G, Ponzo A, and Baker D M 2018 Whale Shark Tourism: Impacts on Coral Reefs in the Philippines *Environmental Management* 1-10.
[28] Zielger J, Dearden P, and Rollins R 2012 But are tourists satisfied? Importance-performance analysis of the whale shark tourism industry on Isla Holbox, Mexico *Tourism Management* **33** 692-701.
[29] Darilaut.co 2024 *Identitas 60 Hiu Paus di Botubarani, Gorontalo: Semua Berjenis Kelamin Jantan* [Internet]. Available from: https://darilaut.id/berita/identitas-60-hiu-paus-di-botubarani-gorontalo-semua-berjenis-kelamin-jantan/2.
[30] Darilaut.id 2024 *Tak Sesuai SOP, Lebih 25 Turis Mengerubuti Seekor Hiu Paus di Botubarani, Gorontalo* [Internet]. Available from: https://darilaut.id/konservasi/tak-sesuai-sop-lebih-25-turis-mengerubuti-seekor-hiu-paus-di-botubarani-gorontalo.

Implementation of AIS Data for Estimation of Fuel Oil Consumption and Energy Efficiency Operational Index (EEOI)

H Prastowo[1], T Pitana[1], K B Artana[1], I M Ariana[1], S N Hibatullah[1]

[1]Department of Marine Engineering, Institut Teknologi Sepuluh Nopember (ITS), Surabaya, Indonesia

corresponding author's e-mail: h-prastowo@its.ac.id

Abstract. Since the development of diesel engine, maritime industry shows an upward trend. As the consequence there is an increasing amount of air pollutant, specifically CO2. As the global authorities, IMO take an action written in MEPC 304(72) to reduce greenhouse gas emission. This research aims to find out Energy Efficiency Operational Indicator (EEOI), in Indonesian flagged containership MV 01 and MV 02 and benchmark the result. Method proposed by Trozzi, Jalkanen, Wang and Mersin were used to estimate the fuel oil consumption then the obtained results were compared with fuel monitoring from ship's bunker report. The resulted error for each method consecutively are 100,53%, 24,3%, 18,44%, and 75889%. Mersin method then modified and resulted error 36,22%. Average EEOI index for MV 01 is 0,000473 tonCO2/TEUS.nm and MV 02 0,000630 tonCO2/TEUS.nm. The result from benchmarking using EEOI variables indicated that in MV 02 has lower operational efficiency than MV 01. MV 01 carried more average cargo in voyage route Surabaya-Semarang and Surabaya-Sampit. From average total fuel oil consumption, MV 02 consumed more fuel oil in voyage route Surabaya-Sampit, Surabaya-Kumai, and Semarang-Kumai. The proposed improvement in EEOI for both vessels is by improving ship cargo management for shipping efficiency. While for MV 02, checking for technical operational aspect is needed especially on machinery system and hull conditions.

Keywords: EEOI, efficiency benchmarking, fuel oil consumption method comparison

1. Introduction

The interaction between human and maritime world show upward trend since the replacement of sails by steam in 19th century and diesel engine in 1912. As the impact of industrial revolution, between 1950 to 2005 shows dramatically increase of cargo shipment from 500 million tons to 7 billion tons [1]. The increase in number of ships using combustible engine as prime mover has consequence of increasing amount of air pollution. International Maritime Organization (IMO) release research about greenhouse gas by ship from international shipping and conclude the amount of CO2 produce by ship rise by 34.793 to 36.573 million tonnes from 2012 to 2018. Emissions are projected to increase from about 90% of 2008 emissions in 2018 to 90-130% of 2008 emissions by 2050 for a range of plausible long-term economic and energy scenarios [2].

IMO established Marine Environment Protection Committee (MEPC) in purpose to control and prevent pollution from ships. Resolution MEPC 304(72) writes that IMO have a vision to reduce greenhouse gas emission in 2050 at least 50% compared to 2008, following Paris Agreement in 2015 [3]. Based on MARPOL Annex VI, in general operation ships create carbon dioxide (CO2), carbon monoxide (CO), sulphur oxides (SOx), nitrogen oxide (NOx), hydrocarbon (HC), and particulates [4][5].

Following the big projection, IMO established the guidelines for the development of Ship Energy Efficiency Management Plan (SEEMP) [6] [7]. This guideline has been developed to assist the preparation of SEEMP required by regulation 22 or MARPOL Annex VI. SEEMP guidelines part I develop a framework and structure of SEEMP. There are planning, implementation, monitoring, and self-evaluation & improvement. For monitoring, there is a specific tool developed by MEPC: EEOI (Energy Efficiency Operational Indicator [8].

This research conducted in container ship MV 01 and MV 02. Both ships have identical design and shipping voyage route in domestic Indonesian area. Using data from shipping company and AIS data, the objective of this study is to calculate EEOI index from both ship and benchmark the results with different operational parameters. The second objective of this research is to compare ship fuel oil estimation method proposed by Trozzi, Jalkanen, Wang, and Mersin with reported fuel oil consumption data from shipping company. Lastly, this research aims to propose possible improvement to increase efficiency based on EEOI index.

2. Methods

This research was conducted using methodology as shown in Figure 1. It began with collecting data from AIS database and ship company's report for EEOI calculation. Thus, all data need to be checked for completeness. All data were processed, sorted then compiled to the database for calculation. The obtained results of EEOI calculation were used to benchmark between two ships. For fuel oil consumption estimation method, obtained data were calculated with four estimation methods and the result compared with reported fuel oil consumption.

2.1. Overview pf Shipping Company Data

The main data source of this research were obtained from shipping company. Data used for EEOI calculation such as amount of fuel oil consumption, amount of cargo carried, and operational time. Those data were available in ship particular, bunker report, and departure report. Information about ship particular is presented in Table 1.

Table 1 Ship Particular

Vessel Name	MV 01	MV 02
Type	Container Ship	
Flag	Indonesia	
Gross Tonnage	3668	
DWT	5161	
LOA	107 meters	
LPP	99 meters	
Breadth	20 meters	
Draught	4,2 meters	
Main Engine	Yanmar 6EY26 – 2 x 1440 kW / 681 RPM	
Propeller	Fixed 4 blades 2 x 2,7 m pitch 0,61	

2.2 Overview of AIS Data

AIS data were supported by FleetMon as AIS data provider [9]. AIS file data formatted in .kml and .csv. Kml file expressing geographic annotation and visualization within Internet-based. File in csv expressing data in table format. Information that can be obtained from AIS data are time in timestamp format, latitude & longitude, which represent ship's position, speed in knot, course & heading in degree, and navigation status.

2.3. Analysis of AIS Data

Kml file was accessed with Google Earth Pro and ArcGIS for analysis, while csv was file accessed with Microsoft Excel. With ArcGIS, kml file was analyzed in each vessel trip. In every voyage, AIS collected ship's condition, such as position, speed, course and heading, from port of departure to the port of arrival. In every epoch (point) of AIS data, connection was made by lines and create continuous voyage line. From voyage line distance can be measured between epoch of AIS location transmission data.

2.4. EEOI Formula

As explained in MEPC.1/Circ.684, EEOI is defined as the ratio of mass of CO2 emitted per unit of transport work [8]. Detailed formula given as follows:

$$EEOI = \frac{\sum_j FC_j \times C_{F\,j}}{m_{cargo} \times D} \tag{1}$$

For average EEOI

$$Average\ EEOI = \frac{\sum_i \sum_j (FC_{ij} \times C_{F\,j})}{\sum_i (m_{cargo} \times D)} \tag{2}$$

Where;

j	= fuel type
i	= voyage number
F_{Cj}	= mass of consumed fuel j at voyage i
C_{Fj}	= fuel mass to CO_2 mass conversion factor of fuel j
m_{cargo}	= cargo carried
D	= distance travelled

2.5. Fuel Oil to CO_2 Conversion Factor

Fuel-mass to CO_2-mass conversion factor is used to calculate the mass of CO_2 released by fuel burning. Conversion factor is non-dimensional factor to obtain the specific amount of CO_2 released by combustion in a specific volume of fuel burning. Each type of fuel has specific carbon chain and other chemical properties. The IMO in EEOI guidelines release the amount of conversion factor for the CO_2 produced as Table *2* [10].

Table 2 Fuel Mass to CO2 Mass Conversion Factors (Ton-CO_2/Ton-Fuel)

Type of fuel	Reference	Carbon content	Cf (ton-CO_2/ton-Fuel)
Diesel/Gas Oil	ISO 8217 Grades DMX through DMC	0,875	3,206000
Light Fuel Oil (LFO)	ISO 8217 Grades RMA through RMD	0,86	3,151040
Heavy Fuel Oil (HFO)	ISO 8217 Grades RME through RMK	0,85	3,114400
Liquified Petroleum Gas (LPG)	Propane	0,819	3,000000
	Butane	0,827	3,030000
Liquefied Natural Gas (LNG)		0,75	2,750000

2.6. Fuel Oil Consumtion Estimation Methods

2.6.1. *Trozzi's Method*

Trozzi proposed method for estimating fuel oil consumption with gross tonnage and operational mode variable [11]. Previous study already implemented this method to estimate fuel oil consumption in Indonesian sea area [12]. Trozzi derived shipping activity in one voyage as:

1. Approaching and docking: this works starts when ship deceleration begins and ends at the moment of docking.
2. Hoteling: this work starts when ship is berthing until the ship leaves the berth.
3. Departing: this work starts with the departure of ship from berth and ends when cruising work reached.
4. Cruising: this work starts after speed of cruising achieves by ship and ends if the ship starts for approaching and docking work.

Both MV 01 and MV 02 using only MFO for main engine in cruising mode, while in maneuvering using only MDO. Therefore, this study use only the cruising data reference to compare the results of the FOC based on the estimation methods. From figure 2, starting point of beginning cruising mode are determined using map and graph. Because of that, docking, hoteling and maneuvering were not included in fuel oil estimation calculation. Trozzi's explained formula for fuel consumption given as follows:

$$S_{jkm}(GT) = C_{jk}(GT) \times P_m \tag{3}$$

Where;

S_{jkm} = daily fuel oil consumption of fuel j in ship class k in operational mode m using gross tonnage function

C_{jk} = daily fuel oil consumption at full power of fuel j in ship class k as a function of gross tonnage (descripted in table 3)

P_m = a fraction of maximum fuel consumption in mode m

Table 3 Trozzi Ship Class and Fuel Consumption at Full Power (Ton/day)

Ship Types	Consumption (Cjk)
Solid bulk	12.0724+0.0012*GT-1.1501E-8*GT^2+4.6484E-14*GT^3
Liquid bulk	7.2194+0.0015*GT-9.1885E-9*GT^2+2.6803E-14*GT^3
General Cargo	-2.2602+0.0049*GT-1.6401E-7*GT^2+1.7394E-12*GT^3
Container	0.0919+0.0038*GT-6.1565E-8*GT^2+6.7917E-13*GT^3
Passenger/ RoRo/Cargo	6.3501+0.0013*GT+1.6852E-7*GT^2-6.2691E-12*GT^3+5.699E-17*GT^4
Passenger	5.2159+0.0034*GT-3.373E-8*GT^2+2.062E-13*GT^3
High Speed Ferry	-9,735+0.0486*GT-4.6603E-6*GT^2+1.3911E-10*GT^3
Inland cargo	-2.2602+0.0049*GT-1.6401E-7*GT^2+1.7394E-12*GT^3
Sail ship	0.42682 + 0.00100 * GT
Tugs	1.0857+0.0259*GT-1.0279E-5*GT^2+1.6828E-9*GT^3
Fishing	1.2744+0.0062*GT-6.4603E-7*GT^2+5.5193E-11*GT^3
Other Ships	1.2744+0.0062*GT-6.4603E-7*GT^2+5.5193E-11*GT^3
All Ships	16.263 + 0.001 * GT

IOP Conf. Series: Earth and Environmental Science **1461** (2025) 012048 doi:10.1088/1755-1315/1461/1/012048

2.6.2. *Jalkanen's Method*

Method proposed by Jalkanen called STEAM2 [13] [14]. There are several data that must be available for STEAM2 calculation, those are:

1. Ship technical data (hull dimension, powering, fuels)
2. Ship activity (speed, distance, sea condition, operational mode)

This method calculates fuel consumption from ship resistance and ship machinery specifications. The total resistance of moving ship in kilo-Newton can be calculated with following formula:

$$R_{Total} \approx R_F + R_R \tag{4}$$

Where;

R_F = frictional resistance acting on wet surface

R_R = residual resistance

Then the frictional resistance coefficient (C_F) and frictional resistance (R_F) can be calculated using

$$C_F = \frac{0,075}{(\log Rn - 2)^2} \tag{5}$$

$$R_F = C_F \frac{\rho}{2} v^2 S \tag{6}$$

Where;

C_F = frictional resistant coefficient

Rn = Reynold's number

ρ = seawater density

v = ship's speed

S = wet surface area

Residual resistance (R_R) is calculated with

$$S_{jkm}(GT) = C_{jk}(GT) \times P_m \tag{7}$$

Where;

C_R = residual resistant coefficient

B = ship's breadth

T = ship's draught

Residual coefficient calculated using Hollenbach method and Schneekluth-Bertram [15][16]. Wetted surface area assumed using Watson formula [17]. After obtaining value of total resistance, propelling power calculated by:

$$P_{Propel} = R_{Total} \cdot v \tag{8}$$

Subsequent calculations are performed to obtaining quasi propulsive constant (η_{qpc}) for describing the efficiency of power transferring from engine to propeller.

$$\eta_{qpc} = 0,84 - \frac{N\sqrt{LBP}}{10000} \tag{9}$$

Where;

N = rpm of the propeller

LBP = length between perpendiculars

Then total required engine power described by:

$$P_{Total} = \frac{P_{Propel}}{\eta_{qpc}} \tag{10}$$

For multi-engine setups, all engine can be assumed as identical. Engine load (EL) can be determined by:

$$n_{OE} = \frac{P_{Total}}{P_E} + 1 \tag{11}$$

$$EL = \frac{P_{Total}}{P_E n_E} \tag{12}$$

Where;

P_E = maximum continuous rating of a single installed engine
n_E = number of installed engines
In STEAM2 method, parabolic function assumed for relative fuel oil consumption as function to relative engine load in all engines. Using regression analysis of the comprehensive SFOC-measured data from Wartsila engine, a second-degree polynomial equation:

$$SFOC_{Relative} = 0{,}455EL^2 - 0{,}71EL + 1{,}28 \tag{13}$$

The absolute fuel consumption is estimated from:

$$FOC = SFOC_{Relative} SFOC_{Base} \tag{14}$$

2.6.3. *Wang's Method*
Wang calibrates daily bunker consumption and relation with the sailing speed [18]. Data obtained from calibration then used for the further calculation. Calibration of bunker consumption and sailing speed function follows the formula:

$$Q = a \times v^b \tag{15}$$

Where;
Q = daily fuel oil consumption
V = sailing speed
a & b = coefficient to be calibrated from data
Formula generated by Wang's method can be different on each ship. Data a and b from calculation are obtained from the collection of fuel oil consumption in each speed and using regression method.

2.6.4. *Mersin's Method*
Mersin developed a method by a formula that show the change of displacement of the ship according to the time [19].

$$\nabla(t) = \left[\sqrt[3]{V(0)} - \frac{\lambda v^3 t}{72}\right] \tag{16}$$

Where:
$\nabla(t)$ = volume displacement of the ship in a time
$\nabla(0)$ = initial displacement of ship
λ = ship engine coefficient
v = ship's speed
t = time of sailing in hour
Explained by Barras, λ value interpreted as [20].

$$\lambda = 1/F_C \tag{17}$$

FC is the fuel coefficient and described as:

$$F_C = \frac{W^{\frac{2}{3}} \times V^3}{Fuel\ Cons/day}$$

Where;
W = ship's displacement (ton)
V = ship's speed (knot)
Then to calculate the amount of remaining fuel on board using:

$$L(t) = \nabla(t) - W_{i,i+1} \tag{18}$$

Where;
L(t) = amount of remaining fuel
$W_{i,i+1}$ = weight of the cargo carried from port i to the next port (i+1)
Calculation of the fuel consumption using formula:

$$FC = \mathrm{L}(0) - L(t) \tag{19}$$

Where:

FC = fuel consumption
L(0) = initial volume of fuel oil in tank

2.7. Fuel Oil to CO_2 Conversion
Benchmarking process was conducted after obtained EEOI index of both ships. The value index used as a comparation between MV 01 and MV 02 by reflecting on the other variables, such as fuel oil consumption, sailing speed, and cargo carried. The purpose of benchmarking process is to evaluate the efficiency and obtain list for chance of optimization for ship in further time. It needs to be underlined that the efficiency is concluded from fuel oil report, cargo carried, and distance sailed in which those are variable used for EEOI calculation. Furthermore, the external factors such as weather and passage plan then technical factors such as the hull/propeller conditions and the engine performance, were excluded in this study.

3. Result and Discussion

3.1. EEOI Results and Analysis
Step by step of EEOI calculation is explained in figure 3. EEOI index calculated with formula (1) and the results can be seen in the table 4 for MV 01 and table 5 for MV 02. Average EEOI value calculated with formula (2).
MV 01 and MV 02 benchmarked with voyage that has same voyage route. Some routes from the data used as a benchmark are as follows:

a. Voyage Route Surabaya Semarang (SBY-SMG)
b. Voyage Route Surabaya Sampit (SBY-SMQ)
c. Voyage Route Surabaya Kumai (SBY-PKX)

Table 4 MV 01 EEOI Results

Voyage Number	Time Window		Port		Trip Number		EEOI Index
	Depart	Arrive	Depart	Arrive			
1901	09/01/2019 22:30	13/01/2019 23:48	Semarang	Pontianak	MV01-	1	0.000517
	13/01/2019 21:54	18/01/2019 8:30	Pontianak	Semarang	MV01-	2	0.000451
1902	18/01/2019 8:36	21/01/2019 20:24	Semarang	Pontianak	MV01-	3	0.000601
	21/01/2019 16:00	26/01/2019 16:28	Pontianak	Semarang	MV01-	4	0.000411
1903	26/01/2019 16:48	30/01/2019 1:00	Semarang	Pontianak	MV01-	5	0.001040
	29/01/2019 20:48	02/02/2019 17:00	Pontianak	Surabaya	MV01-	6	0.000460
1904	03/02/2019 18:18	06/02/2019 11:54	Surabaya	Sampit	MV01-	7	0.000396
	06/02/2019 13:24	09/02/2019 1:00	Sampit	Surabaya	MV01-	8	0.000533
1905	09/02/2019 9:00	12/02/2019 17:12	Surabaya	Kumai	MV01-	9	0.000373
	12/02/2019 17:36	15/02/2019 14:24	Kumai	Semarang	MV01-	10	0.000400

Voyage Number	Time Window		Port		Trip Number		EEOI Index
	Depart	Arrive	Depart	Arrive			
	15/02/2019 9:36	17/02/2019 14:42	Semarang	Surabaya	MV01-	11	0.000457
1906	17/02/2019 18:42	20/02/2019 3:54	Surabaya	Sampit	MV01-	12	0.000421
	20/02/2019 4:00	22/02/2019 9:24	Sampit	Surabaya	MV01-	13	0.000306
1907	22/02/2019 9:18	26/02/2019 0:12	Surabaya	Sampit	MV01-	14	0.000316
	25/02/2019 20:24	28/02/2019 1:12	Sampit	Surabaya	MV01-	15	0.000322
1908	28/02/2019 9:30	03/03/2019 16:42	Surabaya	Kumai	MV01-	16	0.000451
	03/03/2019 16:48	06/03/2019 13:12	Kumai	Surabaya	MV01-	17	0.000615
1909	06/03/2019 13:24	08/03/2019 7:18	Surabaya	Semarang	MV01-	18	0.000618
	08/03/2019 7:24	10/03/2019 11:24	Semarang	Kumai	MV01-	19	0.000368
	10/03/2019 11:30	14/03/2019 7:18	Kumai	Surabaya	MV01-	20	0.000413
1910	15/03/2019 2:00	17/03/2019 9:18	Surabaya	Semarang	MV01-	21	0.000772
	17/03/2019 9:24	19/03/2019 19:54	Semarang	Kumai	MV01-	22	0.000667
	19/03/2019 20:06	23/03/2019 3:12	Kumai	Surabaya	MV01-	23	0.000406
					Average		0.000473

Table 5 MV 02 EEOI Results

Voyage Number	Time (GMT +7)		Port		Trip Number		EEOI Index
	Depart	Arrive	Depart	Arrive			
1901	06/01/2019 20:18	11/01/2019 9:48	Surabaya	Kumai	MV02-	1	0.000626
	11/01/2019 9:36	14/01/2019 2:30	Kumai	Semarang	MV02-	2	0.000593
	14/01/2019 2:12	16/01/2019 12:00	Semarang	Surabaya	MV02-	3	0.000679
1902	16/01/2019 16:54	21/01/2019 8:48	Surabaya	Samarinda	MV02-	4	0.000545
	21/01/2019 8:54	25/01/2019 20:06	Samarinda	Surabaya	MV02-	5	0.000883
1903	26/01/2019 14:00	30/01/2019 5:24	Surabaya	Kumai	MV02-	6	0.000796
	30/01/2019 3:00	02/02/2019 10:12	Kumai	Semarang	MV02-	7	0.000453

Voyage Number	Time (GMT +7)		Port		Trip Number		EEOI Index
	Depart	Arrive	Depart	Arrive			
	02/02/2019 10:18	03/02/2019 1:30	Semarang	Surabaya	MV02-	8	0.000513
1904	04/02/2019 10:42	07/02/2019 8:48	Surabaya	Sampit	MV02-	9	0.000688
	07/02/2019 9:00	09/02/2019 20:00	Sampit	Surabaya	MV02-	10	0.000930
1905	10/02/2019 9:00	12/02/2019 5:18	Surabaya	Semarang	MV02-	11	0.000759
	12/02/2019 5:24	14/02/2019 4:42	Semarang	Kumai	MV02-	12	0.000441
	14/02/2019 5:00	16/02/2019 22:42	Kumai	Surabaya	MV02-	13	0.000552
1906	16/02/2019 21:00	20/02/2019 4:42	Surabaya	Kumai	MV02-	14	0.000503
	20/02/2019 3:36	23/02/2019 1:48	Kumai	Surabaya	MV02-	15	0.000424
1907	22/02/2019 22:18	26/02/2019 9:12	Surabaya	Kumai	MV02-	16	0.000525
	26/02/2019 2:00	02/03/2019 17:00	Kumai	Surabaya	MV02-	17	0.000400
1908	02/03/2019 14:06	06/03/2019 2:42	Surabaya	Kumai	MV02-	18	0.000602
	05/03/2019 20:00	08/03/2019 23:30	Kumai	Semarang	MV02-	19	0.000674
	08/03/2019 23:48	12/03/2019 1:36	Semarang	Surabaya	MV02-	20	0.002516
1909	12/03/2019 1:42	15/03/2019 9:30	Surabaya	Kumai	MV02-	21	0.000749
	15/03/2019 9:48	18/03/2019 7:48	Kumai	Semarang	MV02-	22	0.000621
	18/03/2019 8:30	21/03/2019 22:00	Semarang	Surabaya	MV02-	23	0.000796
1910	21/03/2019 21:00	25/03/2019 4:06	Surabaya	Kumai	MV02-	24	0.000771
	25/03/2019 4:00	27/03/2019 23:30	Kumai	Semarang	MV02-	25	0.000564
	28/03/2019 0:24	30/03/2019 17:48	Semarang	Surabaya	MV02-	26	0.000947
					Average		0.000630

Benchmarking was done using comparative variables obtained from the report about the cargo load, ship fuel consumption from port-to-port which includes MFO and HSD, and ship average service speed. The rolling average for EEOI index is shown in figure 4

From figure 6.(a), it can be analyzed that in Surabaya-Semarang route MV 01 has mean cargo carried 174 TEUs with average EEOI 0.000625 and MV 02 has mean cargo carried 122.5 TEUs with average EEOI 0.000807. For Surabaya-Sampit route, MV 01 has average cargo carried 223.5 TEUs with average EEOI 0.000371 and MV 02 163.5 TEUs, with average EEOI 0.000771. For Surabaya-

Kumai route, MV 01 has average cargo 198 TEUs with average EEOI 0.000437 and MV 02 has average cargo 209.6 TEUs with average EEOI 0.000589. For Semarang-Kumai route, MV 01 has average cargo carried 193 TEUs and average EEOI index 0.000443. MV 02 has average cargo 234.5 TEUs with average EEOI 0.000557.

The highest average EEOI index for MV 02 was at Surabaya-Semarang route and the lowest was at Semarang-Kumai route. At highest average of EEOI index (Surabaya-Semarang), MV 02 carried least of average cargo while at lowest average EEOI index (Semarang-Kumai) it carried most cargo. While MV 01 carried most average cargo in route Surabaya-Sampit and has lowest average EEOI index. Although MV 02 carried more average cargo than MV 01 in route Surabaya-Kumai and Semarang-Kumai, but still has average EEOI index more than MV 01.

From figure 6.(b) shows the relation between average EEOI and Average fuel oil consumption. At route Surabaya-Semarang, MV 01 has average fuel oil consumption 11.2262 ton with average EEOI 0.000625, while MV 02 has average fuel oil consumption 9.0359 ton with average EEOI 0.000807. At route Surabaya-Sampit, MV 01 has average fuel oil consumption 11.9055 ton with average EEOI 0.000371, and MV 02 average fuel oil consumption 16.9852 ton with average EEOI 0.000771. At route Surabaya-Kumai, MV 01 average fuel oil consumption is 12.9376 ton with average EEOI 0.000437, while MV 02 has average fuel oil consumption 17.3680 ton with average EEOI 0.000589. At Semarang-Kumai route, MV 01 has average fuel consumption 10.6527 ton with 0.000443 and MV 02 average fuel oil consumption 16.4194 ton with average EEOI 0.000557.

Fuel oil consumption in MV 02 is higher than MV 01. The lowest average point of fuel oil consumption for MV 02 was at voyage route Surabaya-Semarang, but it has highest average EEOI index. In another route, average fuel oil consumption for MV 02 is higher than MV 01. The average value of all EEOI index for MV 02 is higher than MV 01.

Fuel oil consumption is the function of resistance and speed of a ship. Resistance of ship is affected by many variables from the ship itself and the environment. Ship operational work creates two big variables, there are wetted surface area and speed. Wetted surface area can be represented by the changing of the ship's draught, which caused based on the load of cargo and ballast.

Figure 6.(c) indicates that the speed of MV 01 is relative higher than MV 02. In all route MV 01 has higher average service speed. At Surabaya-Semarang route, Surabaya-Sampit, Surabaya-Kumai, and Semarang-Kumai each has average service speed 7.572 knot, 7.256 knot, 7.798 knot, and 7.664 knot. While MV 01 has average service speed 7.052 knot, 5.957 knot, 7.229 knot, and 6.788 knot.

Figure 6 shows overall analysis on how significant the influence of mentioned variables. As shown in Figure 6.(a), the obtained EEOI Index values from MV 01 were sorted from lowest to highest, with the amount of loaded cargo indicated by line. It showed that EEOI Index and the amount of loaded cargo had opposite relation. Trendline shows linear function with negative gradient from the resulted equation. Meanwhile Figure 6.(d) for MV 02 also shows the same finding. The relation between EEOI Index and the amount of loaded cargo denoted that it was inversely proportional. The shown negative gradient indicates that the amount of cargo fell dramatically with the rise of EEOI index value.

Figure 6.(b) explains the correlation between EEOI index and total fuel oil consumption of MV 01. The graph shows that its fuel oil consumption was proportional with EEOI index. It is shown by trendline with positive gradient. For MV 02, it can be analyzed from Fig 6.(e) that the result was similar with MV 02 in which the trendline increase insignificantly.

Figure 6.(c) and 6.(f) show the correlation between EEOI index and average ship service speed in any trip. Average service speed of MV 01 and MV 02 did not show significant increase correlated with EEOI index that significantly increase. The shown trendline has positive gradient. It indicates that the increase of sailing speed in MV 02 has effect on its EEOI index. From the graph, it can be concluded that the increase of sailing speed does not affect EEOI index significantly.

3.2. Fuel Oil to CO_2 Conversion

The calculations of the estimated fuel oil consumption using several models were done by only considering when the ship is in sailing speed/service speed condition. This condition applied because both MV 01 and MV 02 use different fuels (HSD) when approaching the port (maneuvering and hoteling).

For fuel oil consumption estimation, performed Mersin method show mismatch results. The example results of Mersin method used for MV 01 trip 1 (MV01-1) fuel oil consumption calculation is written below.

$\nabla(0)$ = 7313 ton
λ = 0.000029
t = 55.85 hours
Cargo = 4017 ton
L(0) = 74.22 ton

Calculation of ($\nabla(t)$) using formula (16)

$$\nabla(t) = \left[\sqrt[3]{7313} - \frac{0{,}000029 \times 7{,}57^3 \times 55.850}{72}\right]$$

$$\nabla(t) = 7302{,}16\ Ton$$

The amount of fuel remaining at destined port L(t) (18).

$$L(t) = 7302{,}16 - 4017$$

$$L(t) = 3285{,}16\ Ton$$

Then, calculating fuel oil consumption is (19).

$$F(t) = 74{,}22 - 3285{,}16$$

$$F(t) = -3210{,}94\ Ton$$

The calculation outcome showed negative amount of fuel consumption thus it indicates that the obtained result from Mersin's method cannot be applied. The reason behind this error was Mersin assumes that remaining fuel after reaching port L(t) is obtained after subtracting the weight of the carried cargo from the ship's displacement. In practical, displacement is the total of ship weight and this in general includes fuel oil, lubricating oil, provision, ballast, ship construction etc. Direct subtraction of displacement with cargo weight neglect the weight of others.

It is decided to modify the assumption of Mersin's method formula. Assumption for the initial fuel weight was changed into 'weight of displacement minus cargo'. Other assumption not necessary to be changed.

$$L(0) = \nabla(0) - cargo\ weight \tag{20}$$

With modified Mersin's formula, the calculation of fuel oil consumption of several methods can be calculated. The calculated results are shown in Table 6 and Table 7. Resulted error in MV 01's fuel oil consumption estimation for Trozzi, Jalkanen, Wang, and Modified Mersin's formula were 123.33%, 30.03%, 19.21%, and 44.74%. While MV 02 resulted error for Trozzi, Jalkanen, Wang and modified Mersin: 77.73%, 18.57%, 17.66%, and 26.69%. So far, Wang method resulted lowest error in sequence followed by Jalkanen method, modified Mersin method, and Trozzi.

Table 6 MV 01 Fuel Consumption Estimation Results

Voyage Number	Trip Number	MFO Fuel Oil Consumption (Ton)					% Error			
		Report	Trozzi	Jalkanen	Wang	Modified Mersin	Trozzi	Jalkanen	Wang	Modified Mersin
1901	MV01-1	11.8603	24.3631	13.8798	12.6842	10.8370	105.42%	17.03%	6.95%	8.63%

Voyage Number	Trip Number	MFO Fuel Oil Consumption (Ton)					% Error			
		Report	Trozzi	Jalkanen	Wang	Modified Mersin	Trozzi	Jalkanen	Wang	Modified Mersin
	MV01-2	10.3123	21.3677	12.1733	10.8754	9.046	107.21%	18.05%	5.46%	7.83%
1902	MV01-3	12.5738	27.7802	15.8265	14.5800	10.2580	120.94%	25.87%	15.96%	18.42%
	MV01-4	9.8109	19.8773	11.3242	9.9959	10.1065	102.60%	15.42%	1.89%	3.01%
1903	MV01-5	17.8023	23.2072	13.2212	11.9433	10.6752	30.36%	25.73%	32.91%	40.03%
	MV01-6	11.5987	25.1702	14.3396	12.8113	11.2931	117.01%	23.63%	10.46%	2.63%
1904	MV01-7	6.1046	15.9731	9.0999	8.2797	37.1136	161.66%	49.07%	35.63%	507.97%
	MV01-8	5.3415	12.0980	6.8923	6.2027	4.4445	126.49%	29.03%	16.12%	16.79%
1905	MV01-9	6.5842	18.8667	10.7485	9.6345	11.7615	186.55%	63.25%	46.33%	78.63%
	MV01-10	6.4534	13.3703	7.6171	6.9772	5.4341	107.18%	18.03%	8.12%	15.79%
	MV01-11	3.3793	8.3682	4.7674	4.3053	3.1321	147.63%	41.08%	27.40%	7.32%
1906	MV01-12	5.8865	14.0246	7.9899	7.3953	5.3135	138.25%	35.73%	25.63%	9.73%
	MV01-13	5.1017	11.7054	6.6686	6.0209	4.3055	129.44%	30.71%	18.02%	15.61%
1907	MV01-14	4.2078	11.5600	6.5858	5.9191	6.8720	174.73%	56.51%	40.67%	63.32%
	MV01-15	5.2325	12.5705	7.1615	6.2750	7.0865	140.24%	36.87%	19.92%	35.43%
1908	MV01-16	6.7586	17.5072	9.9739	9.1783	7.2319	159.03%	47.57%	35.80%	7.00%
	MV01-17	5.3415	12.7741	7.2775	6.4583	6.3479	139.15%	36.24%	20.91%	18.84%
1909	MV01-18	4.3822	8.6954	4.9538	4.5916	2.9494	98.43%	13.04%	4.78%	32.69%
	MV01-19	5.3197	12.2288	6.9668	6.2811	6.9719	129.88%	30.96%	18.07%	31.06%
	MV01-20	5.0145	10.4839	5.9727	5.3049	6.2998	109.07%	19.11%	5.79%	25.63%
1910	MV01-21	5.2325	8.5936	4.8958	4.2737	7.9195	64.24%	6.43%	18.32%	51.35%
	MV01-22	5.9519	13.2103	7.5260	6.8239	6.5464	121.95%	26.45%	14.65%	9.99%
	MV01-23	5.9737	13.0940	7.4597	6.6869	7.2445	119.19%	24.88%	11.94%	21.27%
Average							123.33 %	30.03%	19.21 %	44.74%

Table 7 MV 02 Fuel Consumption Estimation Results

Voyage Number	Trip Number	MFO Fuel Oil Consumption (Ton)					% Error			
		Report	Trozzi	Jalkanen	Wang	Modified Mersin	Trozzi	Jalkanen	Wang	Modified Mersin
1901	MV02-1	6.8528	14.6135	8.3254	7.3404	10.1839	113.25%	21.49%	7.12%	48.61%

Voyage Number	Trip Number	MFO Fuel Oil Consumption (Ton)					% Error			
		Report	Trozzi	Jalkanen	Wang	Modified Mersin	Trozzi	Jalkanen	Wang	Modified Mersin
	MV02-2	7.2343	11.4800	6.5402	5.9828	4.8117	58.69%	9.59%	17.30%	33.49%
	MV02-3	5.0878	6.5288	3.7195	3.2733	3.7711	28.32%	26.89%	35.66%	25.88%
1902	MV02-4	13.6708	27.6712	15.7644	14.2235	14.8210	102.41%	15.31%	4.04%	8.41%
	MV02-5	22.8207	27.1623	15.4745	14.3506	8.8674	19.02%	32.19%	37.12%	61.14%
1903	MV02-6	12.9236	22.6401	12.8982	11.7700	10.4968	75.18%	0.20%	8.93%	18.78%
	MV02-7	7.9161	15.5587	8.8639	8.2047	5.6832	96.54%	11.97%	3.65%	28.21%
	MV02-8	4.2018	8.8481	5.0408	4.5228	3.9753	110.58%	19.97%	7.64%	5.39%
1904	MV02-9	10.9476	13.1522	7.4929	6.8823	5.5311	20.14%	31.56%	37.13%	49.48%
	MV02-10	6.8815	12.6432	7.2029	6.5758	4.1805	83.73%	4.67%	4.44%	39.25%
1905	MV02-11	5.1037	10.5494	6.0100	5.3621	6.4119	106.70%	17.76%	5.06%	25.63%
	MV02-12	7.7268	13.3267	7.5923	6.8251	7.7250	72.47%	1.74%	11.67%	0.02%
	MV02-13	6.4940	15.7550	8.9757	8.1404	7.1160	142.61%	38.21%	25.35%	9.58%
1906	MV02-14	7.4721	17.8198	10.1520	9.5160	5.4315	138.48%	35.86%	27.35%	27.31%
	MV02-15	6.5347	14.7953	8.4289	7.5937	6.5595	126.41%	28.99%	16.21%	0.38%
1907	MV02-16	8.1321	17.6380	10.0485	9.3028	6.5368	116.89%	23.56%	14.40%	19.62%
	MV02-17	6.8022	15.1588	8.6360	7.8484	6.9512	122.85%	26.96%	15.38%	2.19%
1908	MV02-18	9.1777	15.9440	9.0834	8.1496	9.5801	73.73%	1.03%	11.20%	4.39%
	MV02-19	11.7870	12.9850	7.3976	6.7375	5.9797	10.16%	37.24%	42.84%	49.27%
	MV02-20	4.5130	8.6809	4.9455	4.5066	3.0082	92.35%	9.58%	0.14%	33.34%
1909	MV02-21	12.0843	13.7483	7.8325	6.9730	9.7606	13.77%	35.18%	42.30%	19.23%
	MV02-22	11.4570	13.9955	7.9733	7.2324	6.1837	22.16%	30.41%	36.87%	46.03%
	MV02-23	4.4932	8.1574	4.6473	4.2084	3.2731	81.55%	3.43%	6.34%	27.15%
1910	MV02-24	10.4759	15.8059	9.0047	8.3997	5.0264	50.88%	14.04%	19.82%	52.02%
	MV02-25	8.0657	13.5375	7.7124	7.2116	3.6567	67.84%	4.38%	10.59%	54.66%
	MV02-26	4.3763	7.6267	4.3449	3.9069	3.0456	74.27%	0.72%	10.72%	30.41%
						Average	77.73%	18.57%	17.66%	27.69%

If viewed from calculation method, Wang resulted lowest error because the initial formula generated from the daily fuel oil consumption from ship's bunker history data. For Jalkanen, it resulted less average error compared with other methods. Operational parameter(s) used in Jalkanen method to calculate the estimation of fuel oil consumption were relative complex and many variables were considered during the process. Using ship resistance calculation for forecasting ship's engine power demands data for ship's hull dimension, propellers, and mechanical efficiency. Method proposed by Mersin considers the alteration of ship's displacement for estimating its fuel oil consumption. Mersin method includes ship engine coefficient in its calculation with displacement, speed and fuel consumption per day. Trozzi method only uses the ship's gross tonnage (GT), operational mode and type of ship in the calculation process. Potential errors that occur in the Trozzi model can be caused by the assumption that the Trozzi model for container ships assumption (hull shape, size, speed, etc.) are incompatible with the design of both MV 1 and MV 2.

For example, typical coefficient block (Cb) for containerships is 0,52 to 0,716 [13]. Both MV 01 and MV 02 has coefficient block 0,8487. Both ships also have lower service speed (6-8 knots) than general containerships design (>16 knots) [18]. From the findings, it can be hypothesized that inter-Indonesian archipelago container shipping fleet has different characteristic compared with global container fleet.

From the result of each models, Wang models represent the lowest error. The result of EEOI index from Wang's model calculated and compared with EEOI error from bunker report showed in Table 8. This result only calculates the EEOI index from sailing speed operation which only use MFO as fuel.

Table 8 MV 01 & MV 02 EEOI Index Result with Report FOC and Estimated FOC calculated using Wang's model

MV 01				MV 02			
Voyage Number	Trip Number	EEOI Report	EEOI Estimated	Voyage Number	Trip Number	EEOI Report	EEOI Estimated
1901	MV01-1	0,000408	0,000436	1901	MV02-1	0,000446	0,000478
	MV01-2	0,000369	0,000390		MV02-2	0,000418	0,000346
1902	MV01-3	0,000516	0,000598		MV02-3	0,000558	0,000359
	MV01-4	0,000345	0,000352	1902	MV02-4	0,000436	0,000454
1903	MV01-5	0,000922	0,000619		MV02-5	0,000749	0,000471
	MV01-6	0,000393	0,000435	1903	MV02-6	0,000619	0,000564
1904	MV01-7	0,000281	0,000381		MV02-7	0,000335	0,000347
	MV01-8	0,000400	0,000464		MV02-8	0,000421	0,000454
1905	MV01-9	0,000274	0,000401	1904	MV02-9	0,000551	0,000347
	MV01-10	0,000288	0,000311		MV02-10	0,000669	0,000639
	MV01-11	0,000339	0,000432	1905	MV02-11	0,000466	0,000490
1906	MV01-12	0,000297	0,000374		MV02-12	0,000393	0,000348

IOP Conf. Series: Earth and Environmental Science **1461** (2025) 012048 doi:10.1088/1755-1315/1461/1/012048

MV 01				MV 02			
Voyage Number	Trip Number	EEOI Report	EEOI Estimated	Voyage Number	Trip Number	EEOI Report	EEOI Estimated
	MV01-13	0,000218	0,000257		MV02-13	0,000397	0,000497
1907	MV01-14	0,000211	0,000297	1906	MV02-14	0,000386	0,000492
	MV01-15	0,000223	0,000268		MV02-15	0,000290	0,000337
1908	MV01-16	0,000330	0,000448	1907	MV02-16	0,000407	0,000466
	MV01-17	0,000422	0,000511		MV02-17	0,000287	0,000331
1909	MV01-18	0,000430	0,000451	1908	MV02-18	0,000450	0,000400
	MV01-19	0,000297	0,000351		MV02-19	0,000539	0,000308
	MV01-20	0,000291	0,000308		MV02-20	0,001975	0,001972
1910	MV01-21	0,000446	0,000364	1909	MV02-21	0,000599	0,000346
	MV01-22	0,000573	0,000657		MV02-22	0,000502	0,000317
	MV01-23	0,000293	0,000328		MV02-23	0,000618	0,000578
				1910	MV02-24	0,000574	0,000460
					MV02-25	0,000433	0,000387
					MV02-26	0,000757	0,000676

4. Conclusion
In conclusion, study above showed that MV 01 has average EEOI index result 0.000473 ton-CO2/TEUS.nm and is lower than MV 02 which index is 0.000630 ton-CO2/TEUS.nm. It means that overall MV 01 EEOI index is 33.2% lower than MV 02. This study also shows that the most significant factor causing the increase in EEOI index is the amount of cargo carried by ship.

The reason underlying MV 01 has lowest index is that MV 02 consumed more fuel oil during voyage route Surabaya-Sampit, Surabaya-Kumai, and Semarang-Kumai. The amount of average carried cargo for MV 02 was less than MV 01 in voyage route Surabaya-Semarang and Surabaya-Sampit.
In terms of four fuel oil estimation methods being compared, Wang's method has lowest average error with overall error is 18.44%. It then followed by Jalkanen with 24.3%, modified Mersin with 36.22%, and Trozzi's method with 100.53% error.

Further action proposed by this study to increase the EEOI index of both MV 01 and MV 01 is by making improvement in its cargo management for shipping efficiency. In addition, for MV 02 checking of operational aspect is needed especially on machinery system and hull conditions. Further research which includes the external and technical factors including hull and propeller condition, weather and sea conditions, and engine performance is strongly recommended to complete conclusions and get better understanding about its impact on the ship's EEOI index.

IOP Conf. Series: Earth and Environmental Science **1461** (2025) 012048 doi:10.1088/1755-1315/1461/1/012048

Acknowledgements
The authors wish to acknowledge AIS data support from FleetMon for Marine Operational and Maintenance Laboratory, Marine Engineering Department, Institut Teknologi Sepuluh Nopember, Surabaya, Indonesia. We also wish to acknowledge one of Indonesian Shipping Company for sharing ship operational data which help the authors to conduct this research

References
[1] M. Stopford. 2009. Maritime Economics, 3rd ed. Oxford, Routledge, 2009, pp. 35-39.
[2] International Maritime Organization. 2021. Fourth IMO GHG Study 2020. London, International Maritime Organization
[3] United Nations Framework Convention on Climate Change. 2015. Paris Agreement. Paris: United Nations Framework Convention on Climate Change.
[4] Deng, J., Wang, X., Wei, Z., Wang, L., Wang, C., & Chen, Z. (2021). A review of NOx and SOx emission reduction technologies for marine diesel engines and the potential evaluation of liquefied natural gas fuelled vessels. Science of the Total Environment, 766, 144319.
[5] Ni, P., Wang, X., & Li, H. (2020). A review on regulations, current status, effects and reduction strategies of emissions for marine diesel engines. Fuel, 279, 118477.
[6] MEPC.282(70). 2016. Guidelines for the Development of a Ship Energy Efficiency Management Plan (SEEMP). London, International Maritime Organization.
[7] Prastowo, H., Pitana, T., Artana, K. B., & Ariana, I. M. (2024). Evaluation of Carbon Intensity Index for Ship with Recommended Actions to Improve CII Rating. International Review of Mechanical Engineering, 18(5), 247-255.
[8] Prastowo, H., Pitana, T., & Kusuma, H. P. (2020, August). Activity-Based Fuel Oil Consumption Estimation for Calculating Energy Efficiency Operational Indicator (EEOI) in an Indonesian Merchant Ship. In IOP Conference Series: Earth and Environmental Science (Vol. 557, No. 1, p. 012053). IOP Publishing.
[9] JAKOTA Cruise Systems GmbH. 2019. FleetMon. [Online], Available at: www.fleetmon.com, Accessed 9 May 2019.
[10] MEPC.1/Circ.684. 2009. Guidelines for Voluntary Use of the Ship Energy Efficiency Operational Indicator (EEOI). London, International Maritime Organization.
[11] Trozzi, C., & Vaccaro, R. (2024). Air pollutant emissions from ships: high Tyrrhenian Sea ports case study. WIT Transactions on The Built Environment, 39.
[12] Irvana, R., Abdullah, K., Uyun, A. S., & Anne, O. (2024). Transformation Conventional Fishing Vessels 30 GT to Hybrid (Diesel-PV) System to Reduce Fuel Oil Consumption and Exhaust Emissions. In BIO Web of Conferences (Vol. 104, p. 00033). EDP Sciences.
[13] Jalkanen, J. P., Johansson, L., Kukkonen, J., Brink, A., Kalli, J., & Stipa, T. (2012). Extension of an assessment model of ship traffic exhaust emissions for particulate matter and carbon monoxide. Atmospheric Chemistry and Physics, 12(5), 2641-2659.
[14] Luo, X., Yan, R., Lang, X., & Wang, S. (2024). Accuracy and applicability of ship's fuel consumption prediction models: A comprehensive comparative analysis. Energy, 133187.
[15] K. U. Hollenbach. 1998. Estimating Resistance and Propulsion for Single Screw and Twin Screw Ships. Ship Technology Research, p. 45/2.
[16] H. Schneekluth, V.Bertram. 1998. Ship Design for Efficiency and Economy, 2nd ed. Oxford, Butterworth-Heinemann. pp. 197-200.
[17] D. G. M. Watson. (1998). Practical Ship Design, vol 1. Oxford, Elsevier. pp. 76, 219.
[18] S. Wang, Q. Meng. (2012). Sailing Speed Optimization for Container Ships In a Liner Shipping Network. Pp. 701-714. DOI: https://doi.org/10.1016/j.tre.2011.12.003.
[19] K. Mersin, G. Alkan, T. Mısırlıoğlu. (2017). A New Method for Calculating Fuel Consumption and Displacement of a Ship in Maritime Transport. DOI: https://doi.org/10.1080/23311916.2017.1415107.

[20] C. B. Barras. (2004). Ship Design and Performance for Masters and Mates, 1st ed. Exford Elsevier, Butterworth-Heinemann, 2004, pp. 133-135.

Utilization of AIS Data and Ship Simulator Integration in Maritime Safety: A Systematic Literature Review

Maulidiah Rahmawati[1,2*], Trika Pitana[1], Dhimas Widhi Handani [1] and Nurhadi Siswantoro [1]

[1] Departement of Marine Engineering, Institut Teknologi Sepuluh Nopember, Surabaya, Indonesia
[2] Sea Transportation, Politeknik Pelayaran Surabaya, Surabaya, Indonesia

*E-mail: 7019222006@student.its.ac.id, maulidiah@poltekpel-sby.ac.id

Abstract. Maritime safety is a crucial issue that has always been a major concern in the global shipping industry. The increase volume in global shipping can have an impact on the density of water traffic which has consequences including on safety aspects, i.e. traffic. Automatic Identification System (AIS), as a communication system in traffic monitoring or traffic management. The consequences related to maritime safety involve human factor, i.e. seafarer. As regulated from International Convention on Standards of Training, Certification and Watchkeeping for Seafarers (STCW) 1978, as amended in 1995/2010 requires the assessment of the fulfilment of training and assessment for seafarers. Means for facilitating the fulfilment of training and assessment for seafarers can be conducted through ship simulator. The paper aims to review the utilization of AIS data and ship simulator integration in maritime safety. It has been identified that research areas involving AIS data and ship simulator in maritime safety are maritime accidents, education & training, navigation related to autonomous ship. The method to analyze data can be classified decision-making, risk management, mathematical/statistical analysis, and digital technology. From the finding, research area can be developed into Maritime Accident, Traffic, Ship Behaviour, Education & Training, Navigation/Autonomous, and Human Factor. Improvement in research area can be conducted to support sustainability and environmental protection.

1. Introduction

Maritime safety is a critical issue that needs essential concern, due to its sustainability, many risks and impacts related to loss of life, property, economic harm, and environmental conditions or pollution [1][2][3]. Environmental condition can be one impact of maritime safety due to fuel or oil spill [3][4], which have consequences in water pollution that can be impact to preserving natural resources.

One component that contributes to maritime safety is traffic. Automatic Identification System, abbreviated as AIS, is a communication system or ship data encounter utilized as a traffic monitoring or traffic management system to monitor the navigation status of the ship in real-time accurately [5]. AIS is an automated tracking system using transceivers installed on ships and used by ship traffic services. At the international level, AIS was utilized in the year of 2000 as the implementation of the Safety of Live At Sea (SOLAS) convention as amended by the International Maritime Organization [6], [7]. Implementation of AIS is mandatory for a vessel with more than 300 GT (Gross Tonnage) for international voyages [8][9][6][10], and a cargo ship with 500 GT, and a passenger ship for all of size [11][7]. AIS is categorized into 2 types, Class A and Class B. AIS Class A is mandatory, and must fitted in a convention vessel with 300 GT or more, and Class B is for non-convention vessels or non-mandatory [9] with lower expense [11]. Based on IMO requirements, AIS has been applied effectively since 31st of December 2004 [9].

However, the IMO provision is applied to vessels in international waters. For territorial waters, each sovereign country has been provided authority to regulate its own compliance according to the needs of each country. For Indonesian territorial waters, vessels are required to install and activate AIS by Transportation Minister Regulation No. PM 7/2019. It has been effective since August 20, 2019. It regulates that Indonesian-flagged ships that fulfill the standards of the Safety of Life at Sea (SOLAS) Convention must install and activate AIS Class A while in Indonesian waters, Non-conventional Indonesian-flagged ships i.e. Passenger and freight ships with a minimum capacity of 35 GT, cross country vessel, and fishing vessels with a minimum size of 60 GT are obliged to install and activate AIS Class B, and foreign-flagged vessels are required to install and activate AIS while in Indonesian waters. There are some differences of AIS Class A and Class B regarding on time interval for data transmission. Time interval for data transmitted by AIS Class A depend on their navigational position, 6 minutes for static and voyage-related information, and every 2-10 second for dynamic information [6], [12], while Class B minimum 5 seconds [12].

AIS is a device that has a function as transmitter [6], [12] fitted on board [7]. It contains information, that consist of static (width, length, draft, type, speed, IMO number, MMSI/Maritime Mobile Service Identity, number and call sign), dynamic (location/ longitude & latitude, speed, course and heading, time, course, rate of turn, speed over ground) and voyage-related data (draft of the ship, destination, and type of cargo, [6], [7], [9], [12]–[14].

One of the factors that play a role in maritime safety is the human factor. Regulations related to human factors, in this case seafarers, are regulated by IMO through International Convention on Standards of Training, Certification and Watchkeeping for Seafarers (STCW) 1978, as amended in 1995/2010. International Convention on Standards of Training, Certification and Watchkeeping for Seafarers stated that there are provisions related to the assessment of the fulfilment of training and assessment requirements for seafarers [15]. Fulfilment of training standards and seafarer certification are implemented through Maritime Education and Training/MET, as a gateway to retrieve seafarer [16]. One of the facilities or means used in maritime education and training is a simulator, i.e. *Full Mission Bridge Simulator*. Through the simulator, seafarers gain experience related to competencies in maritime operations, such as berthing, navigation and pilotage, in a safe and comfortable environment [15], as a virtual setting according to the condition on board. Some effort to improve maritime safety, come from some research or studies that integrate the utilization of AIS data and ship simulators. This paper aims to present systematic review on utilization and methodology of AIS data and maritime simulator integration related to maritime safety.

2. Utilization of AIS

Initially, AIS was used for navigation-oriented [12] and traffic service purposes [6] to reduce the risk of ship collision [9], [11], [13], [14]. The utilization of AIS Data has been discussed by many scholars in their research. The research shows that in its development, AIS is not only used for safety purposes but also the other applications. [12] divided AIS application into 3 categories, i.e. basic application (data mining, navigation safety); extended application (ship behaviour, environmental analysis), and advanced application (trade analysis, arctic shipping, ship, and port performance). While [6] identify that AIS data has been utilized for traffic, logistics and transport economy, monitoring, collision and navigational safety, emissions, oil discharge, the problem of noise and marine mammals, interaction with whales, movement of fishing vessels, and ice condition.

3. AIS Data in Maritime Safety

Research on the utilization of AIS data related to maritime safety has been conducted by some scholars. . Initially, AIS was used for navigation-oriented. [11] have reviewed some research related to Improving Ship Safety Operations at Sea based on AIS data in Indonesia. They identified and analysed issues of AIS data related to danger score, hazard navigation map, ship inspection, and emission monitoring [11]. Initially AIS data was utilized for ship collision avoidance, as intended objective [6]. [6] find that AIS research is not only about maritime safety but also related to environmentally sustainable marine transportation systems. The review shows that two current global issues related to maritime safety and the environment can be a challenge for future research [6]. AIS data can be used to improve maritime safety by providing information for traffic management, collision avoidance, dark vessel detection, maritime spatial planning, and official statistics [14].

The systematic reviews attempt to classify research of utilization of AIS data and ship-simulator integration related to maritime safety. It contains risk assessment, traffic monitoring, ship behaviour and big data.

3.1. AIS Data in Risk Analysis & Evaluation

Some researchers discuss the utilization of AIS data related to risk analysis. [17] propose a fuzzy failure mode and effects analysis (FMEA) model to evaluate the risk of ship collisions in the Malacca Strait using Automatic Identification System (AIS) data, that can be developed and applied to real-world problems. The proposed model considers various factors that affect collision risk, such as vessel speed, distance, and course [18]. By using AIS data, [19] estimate the probability and consequence, indices from a risk matrix, the time and encounter conditions to determine the level of risk and to obtain safety measurements.

Statistical analysis, the combination of Correspondence Analysis and F-test with a multivariate logistic regression model, from AIS data and accident database, are used as risk indicators to predict the likelihood of a navigation-related accident [20]. [21] propose the improvement of safety measures by developing a collision risk model from AIS data to obtain the Collision Risk Index of Ferries (CRIF) by considering some Risk Influencing Factors (RIFs) to enhance navigation safety. [20] apply bivariate and multivariate statistical analysis to identify risk factors and indicators by using a combination of AIS data and data sets related to navigational accidents, e.g. grounding and collision.

3.2. AIS Data in Traffic Analysis

In line with its function as a traffic monitoring system, some researchers point out the utilization of AIS data. [18] use AIS as a data source to calculate the dangerous score to explain the features of ship traffic volume by a geographic information system (GIS) and the analytic hierarchy process (AHP) to assess the degree of safety and weighting variables in the Malacca Strait. Research on safety measurements was conducted by estimating the probability and consequence, indices from a risk matrix, and the time, and encounter conditions determined the level of risk [5].

AIS data can be a data source to assess the performance of ship traffic. [10] propose a new method of evaluating the performance of shipping traffic under current speed limits by utilizing AIS data which involves data acquisition, error elimination, modelling of traffic flow characteristics, and estimation of the correlation between ship speed and congestion level. From the expert judgment, [10] offer insights that can be applied to test the rationality of speed limits in other waterways or shipping channels.

To cope with water conditions. [2] assess and verify the suggested alternative clustering algorithms, utilizing historical AIS data by comparing the effectivity of main modules in the proposed approach's functionality and usefulness. The new methodology shows significant potential to improve the operators' intelligence surveillance skills and to make risk mitigation easier to apply [2].

3.3. AIS Data in Ship Behavior

Some researchers discuss the utilization of AIS data in ship behavior. [7] propose a method by k-means clustering and a naïve Bayes classifier to categorize ships based on their behavior patterns, including speed, direction, and frequency of maneuvering. The clustering results show patterns in ship path and speed over ground, which are dominant behavior attributes in ports, as well as patterns in integral ship behavior through correlation analysis, and ship length and beam are identified as explanatory variables [7].

Proactive measures to prevent collision can be conducted by prediction based on previous AIS data. [14] propose the usage of deep learning model based on AIS data to generate predictions about the ship behaviour in the region. The model extracts long-term correlations from AIS trajectories in an enriched space to reduce the complexity of the data and improve accuracy [14]. From the research [14] find that the challenges of using AIS data for maritime safety include the ability to receive large numbers of messages simultaneously, incomplete or infrequent tracks, false signals, spoofed tracks, and mistaken identity, gaps in transmissions, and privileged access through competent public bodies.

According to maritime navigation, the other function of AIS is to assist with ship behaviour in restricted waterways by U-turn movement as the ship manoeuvres to reduce the possibility of encounter or collision [22]. [22] offer an AIS data-based technique to capture the off-destination-dock U-turn behaviour of large ships in narrow channels rather effectively, that is adaptable and can be applied to other similar narrow waterways. [22].

4. AIS as Big Data in Maritime Safety

The maritime industry generates a vast amount of data from many sources and in various forms [12], [23] recently known as big data. Big data has become a popular subject, which analysis looks for connections between many measurable and unmeasurable characteristics, that can be described from its volume, variety, velocity, and veracity, with the purpose to reveal hidden patterns and trends [17][23], that can provide important progress and influence in shipping operation [12]. In adapting to technological demands, related to maritime safety, big data have some advantages and disadvantages, as opportunities and challenges. In line with the impact of big data technology, such as enhancing performance monitoring, eliminating human errors, and

IOP Conf. Series: Earth and Environmental Science **1461** (2025) 012049 doi:10.1088/1755-1315/1461/1/012049

fostering more significant linkages between components, but there are still some constraints related to data processing, reliability, and ensuring data security [17].

Considering the capacity of data, AIS can be viewed as a source of big data. [12] describe application of AIS data in maritime study into three categorizes e.g. basic application, extended application, and advanced application. While, [24] figure out the application of Big Data from AIS as a part of Big Data and Artificial Intelligence in maritime related to maritime surveillance and sustainability.

5. Ship Simulator in Maritime Safety

5.1. Ship Simulator in Training and Education

Currently, simulation and simulator-based learning techniques have been utilized as one of training tools that can open up new opportunities to present real conditions in the learning environment [25]. Some scholars conducted simulation research using simulators in maritime training and education as an effort to increase maritime safety. The diverse topics in ship simulators was discussed covering some different areas. The areas emphasize on human factor related to seafarer's psychology and physiology, for example, seafarers' emotion on human performance using Self-Assessment Manikin (SAM) method and machine learning [26], mental state and physiology using explorative experiment to find their correlation [27] in ship simulator operation. The assessment performance of pilotage operation by using computer and full scale simulator is conducted by [15], to test the validity and reliability of the proposed tool.

In line with development in computer technology, bridge simulator, as a simulation tool related to skill and experience in navigation learning [28], can be attached with additional application. One of additional application is virtual reality that can provide immersive experince and interaction to present imitation circumstances by using 3 D software [29]. From virtual reality, the certain environment can be created to model the ship bridge simulator [30]. A full mission bridge simulator, inserted with operational scenario, can be modified with virtual reality, to be implemented and tested to users [31]. In maritime training and education, simulator can be used as a tool to conduct the simulation on accident scenario [32].

5.2. Ship Simulator in Maritime Traffic and Risk Assessment

Maritime simulator can be a tool to evaluate performance of pilotage operation to test computer-based reliability and validity [15]. Maritime simulator is improved as evaluation tool not only for pilotage operations [15] and pilotage training [33], but also for competence of watchkeeping officers [28] and maneuvering [33]. Usage of ship simulator and AIS data were discussed in the other areas related to ship risk management such as risk assessment [34] and risk analysis [35].

5.3. Ship Simulator in Navigation and Autonomous Ship

Ship-handling simulator has an important role in navigation safety related to port and water way for ship manoeuvre models [36]. A full mission bridge simulator has been utilized as equipment tools to test the ACCSEAS project's e-Navigation route exchange [37]. In navigation areas, simulator is used to evaluate navigation safety [38], [39] related to the path, sea traffic [38] and trial means algorithm [39]. Discussion related to simulation of navigation is conducted by Aydogdu (2022). He used scenario in Full Mission Ship Handling Simulator as a means to realize the passage environment, to identify the level of risk by using the Environmental Stress model. Ship motion simulator is also used as a tool to test and verify the simulation-based environment before implemented to real ship by combining with the electronic chart platform and scenario design to develop the system related to Autonomous Navigation Decision-making [41]. In line with development of navigation technology in decreasing accident, shiphandling simulator is utilized as a means to compare emotional response to emergency between navigational officers and marine pilots [42]

As increasing role of Ship-bridge simulators for research and innovation, it has wide scope, not only in maritime training, but also related to autonomous controller and pilots recruitment [43]. A ship-bridge simulator-based scenario provide some contributions as an efective tool in evaluation of human performance or operator behaviour involved in developing complex system [44], such as Maritime Autonomous Surface Ship (MASS) [16].

6. Published Research of AIS and Ship Simulator in Maritime Safety

The published papers from some scholars that have discuss AIS data related to maritime safety are shown in Table 2.

Table 1. Legend of Table 2

Means	AIS	Automatic Identification System
	SS	Ship Simulator
Type of Method/ Analysis	DM	Decision Making
	RM	Risk Management
	M/SA	Mathematical/Statistical Analysis
	DT	Digital Technology (AI/VR/AR)
	Q/KA	Qualitative/Kognitif Analysis
Research Area	MA	Maritime Accident
	T	Traffic
	SB	Ship Behaviour
	ET	Education & Training
	N/A	Navigation/Autonomous
	HF	Human Factor

From Table 2, it can be known that the trend of data analysis is statistical approach. Along with development of technology, there are potential research in AIS data related to artificial intelligence. Based on Table 1, it can be shown the distribution of Type of Data Analysis and Research Area as displayed in Figure 1.

Table 2. Published Research of AIS Data and Ship Simulator in Maritime Safety

No.	Author /year	Topic	Means		Type of Method/Analysis					Research Area						Method/Analysis
			AIS	SS	DM	RM	M/SA	DT	Q/KA	MA	MT	SB	MET	N/A	HF	
1	[8]	*Formulation of Dangerous Score*	√		√					√						*Analytic Hierarchy Process (AHP)*
2	[13]	*Formulation of Dangerous Score*	√		√						√					*Analytic Hierarchy Process (AHP)*

No.	Author /year	Topic	*Means*			*Type of Method/Analysis*				*Research Area*						*Method/Analysis*
			AIS	SS	DM	RM	M/SA	DT	Q/KA	MA	MT	SB	MET	N/A	HF	
3	[45]	*Development of ship sinking frequency model*	√			√				√						*A Bayesian Network*
4	[18]	*model for ship collisions risk evaluation*	√			√				√						*Fuzzy FMEA model*
5	[46]	*Simulation of Collision Avoidance*	√			√				√						*a three-degree-of-freedom (DOF) model, servo model*
6	[47]	*Ship Safety Index*	√				√			√						*Binary Logistic Regression*
7	[5]	*Risk of Navigation for Marine Traffic*	√			√					√					*Quantitative risk assessment models*
	[48]	*Ship Accident & Safety Level*					√			√						*Causality, statistical analysis*
8	[49]	*Generating real-time objects for ship simulator*	√	√				√		√				√		*machine learning*
9	[50]	*Developing a Maritime Safety Index using Fuzzy Logics*	√	√			√				√					*Fuzzy Logic*
10	[20]	*Risk Indicators*	√				√			√						*Correspondence analysis and F-tests logistic regression model*
11	[26]	*Seafarers' Emotion On Human Performance*		√				√		√			√		√	*Machine Learning*
12	[34]	*Index of Risk Assessment and Simulation of Collision Avoidance*	√			√	√			√					√	*CRI (collision-risk index)*
13	[7]	*Ship classification*	√				√					√				*Cluster analysis*
15	[51]	*A simulation model for ship navigation*	√				√				√					*A simulation model*
16	[25]	*Usage of simulators*	√	√			√						√			*interview, survey, observation*
17	[10]	*Maritime Autonomous Navigation Decision-Making System*	√				√				√					*estimation of the correlation*

No.	Author /year	Topic	*Means*		*Type of Method/Analysis*					*Research Area*						*Method/Analysis*
			AIS	SS	DM	RM	M/SA	DT	Q/KA	MA	MT	SB	MET	N/A	HF	
18	[52]	*Exploring AIS data*	√					√			√					*combination ship trip semantic object (STSO) and graph theory*
19	[15]	*Performance Assessment In Ship Simulator*		√	√								√		√	*Analytic Hierarchy Process (AHP) and Bayesian Network*
20	[36]	*Ship's Manoeuvre Models*	√	√			√							√		*Regression, ARMAX Model And Semi-Markov Process*
21	[28]	*Evaluation Procedures for Watchkeeping using ship simulator*		√					√				√	√		*Goal-Directed Task Analysis (GDTA)*
22	[38]	*The Use of a Full Mission Bridge Simulator*		√										√		*experiment*
23	[27]	*relationship between mental state and physiology using ship simulator*		√			√								√	*explorative experiments & correlation*
24	[53]	*Information retrieval for navigation*	√				√				√					*Multi linear and Lasso regressions*
25	[14]	*Ship Behavior Prediction*	√					√				√				*Trajectories Simplification Algorithm - deep learning*
26	[21]	*Index of Collision risk analysis*	√			√	√			√						*A collision risk model entropy theory*
27	[35]	*Risk Analysis Of Ship Collisions*	√	√		√				√	√					*Monte Carlo,a parameter sensitivity analysis, a mitigation analysis*
28	[31]	*Operational Scenarios in a Virtual Reality*		√				√	√				√			*tematic analysis*
29	[32]	*Simulator in maritime training and education*		√				√					√			*combination of face-to-face, blended, and online learning*
30	[54]	*Characterization of navigation conflicts*	√				√				√					*Graphical analysis*
31	[22]	*Study on U-turn behavior of vessel*	√				√					√				*corresponding solution algorithms*

No.	Author /year	Topic	Means		Type of Method/Analysis					Research Area						Method/Analysis
			AIS	SS	DM	RM	M/SA	DT	Q/KA	MA	MT	SB	MET	N/A	HF	
32	[39]	*Use Of Maritime Training Simulators*		√										√	√	*decision-making, mixed scenarios, algorithms*
33	[40]	*Navigational Risk Assessment using ship simulator*		√		√								√	√	*simulated scenario, risk analysis,*
34	[2]	*Maritime Traffic Clustering*	√				√				√					*Clustering technique*
35	[55]	*Simulator Fidelity*		√			√	√					√			*experiment, questionnaire*
36	[29]	*Application of virtual reality in ship bridge simulator*		√				√					√			*experiment, questionnaire*
37	[30]	*Modeling The Ship Bridge Simulator*		√			√	√					√			*Simulation, experiment, questionnaire*
38	[44]	*Maritime Safety Assessment*		√			√								√	*experiment, questionnaire, ANOVA, MANOVA*
39	[41]	*Maritime Autonomous Navigation Decision-Making System*	√	√	√									√		*Decision-making System, modeling, integration, real -ship trial*
40	[42]	*Exploring Seafarers' Emotion By Ship Simulator*		√											√	*support vector machine-based*

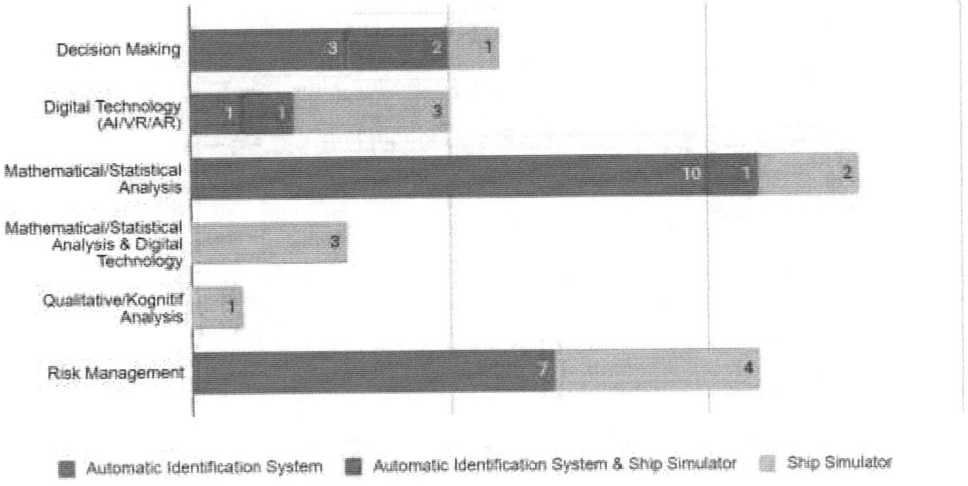

Figure 1. Type of Data Analysis of AIS and Ship Simulator.

Figure 1 shows the most frequence usage of integration of AIS data and ship simulator, that is related to decision making, while the other is associated with Mathematical/Statistical Analysis and Digital Technology. The other type of data analysis such as Risk Management and Qualitative/Kognitif Analysis is utilized for AIS and ship simulator respectively.

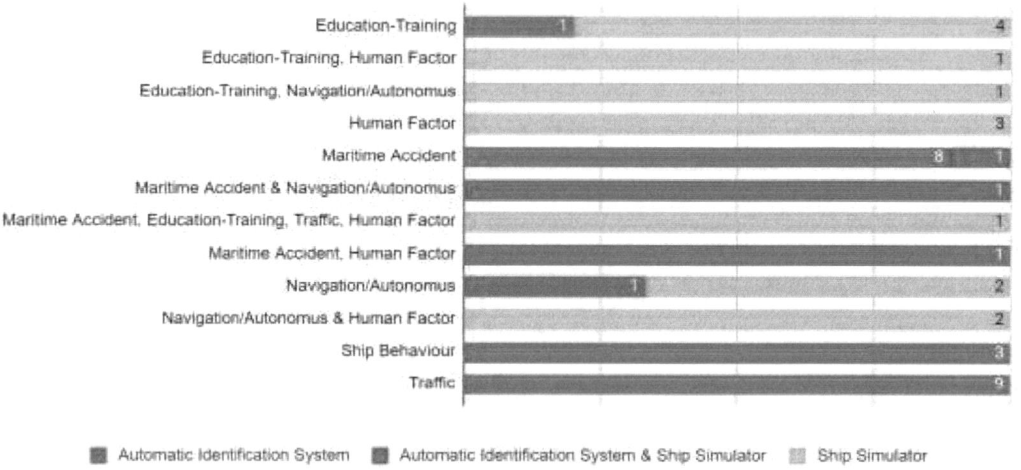

Figure 2. Type of Research Area of AIS and Ship Simulator

According to Figure 2, the highest frequency of research area in AIS data is traffic, the second ia maritime accident, while the most utilization of ship simulator is in relation to education and training respectively. The research area of integration of AIS data and ship simulator is related to navigation/autonomous, maritime accident and education & training.

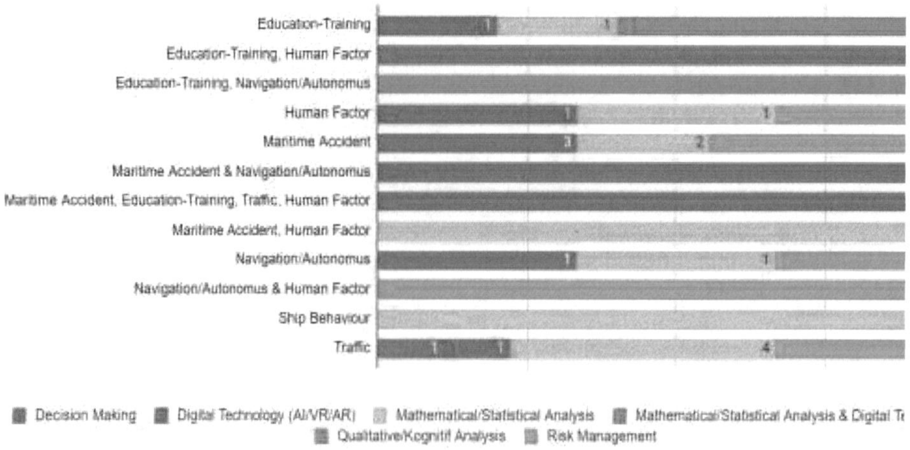

Figure 3. Type of Research Area and Data Analysis of AIS and Ship Simulator

Type of research area and data analysis of the utilization of AIS and ship simulator is shown in Figure 3. Meanwhile, education and statistical analysis is most used in research related to traffic, beside maritime accident and ship behaviour. Usage of Digital Technology is used in research related to education & training and human factor respectively, while Decision making utilized in education & training and human factor simultaneously.

Utilization of AIS data related to maritime safety can be classified into several types that consists of some analysis. Table 1 shows some analysis and models that can be classified into several types, such as multicriteria decision making, risk management, statistical analysis and big data. It involves many types of analysis, either probabilistic or non-probabilistic analysis.

Analysis of AIS Data related to decision making theory, as non-probabilistic or deterministic approach, is Analytical Hierarchy Process (AHP). Analytical Hierarchy Process (AHP) is combined with AIS Data to build navigation map from its weight criteria [8] and to analyse degree of safety from its danger score [13]. In multi criteria decision making, weight of criteria can be identified to identify the rank and to find the most influences variables, that can be used as a hint for further stage.

From previous researches, AIS Data is utilized for risk management to analyze maritime accident, whether ship collision [13], [21], [46] ship sinking [45] or ship traffic [5], [56].

Considering AIS as data base, statistical analysis, as probabilistic approach, is utilized for data analysis, that consist of bivariate and multivariate analysis, i.e. correspondence analysis [20], cluster analysis [2], [7], regression analysis [53], correlation [10]. Research area that use statistical analysis are related to maritime accident, traffic and ship behavior. By using regression analysis, it is possible to find the significance of correlation between variables and the magnitude of contributing factor or variables, between predictor/independent variables and independent variables. Another type of multivariate analysis is cluster analysis, which possible to classify objects with similar characteristics based on its variables.

Recently, big data is known as breakthrough technology related to data mining, processing, storing, and analysing data in diverse forms, where the huge amount or volume of data processed, saved, and evaluated. In maritime safety, AIS data is utilized using big data technology by machine learning algorithms [52] and deep learning frameworks [14], related to Artificial Intelligence in maritime.

7. Conclusion

The purpose of this paper was to provide an overview of identification the utilization of AIS data and ship simulator in maritime safety. The point of view can be different from many characteristics. Research area of AIS data related to maritime safety are maritime accident, traffic and ship behaviour, while type of method for data analysis consists of multicriteria decision making, risk management and statistical analysis. According to the analysis, AIS data should be integrated with another database. Another means that can be integrated with AIS data is ship simulator. It can be combined with some different perspectives such as human factor, risk management, evaluation of operation performance, navigation safety etc. In line with emerging technologies, AIS data can be considered as big data and can be threatened using machine learning or deep learning related to Artificial Intelligence. From literature review, it can be found that research area can be developed into Maritime Accident, Traffic, Ship Behaviour, Education & Training, Navigation/Autonomous, and Human Factor. Based on the research area, future work can be improved related to sustainability and environmental protection in water pollution prevention, preserving natural resources, etc.

References

[1] M. Fallahzadeh, "Evaluating safety control criteria in maritime traffic using formal safety assessment. Case study: Iranian port of bushehr," *J. Marit. Res.*, vol. 12, no. 3, pp. 37–48, 2015, [Online]. Available: https://api.elsevier.com/content/abstract/scopus_id/85028764702

[2] X. Xin, K. Liu, S. Loughney, J. Wang, and Z. Yang, "Maritime traffic clustering to capture high-risk multi-ship encounters in complex waters," *Reliab. Eng. Syst. Saf.*, vol. 230, no. October 2022, p. 108936, 2023, doi: 10.1016/j.ress.2022.108936.

[3] L. Zhang, H. Wang, Q. Meng, and H. Xie, "Ship accident consequences and contributing factors analyses using ship accident investigation reports," *Proc. Inst. Mech. Eng. Part O J. Risk Reliab.*, vol. 233, no. 1, pp. 35–47, 2019, doi: 10.1177/1748006X18768917.

[4] C. Dominguez-Péry, L. N. R. Vuddaraju, I. Corbett-Etchevers, and R. Tassabehji, *Reducing maritime accidents in ships by tackling human error: a bibliometric review and research agenda*, vol. 6, no. 1. Springer Singapore, 2021. doi: 10.1186/s41072-021-00098-y.

[5] M. B. Zaman, E. Kobayashi, N. Wakabayashi, and A. Maimun, "Risk of Navigation for Marine Traffic in the Malacca Strait Using AIS," *Procedia Earth Planet. Sci.*, vol. 14, pp. 33–40, 2015, doi: 10.1016/j.proeps.2015.07.082.

[6] M. Svanberg, V. Santén, A. Hörteborn, H. Holm, and C. Finnsgård, "AIS in maritime research," *Mar. Policy*, vol. 106, no. December 2018, p. 103520, 2019, doi: 10.1016/j.marpol.2019.103520.

[7] Y. Zhou, W. Daamen, T. Vellinga, and S. P. Hoogendoorn, "Ship classification based on ship behavior clustering from AIS data," *Ocean Eng.*, vol. 175, no. February, pp. 176–187, 2019, doi: 10.1016/j.oceaneng.2019.02.005.

[8] T. Pitana, "Development of Hazard Navigation Map by Using AIS data," *J. Marit. Res.*, vol. 1, no. 1, p. 61—85, 2011, doi: https://doi.org/10.24546/81004909.

[9] K. B. Artana, E. Pratiwi, D. Kristianto, D. S. Antara, and R. Time, "Implementation of Automatic Identification System for Safety Operation of Maritime Installation," vol. 74, no. Mastic, pp. 1–9, 2019, doi: 10.23977/mastic.001.

[10] L. Wang *et al.*, "Use of AIS data for performance evaluation of ship traffic with speed control," *Ocean Eng.*, vol. 204, no. October 2019, p. 107259, 2020, doi: 10.1016/j.oceaneng.2020.107259.

[11] A. A. Masroeri, "A Review on Some Research Issues on AIS to Improve the Ship Safety Operat ion at Sea," *J. Marit. Res.*, vol. 33, no. 1, p. 61—85, 2012, doi: https://doi.org/10.24546/81004320.

[12] D. Yang, L. Wu, S. Wang, H. Jia, and K. X. Li, "How big data enriches maritime research–a critical review of Automatic Identification System (AIS) data applications," *Transp. Rev.*, vol. 39, no. 6, pp. 755–773, 2019, doi: 10.1080/01441647.2019.1649315.

[13] Zaman, E. Kobayashi, N. Wakabayashi, T. Pitana, and A. Maimun, "Implementation of Automatic Identification System (AIS) for Evaluation of Marine Traffic Safety in Strait of Malacca using Analytic Hierarchy Process (AHP)," *J. Japan Soc. Nav. Archit. Ocean Eng.*, vol. 16, no. 0, pp. 141–153, 2012, doi: 10.2534/jjasnaoe.16.141.

[14] B. Murray and L. P. Perera, "An AIS-based deep learning framework for regional ship behavior prediction," *Reliab. Eng. Syst. Saf.*, vol. 215, no. May, p. 107819, 2021, doi: 10.1016/j.ress.2021.107819.

[15] J. Ernstsen and S. Nazir, "Performance assessment in full-scale simulators – A case of maritime pilotage operations," *Saf. Sci.*, vol. 129, no. April, p. 104775, 2020, doi: 10.1016/j.ssci.2020.104775.

[16] S. Fan and Z. Yang, "Towards objective human performance measurement for maritime safety: A new psychophysiological data-driven machine learning method," *Reliab. Eng. Syst. Saf.*, vol. 233, no. September 2022, p. 109103, 2023, doi: 10.1016/j.ress.2023.109103.

[17] M. B. Zaman and Semin, "Study on Danger Index for Ship’s Tanker," *Am. J. Eng. Appl. Sci.*, vol. 10, no. 2, pp. 366–372, 2017, doi: 10.3844/ajeassp.2017.366.372.

[18] M. B. Zaman, E. Kobayashi, N. Wakabayashi, S. Khanfir, T. Pitana, and A. Maimun, "Fuzzy FMEA model for risk evaluation of ship collisions in the Malacca Strait: Based on AIS data," *J. Simul.*, vol. 8, no. 1, pp. 91–104, 2014, doi: 10.1057/jos.2013.9.

[19] M. B. Zaman, E. Kobayashi, N. Wakabayashi, and A. Maimun, "Development of Risk Based Collision (RBC) Model for Tanker Ship Using AIS Data in the Malacca Straits," *Procedia Earth Planet. Sci.*, vol. 14, pp. 128–135, 2015, doi: 10.1016/j.proeps.2015.07.093.

[20] R. J. Bye and A. L. Aalberg, "Maritime navigation accidents and risk indicators: An exploratory statistical analysis using AIS data and accident reports," *Reliab. Eng. Syst. Saf.*, vol. 176, no. March, pp. 174–186, 2018, doi: 10.1016/j.ress.2018.03.033.

[21] M. Cai, J. Zhang, D. Zhang, X. Yuan, and C. G. Soares, "Collision risk analysis on ferry ships in Jiangsu Section of the Yangtze River based on AIS data," *Reliab. Eng. Syst. Saf.*, vol. 215, no. July, p. 107901, 2021, doi: 10.1016/j.ress.2021.107901.

[22] M. Kabir, M. J. Kang, X. Wu, and M. Hamidi, "Study on U-turn behavior of vessels in narrow waterways based on AIS data," *Ocean Eng.*, vol. 246, no. January, p. 110608, 2022, doi: 10.1016/j.oceaneng.2022.110608.

[23] I. Zaman, K. Pazouki, R. Norman, S. Younessi, and S. Coleman, "Challenges and opportunities of big data analytics for upcoming regulations and future transformation of the shipping industry," *Procedia Eng.*, vol.

194, pp. 537–544, 2017, doi: 10.1016/j.proeng.2017.08.182.

[24] Z. H. Munim, M. Dushenko, V. J. Jimenez, M. H. Shakil, and M. Imset, "Big data and artificial intelligence in the maritime industry: a bibliometric review and future research directions," *Marit. Policy Manag.*, vol. 47, no. 5, pp. 577–597, 2020, doi: 10.1080/03088839.2020.1788731.

[25] K. Saastamoinen, A. Rissanen, and R. Linnervuo, "Usage of simulators to boost marine corps learning," *Procedia Comput. Sci.*, vol. 159, pp. 1011–1018, 2019, doi: 10.1016/j.procs.2019.09.268.

[26] S. Fan, J. Zhang, E. Blanco-Davis, Z. Yang, J. Wang, and X. Yan, "Effects of seafarers' emotion on human performance using bridge simulation," *Ocean Eng.*, vol. 170, no. August, pp. 111–119, 2018, doi: 10.1016/j.oceaneng.2018.10.021.

[27] A. S. Wulvik, H. Dybvik, and M. Steinert, "Investigating the relationship between mental state (workload and affect) and physiology in a control room setting (ship bridge simulator)," *Cogn. Technol. Work*, vol. 22, no. 1, pp. 95–108, 2020, doi: 10.1007/s10111-019-00553-8.

[28] M. S. Solmaz, B. Özsever, A. Güllü, and C. Meşe, "Development of evaluation procedures for Watchkeeping officers using bridge simulator," *TransNav*, vol. 14, no. 3, pp. 656–571, 2020, doi: 10.12716/1001.14.03.07.

[29] T. Pitana, M. T. Pramudhita, N. Siswantoro, and J. R. C. Napitupulu, "Application of virtual reality as an alternative training media as complement of ship bridge simulator (case study: Port of tanjung emas semarang)," *IOP Conf. Ser. Earth Environ. Sci.*, vol. 1166, no. 1, 2023, doi: 10.1088/1755-1315/1166/1/012049.

[30] N. Siswantoro, D. Haryanto, M. A. Hikmahwan, and T. Pitana, "Implementation of Metaverse for Modeling the Ship Bridge Simulator Based on Virtual Reality As Educational Purposes, Case Study: Tanjung Priok Port Area," *IOP Conf. Ser. Earth Environ. Sci.*, vol. 1166, no. 1, p. 012053, 2023, doi: 10.1088/1755-1315/1166/1/012053.

[31] K. Aylward, J. Dahlman, K. Nordby, and M. Lundh, "education sciences Design Process," *Educ. Sci.*, vol. 11, p. 448, 2021.

[32] T. eun Kim *et al.*, "The continuum of simulator-based maritime training and education," *WMU J. Marit. Aff.*, vol. 20, no. 2, pp. 135–150, 2021, doi: 10.1007/s13437-021-00242-2.

[33] F. Fan, R. Guo, D. Wang, H. Geng, and X. Jiang, "Safety Assessment of Multi-Ships Entering and Berthing with Full Consideration of Real Navigation Environment in Complex Waters," *Am. J. Traffic Transp. Eng.*, vol. 8, no. 4, pp. 88–98, 2023, doi: 10.11648/j.ajtte.20230804.12.

[34] M. C. Nguyen, S. Zhang, and X. Wang, "A novel method for risk assessment and simulation of collision avoidance for vessels based on AIS," *Algorithms*, vol. 11, no. 12, 2018, doi: 10.3390/a11120204.

[35] A. Hörteborn and J. W. Ringsberg, "A method for risk analysis of ship collisions with stationary infrastructure using AIS data and a ship manoeuvring simulator," *Ocean Eng.*, vol. 235, no. October 2020, 2021, doi: 10.1016/j.oceaneng.2021.109396.

[36] N. M. Quy, K. Łazuga, L. Gucma, J. K. Vrijling, and P. H. A. J. M. van Gelder, "Towards generalized ship's manoeuvre models based on real time simulation results in port approach areas," *Ocean Eng.*, vol. 209, 2020, doi: 10.1016/j.oceaneng.2020.107476.

[37] T. Porathe, A. Brodje, R. Weber, D. Camre, and O. Borup, "Supporting Situation Awareness on the Bridge: Testing Route Exchange in a Practical e-Navigation Study," *Information, Commun. Environ.*, vol. 2, pp. 85–92, 2015, doi: 10.1201/b18514-12.

[38] V. Senčila, R. Zažeckis, A. Jankauskas, and R. Eitutis, "The use of a full mission bridge simulator ensuring navigational safety during the klaipeda seaport development," *TransNav*, vol. 14, no. 2, pp. 417–424, 2020, doi: 10.12716/1001.14.02.20.

[39] A. Vagale, O. L. Osen, A. Brandsæter, M. Tannum, C. Hovden, and R. T. Bye, "On the use of maritime training simulators with humans in the loop for understanding and evaluating algorithms for autonomous vessels," *J. Phys. Conf. Ser.*, vol. 2311, no. 1, 2022, doi: 10.1088/1742-6596/2311/1/012026.

[40] Y. V. Aydogdu, "Utilization of Full-Mission Ship-Handling Simulators for Navigational Risk Assessment: A Case Study of Large Vessel Passage through the Istanbul Strait," *J. Mar. Sci. Eng.*, vol. 10, no. 5, 2022, doi: 10.3390/jmse10050659.

[41] S. Wang, Y. Zhang, X. Zhang, and Z. Gao, "A novel maritime autonomous navigation decision-making system: Modeling, integration, and real ship trial," *Expert Syst. Appl.*, vol. 222, no. September 2022, p. 119825, 2023, doi: 10.1016/j.eswa.2023.119825.

[42] K. Shi, J. Weng, S. Fan, Z. Yang, and H. Ding, "Exploring seafarers' emotional responses to emergencies: An empirical study using a shiphandling simulator," *Ocean Coast. Manag.*, vol. 243, no. March, p. 106736, 2023, doi: 10.1016/j.ocecoaman.2023.106736.

[43] R. Zghyer, R. Ostnes, K. Henning Halse, O. Sveinung Hareide, and E. Johnsen, *Applications of Maritime Simulators in Industry and Research*, vol. 7, no. 1. 2022.

[44] S. Fan *et al.*, "Incorporation of seafarer psychological factors into maritime safety assessment," *Ocean Coast. Manag.*, vol. 237, no. December 2022, p. 106515, 2023, doi: 10.1016/j.ocecoaman.2023.106515.

[45] Y. Mulyadi, E. Kobayashi, N. Wakabayashi, T. Pitana, and Wahyudi, "Development of ship sinking frequency model over Subsea Pipeline for Madura Strait using AIS data," *WMU J. Marit. Aff.*, vol. 13, no. 1, pp. 43–59,

2014, doi: 10.1007/s13437-013-0049-2.

[46] I. Putu, S. Asmara, E. Kobayashi, and T. Pitana, "Simulation of Collision Avoidance in Lamong Bay Port by Considering the PAW of Target Ship using MMG Model and AIS Data," *Mar. Eng. Front.*, vol. 2, pp. 31–38, 2014, [Online]. Available: www.seipub.org/mef

[47] K. X. Li, J. Yin, and L. Fan, "Ship safety index," *Transp. Res. Part A Policy Pract.*, vol. 66, no. 1, pp. 75–87, 2014, doi: 10.1016/j.tra.2014.04.016.

[48] E. Eleftheria, P. Apostolos, and V. Markos, "Statistical analysis of ship accidents and review of safety level," *Saf. Sci.*, vol. 85, pp. 282–292, 2016, doi: 10.1016/j.ssci.2016.02.001.

[49] P. Last, M. Kroker, and L. Linsen, "Generating real-time objects for a bridge ship-handling simulator based on automatic identification system data," *Simul. Model. Pract. Theory*, vol. 72, pp. 69–87, 2017, doi: 10.1016/j.simpat.2016.12.011.

[50] F. Olindersson, W. Bruhn, T. Scheidweiler, and A. Andersson, "Developing a Maritime Safety Index using Fuzzy Logics," *TransNav, Int. J. Mar. Navig. Saf. Sea Transp.*, vol. 11, no. 3, pp. 469–475, 2017, doi: 10.12716/1001.11.03.12.

[51] X. Xin, K. Liu, X. Yang, Z. Yuan, and J. Zhang, "A simulation model for ship navigation in the 'Xiazhimen' waterway based on statistical analysis of AIS data," *Ocean Eng.*, vol. 180, no. April, pp. 279–289, 2019, doi: 10.1016/j.oceaneng.2019.03.052.

[52] Z. Yan *et al.*, "Exploring AIS data for intelligent maritime routes extraction," *Appl. Ocean Res.*, vol. 101, no. June, p. 102271, 2020, doi: 10.1016/j.apor.2020.102271.

[53] T. Koyama, T. Nakanowatari, and J. Inoue, "Information retrieval for Northern Sea Route (NSR) navigation: A statistical approach using the AIS and TOPAZ4 data," *Polar Sci.*, vol. 27, no. December 2020, p. 100626, 2021, doi: 10.1016/j.polar.2020.100626.

[54] S. A. Breithaupt, M. Bensi, and A. Copping, "AIS-based characterization of navigation conflicts along the US Atlantic Coast prior to development of wind energy," *Ocean Eng.*, vol. 264, no. February, p. 112235, 2022, doi: 10.1016/j.oceaneng.2022.112235.

[55] R. P. de Oliveira, G. Carim Junior, B. Pereira, D. Hunter, J. Drummond, and M. Andre, "Systematic Literature Review on the Fidelity of Maritime Simulator Training," *Educ. Sci.*, vol. 12, no. 11, 2022, doi: 10.3390/educsci12110817.

[56] M. Fallahzadeh, M. Kiani, and H. Talebnezhad, "Evaluating safety control criteria in maritime traffic using formal safety assessment. Case study: Iranian port of bushehr," *J. Marit. Res.*, vol. 12, no. 3, pp. 37–48, 2015.

IOP Conf. Series: Earth and Environmental Science **1461** (2025) 012050 doi:10.1088/1755-1315/1461/1/012050

Design of the Cyclone Separator for Eco Sandblasting System with CFD Method

F.I. Sitopu[1], I.S. Arief*[1], S. Widjaja[2], V. Kusumadewi[3], D.B. Purwanto[2] A. Amiadji[1]

[1] Department of Marine Engineering, Institut Teknologi Sepuluh Nopember, Surabaya, Indonesia
[2] Department of Naval Architecture, Institut Teknologi Sepuluh Nopember, Surabaya, Indonesia
[3] Department of Mechanical Engineering, Institut Teknologi Sepuluh Nopember, Surabaya, Indonesia
*E-mail: irfansya@.its.ac.id

Abstract. The sandblasting products currently available on the market come in various types and serve multiple purposes, but they often need more environmental friendliness. This has prompted an analysis leading to developing an Eco Sandblasting system to reduce harmful dust emissions. The research focuses on the cyclone separator using Computational Fluid Dynamics (CFD) as a simulation method. In the study, SEM (Scanning Electron Microscopy) and PSA (Particle Size Analysis) laboratory tests were conducted to examine the characteristics of silica sand particles before (pre-) and after (post-) the sandblasting process. The results indicated that the particle diameter of sand before the sandblasting process (pre-) ranged from 192.513 μm to 1068.063 μm with a Polydispersity Index (PDI) of 0.4405. After the sandblasting process (post-), the particle diameter ranged from 80.311 μm to 576.70 μm with a PDI of 0.8257. The study aimed to determine the efficiency of the cyclone separator design using the Stairmand method. Based on the design calculations for particle scales ranging from 80.31 μm to 576.70 μm, with a constant inlet velocity of 23.73 m/s, the cyclone separator achieved an efficiency of 86.58% and a pressure drop of 11.06 Mbar. In the CFD DPM simulation, the cyclone separator attained an efficiency of 83.42% with a pressure drop of 12.1 mbar. Correcting the efficiency discrepancy between these two methods, an error correction of 3.6% was determined.

1. Introduction

Sandblasting is a technique for cleaning and preparing metal surfaces using high-pressure silica sand flow to remove rust, dust, and dirt. This process can also create a uniformly rough surface texture, enhancing the durability and adhesion of paint or coating layers and improving corrosion resistance [1]. Protecting ship hulls from corrosion and biofouling using paint coatings has become a standard practice in the market. To enhance the cohesive properties of the paint coating on ship hulls made of steel plates, it is necessary to clean the surface layer of the steel plate to meet ISO 8501-1 Sa 2.5 standards, which specifies surface cleanliness achieved through abrasive blasting methods [2]. Sandblasting products are available in various types and applications on the market, but these products tend to be environmentally unfriendly. This forms the basis for analysing an eco-friendly sandblasting system aimed at reducing hazardous dust emissions, preventing soil and water contamination from metal particles, and minimising the use and degradation of sand as a blasting medium.

One of the primary reasons for community protests against sandblasting activities in shipyards is air pollution, which results in dust, fine particles, and chemicals that can contaminate the surrounding air with hazardous metal particles such as lead and chromium. These particles can have negative impacts on public health and the surrounding environment. According to Presidential Decree No. 22 of 1993 concerning occupational diseases, one such disease is Silicosis, which carries a 12% increased risk for workers with more than 30 years of exposure to silica particles [3]. Environmental impacts of the particles and chemicals used in sandblasting, such as silica, can also contaminate nearby water bodies if not adequately managed, disrupting marine ecosystems and affecting marine organism health, thereby indirectly impacting public health. Current product development requires innovation to address the pollution issues by integrating additional components into sandblasting equipment, such as vacuum technology with recyclable abrasive materials, thereby creating eco-friendly sandblasting products. The vacuum technology design requires initial pressure parameter calculations within the blasting room. This research focuses on the Cyclone Separator in designing an Eco Sandblasting System using Computational Fluid Dynamics (CFD).

Based on the background above, the problem formulation is as follows: how to achieve efficiency in the Cyclone Separator using the Stairmand method. Based on the above problem formulation, this research aims to achieve efficiency in the Cyclone Separator using the Stairmand method. Based on the problem formulation, it is necessary to define the scope of the study to facilitate and clarify the analysis. The scope of this research is as follows: the research focuses on an eco-sandblasting system that does not harm the environment, excluding innovations in noise reduction for sandblasting. The study only analyses modelling CFD the Cyclone Separator. The results are expected to provide benefits by analysing the eco sandblasting system design to reduce air pollution.

2. Method

2.1 Data Collection

In this research, meticulous attention is given to collecting pertinent data, as it significantly impacts the research process. The data for analysis are obtained from a shipyard. Data is first gathered for laboratory testing and as input for CFD simulation. After obtaining the actual data from the sandblasting system, as mentioned earlier, the identification of silica sand before and after sandblasting was conducted. This data is subsequently used in the calculation of the Eco-Sandblasting System.

2.2 Design Dimension Calculations for Cyclone Separators

After the abrasive particles exit the blasting room, they are directed into the Cyclone Separator component. A cyclone separator is a device that operates by causing gas to flow spirally in a closed chamber, thus directing sand particles towards the vessel walls with centrifugal force. Heavier sand particles move downward due to gravity and exit the cyclone through the outlet hole towards the rotary screen. In comparison, lighter particles and filtered gas rotate through the upper outlet. Various factors, including particle size, inlet gas velocity, particle density, dust loading, and cyclone diameter, influence the cyclone separator's performance.

This design guidance is based on the work published by Bohnet in 1997. The approach applies to standard cyclones with a square tangential inlet and a low dust load of approximately 10 g/m3 or less. Some corrections may be required for different inlet types or higher dust loads. The Bohnet method helps estimate cyclone performance in basic design or troubleshooting scenarios. However, it's important to note that it can produce errors of up to 40% compared to experimental results, depending on the specific conditions. For more detailed calculations, it is recommended to seek assistance from a company specializing in cyclone design that has validated its calculation methods [4].

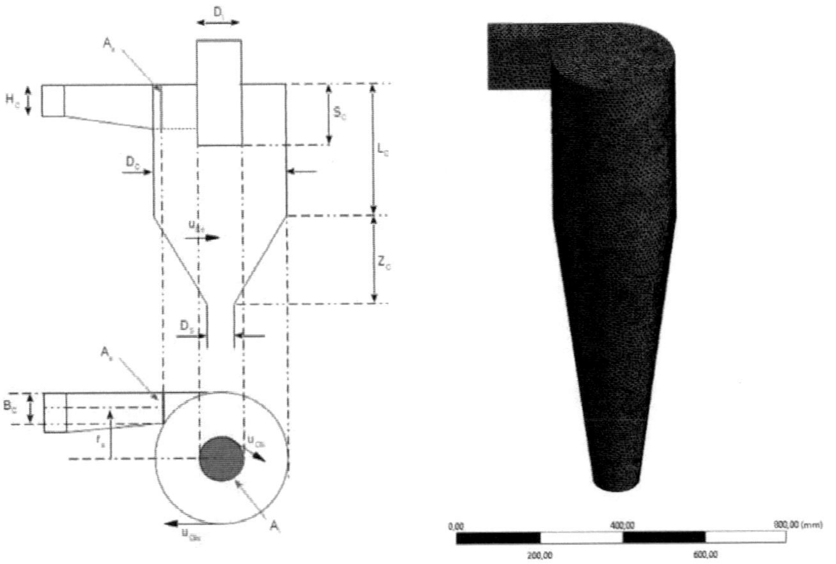

Figure 1. Cyclones and Nomenclature of Geometric Characteristics

2.3 Computational Fluid Dynamics (CFD) Simulation of Cyclone Separator
Based on the complex fluid issues encountered in the simulation, double precision is employed with two solver processes to enhance simulation accuracy, where each floating-point number is represented using 64 bits. The energy equation modelling tools are activated in this stage, and a k-epsilon viscosity model is chosen. This model assumes turbulent flow conditions with the RNG model known for its high accuracy in rotational flows. The next step involves defining the boundary conditions after selecting the input for both discrete and continuous particles. Boundary conditions encompass the flow conditions at inlet, outlet, and repeated settings. The inlet conditions vary with velocities of 7 m/s, 8 m/s, 9 m/s, and 23.73 m/s. The outlet conditions maintain a constant pressure of 4 bar at outlet 2 with an escape condition and at outlet 1 with a trapped condition (no pressure outlet).

3. Result and Discussion

3.1 Efficiency of Cyclone Separators using the Stairmand Method
The following data were obtained using the formula derived in the previous subsection. Since efficiency increases with particle diameter, the table below presents the calculated efficiency based on the variation in particle diameter.

Table 1. Cyclone Separator Efficiency on Particle Diameter.

Diameter Partikel (μm)	Efisiensi Cyclone (%)
576.70	98.87%
355.20	96.43%
290.15	92.09%
217.44	86.74%
166.15	79.25%
116.41	65.21%
110.39	62.77%
87.38	62.03%
80.31	47.15%

One of the key parameters influencing cyclone efficiency is particle diameter. The table below presents the efficiency values obtained for particle diameters ranging from 80.31 μm to 576.70 μm. The inlet velocity was maintained at 23.73 m/s during the calculations. The provided statement accurately highlights the direct correlation between cyclone efficiency and particle diameter. This relationship stems from the fundamental principles of cyclone operation and the inherent characteristics of particle behavior within the cyclone separator.

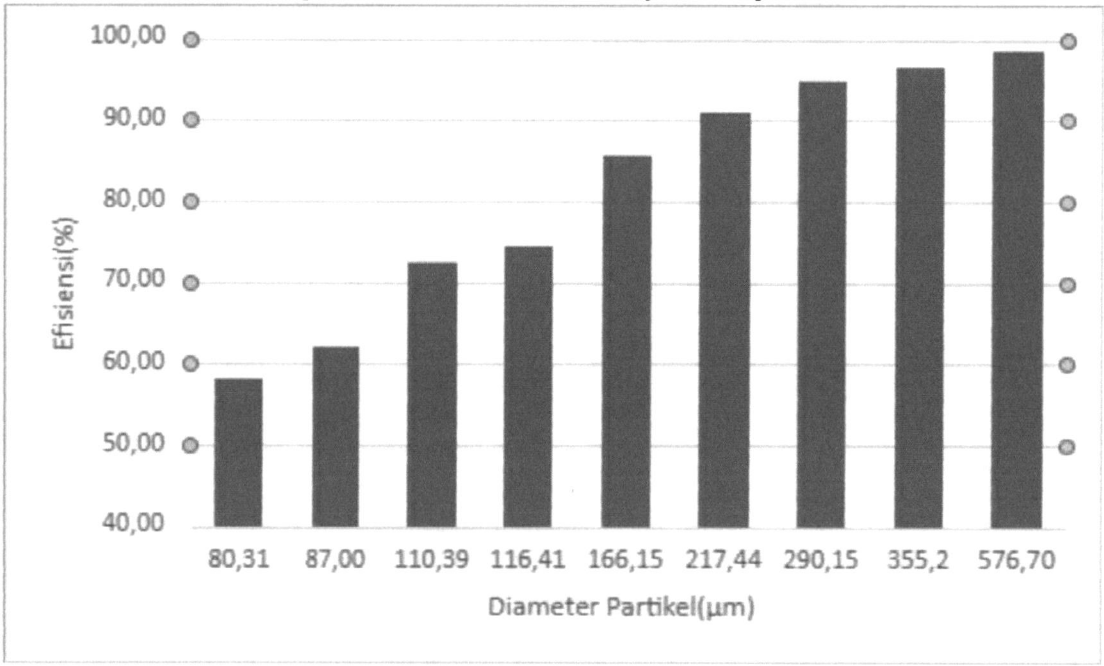

Figure 2. Effect of Particle Diameter on Efficiency

Based on the graph illustrating the influence of particle diameter, it is observed that the largest particle diameter, 576.7μm, exhibits an efficiency of 98.62%, while the smallest particle diameter, 80.31μm, demonstrates an efficiency of 58.04%. Statistical analysis using linear regression reveals a coefficient of determination (R2) value of 0.7178. According to Chin (1998), an R-Square value greater than 0.67 is considered strong, between 0.33 and 0.67 is considered moderate, and between 0.19 and 0.33 is considered weak. The employed method in this study is the Stairmand method, which does not account for flow resistance. The cumulative efficiency achieved using this method is 86.58%.

3.2 CFD Simulation of Cyclone Separators

Figure 3. Streamline Particle's Trajectory

The cyclone separator simulation was conducted using an absolute criterion of 0.001. Particle tracking analysis revealed a total of 356 particles involved in the simulation. Out of the 356 tracked particles, 297 were successfully trapped by the cyclone separator. Based on the particle tracking results, the cyclone separator efficiency was calculated as 83.42%. The efficiency obtained from the simulation (83.42%) exhibits a slight deviation from the efficiency calculated using the Stairmand method (86.58%). This difference is attributed to the inherent limitations of the Stairmand method, which does not account for flow resistance and other factors that can influence actual cyclone performance. To account for the discrepancy between simulated and calculated efficiencies. The correction factor (CF) of 0.036 provides a means to adjust the Stairmand method's efficiency for more accurate predictions.

- Pressure contour

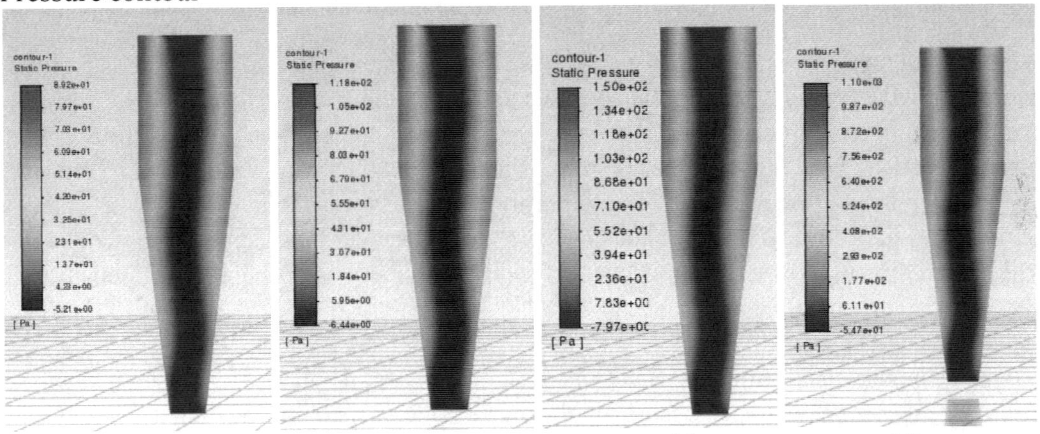

(a) (b) (c) (d)

Figure 4. Pressure contour with cyclone separator speed (a) at a speed of 7 m/s, (b) at a speed of 8 m/s, (c) at a speed of 9 m/s, (d) at a speed of 23.73 m/s

The maximum pressure occurs in the inlet region, reaching a value of 1131.99 Pa. This is attributed to the collision of particles and air in this region due to the tangential inlet design. The inlet region is located on the left side of the cyclone wall. The maximum pressure is visualized in red on the pressure contour plot. The minimum pressure is found in the central region of the cyclone, as depicted in Figure 4. This is because the central region has reduced flow resistance and the gas stream expands, leading to a lower pressure. The minimum pressure is -61.438 Pa. The pressure contour is influenced by the width of the inlet. A larger inlet surface area allows for a greater influx of air, leading to an increase in pressure. This is because a wider inlet increases the momentum of the incoming air, resulting in higher pressure values.

Table 2. Cyclone Separator Pressure Simulation Data

7 m/s					
Surface			Volume		
Average Pressure	51.883	Pa	Average Pressure	-	Pa
Maximum Pressure	91.035	Pa	Maximum Pressure	92.4152	Pa
Minimum Pressure	-3.176	Pa	Minimum Pressure	-5.7056	Pa
8 m/s					
Surface			Volume		
Average Pressure	68.871	Pa	Average Pressure	-	Pa
Maximum Pressure	120.358	Pa	Maximum Pressure	122.179	Pa
Minimum Pressure	-3.248	Pa	Minimum Pressure	-7.261	Pa
9 m/s					
Surface			Volume		
Average Pressure	88.0899	Pa	Average Pressure	-	Pa
Maximum Pressure	153.578	Pa	Maximum Pressure	155.9385	Pa
Minimum Pressure	-3.542	Pa	Minimum Pressure	-8.99851	Pa
23.734 m/s					
Surface			Volume		
Average Pressure	337.822	Pa	Average Pressure	-	Pa
Maximum Pressure	1131.99	Pa	Maximum Pressure	1149	Pa
Minimum Pressure	- 61.4384	Pa	Minimum Pressure	-61.4404	Pa

The performance of a cyclone separator can be assessed based on its pressure drop, which represents the energy loss associated with the separation process. A lower pressure drop indicates more efficient separation. Sure, here is the translation of the Indonesian text. The performance of a cyclone separator can be assessed based on its pressure drop, which represents the energy loss associated with the separation process. Lower pressure drop indicates more efficient separation. The provided text presents two different pressure drop values, from calculated stairmand method pressure drop: 11.061 mbar, and from simulated pressure drop: 12.1 mbar. The observed difference of 1.039 mbar.

- Velocity distribution

The maximum velocity occurs in the inlet region, reaching a value of 40.28 m/s.This is attributed to the high inlet velocity of 23.73 m/s and the tangential inlet design, which imparts a rotational motion to the fluid. The minimum velocity is found in the bottom outlet region, as indicated by the blue color in the velocity contour plot. This is because the bottom outlet serves as the exit point for the captured powder, resulting in a lower flow velocity in that area. The velocity along the cyclone wall is relatively high, as depicted in Figure 5. This is due to the centrifugal force acting on the particles, causing them to move towards the wall.

doi:10.1088/1755-1315/1461/1/012050

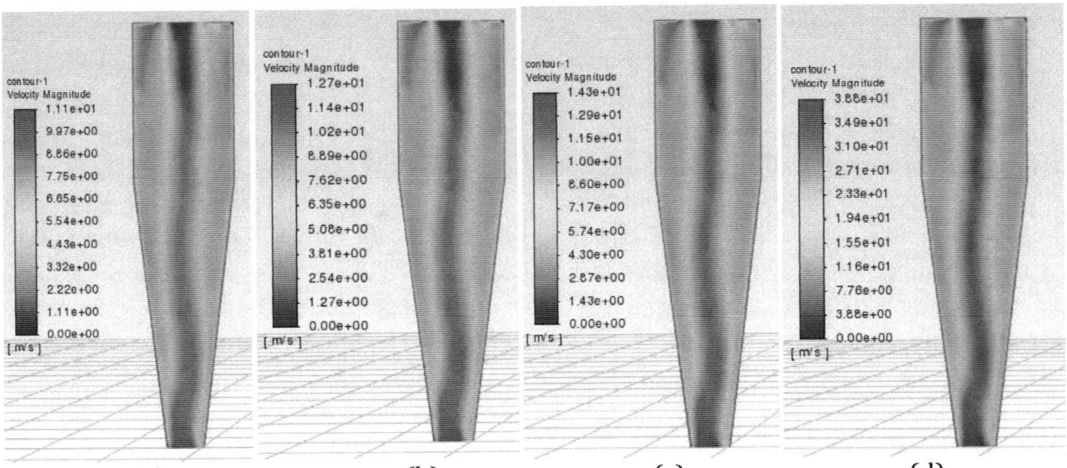

(a) (b) (c) (d)

Figure 5. Distribution of air speed with cyclone separator speed (a) at a speed of 7 m/s, (b) at a speed of 8 m/s, (c) at a speed of 9 m/s, (d) at a speed of 23.73 m/s

The cone-shaped section of the cyclone is designed to generate sufficient rotational velocity for the particles to maintain their movement along the wall. This rotational motion is essential for capturing and separating the particles. A higher inlet velocity increases the likelihood of particles remaining on the cyclone wall, leading to a higher particle collection efficiency. This is because a faster inlet velocity imparts a stronger centrifugal force on the particles, pushing them towards the wall. The height of the cone is typically equal to the height of the cylindrical section of the cyclone. This ratio ensures a balance between the rotational flow and the axial flow within the cyclone, optimizing its performance.

Table 3. Cyclone Separator Speed Simulation Data

7 m/s					
Surface			Volume		
Average Velocity	1.1078	m/s	Average Velocity	-	m/s
Maximum Velocity	11.471	m/s	Maximum Velocity	11.4715	m/s
Minimum Velocity	0	m/s	Minimum Velocity	0.078	m/s
8 m/s					
Surface			Volume		
Average Velocity	1.271	m/s	Average Velocity	-	m/s
Maximum Velocity	13.179	m/s	Maximum Velocity	13.179	m/s
Minimum Velocity	0	m/s	Minimum Velocity	0.1177	m/s
9 m/s					
Surface			Volume		
Average Velocity	1.4347	m/s	Average Velocity	-	m/s
Maximum Velocity	14.8752	m/s	Maximum Velocity	14.8752	m/s
Minimum Velocity	0	m/s	Minimum Velocity	0.111769	m/s
23.734 m/s					
Surface			Volume		
Average Velocity	12.9197	m/s	Average Velocity	-	m/s
Maximum Velocity	40.2895	m/s	Maximum Velocity	40.2895	m/s
Minimum Velocity	0	m/s	Minimum Velocity	0.4466	m/s

- Turbulent Contour

Turbulence is a characteristic flow pattern in fluids that is characterized by chaotic fluctuations and eddies. It occurs when inertial forces, which govern the momentum of the fluid, become dominant over viscous forces, which represent the resistance to flow caused by the fluid's internal friction. This dominance of inertial forces is quantified by a dimensionless parameter called the Reynolds number (Re). When the Reynolds number exceeds a critical value, the flow transitions from laminar, where viscous forces prevail, to turbulent. The provided description indicates that the simulation was conducted using the k-epsilon RNG (Renormalized Group) turbulence model. This model is particularly well-suited for simulating rotating flows, such as the swirling air stream within a cyclone separator. The RNG approach incorporates additional features to account for the effects of rotation on turbulence, leading to improved accuracy in predicting turbulent behavior in swirling.

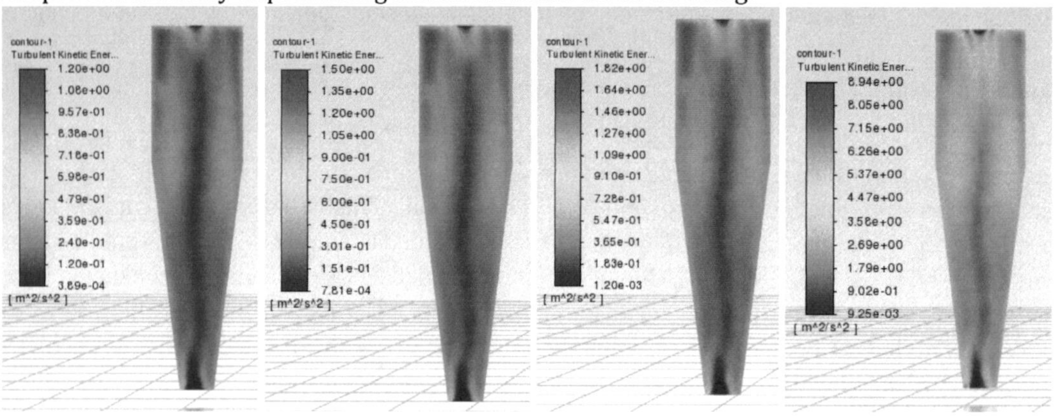

Figure 6. Turbulation contour distribution with cyclone separator speed (a) at a speed of 7 m/s, (b) at a speed of 8 m/s, (c) at a speed of 9 m/s, (d) at a speed of 23.73 m/s

The simulation results reveal that the highest turbulence intensity is concentrated in the central region of the cyclone, encompassing both the cylindrical and cone sections. This is attributed to the intense rotational motion and the interaction of the air stream with the cyclone walls. The maximum turbulence level reaches a value of 11.236 m^2/s^2, indicating the highly turbulent nature of the flow in this region. In contrast, the minimum turbulence level of 0.00343 m^2/s^2 is observed in areas with less energetic flow, such as the bottom outlet.

Table 4. Cyclone Separator Turbulation Simulation Data

7 m/s					
Surface			Volume		
AverageTurbulence	0.3414	m^2/s^2	AverageTurbulence	-	m^2/s^2
Maximum Turbulence	1.9604	m^2/s^2	Maximum Turbulence	1.9606	m^2/s^2
Minimum Turbulence	0.000122	m^2/s^2	Minimum Turbulence	0.000994	m^2/s^2
8 m/s					
Surface			Volume		
AverageTurbulence	0.4393	m^2/s^2	AverageTurbulence	-	m^2/s^2
Maximum Turbulence	2.4501	m^2/s^2	Maximum Turbulence	2.4503	m^2/s^2
Minimum Turbulence	0.00028	m^2/s^2	Minimum Turbulence	0.00146	m^2/s^2
9 m/s					
Surface			Volume		

AverageTurbulence	0.54731	m^2/s^2	Average Pressure	-	m^2/s^2
Maximum Turbulence	2.9543	m^2/s^2	Maximum Pressure	2.9545	m^2/s^2
Minimum Turbulence	0.000404	m^2/s^2	Minimum Pressure	0.001794	m^2/s^2
		23.734 m/s			
Surface			Volume		
AverageTurbulence	3.3780	m^2/s^2	Average Pressure	-	m^2/s^2
Maximum Turbulence	11.2361	m^2/s^2	Maximum Pressure	16.1877	m^2/s^2
Minimum Turbulence	0.00343	m^2/s^2	Minimum Pressure	0.01891	m^2/s^2

4. Conclusion

Conclusions of CFD Analysis for Eco Sandblasting System Based on the simulation results of blasting room and cyclone separator component analysis in the design of an eco sandblasting system using the CFD method, it can be concluded that:

1. Cyclone separator efficiency was determined by comparing the Stairmand method and CFD DPM simulations. The Stairmand method uses two inputs to calculate efficiency: particle diameter and inlet velocity. For particle sizes ranging from 80.31 μm to 576.70 μm and a constant inlet velocity of 23.73 m/s, the Stairmand method yielded an efficiency of 86.58% and a pressure drop of 11.061 mbar. CFD DPM simulations resulted in a cyclone efficiency of 83.42% with an inlet velocity of 23.73 m/s and a pressure drop of 12.1 mbar. The error correction for efficiency was 3.6%. The simulation results indicate that larger sand particles spend more time rotating in the cyclone, maximising the separation process. Similarly, a higher inlet velocity increases the likelihood of particles remaining on the cyclone walls, leading to higher separation rates. However, excessive particle residence time in the cyclone can prolong the separation process and increase the risk of slugging at the inlet and outlet.

References

[1] N. Donthu, S. Kumar, D. Mukherjee, N. Pandey, and W. M. Lim, “How to conduct a bibliometric analysis: An overview and guidelines,” J Bus Res, vol. 133, pp. 285–296, Sep. 2021, doi: 10.1016/j.jbusres.2021.04.070.

[2] R. Heersmink, J. van den Hoven, N. J. van Eck, and J. den van Berg, “Bibliometric mapping of computer and information ethics,” Ethics Inf Technol, vol. 13, no. 3, pp. 241–249, Sep. 2011, doi: 10.1007/s10676-011-9273-7.

[3] X. Yu, X. Wei, X. Huang, and L. Zhang, “Study on the Modeling Method and Influencing Parameters of Sandblasting Process for Blade Grinding,” Advances in Materials Science and Engineering, vol. 2022, 2022, doi: 10.1155/2022/7905927.

[4] M. A. El-Emam, L. Zhou, W. Shi, and C. Han, “Performance evaluation of standard cyclone separators by using CFD–DEM simulation with realistic bio-particulate matter,” Powder Technol, vol. 385, pp. 357–374, Jun. 2021, doi: 10.1016/j.powtec.2021.03.006.

[5] A. Windyandari, H. S. Huboyo, and A. F. Zakki, “INVESTIGATION OF STRUCTURAL RESPONSE OF FLOOR PANEL UNIT ON THE PORTABLE BLAST ROOM USING FINITE ELEMENT METHOD,” vol. 15, no. 3, 2020, [Online]. Available: www.arpnjournals.com

[6] M. Multigner, E. Frutos, J. L. González-Carrasco, J. JimÉNEZ, P. Marin, and J. Ibañez, “Influence of the sandblasting on the subsurface microstructure of 316LVM stainless steel: Implications on the

magnetic and mechanical properties," Materials Science and Engineering C, vol. 29, pp. 1357–1360, May 2009, doi: 10.1016/j.msec.2008.11.002.

[7] A. Rosidah, "ANALISIS KEKASARAN PERMUKAAN PADA PROSES SAND BLASTING DENGAN VARIASI JARAK, TEKANAN, DAN SUDUT PADA PELAT A 36 MENGGUNAKAN METODE BOX BEHNKEN," 2015. [Online]. Available: https://api.semanticscholar.org/CorpusID:140032928

[8] M. Bohnet, O. Gottschalk, and M. Morweiser, "Modern design of aerocyclones," Advanced Powder Technology, vol. 8, no. 2, pp. 137–161, 1997, doi: https://doi.org/10.1016/S0921-8831(08)60472-0.

IOP Conf. Series: Earth and Environmental Science **1461** (2025) 012051 doi:10.1088/1755-1315/1461/1/012051

Assessing the Comfort of Fast Boat Passengers in the Jepara - Karimunjawa Operation Area with Direct Measurement Method

Faizareta Nur Rosiyah[1], Aries Sulisetyono[1*], Wasis Dwi Aryawan[1], Dony Setiawan[1], and Muhammad Nurul Misbach[1]

[1] Department of Naval Architecture, Institut Teknologi Sepuluh Nopember, Indonesia

*E-mail: sulisea@na.its.ac.id

Abstract. Karimunjawa Island, a national park, draws a significant number of tourists. The relatively high seawater conditions prevent tourists from moving around the island, which is due to the comfortability of ship transportation. This study intends to examine the comfort level of passengers on the Express Bahari 1C Fast Motorboat operating on the Jepara-Karimunjawa route. The study uses a direct measurement approach of the ship's sailing motion with the ADIS 16364 sensor device, which includes an accelerometer component that acts to measure translational motions (heave, surge, sway) and a gyroscope for rotational motions (roll, pitch, yaw). The measurement findings are in the form of acceleration and velocity motion signals in time domain, which are subsequently processed using filtering, derivative, and integration algorithms to reduce noise and obtain motion displacement. According to the study results, the highest roll velocity occurs at 0.37 deg/s, and the highest roll displacement reaches 0.25 degrees. Meanwhile, the maximum pitch velocity and displacement occurred at 0.36 deg/s and 0.15 degrees, respectively. The maximum heave acceleration measured is about 0.66 m/s^2. The analysis of passenger comfort was conducted using the international comfort criteria outlined in NATO STANAG 4154 and ISO 2631-1:1997. The equation for motion sickness incidence (MSI) also provided a prediction for passenger comfort, resulting in an average of 11.75% of all passengers reporting discomfort.

1. Introduction

Karimunjawa is an island located in the Java Sea that has been transformed into a prime tourist attraction. Karimunjawa National Park was officially established as a National Park in 1999 by the Minister of Forestry and Plantation Decree no. 78/Kpts-II/1999. The park comprises a water area of around 110.117 hectares [1]. Given its strong allure, both domestic and foreign tourists often opt for Karimunjawa as their travel destination.

Maritime tourism destinations undoubtedly rely on maritime transportation for the movement of people and commodities. One means of transportation for crossing the Jepara - Karimunjawa strait is the Express Bahari Fast Motorboat (KMC). Compact, high-speed motorboats are highly susceptible to the effects of wave action in the open ocean. Ocean waves have the potential to present hazards to the ship's architectural framework and can impact human elements. Thus, it is crucial to prioritise the assurance of safety and comfort for passengers during the journey [2].

Ship motion acceleration during wave navigation is the primary determinant of comfort. To quantify the acceleration of motion, various techniques can be employed, one of which is the direct measurement approach. This approach utilises a sensor device to measure the acceleration and velocity of the ship's movement during its transit. An analysis and identification of the results obtained from these measures is conducted by reference several worldwide comfort standards. The cruise criteria adopted from the international seakeeping criteria serve as the benchmark for assessing ship comfort. Nevertheless, it is necessary to do research and analysis in order to determine the level of comfort experienced by the passengers on the ship, which should align with the established comfort standards. Commonly used international seakeeping standards include the North Atlantic Treaty Organization (NATO) STANAG 4154 Standardization Agreement, the Nordic Co-operative Organization for Applied Research (NORDFOSK 1987), the US Coast Guard Cutter Certification Plan, and ISO 2631-1:1997. The seakeeping criteria encompass specific parameters such as the maximum roll amplitude, pitch amplitude, and acceleration of vertical and lateral motion [3].

The objective of this study is to evaluate the level of comfort experienced by passengers on the fast ship route between Jepara and Karimunjawa. This will be achieved by monitoring the ship's motion response using the ADIS 16344 which is a validated device for the measurement of the ship motion in wave [4].

2. Theoretical of ship motion measurement

2.1 Ship Motion in Irregular Wave

Seakeeping refers to the study of ship motion and its effects on people, systems, and ship operations, specifically focusing on the analysis of ship motion in waves. Vessels exhibit six degrees of freedom (6 DOF) in response to external forces. These degrees include translational motions such as surge, sway, and heave, as well as rotational motions including roll, pitch, and yaw, as depicted in Figure 1 [5].

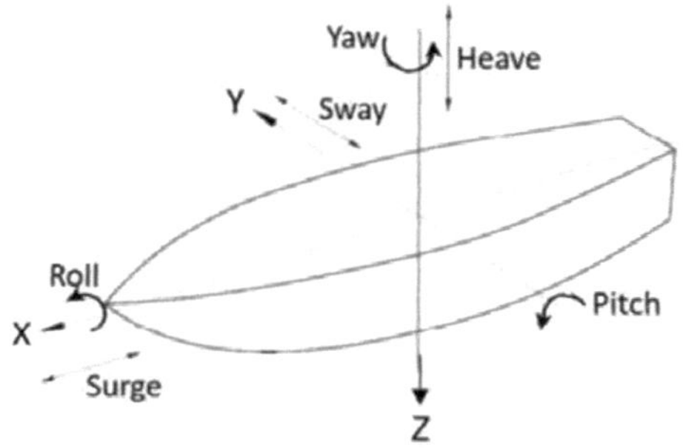

Figure 1. The six Degrees of Freedom (6 DoF)

2.2 Sensor Devices for Direct Measurement

In this study, the ADIS16364 sensor is employed for direct measurement. The ADIS16364 sensor is a physical device that employs an Analogue to Digital Converter (ADC) to process and record data, see Figure 2. This sensor comprises a gyroscope sensor for measuring rotational motion in degrees per second and an accelerometer sensor for measuring translational motion in meter per second quadrat.

Figure 2. ADIS16364 Sensor Device

The specs of the ADIS16364 sensor include exceptional accuracy and precision. The gyroscope sensor achieves an accuracy of ±350º/sec and a precision of 0.05º/sec/LSB. The accelerometer sensor achieves an accuracy of ±5.25 g and a precision of 1.00 mg/LSB. Accuracy is the degree of proximity between a measurement result and the actual value, whereas precision is the degree of consistency across measurements performed on different samples [6].

2.3 Wave Spectrum Generation

Information pertaining to the response energy spectrum is generated from the motion measurement of the ship in time histories. The generation of a response spectrum curve involves the application of mathematical procedures, namely Fast Fourier Transforms (FFT) and Power Spectral Density (PSD), to convert a random motion recording in the time domain into a response energy spectrum curve in the frequency domain. In general, the proceed data is described as following as,

- The Moving Average Filter (MAF) is a commonly employed filtering technique for velocity and accelerometer data recorded. This filter technique performs signal processing to calculate the average of the data within a specified time interval, with the aim of minimizing the noise present in the signal. The spectral window is a smoothing technique employed to eliminate the residual noise response present in the spectrum curve. The outcomes of the procedure of transformation are graphically represented, with the encountering wave frequency serving as the horizontal axis and the spectrum motion responses as the vertical axis [7].
- Random phenomena can be characterized as stochastic signals within recurring oscillations in mechanical systems. The amplitude of motions and forces can be resolved by representing the signal in the form of actual oscillation frequency data and amplitude at that frequency using Fast Fourier Transforms (FFT) [7].
- Power Spectral Density (PSD) is a statistical measure employed to evaluate signals by quantifying the intensity of a signal or time series spread across different frequencies. Within the realm of ship signal analysis and its motions, Physical Signal Decomposition (PSD) can be employed to assess a ship's reaction to ocean waves and forecast its dynamic characteristics. Power spectral density (PSD) computations usually include applying a Fourier transform to convert the temporal signal into a frequency representation, followed by the computation of power density at each of those frequencies [8].

IOP Conf. Series: Earth and Environmental Science **1461** (2025) 012051 doi:10.1088/1755-1315/1461/1/012051

2.4 Seakeeping Criteria

Ship passenger discomfort research investigates the parameters for restricting ship movement in order to prevent passenger discomfort. The criteria employed are based on many international standards, which include the following:

1. The criterion of NATO STANAG 4154 governs comfort by considering several aspects, including Motion Induced Interruptions (MII), Motion Induced Fatigue (MIF), Cognitive performance, Motion Sickness Incidence (MSI), and Habituation. Table 1 describes the seakeeping criteria of NATO STANAG 4154 [9].

Table 1. Seakeeping criteria of NATO STANAG 4154

Recommended Criteria	Limit	Location
Motion sickness incidence (MSI)	20% of crew @ 4 hrs	Task location
Motion induced interruption (MII)	1/m	Task location
Default Criteria	Root mean square (RMS)	
Roll	4 degree	
Pitch	1.5 degree	
Vertical acceleration	0.2 g	Bridge
Lateral acceleration	0.1 g	Bridge

2. The requirements established by the Nordic Co-operative Organization for Applied Research (NORDFORSK) centre on the limitation of ship operations by imposing roll and acceleration restrictions in order to ensure the comfort of passengers [10]. The standards for seakeeping that NORDFORSK adheres to are outlined in Table 2.

Table 2. Seakeeping criteria of NORDFORSK

General Operability Limiting Criteria for Ships	Marchant Ship	Navy Vessels	Fast Small Craft
RMS of vertical acceleration at FP	0.275 g (L ≤ 100 m) 0.050 g (L ≥ 330 m)	0.275 g	0.65 g
RMS of vertical acceleration at Bridge	0.15 g	0.20 g	0.275 g
RMS of lateral acceleration at Bridge	0.12 g	0.10 g	0.10 g
RMS of Roll	6.0 deg.	4.0 deg.	4.0 deg.
Probability of slamming	0.03 (L ≤ 100 m) 0.01 (L ≥ 330 m)	0.03	0.03
Probability of deck wetness	0.05	0.05	0.05
Criteria regard to acceleration and roll	Vert. Acc.	Lat. Acc.	Roll
Light manual work (RMS)	0,20 g	0.10 g 0,07	$6,0^{0}$
Heavy manual work (RMS)	0,15 g	g	$4,0^{0}$
Intellectual work (RMS)	0,10 g	0,05 g	$3,0^{0}$
Transit passengers (RMS)	0,05 g	0,04 g	$2,5^{0}$
Cruise liner (RMS)	0,02 g	0,03 g	$2,0^{0}$

3. The United States Coast Guard Cutter Certification Plan is a set of criteria that is utilized by both the United States Navy and the United States Coast Guard. This plan regulates roll/pitch movements by utilizing significant single amplitude (SSA) units [11]. In Table 3, the United States Coast Guard explains the seakeeping standards.

IOP Conf. Series: Earth and Environmental Science **1461** (2025) 012051 doi:10.1088/1755-1315/1461/1/012051

Table 3. Seakeeping Criteria of USCG

Criterion	Limit	Location
Vertical acceleration	0.4g SSA	forward perpendicular
Vertical acceleration	0.2g SSA	bridge
Lateral acceleration	0.2g SSA	bridge
Motion Sickness Incidence (MSI)	5% in a 30-minute exposure	task location
Motion Induced Interruption (MII)	2.1 tip per minute	task location
Roll	8.0° SSA	
Pitch	3.0° SSA	

4. The International Organization for Standardization (ISO) 2631-1:1997 and ISO 2631-3:1997, which limit the value of vibration magnitude that is permissible for comfort and focus on the vertical acceleration of the ship [12]. ISO 2631-1:1997 and ISO 2631-3:1997 each have their own set of seakeeping requirements, which are displayed in Table 4 and Figure 3, respectively.

Table 4. Seakeeping criteria of ISO 2631-1:1197

Acceleration	Criteria
Less than 0.315 m/s	Not uncomfortable
0.315 m/s^2 - 0.63 m/s^2	A little uncomfortable
0.50 m/s^2 - 1.00 m/s^2	Fairly uncomfortable
0.80 m/s^2 - 1.60 m/s^2	Uncomfortable
1.25 m/s^2 - 2.50 m/s^2	Very uncomfortable
Greater than 2 m/s^2	Extremely uncomfortable

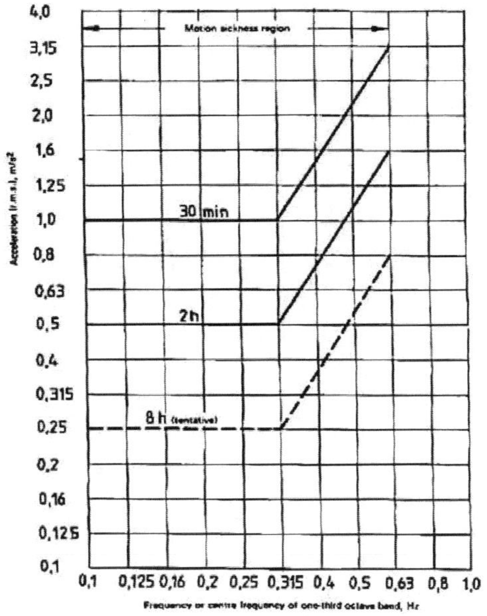

Figure 3. Seakeeping criteria of ISO 2631-3:1197

2.5 Motion Sickness Incidence (MSI)

Both the safety and comfort of passengers are essential aspects of ship design, and they are often regarded as the primary requirements for achieving the highest possible seakeeping performance. The Motion Sickness Incidence (MSI) equation was developed as a consequence of an experiment that was carried out in 1974 by McCauley and O'Hanlon. The experiment involved approximately two hours of participation from three hundred volunteers and was named after the equation.

Motion Sickness Incidence (MSI) is the outcome of the percentage of passengers who feel seasickness as a result of movement or motion sickness while they are aboard the ship. The MSI value can be defined using a number of different formulas, such as the one developed by Scamardella and Piscopo in [13]. The MSI measurement is carried out by utilizing the following equation, which is based on the Scamardella and Piscopo's formula [13]. This formula is a development of the formula that was originally proposed by McCauley and O'Hanlon in1974.

$$MSI = 100\left[0.5 + erf\left(\frac{log_{10}(0.798\sqrt{m_4}/g) - \mu_{MSI}}{0.4}\right)\right] \quad (1)$$

Where *erf* is the error function, μ_{MSI} is the average MSI, and m_4 is the area under the acceleration spectrum.

In addition, Aranda, de la Cruz, and Díaz in 2005 extended this experiment to include 500 participants [14]. Participants were seated and maintained an open gaze for a duration of 2 hours, employing 25 different combinations of motions corresponding to 10 distinct frequencies. MSI was subsequently calculated using the following formula.

$$MSI = 100\left[0.5 + erf\left(\frac{log_{10}\left(\frac{WVA}{g}\right) - \mu_{MSI}}{0.4}\right)\right] \quad (2)$$

Where, wave vertical acceleration (WVA) is the mean value of vertical acceleration (heave) as defined in reference [14].

3. Ship motion measurement

The process of research preparation involves the preparation of suitable measuring tools. The measurements were conducted using the ADIS16364 sensor device, commonly known as the 6 DOF sensor device, which has accelerometer and gyroscope sensors. The vessel employed in the study cases is the Express Bahari 1C Fast Motorboat (KMP), which is operational between Karimunjawa harbour and Kartini harbour, Jepara. The primary dimensions of the vessel are outlined in Table 5. The positioning of measuring hardware on the ship is seen in the ship layout diagram shown in Figure 4.

Table 5. Main Dimension of Express Bahari 1C Ship

Ship's name	Express Bahari 1C
LOA	37.80 m
Breadth	6.75 m
Depth	2.55 m
Draught	1.4 m

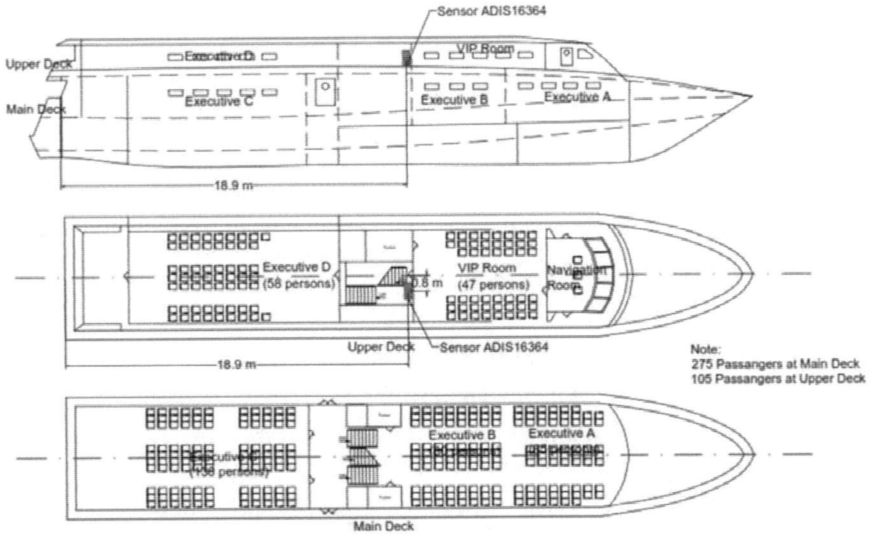

Figure 4. General Arrangement of Express Bahari 1C Ship and ADIS16364 Sensor Placement Points

The ADIS 16364 sensor device is put at the designated location. In this study, one location point is chosen, which is in the middle of the top deck. The sensor device is insured not to move and move places. As for the operation of the tool, numerous supporting tools are needed such as batteries, routers, data cables, and mini-PCs depicted in Figure 5. The battery has a function as a supplier of electrical resources so that the sensor gadget can be employed. The router works as a signal transmitter received from the sensor so that it can be connected to a laptop or smartphone. The data cable serves to link the sensor and/or GPS device to the mini-PC. Mini-PC is utilized as a receiver and storage of data captured by sensors and/or GPS [15].

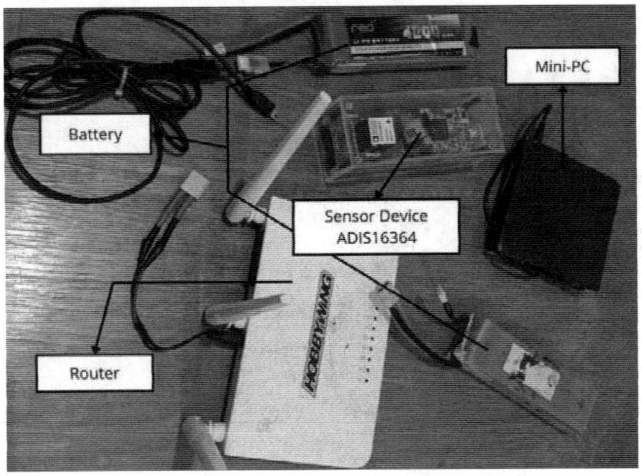

Figure 5. Sensor Device Support Tools

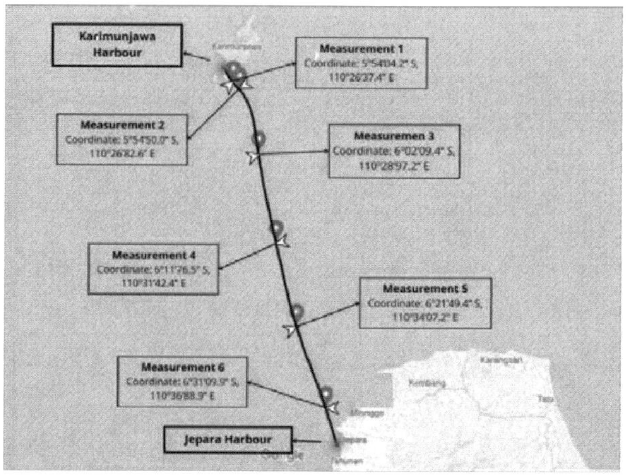

Figure 6. The coordinates of the ship motion measurement location

The data collection measurement was conducted several times at different locations during the ship's voyage. The measurement locations were pre-designed before the ship operated, as they were predicted consider to ISO standard of time measurements in the water area between Jepara and Karimunjawa. The six locations of measurement points were selected, as shown in Figure 6.

Following the conversion of the data acquired by the ADIS16364 device, which was initially received in g units, into m/s units, the data was multiplied by 9.81 m/s^2. Following that, the next step was to filter the data by employing the Moving Average Filter (MAF) technique with the assistance of Matlab. Considering that the data provided by the sensor contains some noise that is rather complicated, this filtering is carried out in order to facilitate the production of data that is more straightforward. The operation of this filter technique involves calculating the average of a number of data points that are next to one another during a specific time frame. This results in the generation of a new data point whose average value is identical to that of the data point [16]. The data that is collected from the sensor is in the form of acceleration and velocity data; therefore, integration is required in order to transform the data into motion data. It is the trapezoidal approach that is utilized for the integration process. Following the integration process, this integral may be utilized to convert acceleration motion into velocity motion. Additionally, it can be utilized to transform velocity motion into movement data after the integration process has been completed.

During this step, more data processing is carried out in order to identify the amount of energy that is included in the motion spectrum that is occurring. This occurs after the data that was acquired has been filtered and integrated. In order to complete this stage, the data that is included inside the time function is transformed into a frequency function. Power Spectral Density (PSD) is the technique that was utilized in order to transform the data. After the results of the calculation of the ship's motion response spectrum have been acquired, the following step is to determine the Motion Sickness Incidence (MSI) value. This is done based on the findings of the computation. In order to do this calculation, it is necessary to consult the calculations that were performed by Equations (1) and (2). Heaving, rolling, and pitching are the three oscillatory motions that are utilized to determine the operability values that are derived from the data. When referring to the evaluation of ship operability, the seakeeping requirements that were developed by NORDFORSK 1987, STANAG 4154, and the United States Coast Guard (USCG) are taken into consideration. This set of seakeeping criteria was chosen after taking into consideration various factors of the kind and type of ship that is being evaluated.

4. Results and Discussion

4.1 Motion Data Measurement

Mobility data of the Express Bahari 1C Ship is generated using the readings acquired from ADIS 16364. The obtained data comprises measured roll, pitch, and heave, which are then filtered using MAF. To illustrate, Figures 7 and 8 display the measurements results of heave acceleration, roll velocity, and pitch velocity, respectively, for the first and second measurements.

According to Figure 7, the initial measurement displays a heave acceleration value of 0.011 m/s^2, along with roll and pitch velocity values of 0.141 deg/s and 0.062 deg/s, respectively. Meanwhile, in Figure 8, the heave acceleration value recorded at the second measurement is 0.111 m/s^2, while the roll and pitch velocity values are 0.249 deg/s and 0.153 deg/s, respectively.

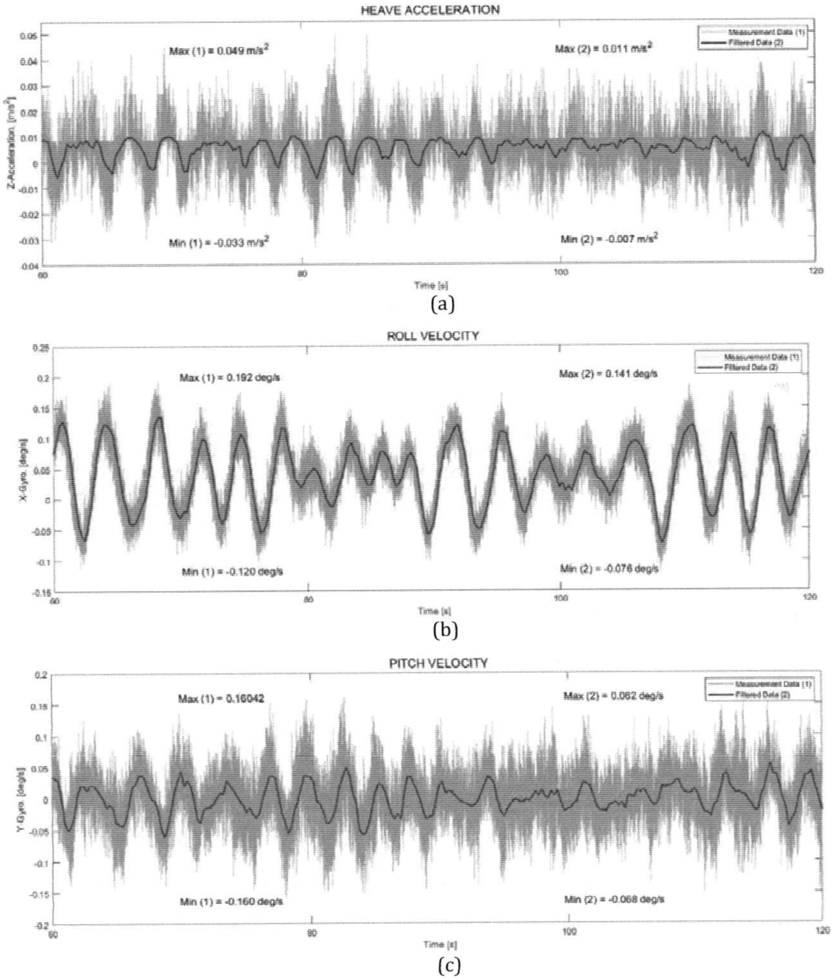

Figure 7. (a) heave acceleration, (b) roll velocity, and (c) pitch velocity at the first measurement

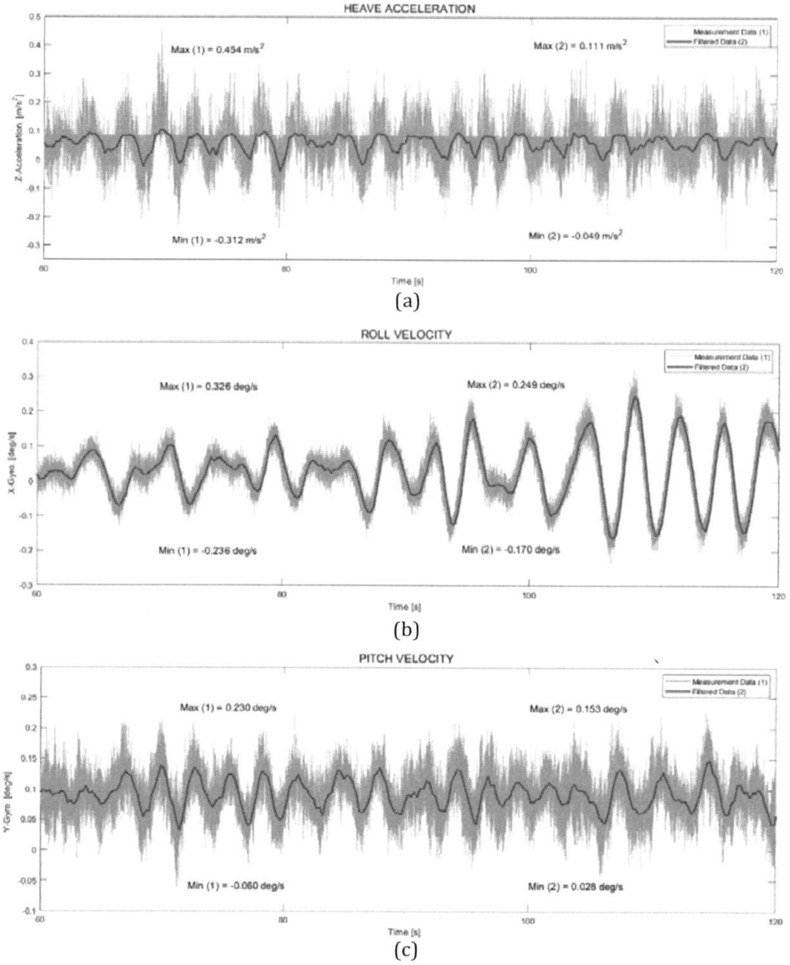

Figure 8. (a) heave acceleration, (b) roll velocity, and (c) pitch velocity at the second measurement

4.2 Seakeeping Criteria Analysis

As the measurement values do not exceed the value limits provided by the three comparison criteria (STANAG 4153, NORDFORSK 1978 and U.S. Coast Guard), the comparison results in Table 6 indicate that the first to the sixth measurement meet the comparison criteria for heave acceleration, roll movement, and pitch movement. Based on the examination of seakeeping criteria, the Express Bahari 1C ship is deemed suitable for operation on the shipping channel as it does not beyond the permitted limits for heave, roll, and pitch motion determined by the seakeeping criteria.

Furthermore, Table 7 presents the study of passenger comfort level according to ISO 2631-1:1997. The findings indicate that the first measurement, with a value of 0.043 m/s^2, falls into the pleasant category (that is, less than 0.315 m/s^2). The second and third recorded measures, measuring 0.325 m/s^2 and 0.423 m/s^2 respectively, are classified as rather unpleasant, falling within the range of 0.315 m/s^2 to 0.63 m/s^2. Furthermore, the readings of 0.215 m/s^2, 0.163 m/s^2, and 0.028 m/s^2 for the fourth, fifth, and sixth time points fall within the comfortable range, which is defined as values below 0.315 m/s^2. Based on these findings, it may be inferred that the ship is pleasant during the initial 15 minutes, somewhat uncomfortable from the 30th to the 60th minutes, and then completely comfortable from the 90th minute till it reaches its destination.

Table 6. Comparison of Measurement Results with Seakeeping Criteria

Criteria	*Heave*	*Roll*	*Pitch*
The direct measurement results on the Express Bahari 1C [RMS]			
The First Measurement	0.043 m/s^2	0.577 deg	0.267 deg
The Second Measurement	0.325 m/s^2	0.912 deg	0.289 deg
The Third Measurement	0.423 m/s^2	0.792 deg	0.165 deg
The Fourth Measurement	0.215 m/s^2	0.494 deg	0.258 deg
The Fifth Measurement	0.163 m/s^2	0.289 deg	0.278 deg
The Sixth Measurement	0.028 m/s^2	0.135 deg	0.063 deg
Comparison [RMS]			
STANAG 4154	1.962 m/s^2	4.0 deg	1.5 deg
NORDFORSK, 1987	1.962 m/s^2	4.0 deg	-
U.S Coast Guard (USG)	0.981 m/s^2	4.0 deg	1.5 deg

Table 7. Comparison of Heave Acceleration of Express Bahari 1C Ship with ISO 2631-1:1997 Comfort Criteria

Direct Measurement Results	ISO 2631-1:1997	
Acceleration RMS [m/s^2]	Acceleration RMS [m/s^2]	Criteria
The 1st measurement = 0.043	< 0.315	Not uncomfortable
The 2nd measurement = 0.325	0.315 – 0.63	A little uncomfortable
The 3rd measurement = 0.423	0.50 – 1.0	Fairly uncomfortable
The 4th measurement = 0.215	0.8 – 1.6	Uncomfortable
The 5th measurement = 0.163	1.25 – 2.5	Very uncomfortable
The 6th measurement = 0.028	> 2 m/s2	Extremely uncomfortable

Apart from consulting ISO 2631-1:1997, ISO 2631-3:1997 offers instructions on motion sickness zones determined by the heave acceleration encountered by the ship. Figure 9 reveals that the first measurement exhibits a root mean square (RMS) vertical acceleration that remains beyond the motion sickness zone for a duration of 8 hours. Furthermore, the second and third readings are classified as motion sickness for journeys lasting less than 2 hours. Furthermore, the fourth to sixth readings occur beyond the motion sickness zone during a 2-hour period of travel. Based on the obtained data, it can be inferred that the Express Bahari 1C ship fails to satisfy the comfort criteria for journeys lasting over 2 hours at the 30th and 60th minutes, since it falls inside the motion sickness zone. Nevertheless, this issue may be surmounted simply reducing the velocity of the ship. The ship satisfies the comfort criteria during the portion of the voyage lasting from the 90th minute until it docks, as it is beyond the motion sickness zone.

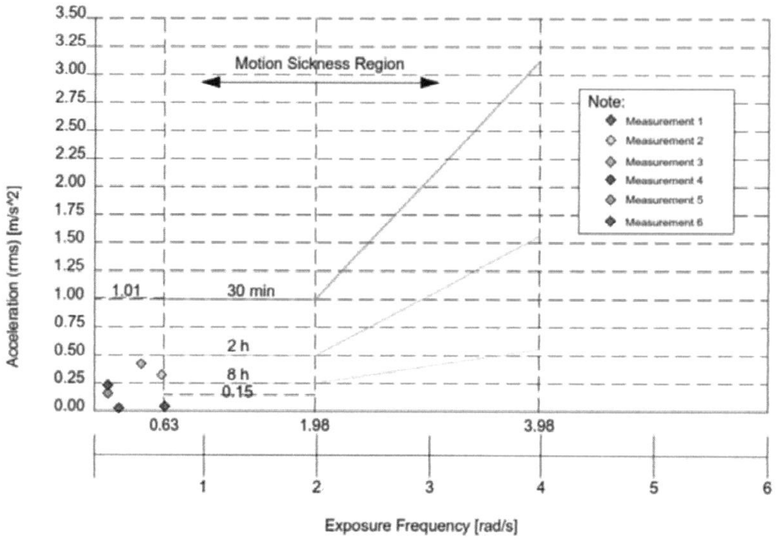

Figure 9. Ship Comfort Points at the 1st to 6th Measurement according to the ISO 2631 Chart

4.3 Motion Sickness Incindence (MSI) Analysis

The MSI value in the measurement is determined by using Equations (1) and (2) derived from the average heave acceleration data and the corresponding values of m^2 and m^4 in each measurement. The calculated findings are subsequently compared to the predetermined benchmark in order to ascertain the degree of comfort experienced by the ship. The MSI scale standard is based on ISO 263-1: 1997, which specifically addresses the categorization of human comfort level responses induced by vibration. Therefore, the comparative results are derived in Table 8.

Table 8. Comparison of MSI Calculation Results with ISO 2631-1:1997 Standard

Measurements	MSI Standard ISO 2631-1:1997 [%]	MSI Formula 1 [%]	Criteria	MSI Formula 2 [%]	Criteria
The 1st	15.27	10.11	Not uncomfortable	10.11	Not uncomfortable
The 2nd	36.69	15.07	Not uncomfortable	13.12	Not uncomfortable
The 3rd	73.22	14.89	Not uncomfortable	15.63	A little uncomfortable
The 4th	87.02	10.11	Not uncomfortable	11.07	Not uncomfortable
The 5th	91.74	10.11	Not uncomfortable	10.52	Not uncomfortable
The 6th		10.11	Not uncomfortable	10.11	Not uncomfortable

According to Table 8, the Express Bahari 1C ship falls into the pleasant category in all parameters since the MSI value does not beyond the 15.27% limit mandated by ISO 2631-1:1997, as calculated using formula 1. Nevertheless, using formula 2, all measures fall within the pleasant range, with the exception of the third measurement (60th minute) which falls into the rather uncomfortable group. The highest Motion Sickness Index (MSI) value obtained using formula 1 was 15.07% on the second assessment, suggesting the presence of motion sickness at 30 minutes. Conversely, the highest MSI value obtained using Formula 2 was 15.63% on the third measurement, suggesting the presence of potential motion sickness at 60 minutes. Given an average MSI score of 11.75%, the ship meets the ISO 2631-1:1997 requirements and may be classified as comfortable.

5. Conclusions

During the third measurement, the Express Bahari 1C ship motion had the greatest heave acceleration, measuring 0.664 m/s^2. The maximum roll speed was recorded in the third measurement, reaching 0.371 degrees per second. The highest roll motion was seen in the second measurement, with a roll angle of 0.245 degrees. Furthermore, the sixth measurement had the greatest pitch velocity and motion, measuring 0.356 deg/s with a pitch angle of 0.151 degrees.

The measurement findings indicate that the Express Bahari 1C ship is deemed suitable for operation on the Karimunjawa - Jepara shipping lane as it does not surpass the boundaries set by seakeeping standards for heave, roll, and pitch movements. Up until the 15th minute of sailing, the ship remains classified as sufficiently comfortable. Nevertheless, from the 30th minute to the 60th minute of cruise, the ship will experience some discomfort due to the presence of motion sickness during the 2-hour journey. Moreover, at the 90th minute of journey till the ship reaches its destination, the ship is once again classified as pleasant.

The calculation results indicate that the average MSI value aboard the Express Bahari 1C ship is 11.75%. Therefore, the Express Bahari 1C ship may be classified as having a pleasant response based on the ISO 2631-1: 1997 standards.

References

[1] Balai Taman Nasional Karimunjawa 2024 Statistik Balai Taman Nasional Karimunjawa Tahun 2023 [Online] Available: https://tnkarimunjawa.id/assets/filepublikasi/3/dokpublik_1709015910.pdf
[2] Jamal, Sulisetyono A, and Aryawan W D 2022 Direct Measurement of Motion on Fast Ferries *IOP Conf. Ser. Earth Environ. Sci.* **1081** 012036 doi: 10.1088/1755-1315/1081/1/012036.
[3] Jamal, Sulisetyono A, and Aryawan W D 2020 Review of the seakeeping criteria for the study of a passenger ship criteria in Indonesian water *IOP Conf. Ser. Mater. Sci. Eng.* **982** 012041 doi: 10.1088/1757-899X/982/1/012041.
[4] Jamal, Sulisetyono A, Aryawan W D, Hafiz M A, and Ali B 2025 The Assessment of ADIS 16364 for the Examination of Ship Motions in the Free-Running Model *Journal of Advanced Research in Applied Sciences and Engineering Technology* vol. **52** no. 2 281-302 doi: 10.37934/araset.52.2.281302
[5] Judg D C Q 2018 Seakeeping and Maneuvering *USNA EN455 - Seakeeping and Manoeuvering* [Online]. Available: https://www.usna.edu/NAOE/academics/en455.php
[6] Devices A 2015 Four Degrees of Freedom Inertial Sensor [Online]. Available: https://www.analog.com/en/products/adis16364.html
[7] Djatmiko E B 2012 Perilaku dan Operabilitas Bangunan Laut di Atas Gelombang Acak *ITS Press.*
[8] Stoica P and Moses M 2005 Spectral Analysis of Signals *Prentice Hall New Jersey.*
[9] Organization N A T 2000 Standardization agreement (STANAG): common procedures for seakeeping in the ship design process *Edition 3. North Atlantic Treaty Organization, Military Agency for Standardization* [Online] Available: https://worldcat.org/title/58973281
[10] Nordforsk 1987 Assessment of Ship Performance in a Seaway: The Nordic Co-operative Project: Seakeeping Performance of Ships *Nordforsk* [Online]. Available: https://books.google.co.id/books?id=hdaQOgAACAAJ
[11] Ghaemi M H and Olszewski H 2017 Total Ship Operability-Review, Concept and Criteria *Polish Maritime Research* vol 24 no s1 74–81, 2017, doi: 10.1515/pomr-2017-0024.
[12] ISO 1997 ISO 2631-1:1997 Mechanical vibration and shock Evaluation of human exposure to whole-body vibration Part 1: General requirements *2nd edition ISO* [Online] Available: https://www.iso.org/standard/7612.html

[13] Scamardella A and Piscopo V 2014 Passenger ship seakeeping optimization by the Overall Motion Sickness Incidence *Ocean Engineering* vol 76 86–97 doi: 10.1016/j.oceaneng.2013.12.005.

[14] Aranda J, Cruz J M de la, and Díaz J M 2005 Design of a multivariable robust controller to decrease the motion sickness incidence in fast ferries *Control Engineering Practice* vol 13 no 8 985–999 doi: 10.1016/j.conengprac.2004.11.003.

[15] Rochmad A N and Sulisetyono A 2024 Analysis of the Hulls Floater Distance in the Manoeuvrability of the N219 Floatplane: An Experimental Approach in the Open Water Test *BIO Web Conference* vol 89 1–14 doi:10.1051/bioconf/20248902002

Empowering Indonesian Women in the Port Industry: Initiatives by *Mutiara Pelindo* in 2022 to 2024

Bahana Wiradanti*, Dothy, Ruri Indrasari Rachmaputri, Riza Rosdiana, Rafika Aulia Hasibuan, Indah Kusumaningrum, Irinda Windyanti, Ditta Frieska Vebrian, Tri Wahyunita, Yona Asri Carera

Mutiara Pelindo, PT Pelabuhan Indonesia (Persero), Jakarta, Indonesia

*E-mail: bahana.wiradanti@pelindo.co.id, bahanawdanti@gmail.com

Abstract. Indonesia's current demographic bonus presents a critical opportunity to enhance the quality of human resources. Empowering women and future generations are key to realizing this potential. This includes human capital particularly in the port and maritime sector. *Pelabuhan Indonesia* also known as *Pelindo* is Indonesia's state-owned enterprise providing port services. Since the merger of four entities into one on October 1st 2021, Pelindo has transformed its organization structure and business, which includes the establishment of Mutiara Pelindo on April 2022. Mutiara Pelindo is a woman organization within Pelindo which aims to empower woman and Pelindo employees, also provide contribution for the environment surrounding the ports and nurture future generations in port and maritime transport business. This paper examines the impact of Mutiara Pelindo's initiatives, leadership and environmental sustainability agendas from 2022 to 2024. Mutiara Pelindo has successfully implemented various empowerment programs focused on increasing female participation and leadership within the maritime industry. These initiatives include specialized training, leadership development, and community engagement, all of which have contributed to a more inclusive and diverse workplace. This paper aims to demonstrate that Indonesian women in the port sector are striving to create meaningful change, and highlights how they can contribute by supporting women in the workplace, environment, and the communities surrounding the port. Findings also shows that challenges remain to spark innovation from women in the company.

Keywords: policy, socioeconomics, port, women, Indonesia

1. Introduction

Indonesia is a vast archipelago country, home to approximately 281 million people [1], making it the fourth most populous nation in the world. With a large and growing population, Indonesia is currently experiencing what is known as a "bonus demography," a period where the proportion of the working-age population is significantly higher than the non-working-age population. This demographic advantage presents a unique opportunity for economic growth and development, as a larger workforce can drive productivity and innovation. However, to fully capitalize on this potential and to reach Indonesia's Golden Vision 2045, human capital must be carefully managed and developed. If not, the country risks facing unemployment, underemployment, and social instability, which could undermine economic progress.

The role of women at the workplace and also at home is both equally crucial. In the workplace, empowering women and ensuring their active participation in the economy is essential for achieving gender equality and maximizing the potential of the workforce. At home, women play a vital role in nurturing and educating future generations, ensuring that children grow up healthy, well-educated, and capable of contributing to society. By supporting women in both domains, Indonesia can harness the full benefits of its demographic bonus, driving sustainable development and ensuring a prosperous future for all its citizens.

According to Indonesian Central Bureau of Statistics, the number of employed individuals has grown by 3.55 million compared to the previous year [2, 1]. By gender, the female Labor Force Participation Rate had risen to 55.41%, representing an increase of about 1% from the previous year in February 2024, and this number is more significant among women compared to men [2]. The following Figure 1 describes Labor Force Participation Rate of Women and Men in 2021 to 2023.

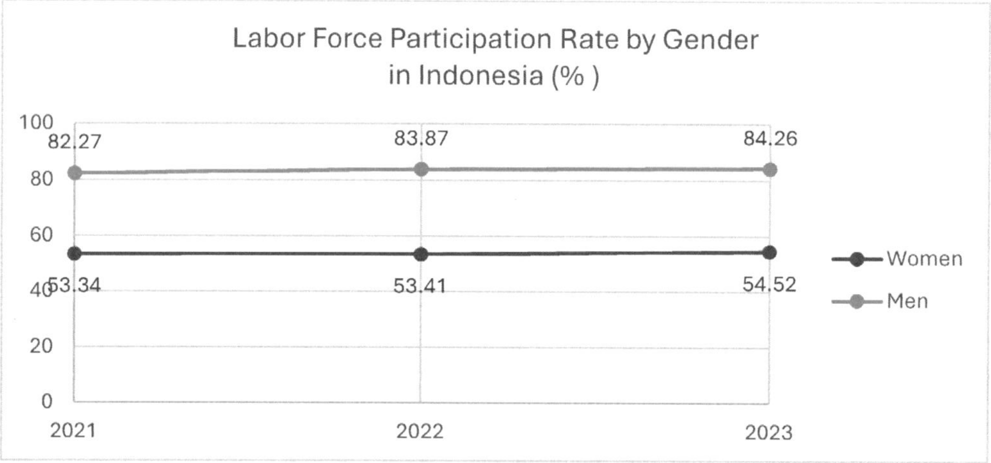

Figure 1. Labor Force Participation Rate by Gender from 2021 to 2023 in Indonesia in Percentage, source: [1]

In the maritime sector, Indonesian seafarer approximately reach a total of 1,168,618 workers [3], which comprises of 95% male and 2,5% female. Meanwhile at the port, *Pelabuhan Indonesia* or Pelindo is the largest state-owned enterprise in Indonesia providing port services. Hence, the number of employees or workers of Pelindo could represent workers of the maritime sector from the port side. In 2023, there are 6,988 organic employees of Pelindo, in which 18,5% of them are women [4]. Meanwhile, there are a large human resource of 16,001 inorganic employees, however, their gender is not recorded. This proportion is quite huge since Pelindo's services covers the whole nation as described in Figure 2.

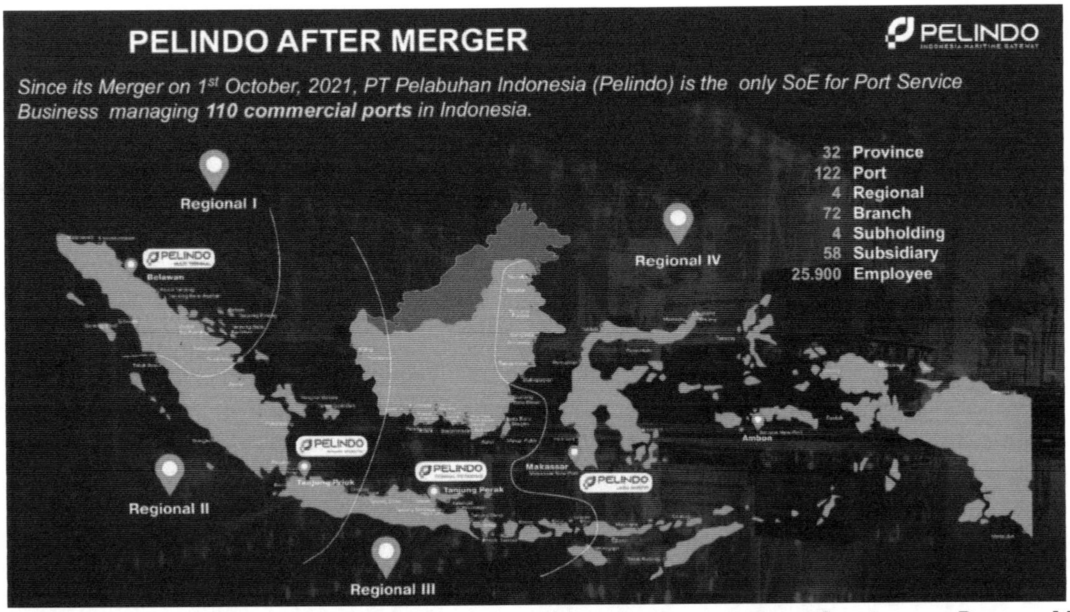

Figure 2. Pelindo's Geographical Scope in Port Services Post Merger, source: [5]

Pelindo post-merger consists of 4 regional areas and 58 subsidiaries. These subsidiaries are clustered to 4 subholdings which are container handling services, non-container handling services, marine services and equipment, and lastly logistics and hinterland development. Pelindo women work in various roles, which dominantly are in administrative staff, where almost 36% of administrative workers are women. 18% of port engineers are female. 25% of cybersecurity and IT workers are female, and 30% of environmental management workers are female. The greater presence in industries like IT and environmental management is indicative of a larger trend of women entering these fields at an increasing rate. Detailed women workers in Pelindo by their job functions are described in the following Figure 3.

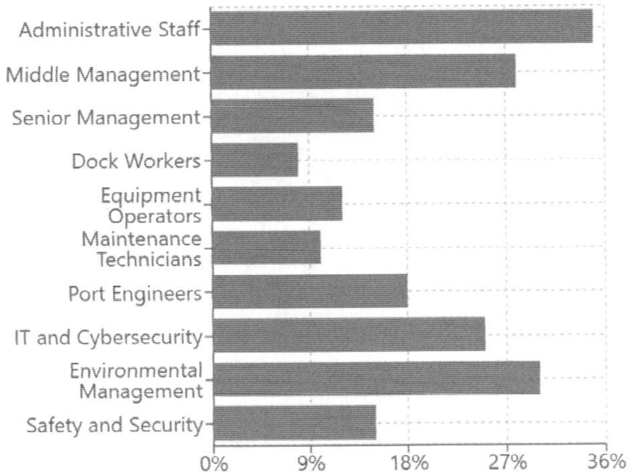

Figure 3. Representation of Women in Pelindo by Job Functions in the Company, source: [4]

There are international women organizations which support women at workplace such as the International Federation of Business and Professional Women (BPW International), Lean In, National Association of Women Business Owners (NAWBO), The Global Women's Leadership Network (GWLN), Women Who Code, Catalyst, Women in Leadership (WIL), Society of Women Engineers (SWE), and Ellevate Network. Particularly in the maritime sector, there are women organizations such as Women's International Shipping and Trading Association (WISTA), International Maritime Organization (IMO) – Women in Maritime Program, Women in Maritime Association (WIMA), Maritime SheEO, African Women in Maritime (WIMAFRICA), Women Offshore, International Association of Maritime Women (IAMW), Women in Maritime Association Caribbean (WiMAC), Seafarers' Welfare and Assistance Network (SWAN) and Women in Maritime Task Force (UK Maritime). These organizations work globally and regionally to provide platforms for women to access training, mentorship, and networking opportunities, helping them overcome barriers in a traditionally male-dominated field. They advocate for policies that support women's participation in various maritime sectors, from shipping and navigation to marine engineering and administration. By fostering inclusivity and diversity, these organizations contribute to the overall growth and sustainability of the global maritime industry, while also encouraging innovation and leadership from a broader talent pool.

Considering the important role of Indonesian women in the workplace and at home, and in order to achieve the vision of Golden Indonesia 2045 through the development of high-quality human resources, especially in the maritime sector, Pelindo post-merger, is striving to contribute by the establishment of Mutiara Pelindo on April 2022. This women organization in Indonesia's port sector to empower woman and Pelindo employees, also provide contribution for the environment surrounding the ports and nurture future generations in port and maritime transport business. This paper examines the impact of Mutiara Pelindo's initiatives on company performance and environmental sustainability from 2022 to 2024. Mutiara Pelindo has successfully implemented various empowerment programs focused on increasing female participation and leadership within the maritime industry. These initiatives include specialized training, leadership development, and community engagement, all of which have contributed to a more inclusive and diverse workplace. This paper also aims to demonstrate that Indonesian women in the port sector are striving to create meaningful change, and highlights how we can contribute by supporting women in our workplaces, our environment, and the communities surrounding the port.

2. Literature on Woman in Workplace and in Port Sector

2.1 Woman in Workplace and Challenges They are Facing Today

Women's participation in the formal workforce began to shift significantly in the 19th century, with key changes occurring during and after the Industrial Revolution and into the 20th century [6]. As factories proliferated, particularly in industries such as textiles, clothing, and manufacturing, women were hired in large numbers. Women were seen as a cheap source of labor, and factory owners capitalized on their willingness to work for lower wages compared to men. Many women, especially young, unmarried women, migrated from rural areas to urban centers to work in these factories. The textile industry, in particular, became a major employer of women. Low Wages and poor working conditions were the main issue at that time [6].

Today, women's participation in the workforce continues to grow across many sectors, with women occupying roles ranging from entry-level jobs to executive positions. However, despite the progress made since the Industrial Revolution, challenges such as the gender pay gap, underrepresentation in leadership, and work-life balance remain [7]. Moreover nowadays, research shows that the role of women in the workplace brings greater benefits for the company. The larger representation of women on their management committees, then the more tendency that the company demonstrates superior performance [7, 8]. Therefore, in order to increase women involvement in senior corporate management, creating social environments that better supports women in the workforce is needed as example in these three different countries as case studies [7, 9]:

- France : A Pro-family policy exist to help balance family and professional life, with various public facilities.
- United Kingdom : Supporting Day-care and state aid for childcare to support working women and a strong tradition of workplace equality.
- Sweden : Gender-neutral parental leave of 480 days and afforadble childcare to create work and family balance, also 2009 Discrimination Act which promotes equality between men and women in workplace.

Women are challenged in the workplace, and the challenge gets more difficult as they climb up the leadership ladder. In research by [10] where they interviewed many working women, findings show that their organizations mostly define leadership in a traditional, stereotypically masculine style demonstrated by most of their senior male colleagues. This is in line with research by [11] which details all types of gender bias and stereotypes in order to break the glass ceiling and reach leadership positions in organizations. One of the stereotypes is the 'queen bee', where a successful female leaders must show the type of masculine leadership characteristics as a way to keep its position in the organization, they must distance themselves from other women, be very competitive and assertive [11]. A model of challenges women experiences to reach high leadership role in organization or companies are described by [11] shown in Figure 4 as follows.

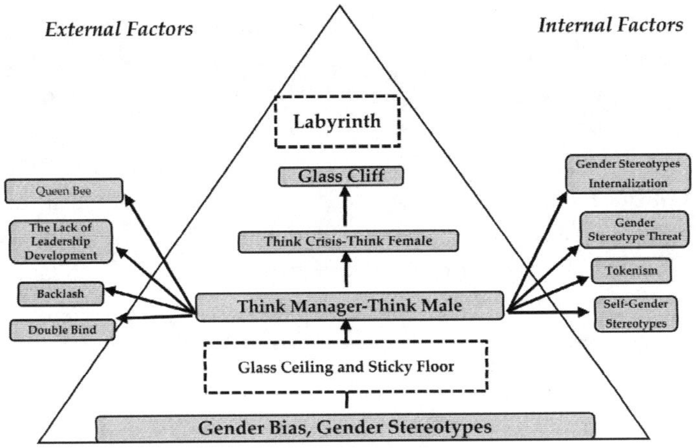

Figure 4. A Model of Challenges Women Experience in Organizations, source: [11]

IOP Conf. Series: Earth and Environmental Science **1461** (2025) 012052 doi:10.1088/1755-1315/1461/1/012052

2.2 Challenges for Women in the Port Sector

The port and maritime sector has traditionally been male-dominated, with women often facing significant barriers to entry and progression in this field. However, over the years, women have slowly made inroads into the industry, although their representation remains relatively low and not many has changed year on year [12]. Men are dominating significantly, both in the office and the yard, employed by port operators and the freight handlers [13]. Women's role in port industry from 2016 to 2022 are mostly in management, administration and corporate office as described in Figure 5.

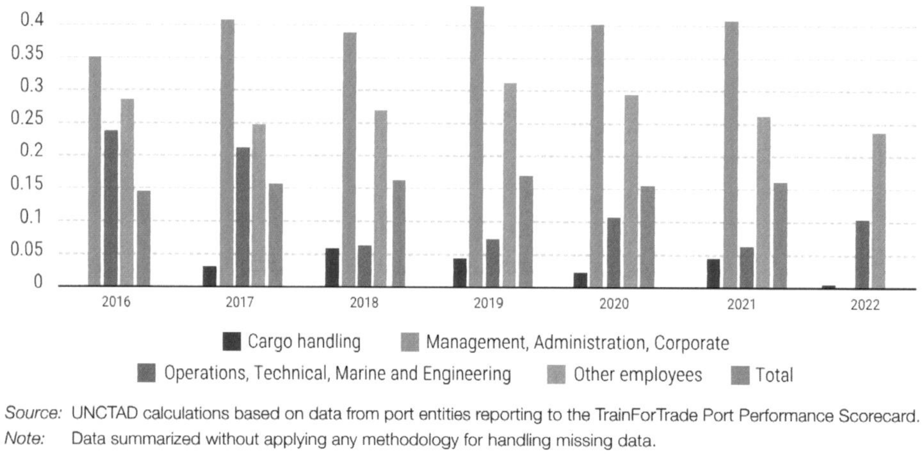

Figure 5. Women's Participation in Port Workforces, Median Across all Ports in 2016 to 2022, source: [12]

Historically in maritime work, whether on ships or in port operations, it was considered too physically demanding and dangerous for women. As a result, women were largely excluded from key roles in this sector, which was dominated by male labourers, engineers, and sailors. In many cultures, there were even superstitions against having women on board ships. In recent decades, societal changes, technological advancements, and efforts toward gender equality have opened more opportunities for women in the maritime and port industries. Automation and technology open opportunities so that many roles in ports and maritime transport have become less physically demanding due to automation and advancements in technology. Tasks that once required significant manual labour such as cargo handling planning and operations, navigation, and machinery operation, which are now increasingly managed with modern equipment and digital systems.

Research by [14] identified challenges women face and also what expectations women have to fulfil in the port sector in their research. They found that there are two main challenges which are cliché-riden environment and physical competence. The first cliché-riden environment consist of: Limited vacancies for women at seaports; Male dominated the industry; Lack of awareness and information about the sector; No or less support from family; and Lack of workshop and seminar for women [14]. Meanwhile, physical competence consists of: Less security assurance for women in this field; The different gender makes it difficult to generalise; Lack of skill and strength; Working in a non-traditional job [14].

There are three main components of expectations or characteristics a women working in ports should have which are patience, professionalism and courage (Jeevan, et al, 2020). These are the details: Patience includes being persistent in doing work and has better understanding in problem solving. Professionalism means Woman are well represented in leadership and workforce, woman is able to manage the tasks professionally, has the critical thinking in problem solving and woman advantages such as positives vibes needs to put into the work environment. Lastly, courage is defined by woman standing well in the seaport sector because of her bravery and her ability in the management of work and operation.

2.3 Indonesian Woman in Workplace
According to [15], only 6% of women in Indonesia who holds the role of Directors from the entire women workers. In their research on women in Indonesian State-Owned Enterprises' (SOEs), [15] did many interviews to women CEOs, women Directors, women EVPs, women VPs and women managers, and found that respondents mostly built up their leadership experience by their own, in which very limited support were given by their respective organisation. They also found out that significant others of the participants, such as children and family, are very critical in supporting their career and that they very much needed coaching and mentoring to improve their work [15].

The development on this issue today has been improving. A major turning point was reached in 2018 with the establishment of Indonesia's Ministry of State-Owned Enterprises' policy which mandates a minimum of 25% female representation in senior roles [16]. In 2021, the President of the Republic of Indonesia, Mr. Joko Widodo, issued a call that led to the establishment of the SOE women community named *Srikandi BUMN* which is additionally an implementation for the UN's HeForShe initiative [17]. According to President Widodo, Indonesia is determined to see 30% more women represented in the legislature and other policy-making organizations. The directive from Mr. Erick Thohir, the Minister of SOE further reaffirmed this commitment. It stated that in line with Southeast Asian standards, SOEs have a future program to increase the number of women, with an aim of 15% at the board of directors level [18].

3. Methodology

The methodology employed in this paper adopts a mixed-methods approach, which integrates both qualitative and quantitative data to provide a comprehensive analysis of the research topic. Qualitative data was gathered through an examination of company documents and relevant publications, allowing for an in-depth exploration of the context and perspectives surrounding the subject matter. This qualitative analysis provided rich, contextual insights that informed the broader understanding of the issues at hand. In parallel, quantitative data was collected through a survey conducted by Mutiara Pelindo, alongside interactions and engagements during various events at the period of April 2022 to April 2024. This dual approach facilitated a triangulation of data, enhancing the validity and reliability of the findings by merging numerical data with contextual qualitative insights. By leveraging both methodologies, the study aims to present a more nuanced and holistic view of the research objectives.

4. Mutiara Pelindo: History and Objective of the Organization

Mutiara Pelindo is a organization within Pelindo for women workers to support each other in their work and gain achievements across numerous roles, whether as moms, wives, or professionals. It was created in 2022 by PT Pelabuhan Indonesia (Persero) / Pelindo. This community, upholds a slogan "Maritime Sisterhood." It is also a component of the SOE's *Srikandi BUMN* program and seeks to give women employed by Pelindo a forum for mutual support, capacity building, adaptability, and role balancing. The phrase "*Muda, Tangguh dan Pemelihara*," which translates to "Young, Resilient, and Nurturing," is the acronym for the word "Mutiara."

Formally, the Management of Pelindo officially launched Mutiara Pelindo in Makassar on April 21, 2022. Mutiara Pelindo is fully supported by the Pelindo Board of Directors and the Deputy for Human Resources, Technology, and Information at the Ministry of State-Owned Enterprises. The primary objective of Mutiara Pelindo is to provide a platform for women at Pelindo to support each other, enhance their capabilities, adapt to change, and maintain a work-life balance. Mutiara Pelindo also provides contribution for the environment surrounding the ports and nurture future generations in port and maritime transport business through its activities and programs.

The organizational structure of Mutiara Pelindo is under Pelindo's President Director and HR Director. It is supported by Group Head of Corporate Secretary and Group Head of HR Management as advisors. The Chairwoman of Mutiara Pelindo is one of the Directors in the Pelindo Subholding Company. Under the Chairwoman of Mutiara Pelindo there are two Vice Chairwoman. Secretary, Treasurer, five Division Heads and team members. In total there are 36 women from across Indonesia in the Mutiara Pelindo 2022-2024 administration. These divisions are as follows: Strategic and Organization Division; Leadership and Development Division; Cooperation and Communication Division; Environment and Sustainability Division; and lastly Welfare and Health Division. Initiatives and programs held by Mutiara Pelindo is executed by the division heads which includes specialized training, leadership development, and community engagements.

The Mutiara Pelindo Survey was conducted in the beginning of the year 2024. A total of 203 respondents' data were gathered, which consists of 79 men employees and 124 women employees. About 54% of respondents are officers or staff, while the rest holds structural or functional positions in Pelindo. Overall, a total of 90% of respondents are aware of Mutiara Pelindo, and among them, 92% are familiar with the identity of its chairwoman. Figure 6 below describes the meaning behind Mutiara Pelindo's logo. Currently the new administration of Mutiara Pelindo 2024-2026 has been appointed, in which there are 23 women in total.

5. Mutiara Pelindo Programs in year 2022-2024 and Its Contribution

In the last 2 years, there has been many initiatives, activities and programs conducted in various venues, or conducted across all Pelindo locations simultaneously, and even online via zoom which could engage many Pelindo employees and external staekholders of Pelindo. Detailed programs are are as follows, with a picture described in Figure 7.

Meaning:

1-The pearl, the first element, is a timeless symbol of elegance. Recognized as a living gemstone, pearls have been treasured throughout history for their beauty and allure;

2-The Pelindo logo, representing PT Pelabuhan Indonesia (Persero), holds significant historical value as the core of Mutiara Pelindo;

3-The Shell, signifies strength and solidity, which is also reflected in the company's ethics. It symbolizes an openness to mutual help and respect, represented by the open-handed gesture in the brand's logo;

4-The shape of a ship, symbolizes a harbor as the destination where ships from all over the world dock;

5-Young, resilient, and nurturing, these are the core values and tagline of Mutiara Pelindo;

6-the "Mutiara" logotype employs a sans-serif font that exudes luxury, elegance, and symbolizes beauty, femininity, and authority;

7-The shape of the Pelindo logo is inspired by a busy network, symbolizing the company's complex and interconnected operations.

Figure 6. Mutiara Pelindo Logo, source: Authors

Figure 7. Mutiara Pelindo Programs in 2022-2024, source: Authors

- Women empowerment, leadership and training programs: sharing sessions hybrid (online and offline), Respectful Workplace Policy, Mutiara Pelindo Teaching in schools, financial planning and management, Women Take Over, capacity building, public speaking trainning, leadership trainning, mentoring, webinars, recorded video as employee learning hours, international networkings, writing and poster competitions, mentorship programs, etc.
- CSR and environment programs: Clothes and books donation, Environment Social Responsibility events, Fundraising with Charity Auction, Stunting Prevention, coral reef cultivation/conservation and ocean diving, waste processing support programs, small medium enterprise support programs, donation collection from Pelindo employees, scholarship for children in need at communities surrounding the port, nutritional support initiatives that treat and/or prevent children from stunting, eco lifestyle, SOE volunteering programs, training seminars for Small Medium Enterprises collaborating with IMO and WIMA, preserving culture events, blood donation events, etc.
- Health and Wellness Programs: running and biking competitions, health and mental health sharing sessions, wellness campaign and awareness, HIIT workout, running buddies, work life balance seminars, employee wellness fair, bring children to work day, parenting and beauty class, safety campaign, HSSE culture campaign, etc.
- Celebration events and Social Media campaigns: Indonesian Women Day (Kartini Day), Indonesian Mothers Day, Environtment Day, Mental Health Day, International Women in Maritime Day, Ramadhan-Eid Fitr and Eid-Qurban or other religion events in paralel with charity programs, social media channels, women profile and stories through media publish articles, etc.

These programs were executed in paralel with the establishment of new supporting policies in the Pelindo, inline with programs from Ministry of SOE such as 3 months maternity leave, work life balance program, respective workplace policy, employee wellbeing policy, flexible working hours, etc [19, 5]. These policy which supports women and employee welfare and wellbeing, makes women workers in Pelindo feel safe, supported and motivated to work, which eventually brings positive impacts for themselves and for their home and children at home. Overall results from the Mutiara Pelindo Survey 2024 shows that 94 % of respondents are satisfied with programs held by Mutiara Pelindo. Their top three programs are Webinar Woman Leadership in first place, Charity and Donation program in second place and Employee Wellness Fair in third place.

In terms of leadership and gender discrimination, according to Pelindo Annual Reports, overall between 2017 to 2023, the proportion of women in middle and senior management roles at PT Pelabuhan Indonesia (Persero) grew from 15% on average in 2017 to 22% in 2023. This shows a favorable trend, with a 46.7% growth over the last six years. However, there has been variation between ports since the data shows there are notable difference in the rate of progress between various ports. For example, by 2023, the Port of Tanjung Priok had 24 percent of its leadership team of women. A few smaller regional ports were unable to make it to 18%.

Related to CSR programs, there are benefits brought by Mutiara Pelindo programs not only for the aid recipients, but actually for the women employees themselves and Pelindo culture. As example, there are 500 children of outsourced workers in Pelindo's four regional areas received educational scholarships. Specifically, a number of children in villages at Takalar and Jayapura received nutritional support initiatives that treat stunting. Moreover, 65% of employees on average engaged in at least one CSR activity each year, in which 72% of female employees participated in CSR activities greater than rates for men which is 61%. According to 82% of survey participants, CSR actions enhanced their opinion of their employer. 75% thought these initiatives helped create a more welcoming culture at work. 68% of female employees stated that CSR initiatives made them feel more a part of the working community.

Health and wellness programs brought benefits as follows. 92% of participants said they felt more helped to manage their obligations to their families and their jobs. Female employees with small children who resigned from ports offering these programs decreased by 15%. Approximately 70% of workers engaged in at least one wellness activity every year on average. 80% percent of female employees said that these programs have made them feel more supported at work. Ports with extensive wellness initiatives showed a 20% decrease in the use of sick days. Among working parents, job satisfaction increased by 25%. There were also 70% participants said they were more aware of mental health issues at work. In addition, according to 65% of managers, talking about mental health with their teams is now more comfortable. On the other hand, training on public speaking brought impact in which 85% of participants said they felt more confident in work-related situations.

6. Discussions

6.1 Mutiara Pelindo Programs aligned towhat is Challenged and Expected from Women in Ports
Mutiara Pelindo programs could be matched to tackle challenges for women in port industry also to fulfil the expectations or characteristics of women working in port industry. The following Table 1 plots this in detail.

Table 1. Mutiara Pelindo Programs Aligned to What is Challenged and Expected from Women in Ports

		Challenges for Women in Port Industry (Jeevan, et al, 2020)		**Expectations of Women Working in Port Industry (Jeevan, et al, 2020)**		
		Cliché-riden environment	**Skill & competence**	**Patience**	**Professionalism**	**Courage**
Mutiara Pelindo Programs	**Women empowerment, leadership & training**	v	o	v	o	o
	CSR and environment programs	v	ir	v	v	v
	Health and Wellness Programs	v	o	v	v	v
	Celebration events & Social Media campaigns	v	v	v	v	v

Source: Authors, v: has been done, o: on progress, ir: irrelevant

Overall, Mutiara Pelindo programs has been well executed to fulfil the needs of Pelindo women employees in terms of preparing them to face challenges and fulfil expectations from the industry. However, there are still things that are on progress such as to build up skill, competence, professionalism and courage. The survey might not be able to detect this. Programs held by Mutiara Pelindo are mostly soft skills or general managerial skills, which are not yet specific professions in port such as tally operators, tug boat operators, security or craine operators for women. However, there are recently new women training programs specific for planning and control in container handling operations held in "*Ladies Forum*" event. Specific business subsidiaries which provide these services run trainings by themselves and do not tailor their training specifically for women. Moreover, Mutiara Pelindo initiatives are still on progress to improve professionalism, courage, also health and wellness. Hence, in the future Mutiara Pelindo aims to contribute by giving more gender specific professional training on these themes.

6.2 Mutiara Pelindo Contributing to Critical Issues Post-Merger
The organizational structure of Mutiara Pelindo and its programs in combination has been well established to tackle critical issues needed in Pelindo, especially Pelindo post merger. There has been issues on inclusivity and diversity because employees somehow relate their work and culture with their previous ex-Pelindo entity before the merger. There are also issues where individual perceives their previous entitiy much stronger or greater than the others. This brings an unhealthy culture for the workplace and hinders organization growth. Through various events, Pelindo employees could mingle and get along together. Mutiara Pelindo also helps the HR directorate to break-down silos between departments and entities within Pelindo Group.

Moreover, having Pelindo Chairwomen and women leaders in the Pelindo Group, brings impact on young women employees to have mentor figures they admire and could look up to. They get insights on many aspects in work such as the ability of women leaders to bring diverse perspectives, improve and nurture team dynamics, good emotional intelligence, nurture strong stakeholder and customer relationships, create resilience and adaptability, also having a growth mindset and open to diversity in workplace.

The women seminar and empowerment programs by Mutiara Pelindo have successfully fostered an image of women leaders who are not "*queen bees*." These leaders are approachable and open to discussions and mentoring, actively engaging with other employees rather than distancing themselves. Their friendly and inclusive approach has built strong connections with both young women and men in the company, creating a supportive environment where everyone feels encouraged to grow and collaborate. All of this are critical for Pelindo post-merger. However, to reach the women leadership percentage target in the company is still a work-in-progress.

6.3 Future Challenges for Indonesian Women in Maritime Sector
Mutiara Pelindo has provided a great foundation for Pelindo Human Capital post-merger, especially for women employees, in the aspect of empowerment, leadership, knowledge and training, CSR and environment activities, as well as improving health and employee wellbeings. Gender equalilty in the port industry where it is a male dominated workplace has been nicely put in place. However, there is a bigger challenge ahead. The next mission Mutiara Pelindo is struggling to do is in the near future is having women to contribute more for the company through operational efficiency and innovation. Innovation and support for young generations and women innovators is needed. This can be seen from the Innovation Competitions held by Pelindo each year, the winners are mostly male employees. Only 6 are women out of total 36 employees winner of the innovation competition in the last 5 years. Mutiara Pelindo current administration must think of creative ways to nurture innovation and learn from best practices in world class ports and companies in growing creativity and new approaches, driving progress and breakthroughs in various fields. Mutiara Pelindo needs to empower women to bring unique insights and problem-solving skills that can lead to innovative solutions and advancements.

7. Conclusions

This paper highlights the efforts of Indonesian women in the port sector to drive positive change and emphasizes the importance of supporting women in the workplace, the surrounding environment, and the communities around the port. In conclusion, Mutiara Pelindo has played a vital role in empowering women at PT Pelabuhan Indonesia (Persero) since its inception in 2022. Through its various programs—ranging from leadership training and capacity building to community engagement and health and wellness initiatives—the organization has created a strong foundation for women to thrive professionally while balancing their personal roles. The 2024 Mutiara Pelindo Survey shows overwhelming satisfaction with these programs, with 94% of respondents expressing positive feedback. Mutiara Pelindo's efforts have also contributed to fostering inclusivity, breaking down silos post-merger, and promoting a supportive environment where women leaders are approachable, engaging, and open to mentoring, thus improving the overall culture within the organization.

However, there are still challenges to address, particularly in enhancing women's participation in operational innovation and efficiency. Despite the strong foundation laid by Mutiara Pelindo, the organization must continue to push for more gender-specific professional training and create avenues for women to contribute to innovation. With the innovation competitions showing a lower percentage of women winners, future efforts must focus on nurturing creativity and fostering a culture of innovation among women employees. By learning from best practices in world-class companies and ports, Mutiara Pelindo can further empower women to become key contributors to the company's success, driving progress and bringing fresh insights into the maritime sector. Empowering women in the maritime workforce has the potential to bring diversity, innovation, and new perspectives to an industry that is crucial to global trade and economic development, hence, it should be fully supported.

Aknowledgement

The team of authors would like to thank the management of PT Pelabuhan Indonesia (Persero) and Mutiara Pelindo administration who supported the writing of this research.

References

[1] BPS. 2024. Mid Year Population. https://www.bps.go.id/en/statistics-table/2/MTk3NSMy/mid-year-population--thousand-people-.html. [Accessed 3 September 2024].
[2] Rainer, P. 2024. February 2024 Woman Workforce Increases. Statistics. https://data.goodstats.id/statistic/februari-2024-angkatan-kerja-perempuan-meningkat-cpeoL#:~:text=Tingkat%20Partisipasi%20Angkatan%20Kerja%20. [Accessed 3 September 2024].
[3] DGST. 2022. Directorate General of Sea Transportation. Ministry of Transportation of Indonesia. Percentage Number of Seafarer by Gender, via http://www.pelautdephubgoid.com/. [Accessed 8 November 2023].
[4] Pelindo I, II, III, IV. 2017-2023. Annual Reports. Retrieved from respective corporate websites.
[5] Pelindo. 2024. Company Profile. Jakarta.
[6] Bythell D. 1993. Women in the workforce. In: O'Brien P, Quinault R, eds. The Industrial Revolution and British Society. Cambridge University Press:31-53.
[7] McKinsey & Company. 2007. Women Matter: Gender diversity, a corporate performance driver. https://www.mckinsey.com/~/media/McKinsey. Accessed July 2024.
[8] Triana & Asri, M. 2017. The Impact of Female Directors on Firm Performance: Evidence from Indonesia. Journal of Indonesian Economy and Business, 32: 1: 19-32.
[9] Swedish Institute. 2023. 5 examples of gender equality in Sweden. https://sharingsweden.se/materials/5-examples-of-gender-equality-in-sweden. Accessed in August 2024.

[10] Catalino, N., & Marnane, K. 2019. When women lead, workplaces should listen. McKinsey Quarterly, December 2019.
[11] Galsanjigmed, E., & Sekiguchi T. 2023. Challenges Women Experience in Leadership Careers: An Integrative Review. Merits: 3: 366-389. https://doi.org/10.3390/ merits3020021.
[12] UNCTAD. 2023. Review of Maritime Transport. https://unctad.org/publication/review-maritime-transport-2023. Accessed July 2024.
[13] Karatas-Cetin, C., Baran-Kasapoğlu, E., & Ayaz, I. S. 2022. An Exploratory Study of Recruitment of Women in the Port Industry. Handbook of Research on the Future of the Maritime Industry. DOI: 10.4018/978-1-7998-9039-3.ch006.
[14] Jeevan, J., Menhat, M., Ruslan, S. M. M., & Cetin, C. K. 2020. Gender inequality: an outlook from a seaport sector. Australian Journal of Maritime & Ocean Affairs, DOI: 10.1080/18366503.2020.1810375.
[15] Dewi, Y. & Rachmawati, R. 2014. Leadership Development Experiences of Women Leaders in State-Owned Enterprises in Indonesia. The South East Asian Journal of Management, 8:2:5. DOI: 10.21002/seam.v8i2.3940.
[16] Ministry of State-Owned Enterprises. 2018. Circular Letter on Women's Representation in Leadership Positions. Jakarta: Government of Indonesia.
[17] Ministry of State-Owned Enterprises. 2022. Women and Milenial Representatives. Ministry of State Owned Enterprise Republic of Indonesia. https://bumn.go.id/prioritas/prioritas/Keterwakilan%20Perempuan%20dan%20Milenial. Accessed August 2024.
[18] Puspadini, M. 2024. Erick Thohir targets 15% Leaders in SOEs are Women. https://www.cnbcindonesia.com/lifestyle/20240308143659-33-520765/erick-thohir-target-15-pimpinan-bumn-diisi-perempuan. CNBC Indonesia March. Accessed August 2024.
[19] Srikandi BUMN. 2023. Respectful Workplace Program Implementation Guide. Jakarta: Srikandi BUMN.

IOP Conf. Series: Earth and Environmental Science **1461** (2025) 012053 doi:10.1088/1755-1315/1461/1/012053

Collision Risk Assessment in a Narrow Channel with Gini Coefficient Approach

F I Prastyasari[1*], AAB Dinariyana[1], and F H Javanica[2]

[1] Department of Marine Engineering, Institut Teknologi Sepuluh Nopember, Surabaya, Indonesia
[2] Undergraduate Student of Department of Marine Engineering, Institut Teknologi Sepuluh Nopember, Surabaya, Indonesia

*E-mail: fadilla@its.ac.id

Abstract. Tanjung Perak Port in Surabaya, Indonesia, serves as a primary hub for trade and economic activities in the eastern region. It makes this port has a great traffic density and prone to a collision accident. Gini coefficient was employed to utilize the Automatic Identification System (AIS) data for ship collision frequency analysis in the port area. AIS data contained voyage related data, one of which is Course Over Ground (COG), which were then subjected to Gini coefficient calculations, measuring the dispersion of COG values. Hierarchical clustering grouped COG values into four clusters: the first comprised of Leg 1, 2, 3, 4, and 6, the second included Leg 8 and 7, the third involved Leg 5 and 10, and the fourth comprised of Leg 9. Cluster proximity validation, indicated by a cophenetic correlation of 0.90484, signified excellent relationships between clusters. Subsequent risk characterization classified clusters 1, 2, and 4 as "extremely remote" risk levels, while cluster 3 was deemed "remote," based on IMO frequency index assessments. IWRAP software facilitated accident frequency calculations, revealing discrepancies in risk levels across different areas in the port. Notably, while IWRAP assigned a "remote" risk level to leg 4, the Gini coefficient characterization indicated "remote" risk levels for legs 10 and 5. Recommendations for reducing ship collision frequency through variable geometry and causal probability values in the Western Shipping Lane of Surabaya should be provided if the ship collision frequency exceeds ALARP level according to the UK HSE standards.

1. Introduction

The Tanjung Perak Port in Surabaya, East Java, stands as a cornerstone of Indonesian maritime infrastructure, commanding a prominent status in the nation maritime domain. Renowned as the second busiest port nationwide, Tanjung Perak Port serves as an indispensable nexus for the facilitation of national trade, import, and export operations. Its strategic location and robust infrastructure underscore its pivotal role as a focal point for commercial activities, further enhancing its significance as a vital cog in the nation's economic machinery. This port exemplifies the critical role that maritime routes play in fostering seamless trade and economic exchanges, particularly in bolstering the commercial activities of the eastern region of Indonesia. Efficient shipping routes are crucial around Tanjung Perak, ensuring smooth vessel transit and sustaining uninterrupted goods flow, reinforcing its pivotal role in Indonesian maritime trade. The management of shipping routes is the responsibility of the government and must be carried out effectively to support the overall smoothness and efficiency of port operations. The Surabaya West Access Channel is a vital route for the traffic of the Tanjung Perak Port in Surabaya and other nearby ports. The number of ships passing through the Surabaya West Access Channel (SWAC) in 2016 was 1,105 units, equivalent to a total ship weight of more than 27 million gross tons. In 2017, there was an increase, with the number of ships reaching 2,857 units, equivalent to a total ship weight of over 53 million gross tons. This represents a significant year-on-year increase of 158.5% for ship traffic and 96.3% for ship weight.

Figure 1 below illustrates the intensity of shipping activities to and from the Tanjung Perak Port in Surabaya, with at least 30 shipping routes [1]. According to the explanation from the Head of Tanjung Perak Main Shipping Office, on April 26, 2017, SWAC is a shipping route that connects ships heading to Tanjung Perak Port from the North Java Sea.

The existence of Surabaya West Access Channel (SWAC) does not automatically eliminate the risk of ship collisions. A ship accident occurred between the Motor Vessel (MV) Tanto Bersinar and the tugboat (TB) Mitra Jaya XIX, which was towing the barge Makmur Abadi V. It was reported that 5 crew members were missing. Another accident involving the MV. Wihan Sejahtera was a result of the busy traffic in the Surabaya West Access Channel (SWAC), leading to an oil spill that polluted the sea. Then, on December 6, 2021, around 05:07 AM local time, a ship collision occurred in the waters near Buoy 5 of the Surabaya West Access Shipping between the TSHD HAM 311 and the MT Crane Vesta [2].

As an effective control measure, ship routes significantly contribute to improving navigational safety and enhancing maritime traffic efficiency. However, ship routes cannot eliminate conflicts in maritime traffic. For example, the enforcement of ship routes can lead to traffic congestion in Precautionary Areas (PA), increasing the complexity of maritime traffic [3]; there are still instances of ships crossing traffic separation zones or lines. With the growth of global trade and the number of ships, traffic density increases along with its associated risks [4].

The determination of ship routes and traffic separation schemes is formulated after surveys and consultations with ship captains familiar with the area [5]. Subjective expert assessments

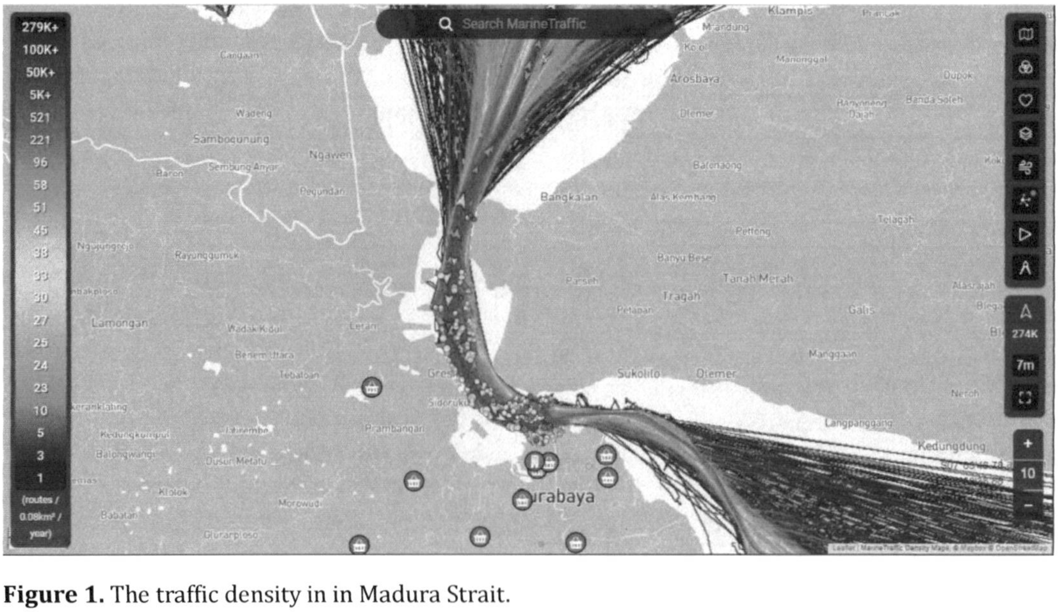

Figure 1. The traffic density in in Madura Strait.

cannot effectively warn incoming ships about high-risk routes or guide authorities on risk reduction measures. Quantitatively assessing ship collision risk in shipping routes is essential [3].

In this research, the Gini coefficient method based on AIS data is proposed as a new perspective for quantifying collision risk in maritime shipping. The Gini coefficient, with values ranging from 0 to 1, indicates the extent of data dispersion within a system. The closer the value is to 0, the more evenly distributed the data is considered, while approaching a value of 1 indicates uneven distribution. The availability of Automatic Identification System (AIS) data is the primary driver for addressing the shortage of ship position data. Static data on AIS includes information about the ship's identity, such as its name, type, size, maximum speed, and destination, while dynamic data on AIS includes real-time information such as the ship's position, speed, and direction [6]. The data obtained from AIS provides a more comprehensive and detailed overview of ship activities, enabling better monitoring of navigation, risk management, and maritime safety in general. The AIS data used consists of Course Over Ground (COG) data. According to my perspective, COG is the direction of a ship between two points relative to the Earth's surface. According to a study by Lin et al. [7], COG can represent the orderliness in shipping lanes and can also be used to estimate the potential for ship collisions.

Compared to other methods and dimensions (such as distance, flow, density, etc.), this method has the advantages of simplicity, accuracy, and robustness in evaluating the risk of ship collisions from overall AIS data in a specific area, without considering differences between different periods [7]. In a study conducted by Okunev [8], he empirically compared the Extended Mean Gini (EMG) model with the average variance model related to the formation of efficient frontiers, asset pricing, portfolio performance, and agricultural planning under uncertainty. The main result of the study is that the extended Gini coefficient is a viable alternative as a measure of risk.

2. Methodology

2.1 Gini Coefficient

The Gini coefficient is the most commonly used approach to measure inconsistency. This coefficient was developed by Italian statistician Corrado Gini (1884-1965) and named after him. This approach is often used as a method for assessing income inequality but can also be used to measure inequality in any type of distribution [9]. The Gini coefficient, with values ranging from 0 to 1, indicates the extent of data dispersion within a system. The closer the value is to 0, the more evenly distributed the data is considered, while approaching a value of 1 indicates uneven distribution. This method is often used not only in the engineering field, but also in the socio-economic topics [9], [10], [11], [12]. In the context of ship traffic in shipping lanes, a research showed that the Gini coefficient can be used to analyze the regularity or chaos in ship traffic patterns [7]. It is important to note that the regularity or irregularity of ship traffic can be a significant factor in the risk of ship accidents. If the Gini value is low, indicating low regularity, it can be interpreted that ship traffic is less orderly and may tend to be "chaotic." In this context, the risk of ship collisions is likely to increase due to the lack of regularity and predictability in shipping lanes. Conversely, if the Gini value approaches 1, indicating high regularity in ship traffic, it may reflect a more organized structure of ship traffic and can help reduce the risk of collisions. Therefore, the Gini coefficient can be a useful tool in understanding and evaluating the risk of ship accidents related to ship traffic patterns in a shipping lane.

2.2 Hierarchical Clustering

The hierarchical clustering method works by grouping data objects into a hierarchy or "tree" cluster. This method can represent data objects in the form of a useful hierarchy for summarizing and visualizing data. Hierarchical clustering algorithms do not require the input of a specific number of clusters (K-cluster), meaning they allow grouping without determining the number of clusters beforehand. This method has been widely used in various fields, such as air transportation, maritime route extraction, and even astronomy. For example, in a study by Hwang and Youn, that explored how to identify and systematize objective navigation scenarios for validating autonomous ship collision avoidance algorithms [13]. In this study, a data-driven approach was used to collect AIS data for 12 months in the waters off the west coast of Korea, to extract ship trajectories, and to hierarchically cluster data based on navigation situations.

Moreover, in several recent studies on ship collision risk assessment, notable developments have been found. For instance, a study by Marino, et al., focused on emerging new technologies, identifying current major trends, and discussing perspectives and challenges for the future [14]. This review reveals a wide and diverse range of methods, including machine learning, hierarchical clustering, swarm intelligence algorithms, and more.

The algorithmic hierarchical clustering method is divided into two, namely the Agglomerative Nesting (AGNES) method [15] and the Divisive Analysis (DIANA) method. In this research, the method chosen for hierarchical clustering of the Gini coefficient COG values is the AGNES method.

3. Results and Discussion

3.1 Data Collection

Data collection is the initial step that must be carried out in this research, and the availability of data is essential as it will be used to analyze frequency using the Gini Coefficient and IWRAP.
The quantity of data in the data frame represents the evaluated traffic.

Raw AIS messages are the primary data that will be used in this research. The mentioned data consists of specific recorded data transmitted by all ships in the area and received by Vessel Traffic Services (VTS), where the recorded data encompasses at least all datasets from August to September. Raw AIS messages are directly retrieved from the server located in the RAMS (Reliability, Availability, Management, and Safety Laboratory). The extracted raw data has beUBen compiled into a text file over a specific period, with all information encrypted in a set of NMEA data packets, and the timestamp of each dataset indicates when the record was transmitted specifically. The content of the text file is listed on separate lines for each raw AIS message, as illustrated in Figure 2 the text file on August 1, 2023.

Every second, we can observe that data from each ship is being recorded, as shown in Figure 3 within this packet dataset. However, it is important to note that the quantity of recorded data may vary from day to day. Furthermore, there may be instances where VTS is unable to record any data at all during certain times. This could be due to various factors such as technical issues, signal interruptions, or other operational constraints. Therefore, it is essential to consider potential fluctuations and gaps in the data when analyzing the dataset.

IOP Conf. Series: Earth and Environmental Science **1461** (2025) 012053 doi:10.1088/1755-1315/1461/1/012053

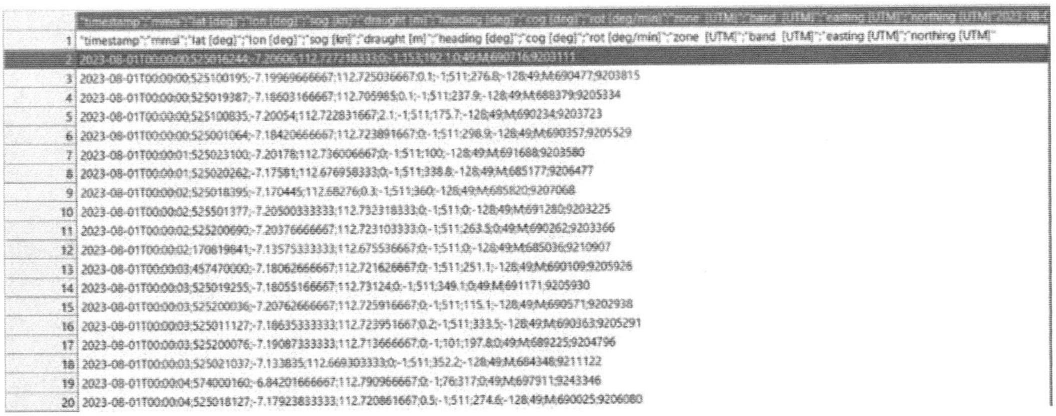

Figure 2. Compiled raw AIS messages

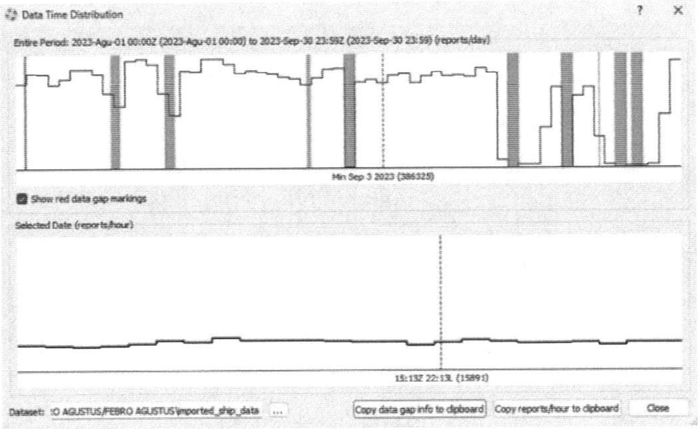

Figure 3. AIS data distribution

Data not recorded during the red-colored periods can occur due to several technical issues, such as disruptions to the AISITS server caused by cable problems, power outages, or signal interruptions. Additionally, other problems may arise due to incomplete data that has not been successfully understood or processed, including errors in transmission or reception and system malfunctions.

3.2 Traffic Analysis in IALA Waterway Risk Assessment Program (IWRAP)

Before calculating the frequency, it is mandatory to divide the track into legs based on the density plot. Creating legs is crucial for conducting frequency calculations because in IWRAP, the frequency of ship collisions will be calculated for different types of collisions, and one of the determining factors for the collisions to be calculated is the leg created. Considerations in creating legs include the density plot, channel width that also determines the maximum leg width, existing ship traffic on the shipping lane, and others. After successfully dividing the legs based on the considerations mentioned in the paragraph above, the frequency calculation is then carried out with the following results on Table 1. The calculations performed using the Gini Coefficient will later refer to each leg; therefore, the frequency calculation results for ship collisions in each leg are needed, as shown on Figure 4.

Table 1. Frequency of each type of collision.

Collision Type	Frequency Value	Unit
Overtaking	0.001131	Incident/Year
Head-on	0.006676	Incident/Year
Crossing	7.29E-05	Incident/Year
Merging	1.83E-08	Incident/Year
Bend	4.45E-05	Incident/Year
Total Collisions	0.007819	Incident/Year

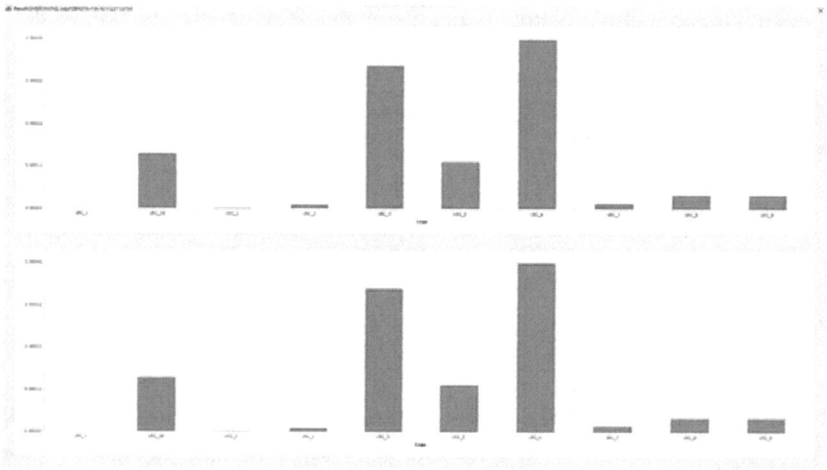

Figure 4. Results of frequency using IWRAP

3.3 Data Processing

At this stage, the raw AIS message is processed and filtered to ensure that the calculation results to be processed in IWRAP software and as input for the Gini coefficient yield accurate values. Based on Figure 5 the group of red-colored points indicates that the data is all successfully read Comma Separated Value (CSV) data, while the black-colored area represents the data selected based on the leg specified in IWRAP according to density mapping. The group of red-colored points indicates that the data is all successfully read the CSV data, while the black-colored area represents the data selected based on the leg specified in IWRAP according to density mapping.

a. Data filtration and conversion.
   During the data processing, Python was utilized as a programming language to do the filtration and conversion. The first step is to remove rows of data containing NaN (Not a Number) values from the DataFrame using the "pandas" module with the 'dropna()' function. Then, the values in the columns within the DataFrame are converted to the 'int' class, which makes Python read the data in the specified column as integers, as well as the 'float' class, which will be read as decimal numbers in Python.

b. Data frame sorting
The data used is sorted based on time and MMSI. The code "ais_19_juli.sort_values(by = ['timestamp', 'mmsi'], inplace = True)" is used to sort the ais_19_juli Data Frame based on the 'timestamp' and 'mmsi' columns. This will sort the data primarily based on the timestamp column and secondarily on the MMSI column.

c. Data interpolation
Data interpolation is necessary due to the frequent occurrence of missing timestamp records. Complete timestamp records for each ship are required for accurate Gini coefficient calculations because when calculating using the Gini coefficient, the COG values of each ship are compared at the same timestamp. The limitation is imposed if the difference between the COG values (COG_2 and COG_1) is greater than 180°. If the difference is less than -180°, it can be formulated as follows:
$COG = COG_1 + Weight \times (COG_1 + 360 - COG_2)$
Weight is a variable used as a weighting factor to calculate interpolation points between two initial points.

$$Weight = \frac{i}{num\_points + 1} \tag{1}$$

where *i* is the current iteration number in the loop and *num_points* is the desired number of interpolation points.
Weight ranges from 0 (in the first iteration) to 1 (in the last iteration), and its value is used to calculate interpolation points between two initial points (lon_1, lat_1) and (lon_2, lat_2) by assigning appropriate weights to these coordinates.

d. Leg distribution
Designing a code to create a Polygon object using coordinates as corner points. In this case, the polygon is formed by four points, each labeled as Leg 1.1, Leg 1.2, Leg 1.3, and Leg 1.4. These coordinates depict the angles that form the polygon on the map, where each angle is represented by a pair of longitude and latitude values.
After that, the exterior coordinates of the polygon are obtained using the .exterior.xy method and plotted using the plt.plot() function from the Matplotlib module with a line color set to black ('black'). The result is a graphical representation of the polygon formed by the coordinates of the Leg 1.1 to Leg 1.4 angles.
The code reads data stored in small blocks with a size of 10000 rows using the pandas library. Each data block is transformed into a GeoDataFrame using the GeoPandas library, adding a geometry column based on latitude and longitude coordinates. Furthermore, a new column 'berada_dalam_LEG1' is generated using the within method from GeoPandas, indicating whether each point in the block is within the specified geometry region.

3.4 Calculation of Gini coefficient

The Gini coefficient has values between 0 and 1, where 0 indicates a perfectly equal distribution, while 1 indicates the opposite. The Gini coefficient is calculated as the ratio of the area under the Lorenz curve to the total area of the triangle formed by the perfectly equal distribution line. The formula for calculating the Gini coefficient for the COG of ships in grouped situations is as follows:

$$Gini\ coefficient = \frac{\sum_{i=1}^{n}\sum_{j=1}^{n}|xi-xj|}{2n\sum_{i=1}^{n}xi} \tag{2}$$

Where |xi-xj| is the absolute value of the difference between each pair of sample distances.

If the distribution of COG in SWAC is homogeneous, it means that the shipping lane has a good level of regularity. Conversely, if the COG of ships is distributed more evenly, indicating irregularities in the shipping lane, the Gini coefficient value will tend to be smaller [7]. The Gini coefficient calculation is performed on groups of Course Over Ground (COG) of ships within specified segments (legs). This process is carried out using Equation 2. Data is derived from the shipping route during a pre-determined three-month period. The Gini coefficient analysis helps understand the distribution of ship COG and the level of regularity in the shipping channel in the SWAC region during a specific period.

3.5 Hierarchical Clustering

In implementing this method to group the Gini coefficients of COG in the SWAC, the specific steps to be followed are as follows:

1) Group all sample points in dataset D into separate categories
2) Calculate the average distance d_{avg} between every two classes and find two class groups Ci and Cj with the smallest distance
3) Merge class groups Ci and Cj into one class group
4) Repeat steps (2) and (3) until all samples are grouped into one category.

There are three ways to calculate distance: minimum distance (d_{min}), maximum distance (d_{max}), and average distance (d_{avg}):

1) Minimum distance (d_{min})
   The steps for calculating the minimum distance between two groups involve finding the smallest distance between each pair of points in Ci and Cj. This process compares each point in Ci with every point in Cj, and the smallest distance from all pairs of points is taken as the minimum distance between the two groups.
2) Minimum distance (d_{max})
   The step in calculating the maximum distance between two groups involves finding the largest distance between each pair of points in Ci and Cj. This process compares each point in Ci with each point in Cj, and the largest distance among all pairs of points is taken as the maximum distance between the two groups.
3) Average distance (d_{avg})
   In calculating the average distance between two groups, Ci and Cj, the process involves computing the average of all distances between each pair of points in Ci and Cj. Specifically, pairs of points in Ci are connected to every pair of points in Cj, the distance between these points is calculated, and the average of all these distances is taken as the average distance between Ci and Cj. The formula for average is:

$$d_{avg}(Ci, Cj) = \frac{1}{|Ci||Cj|} \sum_{x \in Ci} \sum_{z \in Cj} dis(x, z) \qquad (3)$$

3.6 Cophenetic Coefficient

The use of the cophenetic correlation coefficient is crucial in assessing the accuracy and reliability of hierarchical clustering. This coefficient measures the extent to which the relationships among objects in the dendrogram reflect the actual distances between objects in the original data. The cophenetic correlation coefficient provides an overview of how well hierarchical clustering reflects the true structure of COG data.

The results of the cophenetic correlation coefficient can offer a clearer perspective on the validity and accuracy of clustering outcomes. A high value indicates that the dendrogram produced by the algorithm effectively mirrors the structure of COG data, while a low value may suggest potential inaccuracies or biases in clustering. With this approach, research can gain higher confidence in the classification outcomes and interpretation of COG data in the context of assessing collision risk in SWAC. The formula for the cophenetic correlation coefficient is:

$$c = \frac{\sum_{i<j}(x(i,j)-x)(t(i,j)-t)}{\sqrt{\left[\sum_{i<j}(x(i,j)-x)^2\right]\left[\sum_{i<j}(t(i,j)-t)^2\right]}} \quad (4)$$

3.7 Charaterization

Based on the results of hierarchical clustering, several groups are formed with closely related Gini coefficient values. The Gini coefficient values of each group are sorted from smallest to largest. Groups with values close to 0 indicate higher risks compared to other groups. This characterization process is crucial for understanding and comparing the risk of ship collisions in SWAC. Furthermore, the characterization results are compared with findings and interpretations from relevant journals. Through this comparison, the research is expected to contribute to understanding the risk factors affecting the SWAC, aiming to enhance maritime safety and reduce the risk of ship collisions in the region.

3.8 Validation with IWRAP

In this research, the IWRAP method was utilized as a tool for detailed calculations. Three main scenarios explored include head-on situations, crossing situations, and overtaking situations. Ships are categorized based on type and length to identify patterns and characteristics of each group.

After collecting the data, the next steps involved the data processing and interpretation using IWRAP MK II. Data grouped by ship type and length serves as the primary basis for determining the probability of collisions within each ship category, particularly within specific length ranges. The use of LEG (lines) and WAY POINT (points) in IWRAP was employed to define the areas where collision probabilities will be calculated. After designing the leg, the next step involves configuring the leg creation using the editor menu. This menu provides facilities to adjust the parameters and characteristics of the created leg. The editor menu facilitates users in adjusting WAY POINTS and other parameters.

The next step is to compare the risk of ship collisions based on their frequencies using IWRAP. After obtaining the frequency values of ship collisions, the subsequent step involves characterizing these results based on their risk levels and then comparing them with the characterization results obtained through the Gini coefficient method.

Analysis of ship collision and grounding risks is a crucial step in ensuring safety and security around SWAC. If the analysis results indicate that the frequency values exceed acceptable limits, especially when the characterization of the Gini Coefficient-COG reaches Medium Risk to High Risk levels, mitigation measures need to be taken promptly. After matching the analysis results with selected standards, areas (legs) in SWAC are marked according to their frequency levels. Mitigation recommendations can then be focused on areas with high collision frequency levels.

Mitigation recommendations may include changes to ship traffic regulations, increased supervision and patrols in high-risk areas, and educational campaigns to raise ship awareness regarding collision prevention. Additionally, improving navigation infrastructure, such as installing more buoys and navigation aids, may also be part of these recommendations.

Based on Figure 5, the clustering results generated using the hierarchical clustering method can be observed. The results show the formation of several clusters that reflect hierarchical relationships between legs. The first cluster consists of Leg 3, 2, 6, 4, and 1, indicated a strong similarity among these three legs.

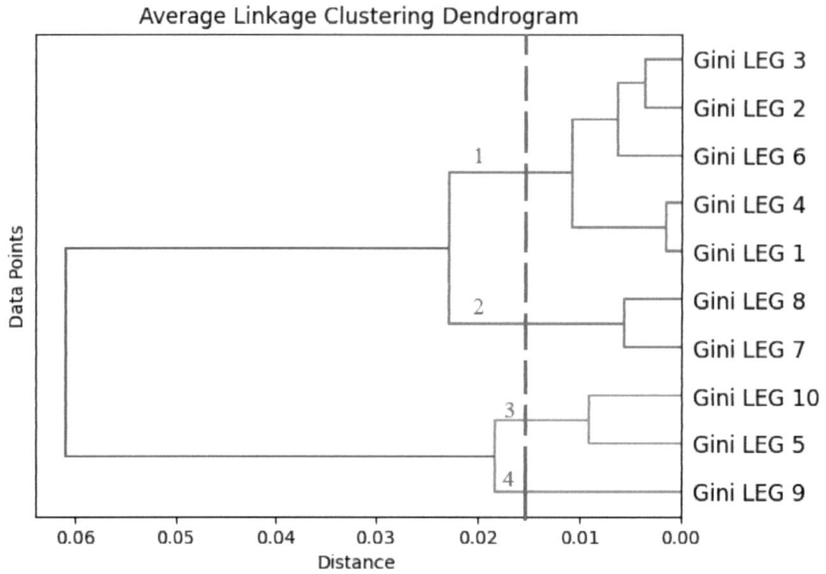

Figure 5. Dendrogram of hierarchical clustering based on ship COG values in each leg.

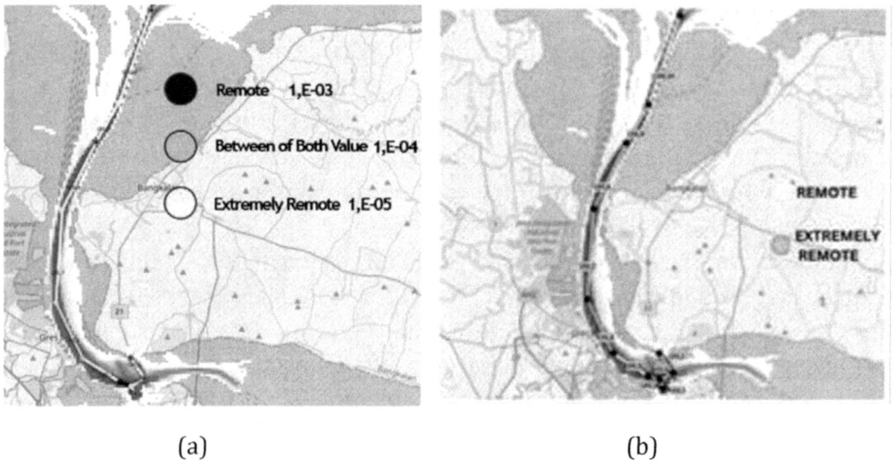

Figure 6. Representation of collision frequency in (a) IWRAP and (b) Gini Coefficient

Meanwhile, the second cluster consists of Leg 4 and 1, indicating similarity in characteristics between these two legs. The third cluster draws attention by combining Leg 10 and 5. This suggests that both legs have similar properties or attributes, leading to their grouping in the same cluster. Lastly, the fourth cluster only contains Leg 9, indicating that this leg has significant differences from the other legs and thus isolated in its cluster.

The utilization of the hierarchical clustering method enables us to meticulously dissect the intricate hierarchical organization deeply ingrained within the dataset. Through this approach, clusters manifest based on the nuanced degrees of similarity among constituent elements, unravelling latent patterns and structures that lie beneath the surface. Consequently, hierarchical

IOP Conf. Series: Earth and Environmental Science **1461** (2025) 012053 doi:10.1088/1755-1315/1461/1/012053

clustering serves as a catalyst for unveiling profound insights into the underlying relationships and characteristics, shedding light on complexities that might otherwise remain obscured when examining individual observations in isolation. In the process of calculating collision frequencies using IWRAP the analysis depicted in Figure 6a reveals that only Leg 4 is categorized as having a "remote" risk level. The collision frequency recorded on leg 4 stands at approximately 1×10E-5, indicating a comparatively low level of risk. In contrast, the remaining legs exhibit significantly higher collision frequency values, typically hovering around 1×10E-3, thus justifying their classification within the "extremely remote" risk category. Figure 6b complements these findings by illustrating the collision frequency values during overtaking for each leg.

4. Conclusion

The process for analyzing the risk of ship collisions in the SWAC using the Gini coefficient approach with AIS data is as follows:

a. Retrieval of raw AIS messages through the AISITS server.
b. Data pre-processing involves decoding raw AIS data, interpolating missing timestamp data, and dividing legs based on the density plot in IWRAP using Python syntax.
c. Gini coefficient calculation using Python to obtain Gini values for each leg.
d. Clustering of the 10 Gini coefficients for each leg to identify clusters with closely spaced values, allowing each cluster to be grouped into the same risk level.
e. Characterization of risk levels based on the IMO frequency index with validation from graphs plotting the COG degree against its frequency of occurrence.
f. Comparison of the risk analysis results for each area between IWRAP calculations and the Gini coefficient. The Tanjung Perak Port in Surabaya, East Java, stands as a cornerstone of Indonesia's maritime

The clustering process using the hierarchical clustering method is carried out by searching for the closest distance between each cluster, utilizing the distance formula with the Euclidean distance measurement method. In this context, the approach applied involves using the "davg" distance formula, referring to the average distance between elements in two clusters to be merged. The steps are implemented iteratively through a looping function in the Python programming language. This process continues iteratively until the formation of the last cluster that includes the entire data. In each iteration, the clusters with the closest distance are calculated and then merged to form a larger cluster.

The comparison between risk levels obtained from IWRAP calculations and those derived from the Gini coefficient generally yields similar outcomes, particularly when examined at the level of individual legs. Both analyses indicate that the collision risk levels do not exceed the standard frequency index set by IMO. As a result, the research suggests that no immediate mitigation measures are necessary. This consistency in findings highlights the reliability of both methodologies in assessing collision risks within the study area. However, ongoing monitoring and vigilance remain essential to ensure navigation safety in the future. Additionally, these aligned results provide valuable insights for refining risk assessment methods and informing decision-making processes in maritime safety management.

References

[1] Marine Traffic, "Density Maps." [Online]. Available: https://www.marinetraffic.com/en/ais/home/centerx:115.3/centery:-8.6/zoom:10

[2] Mahkamah Pelayaran, "Sidang Pembacaan Keputusan Mahkamah Pelayaran tentang kasus Kecelakaan Kapal tubrukan antara TSHD HAM 311 dan MT Crane Vesta pada tanggal 06 Desember 2021 sekira pukul 05.07 WIB di perairan sekitar Buoy 5 Alur Pelayaran Barat Surabaya pada koordinat 06° 52.04' S / 112° 44.53' T," *Mahkamah Pelayaran*, Aug. 2023.
[3] S. Li, J. Zhou, and Z. Yuanqiang, "Research of Vessel Traffic Safety in Ship Routeing Precautionary Areas Based on Navigational Traffic Conflict Technique," *J. Navig.*, vol. 68, pp. 1–13, May 2015, doi: 10.1017/S0373463314000939.
[4] A. Zhang, "A study on ship collision and grounding using nonlinear finite element method," *Univ. Tokyo*, no. Query date: 2022-10-14 07:30:50, 2007, [Online]. Available: https://repository.dl.itc.u-tokyo.ac.jp/record/2265/file_preview/K-295-1.pdf?allow_aggs=False
[5] X. Qu, Q. Meng, and L. Suyi, "Ship collision risk assessment for the Singapore Strait," *Accid. Anal. Prev.*, vol. 43, no. 6, pp. 2030–2036, Nov. 2011, doi: 10.1016/j.aap.2011.05.022.
[6] F. I. Prastyasari and T. Shinoda, "Study on Probabilistic Risk Assessment Model for Crossing Situation in Sunda Strait," *IOP Conf. Ser. Earth Environ. Sci.*, vol. 972, no. 1, p. 012043, Jan. 2022, doi: 10.1088/1755-1315/972/1/012043.
[7] Q. Lin, B. Yin, X. Zhang, M. Grifoll, and H. Feng, "Evaluation of ship collision risk in ships' routeing waters: A Gini coefficient approach using AIS data," *Phys. Stat. Mech. ...*, no. Query date: 2024-06-04 08:31:10, 2023, [Online]. Available: https://www.sciencedirect.com/science/article/pii/S0378437123004910
[8] J. Okunev, *An Analysis of the Extended Mean Gini Coefficient as an Alternative Measure of Risk in Investment Decision Making Under Uncertainty*. search.proquest.com, 1989. [Online]. Available: https://search.proquest.com/openview/7c225e2c38a5e6ed7fc2791acd8278d9/1?pq-origsite=gscholar&cbl=2026366&diss=y
[9] J. Hasell and M. Roser, "Measuring inequality: What is the Gini coefficient?," *Our World Data*, no. Query date: 2024-06-04 08:07:41, 2023, [Online]. Available: https://ourworldindata.org/what-is-the-gini-coefficient
[10] M. S. Amin Megat Ali *et al.*, "Short-term Gini coefficient estimation using nonlinear autoregressive multilayer perceptron model," *Heliyon*, vol. 10, no. 4, p. e26438, Feb. 2024, doi: 10.1016/j.heliyon.2024.e26438.
[11] F. Meng, D. Zhao, and X. Zhang, "A fair consensus adjustment mechanism for large-scale group decision making in term of Gini coefficient," *Eng. Appl. Artif. Intell.*, vol. 126, p. 106962, Nov. 2023, doi: 10.1016/j.engappai.2023.106962.
[12] M. Qin, K. Duan, Z. Ma, Z. Li, X. Ju, and H. Sun, "What makes the provincial inequality of natural gas consumption in China? Evidence from Gini-coefficient approach," *Energy Convers. Manag. X*, vol. 22, p. 100601, Apr. 2024, doi: 10.1016/j.ecmx.2024.100601.
[13] T. Hwang and I. Youn, "Navigation Situation Clustering Model of Human-Operated Ships for Maritime Autonomous Surface Ship Collision Avoidance Tests," *J. Mar. Sci. Eng.*, no. Query date: 2022-10-14 07:30:50, 2021, [Online]. Available: https://www.mdpi.com/1413614
[14] M. Marino, L. Cavallaro, E. Castro, R. Musumeci, and ..., "New frontiers in the risk assessment of ship collision," *Ocean ...*, no. Query date: 2024-06-04 08:41:15, 2023, [Online]. Available: https://www.sciencedirect.com/science/article/pii/S0029801823003839
[15] Q. Gao, Z. Li, W. Zhang, X. Bao, J. Xia, and Z. Zheng, "Channel reconstruction for mmWave Massive MIMO systems based on Channel Path Map," *Phys. Commun.*, vol. 61, p. 102232, Dec. 2023, doi: 10.1016/j.phycom.2023.102232.

AUTHOR INDEX